Microbes Based Approaches for the Management
of Hazardous Contaminants

# Microbes Based Approaches for the Management of Hazardous Contaminants

*Edited by*

*Ajay Kumar*
*Amity Institute of Biotechnology*
*Amity University*
*Noida, India*

*Livleen Shukla*
*ICAR-Indian Agricultural Research Institute*
*New Delhi, India*

*Joginder Singh*
*Nagaland University*
*Nagaland, India*

*Luiz Fernando R. Ferreira*
*Catholic University of Brasília*
*Brasília, Brazil*

WILEY

Published by John Wiley & Sons, Inc., Hoboken, New Jersey.
Published simultaneously in Canada.



For general information on our other products and services or for technical support, please contact our Customer Care Department within the United States at (800) 762-2974, outside the United States at (317) 572-3993 or fax (317) 572-4002.

Wiley also publishes its books in a variety of electronic formats. Some content that appears in print may not be available in electronic formats. For more information about Wiley products, visit our web site at www.wiley.com.

***Library of Congress Cataloging-in-Publication Data applied for:***
Hardback ISBN: 9781119851127

Cover Design: Wiley
Cover Image: © Jeffery Kent/Alamy Stock Photo

Set in 9.5/12.5pt STIXTwoText by Straive, Pondicherry, India

# Contents

## List of Contributors

***Mariana Miranda de Abreu***
Laboratory of Physiology and Biochemistry of Microorganisms, Centro de Biociências e Biotecnologia
Universidade Estadual do Norte Fluminense Darcy Ribeiro (UENF)
RJ, Brazil

Biotechnology and Environmental Microbiology Laboratory
Universidade Vila Velha (UVV)
ES, Brazil

***Ankita Agrawal***
Department of Biotechnology and Bioinformatics
Sambalpur University
Burla, Odisha
India

***Saif Alharbi***
King Abdulaziz City for Science and Technology
Riyadh
Saudi Arabia

***Fahad S. Alotaibi***
King Abdulaziz City for Science and Technology
Riyadh
Saudi Arabia

***Abdullah Alrajhi***
King Abdulaziz City for Science and Technology
Riyadh
Saudi Arabia

***S.T.M. Aravindharajan***
Division of Microbiology
ICAR-Indian Agricultural Research Institute
New Delhi
India

***Sidra Ashraf***
Department of Biochemistry
University of Agriculture
Faisalabad
Pakistan

***Aldona Balčiūnaitė***
Department of Catalysis
Center for Physical Sciences and Technology
Vilnius
Lithuania

***Ram Naresh Bharagava***
Laboratory of Bioremediation and Metagenomics Research (LBMR), Department of Environmental Microbiology (DEM)
Babasaheb Bhimrao Ambedkar University (A Central University)
Lucknow
India

***Priyank Chaturvedi***
Department of Environmental Science
ITM University
Gwalior, Madhya Pradesh
India

***T. Chethan***
Department of Agricultural Entomology
University of Agricultural Sciences, GKVK
Bengaluru
India

***Ashok Chhetri***
Department of Plant Protection & Department of Horticulture, Multi-Technology Testing Centre & Vocational Training Centre
Central Agricultural University (Imphal Manipur)
Agartala, Tripura
India

***Emilane Pinheiro da Cruz Lima***
Laboratory of Physiology and Biochemistry of Microorganisms, Centro de Biociências e Biotecnologia
Universidade Estadual do Norte Fluminense Darcy Ribeiro (UENF)
RJ, Brazil

***Saprativ P. Das***
Department of Chemical Engineering
Indian Institute of Technology Bombay
Mumbai, Maharashtra
India

***Soham Das***
Amity Institute of Biotechnology
Amity University
Noida
India

***Ashim Debnath***
Department of Genetics and Plant Breeding
Faculty of Agricultural Sciences
Rajiv Gandhi University
Doimukh, Arunachal Pradesh
India

***Anshika Dedha***
Amity Institute of Biotechnology
Amity University
Noida
India

***S.S. Deepti Varsha***
Division of Microbiology
ICAR-Indian Agricultural Research Institute
New Delhi
India

***Anuradha Devi***
Laboratory of Bioremediation and Metagenomics Research (LBMR), Department of Environmental Microbiology (DEM)
Babasaheb Bhimrao Ambedkar University (A Central University)
Lucknow
India

***Joy Kumar Dey***
Krishi Vigyan Kendra-Sepahijala
CAU (I)
Latiacherra, Tripura
India

***Farhana Ehsan***
Center for Applied Molecular Biology (CAMB)
University of the Punjab
Lahore
Pakistan

***Luiz Fernando R. Ferreira***
Post Graduate Program in Genomic Sciences and Biotechnology
Genomic Sciences and Biotechnology Lab
Catholic University of Brasília
Brasília
Brazil

***Poulomi Ghosh***
Department of Biotechnology
Institute of Genetic Engineering (affiliated to MAKAUT)
Kolkata, West Bengal
India

***Babett Greff***
Department of Food Science
Faculty of Agricultural and Food Sciences
Széchenyi István University
Mosonmagyaróvár
Hungary

***Shazaf Gul***
Department of Biochemistry
University of Agriculture
Faisalabad
Pakistan

***Irfan Haidri***
Department of Environmental Sciences
Government College University
Faisalabad
Pakistan

***Hafsa***
Centre for Applied Molecular Biology (CAMB)
University of Punjab
Lahore
Pakistan

***Rachel Ann Hauser-Davis***
Ambiental Environmental Health Assessment and Promotion Laboratory
Instituto Oswaldo Cruz
Fundação Oswaldo Cruz, RJ
Brazil

***Asim Hussain***
Department of Biochemistry
University of Agriculture
Faisalabad
Pakistan

***Nazim Hussain***
Centre for Applied Molecular Biology (CAMB)
University of Punjab
Lahore
Pakistan

***Sabir Hussain***
Department of Environmental Sciences
Government College University
Faisalabad
Pakistan

***Fatima Idress***
Department of Biochemistry
University of Agriculture
Faisalabad
Pakistan

***Aline Chaves Intorne***
Laboratory of Physiology and Biochemistry of Microorganisms, Centro de Biociências e Biotecnologia
Universidade Estadual do Norte Fluminense Darcy Ribeiro (UENF)
RJ, Brazil

Chemistry and Biology Laboratory, Departamento de Ensino
Instituto Federal de Educação
Ciência e Tecnologia do Rio de Janeiro, RJ,
Brazil

***Aneeza Ishfaq***
Department of Environmental Sciences
Government College University
Faisalabad
Pakistan

***Prakriti Jhilta***
Department of Microbiology
Central University of Haryana
Mahendergarh, Haryana
India

***Raj Kishor Kapardar***
Microbial Biotechnology Division
The Energy and Resources Institute
New Delhi
India

***Praveen Karijadar***
Department of Agricultural Microbiology
University of Agricultural Sciences, GKVK
Bengaluru
India

***T. Kavya***
Division of Microbiology
ICAR-Indian Agricultural Research Institute
New Delhi
India

***Ajay Kumar***
Amity Institute of Biotechnology
Amity University
Noida
India

***Haralakal Keerthi Kumari***
Department of Agricultural Microbiology
University of Agricultural Sciences
Bangalore
India

***Tabassum Khan***
Department of Pharmaceutical Chemistry
SVKMs Dr. Bhanuben Nanavati College of Pharmacy
Vile-Parle, Mumbai
Maharashtra
India

***Arun Kumar***
Department of Horticulture
School of Agriculture Lovely Professional University
Phagwara
Jalandhar
Punjab
India

*Pawan Kumar*
Department of Microbiology,
Central University of Haryana
Mahendergarh, Haryana
India

*Sandeep Kumar*
Department of Microbiology
Central University of Haryana
Mahendergarh, Haryana
India

*Vijay Kumar*
Department of Plant Protection & Department of Horticulture, Multi-Technology Testing Centre & Vocational Training Centre
Central Agricultural University (Imphal Manipur)
Agartala, Tripura
India

*Bharati Lap*
Department of Genetics and Plant Breeding
Faculty of Agricultural Sciences
Rajiv Gandhi University
Doimukh, Arunachal Pradesh
India

*Jitesh Kumar Maharana*
Department of Biotechnology and Bioinformatics
Sambalpur University
Burla, Odisha
India

*Faisal Mahmood*
Department of Environmental Sciences
Government College University
Faisalabad
Pakistan

*Hafiza Ayesha Malik*
Center for Applied Molecular Biology (CAMB)
University of the Punjab
Lahore
Pakistan

*V. Mamtha*
Department of Agricultural Microbiology
University of Agricultural Sciences
Bangalore
India

*Menpadi Manoj*
Department of Agricultural Microbiology
University of Agricultural Sciences, GKVK
Bengaluru
India

*Sikandar I. Mulla*
Department of Biochemistry, School of Applied Sciences
REVA University
Bangalore
India

*Mehvish Mumtaz*
Centre for Applied Molecular Biology (CAMB)
University of Punjab
Lahore
Pakistan

*Azha Ufaq Nabi*
Department of Bioresources, School of Biological Sciences
University of Kashmir
Srinagar
India

*Sahaya Nadar*
Department of Pharmaceutical Chemistry
SVKMs Dr. Bhanuben Nanavati College of Pharmacy
Vile-Parle, Mumbai
Maharashtra
India

Department of Pharmaceutical Chemistry
St. John Institute of Pharmacy and Research
Palghar, Maharashtra
India

*Raj Nandini*
Amity Institute of Biotechnology
Amity University
Noida
India

*Sivaprakasam Navarasu*
Department of Plant Pathology
Tamil Nadu Agricultural University
Coimbatore
India

*Konderu Niteesh Varma*
Division of Microbiology
ICAR-Indian Agricultural Research Institute
New Delhi
India

***Amiya Kumar Patel***
Department of Biotechnology and Bioinformatics
Sambalpur University
Burla, Odisha
India

***Warda Perveez***
Centre for Applied Molecular Biology (CAMB)
University of Punjab
Lahore
Pakistan

***Gabriela Petroceli-Mota***
Laboratory of Cellular and Tissue Biology
Centro de Biociências e Biotecnologia
Universidade Estadual do Norte Fluminense Darcy Ribeiro (UENF)
RJ, Brazil

***Vikram Poria***
Department of Microbiology
Central University of Haryana
Mahendergarh, Haryana
India

***Nischal Pradhan***
Amity Institute of Biotechnology (AIB)
Amity University Uttar Pradesh
Noida
India

***Shriniketan Puranik***
Division of Microbiology
ICAR-Indian Agricultural Research Institute
New Delhi
India

***Gabriel Quintanilha-Peixoto***
Laboratory of Chemistry and Function of Proteins and Peptides, Centro de Biociências e Biotecnologia
Universidade Estadual do Norte Fluminense Darcy Ribeiro (UENF)
RJ, Brazil

Laboratory of Molecular and Computational Biology of Fungi
Instituto de Ciências Biológicas, Universidade Federal de Minas Gerais (UFMG)
MG, Brazil

***Arathi Radhakrishnan***
Amity Institute of Biotechnology
Amity University
Noida
India

***Hamza Rafeeq***
Department of Biochemistry
Riphah International University
Faisalabad
Pakistan

Department of Biochemistry
University of Agriculture
Faisalabad
Pakistan

***Alessandro Coutinho Ramos***
Biotechnology and Environmental Microbiology Laboratory
Universidade Vila Velha (UVV)
ES, Brazil

***Sobia Riaz***
Institute of Soil & Environmental Sciences
University of Agriculture Faisalabad
Faisalabad
Pakistan

***Zainab Riaz***
Department of Biochemistry
Riphah International University
Faisalabad
Pakistan

***Sajal Saha***
Department of Genetics and Plant Breeding, SAS, Nagaland University
Medziphema, Nagaland
India

***Rashba Sahar***
Department of Plant Pathology
University of Agriculture
Faisalabad
Pakistan

***Velmurugan Shanmugam***
Division of Plant Pathology
ICAR-Indian Agricultural Research Institute
New Delhi
India

***Shakshi***
Amity Institute of Biotechnology
Amity University
Noida
India

**D. Sai Srinivas**
Department of Microbiology, School of Life Sciences
JSS Academy of Higher Education and Research
Mysore
India

**Christina Saran**
Laboratory of Bioremediation and Metagenomics Research (LBMR), Department of Environmental Microbiology (DEM)
Babasaheb Bhimrao Ambedkar University (A Central University)
Lucknow
India

**Ganesh Dattatraya Saratale**
Department of Food Science and Biotechnology
Dongguk University-Seoul
Goyang-si
Republic of Korea

**Rijuta Ganesh Saratale**
Department of Food Science and Biotechnology
Research Institute of Biotechnology and Medical Converged Science
Dongguk University-Seoul
Goyang-si
Republic of Korea

**Muhammad Shahid**
Department of Bioinformatics & Biotechnology
Government College University
Faisalabad
Pakistan

**Areej Shahbaz**
Center for Applied Molecular Biology (CAMB)
University of the Punjab
Lahore
Pakistan

**Tanvir Shahzad**
Department of Environmental Sciences
Government College University
Faisalabad
Pakistan

**Anum Shahzadi**
Department of Biochemistry
Riphah International University
Faisalabad
Pakistan

**Faamiya Shajar**
Department of Bioresources, School of Biological Sciences
University of Kashmir
Srinagar
India

**Livleen Shukla**
Division of Microbiology
ICAR-Indian Agricultural Research Institute
New Delhi
India

**Anamika Singh**
Amity Institute of Biotechnology
Amity University Uttar Pradesh
Noida
India

**Balkar Singh**
Department of Botany
Arya PG College
Panipat, Haryana
India

**Dilbag Singh**
Division of Microbiology
ICAR-Indian Agricultural Research Institute
New Delhi
India

**Gourav Kumar Singh**
Department of Environmental Science
ITM University
Gwalior, Madhya Pradesh
India

**Jatinder Singh**
Department of Horticulture, School of Agriculture Lovely Professional University Phagwara
Jalandhar
Punjab
India

**Rachana Singh**
Amity Institute of Biotechnology
Amity University Uttar Pradesh
Noida
India

**Rahul Prasad Singh**
Laboratory of Algal Research, Department of Botany
Institute of Science, Banaras Hindu University
Varanasi
India

***Sandeep Kumar Singh***
Division of Microbiology
ICAR-Indian Agricultural Research Institute
New Delhi
India

***Surender Singh***
Department of Microbiology
Central University of Haryana
Mahendergarh, Haryana
India

***Surbhi Sinha***
Amity Institute of Biotechnology,
Amity University Uttar Pradesh
Noida
India

***Muhammad Sohail***
Department of Forestry and Range Management
University of Agriculture Faisalabad
Faisalabad
Pakistan

***K. Sowmiya***
Department of Agricultural Microbiology
University of Agricultural Sciences
Bangalore
India

***José Cleiton Sousa dos Santos***
Instituto de Engenharias e Desenvolvimento Sustentável
Universidade da Integração Internacional da Lusofonia
Afro-Brasileira
Campus das Auroras, Redenção
Brazil

***Glacielen Ribeiro de Souza***
Laboratory of Biotechnology, Centro de Biociências
e Biotecnologia
Universidade Estadual do Norte Fluminense Darcy
Ribeiro (UENF)
RJ, Brazil

***Jussara Tamires de Souza Silva***
Laboratory of Physiology and Biochemistry of
Microorganisms, Centro de Biociências e Biotecnologia
Universidade Estadual do Norte Fluminense Darcy
Ribeiro (UENF)
RJ, Brazil

Phytotechnics Laboratory, Centro de Ciências
e Tecnologias Agropecuárias
Universidade Estadual do Norte Fluminense Darcy
Ribeiro (UENF)
RJ, Brazil

***Rajpal Srivastav***
Amity Institute of Biotechnology
Amity University Uttar Pradesh
Noida
India

***Kallinkal Sobha Sruthy***
Division of Microbiology, ICAR-Indian Agricultural
Research Institute
New Delhi
India

***Swati***
Department of Agricultural Microbiology
University of Agricultural Sciences
Bangalore
India

***Vikanksha Thakur***
Department of Horticulture
School of Agriculture Lovely Professional University
Phagwara
Jalandhar, Punjab
India

***Reiaz Ul Rehman***
Department of Bioresources, School of Biological Sciences
University of Kashmir
Srinagar
India

***V.H.S. Vaishnavee***
Amity Institute of Biotechnology
Amity University
Noida
India

***Konaghatta Vijayakumar Vikram***
Division of Microbiology
ICAR-Indian Agricultural Research Institute
New Delhi
India

***D. Vijaysri***
Division of Microbiology
ICAR-Indian Agricultural Research Institute
New Delhi
India

***Priya Yadav***
Laboratory of Algal Research, Department of Botany
Institute of Science, Banaras Hindu University
Varanasi
India

# Preface

Hazardous contaminants in waste from the industrial, home, and agricultural sectors inflict enormous harm to the ecosystem and the lives of those who live nearby. These hazardous contaminants can exist in a variety of forms that are inclined by climate features like the presence of various types of organic matter, pH, water system hardness, transformation, and bioavailability. Since majority of them are mobile and soluble, they have the potential to bioaccumulate in the food chain and cause serious harm above certain concentrations. Microbe-based approaches are an important tool for removing toxic contaminants from the ecosystem and have piqued the concern of investigators over the centuries. Hence, the use of appropriate microorganisms in the remediation of pollutants is critical to effectively reducing the negative effects of toxic pollutants. Several microorganisms were discovered as promising candidates for bioremediation of hazardous contaminants via biotransformation, bioremediation, bioaccumulation, or biosorption processes. Hazardous contaminants degradation mechanisms exhibited by microorganisms are primarily determined by their degradative plasmids and spores. The book entitled *Microbes Based Approaches for the Management of Hazardous Contaminants* is focused on the most insightful aspects of microbe-based approaches for addressing hazardous contaminants in our ecosystem. This book covers novel and indigenous microbes and microbial products. Biofilms, exopolysaccharides, bio-surfactants, enzymes, metabolites, microbially synthesized nanoparticles, and the latest genetic and other metabolic engineering approaches based on microbial technologies that are utilized in the remediation and management of hazardous contamination from the environment and surroundings are covered. The book details many methods used, including degradation and remediation of hazardous pollutants, genetically engineered microorganisms for the removal of toxic organic compounds, metabolic engineering as alternative strategies for microbial bioremediation, microbial immobilization and use of microbial surfactants, microorganisms as the indicator for pollutants, biochemical approaches in microbial biodegradation, to name a few. The chapters provide readers with rich sources of reference information on important topics in this field. There are several illustrations so that the scholars can grasp the flow of the content and the transition to the next topic.

Chapters are written by active researchers and presented in an accessible way to the public beyond those specialists in the topic dealt with. The content aims to bridge the gap between advanced research and general understanding, providing clear explanations and practical insights that can be appreciated by a broader audience. This book serves as a helpful tool for scientists, industrial professionals, and experts working in diverse aspects to understand the various microbial-based methodologies toward hazardous waste management with proper resource management and technological applications. We are honored that the leading experts with extensive, in-depth experience and expertise in diverse microbial technologies have taken the time and effort to develop outstanding chapters.

We thank team Wiley for their generous assistance, constant support, and patience in initializing the book. The authors are also grateful to our esteemed friends, well-wishers, and faculty colleagues.

***Editors:***
*Dr. Ajay Kumar*
*Dr. Livleen Shukla*
*Prof. Joginder Singh*
*Prof. Dr. Luiz Fernando R. Ferreira*

# 1

# Mycobial Nanotechnology in Bioremediation of Wastewater

*Vikanksha Thakur, Arun Kumar, and Jatinder Singh*

*Department of Horticulture, School of Agriculture Lovely Professional University Phagwara, Jalandhar, Punjab, India*

## 1.1 Fungi

Fungal species represent the earliest known examples of eukaryotic life. The vast majority of fungi are minute in size, but certain varieties of them may grow to be very large (Van den Brule et al. 2020). Examples of these include mushrooms, toadstools, and stinkhorns. In broad terms, they get nourishment by the process of heterotrophy; however, a portion of them are parasitic, while others are saprophytic organisms. The dispersion of fungus spans the whole planet. Fungi are essential to the functioning of the ecosystem because of their roles as food suppliers, decomposing organisms, symbiotic partners, and protectors (Ray et al. 2020). Fungi additionally have a significant part in the cycle of nutrients. Since the beginning of time, people have turned to fungus not only as a supply for nourishment but also as an agent in the fermentation and preservation of foods and drinks. Antibiotics, medications to lower lipids, and immunosuppressants are all examples of how humans have benefited from the study of fungi in the modern age. The 1980s saw the beginning of the commercial application of fungi for the synthesis of amino acids, chemicals, and biosurfactants because of the development of contemporary nanotechnology (Gayathiri et al. 2022). Fungi are able to be quickly separated through a variety of diverse environmental sources. They can additionally be easily cultured on a simple and low-nutrient medium in the testing facility, such as Potato Dextrose Agar. Additionally, maintaining fungus in the laboratory is quite inexpensive. Investigators concluded that fungi are capable of higher enzyme-secreting capacity, and because they are easy to identify and keep alive, fungi serve as the organisms of choice for the creation of silver as well as different nanoparticles (NPs). It is well recognized that fungi constitute the natural world's foremost powerful catalyst for the breakdown or disintegration of recyclables, and in exchange supplies nutrition to different organisms in the natural world (García-Depraect et al. 2021). In fact, although a forest floor is covered with foliage, many of the vitamins and minerals remain trapped within it because twigs are challenging to break down and digest. Fungi are among the most important creatures that can break down leaf material. On both the bark and the leaves, the spores of the fungus develops, and as it does, it produces incredibly strong extrinsic enzymes and acids that possess the capacity to break down lignin and cellulose. Both cellulose and lignin, the two main elements that comprise plant material, are notoriously difficult to degrade. Cellulose is particularly resistant to this process. Fungi are also responsible for the breakdown of plant fibers, which results in the production of humus, which is a substance that is rich in nutrients and beneficial to the health of plants (Krishnaveni et al. 2020). Fungi are among the few known creatures that possess the capability of breaking down wood. Majority of essential organisms in the environment compete with one another for the minerals and vitamins that are necessary for their own survival and growth. These essential elements include nitrogen, potassium, and phosphorus. Because of that, stress related to the environment is developed upon the substrate, which ultimately contributes to a quicker pace of inorganic pollution breakdown. Microorganisms are very effective in breaking down materials that include paper and fabrics such as cotton, jute, and linens. This particular stage is referred to as biodeterioration. It is widely acknowledged that Aspergillus and molds are able to degrade cellulose, starches, hemicellulose, and a variety of carbohydrate polysaccharides very well. Furthermore, Aspergillus possesses the ability to break down seemingly unbreakable compounds, including fatty acids, fats, keratins, and chitosans (Sanchez et al. 2012). It is essential that the appropriate species of fungi be considered in order to perform mycoremediation.

*Microbes Based Approaches for the Management of Hazardous Contaminants*, First Edition. Edited by Ajay Kumar, Livleen Shukla, Joginder Singh, and Luiz Fernando R. Ferreira.

## 1.2 Nanotechnology Aspects

NPs are extremely tiny molecules that range in size between 1 and 100 nm that are starting to adhere more closely to the laws of quantum theory than to those of traditional science. For instance, an identical substance is going to develop distinct visual and electrical characteristics being seen across the nanoscale in comparison to being viewed at the larger scale (Sudha et al. 2018). The research and investigation of various sorts of substances, including NPs, which, owing to their extremely tiny dimensions, have important qualities, processes, and occurrences is commonly referred to the nanoscience. The production of NPs over purposes of improving the well-being of humans is an example of nanotechnology, which involves applying the techniques of nanotechnology within a production or industrial environment.

## 1.3 The Production of Nanoparticles Using an Origin of Fungi

There are two distinct procedures that govern the production of NPs from a fungal origin. These are the external and internal processes. Oxidation of metallic ions occur in the midst of enzymes released by fungal cells, which is necessary for the external biosynthesis to take place (Kotrange et al. 2021).

Intracellular enzymes also prove helpful in the procedure of converting metallic ions to NPs. That is sometimes accomplished by either capturing metallic particles onto the outermost layer of cellular structures and subsequently reducing them in the presence of enzymes. The dimensions of the fragments are unique to both the fungus from which they are derived and the metal that was oxidized (Bhardwaj et al. 2020). Different types of fungi need different amounts of time to germinate (Figure 1.1).

### 1.3.1 Silver Nanoparticles

Silver NPs are produced because the characteristics of the metallic element alter substantially when reduced to small particles in the range of 1–100 nm. This reduction in size of particles results in the formation of SilverNPs (Table 1.1). In comparison to macro-sized silver material, silver NPs have a variety of physicochemical and physiological characteristics that set them apart (Bishayee et al. 2022). The enhancement in chemical resistance also results in a rise in both the electric as well as heating conductance of the substance. Because of its improved qualities, they are becoming more important in the

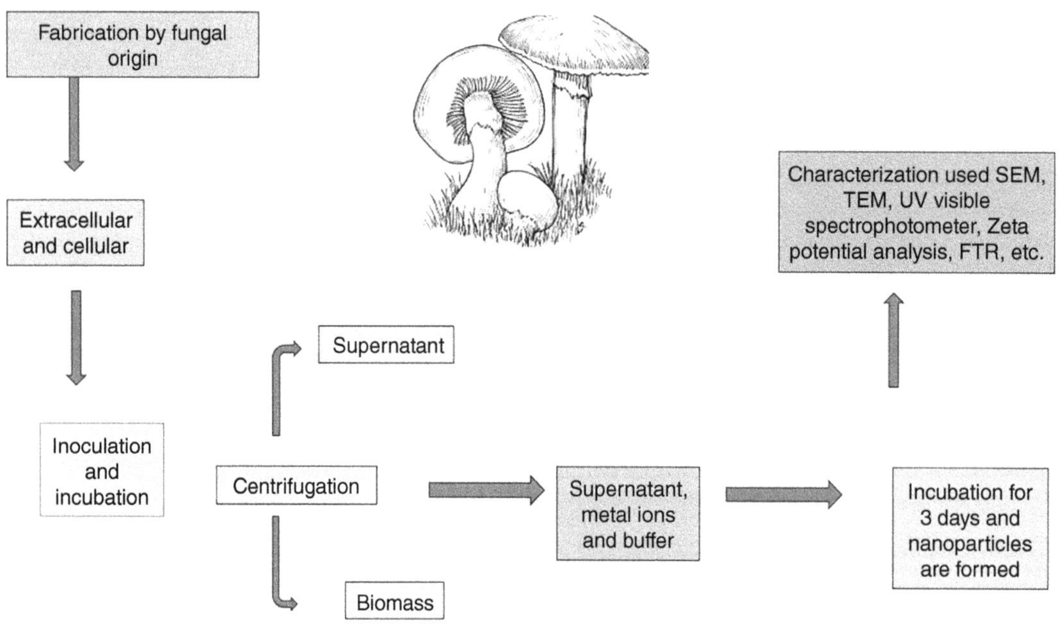


Figure 1.1 Procedure of nanoparticle formulation from a fungal origin. (*Source:* Vikanksha).

**Table 1.1** Below is a compilation of fungus species that were previously used for the synthesis of nanoparticles.

| Fungus used | Type of nanoparticles produced | Size (nm) | Shape | References |
|---|---|---|---|---|
| *A. niger* and *A. terreus* | Silver nanoparticles | 15 ± 3 nm | Rod and cube | Alananbeh et al. (2017) |
| *Aspergillus* spp. | Silver nanoparticles | 6–25 nm | Sphere and hexagon | Gola et al. (2021) |
| *Penicillium citreonigum Dierck* and *Scopulariopsis brumptii* | Silver nanoparticles | 350–450 nm | Sphere | Moustafa (2017) |
| *Aspergillus foetidus* MTCC8876 | Silver nanoparticles | 100–100 nm | Sphere | Roy et al. (2013) |
| *Trichoderma harzianum* | Gold nanoparticles | 70–300 nm | Sphere | do Nascimento et al. (2021) |
| *Trichoderma harzianum* | Gold nanoparticles | 30–45 nm | Sphere | Tripathi et al. (2018) |

(*Source:* Vikanksha)

fields of miniature electronics and health care imaging, in addition to the development of antibacterial medications. India is responsible for publishing 9.8% of the worldwide total scientific papers on SilverNPs.

Synthesis methods include the processes of chemical synthesis, physical manufacturing, photochemical reactions fabrication, and biological production. Chemical synthesis is the most common method. Synthesis of silver NPs is frequently accomplished by chemical production. Although the aforementioned techniques are useful in the manufacturing of Silver NPs on a broadscale, the upfront investment in the necessary equipment could prove prohibitively expensive and might not be feasible (Mazurkiewicz et al. 2020).

In order to complete the reduction procedure, photosynthesis production requires the application for a variety of luminescent resources. The photochemical production of silver NPs has a number of advantages, including its low environmental impact and its adaptability (NPs may be produced in a variety of mediums). The biosynthetic method is not only beneficial to the natural world but also economical. Additionally, the overall pace of NP biogenesis is considerably quicker compared to that of different approaches (Khorrami et al. 2019).

Silver NPs are well-known because of the antibacterial properties they possess. Past studies have revealed that they are effective against a wide variety of harmful microorganisms, including bacterium, fungi, and additional viruses.

According to Salvadori et al. (2014), the dead tissue of *Trichoderma koningiopsis* is capable of producing copper NPs rapidly inside the extracellular environment. The NPs were created extracellularly, with majority of them having a sphere-like form (Zhou et al. 2020). Their median dimension was 87.5 nm, indicating that they had been produced in a lab. It has been established that proteins act as stabilizing substances for the NPs. The external manufacture and absorption of copper NPs utilizing decomposing fungal cells is a cost-effective ecological approach that also serves as bioremediation of nearby damaged regions.

### 1.3.2 Gold Nanoparticles

The formation of gold NPs occurs as a result of reducing $HAuCl_4$ or another aurum salt (such as $AuCl_3$) with the help of enzymatic or similar reducing chemicals (Voliani 2020). This process leads to the formation of NPs with dimensions ranging from 1 to 100 nm and causes shifts in the fundamental characteristics of the metal. During the process Gold NPs exhibit a color that ranges from a reddish-orange to a purplish-red.

The characteristics of Gold NPs change depending on both, the dimensions and its form (Table 1.1). As a result of their advantageous refractive qualities, high surface-to-volume ratios, and electrical characteristics, these particles are capable of serving as a significant method within the field of bio-nanotechnology. Surfaces plasmon resonance, often known as SPR, and the capacity of goldNPs to dampen fluorescence constitute two of the material's most significant mechanical characteristics (Mahrouk et al. 2021). These have applications in immunostaining, monitoring individual particles, and contrasting compounds for X-rays, all with the purpose of identification. As for context of delivery, Au-NPs are appropriate for application in genetic guns and cellular absorption. The ability of GoldNPs to function as thermal producers can be utilized in cancer chemotherapy through the application of thermal effects. This capability enables the disruption of biological interactions, visually activated by photo-induced warmth. In addition to this, gold NPs are safe for human consumption, offering a significant advantage above silver NPs (Kumar and Yadav 2009).

### 1.3.3 Additional Nanoparticles

NPs of silver and gold were the ones that have been frequently researched. In addition to those, NPs may be manufactured using non-metals such as selenium and transition metal oxides such as zinc oxide and tin oxide, as well as metals such as platinum and copper.

The potency of zinc oxide NPs as antibiotics against disease-causing bacteria was discovered, indicating that these NPs are potent antibacterial substances (Pati et al. 2014). Magnetite–NPs have a wide range of uses. Additionally, magnetite NPs have several mechano-electrical functions. Quantum dots (QDs) that include cadmium have a great deal of unexplored possibilities in the domain of healthcare therapy. On the contrary, the presence of poisonous cadmium in nanoparticle state might pose a constraint. When using NPs containing any of those parts, there is a need for an appropriate review of their applications.

## 1.4 Categories and Characteristics of Synthesized Nanoparticles

Various categorization methods are utilized to categorize NPs. Division of NPs according to their various dimension patterns. One-dimensional NPs consist of thin coatings or artificial materials utilized in the fields of living organism detection, catalytic processes, and solar energy production. The two-dimensional NPs consist of 1 to 100 nm width carbon nanotubes (CNTs). Both single-walled (SWCNTs) and multi-walled (MWCNTs) CNTs are possible (Kosynkin et al. 2009).

CNTs are useful because of their metals and semiconducting characteristics, as well as their enhanced electrical conductivity, which allow them to function as semiconductors. The NPs are composed of dendrimers, and QDs. Fullerenes are the shape of hollowed spheres resembling sports balls. They are distinguished by their resistance to high stress as well as the capacity to recover their initial form upon tension reduction. Because of the fact that fullerenes are fluid-filled molecules, their uses involve loading them with different compounds; consequently, they have healthcare uses.

Dendrimers are polymer compounds with a length of 10–100 nm that encompass multiple bioactive atoms, demonstrating the possibility of utilization within pharmaceutical distribution methods (Pramanik et al. 2021). QDs are nanocrystals with dimensions of 2–10 nm that may contain anywhere from one to a million electrons. QDs are frequently utilized in photonic and electrical equipment, quantum computation, and data retention, while color-coded QDs are utilized for rapid examinations of DNA. Recent researches on inorganic NPs, i.e., noble metallic NPs, is gaining prime importance as these find wide application in diverse medical fields (Sindhu et al. 2021). The categorization of NPs according to size and chemical composition is depicted in Figures 1.2 and 1.3.

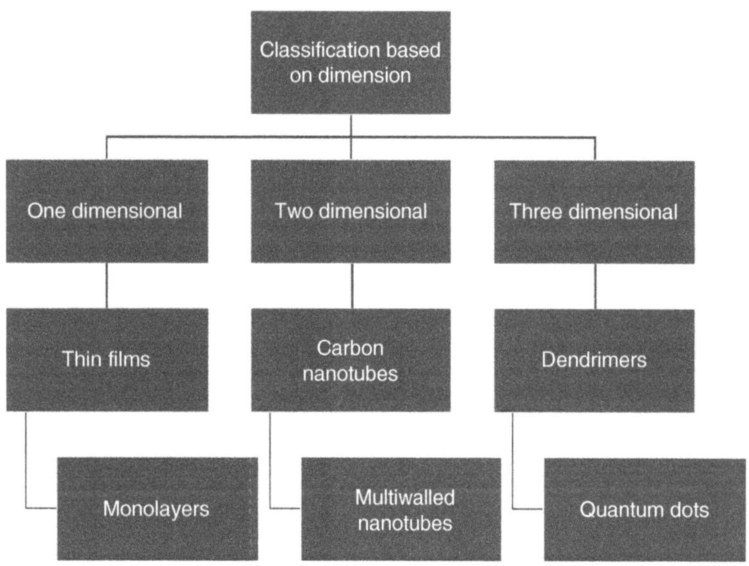


**Figure 1.2** Categorization of nanoparticles based on dimension (*Source:* Vikanksha).

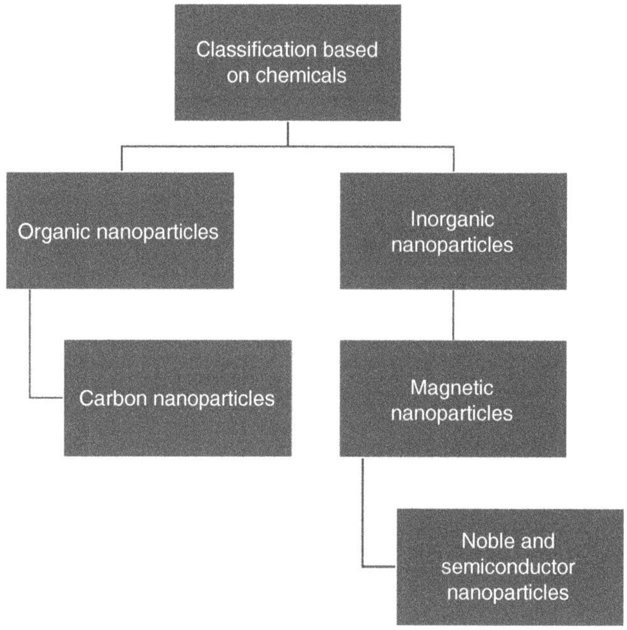


**Figure 1.3** Categorization of nanoparticles based on chemicals (*Source:* Vikanksha).

### 1.4.1 Characteristics on Nanoparticles

NPs are a subcategory of colloidal substances. As a result of the interaction between traditional science and quantum engineering, nanotechnologies discover their primary functions in the manufacture of temporary items linking various fields, consequently laying the groundwork for what is referred to as Nanotechnology. The dimension, framework, and geometry of NPs have a significant effect on their unique or enhanced characteristics (Dekura et al. 2019). Changing each of the parameters allows experts to create compounds with particular purposes. The characteristics of NPs that will be addressed are extremely substantial.

### 1.4.2 Physical Characteristics

NPs are distinct owing to its physical features and significant dimension. Because NPs are incredibly tiny, particles trap additional amounts of sunlight. The ideal instance for such can be seen in the inclusion of zinc oxide (ZnO) NPs in sunblock creams with UV-blocking properties (Kaur et al. 2020).

### 1.4.3 Biological Characteristics

The applications of NPs within the food industry have expanded tremendously as a component of food packaging to limit degradation of food against the natural environment that covers the food, maintaining it secure from hazardous pathogenic microbes that could compromise the food. Claylike–NPs limit water infiltration and restrict gaseous transfer by means of packing. Antibacterial properties of NPs have recently been included with paintwork and wallpaper coatings, rendering goods easier to use and secure for healthcare and clinical laboratory environments (Kumar et al. 2019).

### 1.4.4 Medical Benefits

Physical qualities of NPs have shown a lot of promise in healthcare fields, including disease detection, medication administration, and scanning. NPs can move freely through the organism, yet they can also infiltrate cells. It improves the appearance of tissues and tumors. The anti-inflammatory capabilities of NPs have been heavily exploited in the health care sector because they affect the production of enzymes that break down protein (Abd El-Hack et al. 2021).

### 1.4.5 Mechanical Characteristics

The mechanical characteristics of NPs are determined by what kind of link that connects the atomic particles, which might be covalent, ionic, or metallic. This means that although NPs elements are robust, robust, and flexible, the existence of contaminants affects their quality. Because NPs are more durable, resilient, and more immune to deterioration, these are utilized as ignition connectors in automobiles (Akhil et al. 2023).

### 1.4.6 Optical Characteristics

Because NPs remain tiny in order to trap particles and induce classical impacts, particles frequently display surprising optical features. Au–NPs, which appear as a shade of red to black, are an excellent illustration of this phenomenon (Iqbal et al. 2021).

### 1.4.7 Electrical Characteristics

Conductance represents a particle's characteristic in materials, whereas resistance is the opposite of conductance. Metals have extremely little resistance; however, nanosized particles exhibit elevated conductivity as electrons disperse across particle restrictions, leading to considerable impedance (Wu et al. 2022).

## 1.5 Various Usage of Nanomaterials

The importance of nanomaterials with various functionalities within our natural surroundings has been demonstrated in Figure 1.4. The NPs rapidly remediated refractory organic molecules through microbial activity. Because of its discontinuous framework, zerovalent iron demonstrated promising outcomes for neutralizing heavy metals. Aside from zero-valent metals, numerous NPs are made utilizing microbes, fungi, algal cells, and botanical compounds. The organic elements found in the environment, such as flavonoids, alkaloid compounds, operate as effective reduction factors for metals utilized for the creation of NPs.

The core methods and workings that comprise the "sustainable" synthetic technique, particularly for metal and metal oxide NPs, such as gold, silver NPs (Samrot et al. 2021), have been previously detailed in several publications.

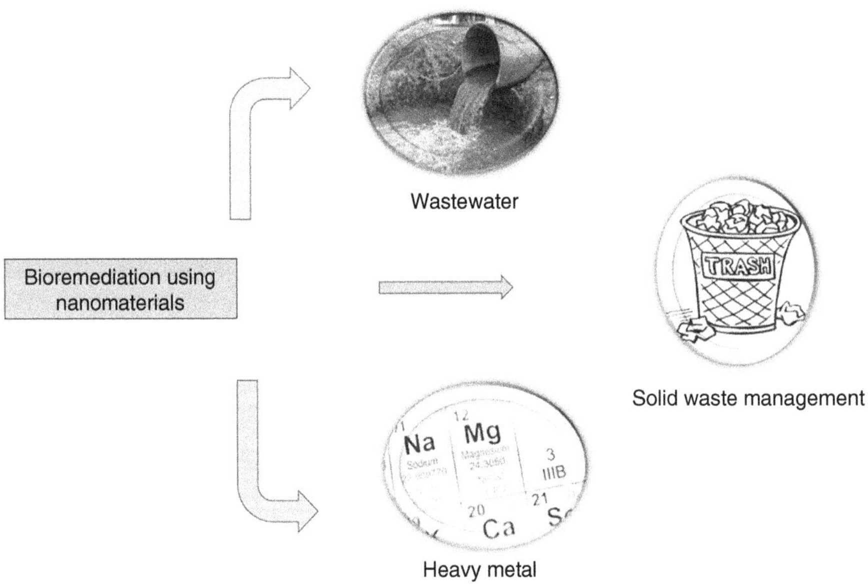


**Figure 1.4** Different applications of nanomaterials in bioremediation (*Source:* Vikanksha).

## 1.6 Mycobial Bioremediation of Heavy Metals from Wastewater

The reason why heavy metals are considered to fall within an assortment of biological contaminants is owing to their tendency to have a harmful effect on many parts of the ecosystem, including people, livestock, and vegetation (Vardhan et al. 2019). Human endeavors, such as farming, metal plating, distillations, metallurgical operations, and excavation, have increased the levels of toxic metals such as Pd, Cd, Co, Cr, As, and Ni in both water and soil to dangerous amounts. The bioremediation strategy for eliminating certain harmful chemicals has proven to be both environmentally and financially feasible.

Fungi play an important part in nanotechnology exploration. Table 1.2 depicts different nano microbial particles produced. Because of its biological metabolism of metals capabilities, NPs have attracted increased interest for the examination of naturally occurring synthesis of metal NPs (Mustapha et al. 2022). Fungi additionally demonstrate effective extracellular enzymatic release, resulting in significant generation of enzymes. Heavy metals are particularly harmful to microbes because they limit the functioning of metal enzymes that aid in the production of reactive oxygen species. Fungi utilize a variety of methods to achieve heavy metal detoxification, including restricting the entry into cells through carriers, intracellular accumulation in vacuoles, and enzyme detoxifying via redox enzymes (Figure 1.5). The eliminated metals are either released in minor amounts or retained in tissues such as metal ions. Whenever employing nanomaterials, it is critical to understand their characteristics (Benhaddouch et al. 2022). Transmission Electron Microscopy (TEM), Scanning Electron Microscopy (SEM), Fourier Transform Infrared spectroscopy (FTIR), Ecological remediation, managing solid waste, metal biosensors, antibacterial, heavy metal removal, and dye decolorization are just a few of the uses for NPs in scientific research. Since NPs possess a large contact/volume proportion, they are being studied for their ability to

**Table 1.2** Different nano microbial nanoparticles synthesized.

| Fungal strain | Nanoparticles |
|---|---|
| *Fusarium oxysporum OSF18* | Copper nanoparticles |
| *Trichoderma asperellum* | |
| *Penicillium chrysogenum* | |
| *Aspergillus niger* | Gold nanoparticles |
| *Fusarium solani* | |
| *Flammulina velutipes* | |
| *Magnusiomyces ingens* | |
| *Trichoderma* sp. | |
| *Pleurotus ostreatus* | |
| *Ganoderma lucidum* | Silver nanoparticles |
| *Penicillium chrysogenum* | |
| *Lentinus tuberregium* | |
| *Penicillium chrysogenum* | |
| *Alternaria tenuissima* | |
| *Ganoderma sessiliforme* | |
| *Periconium* sp. | ZnO nanoparticles |
| *Alternaria tenuissima* | |
| *Penicillium chrysogenum* | |
| *Aspergillus terreus* | |
| *Fusarium keratoplasticum* | |
| *Aspergillus flavus* | $TiO_2$ nanoparticles |
| *Aspergillus flavus* | |

(*Source:* Vikanksha)

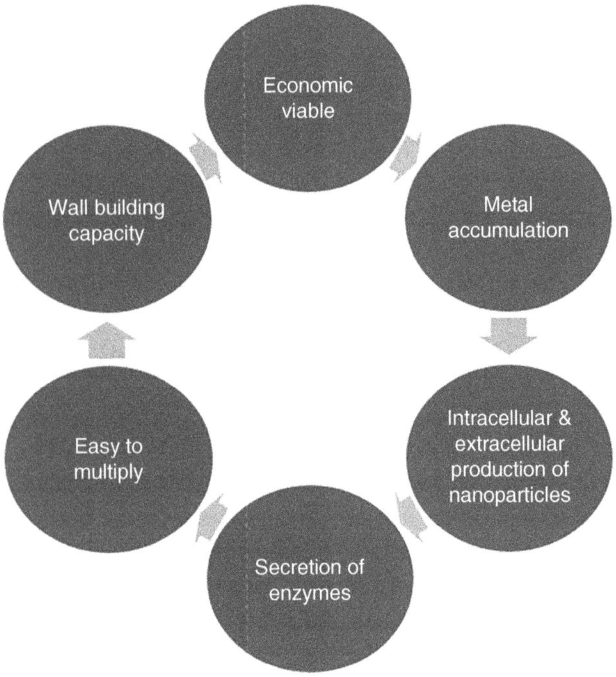


**Figure 1.5** Benefits of fungal nanoparticles (*Source:* Vikanksha).

recognize and eliminate contaminants in diverse environments such as wastewater, soil, and debris (Singh et al. 2022). Current pathogenic disinfection techniques include the development and use of novel antibacterial compounds that include vegetation or microorganisms, novel synthetic compounds, or variants. Metal NPs possess uncommon adsorption characteristics and dimensions because of their differing amounts of receptive surface areas and disorganized exterior areas, resulting in considerably larger exterior regions compared to bulk fragments, with a significant impact on the adsorption for suspended heavy metals. As a result, the current work concentrated on mycobial synthesis and potential important applications for pathogenic disinfecting and heavy metal elimination. Along with contamination prevention and the introduction of unconventional water resources.

## 1.7 Benefits of Mycobial Bioremediation

Bioremediation is the systematic utilization of organisms, such as bacteria, fungus, algae, and different microbes, for pollution removal. The key benefits of bioremediation include environmental durability and long-term viability, as the process lacks poisonous, unstable, or dangerous substances. Bioremediation makes use of biologically active compounds such as catalysts to break down complicated organic contaminants into harmless derivatives (Gaur et al. 2018).

Microorganisms such as fungi extracted from polluted locations use their chemical reactions to remove contaminants. The majority of contaminants, from simple inorganic contaminants and heavy metals, to more complicated volatile hydrocarbons and chemical dyes, are capable of being safely eliminated through the method of bioremediation (Rao et al. 2010). A further aspect on that method is that it is inexpensive as it comes from the use of resources from the environment. When contrasted with alternative ways, this procedure is substantially more effective and requires fewer worker oversight. Myconanotechnology is a rapidly growing sector with applications in multiple nanotechnology disciplines. Mycosynthesis of NPs by fungi caught the interest of experts in microbiology as well as different academics because of their vast potential in a variety of disciplines. Since a result of these genuine characteristics, NPs are useful in numerous subjects as catalysts, energy components, grooming products, fluids, coatings, packaging for food, agricultural chemicals, clothing technology, optical science, electronic devices, ecological detecting, delivery of genes, and biodetection of infectious agents, the application of nanomedicine tumors damage, visualization, resonance imaging, water treatments, farming investigations, and various other usages (Rasmussen et al. 2010).

## 1.8 Constraints of Mycobial Bioremediation

Unquestionably, mycobial bioremediation has advantages in terms of preliminary concentration of contaminants. Nonetheless, a number of drawbacks related to this method should have been addressed. The use of bioremediation is particularly suitable for recyclable contaminants. Nonbiodegradable contaminants, such as heavy metals, are more difficult to decompose or eliminate (Tufail et al. 2022). An essential drawback of the said technique involves smaller applicability toward the direct remediation of contaminants compared to the soils possessing limited permeability. Occasionally, inadequate breakdown of macromolecular contaminants results in the formation of harmful compounds that require additional remediation approaches. Bioremediation constitutes a slow management, which implies it can require a while to achieve the intended consequences (Ambaye et al. 2023).

## 1.9 Conclusion and Future Prospects

In summary, the green synthesis of NPs utilizing fungi especially is widely favored because it benefits the enormous secretory compounds that they produce, the simplicity and ease with which they can be scaled out, the ease with which they can be processed, affordability, and the handling of biomass. Green biosynthesis of NPs includes myconanotechnology, which emphasizes the relationship among mycology and nanotechnology. Since it is a biofactory, fungi can readily produce NPs by resolving their charged particles in suspension using both an extracellular or intracellular method. To enhance the mycosynthesis of NPs and regulate their form and dimensions, various external variables, such as fungal biomass, humidity, optimum amounts, time interaction, pH, and the existence of a specific protein have been thoroughly studied for maximizing the amount of created NPs, and to achieve optimal synthesis of NPs as well. In addition, physicochemical evaluation of mycosynthesized NPs is regarded as one of the most essential variables for studying the properties of NPs, as it reveals the primary characteristics of the manufactured NPs. Because of their distinct biochemical and physical properties, mycosynthesized NPs had an extensive list of potential uses in numerous domains. Potential purposes for mycosynthesized NPs include the farming industry, treatment of effluents, heavy metals remediation, textile manufacturing, food packaging and its security, biological remediation for ecological hazardous compounds, and therapeutic uses. Without any question, the biosynthesis of NPs employing fungus provides all intriguing opportunities for making our everyday lives as environment friendly as attainable. Global support for microbial bioremediation-based strategies to the reduction of widespread environment pollutants has skyrocketed. In conclusion, we discovered that the various bioremediation techniques may serve a substantial part within environmental sanitation. The majority of pollutants are removable through soil and water. It seemed obvious through the referenced material that mycobial bioremediation techniques could degrade harmful organic contaminants to less complicated and beneficial compounds. Observations suggest the method is just as efficient for treating wastewater from industries. With the goal to increase the efficacy of bioremediation, we believe this is necessary to close the existing knowledge deficits among microbial biology, biological functioning, and genetic features. Currently, nanotechnology was recognized as a substantial technique which can be combined with bioremediation to enhance its efficacy.

## References


Abd El-Hack, M.E., El-Saadony, M.T., Swelum, A.A. et al. (2021). Curcumin, the active substance of turmeric: its effects on health and ways to improve its bioavailability. *Journal of the Science of Food and Agriculture* 101 (14): 5747–5762.

Akhil, U.V., Radhika, N., Saleh, B. et al. (2023). A comprehensive review on plant-based natural fiber reinforced polymer composites: fabrication, properties, and applications. *Polymer Composites* 44 (5): 2598–2633.

Alananbeh, K.M., Al-Refaee, W.J., and Al-Qodah, Z. (2017). Antifungal effect of silver nanoparticles on selected fungi isolated from raw and waste water. *Indian Journal of Pharmaceutical Sciences* 79 (4): 559–567.

Ambaye, T.G., Vaccari, M., Franzetti, A. et al. (2023). Microbial electrochemical bioremediation of petroleum hydrocarbons (PHCs) pollution: recent advances and outlook. *Chemical Engineering Journal* 452: 139372.

Benhaddouch, T.E., Pinzon, S.K., Landi, D.M.C. et al. (2022). Micro-fuel cell principal biosensors for monitoring transdermal volatile organic compounds in humans. *ECS Sensors Plus* 1 (4): 041602.

Bhardwaj, K., Sharma, A., Tejwan, N. et al. (2020). Pleurotus macrofungi-assisted nanoparticle synthesis and its potential applications: a review. *Journal of Fungi* 6 (4): 351.

Bishayee, B., Chatterjee, R.P., Ruj, B. et al. (2022). Strategic management of nitrate pollution from contaminated water using viable adsorbents: an economic assessment-based review with possible policy suggestions. *Journal of Environmental Management* 303: 114081.

Dekura, S., Kobayashi, H., Kusada, K., and Kitagawa, H. (2019). Hydrogen in palladium and storage properties of related nanomaterials: size, shape, alloying, and metal-organic framework coating effects. *ChemPhysChem* 20 (10): 1158–1176.

García-Depraect, O., Bordel, S., Lebrero, R. et al. (2021). Inspired by nature: microbial production, degradation and valorization of biodegradable bioplastics for life-cycle-engineered products. *Biotechnology Advances* 53: 107772.

Gaur, N., Narasimhulu, K., and PydiSetty, Y. (2018). Recent advances in the bio-remediation of persistent organic pollutants and its effect on environment. *Journal of Cleaner Production* 198: 1602–1631.

Gayathiri, E., Prakash, P., Karmegam, N. et al. (2022). Biosurfactants: potential and eco-friendly material for sustainable agriculture and environmental safety—a review. *Agronomy* 12 (3): 662.

Gola, D., Tyagi, P.K., Arya, A. et al. (2021). Antimicrobial and dye degradation application of fungi-assisted silver nanoparticles and utilization of fungal retentate biomass for dye removal. *Water Environment Research* 93 (11): 2727–2739.

Iqbal, M.Z., Ali, I., Khan, W.S. et al. (2021). Reversible self-assembly of gold nanoparticles in response to external stimuli. *Materials & Design* 205: 109694.

Kaur, R., Thakur, N.S., Chandna, S., and Bhaumik, J. (2020). Development of Agri-biomass based lignin derived zinc oxide nanocomposites as promising UV protectant-cum-antimicrobial agents. *Journal of Materials Chemistry B* 8 (2): 260–269.

Khorrami, S., Zarepour, A., and Zarrabi, A. (2019). Green synthesis of silver nanoparticles at low temperature in a fast pace with unique DPPH radical scavenging and selective cytotoxicity against MCF-7 and BT-20 tumor cell lines. *Biotechnology Reports* 24: e00393.

Kosynkin, D.V., Higginbotham, A.L., Sinitskii, A. et al. (2009). Longitudinal unzipping of carbon nanotubes to form graphene nanoribbons. *Nature* 458 (7240): 872–876.

Kotrange, H., Najda, A., Bains, A. et al. (2021). Metal and metal oxide nanoparticle as a novel antibiotic carrier for the direct delivery of antibiotics. *International Journal of Molecular Sciences* 22 (17): 9596.

Krishnaveni, A., Chinnasamy, S., Elumalai, J., and Muthaiyan, P. (2020). Sugar industry wastes as wealth of organic carbon for soil. *Environmental Factors Affecting Human Health*, IntechOpen 6471: 137–160.

Kumar, V. and Yadav, S.K. (2009). Plant-mediated synthesis of silver and gold nanoparticles and their applications. *Journal of Chemical Technology & Biotechnology: International Research in Process, Environmental & Clean Technology* 84 (2): 151–157.

Kumar, P., Huo, P., Zhang, R., and Liu, B. (2019). Antibacterial properties of graphene-based nanomaterials. *Nanomaterials* 9 (5): 737.

Mabrouk, M., Das, D.B., Salem, Z.A., and Beherei, H.H. (2021). Nanomaterials for biomedical applications: production, characterisations, recent trends and difficulties. *Molecules* 26 (4): 1077.

Mazurkiewicz, W., Podrażka, M., Jarosińska, E. et al. (2020). Paper-based electrochemical sensors and how to make them (work). *ChemElectroChem* 7 (14): 2939–2956.

Moustafa, M.T. (2017). Removal of pathogenic bacteria from wastewater using silver nanoparticles synthesized by two fungal species. *Water Science* 31 (2): 164–176.

Mustapha, T., Misni, N., Ithnin, N.R. et al. (2022). A review on plants and microorganisms mediated synthesis of silver nanoparticles, role of plants metabolites and applications. *International Journal of Environmental Research and Public Health* 19 (2): 674.

do Nascimento, J.M., Cruz, N.D., de Oliveira, G.R. et al. (2021). Evaluation of the kinetics of gold biosorption processes and consequent biogenic synthesis of AuNPs mediated by the fungus *Trichoderma harzianum*. *Environmental Technology & Innovation* 21: 101238.

Pati, R., Mehta, R.K., Mohanty, S. et al. (2014). Topical application of zinc oxide nanoparticles reduces bacterial skin infection in mice and exhibits antibacterial activity by inducing oxidative stress response and cell membrane disintegration in macrophages. *Nanomedicine: Nanotechnology, Biology and Medicine* 10 (6): 1195–1208.

Pramanik, S., Mohanto, S., Manne, R. et al. (2021). Nanoparticle-based drug delivery system: the magic bullet for the treatment of chronic pulmonary diseases. *Molecular Pharmaceutics* 18 (10): 3671–3718.

Rao, M.A., Scelza, R., Scotti, R., and Gianfreda, L. (2010). Role of enzymes in the remediation of polluted environments. *Journal of Soil Science and Plant Nutrition* 10 (3): 333–353.

Rasmussen, J.W., Martinez, E., Louka, P., and Wingett, D.G. (2010). Zinc oxide nanoparticles for selective destruction of tumor cells and potential for drug delivery applications. *Expert Opinion on Drug Delivery* 7 (9): 1063–1077.

Ray, P., Lakshmanan, V., Labbé, J.L., and Craven, K.D. (2020). Microbe to microbiome: a paradigm shift in the application of microorganisms for sustainable agriculture. *Frontiers in Microbiology* 11: 622926.

Roy, S., Mukherjee, T., Chakraborty, S., and Das, T.K. (2013). Biosynthesis, characterisation & antifungal activity of silver nanoparticles synthesized by the fungus *Aspergillus foetidus* MTCC8876. *Digest Journal of Nanomaterials and Biostructures* 8 (1): 197–205.

Salvadori, M.R., Ando, R.A., Oller Do Nascimento, C.A., and Correa, B. (2014). Bioremediation from wastewater and extracellular synthesis of copper nanoparticles by the fungus *Trichoderma koningiopsis*. *Journal of Environmental Science and Health, Part A* 49 (11): 1286–1295.

Samrot, A.V., Sahithya, C.S., Selvarani, J. et al. (2021). A review on synthesis, characterization and potential biological applications of superparamagnetic iron oxide nanoparticles. *Current Research in Green and Sustainable Chemistry* 4: 100042.

Sanchez, J.F., Somoza, A.D., Keller, N.P., and Wang, C.C. (2012). Advances in *Aspergillus* secondary metabolite research in the post-genomic era. *Natural Product Reports* 29 (3): 351–371.

Sindhu, R.K., Chitkara, M., and Sandhu, I.S. (2021). *Nanotechnology: Principles and Applications*. CRC Press.

Singh, N.B., Agarwal, A., De, A., and Singh, P. (2022). Coal fly ash: an emerging material for water remediation. *International Journal of Coal Science & Technology* 9 (1): 44.

Sudha, P.N., Sangeetha, K., Vijayalakshmi, K., and Barhoum, A. (2018). Nanomaterials history, classification, unique properties, production and market. In: *Emerging Applications of Nanoparticles and Architecture Nanostructures* (ed. A. Barhoum and A.S.H. Makhlouf), 341–384. Elsevier.

Tripathi, R.M., Shrivastav, B.R., and Shrivastav, A. (2018). Antibacterial and catalytic activity of biogenic gold nanoparticles synthesised by *Trichoderma harzianum*. *IET Nanobiotechnology* 12 (4): 509–513.

Tufail, M.A., Iltaf, J., Zaheer, T. et al. (2022). Recent advances in bioremediation of heavy metals and persistent organic pollutants: a review. *Science of the Total Environment* 850: 157961.

Van den Brule, T., Punt, M., Teertstra, W. et al. (2020). The most heat-resistant conidia observed to date are formed by distinct strains of *Paecilomyces variotii*. *Environmental Microbiology* 22 (3): 986–999.

Vardhan, K.H., Kumar, P.S., and Panda, R.C. (2019). A review on heavy metal pollution, toxicity and remedial measures: current trends and future perspectives. *Journal of Molecular Liquids* 290: 111197.

Voliani, V. (2020). *Gold Nanoparticles: An Introduction to Synthesis, Properties and Applications*. Walter de Gruyter GmbH & Co KG.

Wu, L., Wang, Y., Guo, X. et al. (2022). Interface science in polymer-based composite solid electrolytes in lithium metal batteries. *SusMat* 2 (3): 264–292.

Zhou, X., Hao, Y., He, X. et al. (2020). Protean morphology of waterborne polyurethane dispersion: an overview of nanoparticles from sphere to irregular elongated shape. *Progress in Organic Coatings* 146: 105742.

# 2

# Microbial Enzymes in Biodegradation of Organic Pollutants: Mechanisms and Applications

*Bharati Lap[1], Ashim Debnath[1], Gourav Kumar Singh[2], Priyank Chaturvedi[2], Joy Kumar Dey[3], and Sajal Saha[4]*

[1] *Department of Genetics and Plant Breeding, Faculty of Agricultural Sciences, Rajiv Gandhi University, Doimukh, Arunachal Pradesh, India*
[2] *Department of Environmental Science, ITM University, Gwalior, Madhya Pradesh, India*
[3] *Krishi Vigyan Kendra-Sepahijala, CAU (I), Latiacherra, Tripura, India*
[4] *Department of Genetics and Plant Breeding, SAS, Nagaland University, Medziphema, Nagaland, India*

## 2.1 Introduction

The interdependence of Earth's quality of life and the state of its environment is undeniable. However, the relentless march of science, technology, and industry has ushered in an era where we witness the release of diverse pollutants, from raw sewage to the formidable challenge of nuclear waste, into our precious ecosystems. This unwelcome trend presents a formidable threat not only to the biogeochemical cycle but also to the very survival of humankind. Traditional methods for removing pollutants have its own limitations, including high cost, non-specificity, and the possibility of producing secondary contamination. As a result, there has been increased interest in environmentally friendly and biological treatments, which are collectively referred to as bioremediation (Sokal et al. 2022; Bhadouria et al. 2020; Giri et al. 2021). Bioremediation is the process of transforming pollutants into nonhazardous substances facilitated by microorganisms as the transformative agents (Gond et al. 2021; Gupta et al. 2020). Microbial enzymes produced by a diverse array of microorganisms exhibit impressive potential for efficient bioremediation as they play a crucial role to detoxify and mitigate environmental pollution, particularly in cases involving contaminants such as hydrocarbons, pesticides, dyes, and various organic pollutants (Singh et al. 2021; Mishra et al. 2021). These enzymes facilitate the breakdown and removal of pollutants, making bioremediation a sustainable and environmentally friendly approach to address pollution issues. Genetic engineering has further broadened the horizons of bioremediation, with genetically modified microbes such as *Pseudomonas fluorescens* and *Chlamydomonas reinhardtii* (Rafeeq et al. 2023). This chapter delves into the intricate mechanisms that underpin the catalytic abilities of microbial enzymes in breaking down a broad spectrum of organic compounds. Conventional pollutant removal methods often lead to secondary pollutants, whereas microbial enzymes offer a targeted and efficient approach. These mechanisms encompass oxidation, hydrolysis, reduction, conjugation reactions, bioaccumulation, bioaugmentation, biosorption, biostimulation, bioleaching, and precipitation (Ayilara and Babalola 2023; Pandey et al. 2023). Notable enzymes (cytochrome P450s, laccases, hydrolases, dehalogenases, dehydrogenases, proteases, and lipases) exhibit potential in degrading a wide range of pollutants, including polymers, aromatic hydrocarbons, halogenated compounds, dyes, detergents, and agrochemicals through oxidation, reduction, elimination, and ring-opening mechanisms.

The enzymatic prowess of microorganisms, meticulously target pollutants and convert them into benign forms and therefore, has extensive and impactful applications. However, the effectiveness of bioremediation hinges on environmental conditions conducive to microbial growth and activity, often necessitating the manipulation of various parameters to expedite the degradation process. In the realm of sustainable materials, microbial enzymes are pioneering the development of biodegradable plastics, offering respite from the ecological burden of synthetic polymers. Their contributions extend to oil spill mitigation, the enhancement of phytoremediation practices, and various industrial applications, underscoring their

*Microbes Based Approaches for the Management of Hazardous Contaminants*, First Edition. Edited by Ajay Kumar, Livleen Shukla, Joginder Singh, and Luiz Fernando R. Ferreira.

versatility and significance. By unraveling the biochemical intricacies of these enzymatic processes, our chapter would shed light on the myriad applications of microbial enzymes in addressing environmental pollution and industrial challenges.

### 2.1.1 Mechanism of Microbial Enzymes in Bioremediation of Organic Pollutants

Microorganisms such as bacteria, fungi, archaea, and algae play the central role in breaking down organic compounds into simpler forms. Hydrolases, oxidoreductases, dehalogenases, oxygenases, and transferases are few important classes of microbial enzymes responsible for the degradation of most of the toxic pollutants in the environment (Figure 2.1). These enzymes facilitate the degradation of toxic environmental pollutants by lowering the activation energy. Enzymatic degradation is more specific than others because the process of degradation completely depends on the interaction of enzyme–substrate (Saravanan et al. 2021; Singh et al. 2020). These mechanisms are broadly categorized under immobilization and mobilization. Mobilization process involves, enzymatic oxidation, bioaugmentation, biostimulation, bioleaching, and enzymatic reduction procedure. While, immobilization includes bioaccumulation, biosorption complexation, and precipitation (solidification) (Ayilara and Babalola 2023).

#### 2.1.1.1 Fungi

Fungi are great assets due to their functional capacity of benignly decomposing pollutants from soil and water. Various enzymatic processes are operationalized, making their enzymatic systems highly versatile. Fungi manufacture an assortment of extracellular enzymes that typically degrades broad range of pollutants by the action of extracellular enzymes such as manganese peroxidase, lignin peroxidase, and laccases in the environment under aerobic conditions. Lignin peroxidases (LiPs) are important in the decomposition of lignin, a polymeric constituent of wood and plant matter, Manganese Peroxidases (MnPs) cooperate with LiPs to destroy lignin and other aromatic molecules and versatile peroxidase (VP) that have the high potential to degrade toxic substances in nature (Rovaletti et al. 2023). Fungi also synthesize enzymes, such as cellulases and hemicellulases, to deconstruct cellulose and hemicellulose, respectively, which are organic molecules located in the walls of plant cells. These compounds are crucial for detaching diverse organic pollutants, such as paper, agricultural by-products, and certain synthetic components. Furthermore, polycyclic aromatic hydrocarbons (PAHs) are also susceptible to degradation as fungi transform PAHs by co-metabolizing with enzymes such as cytochrome P450 monooxygenase (Espinosa-Ortiz et al. 2022). Enzyme laccases are extracellular multicopper oxidases that use molecular oxygen to oxidize a wide variety of organic compounds ranging from phenolic and nonphenolic lignin-related compounds to highly recalcitrant

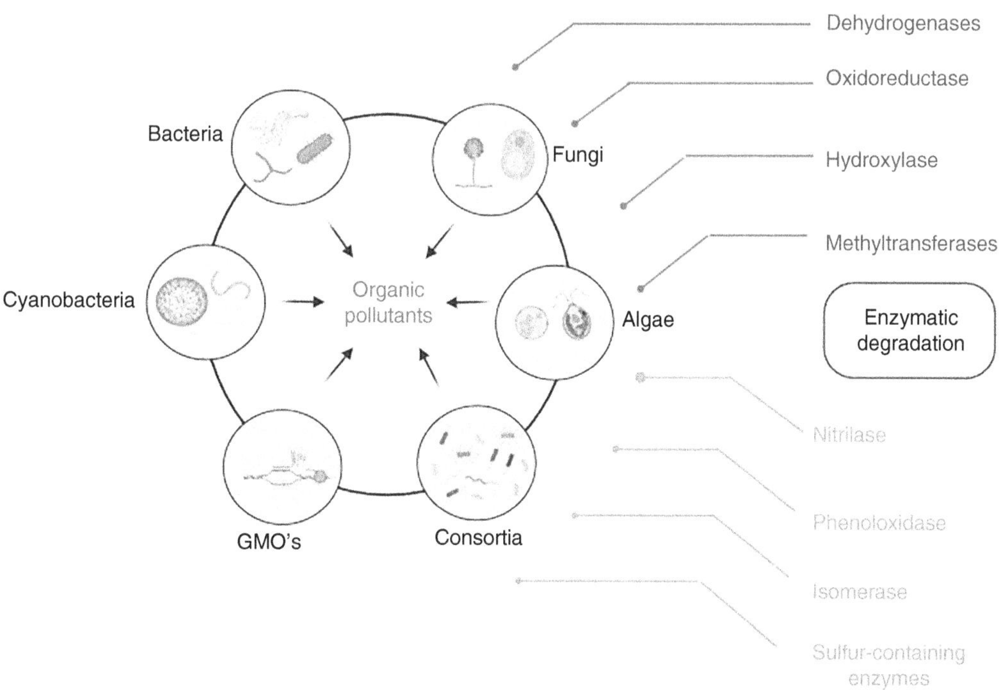


**Figure 2.1** Microbes-mediated enzymatic degradation of organic pollutants.

environmental pollutants, by a radical-catalyzed reaction mechanism (Patel et al. 2020). Laccases from white rot Basidiomycetes (e.g., *Phanerochaete chrysosporium*) and other fungal genera such as *Trichoderma, Candida, Agrocybe, and Aspergillus* constitute great importance in the biodegradation of organic pollutants, particularly the breakdown of complex organic molecules (Sánchez, 2022). Additionally, they have the ability to generate lipases and esterases, which are capable of breaking down ester ties of assorted pollutants, such as certain pesticides and industrial chemicals. Fungi feature a host of redox enzymes, including dehydrogenases and reductases that are crucial for the overall outcome of reduction–oxidation (redox) processes involved in the decomposition of organic compounds (Saravanan et al. 2021).

Yeast is not as functional as some fungi or other microorganisms when it comes to the biodegradation of intricate organic impurities. However, they do contribute in the degradation of some organic pollutants. While yeast may not boast as much enzymatic capability as certain microorganisms, but it still participates in the biodegradation of certain organic substances. They have the ability to carry out both reduction and oxidation responses. Certain types of yeast are capable of reducing nitroaromatic elements, resulting in their transformation to relatively innocuous amines. Yeast also produces esterases and lipases, two kinds of enzymes that breaks down ester and lipid bonds in organic chemicals. *Yarrowia lipolytica, Meyerozyma guilliermondii, Candida utilis, Magnusiomyces capitatus, Trichosporon asahii, Guehomyces pullulans, Thermomyces lanuginosus,* and *Candida rugose* are very good primary producers of lipases (Kumar et al. 2023). Yeast utilizes the catechol pathways for decomposing quite a few phenolic compounds (Filipowicz et al. 2020).

#### 2.1.1.2 Bacteria

The enzymatic methods to breakdown bacteria's organic waste involve a progression of enzyme-initiated reactions that deconstruct complex organic compounds into simpler, less noxious molecules. Different kinds of enzymes are utilized depending on the type of pollutant a bacterium is metabolizing. They can metabolize aliphatic saturated and unsaturated hydrocarbons such as alkanes, cycloalkanes, aldehydes, and ethers by using variety of degrading pathways, including catechol, protocatechuate, gentisate, benzoate, and biphenyl, to remove aromatic organic contaminants such as polychlorinated biphenyls (PCBs), PAHs, and pesticides from the environment (Kumari and Das 2023).

Bacteria's produce intra and extracellular oxidoreductive enzymes (e.g., azoreductases, laccases, and dichlorophenolindophenol-reductases) for the degradation of pollutant under aerobic and anaerobic condition (Espinosa-Ortiz et al. 2022). Aerobic bacteria harbor numerous oxidative enzymes, such as oxygenases and peroxidases, e.g., methane monooxygenase, ammonia monooxygenase, toluene monooxygenase, and toluene dioxygenase. Among all the other species, *Pseudomonas* spp. stands in great fame for decomposing diverse organic substances, ranging from fuel-based hydrocarbons to pesticidal chemicals, and factory-made materials. *Pseudomonas aeruginosa*, *Pseudomonas fluorescens*, and *Pseudomonas putida* have generously been utilized in bioremediation technology (Kumari and Das 2023). Other bacterial species include *Comamonas, Klebsiella, Burkholderia, Dehalococcoides, Meyerozyma, Xanthomonas, Bacillus, Staphylococcus, Achromobacter, Rhodococcus* and *Acinetobacter* that have the unique genetic makeup to metabolically degrade toxic pollutants (Sánchez 2022). While, *Desulfotomaculum*, *Desulfovibrio*, *Methanospirillum*, and *Methanococcus* are anaerobic bacteria that break down organic contaminants.

Many bacteria employ oxygenases or hydroxylases to incorporate oxygen into organic molecules, increasing their capacity for degradation. For instance, when degrading aromatic compounds such as benzene, toluene, and xylene, dioxygenases are capable of introducing molecular oxygen. Whereas, monooxygenases enzymes incorporate single oxygen atoms into the organic pollutant. An illustrative example is cytochrome P450, which is involved in metabolism of several hydrocarbons, e.g., PAHs. Additionally, *Rhodococcus* spp. can be employed to separate assortments of hydrocarbons, in particular PAHs and chlorinated solvents. Bacterial systems make use of dehalogenase enzymes for extracting halogen atoms (e.g., chlorine and bromine) from organic compounds, a seminal instance being the biodegradation of trichloroethylene (TCE). Meanwhile, *Dehalococcoides* spp. takes part in an exceptionally dearest process, which is none other than the bioremediation of chlorinated substances, some of which are TCE and perchloroethylene (PCE). Additionally, nitroreductases have been observed in diminishing nitro groups ($NO_2$) arising from organic molecules, highlighted by degradation of nitroaromatic compounds. Finally, hydrolysis reactions constitute another important role in this process. Bacteria utilizes esterases and hydrolases to dismantle ester or amide linkages in organic toxins such as pesticides and drugs. And lipases hydrolyze lipids and fat, which are organically based components found in some contaminants. Glutathione S-transferases facilitates the combination of contaminants with glutathione, a tripeptide. This transformation renders the pollutant more permeable to water and facilitates its removal. Bacterial species participating in the breakdown of aromatic compounds such as PAHs employ ring-cleaving enzymes for splitting the aromatic ring of the molecule, dividing it into smaller fragments that can be metabolized even more. Moreover, these microorganisms possess cofactor recycling enzymes that permit the reuse of these compounds. Nicotinamide adenine dinucleotide phosphate hydrogen (NADPH)-dependent dehydrogenases are necessary enzymes for recycling the crucial cofactors Nicotinamide adenine dinucleotide hydrogen (NADH) and NADPH. These factors are vital for many enzymatic procedures utilized in the degradation of pollutants. Different bacteria have adapted to be able to metabolize particular types of organic compounds through biodegradation, a process that is exclusive to each species. As a result, organic contaminants are broken down and converted into by-products, such as carbon dioxide, water, and biomass, which are fitted into the natural biogeochemical cycles.

#### 2.1.1.3 Algae

Algae, known chiefly as the foundation of photosynthesis and a key element in aquatic ecosystems, can likewise participate in the biodegradation of specific organic contaminants by means of varied enzymatic procedures, known as phycoremediation. While these organisms may not be as famous for biodegradation as some bacteria and fungi, they can be crucial components of organic remediation systems in aqueous places. Algal biodegradation occurs mainly in different processes such as corrosion, hydrolysis, penetration, and fouling.

Algal processes and enzymatic actions for biodegrading involves photosynthesis and oxygen production. Algae carry out photosynthesis, which allows them to take in carbon dioxide and emit oxygen. The boost in oxygen levels in aquatic environments enhances the proliferation of aerobic microbes, such as bacteria and fungi, normally associated with the decomposition of organic pollutants. Excretion of enzymes and metabolic products: Algae secrete diverse organic substances, comprising of enzymes, in their local environment. These enzymes can promote the decomposition of organic matter, which can include certain organic pollutants, carried out with techniques such as oxidation and hydrolysis. Bioaccumulation and biosorption: Algae can comprehend and absorb organic pollutants from the encircling water. Though this is not exact enzymatic decomposition, it can be a significant process for amassing and sheltering organic pollutants, making them less accessible in the aquatic environment. Biosorption is a physicochemical process that is rather dependent on different mechanisms, including absorption, adsorption, surface complexation, ion exchange, and precipitation. Emission of secondary metabolites: Some algae produce secondary metabolites, which can display allelopathic characteristics. These compounds are capable of impeding the growth of other microorganisms, which may affect the decomposition of organic waste by competing microorganisms. Moreover, these substances may also influence nitrogen and phosphorus uptake. Algae need essential elements, such as nitrogen and phosphorus, for their development. The uptake of these nutrients by algae can influence the accessibility of these constituents for other microorganisms engaged in decay. *Chlorella vulgaris, Scenedesmus quadricauda, Skeletonema costatum, Selenastrum capricornutum, Prototheca zopfii, Chlamydomonas moewusii, Ankistrodesmus braunii and S. quadricauda, Chlamydomonas reinhardtii* (Touliabah et al. 2022). Algae can develop symbiotic affiliations with bacteria and fungi. These microbial aggregations can assist in the decomposition of organic contaminants, with each organism taking on a particular task in the technique. It is essential to be aware of that the biodegradation power of algae is contingent on the context and can differ influenced by the species of algae, the pollutant being studied, and the environmental factors.

Algae and Cyanobacteria are able to act as biodegrades owing to their capability of assimilating and eliminating organic waste, by developing several detoxifying mechanisms, including biosorption, bioaccumulation, biotransformation, biomineralization, and *in situ* and/or *ex situ* biodegradation. Laccase and polyphenol oxidase of the cyanobacterium *Phormidium valderianum* have been identified as the enzymes responsible for most phenol biodegradation. Other cyanobacteria include *Westiellopsis* sp., *Spirulina* sp., *Oscillatoria* sp., *Nostoc linckia, Anabaena flos-aquae, Aphanothece conferta, Phormidium* sp., *Nostoc* sp., *Synechocystis aquatilis, Spirulina maxima*, and *Aulosira fertilissima* (Touliabah et al. 2022).

#### 2.1.1.4 Other Microbes

***Archaeal Species*** Archaeal species though not so widely recognized as bacteria and fungi, can also be significant contributors to the breakdown of organic pollutants in their particular habitats. Their unique metabolic characteristics allows them to flourish in extreme conditions, and some of their enzymatic modes can be beneficial to biodegradation. Important one being the methanogenic archaea that assist in the decomposition of organic debris in anaerobic conditions. These Archaea, referred to as "methanogens," possess the propensity to manufacture methane gas ($CH_4$) as a resultant of their metabolic processes. Archaea's have the ability to break down various organic pollutants, i.e., hydrocarbons and volatile organic compounds (VOCs), by utilizing them as a carbon source; methane and carbon dioxide are produced as the two by-products in the course of their functioning. Anaerobic oxidation of methane (AOM) microorganisms is a group of methanotrophs that utilize high-valence compounds as electron acceptors to oxidize $CH_4$ and generate $CO_2$. As typic AOM microorganisms, Candidatus Methylomirabilis oxyfera (*M. oxyfera*) bacteria metabolize methane via an intra-aerobic pathway and Candidatus Methanoperedens nitroreducens (*M. nitroreducens*) archaea metabolize methane via a reverse methanogenesis pathway (Fu et al. 2019).

Astonishingly, certain species of halophilic archaea (e.g., *Marinobacter hydrocarbonoclasticus*) have demonstrated the capacity to dehalogenate halogenated organic compounds. This biochemical phenomenon involves the subtraction of halogen substances, such as chlorine or bromine, from pollutant particles, consequently reducing their damaging properties. Additionally, hydrolyzing and fermenting components have been indicated to influence the process. Certain archaea have the aptitude to catalyze the hydrolysis of organic molecules, resulting in the dissolution of their integral components into easier compounds. Under anaerobic fermentation episodes, they can get involved in the catabolism of organic matter, and convert it to a number of products such as organic acids, alcohols, and gases. Additionally, archaea can also be implicated in sulfate reduction. It is noteworthy that sulfate-reducing archaea (SRAs) play a specialized role in the anaerobic biodegradation of organic contamination in sulfate-rich conditions. These microorganisms can use organic compounds as their source of carbon while reducing sulfate to sulfide. Their contribution, when compared to those of other microbial components such as bacteria and fungi, is typically highly specific to certain environment factors and pollutant substrates. In some cases, cooperative action between these diverse organisms in microbial consortia may be the best approach for rapid and expeditious pollutant degradation.

***Microbial Consortia*** Microbial consortia consist of different types of microorganisms (bacteria, fungi, archaea, and sometimes even algae) that often work together to efficiently biodegrade complex organic pollutants. These consortia can harness a wide range of enzymatic mechanisms and metabolic pathways to break down organic pollutants. Microbial degradation may be most successful when small organisms act together in consortia, as distinct species can manufacture adjunctive enzymes and metabolic abilities. Synergistic relationships can enable comprehensive abatement of dense pollutants. In bioremediation and wastewater remediation practices, multiple microorganisms such as bacteria, fungi, and yeast are combined to achieve stronger degradation of a range of organic pollutants through the collaborative enzyme activity of the organisms. The cocultivation of the microbial consortium is more effective than single microbe as it degrades pollutants faster, and it significantly enhances the biodegradation of pollutants.

Several studies have reported increasing rate of biodegradation by incorporating a microbial consortium of many enzymes such as lipase, esterase, alcalase, proteinase K, and cutinases are involved in hydrolysis of organic materials. Combinations of different bacteria such as *Clostridium, Lactobacilli, Salmonella, Actinobacteria, Pseudomonas*, and *Escherichia coli* have been found to have the inherent ability to degrade pollutants (Bhatt et al. 2021). Fungal-bacterial cocultures, mostly relying on commensalism and co-metabolism degradation of chlorobenzene (volatile organic compound) using a mixed culture of *Trichoderma viride* and *Ralstonia pickettii* (Espinosa-Ortiz et al. 2022). Fungi are able to form symbiotic affiliations with bacteria, creating microbial assemblages, which degrade complex organic contaminants with extreme efficacy. Effectiveness of other consortia-like coculture strains of *Burkholderiales, Sphingobacteriales, Rhodobacterales, Actinomycetals,* and *Rhizobiales* against plastic (Gaytán et al. 2020) and strains of *Meyerozyma guilliermondii, Meyerozyma caribbica, Debaryomyces hansenii,* and *Vanrija humicola* against azo dyes (Ali et al. 2022) has been widely studied. Such microbial collaborations result in an increased total efficiency in biodegradation. The enzymatic processes employed by fungi in biodegradation are often modified and specialized to certain substrates. As a consequence, fungi are essential in bioremediation efforts, particularly in the purification of sullied soils, liquids, and other surroundings comprising intricate organic pollutants.

***GMO's*** Genetically engineered or specialized microorganisms are introduced into polluted locations to speed up biodegradation by a process known as bioaugmentation. Such strains are created to break down particular pollutants effectively. Genetically modified microorganisms (GMO's) are employed in bioremediation as a means of enhancing the decomposition of organic residues. These engineered microbes are tailored using biotechnological tools of Clustered regularly interspaced short palindromic repeats- CRISPR associated protein (CRIPSR-Cas9), Transcription activator-like effector (TALENs), and Zinc-finger nucleases (ZFNs) to systematically deconstruct certain types of contaminants. Many microorganisms have been genetically modified and engineered to detoxify organic contaminants that contain chlorine or other halogens, for example, *Dehalococcoides* spp. have been manipulated with the intent of increasing their potency to degrade chlorinated solvents such as TCE and PCE. Two pesticide degrading genes (*linA* and *mpd*) have been integrated with *E. coli* for the degradation of organochlorine and organophosphates (Mishra et al. 2020). Genetic engineering has been implemented for amplifying the potency of microbes in the management of oil spills. *P. putida* has been genetically altered so as to degrade hydrocarbons, comprising of aromatic compounds and petroleum-based contaminants, in a more effectual manner; *Alcanivorax borkumensis* has been used to refine the hydrocarbon-degradation abilities; Naphthalene-degrading bacteria has also been modified to reinforce the cleavage of naphthalene, an ordinarily occurring polycyclic aromatic hydrocarbon (PAH), in *P. fluorescens*; *Acinetobacter* spp. possess certain strains that have been manipulated genetically in order to optimally metabolize phenol and related compounds, which are usually detected as pollutants in industrial wastewater (Rafeeq et al. 2023). *Methylobacterium* spp. have been altered genetically so as to efficiently disintegrate methanol, which is utilized as a commercial solvent and can be a groundwater contaminant. Certain yeasts such as *Saccharomyces cerevisiae* have been overridden to scope out precise catalysts, such as lignin-removal peroxidases, that can lend a hand to the dissipation of lignin in flora tissues and certain chemical pollutants. Certain strains of algae have been modified to fortify their capacity to absorb and break down organic impurities from aquatic ecosystems, including types of industrial chemicals and nutrients such as nitrates and phosphates, for example *C. reinhardtii,* which was genetically modified to produce polyethylene terephthalate hydrolase to degrade polyethylene terephthalate films and terephthalic acid (Anand et al. 2023). Specifically, cyanobacteria have been shown to be particularly effective for the removal of nitrate. Cyanobacteria have been purposefully altered to extract nitrates from aquatic systems by transforming them into innocuous nitrogen gas. At times, engineered microbial collectives have been created, integrating a diversity of microbes with mutual functionalities to effectively process complex amalgamations of organic contaminants. The bioengineered strains of *Pseudomonas*, *Alcaligenes*, *Rhodococcus*, *Burkholderia*, *Ralstonia*, *Achromobacter*, *Dehalococcoides*, *Comamonas*, and *Sphingomonas* are a few examples that have reportedly been actively participating in bioremediation processes (Rafeeq et al. 2023). These bioengineered microorganisms and collaboratives present attractive resolutions for bioremediation endeavors by enlarging the efficiency and exactness of pollutant destruction processes. However, when employing genetically transformed organisms for ecological renewal undertakings, it is essential to contemplate the environmental and administrative effects.

### 2.1.2 Applications of Microbial Enzymes Mediated Bioremediation

Microbial enzyme-based bioremediation is a promising approach due to its specificity, efficiency, and environmental sustainability. Researchers continue to explore new enzymes and microbial strains for its application for improving the effectiveness of bioremediation processes for a wide range of pollutants.

- **Biodegradation of hydrocarbons:** Microbial enzymes, such as hydrocarbon-degrading enzymes (e.g., alkane hydroxylases and aromatic ring-cleaving dioxygenases), help break down hydrocarbon-based pollutants such as oil spills. Bacteria such as *Pseudomonas* and *Bacillus* species are well-known for their ability to produce these enzymes.
- **Organic pollutant removal:** Microbial enzymes, particularly peroxidases and laccases, are involved in the degradation of various organic pollutants, including pesticides, dyes, and pharmaceuticals. Fungi such as *Trametes versicolor* produce laccases that can break down a wide range of organic compounds.
- **Nitrogen and phosphorus removal:** Microorganisms with enzymes such as nitrate reductases and phosphatases are employed in wastewater treatment to remove excess nitrogen and phosphorus, preventing water eutrophication and contamination of aquatic ecosystems.
- **Bioremediation of chlorinated compounds:** Certain bacteria, such as *Dehalococcoides*, produce enzymes called dehalogenases that can break down chlorinated compounds, such as PCBs and chlorinated solvents.
- **Aromatic compound degradation:** Microbial enzymes such as catechol dioxygenases and catechol 1,2-dioxygenases are involved in the degradation of aromatic compounds found in pesticides, herbicides, and other pollutants.
- **Bioremediation of industrial effluents:** Enzymes can be used in the treatment of industrial wastewater, such as in the pulp and paper industry, where enzymes such as xylanases and ligninases are used to break down lignocellulosic compounds and reduce the environmental impact of effluents.
- **Bioremediation of landfills**: Microbial enzymes can enhance the degradation of organic matter in landfills, reducing the production of $CH_4$ and minimizing the leaching of harmful substances into the groundwater.
- **Oilfield bioremediation:** In oilfields, microbial enzymes can be used to treat produced water contaminated with hydrocarbons, ensuring compliance with environmental regulations.
- **Bioremediation of radioactive contaminants:** Enzymes produced by extremophilic microorganisms can be used to remediate radioactive waste by facilitating the immobilization or reduction of radioactive ions.

### 2.1.3 Factors Affecting Enzymatic Biodegradation

Environmental factors such as pH, temperature, oxygen levels, soil structure, moisture content, nutrient availability, soil type, the bioavailability of contaminants, and the presence of microbial community limits and influences the overall process of biodegradation. Diverse environmental variables have profound effect on the rate and level of microbial breakdown despite having numerous biodegrading microbes present in nature. The environmental conditions must be such that microorganisms may thrive and actively destroy contaminants for bioremediation to be effective. The optimum pH for biodegrading is 6.5–8.5, bacteria favor a basic pH, while fungi prefer an acidic pH. Oil degrading bacteria such as *P. aeruginosa* work best at a pH of 8.0 whereas pH values of 6.5–7.0 for hydrocarbon biodegradation by *Burkholderia cocovenenans* (Bacosa et al. 2022). Temperature has profound effect on the enzyme activity, survival rate of microbes as they adapt to range of temperature. Low temperature crystallizes oil components enabling their degradation and each 10°C decrease in temperature decreases the rate of biodegradation by one-half (Mbachu et al. 2020). Moisture content also influences the rate of contaminant metabolism because it influences the kind and number of soluble materials that are available as well as the osmotic pressure and pH of terrestrial and aquatic systems. Optimum nutrient availability (N, P, K) enhances microbial growth and activity while higher concentration is toxic to the cells. The microbial consortium has a higher degradation ability than single microbe isolates as they can transform more components. Soil conditions such as salinity adversely affect the bioremediation process as microbe's efficacy is direct inhibitory due to the unfavorable high osmotic potential, as well as altered solubility or sorption of toxic compounds.

## 2.2 Conclusion

In an era marked by unprecedented challenges arising from pollution and environmental degradation, the significance of microbial enzymes in bioremediation cannot be overstated. These enzymes, produced by a diverse array of microorganisms, have emerged as indispensable tools for mitigating the adverse impacts of organic pollutants. Understanding the mechanisms that drive microbial enzymes to transmute complex organic pollutants into less harmful, more manageable forms, and illuminating the diverse applications of microbial enzymes, from cleansing contaminated sites through bioremediation to purifying wastewater, and aiding in the development of eco-friendly, biodegradable plastics is critical for mitigating the impact of organic pollutants on ecosystems and human health. These enzymes also contribute to oil spill, plastic, dyes, pesticides, and other organic pollutant mitigation, and find its use in various industrial applications. Further research and innovation need emphasis for fostering a future where the catalytic power of nature's enzymes plays a pivotal role in building a cleaner, more sustainable world.

## References


Ali, S.S., Al-Tohamy, R., Khalil, M.A. et al. (2022). Exploring the potential of a newly constructed manganese peroxidase-producing yeast consortium for tolerating lignin degradation inhibitors while simultaneously decolorizing and detoxifying textile azo dye wastewater. *Bioresource Technology* 351: 126861.

Anand, U., Dey, S., Bontempi, E. et al. (2023). Biotechnological methods to remove microplastics: a review. *Environmental Chemistry Letters* 21: 1787–1810.

Ayilara, M.S. and Babalola, O.O. (2023). Bioremediation of environmental wastes: the role of microorganisms. *Frontiers in Agronomy* 5: https://doi.org/10.3389/fagro.2023.1183691.

Bacosa, H.P., Ancla, S.M.B., Arcadio, C.G.L.A. et al. (2022). From surface water to the deep sea: a review on factors affecting the biodegradation of spilled oil in marine environment. *Journal of Marine Science and Engineering* 10 (3): 426. https://doi.org/10.3390/jmse10030426.

Bhadouria, R., Das, S., Kumar, A. et al. (2020). Mycoremediation of agrochemicals. In: *Agrochemicals Detection, Treatment and Remediation* (ed. M. Narasimha Vara Prasad), 593–620. Oxford: Butterworth-heinemann, Oxford https://doi.org/10.1016/B978-0-08-103017-2.00022-2.

Bhatt, P., Gangola, S., Bhandari, G. et al. (2021). New insights into the degradation of synthetic pollutants in contaminated environments. *Chemosphere* 268: 128827.

Espinosa-Ortiz, E.J., Rene, E.R., and Gerlach, R. (2022). Potential use of fungal-bacterial co-cultures for the removal of organic pollutants. *Critical Reviews in Biotechnology* 42: 361–383.

Filipowicz, N., Momotko, M., Boczkaj, G., and Cieśliński, H. (2020). Determination of phenol biodegradation pathways in three psychrotolerant yeasts, *Candida subhashii* A01(1), *Candida oregonensis* B02(1) and *Schizoblastosporion starkeyi-henricii* L01(2), isolated from Rucianka peatland. *Enzyme and Microbial Technology* 141: 109663.

Fu, L., Bai, Y.-N., Lu, Y.-Z. et al. (2019). Degradation of organic pollutants by anaerobic methane-oxidizing microorganisms using methyl orange as example. *Journal of Hazardous Materials* 364: 264–271.

Gaytán, I., Sánchez-Reyes, A., Burelo, M. et al. (2020). Degradation of recalcitrant polyurethane and xenobiotic additives by a selected landfill microbial community and its biodegradative potential revealed by proximity ligation-based metagenomic analysis. *Frontiers in Microbiology* 10: 2986. https://doi.org/10.3389/fmicb.2019.02986.

Giri, D.D., Singh, S.K., Giri, A. et al. (2021). Bioremediation potential of methylotrophic bacteria. In: *Microbe Mediated Remediation of Environmental Contaminants* (ed. A. Kumar, V.K. Singh, P. Singh, and V.K. Mishra), 199–207. MA: Woodhead Publishing Cambridge https://doi.org/10.1016/B978-0-12-821199-1.00017-1.

Gond, D.P., Jha, S.S., Kumar, A., and Singh, S.K. (2021). Plant growth promoting bacteria and its role in green remediation. In: *Sustainable Environmental Clean-up* (ed. V.K. Mishra and A. Kumar), 149–163. Cambridge, MA, USA: Elsevier https://doi.org/10.1016/B978-0-12-823828-8.00007-4.

Gupta, A., Singh, S.K., Singh, V.K. et al. (2020). Endophytic microbe approaches in bioremediation of organic pollutants. In: *Microbial Endophytes: Prospects of Sustainable Agriculture* (ed. A. Kumar and V.K. Singh), 157–174. Cambridge, MA: Elsevier https://doi.org/10.1016/B978-0-12-818734-0.00007-3.

Kumar, A., Verma, V., Dubey, V.K. et al. (2023). Industrial applications of fungal lipases: a review. *Frontiers in Microbiology* 14: 1142536. https://doi.org/10.3389/fmicb.2023.1142536.

Kumari, S., Das, S. (2023). Bacterial enzymatic degradation of recalcitrant organic pollutants: catabolic pathways and genetic regulations. *Environmental Science and Pollution Research*, 30: 79676–79705. https://doi.org/10.1007/s11356-023-28130-7.

Mbachu, A.E., Chukwura, E.I., and Mbachu, N.A. (2020). Role of microorganisms in the degradation of organic pollutants: a review. *Energy and Environmental Engineering* 7 (1): 1–11.

Mishra, B., Varjani, S., Pradhan, I. et al. (2020). Insights into interdisciplinary approaches for bioremediation of organic pollutants: innovations, challenges and perspectives. *Proceedings of the National Academy of Sciences, India Section B: Biological Sciences* 90: 951–958.

Mishra, M., Singh, S.K., and Kumar, A. (2021). Environmental factors affecting the bioremediation potential of microbes. In: *Woodhead Publishing Series in Food Science, Technology and Nutrition, Microbe Mediated Remediation of Environmental Contaminants* (ed. A. Kumar, V.K. Singh, P. Singh, and V.K. Mishra), 47–58. Cambridge, MA: Elsevier, Woodhead Publishing https://doi.org/10.1016/B978-0-12-821199-1.00005-5.

Pandey, K.D., Singh, S.K., Shukla, L. et al. (2023). Mechanistic approaches and factors regulating microalgae mediated heavy metal remediation from the aquatic ecosystem. In: *Advances in Chemical Pollution, Environmental Management and Protection* (ed. L.F.R. Ferreira, A. Kumar, and M. Bilal), 285–299. Cambridge: Elsevier, Academic Press Volume 9. https://doi.org/10.1016/bs.apmp.2022.10.011.

Patel, A., Patel, V., Patel, R. et al. (2020). Fungal laccases: versatile green catalyst for bioremediation of organopollutants. In: *Emerging Technologies in Environmental Bioremediation* (ed. M.P. Shah, S. Rodriguez-Couto, and S.S. Şengör), 85–129. Elsevier.

Rafeeq, H., Afsheen, N., Rafique, S. et al. (2023). Genetically engineered microorganisms for environmental remediation. *Chemosphere* 310: 136751.

Rovaletti, A., De Gioia, L., Fantucci, P. et al. (2023). Recent theoretical insights into the oxidative degradation of biopolymers and plastics by metalloenzymes. *International Journal of Molecular Sciences* 24: 6368.

Sánchez, C. (2022). A review of the role of biosurfactants in the biodegradation of hydrophobic organopollutants: production, mode of action, biosynthesis and applications. *World Journal of Microbiology and Biotechnology* 38: 216.

Saravanan, A., Kumar, P.S., Vo, D.V.N. et al. (2021). A review on catalytic-enzyme degradation of toxic environmental pollutants: microbial enzymes. *Journal of Hazardous Materials* 419: 126451.

Singh, D., Singh, S.K., Singh, V.K. et al. (2020). Plant growth–promoting bacteria and their role in environmental management. In: *Abatement of Environmental Pollutants* (ed. P. Singh, A. Kumar, and A. Borthakur), 161–175. Cambridge, MA: Elsevier https://doi.org/10.1016/B978-0-12-818095-2.00008-4.

Singh, S.K., Singh, M.K., Singh, V.K. et al. (2021). Microbial enzymes and their exploitation in remediation of environmental contaminants. In: *Microbe Mediated Remediation of Environmental Contaminants* (ed. A. Kumar, V.K. Singh, P. Singh, and V.K. Mishra), 59–71. Cambridge, MA: Woodhead Publishing https://doi.org/10.1016/B978-0-12-821199-1.00006-7.

Sokal, S., Palsania, P., and Kaushik, G. (2022). Bioremediation and functional metagenomics: advances, challenges, and opportunities. In: *Omics Insights in Environmental Bioremediation* (ed. V. Kumar and I.S. Thakur), 3–6. Singapore: Springer https://doi.org/10.1007/978-981-19-4320-1_1.

Touliabah, H.E.-S., El-Sheekh, M.M., Ismail, M.M., and El-Kassas, H. (2022). A review of microalgae- and cyanobacteria-based biodegradation of organic pollutants. *Molecules* 27: 1141.

# 3

# Microbe Assisted Remediation of Xenobiotics: A Sustainable Solution

*Azha Ufaq Nabi, Faamiya Shajar, and Reiaz Ul Rehman*

*Department of Bioresources, School of Biological Sciences, University of Kashmir, Srinagar, India*

## 3.1 Introduction

According to Atashgahi et al. (2018), xenobiotics are chemical substances that are chemically synthesized and are not produced naturally. They are extensively used in huge quantities for industrial, agrochemicals, and residential usage. According to Schwab (2008), xenobiotics are unusual foreign anthropogenic substances that are harmful in nature. The word "xenobiotics" comes from the Greek words "Xenos" which means "foreign" and "bios" which means life. These environmental pollutants are frequently used in the production of dyes, paints, pesticides, herbicides, insecticides, explosives, and other industrial chemicals (Lellis et al. 2019; Singh et al. 2017). Most common xenobiotic substances comprise of fuels, solvents, herbicides, antibiotics, alkanes, polycyclic aromatic hydrocarbons (PAH), synthetic azo dyes, polyaromatic chemicals, polychlorinated biphenyls (PCBs), dioxins, and chlorinated compounds (Atashgahi et al. 2018; Goedtke et al. 2020; Sarkar et al. 2017). Numerous xenobiotic pollution sources have the potential to directly or indirectly affect humans, animals, and plants. These poisonous substances are discharged into the surroundings through incorrect waste disposal and human practices, severely harming aquatic and terrestrial ecosystems, and posing a threat to all other life forms (Zaki et al. 2014; Zhu et al. 2017). Since majority of the xenobiotic compounds are durable and possess moderate to prolonged stability in soil materials, they have a major influence on the soil ecosystem (Liu et al. 2020). This also impacts the overall nutrient cycling as well as the decomposition process (Litskas et al. 2019). Such toxicity causes serious health risks and presents a significant risk to the security and safety of the environment. The enzymatic microbial breakdown of these xenobiotics offers an intriguing approach for bioremediating hazardous environmental contaminants into nonhazardous and harmless forms.

### 3.1.1 Sources of Xenobiotics

There are both direct and indirect sources of xenobiotics (Figure 3.1). In addition to wastewater, other effluents including hydrocarbons, phenols, pesticides, insecticides, paint effluents, dyes, and other solid wastes discharged by industries such as chemical, agrochemical, plastics, pharmaceutical, paper, and textile mills are the main source of the majority of xenobiotic toxicants (Donner et al. 2010; Mathew et al. 2017). Industries (such as POPs), disinfection by-products (such as DBPs), and residential use pharmaceutical and personal care products (PPCPs) (Agus et al. 2015; Petriello et al. 2014) are the main causes of halogenated xenobiotics pollution. It is well known that the proliferation of halogenated substances in the water cycle is assisted by raw sewage and treated discharge from wastewater treatment plants. Most of these treatment facilities are either unable to eliminate these halogenated toxins or generate transformed byproduct. This is more problematic, as the transformed products prove to be more detrimental to the environment than the original contaminant, as demonstrated by the case of transformed products of pesticides (Menger et al. 2021). In addition, perfluorooctanesulfonic acid, a persistent organic pollutant (POP) that may be produced by transforming various synthetic compounds, such as perfluoroalkyl acids, found in wastewater treatment facilities, is another example (Wang et al. 2019). Similar to this, 4-chlorophenol, which is more hazardous, is created during the biodegradation of clofibric acid (Igbinosa et al. 2013; Salgado et al. 2012).

*Microbes Based Approaches for the Management of Hazardous Contaminants*, First Edition. Edited by Ajay Kumar, Livleen Shukla, Joginder Singh, and Luiz Fernando R. Ferreira.

**Figure 3.1** Various sources of xenobiotics

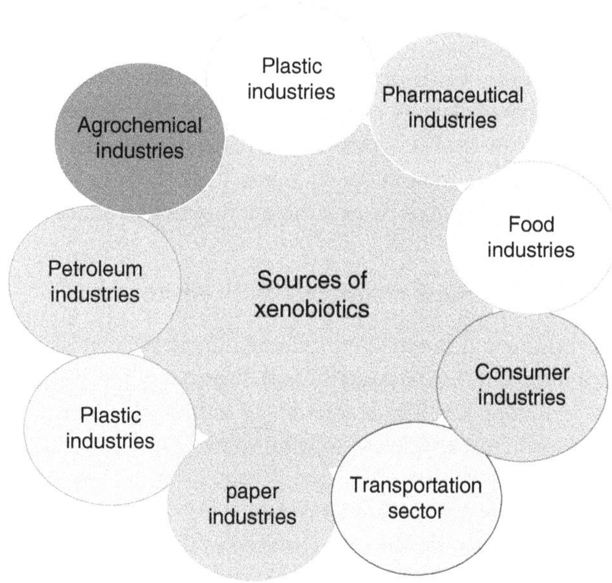


Although in trace concentrations, because of their high persistence and hazardous levels, such extremely poisonous biotransformed products are of major concern since they are easily discovered in water, soil, and in the food for people and other animals. Biosolids, phenolic chemicals, petrochemical products, pigments, and dyes are a few of these direct-origin xenobiotic contaminants. Hydrophobic organohalogens, which are essential industrial compounds in dyes, pigments, paints, medicines, and synthetic polymers are also found in biosolids released from processed sewage sludge. For example, flame-retardant polybrominated diphenyl ethers generated from biocidal chemicals, such as triclocarban and triclosan, are known for their potential to contaminate soil and water (Andrade et al. 2015).

Similarly, toxic biosolids are also produced by cattle, when organohalogen antibiotics, such as fluoroquinolones, are used in feed to promote growth, and prevent diseases. If used as manures or fertilizers, these biosolids may enter the body in a number of ways. Veterinary medications such as florfenicol and enrofloxacin can be consumed up by plants and animals, ultimately entering into the human body (Chen et al. 2016; Hanna et al. 2018). Being toxic even at low concentrations, phenolic compounds are the most common pharmaceutical and chemical contaminants. According to (Bhuyan and Basu 2017; Igbinosa et al. 2013), they facilitate the production of transformed compounds during the oxidation process, that directly influence the environment by depleting the ozone layer, disrupting the earth's thermal balance, increasing the concentration of acidic air pollutants, and reducing visibility. Besides, they are also carcinogenic in nature, which makes their biodegradation difficult since it results in the formation of harmful end products and secondary metabolites. Therefore, before wastewater is discharged, phenol removal from it becomes crucial. Polycyclic saturated hydrocarbons, aromatic hydrocarbons, and various nitrogen, sulfur, and oxygen containing organic components make up petrochemical products. Their aromatic components are more difficult to degrade. As a result, the sensitivity to biodegradation for any molecule used in bioremediation is inversely proportionate to its aromaticity (Fulekar 2017; Sharma et al. 2017). According to El-Sikaily et al. (2012), additional direct sources of xenobiotic pollution include dyes and pigments produced by different industries, such as textile, photographic, and paper manufacturing facilities. Due to the low light penetration caused by dye aggregation in aquatic environments xenobiotic persistence occur (Berradi et al. 2019). Azo dyes such as Black B, Yellow HEM, and Turquoise Blue, as well as substances such as phthalocyanine and anthraquinones, all create aromatic amines after degrading, which causes carcinogenic mutagenesis in a variety of animals (Govindwar et al. 2014). Pharmaceutically active substances such as antibiotics, anesthetics, and hormones are some of the secondary or indirect sources of xenobiotics. These chemicals are easily transported through the food chain and as such have the potential to bioaccumulate in the environment. Biomaterials manufactured from synthetic polymers may undergo degradation, yielding potentially hazardous substances, despite the biodegradability of the parent material. These occurrences can adversely affect the ecological cycle.

Non-steroidal anti-inflammatory drugs (NSAIDS), which are a group of drugs utilized for alleviation of pain, inflammation, and fever represent another class of such indirect sources (Harirforoosh et al. 2013). For instance, a dramatic fall in the number of vultures in Pakistan, which increased from 95% in 2003 to a 99.9% decline in 2008 (Oaks et al. 2004), was attributed to the use of diclofenac salt in animals. The pollution causing plastic compounds are a significant source of xenobiotic contamination (Mondal and Subramaniam 2020). This is due to the fact that they degrade slowly because of their resistant and durable molecular structure, which is made up of powerful molecular connections. The plastics such as polystyrene, polyvinyl chloride (PVC), and polyethylene, and their hazardous by products that are frequently used as fuel source in various industries, contribute significantly to xenobiotic pollution. Their microbial breakdown has garnered increasing attention in recent years, since these usually get degraded into hydrocarbons and other types of fragmented compounds that impose negative environmental effects (Wilkes and Aristilde 2017).

## 3.1.2 The Effects of Xenobiotics on Environment

Environmental xenobiotic contamination is a global issue brought on by human activities such as urbanization and population growth. The ecosystems have been contaminated because of the massive discharge of dangerous substances into the environment. Significant amounts of compounds, such as heavy metal ions, pesticides, fertilizers, PAHs, and oil derivatives can be seen present in the soil and water.

### 3.1.2.1 Effect of Xenobiotics on Soil

The main threat for the soil ecosystems of developed countries comes from xenobiotics such as dioxins, DDT, PCBs, chlordanes, PAHs, and nitroaromatics. Some additional pollutants, including benzene, nitrobenzene, toluene, xylene, aniline, ethylbenzene, trinitrotoluene/dibenzofurans, and chlorinated solvents, may be xenobiotic, particularly in the soil ecosystem, according to some reports (Salem et al. 2017). Due to the presence of aromatic rings and azo linkages, cosmetics and personal care products, particularly parabens, in the soil and air (Juliano and Magrini 2017), and azo dyes in the soil also act as xenobiotic pollutants.

Industrial activities, fuel combustion, military activities, usage of pesticides and fertilizers in agricultural activities are examples of activities that lead to deposition of these chemicals in soil (Alkorta and Garbisu 2001; Samanta et al. 2002). Soil organic matter (SOM) have a significant influence on xenobiotics' bioavailability and distribution in soil, which is also influenced by site conditions and their chemical characteristics (Borpatragohain et al. 2019). Pesticides that are used for management and protection of crops, include herbicides, insecticides, fungicides, bactericides, and algaecides. They have been the predominant environmental toxins over the past years with tons of pesticides manufactured and used annually across the globe (Schwitzguébel et al. 2006). Environmental factors including temperature, soil pH, and moisture influence how POPs present in the soil will behave. Numerous xenobiotics and the by-products of their degradation mimic the predecessors of humus and are commonly utilized in humification. It has been proposed that this natural mechanism can be employed to neutralize pollutants found in soil. Inorganic minerals have favorable xenobiotic interactions and are essential for xenobiotic transformation (Hayat et al. 2010).

### 3.1.2.2 Effect of Xenobiotics on Water

Anthropogenic activities such urbanization, industrialization, transportation, and construction, pollute surface as well as groundwater. Numerous research studies have focused on investigating the occurrence of chemicals and indications of human interference in urban water systems (Strauch et al. 2008). Before their release into aquatic systems, certain widely recognized xenobiotic substances need to undergo processing at municipal wastewater treatment facilities. Diverse water bodies have also been shown to include a number of trace metals, xenobiotics, synthetic organic compounds such as PAHs, phthalates, and pesticides (Mishra et al. 2019). Different sources produce different xenobiotic chemicals that eventually enter the water bodies. They include aerial particle deposition, water flowing from land surfaces, inflows from commercial and sewage effluents, fossil fuel products, and solid waste burning (Essumang 2010). Leaching is another method by which xenobiotic chemicals enter the water table. This process compromises the biological stability of aquatic ecosystems (Fent et al. 2006). Aquatic species experience oxidative stress when xenobiotic contaminants are present. In a recent study (Ibor et al. 2019), it was found that fish in Eleyele lake of Nigeria, significantly increased their oxidative stress response. By boosting the production of reactive oxygen species (ROS) and weakening the antioxidant mechanisms, xenobiotic chemicals cause fish to experience oxidative stress and alter their homeostasis (Burgos-Aceves et al. 2018).

### 3.1.2.3 Effect of Xenobiotics on Plants

The physiological and morphological properties of plants are affected by xenobiotics through different ways. For instance, plant's photosynthetic pigments, proteins, cysteine levels, leaf area, and leaf surface are all altered by particulate matter produced by the automotive sector (Verma and Singh 2006). The enormous range of xenobiotics present in higher plants, each with a unique structure and design, alters gene expression, regulation, and signal transduction. Xenobiotics interact intrinsically with the phytohormone receptors and signaling cascades, which is comparable to phytohormone analogues (Ramel et al. 2012). While nonessential metals such as Pb, Cd, Hg, and As possess detrimental effects even at low quantities, essential metals such as Cu, Zn, Fe, and Mo pose threats at high concentrations (Kumar and Aery 2016). Xenobiotics damage plant DNA through oxidative stress and the production of ROS. Plant signaling pathways are disrupted as a result of xenobiotic toxicity which alters various signaling receptors including the G-Protein coupled receptor and receptor tyrosine kinase (Pande et al. 2022).

#### 3.1.2.4 Effect of Xenobiotics on Marine Life

In particular, the metabolic processes of developing fish embryos are adversely affected by xenobiotics, which causes morphological and functional abnormalities, stunted growth, and final mortality in a variety of marine creatures. Additionally, fish have been known to develop anomalies in their bodies, including changed body shapes, delayed hatching, and mortality (Arya and Haq 2019). Because they obstruct gas exchange and restrict sunlight penetration, dyes and paints are considered xenobiotics even when present in minute quantities (Abdelkader et al. 2011). Pesticides and herbicides are the principal xenobiotic contaminants in aquatic ecosystems. Commonly employed chemicals in agriculture and everyday use include carbamates, nitrophenols, morpholine, synthetic pyrethroids, and organophosphorus. Later, these pollutants enter into aquatic ecosystems such as sea and ocean. Insecticides such as Cypermethrin pose a major threat to the marine ecosystem (Zhang et al. 2011).

#### 3.1.2.5 Effect of Xenobiotics on Terrestrial Animals

Xenobiotic exposure can also manifest when pharmaceutical drugs or other substances are introduced through regular conditioning protocols or experimental methods. The toxicity of xenobiotics is influenced by their consumption and subsequent disposal. Moreover, the mechanical and chemical properties of these xenobiotics have a significant role in determining the toxicity of their components (Perkins and Hankenson 2020). Animals may undergo physiological changes because of xenobiotics and their metabolites as they affect their immunological, cardiovascular, and other organ systems. Ivermectin, a common anthelmintic and acaricide, for example, harms several dog breeds and mouse strains because it lacks p-glycoprotein (Macdonald and Gledhill 2007). It was revealed that Pazufloxacin and Meloxicam damage rabbit's oxidative systems, causing glutathione levels to drop, and significant lipid peroxidation (Khan and Rampal 2014).

#### 3.1.2.6 Effect of Xenobiotics on Human Health

Environmentally harmful xenobiotics have also caused a sharp rise in the absorption of living organisms in recent years. These compounds could cause genetic changes, immune system deterioration, metabolic abnormalities, disruptions of natural ecosystem processes, and an increase in allergic reactions (Kobzev et al. 2020). Humans regularly encounter a diverse range of potentially harmful xenobiotics, such as drugs and nonessential external substances, through food, inhalation, skin contact, and other intravenous exposure routes that maybe harmful for their health (Miglani et al. 2022). Xenobiotics may change the microbiome in the human gut, causing dysbiosis, which is indirectly related to a number of unfavorable health effects. The ongoing biotransformation process, which is developed to neutralize and eliminate body-invading substances, constantly strives to strike a balance between the metabolically activating xenobiotics and detoxifying their harmful by-products. Chronic illnesses and DNA damage can develop in the human body when this equilibrium gets upset. Xenobiotic toxicity varies greatly from person to person. The human body is totally foreign to a number of chemicals. Several human organs and systems are adversely affected by these substances, both directly and indirectly (Kucherenko et al. 2021).

The degradation of such harmful components becomes crucial due to the dangers it presents to both human health and environmental stability. Filtration, coagulation, oxidation, chemical precipitation, adsorption, electrolysis, ozonation, dehalogenation, and hydrolytic processes are only a few of the physical and chemical treatment strategies that are already designed in the past for managing xenobiotics (Mishra et al. 2021). While all of these methods might be effective to some extent, they come with a number of drawbacks (Olak-Kucharczyk et al. 2017). These procedures are often too expensive and difficult to be used on a large scale. Additionally, they are often subject to stringent regulatory constraints and widespread public opposition. These processes produce a lot of sludge, that creates other secondary dangerous pollutants; therefore, they need to be safely disposed of, which involves exceedingly risky transport. Due to these shortcomings, bioremediation was chosen as the preferred approach for xenobiotic degradation right away (Karigar and Rao 2011). The metabolic skills of the microorganisms are used in this biologically friendly method to transform the toxins into harmless by-products. It influences the physiologically mediated modification of pollutants involved in repairing the compromised environment, as well as the regulation of contaminants through mineralization and detoxification. As such bioremediation has grown as the easiest and most practical method for the detoxification of xenobiotics. The microbes which enzymatically degrade the contaminants and transform them into harmless compounds are absolutely necessary for the bioremediation. The vast majority of these microbes may be found in diverse environments, including humus, water, drainage, oil reservoirs, hydrothermal vents, and the human gut (Das et al. 2016). They also flourish at severe temperatures, pH levels, and soil conditions. Microbes have the ability to clean up any polluted site because of their genetic diversity. In order to gain energy and biomass, they produce certain enzymes that digest or convert xenobiotic contaminants from soil and water, including microplastic, agrochemicals, and poly-halogenated substances, leading to the production of nontoxic transformed products. Numerous microorganism have the ability to bioremediate various forms of xenobiotics, including bacteria such as *Microbacterium*, *Pseudomonas*, *Alcaligenes*, *Micrococcus*, *Aeromonas*, *Methanospirillum*, *Cellulosimicrobium*, *Rhodococcus*, B, *Flavobacterium*, and *Sphingobium*, and fungi such as *Penicillium*, *Aspergillus*, *Trichoderma*, *Sporotrichum*, *Ganoderma*, *Talaromyces*, *Fusarium*, *Thermoascus*, *Phanerochaete*, *Cladosporium*, *Trametes*, and *Chaetomium* (Arora 2020; Palleroni and Moore 2010).

## 3.2 Bioremediation

Bioremediation is a carefully managed process that biologically converts organic waste materials into a benign state, often reducing them to dosages lower than the limits set by regulatory agencies (Mueller et al. 1996). Contaminants which are hazardous for the environment or for human health are degraded or detoxified using plants or naturally occurring bacteria, and fungi. The microbes may be isolated from another location and delivered to the contaminated locations, or they may be native to the contaminated site. By means of reactions that take place in the course of their metabolic activities, organisms transform the contaminants. Different organisms often work together to biodegrade a chemical compound. The process of bioaugmentation involves introducing microbes to a contaminated area in order to hasten degradation of the contaminants. For bioremediation to work, pollutants must be attacked enzymatically by microbes and converted into innocuous compounds. Application of bioremediation typically involves altering environmental factors to hasten microbial growth and breakdown since bioremediation is only effective in settings which favors the growth and development of microbes. Like other methods, bioremediation has inevitable limitations. Some pollutants such as highly aromatic hydrocarbons and chlorinated organic substances are often immune to microbial degradation. Due to the difficulty in predicting cleanup rates for bioremediation operations, there are no established standards to determine whether a contaminant can be effectively degraded. Bioremediation methods can remediate some pollutants on-site, reducing risks of exposure for personnel who clean up or probably a wider exposure due transportation mishap. Compared to traditional practices such as incineration, they are more cost-effective. The public views bioremediation as being more acceptable as compared to other solutions due to the fact that it is based on natural degradation. Although most bioremediation systems work in aerobic environments, operating a system in anaerobic settings may enable the microbes to breakdown chemicals that are generally resistant (Colberg and Young 1995).

### 3.2.1 Factors Affecting Bioremediation

Controlling and improving the bioremediation process requires a complicated system with numerous different variables. The existing population of microbes that can degrade the contaminants, the accessibility of pollutants to these microbes, and environmental variables (such as soil type, temperature, pH, the presence of oxygen or other electron acceptors, and nutrients) are some of these factors.

Microbial Populations for Processes of Bioremediation: It is possible to isolate microorganisms practically from any environmental circumstance. Microbes may adapt to different environments and develop in a variety of conditions, including extreme cold, intense heat, arid regions, high oxygen levels, anaerobic conditions, in the presence of harmful materials, and on any waste stream. Due to their versatility, microbes can be used to reduce or eliminate environmental concerns. The different categories of microorganisms have been elaborated upon as follows:

**Aerobes**: *Pseudomonas*, *Alcaligenes*, *Sphingomonas*, *Rhodococcus*, and *Mycobacterium* are few examples of aerobic bacteria known for having immense degenerative capacity. There are numerous reports of these bacteria degrading alkanes and polyaromatic compounds as well as pesticides. Many of these bacteria exclusively derive their energy from the pollutant itself.

**Anaerobes**: Anaerobic microorganisms are not employed as often as aerobic ones. Anaerobic microbes used in the bioremediation of PCBs found in river sediments, for dechlorination of the solvents such as trichloroethylene (TCE) and chloroform, are gaining more and more attention.

**Ligninoytic fungi**: The white rot fungus *Phanerochaete chrysosporium* is one example of a ligninolytic fungus which breakdown a variety of harmful or persistent environmental contaminants. Typical substrates are comprised of straw, sawdust, or maize cobs.

**Methylotrophs**: Aerobic bacteria called methylotrophs thrive by utilizing methane as a source of carbon and energy. The enzyme used in its aerobic degradation pathway is methane monooxygenase that is effective against a variety of substances such as chlorinated aliphatics, TCE, and 1,2-dichloroethane. Bacteria must come in touch with the pollutants for degradation to occur. Due to the uneven distribution of pollutants and bacteria in the soil, this is not an easy feat to accomplish. Some bacteria are chemotactic, meaning they move toward contaminants when they detect them. These bacteria are mobile. In the direction of the contamination, other bacteria, such as fungus, develop in a filamentous shape. Utilizing specific surfactants such as sodium dodecyl sulfate (SDS), it is feasible to increase the mobilization of the pollutant.

## 3.3 Environmental Factors

**Nutrients**: Although there may be microorganisms in contaminated soil, their numbers may not be sufficient to facilitate bioremediation at the site. Their growth and activities must be supported. In most cases, biostimulation necessitates the provision of nutrients and oxygen so as to benefit natural microbes. The nutrients allow bacteria to create the enzymes needed to breakdown pollutants. All of them require nitrogen, phosphorus, and carbon. Carbon is the most essential element for all organisms, more than any other element. Along with hydrogen, oxygen, and iron, it accounts for roughly 95% of the weight of cells.

**Environmental requirements**: The ideal environmental conditions for the degradation of pollutants are pH, temperature, and moisture as they easily influence microbial activity and growth. The majority of microorganisms grow within a certain range, so it is very crucial to create ideal settings, even though they have also been isolated in harsh conditions. If the soil is very acidic, we can add lime so as to alter the pH. Temperature affects biochemical response rates, and several reactions double with every 10°C rise in temperature. But once a specific temperature threshold is reached, the cells degenerate. To increase solar heating in the late spring, summer, and autumn, a plastic cover might be used. Water is a necessity for all living organisms and irrigation is necessary to keep a proper moisture content. Soil structure determines how well air, water, and other nutrients are delivered. To enhance soil structure, substances such as gypsum or organic matter can be utilized. Low soil permeability usually hinders the transfer of water, nutrients, and oxygen which make such soils unsuitable for *in situ* cleanup techniques.

### 3.3.1 Strategies for Bioremediation

Depending on how saturated and aerated an area is, several strategies are used. These strategies can be divided into two categories – *in situ* and *ex situ* (Figure 3.2). The least disruptive methods are those employed *in situ* to preserve the soil and groundwater of the region. Ex situ techniques are employed for treating groundwater and soil that have been pumped or excavated from the site, respectively.

#### 3.3.1.1 *In Situ* Bioremediation Strategies

Since these methods are less costly as well as create less disruption, they are often the best options, as they provide treatment where it is needed without excavating or moving the contaminants. Heightened microbial activity is achieved by establishing an environment that is conducive to the growth of the natural or introduced populations of microbes. The kind and quantity of a particular pollutant present, oxygen supply, water content, temperature, pH, nutritional supply, bioaugmentation with strains showing specific catabolic abilities, and cometabolism all have an impact on how quickly anything degrades or transforms (de Lorenzo 2006). Various in situ bioremediation methods exist, including bioventing, biosparging, biostimulation, and bioaugmentation.

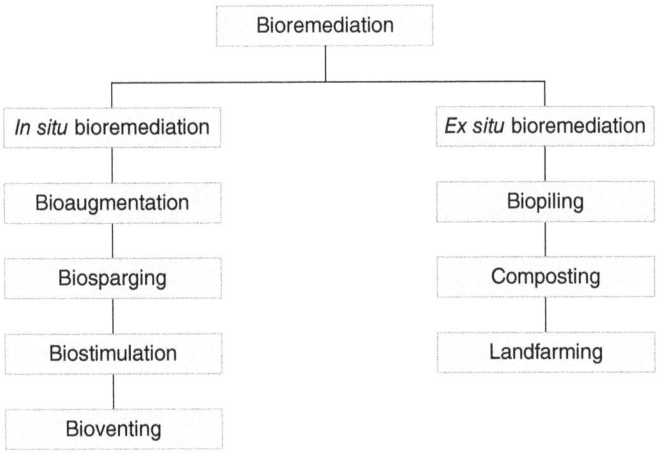


**Figure 3.2** Classification of different bioremediation techniques.

### 3.3.2 Bioventing

It is a technique to breakdown any substance that can be degraded aerobically. Nitrogen, phosphorous, and oxygen are all introduced into the contaminated site during bioventing. The texture of the soil affects how these nutrients are distributed in the soil. Reduced flow rates of air are utilized in bioventing in order to supply the required amount of oxygen needed for the sustenance of the microbes. Direct air injection into soil with persistent pollution via wells is the most common way to supply oxygen. Bioventing becomes efficient when the water table is fairly deep beneath the surface and the environment is warm. It is mostly used to get rid of oil, petrol, and other dangerous pollutants. Different chemicals are eliminated at different rates, varying from one site to another. This is just a result of the various soil textures and hydrocarbon compositions (Rigas et al. 2005). Numerous studies have shown that bioventing has been effective at bioremediating soils that are contaminated with petroleum (Agarry and Latinwo 2015; Lee et al. 2002).

### 3.3.3 Biosparging

In biosparging, air is delivered beneath groundwater under pressure so as to increase the oxygen content for microbes so that they can degrade the pollutants. Biosparging accelerates the aerobic degradation and volatilization (Leahy and Colwell 1990). This technique involves injecting air at the saturated zone, that generate a flow of volatile organic compounds (VOCs) to the unsaturated zone thereby encouraging their biodegradation. Efficacy of biosparging depends upon two fundamental parameters: the permeability of the soil, which directly influences bioavailability of the contaminant to microbes, and the inherent biodegradability of the pollutant (Pongratz and Heumann 1999).

### 3.3.4 Bioaugmentation

To speed up the decomposition of waste, microorganisms (natural, exotic, or created) with specialized metabolic capabilities are introduced to the contaminated region. Thus, bioaugmentation results in an expansion of the gene pool, which raises the genetic diversity of the region. Continuously adding microbes to the bioreactor for the treatment of ground water may boost this process. Commercial inoculants of improved cultures made up of different species of microbes have been used to efficiently colonize the polluted areas whereas the indigenous population of microbes degrade metals successfully. In order to boost natural microbial decomposition, microorganisms present at the remediation site are gathered, grown in isolation, genetically modified, and then reintroduced into the original area. It is most frequently used to reactivate the activated sludge bioreactors used in municipal waste water treatment. Bioaugmentation is employed to make sure that the *in situ* microbes can totally degrade the chlorinated ethane contaminants, such as tetrachloroethylene and TCE to ethylene and chloride, that are naturally harmless, at places in which soil and ground water are polluted (O'Neill et al. 1999). Bioaugmentation is mostly used in areas that are contaminated with oil.

### 3.3.5 Biostimulation

This strategy involves adding particular nutrients such as carbon, nitrogen, phosphorus, oxygen, or other electron donors or acceptors, to the polluted area, so as to stimulate the activity of local microbes (Rigas et al. 2005). These supplements may be administered either in liquid or gaseous form. For the purpose of bioremediation, liquids can be introduced into superficial or deepwater reserves so as to promote the growth of specific microbes. Its key aim is to stimulate the indigenous or naturally existing microbial communities. This is achieved by giving out fertilizers, growth aids, and traces of minerals, and by meeting their other needs, such as pH, temperature, and oxygen, so as to increase the speed of their metabolism (Jilani and Khan 2004; Kumar 2011). A small quantity of pollutants alone can function as a catalyst and cause the activation of genes that are responsible for production of the requisite enzymes which are crucial for bioremediation. In order to aid local microbes, this kind of strategic approach is frequently extended by adding nutrients and oxygen. These nutrients, which act as the core components of life, allowing microbes to generate essential elements including energy, cell biomass, and enzymes for elimination of the pollutant. Petroleum hydrocarbons, sulphates, and polyester polyurethanes are the primary contaminants that biostimulation can eliminate effectively (Marzan et al. 2017; Miao et al. 2012).

## 3.4 *Ex Situ* Bioremediation Strategies

These strategies involve moving of soils to any other place and bioremediated as they cannot be treated on the polluted site due to legal restrictions, lack of available land, risk of ground water or air pollution. *Ex situ* bioremediation procedures include land farming, composting, bioreactors, and biopiling.

### 3.4.1 Landfarming

Land farming is a straightforward method that encompasses the digging up of polluted soil, distributing it over a designated area, and regularly tilling the soil until the contaminants are effectively eliminated. The procedure is restricted to the treatment of the upper 10–35 cm of soil with the aim of stimulating native biodegradative bacteria and facilitating the aerobic degradation of pollutants. Since landfarming has the ability to reduce surveillance and upkeep expenses, it has drawn a lot of attention as a disposal option. Land farming practices have been used to bioremediate spilled oil and wood-preserving pollutants (Head et al. 2006; Maas and Chaudhari 2005). Land farming is crucial for the breakdown of pesticides, and the soil that was dug up is preserved as an intermediary layer between clean soil and clay soil. It permits spontaneous disintegration and also offers oxygen, moisture, and pH maintenance through the use of lime.

### 3.4.2 Composting

An *ex situ* bioremediation technique that dates back thousands of years, composting is still used today on a wide range of scales, from little garden piles to massive industrial operations. Microorganisms that are naturally present in soil work to breakdown organic matter into compost. Under natural circumstances, the mechanical disintegration of organic matter into smaller fragments is done by earthworms, nematodes, and soil insects such as mites, sow bugs, springtails, ants, and beetles. The waste is broken down by microorganisms at a high temperature of 55–65°C. Microbes produce heat during the degradation process, causing a rise in the temperature, which subsequently increases the solubility of trash and promotes increased metabolic activity within the compost. When organic materials are combined and stacked up, composting occurs either aerobically or anaerobically. The fastest way to make finished compost is through aerobic composting, which is also the most effective type of decomposition. Composters use chopping or grinding methods to breakdown big particles under controlled circumstances. The composting process begins when microbes colonize the organic matter under ideal physical conditions. The process of creating finished compost has no set duration. The length of time depends on the feedstocks, the composting technique, and management. It could generally take three months to two years to complete.

### 3.4.3 Biopiling

Compost piles, bio-mounds, bio-cells, and bio-heaps are all interchangeable terms used for the bio-piles. *Ex situ* bioremediation techniques such as these ones are utilized to lower petroleum pollutant concentrations in excavated soils. In bio-piles, dirt is excavated, moved, and piled into heaps. A bottom inert liner creates a protective layer on top of which the soil is put. In this procedure, the bio-pile system is fed with air through a pumping system that either pumps the air at positive pressure into the pile or removes the air from the pile at a negative pressure (Delille et al. 2008). The breakdown of adsorbed petroleum pollutants increases because of the enhanced activity of microbes brought on by microbial respiration (Emami et al. 2012; Karpouzas et al. 2005). The basic bio-pile system consists of a treatment bed, a ventilation system, an irrigation/nutrient system, and a wastewater collection system. To ensure efficient decomposition, pH, oxygen, nutrition, heat, and moisture must all be under proper regulation. Vacuum is used by underground irrigation systems to provide nutrients and air. Plastic has to be used to coat the soil so as to minimize the runoff. It also inhibits evaporation and volatilization and promotes solar heating. The bio-pile treatment procedure can be completed in 20 days to 3 months (O'Neill et al. 1999).

## 3.5 Genetic Engineering Approaches

Genetic engineering has enhanced the usage and eradication of dangerous undesirable by-products in research facilities through the development of genetically engineered organisms (Jain et al. 2011). By using recombinant DNA or by naturally transferring genetic material between species, recombinant organisms can be developed. The potential of genetically engineered microorganisms (GEMs) for bioremediation applications has been seen in soil, groundwater, and activated sludge settings by demonstration of increased degradative capabilities covering a variety of chemical pollutants. There have recently been several opportunities to use genetic engineering techniques to improve degradative performance. For instance, rate-limiting phases in widely recognized metabolic pathways can be genetically altered in order to accelerate the rate of decomposition, or entirely novel metabolic pathways can be introduced into microbial strains to speed up the breakdown of compounds that were formerly resistant. Four approaches need to be carried out in GEMs: (i) Modifying enzyme's specificity and affinity; (ii) Creation and regulation of pathways; (iii) Developing, monitoring, and controlling bioprocesses; and (iv) Using bioaffinity bioreporter sensors for chemical detection, toxicity reduction, and end point assessment. Since GEMs can be successfully used for biodegradation goals, their development signifies or refers to a research area with a significant potential (Kulshreshtha 2013). Recent research has placed a significant emphasis on harnessing the microbial enzymes through the creation of genetically modified microorganisms which produce the required enzymes under optimal settings for better and quicker breakdown of contaminants as maintenance of the microbes under precise conditions can be a protracted and labor-intensive process. Enzymes are biological catalysts that make it easier for different substrates to be converted into diverse products by lowering the activation energy of the reaction. Different enzymes obtained from the microorganisms possess a significant capacity for transforming dangerous xenobiotic chemicals into less hazardous forms.

**Cytochrome P450**: Is a mono-oxygenase enzyme that is commonly present in microbes. It performs different tasks such as the biotransformation of harmful substances in the environment and the creation of sophisticated natural products for medication metabolism in living systems. Through chemical transformation processes and other mechanism-based inactivation associated with bioremediation, it facilitates the biotransformation of several xenobiotics (Esteves et al. 2021).

**Dehydrogenase**: Is a member of the oxidoreductase family and is present in different organisms. In order to verify the polypropylene glycol (PPG) dehydrogenase enzyme's activity in its oxidative route (Hu et al. 2008), conducted an investigation on the *Sphingobium* species that utilize PPG. According to a further investigation by (Hong et al. 2016), the overexpression of the dehydrogenase enzyme and monooxygenases in *Achromobacter* species yields favorable outcomes for the breakdown of hydrocarbons.

**Laccase**: Is a member of the oxidoreductase family and is widely formed by different organisms. Laccases are able of oxidizing a variety of reduced aromatic and phenolic substrates as well as reducing molecular oxygen to water (Lu et al. 2013), demonstrated that employing an alkaline laccase enzyme generated by recombinant *bacillus licheniformis* degenerate synthetic color carmine in just one hour. In addition, they also found that the laccase enzyme could breakdown about 93% of all the dyes assayed at a pH of 9 in just four hours. Recombinant *Trametes (Pycnoporus) sanguineus* from the *Trichoderma atro-viride* species was employed by (Balcazar-Lopez et al. 2016) to express laccase enzyme and assess the organism's capacity to eliminate xenobiotics.

**Hydrolases**: Are hydrolytic enzymes that use water to disassemble the chemical interactions of large harmful compounds and turn them into smaller, harmless substances. For the hydrolysis-based breakdown of xenobiotic pollutants including carbaryl, carbofuran, coumaphos, and diazinon, numerous such hydrolases are isolated from microbes such as *Bacillus cereus, Flavobacterium, Pseudomonas*, and *Nocardia* (Vermassen et al. 2019). A hydrolytic enzyme from

*Bacillus licheniformis* is utilized to bioremediate the soil polluted with malathion because it uses the chemical as a carbon source and breaks it down into nontoxic forms (Khan et al. 2016).

**Cellulase**: Cellulase enzymes such as hemicellulase, extracellular cellulase, and pectinases are generated by a range of bacterial and fungal species. These enzymes catalyze cellulolytic processes and aid in the breakdown of cellulose molecules as well as a number of other polysaccharides. Cellulase is used to breakdown nanofibers of cellulose generated during the washing of cotton clothing and as such are employed in the textile industry to soften and brighten the material (Kamdem Tamo et al. 2020). Cellulases are frequently employed in the paper and pulp sector to aid the ink removal during the process of paper recycling (Pelach et al. 2003).

**Protease**: The protein molecules that are discharged into the environment as industrial waste products by the fisheries, leather, and poultry industries, from deceased animals are hydrolytically cleaved by the protease enzymes. They are widely distributed across bacterial species such as *Amycolatopsis* or *Bacillus* species as well as fungal species such as *Aspergillus* (Souza et al. 2015). Protease enzymes find application across various industries including pharmaceutical, food, leather, and wastewater treatment because of their high productivity, affordable cost, and great potentiality (Souza et al. 2015).

**Lipase**: Various kinds of bacteria, fungi, plants, and animals produce lipases, which are widely distributed enzymes that can hydrolyze ester linkages of triglycerides and transform them into fatty acids and glycerol molecules (Kieliszek et al. 2021). They effectively degrade effluents, oil, and petroleum pollutants, and have a great potential for bioremediation (Dahiya et al. 2010). By decomposing a number of PAHs and n-alkanes, bacteria from the *Mycobacterium, Acinetobacter,* and *Rhodococcus* genera are used to control oil spills. For the bioremediation of wastewater and soil polluted with industrial waste oil, lipase derived from *Pseudomonas aeruginosa* is employed (Amara and Salem 2009).

**Dehalogenase**: Dehalogenase enzymes use hydrolytic, oxidative, and reductive methods to biodegrade a vast range of organic halogenated substances by swapping the halogen atom for a hydroxyl group (Wang et al. 2018). Dehalogenase enzymes are known to be released by a number of bacterial species, including *Pseudomonas umsongensis* YCIT1612, *Ancylobacter aquaticus* strain UV5, and *Rhizobium species* (Xue et al. 2018). These enzymes can bioremediate a range of xenobiotic contaminants. Trichlorobenzenes have been dechlorinated utilizing the dehalogenase enzyme obtained from Desulfovibrionaceae sub-lineage and Proteobacteria (Alfan-Guzman et al. 2017).

**Peroxidase**: Lignin and other phenolic compounds can be oxidized by peroxidases using hydrogen peroxidase, usually in presence of a mediator. Numerous species, including bacteria, fungus, and yeast express the enzyme. These enzymes have substantial potential for bioremediation of different toxic substances due to their ability to withstand extreme temperatures, pH, alkalinity, and salt concentrations. Their applications include the breakdown of various contaminants, decolorization of different dyes, lignin degradation, and the treatment of sewage water (Shrestha et al. 2017).

## 3.6 The Beneficial Role of Microbes in Degradation of Different Pollutants

### 3.6.1 In Heavy Metal Bioremediation

Heavy metals are widely recognized pollutants because of their hazardous effects, stability in the atmosphere, and potential of getting bioaccumulated within the organisms. Most of the hazardous heavy metals are used frequently in different industrial procedures and products. This can be confirmed by presence of heavy metals in different products such as presence of Pb in Herbicides, batteries, insecticides, and aerial emissions from petrol; Hg in Medical waste, coal burning, and mining of precious metals such as gold and silver; Ni in Battery manufacturing, steel alloys, kitchen appliances, and surgical instruments; Cu in Pesticides and fertilizers; Cd in Electroplating, plastic burning, phosphate fertilizer, paints, and pigments; Zn in Priming paints for metals, varnishes, and pigments in aerospace paints; As in Wood storage and pesticides. They have harmed our ecology and contaminated agricultural soils. Heavy metals can be dangerous to ecosystems and have a detrimental effect on biologically mediated processes of soil (Levin et al. 2004). Heavy metals are known to be harmful for humans, animals, plants, and microbes. According to numerous studies (Dheeba et al. 2015; Godlewska-Żyłkiewicz 2006; Hussain et al. 2007; Kamanavalli and Ninnekar 2004; Luptakova and Kusnierova 2005), heavy metal stress severely threatens agricultural crops by limiting plant development parameters and productivity. Metals are more readily available and last longer in soils due to the fact that they are non-biodegradable. Metals that persist for an extended period in soil have mutagenic and carcinogenic effects and enter our food chain (Ahemad and Kibret 2013; Ali et al. 2013). A change in microbial activity and the soil health is another effect of exceeding the threshold limit for heavy metals (Hu et al. 2013). According to many sources, microbes play an important role in the removal of heavy metals. Even at low levels of metal concentration, they are very effective. According to reports, *Vibrio harveyi*, a typical resident of the saline environment, has the capacity to bioaccumulate Cd up to 23.3 mg per gram of dry cells. According to studies, marine bacterium *Enterobactercloaceae*, also have the capacity to chelate heavy metals, thereby eradicating them from polluted environments by releasing exopolysaccharides (EPS) (Jalil et al. 2013). This bacterium can chelate up to 65% of Cd, 20% of Cu, and 8% of Co. In agreement with that, it has also been discovered that some purple non-sulfur bacteria such as *Rhodobium marinum* and *Rhodobacter sphaeroides*, have the ability to eliminate heavy metals from contaminated areas through biosorption or biotransformation. By tolerating and biosorbing heavy metals, bacteria have been specially developed to detect contamination. The genetic pathways for bioremediation of dangerous metals have been simplified for a certain number of bacteria only. For example, Zn and Cu stress has been treated with *Bacillus* spp. and *P. aeruginosa* (Kumar 2011). Furthermore, rhizobacteria and plant symbiosis effectively enhance growth of plants by boosting mineral nutrition and reducing heavy metal toxicity on target plants (Tariq et al. 2006). Plants do, however, use a variety of defenses against heavy metals. By a variety of strategies, including valence transformation, extracellular and intracellular precipitation, and active absorption, fungi are also known for their tolerance and detoxification of heavy metals (Gadd 1994).

## 3.7 Mechanism of Heavy Metal Detoxification by Microbes

In presence of inorganic metals, microorganisms utilize various interactions and coping mechanisms. Biotransformation, the usage of enzymes, the generation of EPS, and synthesis of metallothioneins are certain strategies that are utilized by microbes to withstand metal poisoning. Electrostatic contact, ion exchange, precipitation, redox reaction, and surface complexation are some of the processes that are utilized in these strategies (Soloman et al. 2009). Metal oxidation, metal methylation, metal–organic complexion, metal ligand degradation, metal flux pumps, intracellular, and extracellular metal sequestration, exclusion by permeability barrier, and production of metal chelators are the methods used by microbes to tolerate heavy metals (Kumar et al. 2016).

### 3.7.1 Biosorption Mechanisms

Metal cations are absorbed to negatively charged bacterial cell surfaces during biosorption. Biosorption procedures for the intake of heavy metals by microbes can be divided into metabolism-independent biosorption, that primarily occurs on surface of cells, and metabolism-dependent bioaccumulation, that include sequestration, redox reactions, and species-transformation techniques (Mavura and Wangila 2004; Solomon et al. 2000). Living or dead biomass can passively absorb substances by forming complexes either on the cell wall or on surface layers in order to perform biosorption (Fomina and Gadd 2014).

## 3.8 Intracellular Sequestration

The formation of metal complexes by different substances that are present within the cell's cytosol is known as intracellular sequestration. Metals can engage with surface ligands and then slowly move into microbial cells, where they can accumulate. By making use of bacteria's capacity to store metals intracellularly, different practices have been employed, primarily in the treatment of effluents. Using low molecular weight proteins rich in cysteine, heavy metal ions such as Cu, Cd, and Zn can be sequestered inside cells (Hommel 1990). Chitin, mineral ions, lipids, proteins, polyphosphates, and nitrogen-containing polysaccharides form the stiff fungal cell wall. They are capable of removing metal ions through energy absorption, extracellular and intracellular precipitation, and valence conversion, and as such, various fungi accumulate metals in their mycelium and spores.

## 3.9 Extracellular Sequestration

It occurs when metal ions form insoluble complexes or by the assemblage of metal ions by different cellular constituents within the periplasm. Cu-inducible proteins CopA, CopB, and CopC, that bind to the Cu ions, have been generated by Cu resistant *Pseudomonas syringae* strains (Cha and Cooksey 1991). Metal precipitation as insoluble complexes is another instance of extracellular sequestration. Numerous metal cations precipitate as a result of the high volumes of hydrogen sulfide produced by sulfate-reducing bacteria (Matsumura et al. 1968; Xu 1996). Under anaerobic circumstances, the *Klebsiella planticola* strain generated hydrogen sulfide from thiosulfate and precipitated out the Cd ions as insoluble sulfides (Saratale et al. 2011).

### 3.9.1 Metal Methylation

Methylation enhances the metal toxicity by inducing a rise in lipophilicity, which increases the penetration of metals through cell membranes. Metal cleanup benefits greatly from microbial methylation. Methylated substances are frequently explosive; for example, Hg (II) can be methylated into gaseous methyl mercury by certain bacteria such as *Bacillus* spp., *Escherichia* spp., *Clostridium* spp., and *Pseudomonas* spp. In polluted top soil, bio methylation of Se to volatile dimethyl selenide, As to gaseous arsines, and Pb to dimethyl lead were observed (Ren et al. 2006).

## 3.10 Reduction of Heavy Metal Ions by Microbial Cell

Metal ions can be converted from one oxidation state to another by microbes, thereby decrease their harmfulness (Kamaludeen et al. 2003). Metals and metalloids are used by microbes either as electron acceptors or donors for energy production. Metals in their oxidized state can act as terminal acceptors of electrons when bacteria breathe anaerobically. Mercury and chromium are converted in less harmful forms when metal ions are reduced by enzymatic action (Barkay et al. 2003; Vijayaraghavan and Yun 2008).

### 3.10.1 In Dye Bioremediation

A dye is an organic molecule that has benzene groups coupled to auxochrome and chromophore groups. Azo dyes are the most practical and popular colorants among the different types of dyes, making up 50% of all industrial dyes produced globally. Azo dyes and a number of other dye types are resistant to standard wastewater treatment methods and remain in the surroundings. The textile, pharmaceutical, food, leather, and cosmetic sectors employ a variety of synthetic dyes obtained from different chemical families (such as azo, base, acid, triphenylmethane, and others) extensively (Chang and Lin 2001; Nadeau et al. 1994). More than 80,000 tons of dyes are produced in industries such as the textile, cosmetics, paper, pulp, and bleaching. Untreated wastewater is one of the main sources of consumed phenol, aromatic amines, and metal dyes (Kamei et al. 2011; Sandhya et al. 2005).The metabolites of different synthetic azo dyes are not only poisonous but also mutagenic as well as carcinogenic (Moosvi et al. 2007). Mammals convert azo dyes to cancer-causing aromatics, which are later on oxidized to *N*-hydroxy derivatives, that generate electrophiles which can make covalent bonds with DNA amines. Numerous bacteria are engaged in the bioremediation of several dye classes. Azo dyes such as Congo red, Orange II, and Tropaeolin have been decolorized by *P. chrysosporium* (Cripps et al. 1990). A basidiomycete called *Flavodonflavus*, which was isolated from a maritime environment along the coast, produces laccases, MnPs, and LiPs, which effectively breakdown the colors poly R, poly B, and azure B. According to one report, the commercial dyes such as Reactive Violet 5, Reactive Orange 96, Reactive Red 198, and Reactive Blue 38 and 15 were all decolored by *Pleurotus eryngii* (Heinfling et al. 1998).

## 3.11 The Degradation Mechanism of the Complex Dye Structure by Microbes

Various degrading strategies are used by both aerobic and anaerobic microbes. In these biotransformation pathways, enzymes are very crucial. Textile dyes are widely known to be degraded by oxidizing enzymes such as veratryl alcohol oxidase, laccase, and tyrosinase. However, reducing enzymes that breakdown complicated dye complexes include azo reductase, riboflavin reductases, and dichlorophenolindophenol (DCIP) reductase (Spain and Nishino 1987). Peroxidases particularly catalyze the radical production from phenolic substrates by employing hydrogen peroxide as an electron source. Both phenolic and nonphenolic aromatic substances, including dyes, can be oxidized by versatile peroxidases (VP) (Heinfling et al. 1998; Higham et al. 1986). An enzyme called laccase, which contains copper, uses a one-electron transfer mechanism to oxidize different contaminants including phenols, polyphenols, and anilines (Philp and Atlas 2005; Sharma et al. 2007; Xiong and Jayaswal 1998). To create free radicals, laccase uses oxygen as an electron acceptor to oxidize substituted phenolic and nonphenolic chemical compounds. Demethylation, depolymerization, repolymerization, or quinone formation are other processes these free radicals go through (Sette et al. 2005).

### 3.11.1 In Pesticide Bioremediation

For pest management, pesticides are employed in horticulture, forestry, fish farming, gardens, food storage, and animal husbandry. Due to farming intensification to boost yields, synthetic pesticide use has quickly expanded globally during the past 50 years (Wang et al. 2015). When excess insecticides, herbicides, and fertilizers are carried into rivers by rain, significant health hazards occur. Fertilizers with too much phosphate cause severe eutrophication. Pesticides that are imported into the water bodies are rated as being acutely harmful to fish, crustaceans, and amphibians in 98% of the cases. Around three million people are poisoned by organophosphorus each year, and 200,000 people die as a result (Kumar et al. 2008; Singh and Tripathi 2007). Pesticides are a problem since many of them are lipophilic and are frequently deposited in the fatty tissue of animals. There are various pesticide classes. These include di-nitrophenol, organochlorine, organophosphate, carbamate, pyrethroid, thiocarbamates, and organotin (Salehizadeh and Shojaosadati 2003). Organochlorine chemicals have detrimental effects on the ecosystem, including colony collapse disorder (CCD), pollination reduction, and the death of bees which pollinate the plants. Many organochlorine pesticides are outlawed in most countries, and their usage at international level is governed by Stockholm Convention on POPs. Numerous microbes are capable of degrading pesticides, according to several researches. Pseudomonas, Bacillus, Flavobacterium, Alcaligenes, and Arthrobacter are the most notable bacteria that breakdown pesticides (Jan et al. 2003). Several fungi such as *Agrocybe semiorbicularis*, *Coriolus versicolor*, *Dichomitus squalens*, *Flammulina velupites*, *Pleurotus ostreatus*, and *Stereum hirsutum* have been able to degrade various pesticide groups such as phenylamide, phenylurea, dicarboximide, chlorinated, and organophosphorus compounds (Bending et al. 2002). White-rot fungi have been reported to breakdown a number of pesticides to varying extents, including lindane, atrazine, diuron, terbuthylazine, DDT, dieldrin, aldrin, heptachlor, chlordane, and mirex.

### 3.11.2 In Petroleum Hydrocarbons and Chlorinated Compound Bioremediation

Alkanes, alkenes, as well as heterocyclic, and aromatic components, make up the complex collection of substances known as petroleum hydrocarbons. The PAHs are another significant class of contaminants. The main energy source for industries and everyday life is petroleum-based products. When petroleum and petroleum products are being explored, refined, transported, and stored, leaks and unintentional spills frequently occur. Water and soil pollution usually occurs by the discharge of hydrocarbons into the surroundings, whether accidently or as a result of human practices (Hubbe et al. 2010). Since the accretion of pollutants in the tissues of animals and plants may result in mortality or mutations, soil contamination with hydrocarbons significantly damages the surrounding ecosystem (Alvarez and Vogel 1991). For remediation of soil, mechanical burying, evaporation, dispersion, and washing technologies are frequently used. These technologies however, result in partial breakdown of the contaminants and are usually costly. Using natural populations of microbes for biodegradation is one of the best strategies to eliminate petroleum and other hydrocarbon pollutants from the environment as it is less expensive as compared to other remediation techniques (Nelson et al. 2004; van Beilen et al. 2006) (Figure 3.3).The quantity and type of hydrocarbons present play a crucial part in the intricate procedure of biodegradation of petroleum hydrocarbons. There are four categories into which petroleum hydrocarbons can be subdivided- saturates, aromatics, asphaltenes (phenols, fatty acids, ketones, esters, and porphyrins), and resins (pyridines, quinolines, carbazoles, sulfoxides, and amides) (Colwell et al. 1977). Bacteria are the dominant contributors to the deterioration of petroleum, and they are the main degradation agents for oil spills (Brooijmans et al. 2009; Rajeswari et al. 2013). Even some bacteria exclusively utilize hydrocarbons as an energy source. It was discovered that *Acinetobacter* sp. could only use n-alkanes with a chain length of C10–C8 as a source of carbon (Thangavel and Subbhuraam 2004). The *Gordonia*, *Brevibacterium*, *Aeromicrobium*, *Burkholderia*, and *Mycobacterium* bacterial genera, which were isolated from soil polluted with petroleum, were found to represent prospective organisms for hydrocarbon degradation (Chaillan et al. 2004). Daugulis and McCracken (2003) reported that *Sphingomonas* degraded poly aromatic hydrocarbons. Additionally, some completely novel marine bacterial species have been used to accelerate the breakdown of PAHs in a saline environment, including *Cycloclasticus spirillensus*, *Lutibacteriumanuloederans*, and *Neptuno monasnaphthovorans* (Chung and King 2001). Bacteria from genus *Cycloclasticus* are primarily responsible for the biodegradation of hydrocarbons. It has been discovered that bacteria such as *Sphingomonas paucimobilis* EPA505 use fluoranthene as their

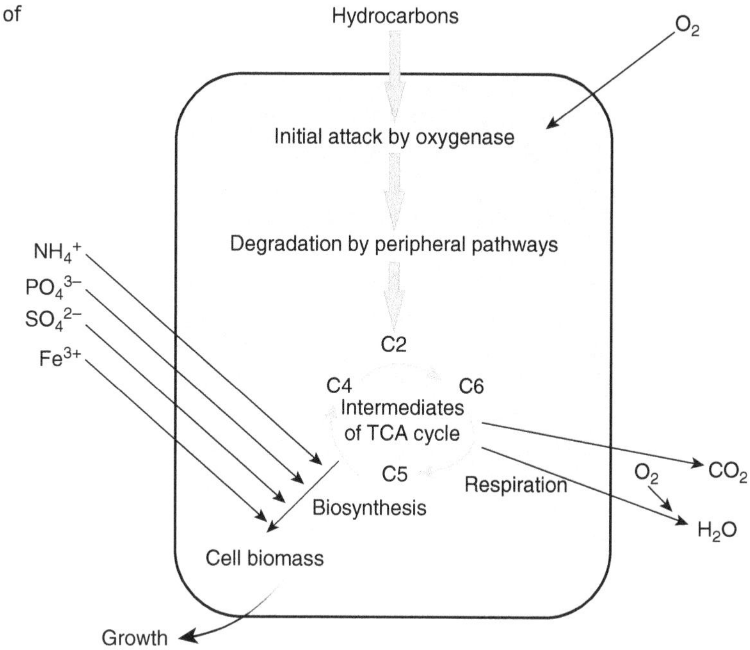


**Figure 3.3** Mechanism of bioremediation of hydrocarbons by aerobic microbes.

exclusive carbon source. *Marinococcus, Methylobacterium*, *Nocardia*, *Planococcus*, and *Rhodococcus* are a few of the significant bacterial genera that can breakdown petroleum oil. For major breakdown of crude oil and its constituent parts, a consortium has been established by (Deppe et al. 2005) employing arctic bacteria such as *Agreia*, *Marinobacter*, *Pseudomonas*, *Psychrobacter*, and *Shewanella*.

## 3.12 In Domestic and Agricultural Lignocellulose Wastes Remediation

Wood, grasses, farm leftovers, forestry wastes, and solid municipal wastes are all sources of lignocellulose. Approximately 40–60% of cellulose, 20–40% hemicellulose, and 10–25% lignin can be found in lignocellulosic complexes. The polluted soil is dug up and combined with natural compounds such as corn cobs, activated sludge, and cow manure. These additions offer the biomass required for the microbial growth. Microorganisms that produce hydrolytic enzyme are crucial for the recycling of agricultural wastes. Particularly significant enzymes required for the degradation of agricultural wastes are the cellulase, laccase, lignin peroxidase, manganese peroxidase, and hemicellulose. One bioremediation technique is composting, and the compost generated through bioconversion of agricultural remnants has a number of advantages over conventional soil amendments including improved soil fertility and soil health, which can increase crop yields, improved biological diversity in the soil, decreased ecological hazards, and an improved environment. The average temperature curve within the piles during aerobic composting displayed the three conventional phases- the mesophilic (moderate temperature phase), the thermophilic (high temperature phase), and the cooling or maturation phase. During the first 25 days within the mesophilic phase the psychrophilic, and mesophilic microbes in trash heaps proliferate. Due to the biodegradation of organic molecules, the temperature during this phase soar to 40–50°C (Alberti 1984). The thermophilic phase, which occurs from 30th day and lasts up to 110th of the composting process, occurs when the temperature is too high encouraging the growth of thermogenic microorganisms. Satish et al. (2017) asserts that temperatures above 55°C are crucial for maximizing sanitization. A temperature range of 35–40°C will boost microbial diversity, while a temperature range of 45–55°C will improve the rate of breakdown. Ventilation and irrigation ensure that the temperature inside the windrow is controlled and kept below 65°C while composting. Microbial activity decreases during the cooling phase, and as the process is about to be finished, the material's temperature approaches to that of ambient air. When compost is finished, it resembles humus, the soil's organic component. The volume, moisture content, and weight of the material are usually reduced by 20–60%. Turning is not as necessary when the product is curing. Though low oxygen levels are still required for microbial activity. The piles should be placed in dry locations, free from moisture, during the curing phase. The curing piles could become anaerobic at this phase if exposed to too much moisture. After the twelfth week, the temperature begins to go down. The C/N ratio in the various piles stabilizes by the end of the fourth month of composting, and the mean temperature inside the various windrows indicate a true dip with value of about 30°C. This decrease leads to the reduction of organic matter. Natural wastes including paddy straw, sugarcane waste, and other agricultural wastes are usually decomposed by composting. During composting, a natural succession of microorganisms occurs. Various fungi play a crucial part in composting the lignocellulosic materials, including *Trichoderma harzianum, Pleurotus ostreatus, Polyporus ostriformis*, and *Phanerochaete chrysosporium* (Bjeldanes and Shibamoto 1993). Although the varied and crucial functions that fungi, bacteria, and actinomycetes play during the process of composting, mixed cultures of microbes hasten the breakdown of lignocellulosic matter as a result of their cooperative activity through the consumption of intermediary products of decomposition (Kim et al. 2005). In perforated trenches, rice straw was composted using a group of four hypercellulolytic fungal cultures namely *Aspergillus nidulans, Trichoderma viride, Phanerochaete chrysosporium,* and *Aspergillus awamori*. Different methods, such as microbe induced mineralization, transformation into harmless compounds, volatilization, the creation of humus and other by-products such as carbon dioxide, water, and salts, may be used to breakdown the pollutants. Process characteristics including pH, temperature, the C:N ratio, moisture content, and the type of the material to be composted, largely impact the crucial parameters and efficacy of compost in bioremediation. Bulking additives such as peat moss, pine wood shavings, and bran flakes speed up the composting by promoting the growth of aerobic microbes (Alburquerque et al. 2006; Naik and Duraphe 2012; Pattus and Abdallah 2000).

## 3.13 Conclusion

There is an ecological crisis, biodiversity loss, and severe climate change as a result of environmental pollution brought on by many pollutant sources from industrial, agriculture, and urbanization activities. Till date, a variety of waste removal techniques have been used to address these global issues. Waste treatment and detoxification methods look for optional and alternative approaches in accordance with cost-effective and environmentally sustainable approach. An incremental technology called bioremediation uses biological agents to clean up hazardous toxic contaminants and eliminate environmental pollutants. It has changed over the past 20 years in terms of commercial use and remedial service use. Through ex situ and *in situ* bioremediation techniques such as biopiling, composting, land farming, bioventing, biosparging, biostimulation, and bioaugmentation, that are used to degrade organic and volatile pollutant, heavy metal waste, petroleum waste, chlorinated compound, and plastic waste, organic and inorganic municipal solid waste, lignocellulosic waste, industry effluent waste, nuclear waste, and various dyestuff, and agrochemical wastes. In the course of bioremediation, enzymes such as oxidoreductases, transferases, hydrolases, lyases, isomerases, and ligases (synthetases) are produced by microorganisms so as to degrade the contaminants. In order to utilize microorganism potential by developing an efficient waste management plan, it is crucial to comprehend the mechanism, mode of action, and role of microorganisms involved in the process of bioremediation. Microbial communities have a significant deal of promise to facilitate the efficient biodegradation of habitats with xenobiotic contamination of soil and water. But all microbes in nature cannot be grown in *in vitro* conditions, as many of them live in viable-but-non-culturable (VBNC) environments, as such the majority of microorganisms that can used in bioremediation are still unexplored. Therefore, modern advanced techniques and sophisticated cutting-edge technologies are extremely needed to comprehend the genetics and molecular biology of microorganisms in order to study the hidden information of these VBNC creatures.

## References

Abdelkader, E., Nadjia, L., and Ahmed, B. (2011). Degradation study of phenazin neutral red from aqueous suspension by paper sludge. *Journal of Chemical Engineering & Process Technology* 2: 109–114.

Agarry, S. and Latinwo, G.K. (2015). Biodegradation of diesel oil in soil and its enhancement by application of bioventing and amendment with brewery waste effluents as biostimulation-bioaugmentation agents. *Journal of Ecological Engineering* 16 (2): 82–91.

Agus, H.H., Sümer, S., and Erkoç, F. (2015). Toxicity and molecular effects of di-*n*-butyl phthalate (DBP) on CYP1A, SOD, and GPx in *Cyprinus carpio* (common carp). *Environmental Monitoring and Assessment* 187: 1–8.

Ahemad, M. and Kibret, M. (2013). Recent trends in microbial biosorption of heavy metals: a review. *Biochemistry and Molecular Biology* 1 (1): 19–26.

Alberti, G. (1984). Aspects bactériologiques du compostage des boues résiduaires de stations d'épuration d'eau. These de 3éme Cycle, Université de Nancy I, France, 200.

Alburquerque, J.A., Gonzálvez, J., García, D., and Cegarra, J. (2006). Composting of a solid olive-mill by-product ("alperujo") and the potential of the resulting compost for cultivating pepper under commercial conditions. *Waste Management* 26 (6): 620–626.

Alfan-Guzman, R., Ertan, H., Manefield, M., and Lee, M. (2017). Isolation and characterization of *Dehalobacter* sp. strain TeCB1 including identification of TcbA: a novel tetra-and trichlorobenzene reductive dehalogenase. *Frontiers in Microbiology* 8: 558.

Ali, H., Khan, E., and Sajad, M.A. (2013). Phytoremediation of heavy metals – concepts and applications. *Chemosphere* 91 (7): 869–881.

Alkorta, I. and Garbisu, C. (2001). Phytoremediation of organic contaminants in soils. *Bioresource Technology* 79 (3): 273–276.

Alvarez, P.J. and Vogel, T.M. (1991). Substrate interactions of benzene, toluene, and para-xylene during microbial degradation by pure cultures and mixed culture aquifer slurries. *Applied and Environmental Microbiology* 57 (10): 2981–2985.

Amara, A.A. and Salem, S.R. (2009). Degradation of castor oil and lipase production by *Pseudomonas aeruginosa*. *American-Eurasian Journal of Agricultural & Environmental Sciences* 5 (4): 556–563.

Andrade, N.A., Lozano, N., McConnell, L.L. et al. (2015). Long-term trends of PBDEs, triclosan, and triclocarban in biosolids from a wastewater treatment plant in the Mid-Atlantic region of the US. *Journal of Hazardous Materials* 282: 68–74.

Arora, P.K. (2020). Bacilli-mediated degradation of xenobiotic compounds and heavy metals. *Frontiers in Bioengineering and Biotechnology* 8: 570307.

Arya, P. and Haq, S.A. (2019). Effects of xenobiotics and their biodegradation in marine life. In: *Smart Bioremediation Technologies* (ed. P. Bhatt), 63–81. Academic Press.

Atashgahi, S., Sánchez-Andrea, I., Heipieper, H.J. et al. (2018). Prospects for harnessing biocide resistance for bioremediation and detoxification. *Science* 360 (6390): 743–746.

Atashgahi, S., Shetty, S.A., Smidt, H., and De Vos, W.M. (2018). Flux, impact, and fate of halogenated xenobiotic compounds in the gut. *Frontiers in Physiology* 9: 888.

Balcazar-Lopez, E., Méndez-Lorenzo, L.H., Batista-García, R.A. et al. (2016). Xenobiotic compounds degradation by heterologous expression of a *Trametes sanguineus* laccase in *Trichoderma atroviride*. *PLoS One* 11 (2): e0147997.

Barkay, T., Miller, S.M., and Summers, A.O. (2003). Bacterial mercury resistance from atoms to ecosystems. *FEMS Microbiology Reviews* 27 (2–3): 355–384.

Bending, G.D., Friloux, M., and Walker, A. (2002). Degradation of contrasting pesticides by white rot fungi and its relationship with ligninolytic potential. *FEMS Microbiology Letters* 212 (1): 59–63.

Berradi, M., Hsissou, R., Khudhair, M. et al. (2019). Textile finishing dyes and their impact on aquatic environs. *Heliyon* 5 (11): e02711.

Bhuyan, D.J. and Basu, A. (2017). Phenolic compounds potential health benefits and toxicity. In: *Utilisation of Bioactive Compounds from Agricultural and Food Production Waste* (ed. Q.V. Vuong), 27–59. Boca Raton, FL, USA: CRC Press.

Bjeldanes, L.F. and Shibamoto, T. (1993). *Introduction to Food Toxicology*. Academic Press.

Borpatragohain, B., Sahoo, S., and Rai, A. (2019). An overview on the impact, interaction and fate of xenobiotic-soil organic matter complexes. *International Journal of Communication Systems* 7: 4935–4941.

Brooijmans, R.J., Pastink, M.I., and Siezen, R.J. (2009). Hydrocarbon-degrading bacteria: the oil-spill clean-up crew. *Microbial Biotechnology* 2 (6): 587.

Burgos-Aceves, M.A., Cohen, A., Smith, Y., and Faggio, C. (2018). MicroRNAs and their role on fish oxidative stress during xenobiotic environmental exposures. *Ecotoxicology and Environmental Safety* 148: 995–1000.

Casas-Godoy, L., Duquesne, S., Bordes, F. et al. (2012). Lipases: an overview. In: *Lipases and Phospholipases: Methods and Protocols* (ed. G. Sandoval), 3–30. New York, NY: Humana.

Cha, J.S. and Cooksey, D.A. (1991). Copper resistance in *Pseudomonas syringae* mediated by periplasmic and outer membrane proteins. *Proceedings of the National Academy of Sciences of the United States of America* 88 (20): 8915–8919.

Chaillan, F., Le Flèche, A., Bury, E. et al. (2004). Identification and biodegradation potential of tropical aerobic hydrocarbon-degrading microorganisms. *Research in Microbiology* 155 (7): 587–595.

Chang, J.S. and Lin, C.Y. (2001). Decolorization kinetics of a recombinant *Escherichia coli* strain harboring azo-dye-decolorizing determinants from *Rhodococcus sp. Biotechnology Letters* 23: 631–636.

Chen, Y., Sun, J., Liao, X.P. et al. (2016). Impact of enrofloxacin and florfenicol therapy on the spread of OqxAB gene and intestinal microbiota in chickens. *Veterinary Microbiology* 192: 1–9.

Chung, W.K. and King, G.M. (2001). Isolation, characterization, and polyaromatic hydrocarbon degradation potential of aerobic bacteria from marine macrofaunal burrow sediments and description of *Lutibacterium anuloederans* gen. nov., sp. nov., and *Cycloclasticus spirillensus* sp. nov. *Applied and Environmental Microbiology* 67 (12): 5585–5592.

Colberg, P.J. and Young, L.Y. (1995). Anaerobic degradation of nonhalogenated homocyclic aromatic compounds coupled with nitrate, iron, or sulfate reduction. *Microbial Transformation and Degradation of Toxic Organic Chemicals* 307330.

Colwell, R.R., Walker, J.D., and Cooney, J.J. (1977). Ecological aspects of microbial degradation of petroleum in the marine environment. *CRC Critical Reviews in Microbiology* 5 (4): 423–445.

Cripps, C., Bumpus, J.A., and Aust, S.D. (1990). Biodegradation of azo and heterocyclic dyes by *Phanerochaete chrysosporium*. *Applied and Environmental Microbiology* 56 (4): 1114–1118.

Dahiya, P., Arora, P., Chaudhury, A. et al. (2010). Characterization of an extracellular alkaline lipase from *Pseudomonas mendocina* M-37. *Journal of Basic Microbiology* 50 (5): 420–426.

Das, A., Srinivasan, M., Ghosh, T.S., and Mande, S.S. (2016). Xenobiotic metabolism and gut microbiomes. *PLoS One* 11 (10): e0163099.

Daugulis, A.J. and McCracken, C.M. (2003). Microbial degradation of high and low molecular weight polyaromatic hydrocarbons in a two-phase partitioning bioreactor by two strains of *Sphingomonas* sp. *Biotechnology Letters* 25: 1441–1444.

Delille, D., Duval, A., and Pelletier, E. (2008). Highly efficient pilot biopiles for on-site fertilization treatment of diesel oil-contaminated sub-Antarctic soil. *Cold Regions Science and Technology* 54 (1): 7–18.

Deppe, U., Richnow, H.H., Michaelis, W., and Antranikian, G. (2005). Degradation of crude oil by an arctic microbial consortium. *Extremophiles* 9: 461–470.

Dheeba, B., Sampathkumar, P., and Kannan, K. (2015). Fertilizers and mixed crop cultivation of chromium tolerant and sensitive plants under chromium toxicity. *Journal of Toxicology* 2015.

Donner, E., Eriksson, E., Holten-Lützhøft, H. C., et al. (2010). Identifying and classifying the sources and uses of xenobiotics in urban environments. Fatta-Kassinos, D., Bester, K., Kümmerer, K. (eds) Xenobiotics in the Urban Water Cycle. Environmental Pollution, vol 16. Springer, Dordrecht. https://doi.org/10.1007/978-90-481-3509-7_2.

El-Sikaily, A., Khaled, A., and El Nemr, A. (2012). Textile dyes xenobiotic and their harmful effect. Nonconventional textile waste water treatment. In: (ed. A. El Nemr), 31–64. New York: Nova Science Publishers Inc.

Emami, S., Pourbabaee, A.A., and Alikhani, H.A. (2012). Bioremediation principles and techniques on petroleum hydrocarbon contaminated soil. *Technical Journal of Engineering and Applied Sciences* 2 (10): 320–323.

Essumang, D.K. (2010). Distribution, levels, and risk assessment of polycyclic aromatic hydrocarbons (PAHs) in some water bodies along the coastal belt of Ghana. *TheScientificWorldJOURNAL* 10: 972–985.

Esteves, F., Rueff, J., and Kranendonk, M. (2021). The central role of cytochrome P450 in xenobiotic metabolism – a brief review on a fascinating enzyme family. *Journal of Xenobiotics* 11 (3): 94–114.

Fent, K., Weston, A.A., and Caminada, D. (2006). Ecotoxicology of human pharmaceuticals. *Aquatic Toxicology* 76 (2): 122–159.

Fomina, M. and Gadd, G.M. (2014). Biosorption: current perspectives on concept, definition and application. *Bioresource Technology* 160: 3–14.

Fulekar, M.H. (2017). Microbial degradation of petrochemical waste-polycyclic aromatic hydrocarbons. *Bioresources and Bioprocessing* 4: 1–16.

Gadd, G.M. (1994). Interactions of fungi with toxic metals. In: *The Genus Aspergillus: From Taxonomy and Genetics to Industrial Application* (ed. K.A. Powell, A. Renwick, and J.F. Peberdy), 361–374. Boston, MA: Springer US.

Godlewska-Żyłkiewicz, B. (2006). Microorganisms in inorganic chemical analysis. *Analytical and Bioanalytical Chemistry* 384: 114–123.

Goedtke, L., Sprenger, H., Hofmann, U. et al. (2020). Polycyclic aromatic hydrocarbons activate the aryl hydrocarbon receptor and the constitutive androstane receptor to regulate xenobiotic metabolism in human liver cells. *International Journal of Molecular Sciences* 22 (1): 372.

Govindwar, S.P., Kurade, M.B., Tamboli, D.P. et al. (2014). Decolorization and degradation of xenobiotic azo dye reactive yellow-84A and textile effluent by *Galactomyces geotrichum*. *Chemosphere* 109: 234–238.

Hanna, N., Sun, P., Sun, Q. et al. (2018). Presence of antibiotic residues in various environmental compartments of Shandong province in eastern China: its potential for resistance development and ecological and human risk. *Environment International* 114: 131–142.

Harirforoosh, S., Asghar, W., and Jamali, F. (2013). Adverse effects of nonsteroidal antiinflammatory drugs: an update of gastrointestinal, cardiovascular and renal complications. *Journal of Pharmacy & Pharmaceutical Sciences* 16 (5): 821–847.

Hayat, M.T., Xu, J., Ding, N., and Mahmood, T. (2010). Dynamic behavior of persistent organic pollutants in soil and their interaction with organic matter. In: *Molecular Environmental Soil Science at the Interfaces in the Earth's Critical Zone* (ed. J. Xu and P.M. Huang), 217–222. Berlin, Heidelberg: Springer.

Head, I.M., Jones, D.M., and Röling, W.F. (2006). Marine microorganisms make a meal of oil. *Nature Reviews Microbiology* 4 (3): 173–182.

Heinfling, A., Ruiz-Dueñas, F.J., Martínez, M.J. et al. (1998). A study on reducing substrates of manganese-oxidizing peroxidases from *Pleurotus eryngii* and *Bjerkandera adusta*. *FEBS Letters* 428 (3): 141–146.

Higham, D.P., Sadler, P.J., and Scawen, M.D. (1986). Cadmium-binding proteins in *Pseudomonas putida: pseudothioneins*. *Environmental Health Perspectives* 65: 5–11.

Hommel, R.K. (1990). Formation and phylogenetic role of biosurfactants. *Journal of Applied Microbiology* 89 (1): 158–119.

Hong, Y.H., Deng, M.C., Xu, X.M. et al. (2016). Characterization of the transcriptome of *Achromobacter* sp. HZ01 with the outstanding hydrocarbon-degrading ability. *Gene* 584 (2): 185–194.

Hu, X., Liu, X., Tani, A. et al. (2008). Proposed oxidative metabolic pathway for polypropylene glycol in *Sphingobium* sp. strain PW-1. *Bioscience, Biotechnology, and Biochemistry* 72 (4): 1115–1118.

Hu, Y., Liu, X., Bai, J. et al. (2013). Assessing heavy metal pollution in the surface soils of a region that had undergone three decades of intense industrialization and urbanization. *Environmental Science and Pollution Research* 20: 6150–6159.

Hubbe, M.A., Nazhad, M., and Sánchez, C. (2010). Composting as a way to convert cellulosic biomass and organic waste into high-value soil amendments: a review. *BioResources* 5 (4): 2808–2854.

Hussain, S., Arshad, M., Saleem, M., and Zahir, Z.A. (2007). Screening of soil fungi for in vitro degradation of endosulfan. *World Journal of Microbiology and Biotechnology* 23: 939–945.

Ibor, O.R., Adeogun, A.O., Regoli, F., and Arukwe, A. (2019). Xenobiotic biotransformation, oxidative stress and obesogenic molecular biomarker responses in Tilapia guineensis from Eleyele Lake, Nigeria. *Ecotoxicology and Environmental Safety* 169: 255–265.

Igbinosa, E.O., Odjadjare, E.E., Chigor, V.N. et al. (2013). Toxicological profile of chlorophenols and their derivatives in the environment: the public health perspective. *The Scientific World Journal* 2013.

Jain, P.K., Gupta, V.K., Bajpai, V. et al. (2011). *GMO's: Perspective of Bioremediation. Recent Advances in Environmental Biotechnology*, 6–23. Germany: LAP Lambert Academic Publishing AG and Co. KG.

Jalil, M.A., Mian, M.N., and Rahman, M.K. (2013). Using plastic bags and its damaging impact on environment and agriculture: an alternative proposal. *International Journal of Learning and Development* 3 (4): 1–14.

Jan, B., Beilen, V., and Neuenschwunder, M. (2003). Rubredoxins involved in alkane degradation. *The Journal of Bacteriology* 184: 1722–1732.

Jilani, S. and Khan, M.A. (2004). Isolation, characterization and growth response of pesticides degrading bacteria. *Journal of Biological Sciences* 4 (1): 15–20.

Juliano, C. and Magrini, G.A. (2017). Cosmetic ingredients as emerging pollutants of environmental and health concern. A mini-review. *Cosmetics* 4 (2): 11.

Kamaludeen, S.P., Megharaj, M., Juhasz, A.L. et al. (2003). Chromium-microorganism interactions in soils: remediation implications. *Reviews of Environmental Contamination and Toxicology* 93–164.

Kamanavalli, C.M. and Ninnekar, H.Z. (2004). Biodegradation of DDT by a *Pseudomonas* species. *Current Microbiology* 48: 10–13.

Kamdem Tamo, A., Doench, I., Morales Helguera, A. et al. (2020). Biodegradation of crystalline cellulose nanofibers by means of enzyme immobilized-alginate beads and microparticles. *Polymers* 12 (7): 1522.

Kamei, I., Takagi, K., and Kondo, R. (2011). Degradation of endosulfan and endosulfan sulfate by white-rot fungus *Trametes hirsuta*. *Journal of Wood Science* 57: 317–322.

Karigar, C.S. and Rao, S.S. (2011). Role of microbial enzymes in the bioremediation of pollutants: a review. *Enzyme Research* 2011.

Karpouzas, D.G., Fotopoulou, A., Menkissoglu-Spiroudi, U., and Singh, B.K. (2005). Non-specific biodegradation of the organophosphorus pesticides, cadusafos and ethoprophos, by two bacterial isolates. *FEMS Microbiology Ecology* 53 (3): 369–378.

Khan, A.M. and Rampal, S. (2014). Effects of repeated oral administration of pazufloxacin mesylate and meloxicam on the antioxidant status in rabbits. *Journal of the American Association for Laboratory Animal Science* 53 (4): 399–403.

Khan, S., Zaffar, H., Irshad, U. et al. (2016). Biodegradation of malathion by Bacillus licheniformis strain ML-1. *Archives of Biological Sciences* 68 (1): 51–59.

Kieliszek, M., Pobiega, K., Piwowarek, K., and Kot, A.M. (2021). Characteristics of the proteolytic enzymes produced by lactic acid bacteria. *Molecules* 26 (7): 1858.

Kim, Y.H., Ahn, J.Y., Moon, S.H., and Lee, J. (2005). Biodegradation and detoxification of organophosphate insecticide, malathion by *Fusarium oxysporum* f. sp. pisi cutinase. *Chemosphere* 60 (10): 1349–1355.

Kobzev, K., Kobzeva, N., Chegge, V. et al. (2020). Ecological problems in the Russian Federation. Impact on the health of people and the country's economy. In: *E3S Web of Conferences*, vol. 217, p. 11001 (ed. V. Breskich). EDP Sciences.

Kucherenko, S.V., Ovcharenko, A.M., and Pushenko, S.L. (2021). Xenobiotics: a threat to the health of living organisms. In: *E3S Web of Conferences*, vol. 285, p. 03006 (ed. S. Belousov and S. Roshchupkin). EDP Sciences.

Kulshreshtha, S. (2013). Genetically engineered microorganisms: a problem solving approach for bioremediation. *Journal of Bioremediation & Biodegradation* 4 (4): 1–2.

Kumar, S. (2011). Composting of municipal solid waste. *Critical Reviews in Biotechnology* 31 (2): 112–136.

Kumar, A. and Aery, N.C. (2016). Impact, metabolism, and toxicity of heavy metals in plants. In: *Plant Responses to Xenobiotics* (ed. A. Singh, S. Prasad, and R. Singh), 141–176. Singapore: Springer.

Kumar, A., Gaind, S., and Nain, L. (2008). Evaluation of thermophilic fungal consortium for paddy straw composting. *Biodegradation* 19: 395–402.

Kumar, S., Chaurasia, P., and Kumar, A. (2016). Isolation and characterization of microbial strains from textile industry effluents of Bhilwara, India: analysis with bioremediation. *Journal of Chemical and Pharmaceutical Research* 8 (4): 143–150.

Leahy, J.G. and Colwell, R.R. (1990). Microbial degradation of hydrocarbons in the environment. *Microbiological Reviews* 54 (3): 305–315.

Lee, I.S., Kim, O.K., Chang, Y.Y. et al. (2002). Heavy metal concentrations and enzyme activities in soil from a contaminated Korean shooting range. *Journal of Bioscience and Bioengineering* 94 (5): 406–411.

Lellis, B., Fávaro-Polonio, C.Z., Pamphile, J.A., and Polonio, J.C. (2019). Effects of textile dyes on health and the environment and bioremediation potential of living organisms. *Biotechnology Research and Innovation* 3 (2): 275–290.

Levin, L., Papinutti, L., and Forchiassin, F. (2004). Evaluation of Argentinean white rot fungi for their ability to produce lignin-modifying enzymes and decolorize industrial dyes. *Bioresource Technology* 94 (2): 169–176.

Litskas, V.D., Karamanlis, X.N., Prousali, S.P., and Koveos, D.S. (2019). The xenobiotic doxycycline affects nitrogen transformations in soil and impacts earthworms and cultivated plants. *Journal of Environmental Science and Health, Part A* 54 (14): 1441–1447.

de Lorenzo, V. (2006). Blueprint of an oil-eating bacterium. *Nature Biotechnology* 24 (8): 952–953.

Liu, J., Song, Y., Tang, M. et al. (2020). Enhanced dissipation of xenobiotic agrochemicals harnessing soil microbiome in the tillage-reduced rice-dominated agroecosystem. *Journal of Hazardous Materials* 398: 122954.

Lu, L., Wang, T.N., Xu, T.F. et al. (2013). Cloning and expression of thermo-alkali-stable laccase of Bacillus licheniformis in Pichia pastoris and its characterization. *Bioresource Technology* 134: 81–86.

Luptakova, A. and Kusnierova, M. (2005). Bioremediation of acid mine drainage contaminated by SRB. *Hydrometallurgy* 77 (1–2, 102): 97.

Maas, R. and Chaudhari, S. (2005). Adsorption and biological decolourization of azo dye Reactive Red 2 in semicontinuous anaerobic reactors. *Process Biochemistry* 40 (2): 699–705.

Macdonald, N. and Gledhill, A. (2007). Potential impact of ABCB1 (p-glycoprotein) polymorphisms on avermectin toxicity in humans. *Archives of Toxicology* 81: 553–563.

Marzan, L.W., Hossain, M., Mina, S.A. et al. (2017). Isolation and biochemical characterization of heavy-metal resistant bacteria from tannery effluent in Chittagong city, Bangladesh: bioremediation viewpoint. *The Egyptian Journal of Aquatic Research* 43 (1): 65–74.

Mathew, B.B., Singh, H., Biju, V.G., and Krishnamurthy, N.B. (2017). Classification, source, and effect of environmental pollutants and their biodegradation. *Journal of Environmental Pathology, Toxicology and Oncology* 36 (1).

Matsumura, F., Boush, G.M., and Tai, A. (1968). Breakdown of dieldrin in the soil by a micro-organism. *Nature* 219 (5157): 965–967.

Mavura, W.J. and Wangila, P.T. (2004). Distribution of pesticide residues in various lake matrices: water, sediment, fish and algae, the case of Lake Nakuru, Kenya. In: *The African Network for Chemical Analysis of Pesticides (ANCAP)- Inaugural conference Proceedings*, 1–6.

Menger, F., Boström, G., Jonsson, O. et al. (2021). Identification of pesticide transformation products in surface water using suspect screening combined with national monitoring data. *Environmental Science & Technology* 55 (15): 10343–10353.

Miao, Z., Brusseau, M.L., Carroll, K.C. et al. (2012). Sulfate reduction in groundwater: characterization and applications for remediation. *Environmental Geochemistry and Health* 34: 539–550.

Miglani, R., Parveen, N., Kumar, A. et al. (2022). Degradation of xenobiotic pollutants: an environmentally sustainable approach. *Metabolites* 12 (9): 818.

Mishra, V.K., Singh, G., and Shukla, R. (2019). Impact of xenobiotics under a changing climate scenario. In: *Climate Change and Agricultural Ecosystems* (ed. K.K. Choudhary, A. Kumar, and A.K. Singh), 133–151. Woodhead Publishing.

Mishra, S., Lin, Z., Pang, S. et al. (2021). Recent advanced technologies for the characterization of xenobiotic-degrading microorganisms and microbial communities. *Frontiers in Bioengineering and Biotechnology* 9: 632059.

Mondal, S. and Subramaniam, C. (2020). Xenobiotic contamination of water by plastics and pesticides revealed through real-time, ultrasensitive, and reliable surface-enhanced raman scattering. *ACS Sustainable Chemistry & Engineering* 8 (20): 7639–7648.

Moosvi, S., Kher, X., and Madamwar, D. (2007). Isolation, characterization and decolorization of textile dyes by a mixed bacterial consortium JW-2. *Dyes and Pigments* 74 (3): 723–729.

Mueller, J.G., Cerniglia, C.E., and Pritchard, P.H. (1996). Bioremediation of environments contaminated by polycyclic aromatic hydrocarbons. *Biotechnology Research Series* 6: 125–194.

Nadeau, L.J., Menn, F.M., Breen, A., and Sayler, G.S. (1994). Aerobic degradation of 1, 1, 1-trichloro-2, 2-bis (4-chlorophenyl) ethane (DDT) by *Alcaligenes eutrophus* A5. *Applied and Environmental Microbiology* 60 (1): 51–55.

Naik, M.G. and Duraphe, M.D. (2012). Review paper on-parameters affecting bioremediation. *Advance Research in Pharmaceuticals and Biologicals* 2 (3).

Nelson, D.L., Lehninger, A.L., and Cox, M.M. (2004). *Lehninger Principles of Biochemistry Lecture Notebook*. Macmillan.

Oaks, J.L., Gilbert, M., Virani, M.Z. et al. (2004). Diclofenac residues as the cause of vulture population decline in Pakistan. *Nature* 427 (6975): 630–633.

Olak-Kucharczyk, M., Foszpańczyk, M., Gmurek, M., and Ledakowicz, S. (2017). Decomposition of xenobiotics during visible light irradiation in the presence of immobilised photosensitisers: kinetics study. *Water Science and Technology* 75 (1): 69–74.

O'Neill, C., Hawkes, F.R., Hawkes, D.L. et al. (1999). Colour in textile effluents–sources, measurement, discharge consents and simulation: a review. *Journal of Chemical Technology & Biotechnology: International Research in Process, Environmental & Clean Technology* 74 (11): 1009–1018.

Palleroni, N.J. and Moore, D.P.E. (2010). *12 Microbiology of Hydrocarbon-Degrading Pseudomonas*. Berlin Heidelberg: Springer-Verlag.

Pande, V., Pandey, S.C., Sati, D. et al. (2022). Microbial interventions in bioremediation of heavy metal contaminants in agroecosystem. *Frontiers in Microbiology* 13: 824084.

Pattus, F. and Abdallah, M.A. (2000). Siderophores and iron-transport in microorganisms. *Journal of the Chinese Chemical Society* 47 (1): 1–20.

Pelach, M.A., Pastor, F.J., Puig, J. et al. (2003). Enzymic deinking of old newspapers with cellulase. *Process Biochemistry* 38 (7): 1063–1067.

Perkins, S.E. and Hankenson, F.C. (2020). Corrigendum to: nonexperimental xenobiotics: unintended consequences of intentionally administered substances in terrestrial animal models. *ILAR Journal* 61 (1): 103–103.

Petriello, M.C., Newsome, B.J., Dziubla, T.D. et al. (2014). Modulation of persistent organic pollutant toxicity through nutritional intervention: emerging opportunities in biomedicine and environmental remediation. *Science of the Total Environment* 491: 11–16.

Philp, J.C. and Atlas, R.M. (2005). Bioremediation of contaminated soils and aquifers. In: *Bioremediation: Applied Microbial Solutions for Real-World Environmental Cleanup*, 139–236. ASM Press.

Pongratz, R. and Heumann, K.G. (1999). Production of methylated mercury, lead, and cadmium by marine bacteria as a significant natural source for atmospheric heavy metals in polar regions. *Chemosphere* 39 (1): 89–102.

Rajeswari, K., Subashkumar, R., and Vijayaraman, K. (2013). Physico chemical parameters of effluents collected from Tirupur textile dyeing and CETP and analysis of heterotropic bacterial population. *Journal of Microbiology and Biotechnology Research* 3 (5): 37–41.

Ramel, F., Sulmon, C., Serra, A.A. et al. (2012). Xenobiotic sensing and signalling in higher plants. *Journal of Experimental Botany* 63 (11): 3999–4014.

Ren, S., Guo, J., Zeng, G., and Sun, G. (2006). Decolorization of triphenylmethane, azo, and anthraquinone dyes by a newly isolated Aeromonas hydrophila strain. *Applied Microbiology and Biotechnology* 72: 1316–1321.

Rigas, F., Dritsa, V., Marchant, R. et al. (2005). Biodegradation of lindane by *Pleurotus ostreatus* via central composite design. *Environment International* 31 (2): 191–196.

Salehizadeh, H. and Shojaosadati, S.A. (2003). Removal of metal ions from aqueous solution by polysaccharide produced from *Bacillus firmus*. *Water Research* 37 (17): 4231–4235.

Salem, H.M., Abdel-Salam, A., Abdel-Salam, M.A., and Seleiman, M.F. (2017). Soil xenobiotics and their phyto-chemical remediation. In: *Xenobiotics in the Soil Environment: Monitoring, Toxicity and Management* (ed. M. Hashmi, V. Kumar, and A. Varma), 267–280. Cham: Springer.

Salgado, R., Oehmen, A., Carvalho, G. et al. (2012). Biodegradation of clofibric acid and identification of its metabolites. *Journal of Hazardous Materials* 241: 182–189.

Samanta, S.K., Singh, O.V., and Jain, R.K. (2002). Polycyclic aromatic hydrocarbons: environmental pollution and bioremediation. *Trends in Biotechnology* 20 (6): 243–248.

Sandhya, S., Padmavathy, S., Swaminathan, K. et al. (2005). Microaerophilic–aerobic sequential batch reactor for treatment of azo dyes containing simulated wastewater. *Process Biochemistry* 40 (2): 885–890.

Saratale, R.G., Saratale, G.D., Chang, J.S., and Govindwar, S.P. (2011). Bacterial decolorization and degradation of azo dyes: a review. *Journal of the Taiwan Institute of Chemical Engineers* 42 (1): 138–157.

Sarkar, S., Banerjee, A., Halder, U. et al. (2017). Degradation of synthetic azo dyes of textile industry: a sustainable approach using microbial enzymes. *Water Conservation Science and Engineering* 2: 121–131.

Satish, G., Ashokrao, D.M., and Arun, S.K. (2017). Microbial degradation of pesticide: a review. *African Journal of Microbiology Research* 11 (24): 992–1012.

Schwab, M. (ed.) (2008). *Encyclopedia of Cancer*. Springer Science & Business Media.

Schwitzguébel, J.P., Meyer, J., and Kidd, P. (2006). Pesticides removal using plants: phytodegradation versus phytostimulation. In: *Phytoremediation and Rhizoremediation. Theoretical Background*, vol. 9 (ed. M. Mackova, D. Dowling, and T. Macek), 179–198. Springer Dordrecht.

Sette, L.D., Oliveira, V.M.D., and Manfio, G.P. (2005). Isolation and characterization of alachlor-degrading actinomycetes from soil. *Antonie Van Leeuwenhoek* 87: 81–89.

Sharma, S., Suryavathi, V., Singh, P.K., and Sharma, K.P. (2007). Toxicity assessment of textile dye wastewater using swiss albino rats. *Australasian Journal of Ecotoxicology* 13 (2): 35–39.

Sharma, A., Sharma, P., Sharma, A. et al. (2017). Hazardous effects of petrochemical industries: a review. *Petrochemical Technology* 13 (3): 1–7.

Shrestha, R., Huang, G., Meekins, D.A. et al. (2017). Mechanistic insights into dye-decolorizing peroxidase revealed by solvent isotope and viscosity effects. *ACS Catalysis* 7 (9): 6352–6364.

Singh, S.N. and Tripathi, R.D. (ed.) (2007). *Environmental Bioremediation Technologies*. Springer Science & Business Media.

Singh, N., Gupta, V.K., Kumar, A., and Sharma, B. (2017). Synergistic effects of heavy metals and pesticides in living systems. *Frontiers in Chemistry* 5: 70.

Soloman, P.A., Basha, C.A., Velan, M. et al. (2009). Electrochemical degradation of Remazol black B dye effluent. *CLEAN - Soil, Air, Water* 37 (11): 889–900.

Solomon, G., Ogunseitan, O.A., and Kirsch, J. (2000). *Pesticides and Human Health A Resource for Health Care Professionals*. eScholarship Publishing.

Souza, P.M.D., Bittencourt, M.L.D.A., Caprara, C.C. et al. (2015). A biotechnology perspective of fungal proteases. *Brazilian Journal of Microbiology* 46: 337–346.

Spain, J.C. and Nishino, S. (1987). Degradation of 1, 4-dichlorobenzene by a *Pseudomonas* sp. *Applied and Environmental Microbiology* 53 (5): 1010–1019.

Strauch, G., Möder, M., Wennrich, R. et al. (2008). Indicators for assessing anthropogenic impact on urban surface and groundwater. *Journal of Soils and Sediments* 8: 23–33.

Tariq, M., Ali, M., and Shah, Z.J.S.E. (2006). Characteristics of industrial effluents and their possible impacts on quality of underground water. *Soil & Environment* 25 (1): 64–69.

Thangavel, P. and Subbhuraam, C.V. (2004). Phytoextraction: role of hyperaccumulators in metal contaminated soils. *Proceedings-Indian National Science Academy Part B* 70 (1): 109–130.

van Beilen, J.B., Funhoff, E.G., van Loon, A. et al. (2006). Cytochrome P450 alkane hydroxylases of the CYP153 family are common in alkane-degrading eubacteria lacking integral membrane alkane hydroxylases. *Applied and Environmental Microbiology* 72 (1): 59–65.

Verma, A. and Singh, S.N. (2006). Biochemical and ultrastructural changes in plant foliage exposed to auto-pollution. *Environmental Monitoring and Assessment* 120: 585–602.

Vermassen, A., Leroy, S., Talon, R. et al. (2019). Cell wall hydrolases in bacteria: insight on the diversity of cell wall amidases, glycosidases and peptidases toward peptidoglycan. *Frontiers in Microbiology* 10: 331.

Vijayaraghavan, K. and Yun, Y.S. (2008). Bacterial biosorbents and biosorption. *Biotechnology Advances* 26 (3): 266–291.

Wang, G., Yinglan, A., Jiang, H. et al. (2015). Modeling the source contribution of heavy metals in surficial sediment and analysis of their historical changes in the vertical sediments of a drinking water reservoir. *Journal of Hydrology* 520: 37–51.

Wang, D., Li, A., Han, H. et al. (2018). A potent chitinase from Bacillus subtilis for the efficient bioconversion of chitin-containing wastes. *International Journal of Biological Macromolecules* 116: 863–868.

Wang, P., Zhang, M., Lu, Y. et al. (2019). Removal of perfluoalkyl acids (PFAAs) through fluorochemical industrial and domestic wastewater treatment plants and bioaccumulation in aquatic plants in river and artificial wetland. *Environment International* 129: 76–85.

Wilkes, R.A. and Aristilde, L. (2017). Degradation and metabolism of synthetic plastics and associated products by *Pseudomonas* sp.: capabilities and challenges. *Journal of Applied Microbiology* 123 (3): 582–593.

Xiong, A. and Jayaswal, R.K. (1998). Molecular characterization of a chromosomal determinant conferring resistance to zinc and cobalt ions in *Staphylococcus aureus*. *Journal of Bacteriology* 180 (16): 4024–4029.

Xu, F. (1996). Oxidation of phenols, anilines, and benzenethiols by fungal laccases: correlation between activity and redox potentials as well as halide inhibition. *Biochemistry* 35 (23): 7608–7614.

Xue, F., Ya, X., Tong, Q. et al. (2018). Heterologous overexpression of *Pseudomonas umsongensis* halohydrin dehalogenase in *Escherichia coli* and its application in epoxide asymmetric ring opening reactions. *Process Biochemistry* 75: 139–145.

Zaki, M.S., Hammam, A.M., and Nagwa, S. (2014). Ata. Xenobiotics as stressful factors in aquatic system (in fish). *Life Science Journal* 11 (4): 188–197.

Zhang, C., Wang, S., and Yan, Y. (2011). Isomerization and biodegradation of beta-cypermethrin by *Pseudomonas aeruginosa* CH7 with biosurfactant production. *Bioresource Technology* 102 (14): 7139–7146.

Zhu, Y., Boye, A., Body-Malapel, M., and Herkovits, J. (2017). The toxic effects of xenobiotics on the health of humans and animals. *BioMed Research International 2017*: 4627872.

# 4

# Bioremediation Strategies as Sustainable Bio-Tools for Mitigation of Emerging Pollutants

*Hamza Rafeeq*[1,2], *Zainab Riaz*[1], *Anum Shahzadi*[1], *Shazaf Gul*[2], *Fatima Idress*[2], *Sidra Ashraf*[2], *and Asim Hussain*[2]

[1] *Department of Biochemistry, Riphah International University, Faisalabad, Pakistan*
[2] *Department of Biochemistry, University of Agriculture, Faisalabad, Pakistan*

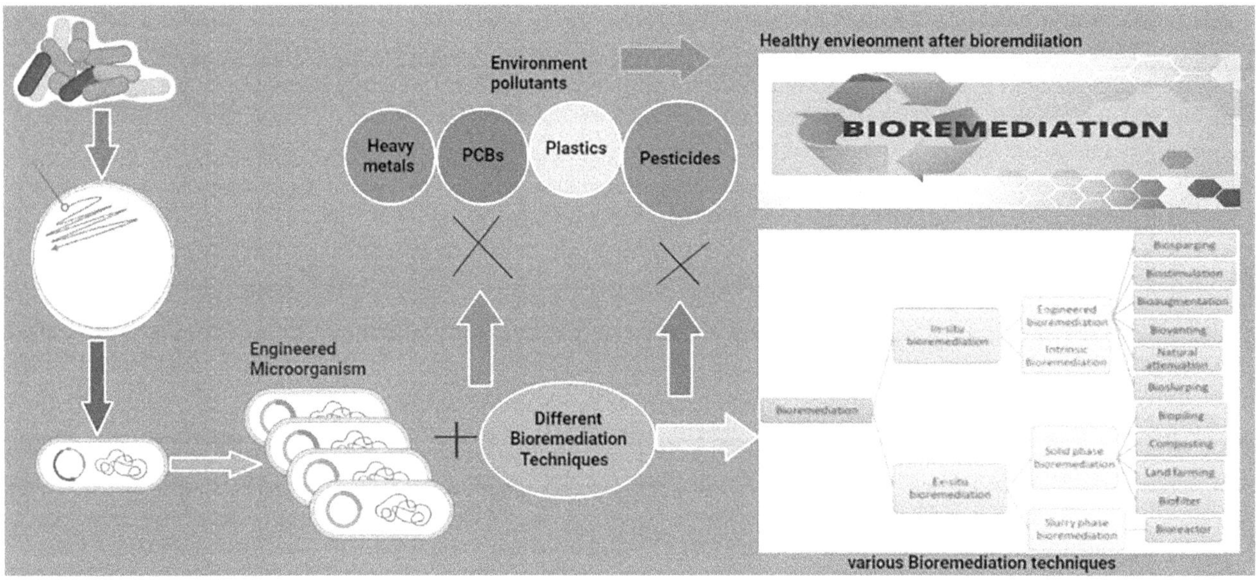


## 4.1 Introduction

Global industrialization has resulted in pollution of the environment, waterways, and topsoil. Water quality has deteriorated because of human activities such as mining and the eventual removal of toxic metal effluents from steel mills, battery firms, and energy generation, raising serious environmental issues. Petroleum, polythene, and trace metals are examples of effluents that impact the ecosystem. Heavy metals are contaminants found in the Earth's crust that are difficult to breakdown. They are found as ores in rocks and recovered as minerals. Heavy metals may be released into the environment because of high-level exposure. They stay hazardous in the environment for much longer. Many of these contaminants are mutagenic to both people and the environment (Alvarez et al. 2017). Heavy metals absorbed accumulate in the brain, liver, and kidney. Cancer, nervous system damage, stunted development, and even death are some of the other impacts on animals. Soil heavy metals diminish food quality and quantity through interfering with nutrient absorption, plant development, and physiological metabolic processes. Metal-contaminated soils are being treated chemically, biologically, and physically. However, since physicochemical approaches generate a lot of waste and pollutants, they are not highly regarded. Bioremediation is a viable and cost-effective method of eliminating environmental toxins (Kalogerakis et al. 2015).

*Microbes Based Approaches for the Management of Hazardous Contaminants*, First Edition. Edited by Ajay Kumar, Livleen Shukla, Joginder Singh, and Luiz Fernando R. Ferreira.

Soil microorganisms have important roles in plant growth promotion, pest control, soil conservation, nutrient recycling, and pollution reduction. In terms of efficiency, affordability, and social acceptability, bioremediation has come a long way. Most of the bioremediation research has concentrated on bacterial processes, which have diverse uses. Archaea are recognized to have a function in bioremediation in several applications involving bacteria. Many adverse environments have deteriorated, necessitating bioremediation (Silva et al. 2018). Microbes may also help with pollutant removal from hyperthermal, acidic, hypersaline, or basic industrial waste. According to a recent study, utilizing more than one live organism improves efficiency and outcomes while allowing for increased microbial variety in bioremediation. Many scientists used the bioremediation method to remove organic and inorganic contaminants (Kumar et al. 2020). Research employed bioremediation technology to remediate several contaminants, including organophosphate pesticides such as chlorpyrifos, methyl parathion, and profenofos by *Aspergillus sydowii* and chloramphenicol by endophytic fungi. Another study found that *Cymbella* sp. can detoxify naproxen-polluted water with 97.1% efficiency (Kaur et al. 2023).

The utilization of microbial enzymes to break down hydrocarbons into less hazardous molecules is required for a bioremediation strategy. The widespread usage of genetically engineered microbes that can also aid in the elimination of petroleum, naphthalene, toluene, benzene, and other xenobiotic compounds is now being investigated. Several variables impact bioremediation for better results, including the surrounding environment's temperature, aerobic or anaerobic conditions, and nutrient availability (Pham et al. 2022). Emerging environmental contaminants, such as persistent organic compounds (OCs), heavy metals, poisons, and air pollutants of synthetic or natural origin, enter ecosystems primarily via human activities and endanger life forms such as plants, animals, and people. Bioremediation is one of the most cost-effective and environmentally friendly biotechnological technologies. Bioremediation is the primary method of waste management. It is capable of removing persistent organic pollutants, which are difficult to degrade and are assumed to be heterologous biological compounds. This study focuses on the most current methodologies and most up-to-date information on bioremediation strategies for eco-friendly detoxification and the efficient degradation of diverse organic and inorganic toxins to manage environmental pollution (Skariyachan et al. 2022).

## 4.2 Bioremediation by Microbial Strains

The addition of microbes to nutritional chains helps to maintain biological homeostasis. Bioremediation is the process of removing polluted materials from the environment utilizing bacteria, algae, fungi, and yeast. Microbes may develop at temperatures as low as 196°F and as high as 1200°F in the presence of hazardous substances or any waste stream. Microbes are an excellent candidate for cleanup due to their versatility and biological systems (Silva et al. 2018). For bacteria, carbon is the most vital nutrient. Bioremediation was carried out using microbes from various settings. Microbes include *Achromobacter, Alcaligenes, Xanthobacter, Arthrobacter, Pseudomonas, Bacillus, Mycobacterium, Corynebacterium, Flavobacterium, Nitrosomonas*, and other bacteria (Jha et al. 2023). Bioremediation by means of Microorganism is given in Figure 4.1.

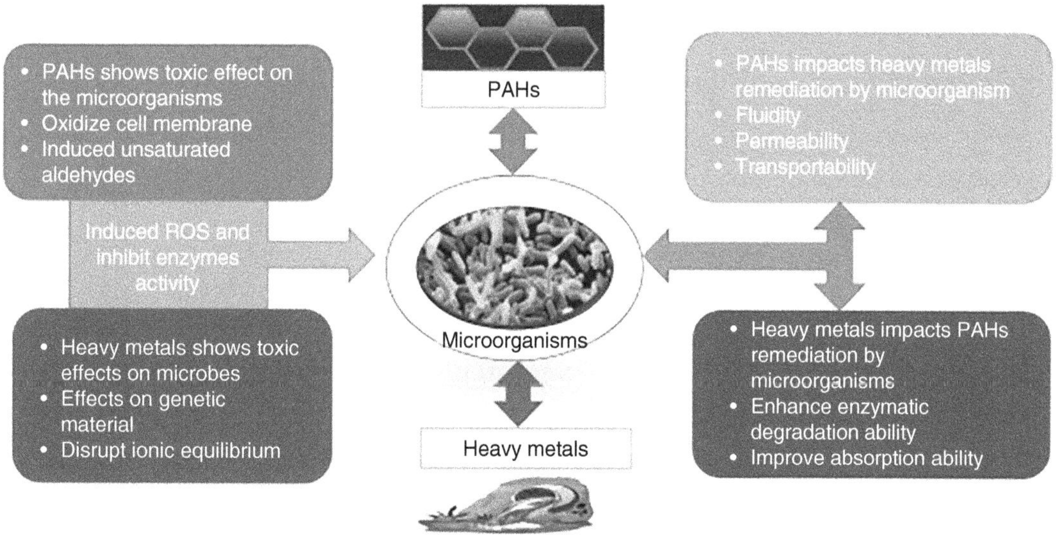


**Figure 4.1** Bioremediation by mean microorganisms.

### 4.2.1 Aerobic

Numerous microorganisms have been shown to bioremediate several distinct classes of environmental contaminants in an aerobic setting. Aerobic bacteria such as *Bacillus, Pseudomonas, Sphingomonas, Flavobacterium, Nocardia, Rhodococcus,* and *Mycobacterium* may breakdown many different types of complex chemical molecules (Bhatt et al. 2023). These microorganisms have been demonstrated to break down a variety of organic pollutants, including pesticides, alkane hydrocarbons, and polyaromatic chemicals. There are microbes that can exploit these pollutants to get the carbon and energy that they need to survive. During aerobic bioremediation, the amount of oxygen limits how many microorganisms can grow (Rafeeq et al. 2023).

### 4.2.2 Anaerobic

The bioremediation of polychlorinated biphenyls (PCBs), chlorine compounds, and the chlorinated solvents, trichlorethylene, and chloroform is gaining popularity due to the use of amphibious bacteria that digest and transform contaminants into less hazardous forms (Liu et al. 2019). Anaerobic bioremediation makes use of a variety of microorganisms, including *Pseudomonas, Aeromonas*, and sulfate-reducing bacteria. In various settings, Garg and Tripathi found that azo dyes, when contaminated by microorganisms, lost their original hue. Through reduction processes that use electrons liberated during the organic substrate's oxidation, azo dyes may break down in the absence of oxygen. The efficiency with which color is removed from a microbe's environment would be significantly influenced by its electrochemical characteristics because of such regulated dye decolorization processes. For industrial processes to obtain such time-variant decolorized metabolites, dyes were anaerobically decolored (Bilal and Iqbal 2019). On the other hand, to conduct more accurate study, it was necessary to supplement the decolorized metabolites collected under controlled settings.

## 4.3 Factors Affecting Microbial Bioremediation

Bacteria, algae, fungus, and plants may be used in bioremediation since these organisms are capable of degrading, transforming, removing, immobilizing, and detoxifying many different types of physical and chemical contaminants. Enzymes in microorganisms' metabolic pathways accelerate biochemical processes that degrade contaminants. To eliminate pollution, microbes need to interact with substances that fuel their growth and provide them with nutrition. Physical, chemical, biological, soil-type characteristics, carbon and nitrogen supplies, kinds of microorganisms (i.e., single or consortium), and other elements all impact the bioremediation process (Bilal and Iqbal, 2020). Figure 4.2 depicts the factors that influence bioremediation. Because diverse species work together to utilize all substrates in the best manner possible, the bioremediation effectiveness of microbial consortiums is often higher than that of single microorganisms, and these consortiums frequently exhibit both multifunctionality and resistance. Researchers found that carbon is one of the most crucial nutrients for *in situ* bioremediation because it boosts the metabolic activity of native microbial populations and expedites the rate at which existing contaminants are degraded. Organic carbon may be the most widely used ingredient in bioremediation. Many microbes do well in places with no oxygen because they can turn organic carbon into hydrogen gas (Mondal et al. 2022).

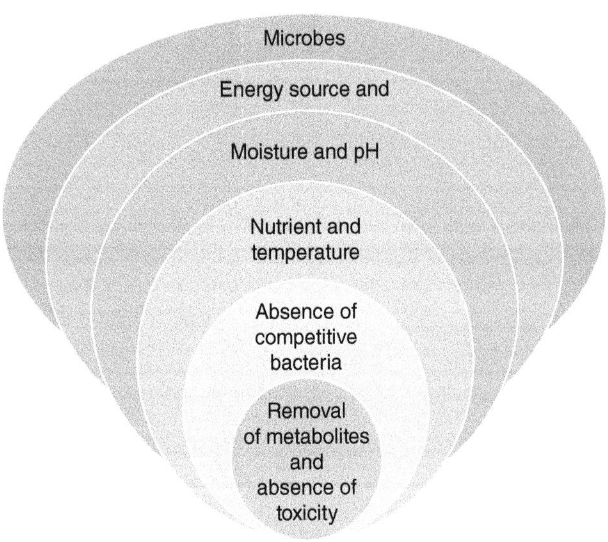


**Figure 4.2** Factors that affect bioremediation.

Results showed that the removal effectiveness of contaminants differed between sandy soil and clay soil, suggesting that soil type has a major impact on bioremediation. Bioremediation is only effective if it makes use of the naturally occurring microorganisms and physicochemical features of the environment. The abundance of pollutant-degrading microbes, contamination accessibility, and the following are all considered. Table 4.1 lists the various factors that influence bioremediation (Hussain et al. 2022a).

**Table 4.1** Factors affecting bioremediation.

| Factors | Remarks | References |
|---|---|---|
| Moisture content | Microorganisms need a sufficient quantity of water in order to grow. When the soil is excessively damp, the biodegradation agents are less effective | Jangir et al. (2019) |
| Oxygen | The use of organisms that do not need oxygen may boost biodegradation rates. Because most living creatures need oxygen to thrive, anaerobic breakdown happens. In most circumstances, the addition of oxygen may enhance hydrocarbon metabolism | Mazumder et al. (2022) |
| Biological factors | Soil microorganisms fight for carbon sources, while bacteriophages and protozoa feed on one other, all of which may have an impact on organic compound breakdown. Contaminants and catalyst levels impact derivatization rates. Expressed enzymes may either accelerate or decrease contaminant breakdown. Enzymes must also be engaged in contaminant metabolism in order for the contaminant to have affinity and availability. The following are the most important biological factors: interaction (competition, predation, and succession), population size, and composition | Alegbeleye et al. (2017) and Sodhi et al. (2021) |
| Temperature | Temperature is the most critical physical element regulating microbe life and hydrocarbon composition. Natural oil deterioration is gradual in freezing areas like the Arctic, placing extra strain on microorganisms to clean up spilled oil. The subzero water here freezes the microbial transport channels, preventing them from performing their metabolic tasks. The metabolic turnover of enzymes involved in degradation is affected by temperature. Furthermore, each compound's breakdown requires a certain temperature<br>Temperature influences microbial physiological parameters, which either accelerates or retards bioremediation. At warmer temperatures, microbial activity increases. It begins to drop abruptly when the temperature rises or falls, and then gradually stops | Sharma et al. (2022) and Ren et al. (2023) |
| Nutrients | Nutrients may impact microbial growth and reproduction, as well as the pace and efficacy of biodegradation. Optimizing the bacterial C:N:P ratio may increase biodegradation efficiency, particularly when important nutrients like N and P are present. Microorganisms need a variety of nutrients to live, including carbon, phosphorus, and nitrogen. Hydrocarbon decomposition is likewise inhibited at low concentrations. Adding nutrients to cold conditions may boost the metabolic activity of microorganisms and consequently the pace of biodegradation.<br>The availability of nutrients limits aquatic biodegradation. Microbes that consume oil need resources to grow. Natural sources of these critical elements are limited | Kebede et al. (2021) and Mupambwa et al. (2022) |
| pH | The acidity, alkalinity, and basicity of a substance influence microbial metabolism and the elimination process. The pH of the soil may predict microbial development. Even slight pH changes have a large influence on metabolic activities | Rajkumar and Kurinjimalar (2021) |
| Metal ions | Metals are required by bacteria and fungi, but in excess, they impair cell metabolism. Metal compounds impact degradation rates both directly and indirectly | Padhan et al. (2021) |
| Site characterization and selection | Before proposing a bioremediation solution, appropriate remedial study work is required to assess the degree of the pollution. Determining the horizontal and vertical extent of contamination, defining criteria, and sample sites, and describing sampling and analysis methodologies are all part of the site selection processes | Sangwan and Dukare (2018) |
| Microorganisms | Toxic substances in high amounts may kill microorganisms and impede the cleaning process. The toxicant, concentration, and bacteria exposed all influence toxicity | Boopathy (2000) |

### 4.3.1 Principle of Bioremediation

Bioremediation refers to the process of biologically degrading organic wastes in a contained setting. By providing the organisms with the nutrients and other chemicals they require to operate efficiently, hazardous compounds may be destroyed or detoxified through bioremediation. Enzymes are necessary in every step of metabolism. These enzymes are related to oxidoreductases, lyases, transferases, and hydrolases. Many enzymes have the ability to degrade a large variety of substrates because they have both broad and narrow substrate specificities (Hussain et al. 2022b). In order for bioremediation to work, the contaminants must be broken down by enzymes. During bioremediation, environmental factors are routinely adjusted to hasten microbial growth and breakdown. This is because beneficial bacteria need favorable conditions in which to thrive and spread before bioremediation may be effective. Bioremediation is a natural process that may be aided by adding living organisms and nutrients. Bioremediation techniques rely heavily on the biodegradation process. It is the transformation of $CO_2$ and water, two common organic pollutants, into inorganic chemicals that are harmless to all living things, including people, plants, animals, and fish (Perera and Hemamali 2022).

## 4.4 Classification of Bioremediations

Bioremediation may be used in a variety of ways, and some of the more prevalent applications are described below and the various strategies of Bioremediation shown in Figure 4.2.

### 4.4.1 Land Farming

Due to its low operating expenses and absence of specialist equipment, land farming is the most effective and straightforward bioremediation approach. The most common approach is *ex situ* bioremediation; however, *in situ* bioremediation is also possible (Mohanty and Paul 2022). This is because of the placement of the therapy. Regularly removing and tilling toxic soils is standard agricultural practice, and the location of treatment determines the kind of bioremediation. *Ex situ* bioremediation techniques, as opposed to *in situ* bioremediation, are used to clean up polluted soil. Typically, extracted contaminated soils are deposited on a permanent layer of substrate far above the surface of the earth to allow natural microorganisms to breakdown pollutants aerobically (Xue et al. 2022a). Land farming bioremediation technology is a simple way to clean up contaminated soil on land. It costs little money, has a small impact on the environment, and uses little energy (Bajpai et al. 2022).

### 4.4.2 Biopile

Bioremediation is the use of aeration and nutrient supplementation to boost the metabolic activity of microorganisms in piled-up, contaminated soil above ground. This technique includes aeration, nutrients, irrigation, leachate collection, and treatment bed systems. This approach is gaining popularity for *ex situ* biodegradation due to its cost-effectiveness and helpful characteristics, such as pH and nutrient management. It is possible to use the biopile to decontaminate contaminated cold settings and remediate low-molecular-weight volatile contaminants (Arunraja et al. 2023). The flexibility of the biopile enables a decrease in remediation time by boosting microbial activity, pollutant availability, and biodegradation rate. When warm air is supplied into the biopile system to simultaneously deliver air and heat, bioremediation is enhanced. The cleanup process of the biopile has been aided by the inclusion of bulking agents such as straw, sawdust, and wood chips. *Ex situ* bioremediation methods, such as land farming, biosparging, and bioventing, may be used to refill the air supply to contaminated heaped soil in biopiles. However, these approaches are costly to install and require distant power sources. Extreme air temperatures that dry soil and make it more likely to evaporate than to be broken down by living organisms may impede bioremediation. Bio-available organic carbon (BOC) is crucial for bioremediation while using the biopile technique (Maqsood et al. 2023). Utilizing alpha, beta, and gamma proteobacteria, petroleum-contaminated soil was bioremediated using mesophilic conditions (30–40°C) and a modest aeration rate for the removal of total petroleum hydrocarbons (TPH). Biopile systems have also been deployed for the remediation of diesel-contaminated subantarctic soil. Within one year, 93% of the TPH were eliminated utilizing the biopile technology (Gupta et al. 2022).

### 4.4.3 Bioreactor

Bioreactors turn basic materials into particular products via a sequence of biological processes. Bioremediation flourishes in bioreactors, which offer optimal growth conditions. The samples of remediation are put in a bioreactor (Bora et al. 2023). Using a bioreactor to remediate polluted soil has several advantages over *ex situ* bioremediation methods. An effective bioremediation procedure based on bioreactors that can accurately control pH, agitation, temperature, aeration, substrate concentration, and inoculum concentration reduces bioremediation time dramatically. Biological responses may occur when the bioreactor can be altered and controlled. The different strategies of bioremediation has been shown in Figure 4.3.

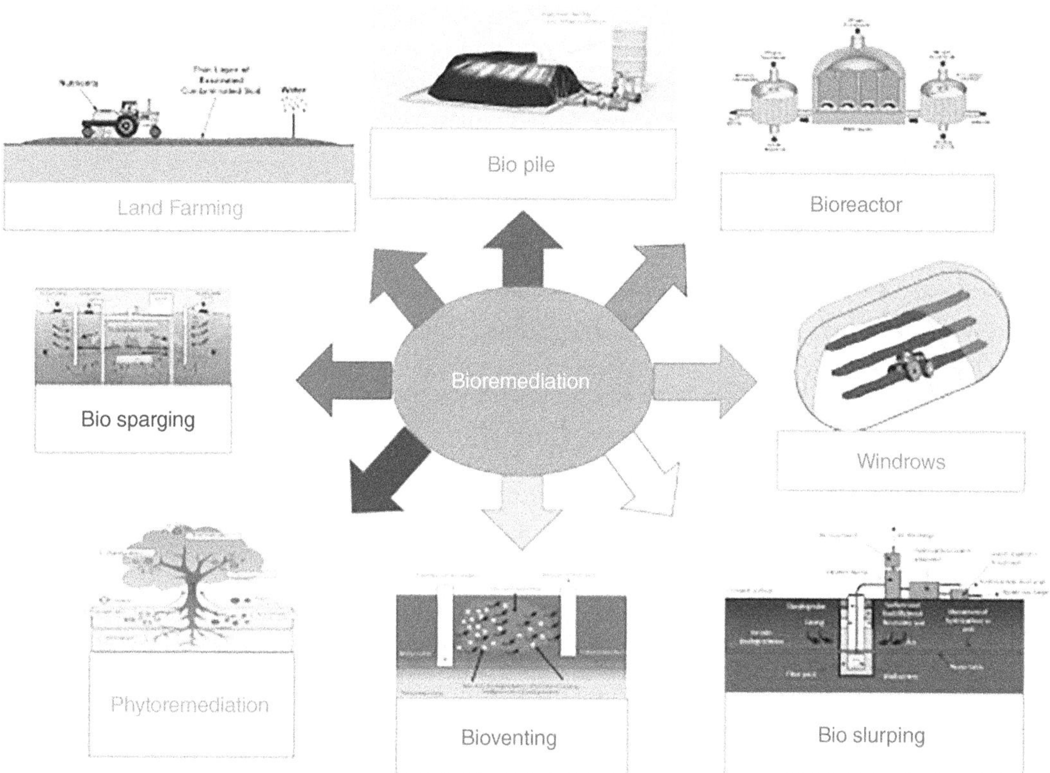


**Figure 4.3** Strategies for bioremediation.

#### 4.4.3.1 *In Situ* Bioremediation Techniques

These techniques include cleaning up contaminated chemicals at their source. It requires no excavation or disruption of the surrounding soil. *Ex situ* bioremediation methods should be less expensive than these methods. *In situ* bioremediation strategies that can be enhanced include bioventing, phytoremediation, and biosparging. However, intrinsic bioremediation and natural attenuation cannot be enhanced (Ahmad et al. 2023a). Chlorine, paints, toxic metals, and hydrocarbon-contaminated sites have been satisfactorily remedied using *in situ* bioremediation methods. Two main forms of *in situ* bioremediation may be distinguished: intrinsic and engineered.

#### 4.4.3.2 Intrinsic *In Situ* Bioremediation

*In situ* bioremediation is often referred to as "natural reduction." Intrinsic bioremediation uses contaminated sites in a noninvasive (human intervention) manner. This method is intended to promote an existing microbial population. Microorganisms rely on aerobic and anaerobic processes for the biodegradation of harmful compounds, including those that are resistant to breakdown (Meyer et al. 2023). This approach is less expensive since there is not much use of power behind it. Utilizing anaerobic reductive dechlorination, aerobic treatment, amendment administration, biosparging, and bioslurping intrinsic *in situ* bioremediation is possible. *In situ* bioremediation has been employed as a solution for the biological treatment of blocked groundwater using a stimulation–optimization strategy that is driven by Particle Swarm Optimization Algorithm-Extreme Learning Machine (ELM–PSO) approaches. *In situ* bioremediation was used to perform this approach. This results in a less expensive technology for the pumping system and reduces the amount of capital required for the whole operation. Using *in situ* bioremediation, the concentration of pollutants was decreased from 10 to 5 ppm (within the allowable limit) in 3 year (Thacharodi et al. 2023a). *In situ* remediation has also been investigated for the removal of Cr(VI) from shallow, unsaturated soil. Microorganisms are able to thrive in soils with high concentrations of Cr(VI), and their sub-cellular machinery is used to interact with heavy metals. Microorganism inoculants could be used to treat heavy metals *in situ*. Also, Cr(VI) interacts with Fe(II) ions through redox processes, and the release of soluble iron speeds up reductive reactions (Ojha et al. 2023).

#### 4.4.3.3 Engineered *In Situ* Bioremediation

In the second procedure, a particular microbe is introduced into the contaminated region to clean it up. *In situ* bioremediation is a technology that uses genetically modified microbes to accelerate the degradation process. This is performed by improving the physicochemical conditions that promote microbial growth. As shown in Table 4.2, bioremediation techniques have numerous limitations.

### 4.4.4 Windrows

The spinning of polluted soils in windrows improves bioremediation by hastening the biodegradation of native and transient hydrocarbon plastics in the soils. Accelerating bioremediation is possible through acclimation, biological treatment, and mineralization of toxic soil, all of which increase soil aeration, mineralization, and biotransformation (Vulpe et al. 2023). When it comes to extracting hydrocarbons from soil, the biopile method outperforms the windrow treatment. For bioremediation of soil contaminated by toxic volatile compounds, the periodic rotation associated with windrow remediation is not a superior selection technique. The anaerobic system that develops inside the piled polluted soil produces greenhouse gas ($CH_4$) during windrow treatment. The Gurugram–Faridabad dumpsite in Bandhwari, India, has been bioremediated using the windrow technique by constructing terraces and windrows and using bioculture, and the results have shown a reduction in pollution (Rani et al. 2022).

### 4.4.5 Bioslurping

Vacuum-assisted pumping, bioventing, and soil vapor extraction (SVE) are utilized to reach soil and groundwater levels for restoration by providing a direct oxygen supply and stimulating pollutant biodegradation. Light non-aqueous phase liquids (LNAPLs), including those in unsaturated and saturated zones, may be recovered using this method. Soils polluted with

**Table 4.2** Limitations of various bioremediation techniques.

| Techniques | Limitations | References |
|---|---|---|
| Land Farming | The disadvantage of this strategy is that the goals defined in the constraint set must be properly followed; else, the problem seems unsolvable. Fresh organic waste may be problematic because it can sometimes cause anoxic conditions that are harmful to plant growth. To retain the quality of preexisting soils, it is best to avoid adding additional organic material over time | Deeb et al. (2020) |
| Biopile | The possibility for biodegradation using the biopiling approach is reduced if the materials become hydrophobic due to the level of weathering | |
| Windrows | The most significant barrier in researching windrows *in situ* is probably not knowing where and when they will appear. Although certain sub-mesoscale convergences may be predicted, it is still difficult to anticipate where and when litter windrows will occur due to the added uncertainty introduced by the litter loading dependence | |
| Bioventing | The downside of this strategy is that it only operates at the lowest depths of the polluted soil environment | Yaashikaa et al. (2022) |
| Bioreactor | The fundamental restriction of utilizing membrane bioreactors (MBR) at such high Mixed liquor suspended solids (MLSS) concentrations seems to be the extremely poor to nil oxygen transfer efficiency recorded when using standard diffused aeration methods (such as fine and coarse bubble diffusers). This implies that a better knowledge of the limits imposed by classic bubble diffusers (measured in terms of the alpha factor) under that particular set of operating parameters is necessary (high MLSS) | Kim et al. (2019) |
| Intrinsic *in situ* bioremediation | The fundamental restriction of *in situ* bioremediation is the need for extremely porous soil at the location | Kumar et al. (2018) |
| Phytoremediation | Phytoremediation, which includes phytoextraction and rhizodegradation, is used to treat contaminated soil in the soil's surface layers. This method may be time-consuming and may not be capable of removing all pollutants | Capuana (2020) |

volatile and semivolatile OCs may be cleaned up with the use of this method. A "slurp" that penetrates the free product layer extracts liquid from it (Wesseler et al. 2023). The pumping mechanism brings light non-aqueous-phase liquids to the surface, where they are isolated from the atmosphere and water. Moisture in the soil is employed in this method to decrease air permeability and oxygen transfer rate, both of which contribute to microbial activity. Although it is not optimal for remediation in low permeable soils, this approach saves money on storage, disposal, and treatment since it consumes less groundwater (Iravani and Varma 2022). Bioslurping calls for digging down 25 ft, but it's worth it to get rid of the pollutants in the water. Methods of bioventing, including aerobic bioremediation of polluted soil *in situ*, are combined in this method. A vacuum-enhanced system is used to recover free product from the capillary fringe using LNAPLs. Once the bioslurping tube has been vacuumed, any unbound product is "sucked" up the tube and into a trap or oil–water separator for further processing. Removing LNAPL causes its level to decline, which in turn attracts more LNAPLs from farther away and draws them into the bioslurping well (Narayanan et al. 2023). When the fluid level in the bioslurping well lowers due to vacuum extraction of LNAPL, the bioslurping tube begins to remove vapors from the unsaturated zone. By increasing the circulation of soil gases, vapor extraction promotes aerobic biodegradation and aeration.

### 4.4.6 Bioventing

Bioventing is a method of increasing the activity of native bacteria for bioremediation by introducing oxygen into the unsaturated zone via the use of controlled airflow. During bioventing, nutrients and moisture are added to promote the bioremediation process. As a result, harmful contaminants will be broken down by microbes into less dangerous ones (Abo-Alkasem et al. 2023). In recent years, this technique has piqued the interest of other *in situ* bioremediation strategies. Bioventing is a way to make biodegradation work better by giving air to and encouraging the native microflora. This makes it more likely that heavy metal pollutants will be taken out of the environment by rain (Novello et al. 2023).

### 4.4.7 Phytoremediation

Phytoremediation is a method for decontaminating contaminated soils. This approach leverages plant interactions at the physical, biological, chemical, biochemical, and microbiological levels to mitigate the harmful effects of pollution in polluted environments (Shahbaz et al. 2023). Phytoremediation uses a wide range of techniques, which change according to the amount and kind of contaminant being removed. Removal of contaminants, such as heavy metals, is often accomplished by extraction, sequestration, and transformation. Organic pollutants such as oils and chloro-compounds may be removed by decomposition, immobilization, rhizoremediation, and evaporation using plants such as willow and alfalfa (Xue et al. 2022b). Important factors in plants that serve as phytoremediators include their tap root system or fibrous root system, penetration, toxicity levels, adaptability to the harsh environmental conditions of the contaminants, plant annual growth, supervision, and, most notably, the time needed to reach a standard of cleanliness. Moreover, the plant cannot easily be damaged by pests or weakened by disease. The extraction of contaminants from plant tissues, including the roots and the shoots, is a crucial step in phytoremediation (Ainur and Sujatha, 2023). Transpiration and partitioning play an important role in the transport of water and nutrients. It is possible to modify this procedure with respect to pollutants and plant nature. Most plants growing in a contaminated area can aid in the cleanup process called phytoremediation. Plants, either natural or manmade, or a mix of the two, may be used to bio-enhance native plants in contaminated areas. One of these is phytomining, which uses plants to get valuable metals out of places where they have been polluted (Mahamuni-Badiger et al. 2023).

More than 300 plant species have been identified as superior Cu, Zn, and Ni sorbers, making them good candidates for phytoremediation. Phytostabilization, also known as *in situ* inactivation or immobilization of heavy metals, reduces both heavy metal bioavailability and off-site transfer (Rathore et al. 2022). It replenishes metals by absorbing them at the plant's roots. Some plant species are able to tolerate cadmium in their environment, and these include *Acanthus ilicifolius* and *Virola surinamensis*. As a result of their ability to phytostabilize CD, the following five ornamental plants were selected: *Cinnamomum camphora, Osmanthus fragrans, Euonymus japonicus, Ligustrum vicaryi,* and *Loropetalum chinense*. Bacterially assisted phytoremediation is an effective method for cleaning up metal-contaminated water in a variety of settings (Fouad et al. 2023). The phytoremediation method of reducing metals in wastewater by using plants can use groups of rhizobacteria that help plants grow, bacteria that break down metals, and bacteria that live inside plants.

### 4.4.8 Biosparging

Just like in bioventing, air is delivered deep into the soil to stimulate microbial activity, which in turn eliminates contaminants. Bioventing is a technique that may be used instead of traditional biodegradation procedures. It entails pumping air into a saturated zone to propel volatile OCs into an adjacent unsaturated zone (Rosca et al. 2019). Biospreading works best when the soil is porous and the contaminants can be broken down naturally. Bioventing and SVE rely on *in situ* Air Sparging (IAS) to volatilize pollutants with high air flow rates and biosparging to promote microbial degradation. Biosparging is often used to get rid of fuel and kerosene in water systems (Muter 2023). During enhanced bioremediation, microorganisms are fed oxygen in order to speed up the biodegradation processes. Adsorption, microbial degradation, biosparging, PRBs, and the usage of modified or synthetic zeolites are only some of the methods that may be used to remove organic contaminants benzene, toluene, ethylbenzene, and xylene (BTEX). Natural zeolite is a cheap and commonly accessible material that might be used for BTEX adsorption, although there have not been many studies on it (Ghosh et al. 2023a).

## 4.5 Bioremediation of Various Pollutants

### 4.5.1 Bioremediation for Inorganic Pollutants

Common inorganic pollutants include toxic heavy metals and their compounds that originate in the mining, power generation, metalworking, and chemical production industries. The discharge of hazardous metals into soils and rivers on or below the surface poses unacceptable health hazards, making this a major worry for environmentalists (Kausar et al. 2022; Khan et al. 2022; Rakib et al. 2022; Rasheed et al. 2022; Rizwan and Bilal 2022; Sarker et al. 2021). Know that microbes can only alter the oxidation states of metals to stabilize them, not destroy them. Metals are just another nutrient that may be metabolized and detoxified by them (Rather and Raina 2022). Bioremediation of organic and inorganic contaminants has been credited to a number of different microorganisms. The bioavailability of metal ions may be significantly altered by microbes that emit chelating agents and acids, as well as those that affect physicochemical features such as redox potential in their environment (Thai et al. 2023). Figure 4.4 shows degradation of organic and inorganic pollutants. The first phase in the interaction between metals and microbial cells include physical adsorption, biosorption, and ion complexation. Metal ions are biochemically transformed by microbes by the action of enzymes for oxidation, methylation, reduction, precipitation, and dealkylation.

The multidrug-resistant *Pseudomonas aeruginosa* T-3 isolate from tannery effluent demonstrates the potential of microorganisms to adapt to heavy metals including iron, zinc, chrome, magnesium, mercury, and barium present in textile

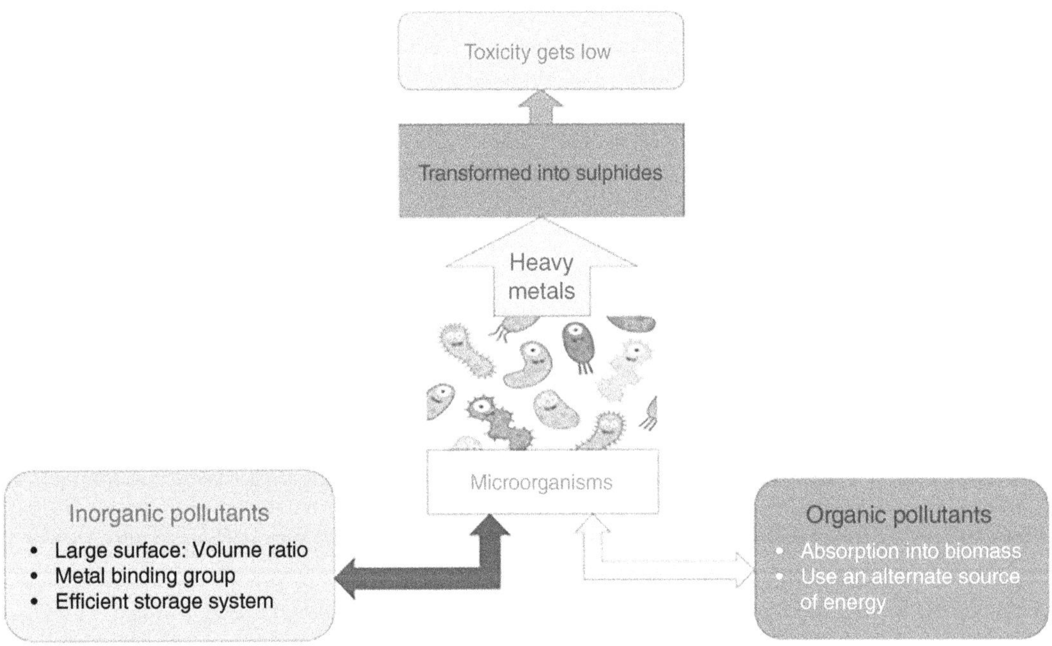


**Figure 4.4** Degradation of organic and inorganic pollutants.

waste. That microorganisms can adjust to new surroundings, demonstrates their resilience. The plasmid-encoded PhCN gene in *Pseudomonas putida* provides the bacterium with resistance to Copper and cadmium (Preetha et al. 2023). Recombinant *Escherichia coli* expressing the metallothionein gene (Neurospora crasa) for Cd absorption was created using plasmid-encoded biochemical information and genetic engineering approaches, resulting in substantially quicker Cd uptake than the donor microbe. Both *Staphylococcus xylosus* and *Staphylococcus carnosus* were genetically modified to bind nickel by introducing a poly-histidyl peptide that contained genes for this ability (Yadav and Sharma 2023). The bioremediation of various organic and inorganic pollutants by means of various microorganisms is given in Table 4.3.

**Table 4.3** Potentially active microorganism and their organic and inorganic target.

| Microorganisms | Organic and inorganic Compounds | References |
|---|---|---|
| *Alcaligenes odorans, Corynebacterium, propinquum, Fusarium* sp. and *P. aeruginosa* | Oils | Pande et al. (2020) |
| *B. coagulans, B. cereus, B. cereus A, Serratia ficaria* and *P. cepacia* | Diesel oil | Miri et al. (2022) |
| *Saccharomyces cerevisiae, Candida krusei, C. glabrata,* and *Aspergillus niger* | Crude oil | Sangkharak et al. (2020) |
| *B. subtilis* strain NAP1, NAP2, NAP4 | Oil based paints | Phulpoto et al. (2016) |
| *Staphylococcus aureus, B. macerans, K. oxytoca* and *B. firmus* | Vat dyes | Sangkharak et al. (2022) |
| *B. cereus, E. aurantiacums* and *A. baumanii* and *Exiguobacterium indicum* | Azo dye effluents | Raquel et al. (2013) |
| *Staphylococcus* and *Bacillus* | Endosulfans | Liu et al. (2018) |
| *Pseudomonas* sp., *Acinetobactor* sp., *Photobacterium* sp. and *Enterobacter* sp. | Methyl parathion and chlorpyrifos | Halak et al. (2006) |
| *P. putida* (F1), *Penicillium chrysogenum* | Toluene and its derivatives | Bustard et al. (1998) |
| *Corynebacterium propinquum, Alcaligenes odorans, P. aeruginosa* and *B. subtilis* | Phenol | Gaur et al. (2018) |
| *Microbacterium* sp., *Acinetobacter* sp., *Pseudomonas* sp. and *Ralstonia* sp. | Aromatic hydrocarbons | Basu et al. (2018) |
| *Phanerochaete chrysosporium* | 4,4-Dibromodiphenyl ether | Sen et al. (2019) |
| *A. radiobacter* (S2) | Sulfonate benzene | Tripathi and Garg (2014) |
| *Kluyveromyces marxianus (IMB3)* | Remazole Black B | Meehan et al. (2000) |
| *Myceliophthorathermophila* and *Burkholderia* sp. | Polycyclic aromatic hydrocarbons | Mohapatra and Phale (2021) |
| *Klebsiella pneumonia* (RS-13) | *N,N*-dimethylpphenylenediamine | Giriyan et al. (2021) |
| *P. putida* (GJ31) | Chlorobenzenes | Kunze et al. (2009) |
| *Arthrospira platensis, Oscillatoria laete-virens, Spirogyra insignis* and *Pseudochlorococcum typicum* | $Pb_{21}$ | Song et al. (2019) |
| *Ulothrix tenuissima, Spirulina* sp., and *C. reinhardtii* | $Sr_{51}$ | Aregbesola et al. (2020) |
| *Chlorella* spp. *Spirulina* sp. (HD-104) | $Cr_2O_{722}$ | Geetha et al. (2021) |
| *Pseudochlorococcum typicum, Cyclotella cryptica, Chlamydomonas reinhardtii, Spirogyra hyaline* | $Hg^{2+}$ | Shah and Jain (2020) |
| *Rhodopseudomonas palustris Aerococcus* sp. | Lead, chromium, and cadmium | Sravya and Sangeetha (2022) |
| *Aeromonas* sp., *P. aeruginosa* | Uranium, copper, nickel, and chromium | Gaur et al. (2022) |
| *Paecilomyces* sp. *A. versicolor, A. fumigatus, Terichoderma* sp. *Paecilomyces* sp., and *Cladosporium* sp. | Cadmium | Unuofin et al. (2021) |
| *Lysinibacillus sphaericus* CBAM5 | Cobalt, chromium, copper, and lead | Kharangate-Lad and D'Souza (2021) |
| *Pseudomonas putida* | $Cr^{6+}$ | Garg et al. (2022) |
| *Cunninghamella elegans* and *S. cerevisiae* | Heavy metals, mercury, nickel, and lead | Duc et al. (2021) |

### 4.5.2 Bioremediation for Organic Pollutants

Because of the pervasive and extensive usage of these chemicals in the environment, OCs such as biocides and flame retardants are now regarded as a hazard to practically all kinds of life on the planet. Microbes can breakdown most OCs in the environment, including PCBs, polybrominated biphenyl ethers (PBEs), and polycyclic aromatic hydrocarbons (PAHs). The process by which bacteria break down organic substances into less hazardous or wholly nontoxic remains is known as biodegradation. The microorganisms eat the organic substrate to obtain organic carbon and energy (Vijayanand et al. 2023). Individual microbial species that are isolated from other microorganisms generally do not breakdown organic substrates and perform well in a community. Resistance, chemical degrading capacity, and tolerance are all provided via the exchange of genetic information across microbial species as a consequence of community microbe interactions (Das et al. 2023). Many OC-degrading bacteria are misdiagnosed because of a lack of universally agreed upon microbial identification methodologies and processes. This highlights the need for research into microbial consortiums employing metagenomics techniques and traditional genetic engineering procedures (Nisa et al. 2022). Bacteria and other microbes may destroy a large variety of chemical substances, depending on chromosomal genes and extracellular enzymatic activity (in the case of bacteria) or the fungal degradation process. Changes in the environment that affect how microbes grow make these steps even more complicated.

Identification of the suitable species and strains for each substrate is required for a successful bioengineered microorganism. Biodegradation by bacteria utilizing easily accessible organic carbon and energy sources in the surrounding environment is a feasible alternative to the recombinant breakdown of refractory organic molecules. Microbes exploit changes in chemical gradients in their surroundings to find the best circumstances for development. This enables them to survive in a favorable environment (Ansari et al. 2023). Microbial consortia and microbial fuel cells (MFCs and bioreactors) are two novel advances in microbiological bioremediation utilized to breakdown resistant organic molecules. Because bacteria cannot grow in large quantities of toxic organics, fungus may be used to remediate toxic organics more efficiently (Sarker et al. 2023). For example, the enzymes laccase (LAC), lignin peroxidase (LIP), and manganese decarboxylase (MDA) are involved in the white rot fungus *Phanerochaete chrysosporium's* metabolism of lignocellulosic substances.

As a clean energy source and a feedstock for certain industries, hydrogen is a useful gas. As a result, the need for hydrogen generation has grown significantly in recent years. The well-known technologies for producing hydrogen gas – electrolysis of water, steam reforming of hydrocarbons, and autothermal processes – are not economically viable owing to their large energy demands (Kugarajah et al. 2023). In comparison to chemical processes, the biological synthesis of hydrogen gas offers a number of benefits. The main biological processes used to produce hydrogen gas include biophotolysis of water by algae, dark fermentation of organic materials, often carbohydrates, by bacteria, and photo-fermentation of organic materials. A relatively recent method of producing biohydrogen is the sequential dark and photo-fermentation process. The cost of raw materials is one of the main issues with the manufacture of dark and photo-fermentative hydrogen. Agricultural and food industry wastes that are high in carbohydrates but low in nitrogen, as well as various food industry wastewaters such as cheese whey, olive mill, and baker's yeast business wastewaters, may all be utilized for the generation of hydrogen when employing the right bioprocess technologies. The aforementioned wastes may be used to create hydrogen, which allows for cheap energy production and waste treatment at the same time (Preeti et al. 2023).

Globally, the health of people and aquatic ecosystems have been threatened by the presence of various contaminants in water. For more than a century, chemical oxidation using highly oxidizing species has been one of the most efficient methods for destroying contaminants and ensuring a safe and clean water supply. However, the removal of growing quantities of organic pollutants with more complex structural makeup – particularly the emerging micropollutants with trace concentrations in the complex water matrix – requires excessive dosages of oxidants and/or energy input, which has a negative impact on cost-effectiveness and may lead to secondary pollution (Raturi et al. 2023). Consequently, achieving selective oxidation of important organic contaminants is both important practically and difficult technically. There are several instances of selective oxidation of organic contaminants nowadays. The logical design of a catalyst for selective catalytic oxidation and a systematic knowledge of the link between the genesis of selectivity and its relevant water treatment scenarios still need to be developed. In this critical review, we summarize the most recent selective oxidation methods used to clean up water and investigate the causes of selectivity, including that which results from the reactivity of the oxidants or the target pollutants, that which results from the accessibility of the pollutants to the oxidants through adsorption and size exclusion, that which results from the interfacial electron transfer process, and that which results from enzymatic oxidation (Saxena et al. 2020).

A novel interface for a powerful biocatalytic system that can successfully remove pollutants from the environment is made possible by the synergistic coupling of biocatalysts with nanomaterials. The effective conversion of $CO_2$ into other, more reduced forms of carbon by biocatalysts provides environmentally friendly alternative to the oxidative processes now used (Zhou et al. 2023).

## 4.6 Recent Advancement and Challenges in Bioremediation

### 4.6.1 Bioinformatics Approaches in Bioremediation

Bioremediation is a valuable waste management technology that may be used to remove trash from polluted regions and places. It is especially interested in the use of organisms to consume or neutralize contaminants. The goal of bioremediation is to interpret the underlying degradation mechanism carried out by a specific organism for a specific pollutant using data from various biological databases, such as databases of chemical structure and composition, RNA/protein expression, OCs, catalytic enzymes, microbial degradation pathways, and comparative genomics (Tufail et al. 2022). To evaluate all of these sources in order to investigate bioremediation and create more efficient environmental cleaning technologies, a range of bioinformatics techniques are utilized. A lack of data on the parameters that affect the development and metabolism of microorganisms with bioremediation potential has resulted in a small number of bioremediation applications. These bioremediation-capable bacteria have been characterized, and their mineralization routes and processes have been mapped out using bioinformatics. The research of bioremediation techniques and technologies also requires the application of proteomic tools such as two-dimensional polyacrylamide gel electrophoresis, microarrays, and mass spectrometry. According to the researchers, it considerably enhances the structural characterization of microbial proteins with contaminant-degradable characteristics. The structural characterization of microbial proteins capable of contaminant degradation has vastly improved. This research straddles the divide between computer science and biology. Computers are used, for example, to store, change, and get information about the DNA, RNA, and proteins in the genome (Thacharodi et al. 2023b).

### 4.6.2 Bioremediation Tools Based on Omics

The fields of genomes, transcriptomics, metabolomics, and proteomics may all contribute to the study of bioremediation. This technological advancement helps in the assessment of the *in situ* bioremediation process by linking DNA sequences to the amounts of metabolites, proteins, and messenger RNA. An omics approach for bioremediation is given in Figure 4.5.

#### 4.6.2.1 Transcriptomics and Metatranscriptomics

The transcriptome reflects the collection of genes that are being transcribed at a certain time and circumstance, making it a vital connection between the cellular phenotype, interactome, genome, and proteome. The capacity to regulate gene expression is crucial for survival in a changing environment. The field of transcriptomics has allowed researchers to examine this phenomenon across the whole human genome. When it comes to transcriptomics, DNA microarray analysis is a very effective method for measuring mRNA expression levels. Transcriptomic analysis includes isolating and enriching total mRNA, synthesizing complementary DNA (cDNA), and sequencing the cDNA transcriptome (Swapnil et al. 2023). The mRNA expression of almost every gene in an organism may be analyzed using a DNA microarray as a transcriptomics tool. Transcriptomics and metatranscriptomics, the study of mRNA transcription patterns, are essential for understanding the roles that microorganisms play in the environment. Throughout the biodegradation process, metagenomics, genome binning, and metatranscriptomics may be used to uncover syntrophism between bacteria, and complementing metabolic processes. In order to study how genes are expressed, scientists may employ a technique called metatranscriptomics (Davis, 2023).

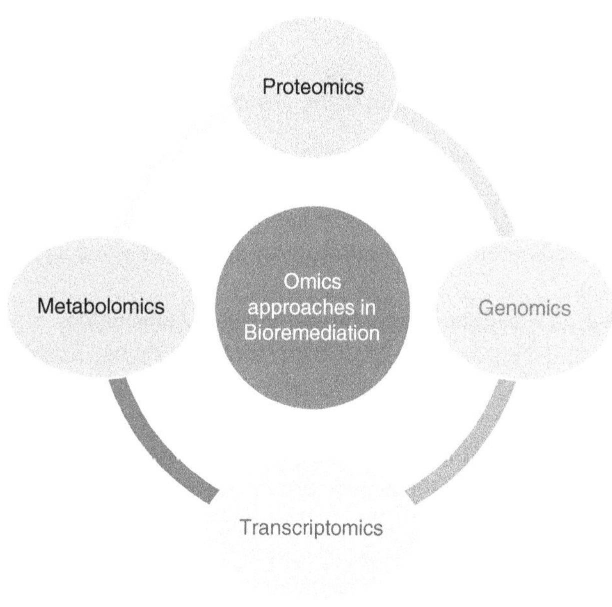


**Figure 4.5** Omics approach for bioremediation.

#### 4.6.2.2 Genomics

The study of bacteria used in bioremediation is a relatively recent area of genomics. This strategy is based on the fact that bacteria can perform all necessary analyses of their genetic material within the cell. Many different kinds of microorganisms are used in bioremediation. Genomic methods including polymerase chain reaction (PCR), isotope ratio analysis (IR), DNA hybridization (HybriDization), molecular connectivity (Molecular Networks), metabolic foot printing (Metabolic Engineering), and metabolic engineering are utilized to get insight into the biodegradation process (Ghosh et al. 2023b). PCR-based methods such as amplified fragment length polymorphisms (AFLP), amplified ribosomal DNA restriction analysis (ARDRA), automated ribosomal intergenic spacer analysis (ARISA), terminal-restriction fragment length polymorphism (T-RFLP), randomly amplified polymorphic DNA analysis (RAPD), single-strand conformation polymorphism (SSCP), and length heterogeneity can be used for genotypic fingerprinting. RAPD may be used to generate genetic fingerprints, build functional structural models, and evaluate closely related bacterial species while examining soil microbial communities. Using length heterogeneity PCR, different small subunit ribosomal ribonucleic acid (SSU rRNA) gene length variations can be detected in microbial communities. T-RFLP allows for the simultaneous profiling of a wide variety of microbial taxa. Genetic fingerprinting, microradiography, Fluorescence *in situ* hybridization (FISH), stable isotope probing, and quantitative PCR are only some of the molecular methods that may be used in studies of soil microbial interactions with environmental variables. The prevalence and distribution of soil taxonomic and functional gene markers may be studied quantitatively by PCR. Direct examination of certain molecular biomarker genes begins with amplified PCR products in DNA analysis methods. Cluster-assisted analysis, which analyses fingerprints from multiple samples, might be utilized to get insight into the interconnectedness of various microbial populations (Hlihor and Cozma 2023).

#### 4.6.2.3 Proteomics and Metabolomics

Total proteins expressed in a cell at a certain time and place are the focus of proteomics, whereas total metabolites generated by an organism during a specific time period or in a specific environment are the focus of metabolomics. Proteomics has been used to analyze protein abundance and changes in protein composition and to identify important proteins connected to microbes. Microbial community functional analysis has more potential and uses than genomics. Metabolomics research has two main applications in the study of living organisms. The first kind of research may be carried out without any previous familiarity with the metabolic pathways of the biological system being studied. This method creates massive quantities of data that may be utilized to determine the interconnection of different samples along metabolic pathways by identifying and recovering multiple metabolites in the sample. Alternatively, you might use the results of previous studies as the basis for a focused investigation to determine which metabolic pathways or metabolites are at play (Johnson 2023). Microbial metabolomics provides a wide range of methods for identifying and quantifying the wide variety of cellular waste products found in live organisms, including metabolite profiling, foot printing, and target analysis. Cell-free bioremediation will benefit from proteome and metabolome data.

### 4.6.3 Bioremediation Using Nanotechnological Methods

The smallest unit of measurement in nanotechnology is the nanometer. Because of their specific properties against numerous resistant pollutants, they may assist in the removal of many harmful compounds. Nanotechnology has provided technologies such as water treatment with a fresh viewpoint. Environmentally friendly techniques are now classified as nanofiltration. The nanotechnology and the bioremediation technology are given in Figure 4.6.

#### 4.6.3.1 Designing the Synthetic Microbial Communities

Recent developments in synthetic biology have had a major influence on ecological problems. Using genetically modified organisms (GMOs), we can clean the environment of harmful chemicals, pesticides, and xenobiotic. A synthetic microbial community requires knowledge of natural microbial communities. For this reason, it may be difficult to determine which species in a natural ecosystem are contributing to bioremediation. Building a synthetic microbial community allows for the creation of an artificial microbiome with purpose-built species (Anand et al. 2023). These neighborhoods serve as natural laboratories for the study of structural and functional aspects. By coculturing two different microbes under carefully controlled settings based on their interactions and roles, synthetic communities were developed. The discovery of bacterial processes and behaviors underpins the community's dynamics and structure. These patterns of microbial interaction are driven by metabolism, which in turn makes it easier for microbes to communicate with one another. Figure 4.7 shows how synthetic biology is used to create various synthetic microbial communities for bioremediation. Two microbial communities may socially interact with one another (such as through mutualism, competition, and cooperation).

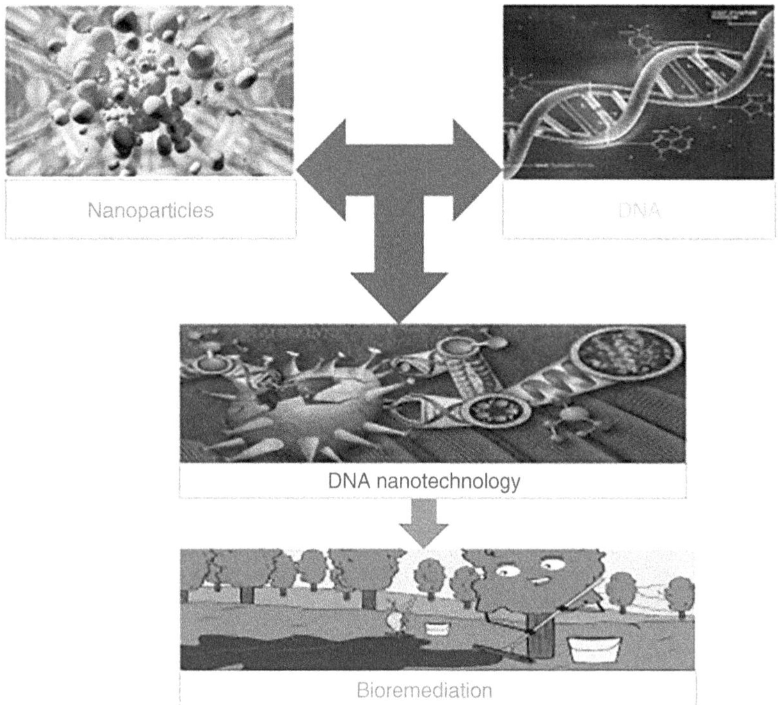


**Figure 4.6** Bioremediation and nanotechnology.

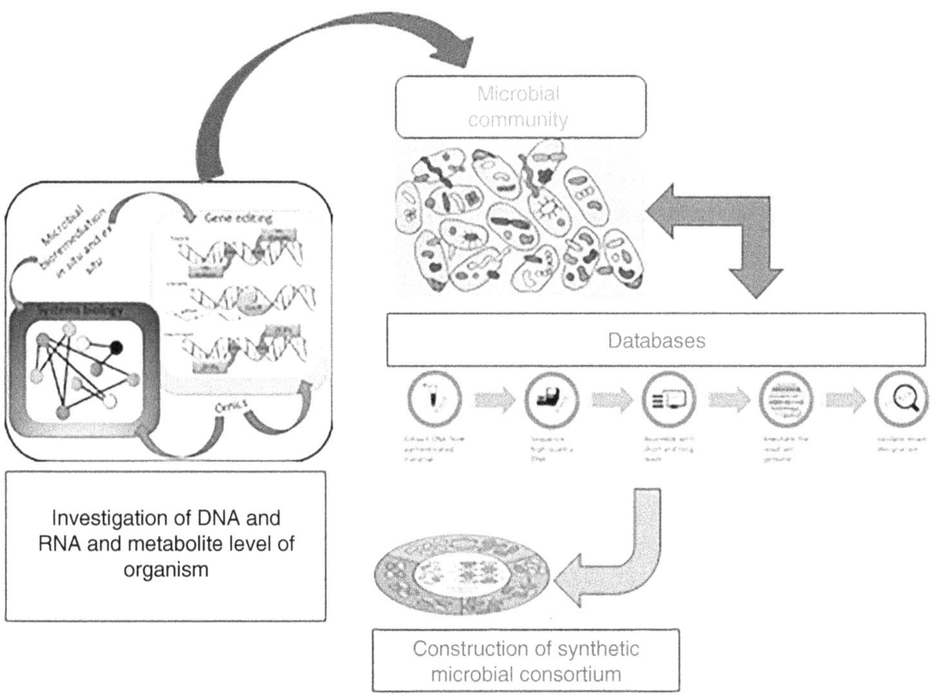


**Figure 4.7** Designing of synthetic microbial communities.

Synthetic communities have been shown to have an effect on community dynamics, and it has been shown that environmental circumstances may be modified – for example, by deleting genes – to engineer cooperation between two microbial strains. It should be noted that the created microbial species included in the synthetic community have also been analyzed for various forms of interaction. This form of designed interaction is often used in bioremediation efforts. Synthetic biology may be used to keep a vast population of microorganisms alive and well (Liaqat et al. 2023).

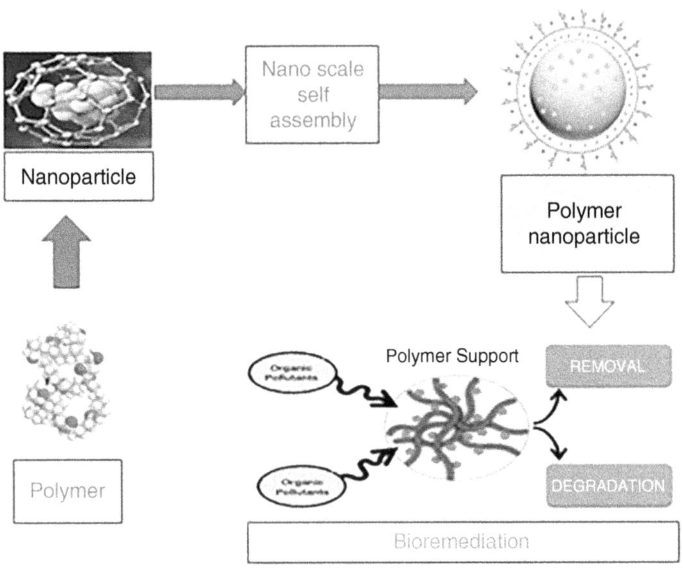


**Figure 4.8** Engineered nanoparticle polymer for bioremediation.

#### 4.6.3.2 Engineered Polymeric Nanoparticles for Hydrophobic Contaminant Bioremediation

Organic pollutants, such as petroleum hydrocarbons and PAHs, are absorbed in soil, which diminishes their solubility and mobility, lowering their environmental effect. Polymeric nano-network particles increase both phenanthrene solubility and phenanthrene release from polluted aquifer material. Polymeric nanoparticles are made using precursor chains of poly(ethylene) glycol-modified urethane acrylate (PMUA) (Bilal et al. 2022). PMUA nanoparticles are engineered to retain their characteristics in the presence of a wide variety of bacterial populations. The engineered nanoparticle polymers are shown in Figure 4.8 and are used for bioremediation.

#### 4.6.3.3 Nanotechnology and Microbes

Water that has been purified using effective microorganisms (EM) technology may be reused for irrigation after being used to clean wastewater. Water treatment with the use of nanotechnology and EM technologies is possible. Recalcitrant organic pollutants, such as PAHs containing numerous benzene rings, cause a plethora of far-reaching environmental problems. Mutagenic and nonbiodegradable, PAHs are dangerous pollutants (Rajhi and Bardi 2023).

#### 4.6.3.4 Genetic and Metabolic Engineering

Gene editing refers to the use of cutting-edge scientific and technological advances that allow for the precise insertion, deletion, or replacement of sections of DNA in a genome using rationally produced genetic fragments. Numerous gene editing techniques, such as Transcription activator-like effector nucleases (TALENs), Zinc finger nucleases, and clustered regularly interspaced short palindromic repeats (CRISPRs), depend on transcription activators to accomplish their goals. Among all gene editing methods, CRISPR-Cas is widely considered the most user-friendly and effective. TALEN uses a DNA-binding element that is specific to the host's DNA sequence. Double-stranded breaks (DSBs) are made in the DNA when TALEN binds to the strand and exposes the sticky ends used for stabilization. The DNA-binding domain of ZFNs is comprised of 30 amino acids (Demarco et al. 2023). The Fok1 cleavage domain generates DSBs at the host DNA site of action. To address molecular issues, researchers have to take a fresh look at composite endonucleases such as TALENs and ZFN nucleases. Unique features of the CRISPR-Cas system include sequence similarity, complementarity, and simultaneous gene editing. This special capability serves as a kind of resistance against viruses, and it is provided by the bacterium *Streptococcus pyogenes*. Crisper-derived RNA (crRNA) and *trans*-acting antisense RNA are linked through guide RNA in the CRISPR-Cas system (trcRNA). When gRNA identifies the desired DNA sequence, the Cas9 enzyme is able to make the necessary DSB. The knock-in and knock-out effects of these gene editing tools are being studied for potential use in bioremediation studies. The CRISPR-Cas system is commonly used in studies of model organisms such as *Escherichia coli* and *Pseudomonas aeruginosa*. Bioremediation researchers are also learning more about how to use CRISPR toolkits an createng gRNA to produce remediation-specific genes in non-model specie (such as *Rhodococcus ruber* TH, *Achromobacter* sp. HZ01, and *Comamonas testosteroni*) (Khan et al. 2023b).

Since they are used to storing a range of hazardous, refractory, and nondegradable xenobiotic substances under challenging conditions, pollutant-tolerant bacteria are the best candidates for genetic engineering and biochemical processes. Additionally, understanding biochemical processes is essential for analyzing microbiological bioremediation. For instance, *Bacillus* sp. DG-02 was used to study the anaerobic decomposition of fenpropathrin, which is used in the bioremediation of harmful pollutants through the production of haloalkane dehalogenases. By changing the current route, metabolic engineering may enhance the bioremediation process (Ahmad et al. 2023b). A genetic approach considerably enhances the possibility of generating recombinant enzymes. According to several investigations, some extracellular enzymes have been discovered to play a part in enzymatic bioremediation. The LiPs from *P. chrysosporium*, which encodes hemoproteins, breaks down PAHs. Although pollutants may serve as substrates or intermediates for bacteria in biological processes, inadequate or partial breakdown results in simpler, non-toxic, degradable chemicals. For instance, LiP can break down benzopyrene to produce 1,6-quinol, 3,6-quinine, and 6,12-quinine, three different quinine molecules. In the presence of MnP, MnP (Mn(II) peroxidase) may also oxidize organic molecules (Manganese peroxidase). Additionally, cytochrome P450, glutathione S-transferase, and LAC all contribute to the biodegradation of refractory substances. It has been shown that the activity and stability of enzymes are all greatly increased by immobilization. Microorganisms may quickly and easily remove and break down stubborn xenobiotic chemicals via the process of enzyme-based bioremediation. Enzyme-producing bacteria have been identified and described despite their low productivity. Organophosphates (OP) and organochlorines (OC), the primary components of insecticides, are present in agricultural soil and flow off into rivers (Petrov et al. 2023).

Hexachlorocyclohexane and methyl parathion have been successfully bioremediated by genetically modified bacteria. Experiments on organophosphate and pyrethroid bioremediation using genetically altered *P. putida* KT2440. Since the invention of metabolic engineering, the breakdown and catabolism of a range of persistent substances have been reported. Bioremediation of the breakdown routes for methyl parathion and hexachlorocyclohexane was shown by *Sphingobium japonicum* and *Pseudomonas* sp. WBC-3. Heterologous catabolism occurs when three enzymes from two distinct bacteria are joined in *E. coli*, resulting in the release of the persistent fumigants 1-, 2-, and 3-trichloropropane into the environment. Microbes may be utilized to convert persistent chemicals into minerals in order to achieve this (Khan et al. 2023c).

## 4.7 Advantages and Disadvantages

Pollution in the environment is a serious public health issue because it harms both people and other living things. It is expensive to completely remove pollutants using chemical and physical cleanup techniques. Both strategies may also cause more pollution and site disturbance, which might have a negative impact on local people and other biota. Therefore, restoration methods that rely on chemicals and physical force are not thought of as environmentally sustainable. In contrast to these methods, using biological processes known as "bioremediation" is offered as a way to get rid of a variety of persistent pollutants (mediated by various types of living organisms) (Singh et al. 2023). However, because of their unique uses, each bioremediation technology has its own benefits and drawbacks. The various advantages and disadvantages are given in Figure 4.9.

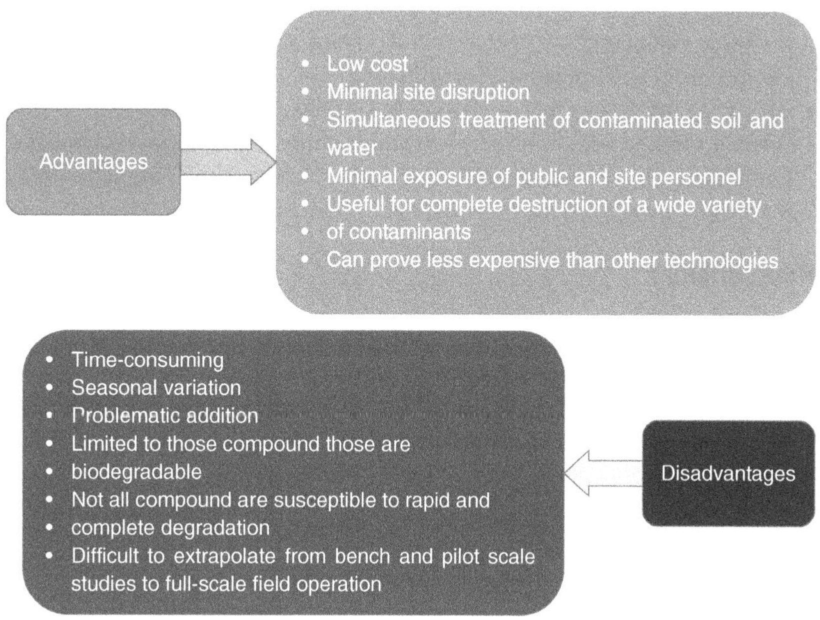


**Figure 4.9** Advantages and disadvantages of bioremediation.

## 4.8 Conclusion

The field of microbial remediation for the pulp and paper industry, textile industry, food industry, dairy industry, wood industry, fisheries, water and soil treatment industry, remediation of solid waste, remediation of heavy metal pollution, and remediation of hydrocarbons has seen growth in the use of omics. Bioremediation data must be mined in order to better understand degradative pathways, and new algorithms can be used to easily incorporate these data into simulation and numerical modeling, along with data assemblage, repositioning, exploration, and transmission, all of which call for standardized protocols. New biomarkers may help in the understanding of bioremediation processes. A complete understanding of the microbial remediation process might be obtained by combining all the omics data with genetically modified tools. It is also possible to research how phytoremediation helps to lessen environmental contamination. Compared to other remediation methods, the phytoremediation process offers a variety of benefits, including reduced costs, higher public acceptability, and a greater potential to degrade contaminants. The production of hazardous waste as a byproduct of semiconductor manufacture, such as glycol ethers, hydrochloric acid (HCl), xylene, hydrogen fluoride (HF), and methanol, pollutes groundwater and the air and causes the formation of toxic waste. Pharmaceuticals that are designed to survive for a long time or perhaps never decompose provide a special environmental risk. Pharmaceutical contaminants are persistent in the environment. Many urban and rural groundwater sources include trace levels of pharmaceutical compounds such as birth control pills, anti-epileptics, painkillers, and antidepressants. When in operation, solar power production facilities emit fewer greenhouse gas emissions than traditional fossil fuel-based power generation facilities do. This includes air pollutants including carbon monoxide, volatile OCs, nitrogen oxides, and carbon dioxide. Through newly found metabolic pathways, enzymes, genes, or operons, genetically modified plants may be able to bio remediate certain contaminants. In the area of bioremediation for a variety of pollutants, certain genes and enzymes that may be used with synthetic biology techniques have contributed to significant advancements. Therefore, it can be stated that GMOs remediation solutions not only boost the efficacy of microbial bioremediation procedures for a specific contamination, but also equip researchers with the most effective approaches.

## 4.9 Future Perspective

- Furthermore, genome reconstruction techniques and technology provide a strong foundation for bioremediation research.
- The next phase in bioremediation research is to find and compare effective gene and protein sequences for eliminating pollutants.
- Through the discovery of metabolic pathways, enzymes, genes, and operons, GMOs may bioremediate certain contaminants.
- Although genomics, metabolomics, and proteomics in bioremediation help in the discovery of potential solutions to particular pollutants, the next step in bioremediation research is to find and compare gene and protein sequences that are successful at eliminating contaminants.
- GMOs can purify a variety of waste effluents and degraded soil.
- When combined with other physical and chemical procedures, bioremediation may give an all-encompassing strategy for reducing environmental pollutants. Since it looks to be a long-term treatment, greater study is required in this area.

## References

Abo-Alkasem, M.I., Hassan, N.'m.H., and Elsoud, M.M.A. (2023). Microbial bioremediation as a tool for the removal of heavy metals. *Bulletin of the National Research Centre* 47: 1–15.

Ahmad, A., Mustafa, G., Rana, A., and Zia, A.R. (2023a). Improvements in bioremediation agents and their modified strains in mediating environmental pollution. *Current Microbiology* 80: 208.

Ahmad, Z., Khan, S.M., Page, S.E. et al. (2023b). Environmental sustainability and resilience in a polluted ecosystem via phytoremediation of heavy metals and plant physiological adaptations. *Journal of Cleaner Production* 385: 135733.

Ainur, N. and Sujatha, M. (2023). Genetically modified organisms: current situation and recommendations for the environment. *Applications of Environmental Biotechnology for Global Sustainability* 1–35.

Alegbeleye, O.O., Opeolu, B.O., and Jackson, V.A. (2017). Polycyclic aromatic hydrocarbons: a critical review of environmental occurrence and bioremediation. *Environmental Management* 60: 758–783.

Alvarez, A., Saez, J.M., Costa, J.S.D. et al. (2017). Actinobacteria: current research and perspectives for bioremediation of pesticides and heavy metals. *Chemosphere* 166: 41–62.

Anand, U., Dey, S., Bontempi, E. et al. (2023). Biotechnological methods to remove microplastics: a review. *Environmental Chemistry Letters* 1–24.

Ansari, F., Ahmad, A., and Rafatullah, M. (2023). Review on bioremediation technologies of polycyclic aromatic hydrocarbons (PAHs) from soil: mechanisms and future perspective. *International Biodeterioration & Biodegradation* 179: 105582.

Aregbesola, O.A., Kumar, A., Mokoena, M.P., and Olaniran, A.O. (2020). Cloning, overexpression, purification, characterization and structural modelling of a metabolically active $Fe^{2+}$ dependent 2,6-dichloro-p-hydroquinone 1,2-dioxygenase (CpsA) from *Bacillus cereus* strain AOA-CPS_1. *International Journal of Biological Macromolecules* 161: 247–257.

Arunraja, D., Romauld, S.I., Devi, P.B. et al. (2023). Genetically engineered microbes for bioremediation and phytoremediation of contaminated environment. In: *Metagenomics to Bioremediation* (ed. V. Tripathi), 709–721. Elsevier.

Bajpai, A., Atoliya, N., and Prakash, A. (2022). Genetically engineered microbes in micro-remediation of metals from contaminated sites. In: *Current Developments in Biotechnology and Bioengineering* (ed. R.D. Tyagi, A. Pandey, D.-J. Lee, R.R. Nagendran, and H.K. Parameswaran), 397–416. Elsevier.

Basu, S., Rabara, R.C., Negi, S., and Shukla, P. (2018). Engineering PGPMOs through gene editing and systems biology: a solution for phytoremediation? *Trends in Biotechnology* 36: 499–510.

Bhatt, A., Shandilya, J., Singal, S., and Prajapati, S.K. (2023). Application of genetically modified microorganisms for bioremediation of heavy metals from wastewater. In: *Genomics Approach to Bioremediation: Principles, Tools, and Emerging Technologies* (ed. V. Kumar, M. Bilal, L.F.R. Ferreira, and H.M.N. Iqbal), 295–320. Wiley.

Bilal, M. and Iqbal, H.M. (2019). Microbial-derived biosensors for monitoring environmental contaminants: recent advances and future outlook. *Process Safety and Environmental Protection* 124: 8–17.

Bilal, M. and Iqbal, H.M. (2020). Microbial bioremediation as a robust process to mitigate pollutants of environmental concern. *Case Studies in Chemical and Environmental Engineering* 2: 100011.

Bilal, M., Lam, S.S., and Iqbal, H.M. (2022). Biocatalytic remediation of pharmaceutically active micropollutants for environmental sustainability. *Environmental Pollution* 293: 118582.

Boopathy, R. (2000). Factors limiting bioremediation technologies. *Bioresource Technology* 74: 63–67.

Bora, J., Imam, S., Vaibhav, V., and Malik, S. (2023). Use of genetic engineering approach in bioremediation of wastewater. In: *Modern Approaches in Waste Bioremediation: Environmental Microbiology* (ed. M.P. Shah), 485–513. Springer.

Bustard, M., McMullan, G., and McHale, A.P. (1998). Biosorption of textile dyes by biomass derived from *Kluyveromyces marxianus* IMB3. *Bioprocess Engineering* 19: 427–430.

Capuana, M. (2020). A review of the performance of woody and herbaceous ornamental plants for phytoremediation in urban areas. *iForest - Biogeosciences and Forestry* 13: 139.

Das, N., Das, A., Das, S. et al. (2023). Petroleum hydrocarbon catabolic pathways as targets for metabolic engineering strategies for enhanced bioremediation of crude-oil-contaminated environments. *Fermentation* 9: 196.

Davis, G.B. (2023). Reviewing the bioremediation of contaminants in groundwater: investigations over 40 years provide insights into what's achievable. *Frontiers in Bioscience-Elite* 15: 16.

Deeb, M., Groffman, P.M., Blouin, M. et al. (2020). Using constructed soils for green infrastructure–challenges and limitations. *The Soil* 6: 413–434.

Demarco, C.F., Quadro, M.S., Selau Carlos, F. et al. (2023). Bioremediation of aquatic environments contaminated with heavy metals: a review of mechanisms, solutions and perspectives. *Sustainability* 15: 1411.

Duc, H.D., Hung, N.V., and Oanh, N.T. (2021). Anaerobic degradation of Endosulfans by a mixed culture of *Pseudomonas* sp. and *Staphylococcus* sp. *Applied Biochemistry and Microbiology* 57: 327–334.

Fouad, F.A., Youssef, D.G., Shahat, F.M., and Abd El-Ghany, M.N. (2023). Role of microorganisms in biodegradation of pollutants. In: *Handbook of Biodegradable Materials* (ed. G.A.M. Ali and A.S.H. Makhlouf), 221–260. Springer.

Garg, S.K. and Tripathi, M. (2017). Microbial strategies for discoloration and detoxification of azo dyes from textile effluents. *Research Journal of Microbiology* 12 (1): 1–19.

Garg, A., Yadav, B.K., Das, D.B., and Wood, P.J. (2022). Biomonitoring and phytoremediation of Cr (VI) contaminated water: Study on bioaccumulation and morphological responses in Eichhornia crassipes. *Journal of Environmental Engineering* 148 (12): 04022075.

Gaur, N., Narasimhulu, K., and PydiSetty, Y. (2018). Recent advances in the bio-remediation of persistent organic pollutants and its effect on environment. *Journal of Cleaner Production* 198: 1602–1631.

Gaur, V.K., Tripathi, V., and Manickam, N. (2022). Bacterial-and fungal-mediated biodegradation of petroleum hydrocarbons in soil. In: *Development in Wastewater Treatment Research and Processes* (ed. M.P. Shah and S. Rodriguez-Couto), 407–427. Amsterdam: Elsevier.

Geetha, N., Bhavya, G., Abhijith, P. et al. (2021). Insights into nanomycoremediation: secretomics and mycogenic biopolymer nanocomposites for heavy metal detoxification. *Journal of Hazardous Materials* 409: 124541.

Ghosh, D., Ghorai, P., Sarkar, S. et al. (2023a). Microbial assemblage for solid waste bioremediation and valorization with an essence of bioengineering. *Environmental Science and Pollution Research* 30: 16797–16816.

Ghosh, S., Sarkar, B., Khumphon, J., and Thongmee, S. (2023b). Genetically modified microbe mediated metal bioaccumulation: a sustainable effluent treatment strategy. In: *Industrial Wastewater Reuse: Applications, Prospects and Challenges* (ed. M.P. Shah), 215–229. Springer.

Giriyan, A.L., Berde, V.B., Pereira, E.J., and Parulekar-Berde, C.V. (2021). Microbial bioremediation of heavy metals: a genetic and omics approach. In: *Handbook of Research on Microbial Remediation and Microbial Biotechnology for Sustainable Soil* (ed. J.A. Malik and M. Shahid), 417–439. Hershey, PA: IGI Global.

Gupta, N., Sundar, R.D.V., and Arunachalam, S. (2022). A review on the genetically engineered microbes for bioremediation of THMs namely Hg and Cr. *ECS Transactions* 107: 11509.

Halak, S., Basta, T., Bürger, S. et al. (2006). Characterization of the genes encoding the 3-carboxy-cis, cis-muconatelactonizing enzymes from the 4-sulfocatechol degradative pathways of *Hydrogenophaga intermedia* S1 and *Agrobacterium radiobacter* S2. *Microbiology* 152: 3207–3216.

Hlihor, R.M. and Cozma, P. (2023). Microbial bioremediation of environmental pollution. *MDPI* 11: 1543.

Hussain, A., Rafeeq, H., Afsheen, N. et al. (2022a). Urease-based biocatalytic platforms – a modern view of a classic enzyme with applied perspectives. *Catalysis Letters* 152: 414–437.

Hussain, A., Rehman, F., Rafeeq, H. et al. (2022b). *In-situ*, *ex-situ*, and nano-remediation strategies to treat polluted soil, water, and air – a review. *Chemosphere* 289: 133252.

Iravani, S. and Varma, R.S. (2022). Genetically engineered organisms: possibilities and challenges of heavy metal removal and nanoparticle synthesis. *Clean Technologies* 4: 502–511.

Jangir, C.K., Kumar, S., and Meena, R.S. (2019). Significance of soil organic matter to soil quality and evaluation of sustainability. In: *Sustainable Agriculture* (ed. R.S. Meena), 357–381. Jodhpur: Scientific Publisher.

Jha, S., Shukla, R., Singh, R. et al. (2023). Phyco-remediation: role of microalgae in remediation of emerging contaminants. In: *Emerging Contaminants and Plants: Interactions, Adaptations and Remediation Technologies* (ed. T. Aftab and M. Hasanuzzaman), 163–192. Springer.

Johnson, R. (2023). Biotransformations for bioremediation. *Nature Chemical Biology* 19: 1287–1289.

Kalogerakis, N., Arff, J., Banat, I.M. et al. (2015). The role of environmental biotechnology in exploring, exploiting, monitoring, preserving, protecting and decontaminating the marine environment. *New Biotechnology* 32: 157–167.

Kaur, S., Midha, T., Verma, H. et al. (2023). Bioremediation: a favorable perspective to eliminate heavy metals from polluted soil. In: *Metagenomics to Bioremediation*, 209–230. Elsevier.

Kausar, F., Bagheri, A.R., Rasheed, T. et al. (2022). Nanomaterials for removal of heavy metals from wastewater. In: *Nano-Biosorbents for Decontamination of Water, Air, and Soil Pollution* (ed. M. Bilal) 135–161.

Kebede, G., Tafese, T., Abda, E.M. et al. (2021). Factors influencing the bacterial bioremediation of hydrocarbon contaminants in the soil: mechanisms and impacts. *Journal of Chemistry* 2021: 1–17.

Khan, M., Ijaz, M., Chotana, G.A. et al. (2022). Bacillus altitudinis MT422188: A potential agent for zinc bioremediation. *Bioremediation Journal* 26 (3): 228–248.

Khan, I.U., Qi, S.-S., Gul, F. et al. (2023a). A green approach used for heavy metals 'Phytoremediation' via invasive plant species to mitigate environmental pollution: a review. *Plants* 12: 725.

Khan, S., Masoodi, T.H., Pala, N.A. et al. (2023b). Phytoremediation prospects for restoration of contamination in the natural ecosystems. *Water* 15: 1498.

Khan, A., Malik, S., Ali, N. et al. (2023c). Metal-organic frameworks for reduction of heavy metals. In: *Emerging Techniques for Treatment of Toxic Metals from Wastewater* (ed. A. Ahmad, R. Kumar, and M. Jawaid), 457–474. Elsevier.

Kharangate-Lad, A. and D'Souza, N.C. (2021). Current approaches in bioremediation of toxic contaminants by application of microbial cells; biosurfactants and bioemulsifiers of microbial origin. In: *Rhizobiont in Bioremediation of Hazardous Waste* (ed. V. Kumar, R. Prasad, and M. Kumar), 217–263. Singapore: Springer.

Kim, S.Y., Garcia, H.A., Lopez-Vazquez, C.M. et al. (2019). Limitations imposed by conventional fine bubble diffusers on the design of a high-loaded membrane bioreactor (HL-MBR). *Environmental Science and Pollution Research* 26: 34285–34300.

Kugarajah, V., Nisha, K.N., Jayakumar, R. et al. (2023). Significance of microbial genome in environmental remediation. *Microbiological Research* 271: 127360.

Kumar, V., Shahi, S.K., and Singh, S. (2018). Bioremediation: an eco-sustainable approach for restoration of contaminated sites. In: *Microbial Bioprospecting for Sustainable Development* (ed. J. Singh, D. Sharma, G. Kumar, and N. Sharma), 115–136. Singapore: Springer.

Kumar, A., Gudiukaite, R., Gricajeva, A. et al. (2020). Microbial lipolytic enzymes–promising energy-efficient biocatalysts in bioremediation. *Energy* 192: 116674.

Kunze, M., Zerlin, K.F., Retzlaff, A. et al. (2009). Degradation of chloroaromatics by *Pseudomonas putida* GJ31: assembled route for chlorobenzene degradation encoded by clusters on plasmid pKW1 and the chromosome. *Microbiology* 155: 4069–4083.

Liaqat, I., Muhammad, N., Ara, C. et al. (2023). Bioremediation of heavy metals polluted environment and decolourization of black liquor using microbial biofilms. *Molecular Biology Reports* 50: 3985–3997.

Liu, Z., Shao, B., Zeng, G. et al. (2018). Effects of rhamnolipids on the removal of 2,4,2,4-tetrabrominated biphenyl ether (BDE-47) by *Phanerochaete chrysosporium* analyzed with a combined approach of experiments and molecular docking. *Chemosphere* 210: 922–930.

Liu, L., Bilal, M., Duan, X., and Iqbal, H.M. (2019). Mitigation of environmental pollution by genetically engineered bacteria – current challenges and future perspectives. *Science of the Total Environment* 667: 444–454.

Mahamuni-Badiger, P., Patel, P.R., Patil, P.M. et al. (2023). Bioremediation of industrial wastewater using microorganisms: an overview with recent developments. In: *Advanced and Innovative Approaches of Environmental Biotechnology in Industrial Wastewater Treatment* (ed. R. Chandra), 333–359. Singapore: Springer.

Maqsood, Q., Sumrin, A., Waseem, R. et al. (2023). Bioengineered microbial strains for detoxification of toxic environmental pollutants. *Environmental Research* 227: 115665.

Mazumder, M.A.R., Jubayer, M.F., and Ranganathan, T.V. (2022). Biodegradation of plastics by microorganisms. In: *Biotechnology for Zero Waste: Emerging Waste Management Techniques* (ed. A. Pandey), 123–141. Hoboken, NJ: John Wiley & Sons.

Meehan, C., Banat, I.M., Mcmullan, G. et al. (2000). Decolorization of REMAZOL BLACK-B using a thermotolerant yeast, *Kluyveromyces marxianus* IMB3. *Environment International* 26: 75–79.

Meyer, L., Guyot, S., Chalot, M., and Capelli, N. (2023). The potential of microorganisms as biomonitoring and bioremediation tools for mercury-contaminated soils. *Ecotoxicology and Environmental Safety* 262: 115185.

Miri, S., Rasooli, A., Brar, S.K. et al. (2022). Biodegradation of p-xylene – a comparison of three psychrophilic *Pseudomonas* strains through the lens of gene expression. *Environmental Science and Pollution Research* 29: 21465–21479.

Mohanty, S. and Paul, S. (2022). Bioremediation of black tides: strategies involving genetically modified organisms. In: *Advances in Oil-Water Separation* (ed. P. Das, S. Manna, and J.K. Pandey), 375–391. Elsevier.

Mohapatra, B. and Phale, P.S. (2021). Microbial degradation of naphthalene and substituted naphthalenes: metabolic diversity and genomic insight for bioremediation. *Frontiers in Bioengineering and Biotechnology* 9: 144.

Mondal, S., Bar, S., Roy, D. et al. (2022). Genetic engineering strategies and degradation of pollutants using genetically engineered microorganisms (GEMs). In: *Biotechnology for Environmental Protection* (ed. R. Selvasembian, E.D. van Hullebusch, and J. Mal), 209–232. Springer.

Mupambwa, H.A., Nciizah, A.D., Nyambo, P. et al. (2022). Potential applications of nanomaterials in agronomy: an African insight. In: *Emerging Nanomaterials for Advanced Technologies*, 581–600. Cham: Springer International Publishing.

Muter, O. (2023). Current trends in bioaugmentation tools for bioremediation: a critical review of advances and knowledge gaps. *Microorganisms* 11: 710.

Narayanan, M., Ali, S.S., and El-Sheekh, M. (2023). A comprehensive review on the potential of microbial enzymes in multipollutant bioremediation: mechanisms, challenges, and future prospects. *Journal of Environmental Management* 334: 117532.

Nisa, R.U., Maqbool, S., and Nisa, A.U. (2022). Bioremediation on the crossroads of technology for environmental clean-up: an overview. In: *Microbial and biotechnological interventions in bioremediation and phytoremediation* (ed. J.A. Malik), 3–25. Cham: Springer.

Novello, B.D., Zanatta, M.M., Albrecht, A.P., and Barroso, A.A.M. (2023). Bioremediation, drought tolerance and biofortification in biotechnological uses. *Acta Scientiarum. Biological Sciences* 45: e64163.

Ojha, A., Jaiswal, S., Thakur, P., and Mishra, S. (2023). Bioremediation techniques for heavy metal and metalloid removal from polluted lands: a review. *International journal of Environmental Science and Technology* 20: 10591–10612.

Padhan, D., Rout, P.P., Kundu, R. et al. (2021). Bioremediation of heavy metals and other toxic substances by microorganisms. In: *Soil Bioremediation: An Approach Towards Sustainable Technology* (ed. R. Chandra), 285–329. Hoboken, NJ: John Wiley & Sons.

Pande, V., Pandey, S.C., Sati, D. et al. (2020). Bioremediation: an emerging effective approach towards environment restoration. *Environmental Sustainability* 3: 91–103.

Perera, I.C. and Hemamali, E.H. (2022). Genetically modified organisms for bioremediation: current research and advancements. In: *Bioremediation of Environmental Pollutants: Emerging Trends and Strategies* (ed. D.C. Suyal and R. Soni), 163–186. Cham: Springer.

Petrov, D.S., Korotaeva, A.E., Pashkevich, M.A., and Chukaeva, M.A. (2023). Assessment of heavy metal accumulation potential of aquatic plants for bioindication and bioremediation of aquatic environment. *Environmental Monitoring and Assessment* 195: 122.

Pham, V.H.T., Kim, J., Chang, S., and Chung, W. (2022). Bacterial biosorbents, an efficient heavy metals green clean-up strategy: prospects, challenges, and opportunities. *Microorganisms* 10: 610.

Phulpoto, A.H., Qazi, M.A., Mangi, S. et al. (2016). Biodegradation of oil-based paint by *Bacillus* species monocultures isolated from the paint warehouses. *International journal of Environmental Science and Technology* 13: 125–134.

Preetha, J.S.Y., Arun, M., Vidya, N. et al. (2023). Biotechnology advances in bioremediation of arsenic: a review. *Molecules* 28: 1474.

Preeti, A., Raj, N., Joshi, J. et al. (2023). Genetically modified microbes in E-waste management: a perspective. In: *Microbial Technology for Sustainable E-waste Management* (ed. D. Das), 259–271. Springer.

Rafeeq, H., Afsheen, N., Rafique, S. et al. (2023). Genetically engineered microorganisms for environmental remediation. *Chemosphere* 310: 136751.

Rajhi, H. and Bardi, A. (2023). Phytoremediation of endocrine disrupting pollutants in industrial wastewater. In: *Current Developments in Biotechnology and Bioengineering* (ed. V.K. Thakur), 55–84. Elsevier.

Rajkumar, R. and Kurinjimalar, C. (2021). Microbes and plant mineral nutrition. In: *Microbiological Activity for Soil and Plant Health Management* (ed. R. Soni, D.C. Suyal, P. Bhargava, and R. Goel), 111–132. Singapore: Springer.

Rakib, M.R.J., Rahman, M., Onyena, A.P. et al. (2022). A comprehensive review of heavy metal pollution in the coastal areas of Bangladesh: abundance, bioaccumulation, health implications, and challenges. *Environmental Science and Pollution Research* 29: 67532–67558.

Rani, K., Singh, P., Agarwal, R., and Kushagra, A. (2022). Concomitant degradation of petroleum products and microplastics in industrial wastewater using genetically modified microorganisms. In: *Petroleum Industry Wastewater* (ed. G. Boczkaj), 19–34. Elsevier.

Raquel, S., Natalia, G., Luis Fernando, B., and Maria Carmen, M. (2013). Biodegradation of high-molecular-weight polycyclic aromatic hydrocarbons by a wood-degrading consortium at low temperatures. *FEMS Microbiology Ecology* 83: 438–449.

Rasheed, T., Ahmad, N., Nabeel, F. et al. (2022). Metal-organic frameworks for removal of heavy metals. In: *Nano-Bioremediation: Fundamentals and Applications* (ed. M.P. Shah), 455–476. Elsevier.

Rather, G.A. and Raina, M. (2022). Microbial modifications and biochemical pathway: mechanism for ecosystem decontamination. In: *Microbial Bioremediation: Sustainable Management of Environmental Contamination* (ed. R.A. Bhat, M. Butnariu, G.H. Dar, and K.R. Hakeem), 109–124. Springer.

Rathore, S., Varshney, A., Mohan, S., and Dahiya, P. (2022). An innovative approach of bioremediation in enzymatic degradation of xenobiotics. *Biotechnology and Genetic Engineering Reviews* 38: 1–32.

Raturi, G., Chaudhary, A., Rana, V. et al. (2023). Microbial remediation and plant-microbe interaction under arsenic pollution. *Science of the Total Environment* 864: 160972.

Ren, L., Weng, L., Chen, D. et al. (2023). Bioremediation of PAEs-contaminated saline soil: the application of a marine bacterial strain isolated from mangrove sediment. *Marine Pollution Bulletin* 192: 115071.

Rizwan, K. and Bilal, M. (2022). Environmental sensing and detection of toxic heavy metals by metal organic frameworks-based electrochemical sensors. In: *Environmental Management Technologies* (ed. M.M. Rahman), 215–227. CRC Press.

Rosca, M., Hlihor, R.-M., and Gavrilescu, M. (2019). Bioremediation of persistent toxic substances: from conventional to new approaches in using microorganisms and plants. In: *Microbial Technology for the Welfare of Society* (ed. P. Arora), 289–312. Singapore: Springer.

Sangkharak, K., Choonut, A., Rakkan, T., and Prasertsan, P. (2020). The degradation of phenanthrene, pyrene, and fluoranthene and its conversion into medium-chain-length polyhydroxyalkanoate by novel polycyclic aromatic hydrocarbon-degrading bacteria. *Current Microbiology* 77: 897–909.

Sangkharak, K., Paichid, N., Yunu, T., and Prasertsan, P. (2020). Enhancing the degradation of mixed polycyclic aromatic hydrocarbon and medium-chain-length polyhydroxyalkanoate production by mixed bacterial cultures using modified repeated batch fermentation. *Journal of Applied Microbiology* 129 (3): 554–564.

Sangwan, S. and Dukare, A. (2018). Microbe-mediated bioremediation: an eco-friendly sustainable approach for environmental clean-up. In: *Advances in Soil Microbiology: Recent Trends and Future Prospects* (ed. T. Adhya, B. Lal, B. Mohapatra, et al.), 145–163. Springer Nature: Singapore.

Sarker, A., Kim, J.E., Islam, A.R.M. et al. (2021). Heavy metals contamination and associated health risks in food webs – a review focuses on food safety and environmental sustainability in Bangladesh. *Environmental Science and Pollution Research* 29: 3230–3245.

Sarker, A., Al Masud, M.A., Deepo, D.M. et al. (2023). Biological and green remediation of heavy metal contaminated water and soils: a state-of-the-art review. *Chemosphere* 332: 138861.

Saxena, G., Kishor, R., Saratale, G.D., and Bharagava, R.N. (2020). Genetically modified organisms (GMOs) and their potential in environmental management: constraints, prospects, and challenges. In: *Bioremediation of Industrial Waste for Environmental Safety: Volume II: Biological Agents and Methods for Industrial Waste Management* (ed. R. Bharagava and G. Saxena), 1–19. Springer.

Sen, S.K., Patra, P., Das, C.R. et al. (2019). Pilot-scale evaluation of bio-decolorization and biodegradation of reactive textile wastewater: an impact on its use in irrigation of wheat crop. *Water Resources and Industry* 21: 100106.

Shah, H. and Jain, S. (2020). Bioremediation: an approach for environmental pollutants detoxification. In: *Waste to Energy: Prospects and Applications* (ed. B.K. Kashyap, M.K. Solanki, D.V. Kamboj, and A.K. Pandey), 121–142. Singapore: Springer.

Shahbaz, A., Hussain, N., Mahmood, T. et al. (2023). Environment-friendly microbial bioremediation. *Microbial Bioreactors for Industrial Molecules* 5: 397–419.

Sharma, P., Singh, S.P., Parakh, S.K., and Tong, Y.W. (2022). Health hazards of hexavalent chromium (Cr(VI)) and its microbial reduction. *Bioengineered* 13: 4923–4938.

Silva, A.B., Costa, M.F., and Duarte, A.C. (2018). Biotechnology advances for dealing with environmental pollution by micro (nano) plastics: lessons on theory and practices. *Current Opinion in Environmental Science & Health* 1: 30–35.

Singh, K., Tripathi, S., and Chandra, R. (2023). Bacterial assisted phytoremediation of heavy metals and organic pollutants by *Cannabis sativa* as accumulator plants growing on distillery sludge for ecorestoration of polluted site. *Journal of Environmental Management* 332: 117294.

Skariyachan, S., Taskeen, N., Kishore, A.P., and Krishna, B.V. (2022). Recent advances in plastic degradation – from microbial consortia-based methods to data sciences and computational biology driven approaches. *Journal of Hazardous Materials* 426: 128086.

Sodhi, K.K., Mishra, L.C., Singh, C.K., and Kumar, M. (2022). Perspective on the heavy metal pollution and recent remediation strategies. *Current Research in Microbial Sciences* 100166.

Song, J., Zhang, S., Xie, Y., and Li, Q. (2019). Purification and characteristics of an aflatoxin B1 degradation enzyme isolated from *Pseudomonas aeruginosa*. *FEMS Microbiology Letters* 366: fnz034.

Sravya, K. and Sangeetha, S. (2022). Feasibility study on bioremediation techniques to contaminated soils. *Materials Today Proceedings* 51: 2556–2560.

Swapnil, P., Singh, L.A., Mandal, C. et al. (2023). Functional characterization of microbes and their association with unwanted substance for wastewater treatment processes. *Journal of Water Process Engineering* 54: 103983.

Thacharodi, A., Hassan, S., Hegde, T.A. et al. (2023a). Water a major source of endocrine-disrupting chemicals: an overview on the occurrence, implications on human health and bioremediation strategies. *Environmental Research* 231: 116097.

Thacharodi, A., Hassan, S., Singh, T. et al. (2023b). Bioremediation of polycyclic aromatic hydrocarbons: an updated microbiological review. *Chemosphere* 328: 138498.

Thai, T.D., Lim, W., and Na, D. (2023). Synthetic bacteria for the detection and bioremediation of heavy metals. *Frontiers in Bioengineering and Biotechnology* 11: 1178680.

Tripathi, M. and Garg, S.K. (2014). Response surface modeling for co-remediation of $Cr^{6+}$ and pentachlorophenol by *Bacillus cereus* RMLAU1: bioreactor trial, structural and functional characterization by SEM-EDS and FT-IR analyses. *Bioremediation Journal* 18: 328–344.

Tufail, M.A., Iltaf, J., Zaheer, T. et al. (2022). Recent advances in bioremediation of heavy metals and persistent organic pollutants: a review. *Science of the Total Environment* 850: 157961.

Unuofin, J.O., Falade, A.O., and Aladekoyi, O.J. (2021). Applications of microbial laccases in bioremediation of environmental pollutants: potential issues, challenges, and prospects. In: *Bioremediation for Environmental Sustainability* (ed. A.K. Mishra), 519–540. Amsterdam: Elsevier.

Vijayanand, M., Ramakrishnan, A., Subramanian, R. et al. (2023). Polyaromatic hydrocarbons (PAHs) in the water environment: a review on toxicity, microbial biodegradation, systematic biological advancements, and environmental fate. *Environmental Research* 227: 115716.

Vulpe, C.B., Matica, M.A., Kovačević, R. et al. (2023). Copper accumulation efficiency in different recombinant microorganism strains available for bioremediation of heavy metal-polluted waters. *International Journal of Molecular Sciences* 24: 7575.

Wesseler, J., Kleter, G., Meulenbroek, M., and Purnhagen, K.P. (2023). EU regulation of genetically modified microorganisms in light of new policy developments: possible implications for EU bioeconomy investments. *Applied Economic Perspectives and Policy* 45: 839–859.

Xue, Y., Du, P., Shendi, A.A.I., and Yu, B. (2022a). Mercury bioremediation in aquatic environment by genetically modified bacteria with self-controlled biosecurity circuit. *Journal of Cleaner Production* 337: 130524.

Xue, Y., Qiu, T., Sun, Z. et al. (2022b). Mercury bioremediation by engineered *Pseudomonas putida* KT2440 with adaptationally optimized biosecurity circuit. *Environmental Microbiology* 24: 3022–3036.

Yaashikaa, P.R., Devi, M.K., and Kumar, P.S. (2022). Engineering microbes for enhancing the degradation of environmental pollutants: a detailed review on synthetic biology. *Environmental Research* 214: 113868.

Yadav, M. and Sharma, P. (2023). New title: current eco-friendly and sustainable methods for heavy metals remediation of contaminated soil and water: special emphasis on use of genetic engineering and nanotechnology. *Pollution* 9: 1028–1048.

Zhou, B., Zhang, T., and Wang, F. (2023). Microbial-based heavy metal bioremediation: toxicity and eco-friendly approaches to heavy metal decontamination. *Applied Sciences* 13: 8439.

# 5

# How Can Plant-microbe Interactions be used for the Bioremediation of Metals in Water Bodies?

*Gabriela Petroceli-Mota*[1,*], *Emilane Pinheiro da Cruz Lima*[2,*], *Mariana Miranda de Abreu*[2,4], *Glacielen Ribeiro de Souza*[3], *Jussara Tamires de Souza Silva*[2,5], *Gabriel Quintanilha-Peixoto*[6,7], *Alessandro Coutinho Ramos*[4], *Rachel Ann Hauser-Davis*[8], *and Aline Chaves Intorne*[2,9]

[1] *Laboratory of Cellular and Tissue Biology, Centro de Biociências e Biotecnologia, Universidade Estadual do Norte Fluminense Darcy Ribeiro (UENF), RJ, Brazil*
[2] *Laboratory of Physiology and Biochemistry of Microorganisms, Centro de Biociências e Biotecnologia, Universidade Estadual do Norte Fluminense Darcy Ribeiro (UENF), RJ, Brazil*
[3] *Laboratory of Biotechnology, Centro de Biociências e Biotecnologia, Universidade Estadual do Norte Fluminense Darcy Ribeiro (UENF), RJ, Brazil*
[4] *Biotechnology and Environmental Microbiology Laboratory, Universidade Vila Velha (UVV), ES, Brazil*
[5] *Phytotechnics Laboratory, Centro de Ciências e Tecnologias Agropecuárias, Universidade Estadual do Norte Fluminense Darcy Ribeiro (UENF), RJ, Brazil*
[6] *Laboratory of Chemistry and Function of Proteins and Peptides, Centro de Biociências e Biotecnologia, Universidade Estadual do Norte Fluminense Darcy Ribeiro (UENF), RJ, Brazil*
[7] *Laboratory of Molecular and Computational Biology of Fungi, Instituto de Ciências Biológicas, Universidade Federal de Minas Gerais (UFMG), MG, Brazil*
[8] *Ambiental Environmental Health Assessment and Promotion Laboratory, Instituto Oswaldo Cruz, Fundação Oswaldo Cruz, RJ, Brazil*
[9] *Chemistry and Biology Laboratory, Departamento de Ensino, Instituto Federal de Educação, Ciência e Tecnologia do Rio de Janeiro, RJ, Brazil*

**These authors contributed equally to this work.*

## 5.1 Water Contamination Issues

Water is an essential resource for life on Earth, and essential for social and economic development. The planets water distribution, however, is highly heterogeneous, with about 27% of the global human population living in drought-affected areas or where water availability is low. In these cases, water becomes a key element of power, causing disputes and conflicts (UN 2018).

The main water conflict areas in the world are (i) Southeast and South Asia, including the Ganges-Brahmaputra-Meghna-Nepal watershed, (ii) Central America, in the San Juan River region in Nicaragua and Costa Rica, (iii) the northern region of South America, in the Mira, Orinoco and Amazonas rivers basins, and (iv) different African continent regions, highlighting the Nile and Congo basins in the central region and Limpopo in the south. As a result, most of the world's river basins (~56%) are under hydropolitical tension (De Stefano et al. 2017).

Brazil, the country with the highest freshwater availability in the world, also displays a significantly heterogeneous water distribution, causing water management problems. For example, water body transposition and overexploitation are both noted at the Paraíba do Sul River, the most important southeastern region hydrological resource (Ovalle et al. 2013). This river has recently experienced one of the worst droughts on record since 1930, resulting in water use disagreements between the states that represent the largest economies in Brazil, namely Rio de Janeiro and São Paulo (ANA 2018).

In addition to conflicts related to multiple water uses, significant loss of biodiversity in aquatic ecosystems and the risk of species extinction due to pollution are also noteworthy (De Stefano et al. 2017). Water quality has deteriorated in most rivers in Latin America, Africa, and Asia since the 1990s, mainly due to population and economic growth and a lack of wastewater management systems (UN 2018). Consequently, several aquatic biota species are now categorized as endangered in these regions (IUCN 2017).

The main factors that contribute to decreased water quality comprise an excessive load of nutrients ($NH_4^+$; $NO_3^-$; $PO_4^{-3}$), pathogens, and thousands of chemical compounds, including heavy metals (Stark and Richards 2008; UN 2018). Pollution sources can be categorized as punctual, such as domestic effluent and industrial discharge points (Calijuri and Cunha 2013),

*Microbes Based Approaches for the Management of Hazardous Contaminants*, First Edition. Edited by Ajay Kumar, Livleen Shukla, Joginder Singh, and Luiz Fernando R. Ferreira.

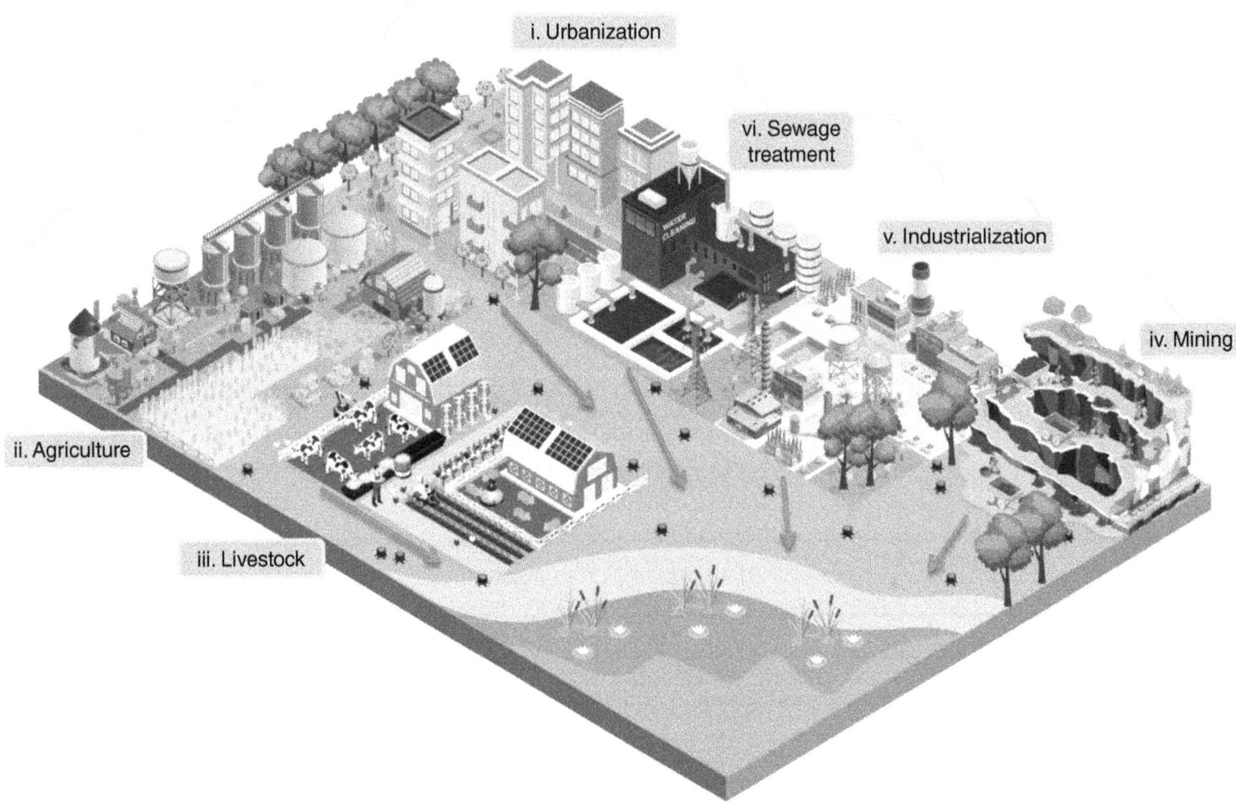


**Figure 5.1** Inappropriate use and management of the soil in waste disposal. The absence of vegetation in the soil increases the speed of surface runoff, facilitating the leach of contaminants generated by (i) urbanization, (ii) agriculture, (iii) livestock, (iv) mining, (v) industrialization, and (vi) wastewater treatment plants for water bodies.

or diffuse, as observed in agricultural activities, where leaching is a notable factor (Kill et al. 2018). Figure 5.1 depicts the main water uses and contamination sources.

The agriculture sector ranks first in water use globally, at around 70%, followed by industrial (20%), and domestic (10%) activities (UN 2018). To meet increasing water needs due to population increases, agricultural production has intensified, causing more environmental damage through the massive application of fertilizers and herbicides (Farooq et al. 2019). In this regard, the responsible use of nitrogen fertilizers may significantly increase cultivation yields, as N is one of the main nutrients required for plant growth (Huang et al. 2017). On the other hand, when overapplied, most of this nutrient is not absorbed by plants and is leached through rainwater and inadequate irrigation, contributing to increased nitrogen levels in water bodies (Stark and Richards 2008). This, in turn, leads to high algae and plant growth, foul smell and turbidity, reducing oxygen availability, and leading to mortality of fauna, reduction of biodiversity, and causing eutrophication (Ansari et al. 2010; Khan and Mohammad 2014; Scherer and Pfister 2015). Because of this, nitrogen fertilizers are considered the most significant cause of surface and groundwater contamination (Khan et al. 2018).

Pesticide use also increases metal concentrations in the soil, altering soil and aquatic biogeochemical cycles (Adriano 2001; Chagnon et al. 2015; Nicholson et al. 2003). Thus, the impact associated with land use (Ayoub 1999; Patnaik 2017) causes water pollution (Norse 2005; Prabakaran et al. 2018), increasing drinking water treatment costs, and upbringing social problems (Norse 2005; Smith and Siciliano 2015).

## 5.2 Metal Contamination Effects

Metals are one of the main classes of environmental contaminants, as they can integrate biological and geological cycles (USEPA 2007) and are not biodegradable (Wu et al. 2016). Water metal contamination may originate from natural sources, such as volcanic activity, weathering, mineral, and rock erosion (USEPA 2007). The bioavailability of these elements in the aquatic environment, however, has reached alarming levels as a direct result of anthropic activities (Sivakumar et al. 2014), including industrial processes, agriculture, mining, and fuel combustion (Sivakumar et al. 2014; USEPA 2007).

Industries involving chloralkali processes and acid, pigment, copper, electronics (including batteries), and metal sulfate production (USEPA 2007), as well as mining activities, have led to significant water metal levels increases for several elements, such as observed for Pb, As, Hg and Cd (Gaur et al. 2014). Concerning mining, the negative impacts of illegal mining in South America are noted mainly, in the Amazon region, where this anthropic activity causes irreversible biodiversity and ecological damages (De Souza Neto et al. 2020). In one study, for example, Capparelli et al. (2020) analyzed metal levels in water and sediments in the Ecuadorian Amazon, close to mining activities, reporting metal concentrations 500-fold times above the maximum limits allowed by local regulatory agencies. The riverside populations that consume the water in that area have also been seriously affected and are at high contamination risk (De Souza Neto et al. 2020).

In Brazil, the recent collapse of mining has led to the release of exorbitant amounts of mining waste into the environment. The rupture that took place in the city of Mariana, in the state of Minas Gerais, southeastern Brazil, was considered the largest environmental disaster in the world due to the mining waste mud volume and covered distance, with 50–60 million $m^3$ of waste discharged, covering 668 km of the Doce River's stretch (Dos Santos Vergilio et al. 2020; Thompson et al. 2020). The rupture of another dam in the city of Brumadinho, also in Minas Gerais, led to the spill of 12 million $m^3$ of mining waste mud, leading to the death of 270 people and contaminating the Paraopeba River (Thompson et al. 2020). Significant metal contamination in Paraopeba waters was reported by Dos Santos Vergilio et al. (2020) for Fe, Al, Mn, Zn, Cu, Pb, Cd, and U, all higher than allowed by Brazilian water legislation. In sediments, Cr, Ni, Cu, and Cd levels were also higher than the established sediment quality guidelines National Oceanic and Atmospheric Administration (TEL-NOAA) for this River.

Excess metals can disturb delicate ecosystem balances and, once available in the environment, be absorbed by plants and animals, entering the food chain (Liu et al. 2011). Chronic and subchronic metal exposure has been reported in several studies (Chowdhury et al. 2016; Liu et al. 2009, 2011; Sivakumar et al. 2014; Wu et al. 2016). Following internalization, metals may impair cellular functions, and compete with essential metals at their protein binding sites, damaging enzymes and resulting in oxidative stress (Wu et al. 2016). Oxidative stress induced by metallic contaminants, in turn, also alters the redox state of cells (He et al. 2014), leading to the production of high amounts of reactive oxygen species (ROS), causing cell death due to membrane degradation and protein and nucleic acid denaturation (Gaur et al. 2014; Wu et al. 2016; Yang et al. 2017).

Metal toxicity can also increase due to bioaccumulation processes in living organisms. The main metal accumulation sites in humans are the adipose tissue and the gastrointestinal tract, which can lead to cancer, among other risks. Few metals, however, can bioaccumulate in the human body (Chowdhury et al. 2016), as bioaccumulation depends on the chemical form of the metal and its affinity for organic tissues and compounds. For example, methylmercury (MeHg) exhibits a high affinity for adipose tissue (Ha et al. 2016), while Cd can be retained in soft tissues, such as liver and kidney, for up to 20 years due to intracellular binding with metallothionein, a metalloprotein (USEPA 2007).

Due to their adverse effects on living organisms (USEPA 2007), metals can significantly decrease biodiversity (Giller et al. 2009). These compounds also accumulate in plant tissues, influencing growth and development and interfering with physiological and biochemical processes, such as photosynthesis and respiration inhibition, main cellular organelle degeneration, anatomical changes, secondary metabolite production, water balance, and nutrient absorption (Sharma and Dubey 2005; Maksymiec 2007; Nagajyoti et al. 2010; Mohanty et al. 2012). Few studies have investigated the toxic effects caused by exposure to metal mixtures, which may lead to synergistic or antagonistic effects (Chowdhury et al. 2016; Wu et al. 2016).

To minimize human health and environmental contamination risks, several regulatory agencies, such as the United States Environmental Protection Agency (USEPA), the National Oceanic and Atmospheric Administration (NOAA), and the World Health Organization (WHO), have established standards for the presence of metals in different environmental compartments (Table 5.1) (Chowdhury et al. 2016; Wu et al. 2016), with the USEPA also creating a list of priority pollutants, which include metals such as As, Be, Cd, Cr, Cu, Pb, Hg, Ni, Se, Ag, and Zn (USEPA 2014).

Some of the metals presented in Table 5.1 are essential micronutrients, such as Co, Cu, Fe, Mn, and Zn (Bielmyer-Fraser et al. 2017; He et al. 2014), although they may become toxic at high concentrations (Clemens 2006). Nonessential metals, on the other hand, such as As, Cd, Pb, and Hg, are toxic even at low concentrations, as they impair vital functions (He et al. 2014).

As discussed previously, the final destination of metals is often water bodies, causing negative aquatic ecosystem and human effects (Wang et al. 2015), and several studies indicate that all water sources intended for human consumption are already, or will soon be, contaminated by these compounds (Chowdhury et al. 2016).

Metal speciation, i.e., the chemical form an element presents, also affects metal toxicity (Baird 2001). Some metals, for example, may present different redox states. Mercury, for example, is found in elemental ($Hg^0$) or oxidized forms, such as

**Table 5.1** Guiding values for metals in sediment, soil, water, plant, and microorganisms. Adapted from (Buchman 2008).

| | Sediment (μg g$^{-1}$) | | | | | Soil (μg g$^{-1}$) | | | Water (mg l$^{-1}$) | | | | |
|---|---|---|---|---|---|---|---|---|---|---|---|---|---|
| | | Freshwater | | Sea water | | | | | | Freshwater | | Sea water | |
| Metal | Background | TEL | PEL | TEL | PEL | Background | Plant | Microorganism | Subterranean | Acute | Chronic | Acute | Chronic |
| Arsenic | 1.1 | 5.9 | 17 | 7.24 | 41.6 | 5.2 | 18 | 100 | 0.01 | 0.34 | 0.15 | 0.069 | 0.036 |
| Barium | 0.7 | – | – | 130.1 | – | 440 | 500 | 3000 | 2 | 0.11 | 0.0039 | 1 | 0.2 |
| Cadmium | 0.1–0.3 | 0.99 | 3.53 | 0.68 | 4.21 | – | 4 | 20 | 0.005 | 0.002 | 0.00025 | 0.04 | 0.0088 |
| Lead | 4–17 | 35 | 91.3 | 30.24 | 112 | 16 | 500 | 900 | 0.015 | 0.065 | 0.0025 | 0.21 | 0.0081 |
| Cobalt | 10 | – | – | – | – | 6.7 | 13 | 1000 | – | 1.5 | 0.003 | – | 0.001 |
| Copper | 10–25 | 31.6 | 197 | 18.7 | 108 | 17 | 70 | 100 | 1.3 | 0.013 | 0.009 | 0.0048 | 0.0031 |
| Chrome | 7–13 | 43.4 | 90 | 52.3 | 160 | 37 | 1 | 10 | 0.1 | – | – | – | – |
| Iron | 0.99–1.18% | – | – | – | – | 1.80% | – | 200 | 0.3 | – | 1 | 0.3 | 0.05 |
| Manganese | 400 | – | – | – | – | 330 | 220 | 100 | 0.05 | 2.3 | 0.08 | – | 0.1 |
| Mercury | 0.004–0.51 | 0.174 | 0.48 | 0.13 | 0.7 | 0.058 | 0.3 | 30 | 0.002 | 0.0014 | 0.00077 | 0.0018 | 0.00094 |
| Nickel | 9.9 | 18 | 36 | 15.9 | 42.8 | 13 | 30 | 90 | 0.02 | 0.47 | 0.052 | 0.074 | 0.0082 |
| Selenium | 0.29 | – | – | – | – | 0.2 | 0.52 | 100 | 0.05 | 0.013–0.186 | 0.005 | 0.29 | 0.071 |
| Zinc | 7–38 | 123 | 459 | 124 | 271 | 48 | 50 | 100 | 5 | 0.12 | 0.12 | 0.09 | 0.081 |

TEL: Concentration below that adverse effects to organisms are rarely expected.
PEL: Concentration above which adverse effects are often expected for organisms.
Background: Relative measure used to distinguish natural concentrations of an element (geogenic and/or biogenic) and the influence of human activities.

mercury ($Hg_2^{2+}$) or mercury ($Hg^{2+}$) (Fu et al. 2016), with MeHg, the organic form of this element, detected mainly in aquatic environments, comprising the most bioavailable and toxic form of this metal (Chen and Driscoll 2018; Genchi et al. 2017). Chromium can be present as $Cr^{3+}$, which, in low concentrations, is an essential micronutrient, or as $Cr^{6+}$, which is carcinogenic (USEPA 2007). Depending on the redox state, metals may also be volatile or become associated with other chemical elements, forming inorganic or organic compounds (Chakraborty et al. 2013; USEPA 1997). In the aquatic environment, metals are found dissolved or associated with particulate matter, which also directly influences their bioavailability and consequent toxicity (Sivakumar et al. 2014; USEPA 2007).

Bioavailability comprises the amount of a compound or chemical element available for use in biological activities, introduced into the body by different routes, such as ingestion, breathing, injection, or direct skin contact. The metal form and environmental temporal and spatial metal distribution will depend on both physical and chemical interactions (Nikinmaa 2014). For example, dissolved organic matter binds strongly to metals in water (Ravichandran 2004), decreasing metal availability, like carbonates, which can co-precipitate Cd, Ni, Pb, and Zn among others (Kabata-Pendias 2010). The presence of fine particles (silt and clay) in the sediment is also noteworthy, leading to metal adsorption due to electrical surface charges (Ferreira et al. 2010). Some substances, however, increase metal availability for absorption by living organisms, such as surfactants, which act by decreasing particle adsorption and increasing their dispersion, facilitating metal transport (Wu et al. 2016).

The excess of bioavailable metals noted today in aquatic ecosystems has led to significant amounts of research on metal damage mitigation of contaminated sites. Several strategies have been evaluated, with bioremediation emerging as a more sustainable and lower-cost alternative compared to traditional technologies (Gaur et al. 2014; Sivakumar et al. 2014).

## 5.3 Metal Bioremediation

Bioremediation consists in the elimination or neutralization of environmental contaminants through natural biological activity (Shukla et al. 2020). In the case of metals, bioremediation acts in neutralizing contaminants, as metals cannot be degraded, as they are already in elemental form. Neutralization can occur through metal conversion to a less toxic form or by immobilization, reducing their bioavailability (Alvarez et al. 2017).

Plants, microorganisms, and even isolated microbial enzymes have been studied concerning metal removal from contaminated areas (Gaur et al. 2014; Sivakumar et al. 2014), and several metal bioremediation methods have been described, such as biosorption (use of dead microbial biomasses), biostimulation (stimulation of native microbial populations), and artificial microorganism introduction (Azubuike et al. 2016; Mani and Kumar 2013; Shukla et al. 2010; Sivakumar et al. 2014). The choice of the bioremediation technique will depend on the contaminated area, including local geological characteristics (soil type and depth), the type and concentration of the metallic contaminant, proximity to human populations, and the peculiarities of each method (performance, costs, and waste generation) (Azubuike et al. 2016).

Different bioremediation approaches based on the employed organisms are noted, comprising phytoremediation, mediated by plants and their associated microorganisms, mycoremediation, which employs live fungi or mycelial ultrafiltration, cyanoremediation – mediated by algae, and genoremediation, which applies genetically modified organisms (Gaur et al. 2014; Sivakumar et al. 2014). A multifaceted approach, employing these processes together, however, has a greater chance of success (Azubuike et al. 2016; Sivakumar et al. 2014).

Phytoremediation has received greater attention concerning metal removal due to its effectiveness and reduced cost (Ahemad 2019). By definition, phytoremediation considers plant physical, chemical, biological, and microbiological interactions that result in contaminant detoxification. Because of this, phytoremediation mechanisms will vary depending on the type of contaminant (elementary or organic) and the applied site (soil, sediments, sludge, wastewater, and groundwater) (Ali et al. 2013; Azubuike et al. 2016; Dixit et al. 2015; Gaur et al. 2014).

Metal phytoremediation processes employing aquatic plants involve the following steps, depicted in Figure 5.2: (i) Phytoremineralization, consisting in the recovery of bioaccumulated metals from the plant, (ii) Phytovolatilization, where metals are converted into volatile forms followed by atmospheric release from plant leaves, (iii) Phytotransformation, where metals are modified by enzymatic action, (iv) Phytofiltration, comprising metal root filtration, (v) Phytostabilization, which limits metal mobility and soil availability by root action, (vi) Uptake, consisting in metal absorption by roots, (vii) Rhizotransformation, where metals are transformed by microorganisms in the rhizosphere, (viii) Translocation, consisting in metal movement toward the aerial portions of the plant, and finally, (ix) Phytoextraction, where metals accumulate throughout the plant.

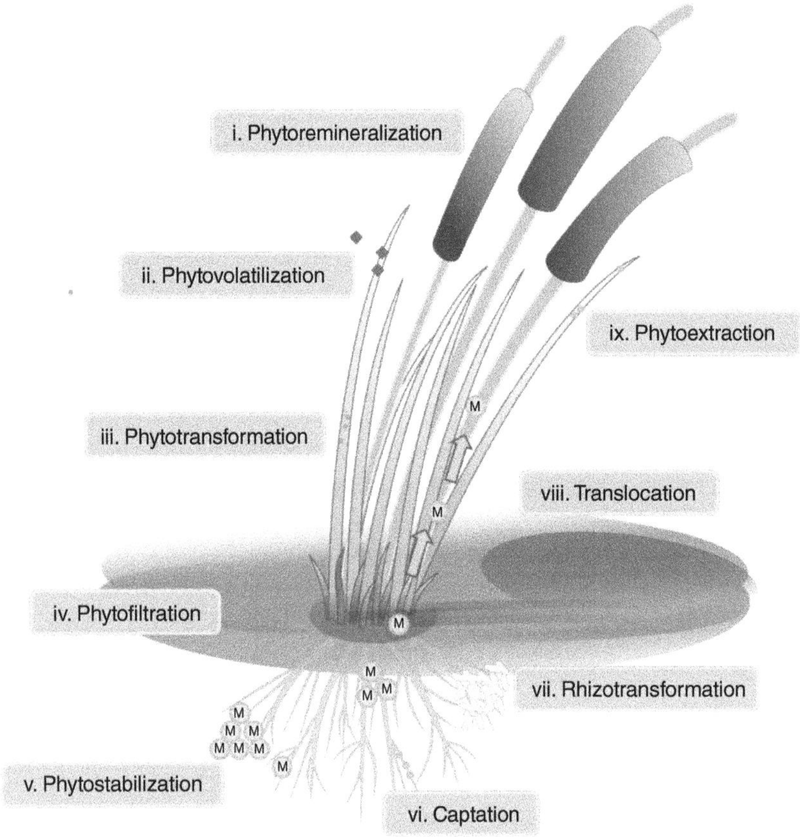


**Figure 5.2** Metal phytoremediation processes employing aquatic plants. M indicates metal ions.

Plants able to grow in metal-contaminated areas are called metallophytes (Ali et al. 2013) and generally exhibit phytoremediation potential. They are also known as hyperaccumulators, due to a higher tolerance to metal exposure and a high ability to accumulate these compounds compared to other plants (Dixit et al. 2015). However, some plant characteristics are crucial for their use in phytoremediation, such as a well-developed root system, aboveground biomass that must not be available for animal consumption, high survival rates, adaptation to environmental conditions, accelerated growth rate, resistance to diseases and pathogens, and metal accumulation ability (Azubuike et al. 2016; Pandey et al. 2015).

The main phytoremediation advantages comprise sustainability, the possibility of process escalation, low installation and maintenance costs, soil structure conservation, increased soil fertility, and erosion and metal leaching prevention (Azubuike et al. 2016), in addition to being sun-powered (Dixit et al. 2015). Disadvantages include a longer period compared to traditional methods and efficiency, which might be reduced depending on contaminant concentrations, toxicity, and bioavailability, resulting in slow growth rates and decreased plant biomass (Azubuike et al. 2016; Dixit et al. 2015). Furthermore, the fate of the plant material following remediation can generate additional costs, while the risk of incorporating metals into the food chain must also be considered (Azubuike et al. 2016; Pandey et al. 2015).

Several plant species displaying high phytoremediation potential have been described, such as *Amaranthus paniculatus*, *Brassica juncea*, *Populus canescens*, and *Mimosa pudica*, among others (Azubuike et al. 2016; Dixit et al. 2015; Ullah et al. 2015). In general, studies usually focus on soil contamination, with aquatic macrophytes more suited to water body contamination treatment (Hadad et al. 2006; Pio et al. 2013).

## 5.4 Aquatic Macrophytes in Metal Phytoremediation Processes

Aquatic macrophytes are herbaceous plants that grow in water, wetlands, or saturated soils. They do not constitute a specific taxonomic group and are classified arbitrarily according to their morphological aspects and life habits (Figure 5.3), from macroalgae to angiosperms (Esteves 2011). These include emerged aquatic macrophytes, rooted in the sediment and with leaves maintained above water (e.g., *Typha domingensis*), emergent with floating leaves, rooted in the sediment and

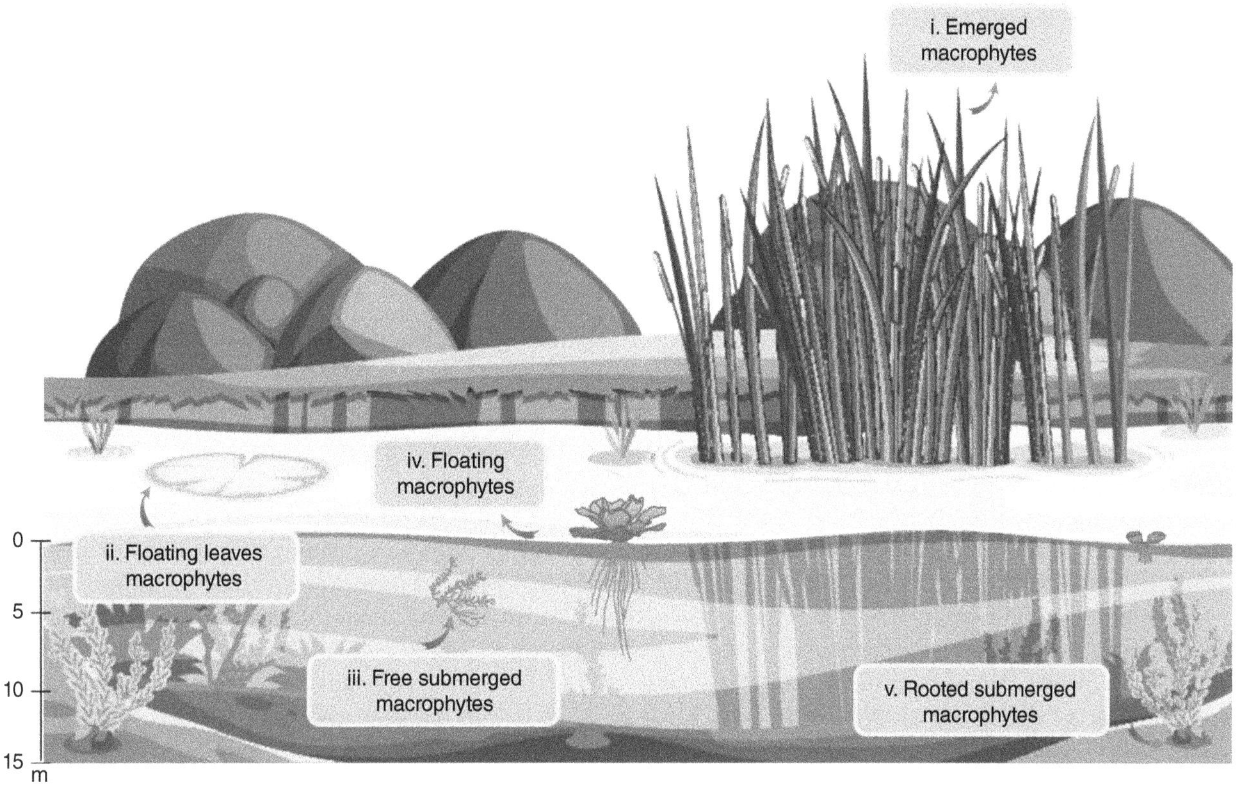


**Figure 5.3** Aquatic macrophyte categorization based on morphological aspects and life habits.

with leaves floating on the water surface (e.g., *Vitoria amazonica*), free submersed aquatic macrophytes, which float underwater (e.g., *Ceratophyllum demersum*), floating macrophytes that float freely on the water surface (e.g., *Pistia stratiotes* and *Salvinia auriculata*), and rooted submersed aquatic macrophyte, that grow entirely underwater (e.g., *Egeria densa*).

Some aquatic macrophytes exhibit fast growth rates, such as *Salvinia auriculata*, which doubles its biomass within 3–5 days (Esteves 2011). This makes the plants able to colonize large water body extensions in a short period, due to their high primary productivity, and they are frequently found in contaminated environments (Dos Santos et al. 2020). Because of this, they are highly relevant in phytoremediation processes and biotechnological research.

Macrophytes function as a biofilter for various contaminants in the phytoremediation approach and can be used in natural wetlands for *in situ* treatments or in wastewater treatment plants for *ex situ* treatment, as both living plants and their dead biomasses display phytoremediation potential (Rai 2009). Some genera are well known for their phytoremediation potential, including *Azolla*, *Eichhornia*, *Lemna*, *Pistia*, *Salvinia*, *Spartina*, and *Typha* (Bonanno 2013; Dhir et al. 2009; Dhir and Srivastava 2011; Malar et al. 2015; Mandal et al. 2016; Mesa et al. 2015; Pio et al. 2013; Shafi et al. 2015).

Studies have reported, for example, that *Eichhornia crassipes* can accumulate Hg in its tissues, resisting high concentrations of this metal (50 $mg\,l^{-1}$) (Malar et al. 2015). It is also able to survive in As-contaminated environments, presenting concentrations of up to 2 $mg\,l^{-1}$, removing up to 20% of this metallo from the environment (Jasrotia et al. 2017). The use of *E. crassipes* in the remediation of other metals has also been described, such as Cd (Zahoor et al. 2018), Cr (Chakrabarty et al. 2017), Cu (Melignani et al. 2015), and Pb (Suryandari et al. 2017). Another macrophyte, *Salvinia natans*, has been reported as capable of accumulating Al, Cr, Cu, Ni, and Zn (Dhir and Srivastava 2011; Mandal et al. 2016), with high Zn- and Cu-removal efficiencies (85 and 74%, respectively) (Dhir and Srivastava 2011). *Salvinia auriculata* also tolerates high Cu concentrations (Bizzo et al. 2014; Panfili et al. 2017). *Typha domingensis* is an emergent macrophyte found in aquatic and terrestrial environment transitions and is indicated as a Hg hyperaccumulator and promising in the As phytostabilization (Bonanno 2013). It is also able to efficiently remove Cr, Ni, and Zn from water at high concentrations (500 $mg\,l^{-1}$) (Mufarrege et al. 2015).

In addition, due to their ability to tolerate and bioaccumulate metals, several aquatic macrophytes may be applied in wastewater treatment, such as in artificially constructed wetlands that simulate natural environments. Several genera are used for this purpose, including *Eichhornia*, *Juncus*, *Phragmites*, *Pistia*, *Salvinia*, *Scirpus*, and *Typha* (Hadad et al. 2018; Sharma et al. 2014; Sricoth et al. 2018).

Among phytoremediation methods, rhizofiltration is the most applied when employing aquatic macrophytes to treat contaminated water, mainly metals. In this type of phytoremediation, contaminants preferentially accumulate in the roots of the employed aquatic plants and not in aerial biomasses (Hadad et al. 2006; Pio et al. 2013). The efficiency of this method is probably due to the presence of beneficial microorganisms at the plant-soil interface, acting to improve plant development and assist in their resistance responses to metals and other water body pollutants.

## 5.5 Microorganisms in Metal Remediation

Like plants, microorganisms play a fundamental role in metal bioremediation (Azubuike et al. 2016), due to their ability to transform natural and synthetic chemicals and raw materials into energy sources for their growth (Sivakumar et al. 2014). Several saprophytic, endophytic, and mycorrhizal microorganisms have been described for this purpose, including fungi, bacteria, and algae.

Fungi belonging to the *Acremonium*, *Aspergillus*, *Mucor*, *Penicillium*, and *Rhizopus* genera are known to tolerate metal exposure (Ma et al. 2014). One yeast, in particular, *Yarrowia lipolytica*, is efficient in accumulating several metals (Cd, Co, Cr, Cu, Hg, Ni, Pb, and Zn) (Asfaram et al. 2016; Dil et al. 2017; Wierzba 2017). Among prokaryotes, *Cupriavidus metallidurans* is noteworthy, which can resist high Ag, Au, Cd, Co, Cr, Cu, Hg, Ni, and Pb concentrations (Mergeay and Van Houdt 2015). Species belonging to the *Achromobacter*, *Bacillus*, *Enterobacter*, *Micrococcus*, *Gemella, Hafnia, Pseudomonas, Aeromonas* and *Mycobacterium* genera are also resistant to metal stress (Anwar et al. 2017; Brito et al. 2015; Marzan et al. 2017; Sharma et al. 2020; Sharma and Shukla 2020; Subudhi et al. 2016). In algae, we can mention some genera such as *Cladophora*, *Hydrodictylon*, *Oedogonium*, *Rhizoclonium*, *Spirogyra*, and *Spirulina* (Lee and Chang 2011; Mane and Bhosle 2012).

Information on the diversity, abundance, and structure of microbial communities is relevant to determine which bioremediation technique is ideal in given contaminated environments (Azubuike et al. 2016). It is also essential to understand microorganism growth and activity control mechanism, their metabolism, and their responses to environmental changes (Dixit et al. 2015). Some factors that may affect microorganism-mediated bioremediation success rates in this regard are nutrient shortage, reduced population size, absence of microorganisms resistant to the contaminant of interest, and contaminant bioavailability (Azubuike et al. 2016).

In cases of contaminated areas with insufficient amounts of nutrients to maintain cellular microorganism metabolism and growth (Sivakumar et al. 2014), certain nutrients, such as nitrogen, phosphorus, and potassium, may be added to the site, in a process called biostimulation, which aims to stimulate the metabolic activities of native microorganisms (Azubuike et al. 2016; Dixit et al. 2015; Sivakumar et al. 2014). Carbon nutritional requirements in comparison with nitrogen (C:N) are usually 10:1 for most microorganisms. However, if the C:N ratio is increased to 25:1, for example, increased microbial soil growth is expected. In addition to nutrients, other supplements can also be added, such as electron acceptors (oxygen) and substrates (methane, phenol, and toluene) (Sivakumar et al. 2014).

Air supplementation can promote the action of certain microbial groups for *in situ* treatments. Oxygen supplementation, for example, is essential to stimulate aerobic microorganisms through a bioremediation process called bioventilation (Mani and Kumar 2013). Another approach comprises bioaugmentation, which consists of introducing selected microbial inocula into the contaminated site (Dixit et al. 2015). This technique, however, increases the risk of competition between native and introduced populations, as well as the accidental introduction of pathogenic microorganisms, and the failure of the introduced population (Azubuike et al. 2016). Furthermore, under ideal environmental conditions, native microorganisms are more efficient in decontaminating affected areas, and should be prioritized (Fodelianakis et al. 2015).

Other factors must also be considered concerning the use of microorganisms in bioremediation, such as pH and temperature. The ideal pH range for the growth of most microorganisms ranged between 5.5 and 8.0 (Sivakumar et al. 2014). However, the original pH of the contaminated site must be first evaluated, in order not to alter natural characteristics and further affect the environment.

A noteworthy aspect that should be highlighted in the case of microorganism bioremediation is the use of microbial consortia, which refers to the use of several microorganisms working together to decontaminate a given environment. This strategy is more effective compared to the use of pure isolates, as it is more similar to what occurs in nature. Each microorganism may have, for example, developed a different resistance mechanism, and when acting together, results in increased bioremediation potential (Carpio et al. 2016; Gupta et al. 2016; Migahed et al. 2017; Mitra and Mukhopadhyay 2016).

### 5.5.1 Microorganism Metal Resistance Mechanisms

Fungi and bacteria exposed to high metal concentrations are expected to suffer cellular function alterations (He et al. 2014). However, microorganisms have developed several strategies to tolerate adverse environmental conditions and still thrive.

As metals cannot be degraded, microorganisms transform these compounds into less harmful forms (Roane et al. 2015). In this regard, several bacterial metal-resistance mechanisms have been described (Figure 5.4) (Das et al. 2016; Dixit et al. 2015; Etesami 2018; Mosa et al. 2016), mainly due to the expression of resistance genes (Das et al. 2016). The mer operon found in the bacterial species *Acidithiobacillus ferrooxidans*, for example, is formed by a gene that regulates resistance to Hg (merR), as well as two other structural genes, merC, which acts as a $Hg^{2+}$ membrane transporter and merA, which encodes the mercuric reductase enzyme (Kusano et al. 1992). Resistance genes to other metals, such as Cd, Cr, Cu, Pb, and Ni, are also known, such as cadB and cadD, chrA, cusF and cueO, pbrD, and NiCoT (Das et al. 2016).

Genetically modified microorganisms have also been described as able to contribute to bioremediation (Azubuike et al. 2016; Dixit et al. 2015). The genetics of metal-accumulating microorganisms can aid in the screening of resistance genes to susceptible species (Dixit et al. 2015; Mosa et al. 2016). For example, the Hg-resistant bacteria, *Cupriavidus metallidurans* MSR33, was genetically modified by the introduction of a plasmid, pTP6, which supplies the merB and merG genes, enabling metal toxicity reduction through the action of organomercurial lysozyme (MerB) proteins and reductase mercury (MerA) (Dixit et al. 2015). This approach, however, requires caution, as horizontal gene transfer and uncontrolled multiplication of these genetically modified microorganisms are a possibility (Azubuike et al. 2016).

Like microorganisms, plants have also been manipulated, receiving bacterial metal resistance genes, and studies have indicated positive results regarding metal resistance and accumulation in different plant species (Dhankher et al. 2012;

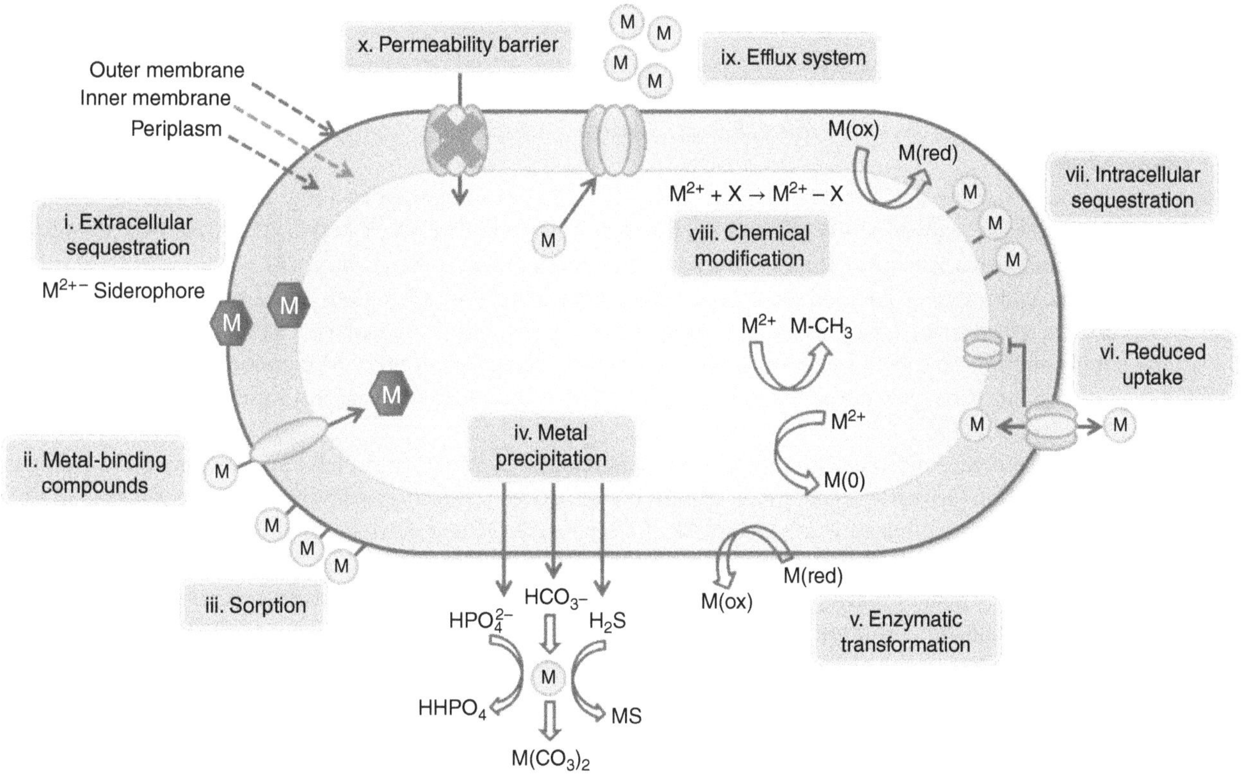


**Figure 5.4** Bacterial metal-resistance mechanisms. (i) Extracellular sequestration is carried out by substances produced by the bacterium, such as siderophores that bind to metal. (ii) Complexation by metal binding to the cell, followed by internalization. (iii) Adsorption, carried out by metal interactions with the cell surface. (iv) Metal precipitation by chemical reactions mediated by bacteria. (v) Enzymatic transformation by redox reactions through the conversion of metallic ions into a less toxic form employing methylation. (vi) Metal entry decreases by blocking a carrier in the bacterium's inner membrane. (vii) Intracellular sequestration with metal accumulation in internal compartments or the production of proteins, such as metallothionein, which can bind to metal ions. (viii) Chemical cell cytoplasm modifications decrease metal toxicity. (ix) Efflux system mediated by proteins that pump metals out of the cell. (x) Membrane permeability prevents excess metal ion cell entry.

Mosa et al. 2016). However, there are other ways to combine the advantageous characteristics of plants and microorganisms to increase metal bioremediation efficiencies. One possibility, for instance, is to inoculate microorganisms in these plants. Thus, a better understanding of plant-microbe interactions is paramount, as, even though they are extensively studied in an agricultural context, are still little explored in bioremediation scenarios, especially concerning the aquatic environment.

## 5.6 Interaction Between Aquatic Macrophytes and Microorganisms

Macrophytes that grow in contaminated and eutrophic areas harbor a diverse group of microorganisms, which can generally tolerate high metal concentrations and provide several plant and environmental benefits (Franchi et al. 2016). This is possible through the synthesis of specific microbial substances that facilitate nutrient absorption (phosphorus, production of phytohormones, and siderophores) and reduce or prevent stresses (Azubuike et al. 2016; Chauhan et al. 2015; Hayat et al. 2010). Considering this, plant-microorganism interactions can prevent contaminant, i.e., metal entry by modulating the transport channels of plant membranes (Srivastava et al. 2017). Rhizosphere microorganisms can also produce metabolites that decrease metal plant absorption, altering metal mobility and bioavailability by altering the pH, releasing chelating agents (organic acids and siderophores), and producing redox reactions (Franchi et al. 2016).

The benefits of plant-microorganism interactions range from increasing plant biomasses to improving plant responses under different stress conditions (Calvo et al. 2014). Microorganisms can promote increased nutrient plant absorption through biological nitrogen fixation (BNF), which is the main plant growth promotion mechanism carried out by microorganisms. This process occurs when fixing bacteria convert atmospheric molecular nitrogen into forms assimilable by plants. These bacteria are also called diazotrophic and contain an enzyme complex named nitrogenase (Burén and Rubio 2018; Merrick 2004; Shcherbak et al. 2014).

Diazotrophic bacteria are usually associated with plant rhizospheres, sometimes establishing even more specific associations, living as endophytics within plant tissues (Baldani et al. 1997; Steenhoudt and Vanderleyden 2000). Several bacteria genera have been described as displaying this potential (*Azoarcus*, *Azospirillum*, *Azotobacter*, *Beijerinckia*, *Burkholderia*, and *Herbaspirillum*).

Plant growth promotion can also occur due to the production of phytohormones by fungi and bacteria. In this case, microorganisms synthesize plant growth regulators, such as indole-3-acetic acid (IAA), making them available to improve root development, and formation and aiding in stress resistance development (Calvo et al. 2014).

Different bacterial communities living in the rhizosphere in association with plants can tolerate high metal concentrations. These rhizobacteria display plant growth-promoting characteristics and can be found in contaminated environments, and several studies indicate that both metal-resistant microorganisms and plant growth-promoting bacteria display the potential to protect plants from toxic metal effects (Braud et al. 2009; Glick 2010; Hansda and Kumar 2017; He et al. 2010; Ma et al. 2011; Zaidi et al. 2006).

Both plants and microorganisms interact for their survival, making local resources available and, consequently, aiding in environmental remediation (Srivastava et al. 2017). This exchange between these organisms characterizes a harmonic ecological relationship, where plants release exudates into the rhizosphere, providing nutrients for microorganisms. The microbial activity of roots is 10- to 100-fold higher when compared to the rest of the soil (Dixit et al. 2015; Franchi et al. 2016).

The interactions between aquatic macrophytes and microorganisms can decrease harmful plant metal effects, depending on the nature of this interaction (Sharma et al. 2013). Studies involving bacteria and aquatic plants have, for example, identified *in vitro* growth promotion characteristics and resistance profiles for Zn, Ni, and Cd in eleven bacteria isolated from the emerged macrophyte *Juncus acutus*. Three of these bacteria were inoculated again in the plant through the bioaugmentation process, and an increased contaminant removal efficiency from the water, mainly in an intercropping treatment by up to 75%, was noted due to microorganism-plant interactions between the microorganisms and the macrophyte (Syranidou et al. 2016).

Mesa et al. (2015) reported a 60% increase in the biomass of the aquatic macrophyte *Spartina maritima* inoculated with *Bacillus strains*, leading to Cu removal of up to 65%. Plant growth was associated with the beneficial effect of this microorganism on the photosynthetic apparatus, with improvements in the functionality of photosystem II (PSII) and chlorophyll concentrations, as well as increased water use efficiency due to decreasing stomatal conductance and an increased root–shoot ratio. In another study, Ejaz et al. (2013) reported that the introduction of a bacterial consortium (*Bacillus pumilus*, *Pseudomonas doudoroffii*, and *Exiguobacterium*) in the cultivation of *P. stratiotes* and *E. crassipes* improved these plants abilities to remove $Cr^{6+}$ from the medium (Ejaz et al. 2013). Table 5.2 depicts a list of macrophyte species, associated microorganisms, and enabled bioremediation of toxic metals due to this interaction.

In nature, plant-microbe interactions do not occur exclusively with a single species, but rather with a highly structured microbial community, which varies with each environment or organism, depending on the nature and availability of organic carbon and the levels of molecular oxygen in the rhizosphere. In the case of rooted macrophytes, structures from the aerial portion to the root facilitate the transfer of $O_2$ to the rhizosphere, supporting aerobic life forms. In other cases, electron acceptors such as $CO_2$, $CH_4$, and $NO_3$ support anaerobic life forms, while optional microorganisms residing close to the rhizosphere in a toxic-anoxic interface are also noted (Srivastava et al. 2017). The main reason for the increased metal removal in the rhizosphere is the presence of high microbial metabolic activity (Dixit et al. 2015).

The use of macrophytes associated with microorganisms has been highlighted in the literature, indicating that this interaction has the potential to be employed in metal bioremediation processes. In this sense, understanding such an ecological relationship can be beneficial to elucidate interaction mechanisms and improve remediation process efficiency. It is clear,

**Table 5.2** Plant-microbe interactions enabling metal remediation.

| Macrophytes | Microorganism | Metal | Concentration(mg $l^{-1}$) | References |
|---|---|---|---|---|
| *Eichhornia crassipes* | *Acinetobacter* spp. | Cu | 20 | Irawati et al. (2017) |
| *Eichhornia crassipes* | *Azobbacter* sp.<br>*Actinomyces* sp.<br>*Bacillus subtilis*<br>*Azospirillum brasiliense*<br>*Pseudomonas aeruginosa* | As | 18 | Kaur et al. (2017) |
| *Eichhornia crassipes* | *Enterobacter* sp. | Cd, Zn, and Pb | 306, 650, and 1.4 respectively | El-deeb et al. (2012) |
| *Eichhornia crassipes* | *Bacillus cereus*<br>*Brevundimonas diminuta*<br>*Nitrobacteria irancium Pseudomonas diminuta Ochrobactrum anthropic* | Cr, Mn, Zn, and Pb | 1 | Abou-Shanab et al. (2007) |
| *Eichhornia crassipes*<br>*Pistia stratiotes* | *Bacillus pumilus*<br>*Exiguobacterium*<br>*Pseudomonas doudoroffii* | Cr | 300 | Ejaz et al. (2013) |
| *Eichhornia crassipes*<br>*Eichhornia azurea*<br>*Hymenachne donacifolia*<br>*Hymenachne rotundifolia*<br>*Ludwigia helminthorrhiza*<br>*Polygonum densiflorum* | Microbial consortium in periphyton | Hg | 0.072 | Correia et al. (2012) |
| *Echinodorus cordifolius* | *Arthrobacter creatinolyticus* | As | 2 | Prum et al. (2018) |
| *Ipomea aquatic* | *Arbuscular mycorrhizal* fungi | Cd | 100 | Bhaduri and Fulekar (2012) |
| *Myriophyllum spicatum*<br>*Potamogeton crispus*<br>*Stuckenia pectinata* | *Acinetobacter calcoaceticus Pseudomonas chlororaphis Pseudomonas putrefaciens Pseudoalteromonas haloplanktis-haloplanktis* | Hg | 0.68 | Caslake et al. (2006) |
| *Phragmites australis* | *Rhizophagus irregularis* | Cd | 20 | Huang et al. (2017) |
| *Pistia stratiotes* | *Bacillus cereus* | Cr | 8 | Chakraborty et al. (2013) |
| *Spartina maritima* | *Bacillus aryabhattai*<br>*Bacillus licheniformis*<br>*Bacillus methylotrophicus* | Cu, Zn, As, Pb, Cd, and Ni | 2453, 1902, 449, 606, 5, and 39 respectively | Mesa et al. (2015) |
| *Typha domingensis* | *Bacillus pumilus*<br>*Microbacterium arborescens* | Ni, Fe, Cr, and Cd | 2.7, 3.3, 0.25, and 0.27 respectively | Shehzadi et al. (2014) |

therefore, that the success of any bioremediation approach will depend on the optimization of the remedial potentials of native plants, which grow in polluted areas, alongside the inoculation of contaminant-resistant microorganisms (Azubuike et al. 2016). These aspects should be considered in future investigations, alongside the impact of different soil properties, on plant responsiveness and the potential of microorganisms as a plant growth promotion tool. However, although bioremediation is recognized as sustainable, plant-microorganism interaction found in nature will not always be ideal for phytoremediation, and human interventions are required to optimize these interactions, increasing contaminant removal as desired (Thijs et al. 2016).

## 5.7 Conclusion

Natural bioremediation is not a very efficient process, due to the increased waste generated by anthropic actions noted in today's world, requiring the application of specific technologies to optimize contaminant removal. In this regard, bioremediation comprises a low-cost and sustainable alternative for environmental contaminant remediation. As discussed herein, this technique can be applied to wastewater decontamination in effluents containing highly toxic metals. Different organisms can be used to this end, such as plants and plant growth-promoting microorganisms. Aquatic macrophytes and microorganisms are noteworthy in this regard, as both display efficient mechanisms and strategies that allow their survival even in eutrophic or severely contaminated environments, while at the same time mitigating the harmful effects of these contaminants or promoting their removal. Thus, bioremediation processes involving the joint action of microorganisms and plants display significant potential, although the intrinsic processes of this strategy remain to be elucidated in full.

## References

Abou-Shanab, R.A.I., Angle, J.S., and Van Berkum, P. (2007). Chromate-tolerant bacteria for enhanced metal uptake by *Eichhornia crassipes* (Mart.). *International Journal of Phytoremediation* 9 (2): 91–105.

Adriano, D.C. (2001). Biogeochemistry, bioavailability and risks of metals. In: *Trace Elements in the Terrestrial Environment*, 2e (ed. D.C. Adriano). New York, NY: Springer-Verlag.

Ahemad, M. (2019). Remediation of metalliferous soils through the heavy metal resistant plant growth promoting bacteria: paradigms and prospects. *Arabian Journal of Chemistry* 12 (7): 1365–1377.

Ali, H., Khan, E., and Sajad, M.A. (2013). Phytoremediation of heavy metals—concepts and applications. *Chemosphere* 91 (7): 869–881.

Alvarez, A., Saez, J.M., Costa, J.S.D. et al. (2017). Actinobacteria: current research and perspectives for bioremediation of pesticides and heavy metals. *Chemosphere* 166: 41–62.

ANA (2018) Paraíba do Sul. Available at: http://www3.ana.gov.br/portal/ANA/sala-de-situacao/paraiba-do-sul/paraiba-do-sul-saiba-mais (accessed in June 28).

Ansari, A.A., Gill, S.S., and Khan, F.A. (2010). Eutrophication: threat to aquatic ecosystems. In: *Eutrophication: Causes, Consequences and Control* (ed. A. Ansari, S. Singh Gill, G. Lanza, and W. Rast), 143–170. Dordrecht: Springer.

Anwar, M., Ali, S., and Asaad, A.T. (2017). Evaluation of heavy metals resistant *Micrococcus* sp. isolated from Rivers in Basra. Iraque. *Journal of Bioremediation & Biodegradation* 8: 1–4.

Asfaram, A., Ghaedi, M., and Ghezelbash, G.R. (2016). Biosorption of $Zn^{2+}$, $Ni^{2+}$ and $Co^{2+}$ from water samples onto *Yarrowia lipolytica* ISF7 using a response surface methodology, and analyzed by inductively coupled plasma optical emission spectrometry (ICP-OES). *The Royal Society of Chemistry* 6 (28): 23599–23610.

Ayoub, A.T. (1999). Fertilizers and the environment. *Nutrient Cycling in Agroecosystems* 55 (2): 117–121.

Azubuike, C.C., Chikere, C.B., and Okpokwasili, G.C. (2016). Bioremediation techniques–classification based on site of application: principles, advantages, limitations and prospects. *World Journal of Microbiology and Biotechnology* 32 (11): 180.

Baird, C. (2001). *Environmental Chemistry*. WH: University of Western Ontario.

Baldani, J., Caruso, L., Baldani, V.L. et al. (1997). Recent advances in BNF with non-legume plants. *Soil Biology and Biochemistry* 29 (5–6): 911–922.

Bhaduri, A.M. and Fulekar, M.H. (2012). Assessment of arbuscular mycorrhizal fungi on the phytoremediation potential of *Ipomoea aquatica* on cadmium uptake. *3 Biotech* 2: 193–198.

Bielmyer-Fraser, G.K., Waters, M.N., Duckworth, C.G. et al. (2017). Assessment of metal contamination in the biota of four rivers experiencing varying degrees of human impact. *Environmental Monitoring and Assessment* 189 (1): 23.

Bizzo, A.L.T., Intorne, A.C., Gomes, P.H. et al. (2014). Short-term physiological responses to copper stress in *Salvinia auriculata* Aubl. *Acta Limnologica Brasiliensia* 26 (3): 268–277.

Bonanno, G. (2013). Comparative performance of trace element bioaccumulation and biomonitoring in the plant species *Typha domingensis, Phragmites australis* and *Arundo donax. Ecotoxicology and Environmental Safety* 97: 124–130.

Braud, A., Jézéquel, K., Bazot, S., and Lebeau, T. (2009). Enhanced phytoextraction of an agricultural Cr- and Pb-contaminated soil by bioaugmentation with siderophore-producing bacteria. *Chemosphere* 74 (2): 280–286.

Brito, E.M.S., De La Cruz, M.B., Caretta, C.A. et al. (2015). Impact of hydrocarbons, PCBs and heavy metals on bacterial communities in Lerma River, Salamanca, Mexico: investigation of hydrocarbon degradation potential. *Science of the Total Environment* 521: 1–10.

Buchman, M.F. (2008). *NOAA Screening Quick Reference Tables, NOAA OR&R Report 08-1*. Seattle, WA: Office of Response and Restoration Division, National Oceanic and Atmospheric Administration 34 p.

Burén, S. and Rubio, L.M. (2018). State of the art in eukaryotic nitrogenase engineering. *FEMS Microbiology Letters* 365 (2): fnx274.

Calijuri, M.C. and Cunha, D.G.F. (2013). *Engenharia Ambiental: conceitos, tecnologia e gestão*. Elsevier Brasil.

Calvo, P., Nelson, L., and Kloepper, J.W. (2014). Agricultural uses of plant biostimulants. *Plant and Soil* 383 (1–2): 3–41.

Capparelli, M.V., Moulatlet, G.M., Abessa, D.M.S. et al. (2020). An integrative approach to identify the impacts of multiple metal contamination sources on the Eastern Andean foothills of the Ecuadorian Amazonia. *Science of the Total Environment* 709: 136088.

Carpio, I.E.M., Franco, D.C., Sato, M.I.Z. et al. (2016). Biostimulation of metal-resistant microbial consortium to remove zinc from contaminated environments. *Science of the Total Environment* 550: 670–675.

Caslake, L.F., Harris, S.S., Williams, C., and Waters, N.M. (2006). Mercury-resistant bacteria associated with macrophytes from a polluted lake. *Water, air, and Soil Pollution* 174: 93–105.

Chagnon, M., Kreutzweiser, D., Mitchell, E.A. et al. (2015). Risks of large-scale use of systemic insecticides to ecosystem functioning and services. *Environmental Science and Pollution Research* 22 (1): 119–134.

Chakrabarty, T., Afrin, R., Mia, M.Y., and Hossen, M.Z. (2017). Phytoremediation of chromium and some chemical parameters from Tannery effluent by using water Hyacinth (*Eichhornia crassipes*). *Research in Agriculture, Livestock and Fisheries* 4 (3): 151–156.

Chakraborty, R., Mitra, A.K., and Mukherjee, S. (2013). Synergistic chromium bioremediation by water lettuce (*Pistia*) and bacteria (*Bacillus cereus* GXBC-1) interaction. *Journal of Biological Chemistry Research* 30 (2): 421–431.

Chauhan, H., Bagyaraj, D.J., Selvakumar, G., and Sundaram, S.P. (2015). Novel plant growth promoting rhizobacteria—prospects and potential. *Applied Soil Ecology* 95: 38–53.

Chen, C.Y. and Driscoll, C.T. (2018). Integrating mercury research and policy in a changing world. *Ambio* 47 (2): 111–115.

Chowdhury, S., Mazumder, M.J., Al-Attas, O., and Husain, T. (2016). Heavy metals in drinking water: occurrences, implications, and future needs in developing countries. *Science of the Total Environment* 569: 476–488.

Clemens, S. (2006). Toxic metal accumulation, responses to exposure and mechanisms of tolerance in plants. *Biochimie* 88 (11): 1707–1719.

Correia, R.R., Miranda, M.R., and Guimarães, J.R. (2012). Mercury methylation and the microbial consortium in periphyton of tropical macrophytes: effect of different inhibitors. *Environmental Research* 112: 86–91.

Das, S., Dash, H.R., and Chakraborty, J. (2016). Genetic basis and importance of metal resistant genes in bacteria for bioremediation of contaminated environments with toxic metal pollutants. *Applied Microbiology and Biotechnology* 100 (7): 2967–2984.

De Souza Neto, H.F., Da Silveira Pereira, W.V., Dias, Y.N. et al. (2020). Environmental and human health risksP of arsenic in gold mining areas in the eastern Amazon. *Environmental Pollution* 114969.

De Stefano, L., Petersen-Perlman, J.D., Sproles, E.A. et al. (2017). Assessment of transboundary river basins for potential hydro-political tensions. *Global Environmental Change* 45: 35–46.

Dhankher, O.P., Pilon-Smits, E.A., Meagher, R.B., and Doty, S. (2012). Biotechnological approaches for phytoremediation. In: *Plant Biotechnology and Agriculture* (ed. A. Altman and P.M. Hasegawa), 309–328.

Dhir, B. and Srivastava, S. (2011). Heavy metal removal from a multi-metal solution and wastewater by *Salvinia natans*. *Ecological Engineering* 37 (6): 893–896.

Dhir, B., Sharmila, P., and Saradhi, P.P. (2009). Potential of aquatic macrophytes for removing contaminants from the environment. *Critical Reviews in Environmental Science and Technology* 39 (9): 754–781.

Dil, E.A., Ghaedi, M., Ghezelbash, G.R. et al. (2017). Highly efficient simultaneous biosorption of $Hg^{2+}$, $Pb^{2+}$ and $Cu^{2+}$ by Live yeast *Yarrowia lipolytica* 70562 following response surface methodology optimization: kinetic and isotherm study. *Journal of Industrial and Engineering Chemistry* 48: 162–172.

Dixit, R., Alaviya, D., Pandiyan, K. et al. (2015). Bioremediation of heavy metals from soil and aquatic environment: an overview of principles and criteria of fundamental processes. *Sustainability* 7 (2): 2189–2212.

Dos Santos Vergilio, C., Lacerda, D., De Oliveira, B.C.V. et al. (2020). Metal concentrations and biological effects from one of the largest mining disasters in the world (Brumadinho, Minas Gerais, Brazil). *Scientific Reports* 10 (1): 1–12.

Dos Santos, N.B.C., De Arruda, E.C.P., Neto, A.G.B., and De Oliveira, A.F.M. (2020). Assessing the effects of water quality on leaf morphoanatomy, ultrastructure and photosynthetic pigment content of *Salvinia auriculata* Aubl. (Salviniaceae). *Ecotoxicology and Environmental Safety* 190: 110061.

Ejaz, S., Rizvi, F.Z., Anwar, S., and Faisal, M. (2013). Biotransformation potential of hexavalent chromium by *Bacillus pumilus*-S4, *Pseudomonas doudoroffii*-S5 and *Exiguobacterium*-S8 in association with hydrophytes. *International Journal of Environmental Science and Technology* 10 (4): 709–718.

El-Deeb, B., Gherbawy, Y., and Hassan, S. (2012). Molecular characterization of endophytic bacteria from metal hyperaccumulator aquatic plant (*Eichhornia crassipes*) and its role in heavy metal removal. *Geomicrobiology Journal* 29 (10): 906–915.

Esteves, F.D.A. (2011). *Fundamentos de Limnologia*, 3e. Interciência, Rio de Janeiro.

Etesami, H. (2018). Bacterial mediated alleviation of heavy metal stress and decreased accumulation of metals in plant tissues: mechanisms and future prospects. *Ecotoxicology and Environmental Safety* 147: 175–191.

Farooq, M., Rehman, A., and Pisante, M. (2019). Sustainable agriculture and food security. In: *Innovations in Sustainable Agriculture* (ed. M. Farooq and M. Pisante), 3–24. Cham: Springer.

Ferreira, T.O., Otero, X.L., De Souza Junior, V.S. et al. (2010). Spatial patterns of soil attributes and components in a mangrove system in Southeast Brazil (São Paulo). *Journal of Soils and Sediments* 10 (6): 995–1006.

Fodelianakis, S., Antoniou, E., Mapelli, F. et al. (2015). Allochthonous bioaugmentation in *ex situ* treatment of crude oil-polluted sediments in the presence of an effective degrading indigenous microbiome. *Journal of Hazardous Materials* 287: 78–86.

Franchi, E., Rolli, E., Marasco, R. et al. (2016). Phytoremediation of a multi contaminated soil: mercury and arsenic phytoextraction assisted by mobilizing agent and plant growth promoting bacteria. *Journal of Soils and Sediments* 17 (5): 1224–1236.

Fu, K.L., Yao, M.Y., Qin, C.G. et al. (2016). Study on the removal of oxidized mercury ($Hg^{2+}$) from flue gas by thiol chelating resin. *Fuel Processing Technology* 148: 28–34.

Gaur, N., Flora, G., Yadav, M., and Tiwari, A. (2014). A review with recent advancements on bioremediation-based abolition of heavy metals. *Environmental Science: Processes & Impacts* 16 (2): 180–193.

Genchi, G., Sinicropi, M.S., Carocci, A. et al. (2017). Mercury exposure and heart diseases. *International Journal of Environmental Research and Public Health* 14 (1): 74.

Giller, K.E., Witter, E., and Mcgrath, S.P. (2009). Heavy metals and soil microbes. *Soil Biology and Biochemistry* 41 (10): 2031–2037.

Glick, B.R. (2010). Using soil bacteria to facilitate phytoremediation. *Biotechnology Advances* 28 (3): 367–374.

Gupta, G., Kumar, V., and Pal, A.K. (2016). Biodegradation of polycyclic aromatic hydrocarbons by microbial consortium: a distinctive approach for decontamination of soil. *Soil and Sediment Contamination: An International Journal* 25 (6): 597–623.

Ha, E., Basu, N., Bose-O'Reilly, S. et al. (2016). Current progress on understanding the impact of mercury on human health. *Environmental Research* 152: 419–433.

Hadad, H.R., Maine, M.A., and Bonetto, C.A. (2006). Macrophyte growth in a pilot-scale constructed wetland for industrial wastewater treatment. *Chemosphere* 63 (10): 1744–1753.

Hadad HR, Mufarrege MM, Di Luca GA, Maine MA (2018). Salinity and pH effects on floating and emergent macrophytes in a constructed wetland. *Water Science and Technology* 2017(1):270–275

Hansda, A. and Kumar, V. (2017). Cu-resistant *Kocuria* sp. CRB15: a potential PGPR isolated from the dry tailing of Rakha copper mine. *3 Biotech* 7 (2): 132.

Hayat, R., Ali, S., Amara, U. et al. (2010). Soil beneficial bacteria and their role in plant growth promotion: a review. *Annales de Microbiologie* 60 (4): 579–598.

He, L.Y., Zhang, Y.F., Ma, H.Y. et al. (2010). Characterization of copper-resistant bacteria and assessment of bacterial communities in rhizosphere soils of copper-tolerant plants. *Applied Soil Ecology* 44 (1): 49–55.

He, D., Ren, L., and Wu, Q.L. (2014). Contrasting diversity of epibiotic bacteria and surrounding bacterioplankton of a common submerged macrophyte, *Potamogeton crispus*, in freshwater lakes. *FEMS Microbiology Ecology* 90 (3): 551–562.

Huang, J., Xu, C.C., Ridoutt, B.G. et al. (2017). Nitrogen and phosphorus losses and eutrophication potential associated with fertilizer application to cropland in China. *Journal of Cleaner Production* 159: 171–179.

Irawati, W., Parhusip, A.J.N., Sopiah, N., and Tnunay, J.A. (2017). The role of heavy metals-resistant bacteria *Acinetobacter* sp. in copper phytoremediation using *Eichhornia crassipes* [(Mart.) Solms]. KnE. *Life Sciences* 208–220.

IUCN, Standards and Petitions Subcommittee (2017) Guidelines for Using the IUCN Red List Categories and Criteria. Version 13. Prepared by the Standards and Petitions Subcommittee. 108 p.

Jasrotia, S., Kansal, A., and Mehra, A. (2017). Performance of aquatic plant species for phytoremediation of arsenic-contaminated water. *Applied Water Science* 7 (2): 889–896.

Kabata-Pendias, A. (2010). *Trace Elements in Soils and Plants*. CRC Press 548 p.

Kaur, P., Singh, S., Kumar, V. et al. (2017). Effect of rhizobacteria on arsenic uptake by macrophyte *Eichhornia crassipes* (Mart.) Solms. *International Journal of Phytoremediation* 20 (2): 114–120.

Khan, M.N. and Mohammad, F. (2014). Eutrophication: challenges and solutions. In: *Eutrophication: Causes, Consequences and Control* (ed. A. Ansari and S. Gill), 1–15. Dordrecht: Springer.

Khan, M.N., Mobin, M., Abbas, Z.K., and Alamri, S.A. (2018). Fertilizers and their contaminants in soils, surface and groundwater. *Encyclopedia of the Anthropocene* 5: 225–240.

Kill, K., Pärn, J., Lust, R. et al. (2018). Treatment efficiency of diffuse agricultural pollution in a constructed wetland impacted by groundwater seepage. *Water* 10 (11): 1601.

Kusano, T., Sugawara, K., Inoue, C. et al. (1992). Electrotransformation of *Thiobacillus ferrooxidans* with plasmids containing a mer determinant. *Journal of Bacteriology* 174 (20): 6617–6623.

Lee, Y.C. and Chang, S.P. (2011). The biosorption of heavy metals from aqueous solution by *Spirogyra* and *Cladophora* filamentous macroalgae. *Bioresource Technology* 102 (9): 5297–5304.

Liu, J., Li, Y., Zhang, B. et al. (2009). Ecological risk of heavy metals in sediments of the Luan River source water. *Ecotoxicology* 18 (6): 748–758.

Liu, J., Zhang, X.H., Tran, H. et al. (2011). Heavy metal contamination and risk assessment in water, paddy soil, and rice around an electroplating plant. *Environmental Science and Pollution Research* 18 (9): 1623.

Ma, Y., Prasad, M.N.V., Rajkumar, M., and Freitas, H. (2011). Plant growth promoting rhizobacteria and endophytes accelerate phytoremediation of metalliferous soils. *Biotechnology Advances* 29 (2): 248–258.

Ma, X.K., Ling, W.U.L., and Fam, H. (2014). Heavy metal ions affecting the removal of polycyclic aromatic hydrocarbons by fungi with heavy-metal resistance. *Applied Microbiology and Biotechnology* 98 (23): 9817–9827.

Maksymiec, W. (2007). Signaling responses in plants to heavy metal stress. *Acta Physiologiae Plantarum* 29 (3): 177.

Malar, S., Sahi, S.V., Favas, P.J., and Venkatachalam, P. (2015). Mercury heavy-metal-induced physiochemical changes and genotoxic alterations in water hyacinths [*Eichhornia crassipes* (Mart.)]. *Environmental Science and Pollution Research* 22 (6): 4597–4608.

Mandal, C., Bera, S., Dey, N., and Adak, M. (2016). Physiological alterations of *Salvinia natans* L. exposed to aluminium stress and its interaction with polyamine. *Plant Science Today* 3 (2): 195–206.

Mane, P.C. and Bhosle, A.B. (2012). Bioremoval of some metals by living algae *Spirogyra* sp. and *Spirullina* sp. from aqueous solution. *International Journal of Environmental Research* 6 (2): 571–576.

Mani, D. and Kumar, C. (2013). Biotechnological advances in bioremediation of heavy metals contaminated ecosystems: an overview with special reference to phytoremediation. *International journal of Environmental Science and Technology* 11 (3): 843–872.

Marzan, L.W., Hossain, M., Mina, S.A. et al. (2017). Isolation and biochemical characterization of heavy-metal resistant bacteria from tannery effluent in Chittagong city, Bangladesh: bioremediation viewpoint. *Egyptian Journal of Aquatic Research* 43 (1): 65–74.

Melignani, E., De Cabo, L.I., and Faggi, A.M. (2015). Copper uptake by *Eichhornia crassipes* exposed at high level concentrations. *Environmental Science and Pollution Research* 22 (11): 8307–8315.

Mergeay, M. and Van Houdt, R. (2015). *Metal Response in Cupriavidus metallidurans: Volume I: From Habitats to Genes and Proteins*. Springer.

Merrick, M.J. (2004). Regulation of nitrogen fixation in free-living diazotrophs. In: *Genetics and Regulation of Nitrogen Fixation in Free-Living Bacteria* (ed. W. Klipp, B. Masepohl, J.R. Galão, and W.E. Newton), 197–223. Dordrecht: Springer.

Mesa, J., Rodríguez-Llorente, I.D., Pajuelo, E. et al. (2015). Moving closer towards restoration of contaminated estuaries: bioaugmentation with autochthonous rhizobacteria improves metal rhizoaccumulation in native *Spartina maritima*. *Journal of Hazardous Materials* 300: 263–271.

Migahed, F., Abdelrazak, A., and Fawzy, G. (2017). Batch and continuous removal of heavy metals from industrial effluents using microbial consortia. *International journal of Environmental Science and Technology* 14 (6): 1169–1180.
Mitra, A. and Mukhopadhyay, S. (2016). Biofilm mediated decontamination of pollutants from the environment. *AIMS Bioengineering* 3 (1): 44.
Mohanty, M., Pattnaik, M.M., Mishra, A.K., and Patra, H.K. (2012). Bio-concentration of chromium—an in situ phytoremediation study at South Kaliapani chromite mining area of Orissa, India. *Environmental Monitoring and Assessment* 184 (2): 1015–1024.
Mosa, K.A., Saadoun, I., Kumar, K. et al. (2016). Potential biotechnological strategies for the cleanup of heavy metals and metalloids. *Frontiers in Plant Science* 7: 303.
Mufarrege, M.M., Hadad, H.R., Di Luca, G.A., and Maine, M.A. (2015). The ability of *Typha domingensis* to accumulate and tolerate high concentrations of Cr, Ni, and Zn. *Environmental Science and Pollution Research* 22 (1): 286–292.
Nagajyoti, P.C., Lee, K.D., and Sreekanth, T.V.M. (2010). Heavy metals, occurrence and toxicity for plants: a review. *Environmental Chemistry Letters* 8 (3): 199–216.
Nicholson, F.A., Smith, S.R., Alloway, B.J. et al. (2003). An inventory of heavy metals inputs to agricultural soils in England and Wales. *Science of the Total Environment* 311 (1–3): 205–219.
Nikinmaa, M. (2014). Chapter 6—factors affecting the bioavailability of chemicals. In: *An introduction to Aquatic Toxicology* (ed. M. Nikinmaa), 65–72. Oxford: Academic.
Norse, D. (2005). Non-point pollution from crop production: global, regional and national issues. *Pedosphere* 15 (4): 499–508.
Ovalle, A.R.C., Silva, C.F., Rezende, C.E. et al. (2013). Long-term trends in hydrochemistry in the Paraíba do Sul River, southeastern Brazil. *Journal of Hydrology* 481: 191–203.
Pandey, V.C., Pandey, D.N., and Singh, N. (2015). Sustainable phytoremediation based on naturally colonizing and economically valuable plants. *Journal of Cleaner Production* 86: 37–39.
Panfili, I., Bartucca, M.L., Ballerini, E., and Del Buono, D. (2017). Combination of aquatic species and safeners improves the remediation of copper polluted water. *Science of the Total Environment* 601: 1263–1270.
Patnaik, P. (2017). *Handbook of Environmental Analysis: Chemical Pollutants in Air, Water, Soil, and Solid Wastes*. CRC Press.
Pio, M.C.S., Souza, K.S., and Santana, G.P. (2013). Ability of *Lemna aequinoctialis* for removing heavy metals from wastewater. *Acta Amazonica* 43 (2): 203–210.
Prabakaran, G., Vaithiyanathan, D., and Ganesan, M. (2018). Fuzzy decision support system for improving the crop productivity and efficient use of fertilizers. *Computers and Electronics in Agriculture* 150: 88–97.
Prum, C., Dolphen, R., and Thiravetyan, P. (2018). Enhancing arsenic removal from arsenic-contaminated water by *Echinodorus cordifolius* endophytic *Arthrobacter creatinolyticus* interactions. *Journal of Environmental Management* 213: 11–19.
Rai, P.K. (2009). Heavy metal phytoremediation from aquatic ecosystems with special reference to macrophytes. *Critical Reviews in Environmental Science and Technology* 39 (9): 697–753.
Ravichandran, M. (2004). Interactions between mercury and dissolved organic matter—a review. *Chemosphere* 55 (3): 319–331.
Roane, T.M., Pepper, I.L., and Gentry, T.J. (2015). Microorganisms and metal pollutants. In: *Environmental Microbiology*, 3e (ed. I.L. Pepper, C.P. Gerba, and T.J. Gentry), 415–439. Academic Press.
Scherer, L. and Pfister, S. (2015). Modelling spatially explicit impacts from phosphorus emissions in agriculture. *International Journal of Life Cycle Assessment* 20 (6): 785–795.
Shafi, N., Pandit, A.K., Kamili, A.N., and Mushtaq, B. (2015). Heavy metal accumulation by *Azolla pinnata* of Dal Lake ecosystem, India. *Development* 1 (1): 8–12.
Sharma, P. and Dubey, R.S. (2005). Drought induces oxidative stress and enhances the activities of antioxidant enzymes in growing rice seedlings. *Plant Growth Regulation* 46 (3): 209–221.
Sharma, B. and Shukla, P. (2020). Lead bioaccumulation mediated by Bacillus cereus BPS-9 from an industrial waste contaminated site encoding heavy metal resistant genes and their transporters. *Journal of Hazardous Materials* 401: 123285.
Sharma, R., Sharma, K., Singh, N., and Kumar, A. (2013). Rhizosphere biology of aquatic microbes in order to access their bioremediation potential along with different aquatic macrophytes. *Recent Research in Science and Technology* 5 (1).
Sharma, S., Singh, B., and Manchanda, V.K. (2014). Phytoremediation: role of terrestrial plants and aquatic macrophytes in the remediation of radionuclides and heavy metal contaminated soil and water. *Environmental Science and Pollution Research* 22 (2): 946–962.
Sharma, R.K., Bartot, K., and Archana, G. (2020). Root colonization by heavy metal resistant *Enterobacter* and its influence on metal induced oxidative stress on *Cajanus cajan*. *Journal of the Science of Food and Agriculture* 100 (4): 1532–1540.
Shcherbak, I., Millar, N., and Robertson, G.P. (2014). Global metaanalysis of the nonlinear response of soil nitrous oxide ($N_2O$) emissions to fertilizer nitrogen. *Proceedings of the National Academy of Sciences* 111 (25): 9199–9204.

Shehzadi, M., Afzal, M., Khan, M.U. et al. (2014). Enhanced degradation of textile effluent in constructed wetland system using *Typha domingensis* and textile effluent-degrading endophytic bacteria. *Water Research* 58: 152–159.
Shukla, K.P., Singh, N.K., and Sharma, S. (2010). Bioremediation: developments, current practices and perspectives. *Journal, Genetic Engineering & Biotechnology* 3 (8): 1–20.
Shukla, S.K., Tripathi, V.K., and Mishra, P.K. (2020). Bioremediation of distillery effluent: present status and future prospects. In: *Bioremediation of Industrial Waste for Environmental Safety* (ed. G. Saxena, R. Kishor, and R.N. Bharagava), 77–97. Singapore: Springer.
Sivakumar, D., Kandaswamy, A.N., Gomathi, V. et al. (2014). Bioremediation studies on reduction of heavy metals toxicity. *Pollution Research* 33: 553–558.
Smith, L.E.D. and Siciliano, G. (2015). A comprehensive review of constraints to improved management of fertilizers in China and mitigation of diffuse water pollution from agriculture. *Agriculture, Ecosystems and Environment* 209: 15–25.
Sricoth, T., Meeinkuirt, W., Pichtel, J. et al. (2018). Synergistic phytoremediation of wastewater by two aquatic plants (*Typha angustifolia* and *Eichhornia crassipes*) and potential as biomass fuel. *Environmental Science and Pollution Research* 25 (6): 5344–5358.
Srivastava, J.K., Chandra, H., Kalra, S.J. et al. (2017). Plant–microbe interaction in aquatic system and their role in the management of water quality: a review. *Applied Water Science* 7 (3): 1079–1090.
Stark, C.H. and Richards, K.G. (2008). The continuing challenge of agricultural nitrogen loss to the environment in the context of global change and advancing research. *Dynamic Soil, Dynamic Plant* 2 (1): 1–12.
Steenhoudt, O. and Vanderleyden, J. (2000). *Azospirillum*, a free- living nitrogen-fixing bacterium closely associated with grasses: genetic, biochemical and ecological aspects. *FEMS Microbiology Reviews* 24 (4): 487–506.
Subudhi, S., Bisht, V., Batta, N. et al. (2016). Purification and characterization of exopolysaccharide bioflocculant produced by heavy metal resistant *Achromobacter xylosoxidans*. *Carbohydrate Polymers* 137: 441–451.
Suryandari, M.K., Hariati, A.M., and Mahmudi, M. (2017). Removal of lead from water by *Eichhornia crassipes* (Mart.) Solms. *Imperial Journal of Interdisciplinary Research* 3 (1): 2387–2392.
Syranidou, E., Christofilopoulos, S., Gkavrou, G. et al. (2016). Exploitation of endophytic bacteria to enhance the phytoremediation potential of the wetland helophyte *Juncus acutus*. *Frontiers in Microbiology* 7: 1016.
Thijs, S., Sillen, W., Rineau, F. et al. (2016). Towards an enhanced understanding of plant–microbiome interactions to improve phytoremediation: engineering the metaorganism. *Frontiers in Microbiology* 7: 341.
Thompson, F., De Oliveira, B.C., Cordeiro, M.C. et al. (2020). Severe impacts of the Brumadinho dam failure (Minas Gerais, Brazil) on the water quality of the Paraopeba River. *Science of the Total Environment* 705: 135914.
Ullah, A., Heng, S., Munis, M.F.H. et al. (2015). Phytoremediation of heavy metals assisted by plant growth promoting (PGP) bacteria: a review. *Environmental and Experimental Botany* 117: 28–40.
UN (2018). *Programa Mundial de las Naciones Unidas de Evaluación de los Recursos Hídricos. Informe Mundial de las Naciones Unidas sobre el Desarrollo de los Recursos Hídricos 2018: Soluciones basadas en la naturaleza para la gestión del agua*. París: UNESCO 168 p.
USEPA (1997) Environmental Protection Agency. Mercury Study Report to Congress: Fate and Transport of Mercury in the Environment. ORD, EPA-452/R-97-003, III
USEPA (2007) Environmental Protection Agency. Framework for Metals Risk Assessment. EPA 120/R-07/001.
USEPA (2014), Protection of Environment. Priority Pollutants. Appendix A to 40 CFR, Part 423-126.
Wang, G., Yinglan, A., Jiang, H. et al. (2015). Modeling the source contribution of heavy metals in surficial sediment and analysis of their historical changes in the vertical sediments of a drinking water reservoir. *Journal of Hydrology* 520: 37–51.
Wierzba, S. (2017). Biosorption of nickel (II) and zinc (II) from aqueous solutions by the biomass of yeast *Yarrowia lipolytica*. *Polish Journal of Chemical Technology* 19 (1): 1–10.
Wu, X., Cobbina, S.J., Mao, G. et al. (2016). review of toxicity and mechanisms of individual and mixtures of heavy metals in the environment. *Environmental Science and Pollution Research* 23 (9): 8244–8259.
Yang, Y., Ma, T., Ding, F. et al. (2017). Interactive zinc, iron, and copper-induced phytotoxicity in wheat roots. *Environmental Science and Pollution Research* 24 (1): 395–404.
Zahoor, A., Ahmad, F., Hameed, M., and Basra, S.M.A. (2018). Structural and functional aspects of photosynthetic response in *Eichhornia crassipes* (Mart.) Solms under cadmium stress. *Pakistan Journal of Botany* 50 (2): 489–493.
Zaidi, S., Usmani, S., Singh, B.R., and Musarrat, J. (2006). Significance of *Bacillus subtilis* strain SJ-101 as a bioinoculant for concurrent plant growth promotion and nickel accumulation in *Brassica juncea*. *Chemosphere* 64 (6): 991–997.

# 6

# Extremophilic Microorganisms for Environmental Bioremediation

*Nazim Hussain, Mehvish Mumtaz, Warda Perveez, and Hafsa*

*Centre for Applied Molecular Biology (CAMB), University of Punjab, Lahore, Pakistan*

## 6.1 Introduction

A massive number of harmful substances accumulated in the surroundings due to fast manufacturing expansion endangers the environment and public health. As a result, preserving the atmosphere from harmful chemicals has developed as a top priority for humanity in recent decades (Farhan et al. 2022; Zhang et al. 2022; Deka et al. 2022). Various solutions have recently become popular for protecting the environment by avoiding dangerous pollutants from diffusing into it. Some physicochemical methods that may be utilized to eradicate harmful compounds include electrical and chemical solutions, excavating process, exchange of ions, precipitated technique, reverse osmosis, evaporation, and phoresis (Muddemann et al. 2019; Rizwan et al. 2022; Liaquat et al. 2022; Javaid et al., 2022; Yang et al. 2022). However, owing to significant limitations like expensive and the potential for contamination from a second source, many of these strategies are not yet widely used in real pollution remediation (Abdelfattah et al. 2020). As an alternative, microbial bioremediation has gained much interest as a potential technique that may overcome the limitations of the presently utilized physicochemical procedures (Zhao et al. 2017a,b; Igiri et al. 2018; Bilal and Iqbal 2020; Hadibarata et al. 2022; Saeed et al. 2021; Rafeeq et al. 2022). Extremophilic microbes, in particular, are the best choice for treating hazardous pollutants (Kiadehi et al. 2018), since they can use microbiological cellular processes to metabolize harmful contaminants and endure very hard environments (Azubuike et al. 2016; Jeong and Choi 2020; Tkavc et al. 2018).

Microbes are one of the few species capable of existing in hostile surroundings. Natural ecosystems where geysers, volcanoes, Antarctic, salt or lakes with high acidity, and hydrothermal vents are all highly harsh environments for practically all sorts of life (Bowman and Nichols 2005). Extremophiles are microorganisms that thrive in specific environments. Acidophiles and alkaliphiles thrive in acidic or alkaline pH environments, halophiles succeed in high concentrations of salt, piezophiles grow under increased pressure, metallophiles prosper in the existence of metals/heavy material, psychrophiles excel at relatively low temperatures, and thermophiles find success at extreme temperatures. Extremophiles dwell in extreme pH, high pressure, and salt, dangerous conditions for humans (Rothschild and Mancinelli 2001).

## 6.2 Extremophiles

The name "extremophiles" (Latin: extremus, Greek: phili) was used to describe these species, which means "extreme lovers." Seasonal variations influence human survival, but extremophiles are organisms adapted to live in an extreme environment. Thermophiles and hyperthermophiles are two heat-loving organisms, depending on which extreme they favor. Temperatures between 80°C and 110°C are ideal for hyperthermophiles (Stetter et al. 1990). Highly extreme temperature values can also be obtained at the bottom of the ocean. Water is pumped through the Earth's interior and comes out through black chimneys with temperatures ranging from 400°C. Other extremes include higher or lower pH values (alkaliphiles or acidophiles), high salt concentrations (halophiles), and high pressure (piezophiles) (Jamieson et al. 2010). They can survive at the

*Microbes Based Approaches for the Management of Hazardous Contaminants*, First Edition. Edited by Ajay Kumar, Livleen Shukla, Joginder Singh, and Luiz Fernando R. Ferreira.

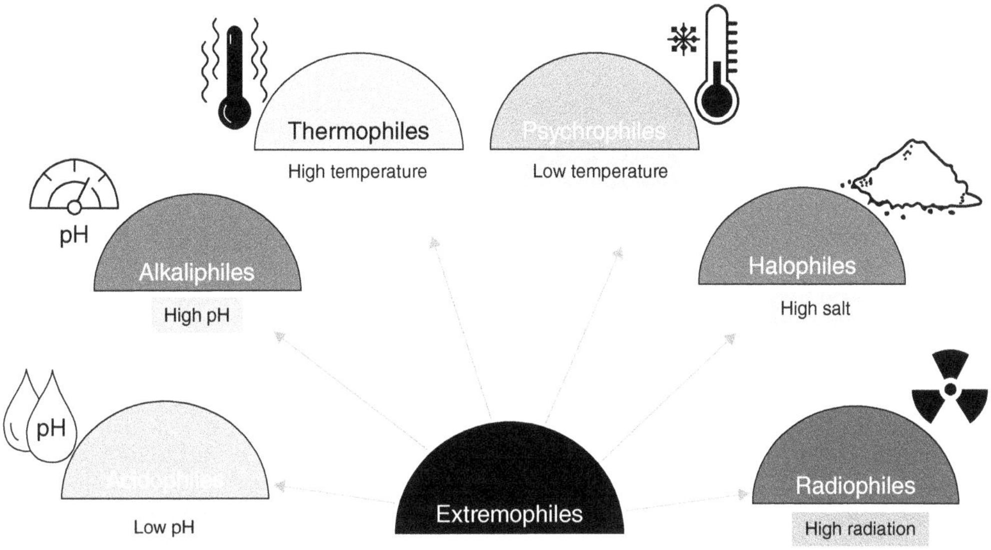


**Figure 6.1** Classification of extremophilic microorganisms.

deepest, about 8200 m or 820 bar's pressure. The classification of extremophilic microorganisms is represented schematically in Figure 6.1. Moreover, tardigrades have been proven to withstand a variety of extremes (Stetter 1998) and are well-known for their hibernating state (Rothschild and Mancinelli 2001).

As a result of human activity, some bacteria have evolved to grow in toxic wastewater released by industries and contaminated soil due to heavy metals, pesticides, and various substances (Liu et al. 2011). Thermozymes considered mostly stable with proteolysis and more durable in hydrocarbon solutions, inactivating chemicals, and excessive saline are all present, mirroring the niches in which they are found. Extremophiles have recently sparked a lot of attention due to their biotechnological and industrial significance. Researchers are trying to develop sustainable remediation technology by studying these odd bacteria, their metabolisms, and catalysts. In this context, exploring further into bio and physiological ecosystems of microbes is vital for the success of white and green bioprocesses in industrial processes and energy generation, pollution prevention through the detective process as well as the elimination of hazardous substances biopolymer formation by renewable sources (Yadav 2021). Thermophilic microbes can clean industrial hot wastewater pollution due to their thermos ability in high temperatures. (Isticato et al. 2020). Thermophiles outperform mesophiles in biorefineries that need high-temperature stages (Gallo et al. 2021). A combination of microbial biotechnology and chemistry is necessary to improve the cleanup methods. Extremophilic bacteria have adapted and developed in several ways to live in severe circumstances throughout the last century due to unique biological processes.

## 6.3 Extremophilic Microorganisms Under Extreme Conditions

### 6.3.1 Acidophilic Microorganisms

Acidophilic bacteria live in very low pH (less than pH 3) environments, managing proton permeability and maintaining pH equilibrium. Acidophiles include Bacteria, or Archaebacteria, and fungi (Sharma et al. 2016). Acidophile bacteria included *Leptospirilum, Hydrogenobacter, Alicyclobacillus, Acidithiobacillus, Acidiphilium,* and *Soosiella* (Sharma et al. 2016), while acidophiles Archaebacteria include *Picrophilus oshimae* and *Picrophilus torridus* (Serour and Antranikian 2002). *Thermoplasma, Ferroplasma*, and *Sulfolobus* are examples of acidophilic bacteria. Acidophilic bacteria have biological membranes that are relatively impermeable, consisting primarily of tetraether triglycerides with a big and heavy isoprenoid center and a wide variety of solvent molecules that can support proton penetration at extremely low pH circumstances. Even though these microbes need a balanced internal pH, barrier insolubility, which regulates proton input inside the unit, influences intracellular pH. Because the lipid on their cell surface has a poor tolerance for proton input. The degradation of organic acids for proton separation and the production of counselors to protect proteins and deoxyribonucleic acid from destruction at low pH are two further adaptation processes (Salwan and Sharma 2020).

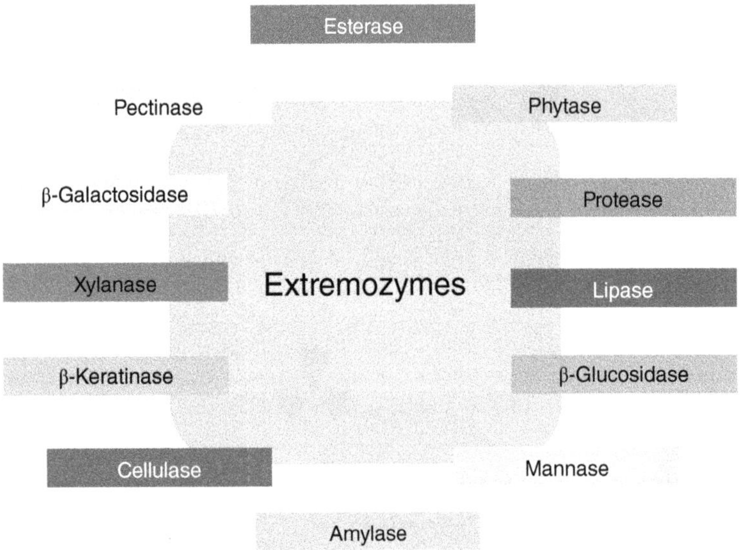


**Figure 6.2** Extremozymes produced by extremophilic microorganisms.

To exist at low pH, protons must be managed into the proton recirculating pump, and proton pump enzymes like $H^+$-ATPase, antiporters, and symporters from *Ferroplasma* type II and *Leptospirillium* group II that are involved in regulating pH (Golyshina et al. 2017). Furthermore, it has been shown that *FOF1*-type ATP synthase is involved in proton absorption in *Thermoplasma acidophilum*, *Bacillus acidocaldarus,* and *Leptospirillium ferriphilum* (Baker-Austin and Dopson 2007). Auxiliary systems like chaperone proteins and cytoplasmic buffering help coping mechanisms in very low pH situations by safeguarding cytoplasmic components, including DNA, RNA, and proteins (Baker-Austin and Dopson 2007; Choi et al. 2019; Jeong et al. 2016). Various molecular strategies of extremophiles for adapting to severe environmental circumstances are depicted in Figure 6.2.

Actinobacteria microorganisms flourish at less than pH 5. High acidophilic bacteria require a pH of 3 or lower for optimum growth, while moderately acidophilic species prefer a pH of 3–5. Acid-tolerant bacteria, on the other hand, flourish at pH levels more than 5, but may still function at lowered pH concentrations (Johnson 2008). Acidophiles may keep their cytoplasmic pH close to normal by establishing a substantial pH gradient, which protects acid-labile biological molecules. Through active proton pumping via proton flux channels is the primary strategy for maintaining pH (Michels and Bakker 1985) and enhancing DNA and peptide-fixing operations by reducing membrane fluidity and preventing protons from entering the cytoplasm (Amaro et al. 1991).

Acidophilic fungus *Bispora* sp. thrived at a pH of 1.5 environments and was able to obstruct a tube meant to transmit hydrochloric acid (0.1 N) (no dissolved nutrients). *Mycobacterium europaeum*, *Mycobacterium parascrofulaceum*, severe *Acidithiobacillus thiooxidans, chemolithotrophic acidophilic* bacteria, and *Sulfobacillus* species (Drienovska et al. 2012), were able to consume *n*-alkane hydrocarbons with pH values of range 1.6–5.5 (Ivanova et al. 2013).

Efforts have been made to optimize some characteristics such as nutrition, oxygen, pH, impurity content, and quantity to speed up dangerous compounds' naturally occurring composting process (Allard and Neilson 1997). Acid mine drainage (AMD) and industrial discharges are examples of environments with low pH. AMD is a severe environmental problem that the mining industry faces globally. Hazardous materials and toxins exist in mining and smelting pollution, producing significant health and environmental issues. AMD enhances cytotoxicity in other aquatic environments by lowering pH in the receiver of water. The solubility of the hazardous element increases, resulting in negative repercussions. Aside from the harmful effects, metal deposition, especially iron, and aluminum, creates agglomeration at the bottom of rivers, interrupting aquatic species' food networks and life cycles by limiting sediment creature reproduction (Hallberg 2010).

The typical treatment for AMD is to increase the pH over various forms needed by microorganisms that oxidize iron, which reduces the likelihood of acidic formations. The functioning expenses of present industrial wastewater treatment systems are too costly. It is, nevertheless, feasible to use acidophilic bacteria to extract harmful metals from contaminated soils. The bacteria can withstand large forms of metals (e.g., metal ion entrapment via metal chaperones or active outflow). Because of genetic recombination, chromosomal islands, and a metal-resistant process based on artificial polyphosphate, acidophiles are excellent participants in phytoremediation (Kambura et al. 2016). Several times, acidic mine runoff and oil spilling due to firms that use carcinogens have made environmental deterioration toxic to both living creatures and the environment. Acidophiles have been reported to destroy a variety of compounds, including aliphatic, heterocyclic, chlorinated solvents, nitrated molecules, and a variety of industrial effluent components (Christen et al. 2012). They are vital to the appropriate cleanup of damaged habitats. It also took DNA from surface petroleum seep and went through community screening, finding that acidophilic bacteria, particularly the proteo-bacteria class, predominated (specifically *Acidiphilium* and *Acidocella*) (Röling et al. 2006).

Aromatic molecules have been demonstrated to be destroyed by acidophiles identified from samples taken from a long coal stack retention basin with a pH of 2.0. However, up to 40% of the parent hydrocarbons (HCs), naphthalenes and toluenes, biodegrade to $CO_2$ and $H_2O$ at such a low pH.

The presence of acyclic alkanes (like *n*-alkanes with chain lengths ranging from C15 to C30, and branched alkanes), especially phytane and pristine, and lower pH values of (2.8–3.8), benefits microbial communities. Researchers observed patterns linked to heterotrophic acidophilic bacteria such as *Acidisphaera*, Proteobacteria *Acidiphilium*, and chemolithotroph *Acidithiobacillus* sp. using 16S rRNA gene cloning libraries. An alkane-degrading extract was also grown, demonstrating one population's alkane degradation potential in acidified compounds infiltrating the soil. A well-acclimatized sample of *Sulfolobus solfataricus*, a thermoacidophilic archaeon, biodegraded in phenol at a temperature of 80°C and pH value of 3.2 (Christen et al. 2012). Phenol, being organic in nature, is found in various industrial effluents, including those from refineries, coking plants, coal treatment plants, and hydrocarbon production plants. For the phytoremediation of phenol-contaminated warm, corrosive environments, acidophiles' ability to flourish and degrade phenol at higher temperatures and lower pH is critical. Because of their capacity to reduce and oxidize iron and sulfur, acidophilic bacteria have long been recognized as effective in remedying acidic pollution (Sharma and Satyanarayana 2012).

### 6.3.2 Alkaliphilic Microorganisms

Alkaliphiles grow at pH 9, whereas severe strains thrive at pH 12–13. The pH range is 10–12, whereas the latter prefers 7–9.5. Alkaliphiles, the majority of which are halophiles, flourish in alkaline conditions such as alkaline lakes and hydrated lime soil. Both the antiporter and substitution mechanisms are engaged in these activities. The electrogenic physiological basis creates an $H^+$ and $Na^+$ ion electrochemical gradient, while the symporter system enables cells to absorb $Na^+$ and other solutes. The involvement of cytochrome c-552 in the respiratory system's electron and hydrogen ion accumulation promotes final oxidation, allowing protons and substances to pass through after an ion redistribution shift (Okumura et al. 2016). The cell's thermodynamic and hydrosaline equilibrium is preserved as a consequence. The carriers seem to be regulated by signals from the epidermal pH sensor. Bioremediation uses archaea, *Natronobacterium*, *Natrialba*, *Halobiforma Natronorubrum*, *Halalkalicoccus*, *Natronococcus, Halorubrum*, and the microbial taxa, *Bacillus*, *Halomonas*, *Aeromonas*, *Corynebacterium*, *Micrococcus*, *Pseudomonas*, and *Methylomicrobium*. 2-Halo acid dehalogenase, found in the bacteria *Paracoccus* sp. *DEH99* is another important alkaline enzyme mineralizing hazardous chemicals (Joshi and Satyanarayana 2013; Moussa and Khalil 2022).

Alkaliphilic microorganisms, unlike acidophilic germs, can withstand high pH. These bacteria have been shown to include three crucial metabolic pathways as survival strategies. Certain alkaliphilic *Bacillus* spp. may increase proton motive force generation at exceptionally high pH balance by producing supplementary acidic cellular membranes composed primarily of peptidoglycan, teichuronic acid, and teichuronopeptide (Aono 1995). Increasing the electrochemical gradient force aids not only in the generation of electricity but also in the maintenance of pH homeostasis. Second, sodium motive force may help to maintain pH balance at very high pH settings (Matsuno et al. 2018). Under elevated $Na^+$ ion circumstances, $Na^+/H^+$ antiporters expel $Na^+$ ions while absorbing more external $H^+$ ions than ejected $Na^+$ ions, thereby starting an energy metabolism mechanism and regulating internal pH. Lastly, it is commonly accepted that organic acid generation, which may be used to calibrate pH, plays a significant metabolic role in pH regulation (Choi et al. 2019; Wernick et al. 2016).

Alkaliphilic microorganisms may include novel compounds and enzymes, including antimicrobials, which get a profitable range from an alkaline environment, like extracting hydrogen sulfide from "sour" hydrocarbons generated in the petroleum chemical sector (Ibrahim et al. 2012). Alkaliphiles may be found in various biological niches, including alkaline soda lakes, insect hindguts, aerobically fermented soils, and high-pH human operation environments. For example, alkaline volcanic eruptions have been considered a way of re-creating conditions that existed before the first living beings appeared (Herschy et al. 2014). Microbiological organisms practice the properties of their environments, alike the niche and the microbial population engage in a developing ballet, and niches affect bacteria. Most of those alkaliphiles are assumed to be in soda lakes (Sorokin et al. 2014), before going on to other alkaline ecologies.

The FOF1-ATP synthase is found in gram-positive *Bacillus* strains such as *Bacillus pseudofirmus* OF4 (Janto et al. 2011) and *Bacillus halodurans* C-125 arealkaliphile. These alkaliphiles use oxidative phosphorylation to assist non-fermentative growth. Proton motive force (PMF) is the electrical and chemical protons gradient between the cytoplasm of bacteria and the mass outside the environment (Krulwich 1995), which promotes ATP synthesis in particular (Preiss et al. 2015).

### 6.3.3 Halophilic

Salt-loving bacteria that can balance the osmolality of their environment are known as halophilic bacteria. They may be found in a wide range of environments, i.e., salt ponds, alkali lakes, and possibly even rock salt crystals. Rather than being a more basic group or branch on the phylogenetic ladder, these prokaryotic species are most likely evolutionary adaptations of more prevalent bacterial forms. Understanding the distinction between "salt tolerance" and "salt demand" is critical. Halophilic bacteria may be aerobic or anaerobic, as well as heterotrophic, phototrophic, or chemoautotrophic. Because of osmolar imbalance and metabolic challenges, halophilic microbes may flourish in a high-salt environment, making it difficult for species to survive. Previous research on halophilic bacteria revealed different adaption mechanisms to high concentration-salt environments. The first one is to utilize the "salting-in" method, in which artificial osmoprotectants like KCl are stored in the cell to preserve osmotic balance in and out of the cell (Gunde-Cimerman et al. 2018). *Halobacterium salinarum,* under elevated conditions, can gather 3.97 and 4.57 M of potassium and chloride ions within the cell to balance the osmotic concentration with the ATP-dependent K channel within the cell to balance the osmotic concentration with transport system (cationic amino acid transporter-3 (Cat3) and KdpFABC complex) and $Na^+$ efflux antiporters, respectively. In addition, halophilic bacteria have developed a large number of negative charges due to glutamate and aspartate residues on protein surfaces, which attach to water, preventing the growth of other bacteria (Jeong and Choi 2020).

Some industrial processes create excessively salty effluent. After centrifugation, hypersaline wastewater is left behind, which has a high proportion of glycerol and is rich in organic material. Pickling plants and tanneries are industries that entangle with higher salt effluents. Wastewater brines have been formed during the production of oil and gas. Several bioremediation procedures have been suggested to remove organic material and harmful compounds from such effluents. Wastewater is treated with activated sludge and reused in various *Dunaliella salina* growth facilities (Santos et al. 2001). For the microbiological elimination of harmful waste waters with a concentration of salts of 10%, like the generation of brines from the pickling industries, extended aeration mechanism basis on oxygenated immersion blenders or discs spinning techniques to promote oxygenation and stirring have been created. Although a colony of *H. salinarum* was added to some experiments to boost breakdown, it is unclear if that species contributed to the system's performance since *Halobacterium* needs relatively 15% salt for survival as well as hydrolyzes at lower concentrations (Hinteregger and Streichsbier 1997). Harmful emissions that permeate liquids and other salty municipal wastewater by close to 6% salt have been filtered using a sequencing batch processor. To treat saline and alkaline sewage, anaerobic decomposition techniques have even been devised. A membrane reactor relying on denitrification by the archaeon *Haloferax denitrificans* was developed to remove nitrate from brines, and *Haloferax mediterranei* has been proposed for nitrate and nitrite phytoremediation in coastal groundwater aquifers. With salinity levels up to 10%, the halophilic fermentative bacteria *Halanaerobium lacusrosei* was successfully employed in an anaerobic fixed bed reactor (Lefebvre et al. 2005).

For industrial wastewater treatment operations, finding halophilic strains that can decompose specific aromatic compounds in wastewater is crucial. Several investigations have found halophilic habitats or halophilic strains that are suitable for bioremediation all over the globe (Oren et al. 1992). In oil-contaminated seawater, an *Achromobacter* strain was found that decomposed PAHs like phenanthrene with a 50.6% efficiency. Similar findings were observed from the other two bacteria (*B. methylotophicus* and *P. sihuiensis*) taken from the sampling location Lagao do Peixe (National Park in Brazil). These strains, under marine conditions, showed a 33% phenanthrene decomposition efficiency (derived from petroleum pollution). Microbes found in organic, non-polluted environments such as uncontaminated saltwater samples have also been examined for their degradation ability toward several contaminants present in various industrial wastewater. For example, microbial sources acquired from uncontaminated seawater eliminated aromatic compounds (PAHs and phenols) from produced water (Lofthus et al. 2018).

The microbial makeup of oil-contaminated seawater was studied in another study. According to the researchers, specific oil-degrading bacteria increased when seawater was contaminated with oil. The degradation of oil molecules has been discovered to be caused by gammaproteobacteria. Within this group, *Cycloclasticus* and *Marinobacter* have been linked to the biological breakdown of aliphatic chemicals (Mainka et al. 2021).

Strains of *Aspergillus flavus* and *Aspergillus penicillioides* were able to effectively secrete extremozymes such as cellulase and xylanase, respectively. According to a study, co-culture of *A. flavus* and *A. penicillioides* as an approach to producing a synergistic mixture of enzymes, including hydrolases (β-xylosidase, xylanase, FPase, and CMCase) under solid-state fermentation (SSF) conditions (Zhao et al. 2021).

### 6.3.4 Thermophiles

Intermediate thermophiles, extreme thermophiles, and hyperthermophiles are microorganisms that thrive in hot conditions, with an optimum growing temperature of 50–60, 60–80, and 80–110°C, accordingly (Gupta et al. 2014). Magmatic or hydrothermal activity, which may be witnessed near volcanoes or deep-sea hydrothermal vents, produces fluid temperatures of up to 100°C when combined with high pressure. Without geothermal impact, 71°C was the greatest known temperature in the Lut Desert (Iran). Hyperthermophiles (hyperthermopiezophiles) are organisms that can live in high-pressure environments when water is liquid at temperatures over 80°C (Merino et al. 2019). Hyperthermopiezophilic archaea strains (*Geogemma barossii* strain 121 and *Methanopyrus kandleri* strain 116), have been discovered. But from the other side, *Thermotoga maritime* and *Aquifex pyrophilus* reached 95°C and 90°C, respectively. Thermophilic fungi are one of the few eukaryotic that survive above 45°C.

Fungal strains like *Aspergillus terreus* TM8 are thermophilic (50°C), which was discovered from hydrothermal vent silt rich in sulfur, and *Penicillium* sp. Y-50-10, which was also discovered in a sulfur-rich hydrothermal vent, has also been studied. The versatility of thermophiles originates from their ability to function at high temperatures and create heat-stable proteins, making them excellent for various biotech applications. Raising the product's temperature affects the absorption and stability of polymeric components, resulting in a faster reaction time, reduced bacterial contamination risk, and more successful bioremediation (Sinha et al. 2012).

In thermophile processors, the interaction of hydrogen bond concentrations increased, resulting in increased hydrophobicity of the active site and decreased exposure of thermolabile amino acids, resulting in more durable thermophilic protein complexes. Because commercial treatments are commonly found at temperatures exceeding 100°C, waste processors containing live microorganisms should be precooled. As a result, hyperthermophiles are ideal since they can thrive in hot conditions and digest a variety of waste pollutants (Vieille and Zeikus 1996). Many studies have shown that thermophiles and associated thermophilic enzymes are efficient in bioremediation. Thermophilic bacteria (*Bacillus licheniformis, Bacillus cereus, Brevibacillus borstelensis,* and *Brevibacillus agri*) can break down fatty food waste and turn it into long-lasting end goods at a far cheaper cost. Antimicrobial metabolites may also be treated using ribonucleases (Horikoshi 1999). The elimination of lead and chromate is mediated by superoxide dismutase and quinone oxidoreductase, respectively (Jardine et al. 2018).

Furthermore, thermophile enzyme systems are unique in their high heat stability, enabling them to function through various adaption mechanisms at high temperatures. Proteins retain their stability through a variety of structural and translational modifications. Thermophiles and related proteins can tolerate high temperatures, making them ideal biotechnological instruments for chemical processing and long-term management of resistant contaminants (Coker 2016). Maximum heat provides other advantages, such as accessibility, easy absorption of organic or nonpolar compounds, and low, and medium viscosity, all being able to form thermophiles and thermozymes. Basic challenges for isolating as well as cultivating thermophiles often by medium elements that restrict growth, such as agar, a lack of population sensing/beneficial contact among population members, or absence of abiotic interactions, which might be required for thermophile existence and development.

Furthermore, media specificity and environmental or cultural requirements are unknown, making developing thermophiles in the lab considerably more difficult. Only 1–10% of the thermophilic community has been reported in the literature. Thermophiles and thermozymes are being studied for their main mechanisms in the creation of protein, bioengineering, medicines, and biodiesel (Kumar et al. 2019).

In conventional mesophilic cells, the maximum temperature induces disintegration or dysfunction of cell components, whereas thermophiles can withstand high-temperature stress. Instead of the glycerol esters lipids seen in mesophilic equivalents, glycerol–ether lipids (R–O–R) sustain the thermophile membrane's elasticity, rigidity, and permeability at extreme temperatures. Archaebacteria have the most ether linkage, however, thermophilic bacteria can also have it. Ether lipid addition increases the resilience of the cell membrane for hydrolysis at high temperatures. In barriers transporting ether lipids, the mobility of carbon molecules is reduced, and penetrability can be further limited by this constraint in hydrocarbon chain movement (Konings et al. 2002). Increased lipid acyl chain length, excessive tetraether lipids in barriers, and isoprenoid chain cyclization, which assists in compact and tight lipid packing, have all been shown to reduce membrane fluidity or restrict $H^+$ permeability. At high temperatures, the permeability of $Na^+$ is low, and some thermophiles rely on this property of $Na^+$ to maintain sodium motif force (SMF) for the generation of ATP. SMF can also be used to transport solute via a secondary transporter in some thermophiles. Alkaliphiles employ SMF to create ATP, hence thermophiles are thought to exhibit polyextremophillic features.

The environments in which thermophiles exist are often nutritionally inadequate. Secondary transporters have a low affinity for solutes, enabling them to effectively overwhelm solutes or nutrients at low temperatures from the environment. The inclusion of a type I DNA topoisomerase (reversed restriction endonucleases), which induces sequences among thermophiles for DNA stabilization, improves the genetic material's (DNA) resilience to extreme temperatures (Lusk 2019). Histone protein's existence provides DNA preservation and stability in archaea.

The genome size reduces dramatically as temperatures and thermophilicity increase. The small size of thermophile genomes allows them to multiply quickly while consuming the least energy for nucleotide synthesis. Regulatory genes may co-regulate these operons, which is an effective strategy for responding quickly to environmental stimuli (Xia et al. 2015). The thermophiles' translational equipment, particularly ribosomes, must remain stable for optimal protein biosynthesis. Upregulation and greater affection for 23S rRNA of ribosomal proteins (LSU, L12A, L10E, and L7AE) aid thermophiles in maintaining their ribosomal structure and membranes at high temperatures. Some of the genes in thermophiles are heat shock (groEL, groES). The glycolytic process provides thermophiles with immediate energy to aid in extreme heat adaptation. Heat-induced oxidative stress in cells may also be reduced by upregulating antioxidative enzymes (peroxidase, catalase) (Bhattacharya and Gupta 2022).

### 6.3.5 Piezophile Microorganism

Approximately 70% of the biosphere is found in the ocean, with an average depth of 3800 m and a pressure of 38 MPa (0.1 megapascals = 1 bar = 0.9869 atmospheres = 1.0197 kg/cm$^2$; for clarity, megapascals [MPa] are used throughout). The pressure is 110 MPa at the Mariana Trench, the deepest known location in the ocean. Even though a large section of the biosphere on Earth exists in a high-pressure environment, deep-sea piezophilic (formerly barophilic) bacteria are significantly less well-studied than soil microorganisms. In deep sea water, the temperature ranges typically between 1°C and 3°C. Hydrothermal vents found on the ocean floor have provided information about environments with high temperatures and pressures. Thereby, marine species live in environments where they may be exposed to temperatures ranging from 1°C to 300°C and pressures ranging from 0.1 to 110 MPa. Deep-sea organisms have caught the attention of marine biologists who want to learn about the survival strategies used in such harsh environments (Abe et al. 1999; Bartlett 1992; Childress 1995). Piezophiles are microorganisms that grow most rapidly at pressures higher than the atmosphere. Piezotolerant microorganisms can grow at atmospheric pressure and higher pressures. However, they can be distinguished from piezophiles because they do not grow as quickly at pressures higher than one atmosphere. Piezotolerant microorganisms can also be differentiated from piezosensitive microbes (whose growth is sensitive to higher pressures) because they can grow at 50 MPa at a rate greater than 30% of their growth rate at atmospheric pressure. *Shewanella benthica* DB172F, for example, exhibits piezotolerant growth at 4°C (identical growth rate at 50 MPa to that at 0.1 MPa), but piezophilic growth at its optimal temperature of 10°C (Kato et al. 1996; Kato and Nogi 2001).

### 6.3.6 Psychrophilic Microorganisms

Temperatures below 0°C are present in places like glaciers, Polar or Antarctic cold areas, as well as high mountain topsoil (Salvador-Castell et al. 2019). Cool soils support the survival of prokaryotes, cyanobacteria, archaea and bacteria, and also eukaryotic fungi, with optimal growing temperatures of less than 5°C (Rambabu et al. 2020). Psychrophiles are microbes that can live at a low level of 15°C and a high level of 20°C (Matsuno et al. 2018). To deal with all these extreme temperatures, they evolved some physiological adaption processes, such as membrane mobility regulation, the action of chaperones, and the creation of antifreeze molecules (De Maayer et al. 2014). Modifications in density, polar/nonpolar carotenoids, the number of polyunsaturated fatty acids, and lipid head size of the group may occur. Polysaccharides and antifreezing proteins present in biofilms, such as mannitol, trehalose, and exopolysaccharides, are examples of cryoprotectants (Białkowska et al. 2020). *Cryobacterium*, *Hymenobacterium*, *Pedobacterium*, *Psychrobacterium*, and *Sphingobacterium* are among the psychrophilic bacteria discovered in Antarctic soils (Zeng et al. 2004).

Nonspore-forming sustainable psychrophilic microorganisms from bacterial genera *Subtercola*, *Arthrobacter*, and *Glaciimonas* have been discovered in Siberia amid tundra. It investigated the biodiversity of microfungi in Antarctica's frozen Lake Fryxell. *Glaciozyma watsonii* was the most prevalent taxon, followed by *Penicillium* spp. *Planococcus halocryophilus* Or1, a severe psychrophile discovered in Arctic permafrost. It multiplies and thrives at 15°C, but metabolic activity is also shown at 25°C in cold tundra microcosms (Connell et al. 2018).

Psychrophiles have evolved characteristics that help them live in cold environments. The presence of pigment and soluble compounds related to osmosis response to stress and the regulation of membrane permeability via membrane phospholipids are two properties. Such cold-adapted microbes generate psychrophilic enzymes with high catalytic efficiency at low temperatures. It has several applications in detergents, textiles, foods, pharmaceuticals, fabric, beer, winemaking, and paper pulp industries (Kuldeep et al. 2011). It is a better alternative to physiochemical approaches in soil and wastewater bioremediation (Singh et al. 2011).

The use of cold-active protease is employed in the pharmaceutical business for environmental remediation and wastewater treatment (Joshi and Satyanarayana 2013). The hydrocarbon-degrading enzyme, biphenyl deoxygenase, taken from psychrotolerant *Pseudomonas* species strain Cam-1, had been used in hydrocarbon bioremediation. The enzyme showed activity at 4°C (Master et al. 2008). *Psychrobacter cryohalolentis* K5 (haloalkane dehalogenase) was also investigated. Because of its more performed activities at low temperatures, an enzyme is suited for subsurface bioremediation (Drienovska et al. 2012). The degrading capability of the indigenous psychrophilic microbial community is responsible for low-temperature biodegradation contaminants in cold settings. Contaminations are transformed or metabolized toward less toxic compounds that are then assimilated through normal biochemical processes. The majority of research on hydrocarbon environmental remediation in cold sites has dealt with the treatment of petroleum hydrocarbons since cold environments are exposed to the public for oil exploration, manufacturing, and transport, all of which causes an increase in the risk of unintentional release of oil (Breaux and Gordon 2013). Such habitats include permafrost, sediments, polar and alpine soils, oceans, temperate subsoils, and groundwaters whose temperatures seldom reach 10°C (Filler et al. 2008).

An increase in the viscosity of liquid hydrocarbons decreases the volatile ability of hazardous chemicals, reduces levels of nutrients, restricts the availability of pollutants, and, in rare situations, salinity and pH extremes are among the issues faced by cold-climate hydrocarbon degraders. Until recently, it was considered that frozen soils constituted a practically impenetrable barrier to contaminants (Yergeau et al. 2009).

Meanwhile, research has demonstrated that even hydrocarbons may penetrate even ice-saturated soils. Microbial activity in colder regions has been detected at temperatures under −10°C and around the freezing point of water, showing that oil-contaminated ice undergoes sluggish hydrocarbon biodegradation (Miyake et al. 2007). Biopiles or landfarming, which are now well enough for frozen and enable reduced remediation of petroleum-contaminated sites, are successful examples of on-site treatment (Filler et al. 2006). Biostimulation of indigenous microorganisms through nutrient supplementation (and optimization with other limiting factors such as pH, oxygen content, and temperature control) is the mostly applied bioremediation procedure in cold soils; however, care must be taken to prevent biodegradation inhibition because of overfertilization (Thomassin-Lacroix et al. 2002). Bioaugmentation using hydrocarbon degraders has been employed to remediate petroleum-contaminated areas, but Inocula like these are much more expensive (Paudyn et al. 2008).

The conjugation of TOL plasmid from (mesophile *Pseudomonas putida*) to the same species of psychrophile resulted in toluene degradation at low temperatures as 0°C, which was initially reported 20 years ago (Filler et al. 2009). The mesophile *Pseudomonas stutzeri* gene, which encodes for the monooxygenase involved in aromatic hydrocarbons breaking, has recently been recombinantly produced in Antarctic *Pseudoalteromonas haloplanktis* (Tin et al. 2009). Fungi and bacteria capable of rapidly degrading large volumes of organic compounds at low temperatures are a possible source of inocula for expedited wastewater treatment (Kolenc et al. 1988). *Arthrobacter psychrolactophilus* strain, which is adapted to a cold environment, cleared a synthetically produced wastewater turbid medium and hydrolyzed starch, proteins, phenol, and lipids at 10°C (Siani et al. 2006). Psychrophile (*Rhodococcus* sp.) and yeasts (*Rhodotorula psychrophenolica*) cleared up to 12.5–15 mM of phenol at 10°C under batch culture; even at very low temperatures as 1°C phenol degradation occurred with certain strains. The ability of psychrophilic yeast strains to immobilize phenol may aid phenol decomposition (Gratia et al. 2009).

Fixed biofilm reactors injected with bacterial consortia were utilized in chlorophenols degradation, and high removal efficiencies were achieved at low temperatures as low as 4°C. Another molecule of current relevance is methyl tert-butyl ether, which has caused considerable surface and groundwater contamination due to its widespread usage as an octane booster and oxygenate in unleaded gasoline during the last few years. An aerobic bacterial culture of various species capable of using methyl tertiary butyl ether (MTBE) and thriving at temperatures ranging from 3°C to 30°C might be influential for cleaning up MTBE-contaminated aquifers. When injected into samples from groundwater, it reduced MTBE and other volatile organic compounds as contaminants (Margesin and Feller 2010).

### 6.3.7 Radiophiles

Microorganisms known as radiophiles are particularly resistant to ionizing and ultraviolet (UV) light. These radiation-resistant microbes offer a lot of promise for dealing with environmental radioactive waste. They were found to withstand high doses of radiation, stress, and even regular DNA damage. Many radiation-resistant bacteria were identified by cultivating the cultures near intense cosmic radiation (Sghaier et al. 2008). The most resistant microbe is *Deinococcus radiodurans*, which can tolerate intense radiation, high temperature, high pressure, and other harsh environments. Radiophiles are usually found in bacteria. However, certain cyanobacteria also contain them (Gabani et al. 2012). It is the only one with a genetic transformation and modification system (Brim et al. 2000).

According to a new study, radiophiles have a sophisticated system of DNA repairing and antioxidation setup that allows them to tolerate the pressure of excess radiation and adapt and thrive under high radiation and oxidative conditions (Ferreira et al. 1999). RecA proteins are vital in repairing damaged DNA when *D. radiodurans* R1, a typical radiophilic bacterium, is exposed to gamma rays (Liu et al. 2017; Srinivasan et al. 2017). When cells are subjected to high amounts of irradiation, they produce many new proteins (PprI, PprA, PprM, and DdrABCDO) and Mutational regulons, which lead to repairing DNA and reconstruction of the damaged genome (Park et al. 2018). Catalase, peroxidase, and superoxide dismutase have been antioxidant enzymes observed in radiophilic bacteria that scavenge reactive oxygen species (ROS) (Tanner et al. 2020).

Radiation-resistant bacteria such as *E. coli* and *S. cerevisiae* have a 30-times less ROS decomposing activity than the superoxide dismutases (SOD) and catalases (CAT) of *D. radiodurans*. Factors that are nonenzymatic, such as internal manganese concentrations, carotenoids, polyphosphate granules, and pyrroloquinoline quinone, help to effectively scavenge various ROSs and prevent protein breakdown (Rajpurohit et al. 2016). Other nonenzymatic mechanisms that protect biomolecules against ionizing radiation include greater intracellular Manganese/Iron concentrations ratio of the polyextremophilic bacteria *H. salinarum*, orthophosphates, huge amounts of free amino acids, and also short peptides (Tanaka et al. 2004).

Microbes could be exposed to excessive radiation in deeper subterranean regions linked with nuclear ore deposits and places with extreme UV radiation (Zahradka et al. 2006). Because of its capacity to cope with greater doses of ionizing radiations, *D. radiodurans* has been widely examined for the possibility of bioremediation of radioactive waste location. Clones that can break down toluene and decrease hazardous elements like chromium and mercury have been created in cesium-137 (Brim et al. 2006). Silicate nanoparticles showed a protective impact on cellular viability when exposed to radiation. High levels of ionizing radiation have been proven to be tolerated by Antarctic microorganisms. These organisms are also excellent candidates for producing commercial enzymes. The survival of species in high-altitude Andean lakes has been studied in search of a good understanding of the principles related to stability. High UV levels, hypersalinity, poor nutrition, high heavy metal concentrations, large temperature variations, and periodic desiccation are all characteristics of these ecosystems. Classification of the extremophiles and various extremozymes produced by them are listed in Table 6.1.

## 6.4 Extremophiles Applications for Environmental Bioremediation

### 6.4.1 Treatment of Radioactive Waste

Extremophilic bacteria and archaea have devised novel strategies for coping with physical and/or geochemical extremes (Watling et al. 2015). Acidophilic or mesophilic bacteria are specific to promising bioremediation substances because of their defensive process. These processes develop extremophilic enzymes such as extremozymes and active and/or stable biomolecules under extreme circumstances, as represented in Figure 6.3 (Rafiq et al. 2018). Because of folded stiffs maintained by ion pair networks, compacted hydrophobic centers, and amino-acid associations, the esterase (EstATII) (Mohamed et al. 2013) is more susceptible to acidic and metal strain (Elleuche et al. 2014). Extremophiles evolved sharpened metal detoxifying pathways as a consequence of rapidly adjusting transcriptional and translational processes that stimulate or inhibit various anti-oxidative stress, metal-transport, metal-binding, and membrane-permeability responses.

**Table 6.1** List of extremozymes produced by extremophiles.

| Extreme environmental parameters | | Class | Habitat conditions | Source | Examples | Extremozymes | References |
|---|---|---|---|---|---|---|---|
| pH | Low | Acidophiles | pH < 5 | Acid mine drainage, volcanic springs | *Thermoplasma acidophilum, Ferroplasma, Sulfobolus, Leptosipirillum ferriphilum, Bacillus acidocaldarius, Picrophilus oshimae Picrophilus torridus, Bispora* sp. | Amylases, glucoamylases, proteases, cellulases, oxidases | Hegde and Kaltenegger (2013), Rasuk et al. (2016), and Wiegel and Canganella (2011) |
| | High | Alkaliphiles | pH > 9 | Alkaline lake, hydrated line soil | *Natronobacterium, Natrialba, Halobiforma Natronorubrum, Halalkalicoccus, Natronococcus, Halorubrum, Bacillus, Halomonas, Aeromonas, Corynebacterium, Micrococcus, Pseudomonas, and Methylomicrobium* | Proteases and cellulases | Hegde and Kaltenegger (2013), Rasuk et al. (2016), and Wiegel and Canganella (2011) |
| Salt | High | Halophiles | 2–5 M NaCl | Salt ponds, alkali lakes, coastal areas | *Halobacterium salinarum, Haloferax mediterranei, Haloferax denitrificans, Halanaerobium lacusrosei* | Protease, dehydrogenase | Hegde and Kaltenegger (2013), Rasuk et al. (2016), and Wiegel and Canganella (2011) |
| Temperature | High | Thermophiles | 60–80°C | Hot springs, hydrothermal vents | *Aquifex pyrophilus, Thermotoga maritime, Aspergillus terreus, Penicillium* sp. | Oxidoreductase, proteases, lipases, catalases, peroxidases chitinases, glucosidases | Hegde and Kaltenegger (2013), Rasuk et al. (2016), and Wiegel and Canganella (2011) |
| | Low | Psychrophiles | <15°C | Snow, ice | *Cryobacterium, Hymenobacterium, Pedobacterium, Psychrobacterium, Sphingobacterium* | Protease, dehydrogenase, cellulases, lipases | Hegde and Kaltenegger (2013), Rasuk et al. (2016), and Wiegel and Canganella (2011) |
| Radiations | High | Radiophiles | High dose of radiation | High UV radiation | *Deinococcus radiodurans* | Catalase, peroxidase, superoxide | Hegde and Kaltenegger (2013), Rasuk et al. (2016), and Wiegel and Canganella (2011) |

Their membranes have a distinct form and composition and are attracted to the positive inner layer, which improves metal transporter function while reducing acidity and M&R toxicity (Dekker et al. 2016; Slonczewski et al. 2009).

Recent advancements in synthetic and separation technologies have created a diverse spectrum of adsorbent systems for radionuclide treatment in soil or water, including nanomaterials with surface modification or hybrid composites (Sadegh et al. 2017). Surface-modified nanoparticles of iron oxide ($Fe_3O_4$), for example, have been used to effectively absorb dangerous heavy metals from aqueous like chromium, cobalt, nickel, Cd(II), and lead (Pb). Gold (Au) nanoparticles with specific functions as adsorbents were created, although, it still requires certain challenges to face when putting these techniques into action (Liu et al. 2017). To start, eliminating unresolved adsorbents following therapies is an increased expense because

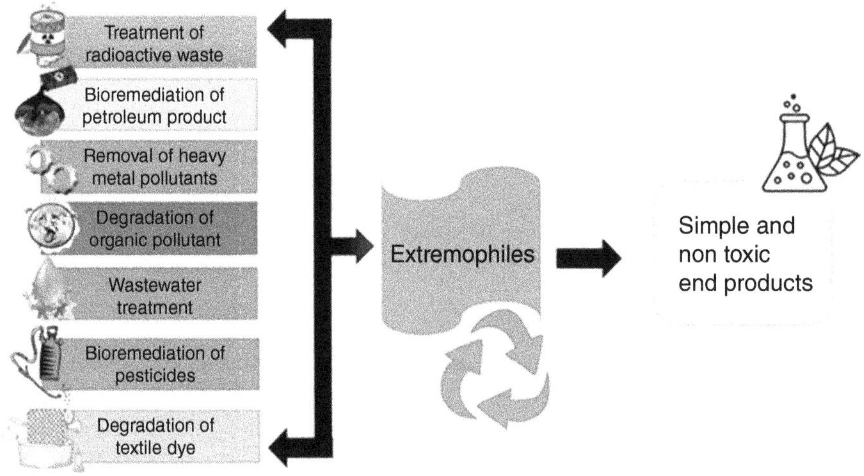


**Figure 6.3** Bioremediation of pollutants by extremophiles.

desalination produces a very substantial amount of secondary radioelement-contaminated adsorbent material. Second, microscopic-sized adsorbents lose their integrity and properties under extreme conditions like high salt concentrations and strong radiation (Choi et al. 2016).

Consequently, utilizing extremophilic bacteria as a live cleaning agent for radioactive waste treatment (Mushtaq et al. 2017) is a realistic option. Biomineralization, biotransformation, and biosorption are interactions between microbes and radioisotopes that might be exploited to treat radioactive waste microbially. The basic technique for radioactive removal from a polluted region has been presented as within a bacterium, and the targeted component is mineralized. *Shewanella* and *Geobacter* bacteria, for example, may be able to break down alpha nuclides like Uranium(VI), Platinum(IV), Am(V), and Thallium(IV) safely (Li et al. 2016). It was proposed how to employ acetate-stimulating *Geobacter* substances to extract Uranium by aqueous medium, with an addition like glucose, ethanol, and acetate acting as electron donors and improving removal efficiency. Since the 1990s, several extremophilic bacteria that can tolerate large doses of ionizing radiation (>15 kGy) have been discovered (Prakash et al. 2013). *D. radiodurans*, one of the most radio-resistant bacteria, has attracted the attention of researchers involved in the remediation of radionuclide-contaminated environments (Gogada et al. 2015).

*D. radiodurans* strain that contains a nonspecific acid phosphatase from (*Salmonella Enterica Serovar Typhi*) or a bacterial Ni/Co transporter may precipitate oxidized uranium contaminants as well as radioactive cobalt (60Co). Combining extremophilic bacteria with nanotechnology is becoming a critical strategy for detoxifying polluted environments. Combining extremophilic microorganisms with nanoparticles such as (nano-adsorbents and oxidizers) would be a great technology for realistic bioremediation with the arrival of nano-biotechnology. For example, a stable method for the abolishment of radioactive iodine (125I) was established with *D. radiodurans* and biogenic Au nanoparticles, which removed 3.7 MBq of 125I (>99%) within 30 minutes. *Sulfolobus tokodaii* 7T (NBRC 100140), a thermo-acidophilic archeon able to produce biogenic Pd(0) nanomaterials with a mean diameter of 8.7 nm.

Kitjanukit et al. (2019) illustrated a four-fold rise in Cr(IV) decreased when 2.0 mg Cr was used (VI). This other study succeeded in reducing Cr in a one-step bacteriological technique utilizing Pd(0) nanoparticles produced by the acidophilic Fe3+-reducing bacteria (Panak and Nitsche 2001).

## 6.5 Bioremediation of Petroleum Product

### 6.5.1 Petroleum Hydrocarbon Microbial Degradation in Hypersaline Environments

High salt concentrations affect microorganisms' physiological qualities and life activities by causing cell membrane disintegration, enzyme denaturation, and reduced oxygen and hydrocarbon solubility. Consequently, halophilic or halotolerant species capable of withstanding high salinity are needed for saline environmental bioremediation even without the expensive diluting of salt-laden soil and water (Liu et al. 2016). To support their osmatic balance halophiles accumulate $K^+$ and $Cl^-$ in the cell for osmotic balance with the atmosphere. Severe halophiles in the Halobacteriaceae family of archea, such as *Haloarcula marismortui*, *H. salinarum,* and *Halococcus morrhuae*, maintain their osmotic balance by accumulating $K^+$ within their cells. *Bacteriorhodopsin*, which includes ATP synthase, membrane-bound proton pumps, and the $Na^+/H^+$ antiporter, work together to generate an electrostatic attraction that stimulates $K^+$ absorption inside the cell via a $K^+$ uniport process.

Because of organic solute accumulation compatibility in a hypersaline environment, the potential of water inside the cell is greater than the water potential outside (Khemili-Talbi et al. 2015). To reduce cell dryness caused by water outflow, the halophile synthesizes low-compatible organic molecules such as glycerol, betaines, glutamine, proline, and glutamate. Furthermore, sugar maintains the integrity of microbial membranes such as sucrose and trehalose. They prevent microbial proteins against conformational changes and preserve cell mechanical strength in salt stress. The halophile protein tissue is notable for its ability to adapt to high salt conditions and for supporting the folding of lipophilic peptide chains into functional proteins, ensuring optimal enzyme activity (Weber and Marahiel 2003).

### 6.5.2 Low-Temperature Environments, Microbial Degradation of Petroleum Hydrocarbons Occurrence

Most studies on bacteria that degrade oil have traditionally been conducted at room temperature. However, indigenous bacteria have been shown to degrade petroleum hydrocarbon components in various terrestrial and marine settings, involving Arctic, Alpine, and Antarctic soil and sediments in recent years. Cold-adapted microorganisms are microorganisms that have adapted to low temperatures. Because of their potential to quickly adapt to a polluted environment, cold-adapted bacteria are becoming increasingly concerning (Xin and Zhou 1998).

By enhancing environmental parameters like temperature and pH, the rate of oil breakdown might be accelerated. Biostimulation and bioaugmentation of polluted cold localities have also been examined as a bioremediation strategy. Biostimulation is a method for increasing enzyme activity in microbes by providing varying levels of dietary components and electrical receptors, such as $O_2$ and $H_2O_2$, resulting in a higher rate of microbe breakdown to petroleum hydrocarbon contaminants (Dias et al. 2015). After 30 days of adding fish meal, total hydrocarbon removal (71%) was discovered. It was revealed that soil tillage increased the clearance rate of petroleum hydrocarbon pollutants in Arctic diesel-contaminated soil to more than 80% after three years (Paudyn et al. 2008). Bioaugmentation is a method of degrading hydrocarbon contaminants by microbes at low temperatures and modifying their accuracy, reliability, and makeup (Jiang et al. 2016; Mao and Guo 2018).

### 6.5.3 In High-Temperature Environments, Microbial Degradation of Petroleum Hydrocarbons

Oil pollution is becoming more of a problem as oil production expands to high-temperature locations and demands for petroleum energy rise. Horikoshi, Morita, and Grant (Morita et al. 1999). Oil-degrading bacteria are well-known medium-temperature microorganisms that perform poorly in high-temperature environments when breaking down petroleum hydrocarbons. As a result, in high-temperature circumstances, thermophilic bacteria are crucial for oil pollution cleaning. *B. licheniformis* WY2 was identified in oil-contaminated soil in China's Daqing Oilfield amid six thermophilic oil-degrading bacteria. The strain WY2 grew best at 52–80°C with 0.5% v/v crude oil and pH 7, as per the study, and the total petroleum hydrocarbon contents were lowered by 52%. In the Daqing Oilfield, scientists discovered a thermophilic oil-degrading bacterium species that can destroy aromatic hydrocarbon compounds, saturated hydrocarbon, and asphaltene as a carbon source at 45°C. Fang, Chang, and Hsu obtained many thermophilic bacteria that use naphthalene as the only carbon source from an oilfield at a high elevation in Lithuania (Fang et al. 1976). These specimens employed anthracene, phenol, benzene-1,3-diol, benzene, and protocatechuic acid as aromatic chemicals.

### 6.5.4 Removal of Heavy Metal Pollutants

Heavy metal toxicity has become a major concern since minor values might harm human health and the environment (Sarker et al. 2021; Kausar et al. 2022; Rizwan and Bilal 2022; Khan et al. 2023). Furthermore, when exposed to high temperatures, dangerous heavy metals are often rendered inaccessible in present chemical treatments (Huang et al. 2011). Scientists have been researching the development of long-term bioremediation systems based on extremophilic bacteria for remedies for heavy metals for decades (Zhang et al. 2020). In the event of severely acidic conditions, acidophilic bacteria that grow in low pH environments have been used as the host strain for heavy metal detoxification via biomining and bio-oxidation (Jameson et al. 2010). Biomining refers to microorganisms' capacity to extract and recover metals from ores and waste concentrates (sometimes referred to as "bioleaching" if metals are solubilized during the process)

(Özdemir et al. 2013). Many investigations have been done on the creation of bioremediation techniques based on Acidothiobacillus strains, the most prevalent acidophilic and chemolithotrophic bacteria. In industrial-scale bioleaching, *Acidothiobacillus ferrooxidans*, for example, has been employed. *At. ferrooxidans* NCIMB 8455 was used in 2009 to study the ability of *At. ferrooxidans* and *Acidothiobacillus ferrivorans* varieties to bioremediate 100 mg/l of U(IV) *ex situ* from contaminated mine water, as well as the versatility of the acidophiles for $H_2S$-assisted copper precipitation (Okibe et al. 2016).

A microbial consortium could be more efficient at eliminating toxic heavy metals, with the added benefit of combining several enzyme systems and metabolic pathways from different bacteria. A bioaugmentation of heavy metals on contaminated port sediment was recently done employing an AMD-isolated acidophilic microbe consortium. Acidophilic microbial communities formed by *A. thiooxidans*, *A. ferrooxidans*, *Acidiphilium cryptum*, and *Leptospirillum ferrooxidans* removed more than 90% of $Cu^{2+}$, $Cd^{2+}$, $Hg^{2+}$, and $Zn^{2+}$ (Beolchini et al. 2009).

Another work, it was investigated *in situ* bioremediation of AMD soil that was very acidic (pH 3.21), sulfate (6285 mg/l), and contaminated with heavy metals. The addition of acidophilic microbial consortia to AMD soil, which included metal-resistant *Chloroflexi* (29%), *Acidobacteria* (21%), *Proteobacteria* (16%), and *Firmicutes* (2%), decreased the dissolved sulfate by 97% and increased the pH to 7.5 (Chakravarty and Banerjee 2012). Halophilic bacteria may aid in the cleaning of hazardous chemicals in high-salt settings. Because marine bacteria can withstand high salt concentrations, bioremediation utilizing marine bacteria, for example, is a viable technique for decontaminating saltwater from hazardous heavy metals. Several experiments employing a variety of marine microorganisms to remove hazardous heavy metals have been reported. *Vibrio harveyi* has a high adsorption capacity (up to 23.3 mg $Cd^{2+}$/g dry cells) and an excellent ability to collect cadmium cations inside the cell. *Enterobacter cloacae*, a marine bacterium, may chelate Cd, Cu, and Co from mixed-salt solutions by up to 65%, 20%, and 8%, respectively (Gupta and Sar 2020).

Some thermophilic bacteria, like *Geobacillus thermantarcticus* and *Anoxybacillus amylolyticus*, have a high potential for heavy metal biosorption, suggesting that they might be used to remove heavy metals from contaminated settings. Lactic acid bacteria, for example, create S-layer proteins, which are potential biomolecules for harmful heavy metal decontamination under extremely low pH (pH 2) settings due to their extraordinary stability and activity against different heavy metals. According to scanning electron microscopy–energy dispersive X-ray analysis, $Pb^{2+}$ ions were efficiently adsorbed and deposited on the cell surface of *Lactobacillus plantarum* YW11 S-layer proteins, which exhibited a 99.9% Pb adsorption capacity. $Cd^{2+}$, $Zn^{2+}$, $Pb^{2+}$, and $Ni^{2+}$ adsorption on S-layer proteins from two *Lactobacillus kefiri* strains (CIDCA 8348 and JCM 5818) was also studied.

Metal biosorption, bioaccumulation, and biomining are important in supplying critical raw materials because they can provide environmentally friendly alternatives to traditional pyrometallurgical and hydrometallurgical processes and in the development of metal biomonitoring and bioremediation strategies (Bondarczuk and Piotrowska-Seget 2013). In this context, thermophiles' capacity to thrive in harsh conditions and eradicate mineral species that are resistant to them offers many benefits. Microorganisms have evolved resistance mechanisms to deal with these dangerous metals, which rely on a balance of absorption and efflux activities (Schelert et al. 2004). Many of these systems are seen in mesophiles as well; however, thermophilic bacteria or archaea may have distinct features. Genome investigations enable finding possible genetic determinants, contributing to the study of heavy metal resistance mechanisms in thermophiles (Zhuang et al. 2015).

The four mechanisms of heavy metal resistance found so far are the extracellular barrier, active metal ion transport (efflux), enzymatic metal ion reduction, and intracellular sequestration. Certain bacteria may generate complexes or chelates with extracellular polymers, which reduces metal permeability. On the other hand, heavy metals may avoid this route and enter the cell via important element absorption systems, such as arsenic, which penetrates the cell through the phosphate or glucose transport systems (Donati et al. 2016). Heavy metal resistance is often generated by the coordinated operation of intracellular enzymatic oxide reduction and heavy metal efflux mechanisms, which require ATP to drive the lethal metal outside the cell. Operons are often used to arrange resistance genes because they assure the expression of a transcription factor that regulates the whole system. Regarding heavy metal sequestration inside cells, the general theory predicts that proteins with many cysteine residues would form complexes with the metals and use the thiol groups (Ilyas et al. 2014).

Microorganisms can recover metals from contaminated environments because of biosorption/bioaccumulation processes. Biosorption refers to passive processes maintaining a kinetic equilibrium, while bioaccumulation refers to energy-driven processes needing active metabolism. Heavy metal metabolism pathways have been intensively studied since many *Geobacillus* species are highly resistant to heavy metals (As, Ag, Cd, Co, Cr, Cu, Fe, Pb, U, Zn). Their biosorption and bioaccumulation processes have been researched and used to remove undesirable metals from environmental samples. Acidophilic chemolithotrophs (autotrophs or mixotrophs) with iron and/or sulfur-oxidizing pathways, such as *Sulfobacillus* sp. and

*Ferroplasma* sp., are recognized for their capacity to solubilize Fe(II) (Hetzer et al. 2006). They are sometimes utilized in microbial consortiums that can bleach many metals simultaneously. Microorganisms capable of extracting and accumulating metals from ores or geothermal sources are of particular interest because of their potential application in the recovery of rare-earth metals such as Europium (Eu), which is widely used in modern devices (solar cells, mobile phones and computers, and biomedical instruments) (Özdemir et al. 2012). *Thermus scotoductus* SA-01 may also be used to monitor metals in the environment at low concentrations. Zdemir and colleagues developed a preconcentration method based on *B. cereus* SO-14 to improve sensitivity in ICP-OES (Inductively Coupled Plasma—Optical Emission Spectrometry) detection of U(VI) and Th(IV). Microorganisms have also been utilized in biomonitoring as whole-cell biosensors. *Thermus thermophilus* HB27 (TtSmtB), an arsenic-responsive transcriptional repressor, controls the expression of the arsenic efflux protein *Tt*ArsX. Antonucci and colleagues used *T. thermophilus* to construct a reporter gene from the *Tt*arsX promoter because *Tt*SmtB reacts to variations in Cd(II), As(III), and As(V) concentrations. It is also feasible to detect a reduction in enzyme activity as a toxicological signal of heavy metals when building metal biomonitoring systems. Poli and colleagues found that heavy metals reduced the amylase activity of *A. amylolyticus* (Yang and Rosen 2016). Shih-Hung and colleagues examined how to prevent an *Acidibacillus ferrooxidans* strain from oxidizing iron. In the domains of heavy metals bioremediation and biomonitoring, many thermophilic oxidoreductases have been discovered. For example, the quinone oxidoreductase, chromate reductase, and superoxide dismutase from different *Anoxybacillus* species have been used for Pb and Cr bio removal. Additionally, the arsenate reductase from *T. thermophilus* HB27 (*Tt*ArsC) has been used as a biological recognition element for the development of various arsenic biosensors.

### 6.5.5 Degradation of Organic Pollutants

Organic pollutants are diverse, chemically distinct organic compounds released as hazardous waste into the environment (Gumulya et al. 2018). Pesticides, fertilizers, hydrocarbons, phenols, plasticizers, biphenyls, detergents, oils, greases, and pharmaceuticals are among the contaminants found in domestic sewage, urban run-off, industrial effluents, and agricultural wastewater (Liu et al. 2019; Kumar et al. 2022; Fernandes et al. 2021). As a result, organic contaminants can take many different forms. The main constituents of persistent organic pollutants (POPs) are organochlorinated pesticides (OCPs), polycyclic aromatic hydrocarbons (PAHs), PCB, dioxins, and dibenzofurans. These are durable since they survive for lengthy periods (years or decades) in the surroundings (Saavedra et al. 2020).

These chemicals are discharged into the air and soil, where they may biomagnify in living organisms and affect human health due to their inability to dissolve in water. Harmful biological contaminants such as petroleum hydrocarbons, aromatic petrochemicals, and other halogenated substances can be converted to harmless molecules by microorganisms. Not only does such a drastic change necessitate a high level of resistance to hazardous organic pollutant exposure, but it also necessitates the ability to utilize hazardous organic pollutants in cellular metabolism. Consequently, extremophilic bacteria have evolved under challenging environments and may be extensively employed to remediate hazardous organic contaminants under the right circumstances. Polycyclic aromatic hydrocarbons (PAHs), including long-chain alkanes (C10–C32), have been found to detoxify by thermophilic *Bacillus*, *Thermus*, and *Geobacillus* culture retrieved from oil-contaminated environments (Waigi et al. 2017).

In a deep oil well isolation, *Geobacillus* SH-1 was found to be very effective in degrading saturated alkanes ranging from C12 to C33, as well as naphthalene. According to another study, C12–C21 *n*-alkanes degrade entirely in eight days, whereas 100 ppm naphthalene decomposes virtually wholly in 72 hours. Biological treatment with *Geobacillus thermoparaffinivorans* IR2, *Geobacillus stearothermophilus* IR4, and *B. licheniformis* increased the detoxification of long alkyl (C32 and C40) chemicals (Elumalai et al. 2017). In the remediation of organic hydrocarbon contaminants, psychrotrophic and halophilic microorganisms and thermophilic bacteria have shown encouraging results. At low temperatures, hydrocarbon mixtures like fuel, military plane fuel, and oil were efficiently weakened by *Pseudoalteromonas* sp. P29 and *Oleispira antarctica* RB-8T. In aquatic systems (e.g., salt lakes, salt marshes, and extremely salty soils), *Marinobacter sedimentalis*, *Marinobacter falvimaris*, and *Marinobacter nanhaiticus* D15-8W proved capable of converting biphenyl, phenanthrene, anthracene, and naphthalene into useable carbon sources (Al-Mailem et al. 2013).

It is impossible to overestimate the importance of EPSs, the following components of halophiles, in removing pollutants from hypersaline environments. Halophile's EPSs may operate as biosurfactants, aggregating oils, and emulsifying hydrocarbons while protecting cells from harmful chemicals. With less than 12 days, the halophilic bacteria *Halobacillus* sp. EG1HP4QL can impair paraffin (34.5%), naphthalene (49.6%), monocyclic and bicyclic aromatic hydrocarbons (51.2%), polycyclic aromatic hydrocarbon (43.5%), and alcohol–benzene resins (25.5%) Bioremediation of the hydrocarbon polluted point of discharge of deepwater horizon was also done with the EPS-producing *Halomonas* strain TG39. Not only did the isolated EPS help to speed up the breakdown of phenanthrene, but it also helped to improve the solubility of hydrocarbons (Gutierrez et al. 2013). As a result, biodegradation with extremophilic bacteria is a promising strategy for treating organic contaminant-polluted areas in extreme conditions because the microorganism can metabolize the organic contaminants.

The most frequent marine environmental pollutants are petroleum hydrocarbons, organic oil spillage, volcanic activities in the ocean depths, and manufactured spillages, like oil ships, oil transportation systems, or oil refining, which contaminate the sea. In most cases, this is the primary source of crude oil pollution in bodies of water. Microorganisms have been excellent hazardous chemical detoxifying catalysts (Mole et al. 2009). Toxic chemicals are often transformed into metabolites that enter important metabolic pathways – oxidation of aliphatic and alicyclic carboxylic acids, coupled – and –oxidation, and aromatization are examples. Members of the multiple antimicrobial susceptibility regulator families (MarR-family) are engaged in eliminating aromatic compounds, including benzaldehyde and salicylate, in *S. solfataricus*. Bacteria degrade aromatic and polyaromatic hydrocarbons, aromatic acids, chlorinated aromatics, and heterocyclic aromatic compounds (Chakravarty and Banerjee 2012), which belong to the Rieske nonheme iron oxygenase family (ROHs). Toluene dioxygenase is an enzyme that converts benzene to benzene *cis*-diol (Parales and Resnick 2004; Yemashova et al. 2007). The employment of thermophiles in crude-oil bioremediation approaches is a fantastic concept; when oil viscosity reduces at higher temperatures, organic compound diffusion rates increase, making them more accessible to microorganisms (da Silva Vilar et al. 2021). Because the degrading activity is often substrate-specific, microbial consortia may accelerate biodegradation while broadening the range of action (Guizhou et al. 2013). For instance, two isolates of *Geobacillus jurassicus* and *Geobacillus subterraneus* observed in China's Dagang high-temperature oil field could also live on benzoate but still not phenol, implying that phenol biodegradation actions must be taken to decompose crude oil. Thermophiles produce large molecules that can be used in biodegradation and natural contamination bioremediation. Low molecular weight peptides, surface-active compounds that could be used as biosurfactants in anaerobic oil/water separation, have now been documented to be produced by *B. licheniformis* and *Anaerophaga thermohalophila* (Feitkenhauer et al. 2001). The surface predictability and consistency of metabolizing thermozymes create them beneficial as biorecognition elements in biosensors, particularly those that require electrical and chemical identification, for instance, the haloacid dehalogenase, I-HADST of *S. tokodaii*, has been immobilized on *N*-hydroxysuccinimidyl Sepharose resin and used for identification of halogenated organic substances, preserving 70% of its initial deployment within a week of six months at 4°C (Kanaly and Harayama 2010).

### 6.5.6 Wastewater Treatment

Nanobiotechnology has potential in wastewater treatment, medicine delivery, biosensors, vaccine development, and genetic editing. Biogenic nanoparticles (NPs), especially biopolymeric metallic nanomaterials, seem to be potential options for bioremediation (Sarkar et al. 2017). Because of the increased enzymatic performance provided by their large surface area/volume, nanoscale adsorbent materials are currently hamming it up in sewage treatment (Pereira et al. 2019). Bioactive metal nanoparticles, on either hand, have several drawbacks, along with a lengthy reducing period, complicated downstream processing, and aggregation formation. To address these concerns, EPS biopolymers could be employed in producing metal nanoparticles as both reducing and capping agents. Because of its higher capacity to synchronize to metal ions, bacterial EPSs could be a preferable option for processing wastewaters, which typically include environmental metal wastes, pharmaceutical wastes, and textile industrial wastes, and treating bactericidal agents. In EPS, there are charged and reducing functional groups (Durán et al. 2011). As a result, bacterial EPS is frequently used to create nanoparticles to reduce the amount of harmful heavy metals in metal-contaminated environment wastes. For example, natural weathering of selenium-rich rock, mining, and irrigation all discharge selenium into the environment in its potentially lethal form, selenite. Reducing groups in bacterial EPSs decreases selenite and eliminates it as elemental selenium in the atmosphere. Many studies show metal degradation and immobilization mediated by bacterial EPS to produce nanoparticles. *Pseudomonas* sp. PAMC 28620, for example, has high metal removal activity (over 99%) against $Fe^{2+}$ ions, making it an excellent biosorbent for treating wastewater. Another research created spherical cadmium sulfide (CdS) nanoparticles with diameters ranging from 20 to 40 nm by using EPS due to aquatic *Pseudomonas aeruginosa* JP-11 (Gahlawat and Choudhury 2019).

CdS NPs incorporated into xanthate functionalized EPS from *P. aeruginosa* JP-11, outperformed bifunctional EPS alone (80.81%) and pristine EPS (88.66%) in solubilizing cadmium ions from metal-contaminated sewage (61.88%). Sulfated mauran (MR), among the most researched extremophilic bacterial EPSs, captivates metal chelate steady nanoparticles due to uronic acid MR isolated from the moderately halophilic bacterium *Halomonas maura* electrospun thin-uniform nanofibers (120 nm) out of a homogeneous poly-vinyl ethanol solution. It has been shown that metal-containing environmental effluent may be rectified by utilizing these MR nanofibers. Co-precipitating $Fe_3O_4$ nanoparticles [iron (III) chloride and iron (II) sulfate with EPS produced out from microalga *C. vulgaris*, resulted in a magnetization nanostructure. Magnetic EPS nanoparticles retrieved 91% of the $PO_4^{3-}$ and 85% of the $NH^{4+}$ as nutrients through sewage in 2017 (Vickers). Because of acute water poisoning, pharmaceutical waste is among the hazardous organic contaminated products in the environment (He et al. 2020). Microbes' reactive sites react with divalent metal ions in waste to generate nanoparticles that are subsequently employed in a cation exchange procedure to treat organic pollutants in pharmaceutical wastewater. *Escherichia coli* BL21 isolated from wastewaters formed biologically active palladium nanoparticles in the context of $H_2$ by transforming Pd(II) to Pd(0). These nano-sized biogenic palladium particles have been proven to reduce and eradicate more than 87% of ciprofloxacin biotic from wastewater due to their significant enzymatic performance (He et al. 2020).

Reactive azo dyes, mostly manufactured by the textile industry, bind functional groups utilizing various forces of attraction (for example, ionic and functional dipolar) and the pH of the medium, making it one of the most polluting chemicals on the planet. Further reduction happens after the creation of an EPS–metal complex. Numerous groups of EPSs surround and support the metal NP anchor. The NPs' exterior charge layer promotes the repulsion of particles, resulting in physically suspect ability in solution and the preservation of the NPs' fundamental phase in the solid state (Pereira et al. 2015). According to studied surveys, EPS can self-assemble into 88 nm spherical nanoparticles. These self-assembled EPS nanoparticles can absorb lead, copper, cadmium, and methylene blue dyes. In addition, bacterial EPS mediates metal nanoparticle production, including its role in isolating and desorption of contaminants from wastewater (Sarkar et al. 2020).

Biogenic metal nanoparticles, as previously stated, displayed outstanding catalytic activity in converting complex aromatic and organic pollutants and metal wastes into non-toxic compounds in a short reaction time. Biologically generated nanoparticles are also predicted to enable multicomponent water treatment (Gahlawat and Choudhury 2019). However, industrial bio-flocculant utilization has been insufficient compared to the well-established use of extremophilic bacterial EPSs in seawater bioremediation. It is yet to be commercialized because of a lack of knowledge that must be bridged among industries and universities. Despite this, bacteria's distinct biological niche in severe conditions and their synthesis capacity structurally varied EPSs suggest that hostile environments across the globe offer a huge source of renewable, multifunctional biomaterials for specialized industrial uses.

### 6.5.7 Textile Dye Degradation

Microbial techniques best fit several dye removal methods for textile effluents. Pathogens, fungi, and bacteria that can remove colorants in sewage water could be used in aerobic, anaerobic, or anaerobic/aerobic systems. Between several wastewater decolorization processes, the adsorption of living or dead microbial biomass plays an important role in dye biorecovery following effluent treatment (McMullan et al. 2001). Bacteria from several taxonomic groups have the potential to decolorize textile colors. Because of their quick development and diverse metabolism, several microorganisms have a high efficacy in wastewater decolorization. The harshness of wastewater effluents due to high salinity, temperature, and alkalinity, on the other hand, is a limiting factor in the ability of mesophilic microbes to remediate wastewater properly. Due to their capacity to thrive in such extreme circumstances, extremophilic microorganisms and their bioactive compounds have an edge over other mesophilic microbes in the bioremediation of wastewater. Most of the research on extremophilic bacteria for biosorption of textile effluent materials has focused on its removal ability, but very little consideration has been devoted to its final products. The discovery of enzyme production and genes responsible for color removal in extremophilic bacteria, as well as their usage in bioremediation, is a future aim (Amoozegar et al. 2015).

Industrialization is a major source of water contamination. Recalcitrant, mutagen, and carcinogenic chemicals may be found in high concentrations in wastewater. Dyes are a kind of highly harmful contaminant found in the effluent of the textile industry. The pH of marine ecosystems, chemical oxygen demand (COD), and biochemical oxygen demand (BOD) are now all affected by such resistant compounds. The dyes are classified based on their chemical structure or desired wide applicability. Acid, synthetic, and direct dyes are the most prevalent dyes. While all dyes are complex chemical compounds, each has its chromophore acid and synthetic azo dyes are distinguished by their azo linkage (—N=N—), anthracene colors are distinguished by their central chromophore, while indigoid dyes are distinguished by indoles (El Harfi and El Harfi 2017).

Colored sewage treatment methods include physical, chemical, mechanical, and biological strategies. Coagulation–flocculation, oxidation, ozonation, Fenton oxidation, photocatalytic oxidation, ion exchange, and electrical and chemical therapies are types of physical treatments; chemical treatments include coagulation–flocculation, oxidation, ozonation, Fenton oxidation, photocatalytic oxidation, ion exchange, and biological methods include aerobic and anaerobic process. Physical–chemical flocculation combined with bioremediation is perhaps the most common technique for wastewater cleaning, and it has various advantages and disadvantages. Intra/extracellular oxidoreductases like dye decoloration peroxidases (DyP), laccases, and azoreductases were found in the bacterial diversity of these streams (Lončar et al. 2016).

Numerous laccases and azoreductases have also been discovered in thermophiles like as *Thermophilus thermophilus*, *Geobacillus*, *Anoxibacillus*, *Thermosediminibacter* species, and many others. Unlike mesophilic bacteria, their enzymes are known to be more resistant to severe temperatures and pH levels. *T. thermophilus* HB27's laccase enzyme can oxidize six distinct colors (the color orange, acid red dye, green dye, naphthol brilliant blue, remazol brilliant blue, and Congo red); the enzyme requires an electron shuffling, which eases the process for certain dyes (Kumari et al. 2018). LacG, a laccase found in *Geobacillus* sp. JS12 decolorizes synthetic substances when heated to 70°C. The UARK 01 laccase of the alkaliphilic and thermophilic bacteria *Anoxybacillus* sp. strain UARK-01 may oxidize the Congo red substrate, among the most toxic dyes. These extremophiles also feature a well-known azoreductase that degrades a variety of colors; for example, *Anoxibacillus* sp. PDR2 destroys straight black (Chen et al. 2020). A thermophilic microflora comprising facultative aerobic bacteria was also discovered (*Tepidiphilus thermophilus* strain JHK30, *Anoxybacillus flavithermus* strain 52-1A, *Tepidiphilus succinatimandens* strain 4). These examples show how thermophilic oxidoreductases are used in biological detoxification.

### 6.5.8 Bioremediation of Pesticides

Pesticides are any compound or mixture of compounds that eradicate or treat insects, weeds, or crop diseases. Pesticides are widely utilized to control agricultural and home pests all over the world. Organophosphorus insecticides seem to be the most extensively employed pesticides, and the very first organophosphorus insecticide, tetraethyl pyrophosphate, was created in 1937 (Maynard and Chilcott 2009), has been used extensively in agriculture since the 1950s. Organophosphate compounds are extremely dangerous due to their ability to alter acetylcholine esterase, a critical protein integral to the development control of the central and peripheral nervous systems. Because organochlorine pesticides such as lindane and dichloro-diphenyl-trichloroethane (DDT) have a longer persistence, a tendency for bioaccumulation, high mammalian toxicity, and potential toxicity to non-target organisms, their use in developed countries has been drastically reduced. They have been replaced by less persistent but more effective and efficient organophosphorus compounds such as chlorpyrifos, glyphosate, and methyl Phosphorus is frequently found as a phosphonate or phosphate ester, which are used in the oxidation, hydrolysis, dealkylation, and alkylation processes. As a result, the most important step in microbiological detox is the hydrolysis of P–O–aryl and P–O–alkyl bonds (Milatovic et al. 2009).

Pesticides serve an important role in crop protection, but their widespread usage has resulted in severe soil contamination and degradation and has polluted a wide range of ecosystems throughout the globe. Pesticide residues have been discovered in soil, water, milk, food, and shellfish from various places. Because these chemicals are very hazardous and constitute considerable health and environmental concern, they must be eliminated from the environment. Many strategies, including physical, chemical, and biological factors, have been studied to establish chemical cleaning methods for these substances; nevertheless, these processes are not appropriate for big environmental remediation and require brutal environments (Jacquet et al. 2016).

Bioremediation has emerged as a viable strategy for eliminating chemical pesticides from damaged soil. *Flavobacterium* sp. was isolated and purified, degrading organophosphorus substances in soil ecosystems and fluid culture. The enzymes organophosphate hydrolase or phosphotriesterase catalyze the degradation of these compounds, which is catalyzed by the genes opd (organophosphate degradation), which was found, sequenced, translated, and altered for improved stability and activity in a variety of species. Extremophile enzymes have recently emerged as a promising choice for sterilizing dangerous chemical compounds cleanly and timely.. SsoPox is a phosphotriesterase-like lactonase produced from the bacterium *S. solfataricus* with bioremediation potential. This enzyme was created and proven to effectively digest a wide range of organophosphorus insecticides (Del Giudice et al. 2016). As a consequence of this enzyme's phosphotriesterase activity on organophosphorus insecticides, the hydrolysis process yields two breakdown products: a phosphodiester and an alcohol. This SsoPox enzyme is potent due to its extremophilic origin, detergent resistance and activity throughout a broad temperature range, and long-term preservation. Because the enzyme is readily immobilized, it might be employed in filtration systems to handle organophosphorus chemicals in effluent (Poirier et al. 2017).

## 6.6 Conclusion and Future Perspective

For the bioremediation of contaminants, microbes serve as the primary catalysts and an alluring substitute to physio-chemical treatment methods. Expanding the taxonomic range of extremophiles and studying the metabolites used as biological defense mechanisms to counter extreme environmental stress requires additional research. For sustainable development, it is essential to clean up a contaminated environment. Therefore, more extremophiles inhabiting ecosystems with extreme circumstances should be isolated, identified, and characterized in future research. The biodegradation of heavy metals and persistent organic pollutants, which will offer a better future for the economy and boost the welfare of people worldwide, should then be investigated for use in biotechnology.

## References

Abdelfattah, A., Hossain, M.I., and Cheng, L. (2020). High-strength wastewater treatment using microbial biofilm reactor: a critical review. *World Journal of Microbiology and Biotechnology* 36 (5): 1–10.

Abe, F., Kato, C., and Horikoshi, K. (1999). Pressure-regulated metabolism in microorganisms. *Trends in Microbiology* 7 (11): 447–453.

Allard, A.-S. and Neilson, A.H. (1997). Bioremediation of organic waste sites: a critical review of microbiological aspects. *International Biodeterioration & Biodegradation* 39 (4): 253–285.

Al-Mailem, D., Eliyas, M., and Radwan, S. (2013). Oil-bioremediation potential of two hydrocarbonoclastic, diazotrophic *Marinobacter* strains from hypersaline areas along the Arabian Gulf coasts. *Extremophiles* 17 (3): 463–470.

Amaro, A., Chamorro, D., Seeger, M. et al. (1991). Effect of external pH perturbations on in vivo protein synthesis by the acidophilic bacterium *Thiobacillus ferrooxidans*. *Journal of Bacteriology* 173 (2): 910–915.

Amoozegar, M., Mehrshad, M., and Akhoondi, H. (2015). Application of extremophilic microorganisms in decolorization and biodegradation of textile wastewater. In: *Microbial Degradation of Synthetic Dyes in Wastewaters* (ed. S.N. Singh), 267–295. Springer.

Aono, R. (1995). Assignment of facultatively alkaliphilic *Bacillus* sp. strain C-125 to *Bacillus lentus* group 3. *International Journal of Systematic and Evolutionary Microbiology* 45 (3): 582–585.

Azubuike, C.C., Chikere, C.B., and Okpokwasili, G.C. (2016). Bioremediation techniques – classification based on site of application: principles, advantages, limitations and prospects. *World Journal of Microbiology and Biotechnology* 32 (11): 1–18.

Baker-Austin, C. and Dopson, M. (2007). Life in acid: pH homeostasis in acidophiles. *Trends in Microbiology* 15 (4): 165–171.

Bartlett, D. (1992). Microbial life at high pressures. *Science Progress (1933-)* 76: 479–496.

Beolchini, F., Dell'Anno, A., De Propris, L. et al. (2009). Auto-and heterotrophic acidophilic bacteria enhance the bioremediation efficiency of sediments contaminated by heavy metals. *Chemosphere* 74 (10): 1321–1326.

Bhattacharya, A. and Gupta, A. (2022). Current trends in applicability of thermophiles and thermozymes in bioremediation of environmental pollutants. In: *Microbial Extremozymes* (ed. M. Kuddus), 161–176. Elsevier.

Białkowska, A., Majewska, E., Olczak, A., and Twarda-Clapa, A. (2020). Ice binding proteins: diverse biological roles and applications in different types of industry. *Biomolecules* 10 (2): 274.

Bilal, M. and Iqbal, H.M. (2020). Microbial bioremediation as a robust process to mitigate pollutants of environmental concern. *Case Studies in Chemical and Environmental Engineering* 2: 100011.

Bondarczuk, K. and Piotrowska-Seget, Z. (2013). Molecular basis of active copper resistance mechanisms in Gram-negative bacteria. *Cell Biology and Toxicology* 29 (6): 397–405.

Bowman, J.P. and Nichols, D.S. (2005). Novel members of the family Flavobacteriaceae from *Antarctic maritime* habitats including *Subsaximicrobium wynnwilliamsii* gen. nov., sp. nov., *Subsaximicrobium saxinquilinus* sp. nov., *Subsaxibacter broadyi* gen. nov., sp. nov., *Lacinutrix copepodicola* gen. nov., sp. nov., and novel species of the genera Bizionia, Gelidibacter and Gillisia. *International Journal of Systematic and Evolutionary Microbiology* 55 (4): 1471–1486.

Breaux, T.D. and Gordon, D.G. (2013). *Regulatory requirements traceability and analysis using semi-formal specifications*. In: *Paper Presented at the International Working Conference on Requirements Engineering: Foundation for Software Quality*.

Brim, H., McFarlan, S.C., Fredrickson, J.K. et al. (2000). Engineering *Deinococcus radiodurans* for metal remediation in radioactive mixed waste environments. *Nature Biotechnology* 18 (1): 85–90.

Brim, H., Osborne, J.P., Kostandarithes, H.M. et al. (2006). *Deinococcus radiodurans* engineered for complete toluene degradation facilitates Cr(VI) reduction. *Microbiology* 152 (8): 2469–2477.

Chakravarty, R. and Banerjee, P.C. (2012). Mechanism of cadmium binding on the cell wall of an acidophilic bacterium. *Bioresource Technology* 108: 176–183.

Chen, G., An, X., Feng, L. et al. (2020). Genome and transcriptome analysis of a newly isolated azo dye degrading thermophilic strain *Anoxybacillus* sp. *Ecotoxicology and Environmental Safety* 203: 111047.

Childress, J.J. (1995). Are there physiological and biochemical adaptations of metabolism in deep-sea animals? *Trends in Ecology & Evolution* 10 (1): 30–36.

Choi, M.H., Shim, H.-E., Yun, S.-J. et al. (2016). Gold-nanoparticle-immobilized desalting columns for highly efficient and specific removal of radioactive iodine in aqueous media. *ACS Applied Materials & Interfaces* 8 (43): 29227–29231.

Choi, J.Y., Lee, K., and Lee, P.C. (2019). Characterization of carotenoid biosynthesis in newly isolated *Deinococcus* sp. AJ005 and investigation of the effects of environmental conditions on cell growth and carotenoid biosynthesis. *Marine Drugs* 17 (12): 705.

Christen, P., Vega, A., Casalot, L. et al. (2012). Kinetics of aerobic phenol biodegradation by the acidophilic and hyperthermophilic archaeon *Sulfolobus solfataricus* 98/2. *Biochemical Engineering Journal* 62: 56–61.

Coker, J. (2016). Extremophiles and biotechnology: current uses and prospects. *F1000Research* 5: 396.

Connell, L., Segee, B., Redman, R. et al. (2018). Biodiversity and abundance of cultured microfungi from the permanently ice-covered Lake Fryxell, Antarctica. *Life* 8 (3): 37.

da Silva Vilar, D., Torres, N.H., Bharagava, R.N. et al. (2021). Emerging contaminants in environment: occurrence, toxicity, and management strategies with emphasis on microbial remediation and advanced oxidation processes. In: *Microbe mediated remediation of environmental contaminants, 1–14*. Woodhead Publishing, New Delhi.

De Maayer, P., Anderson, D., Cary, C., and Cowan, D.A. (2014). Some like it cold: understanding the survival strategies of psychrophiles. *EMBO Reports* 15 (5): 508–517.

Deka, R., Shreya, S., Mourya, M. et al. (2022). A techno-economic approach for eliminating dye pollutants from industrial effluent employing microalgae through microbial fuel cells: barriers and perspectives. *Environmental Research* 212: 113454.

Dekker, L., Arsène-Ploetze, F., and Santini, J.M. (2016). Comparative proteomics of *Acidithiobacillus ferrooxidans* grown in the presence and absence of uranium. *Research in Microbiology* 167 (3): 234–239.

Del Giudice, I., Coppolecchia, R., Merone, L. et al. (2016). An efficient thermostable organophosphate hydrolase and its application in pesticide decontamination. *Biotechnology and Bioengineering* 113 (4): 724–734.

Dias, R.L., Ruberto, L., Calabró, A. et al. (2015). Hydrocarbon removal and bacterial community structure in on-site biostimulated biopile systems designed for bioremediation of diesel-contaminated Antarctic soil. *Polar Biology* 38 (5): 677–687.

Donati, E.R., Castro, C., and Urbieta, M.S. (2016). Thermophilic microorganisms in biomining. *World Journal of Microbiology and Biotechnology* 32 (11): 1–8.

Drienovska, I., Chovancova, E., Koudelakova, T. et al. (2012). Biochemical characterization of a novel haloalkane dehalogenase from a cold-adapted bacterium. *Applied and Environmental Microbiology* 78 (14): 4995–4998.

Durán, N., Marcato, P.D., Durán, M. et al. (2011). Mechanistic aspects in the biogenic synthesis of extracellular metal nanoparticles by peptides, bacteria, fungi, and plants. *Applied Microbiology and Biotechnology* 90 (5): 1609–1624.

El Harfi, S. and El Harfi, A. (2017). Classifications, properties and applications of textile dyes: a review. *Applied Journal of Environmental Engineering Science* 3 (3): 00311–00320.

Elleuche, S., Schröder, C., Sahm, K., and Antranikian, G. (2014). Extremozymes—biocatalysts with unique properties from extremophilic microorganisms. *Current Opinion in Biotechnology* 29: 116–123.

Elumalai, P., Parthipan, P., Karthikeyan, O.P., and Rajasekar, A. (2017). Enzyme-mediated biodegradation of long-chain n-alkanes (C32 and C40) by thermophilic bacteria. *Biotech* 7 (2): 1–10.

Fang, C., Chang, P., and Hsu, T. (1976). The antibacterial components of dan-shen. *Acta Chimica Sinica* 34: 197–209.

Farhan, A., Rashid, E.U., Waqas, M. et al. (2022). Multifunctional graphene-based nanocomposites and nanohybrids for the abatement of agro-industrial pollutants in aqueous environments—a review. *Environmental Pollution* 308: 119557.

Feitkenhauer, H., Schnicke, S., Müller, R., and Märkl, H. (2001). Determination of the kinetic parameters of the phenol-degrading thermophile *Bacillus themoleovorans* sp. A2. *Applied Microbiology and Biotechnology* 57 (5): 744–750.

Fernandes, C.D., Vilar, D.S., Torres, N.H. et al. (2021). Fungal potential for the degradation of synthetic dyes: an overview of renewable alternatives for the production of lignin-modifying enzymes. In: *Microbial Products for Health, Environment and Agriculture* (ed. P.K. Arora), 153–181. Springer.

Ferreira, A.C., Nobre, M.F., Moore, E. et al. (1999). Characterization and radiation resistance of new isolates of *Rubrobacter radiotolerans* and *Rubrobacter xylanophilus*. *Extremophiles* 3 (4): 235–238.

Filler, D.M., Reynolds, C.M., Snape, I. et al. (2006). Advances in engineered remediation for use in the Arctic and Antarctica. *Polar Record* 42 (2): 111–120.

Filler, D.M., Snape, I., and Barnes, D.L. (2008). *Bioremediation of Petroleum Hydrocarbons in Cold Regions*. Cambridge University Press.

Filler, D.M., Stempvoort, D.R.V., and Leigh, M.B. (2009). Remediation of frozen ground contaminated with petroleum hydrocarbons: feasibility and limits. In: *Permafrost Soils* (ed. R. Margesin), 279–301. Springer.

Gabani, P., Copeland, E., Chandel, A.K., and Singh, O.V. (2012). Ultraviolet-radiation-resistant isolates revealed cellulose-degrading species of *Cellulosimicrobium cellulans* (UVP 1) and *Bacillus pumilus* (UVP 4). *Biotechnology and Applied Biochemistry* 59 (5): 395–404.

Gahlawat, G. and Choudhury, A.R. (2019). A review on the biosynthesis of metal and metal salt nanoparticles by microbes. *RSC Advances* 9 (23): 12944–12967.

Gallo, G., Puopolo, R., Carbonaro, M. et al. (2021). Extremophiles, a nifty tool to face environmental pollution: from exploitation of metabolism to genome engineering. *International Journal of Environmental Research and Public Health* 18 (10): 5228.

Gogada, R., Singh, S.S., Lunavat, S.K. et al. (2015). Engineered *Deinococcus radiodurans* R1 with NiCoT genes for bioremoval of trace cobalt from spent decontamination solutions of nuclear power reactors. *Applied Microbiology and Biotechnology* 99 (21): 9203–9213.

Golyshina, O.V., Tran, H., Reva, O.N. et al. (2017). Metabolic and evolutionary patterns in the extremely acidophilic archaeon *Ferroplasma acidiphilum* YT. *Scientific Reports* 7 (1): 1–12.

Gratia, E., Weekers, F., Margesin, R. et al. (2009). Selection of a cold-adapted bacterium for bioremediation of wastewater at low temperatures. *Extremophiles* 13 (5): 763–768.

Guizhou, G., Zheng, L., Dongfeng, Z., and Chaocheng, Z. (2013). Isolation and characterization of a thermophilic oil-degrading bacterial consortium. *China Petroleum Processing & Petrochemical Technology* 15 (2): 82.

Gumulya, Y., Boxall, N.J., Khaleque, H.N. et al. (2018). In a quest for engineering acidophiles for biomining applications: challenges and opportunities. *Genes* 9 (2): 116.

Gunde-Cimerman, N., Plemenitaš, A., and Oren, A. (2018). Strategies of adaptation of microorganisms of the three domains of life to high salt concentrations. *FEMS Microbiology Reviews* 42 (3): 353–375.

Gupta, A. and Sar, P. (2020). Characterization and application of an anaerobic, iron and sulfate reducing bacterial culture in enhanced bioremediation of acid mine drainage impacted soil. *Journal of Environmental Science and Health, Part A* 55 (4): 464–482.

Gupta, G., Srivastava, S., Khare, S., and Prakash, V. (2014). Extremophiles: an overview of microorganism from extreme environment. *International Journal of Agriculture, Environment and Biotechnology* 7 (2): 371.

Gutierrez, T., Berry, D., Yang, T. et al. (2013). Role of bacterial exopolysaccharides (EPS) in the fate of the oil released during the deepwater horizon oil spill. *PLoS One* 8 (6): e67717.

Hadibarata, T., Kristanti, R.A., Bilal, M. et al. (2022). Microbial degradation and transformation of benzo [a] pyrene by using a white-rot fungus *Pleurotus eryngii* F032. *Chemosphere* 307: 136014.

Hallberg, K. (2010). New perspectives in acid mine drainage microbiology. *Hydrometallurgy* 104 (3-4): 448–453.

He, P., Mao, T., Wang, A. et al. (2020). Enhanced reductive removal of ciprofloxacin in pharmaceutical wastewater using biogenic palladium nanoparticles by bubbling H 2. *RSC Advances* 10 (44): 26067–26077.

Hegde, S. and Kaltenegger, L. (2013). Colors of extreme exo—Earth environments. *Astrobiology* 13 (1): 47–56.

Herschy, B., Whicher, A., Camprubi, E. et al. (2014). An origin-of-life reactor to simulate alkaline hydrothermal vents. *Journal of Molecular Evolution* 79 (5): 213–227.

Hetzer, A., Daughney, C.J., and Morgan, H.W. (2006). Cadmium ion biosorption by the thermophilic bacteria *Geobacillus stearothermophilus* and *G. thermocatenulatus*. *Applied and Environmental Microbiology* 72 (6): 4020–4027.

Hinteregger, C. and Streichsbier, F. (1997). Halomona ssp., a moderately halophilic strain, for biotreatment of saline phenolic waste-water. *Biotechnology Letters* 19 (11): 1099–1102.

Horikoshi, K. (1999). Alkaliphiles: some applications of their products for biotechnology. *Microbiology and Molecular Biology Reviews* 63 (4): 735–750.

Huang, Y., Dong, X., Liu, Y. et al. (2011). Graphene-based biosensors for detection of bacteria and their metabolic activities. *Journal of Materials Chemistry* 21 (33): 12358–12362.

Ibrahim, A.S., Al-Salamah, A.A., El-Tayeb, M.A. et al. (2012). A novel cyclodextrin glycosyltransferase from alkaliphilic *Amphibacillus* sp. NPST-10: purification and properties. *International Journal of Molecular Sciences* 13 (8): 10505–10522.

Igiri, B.E., Okoduwa, S.I., Idoko, G.O. et al. (2018). Toxicity and bioremediation of heavy metals contaminated ecosystem from tannery wastewater: a review. *Journal of Toxicology 2018*: 1–16.

Ilyas, S., Lee, J.-C., and Kim, B.-S. (2014). Bioremoval of heavy metals from recycling industry electronic waste by a consortium of moderate thermophiles: process development and optimization. *Journal of Cleaner Production* 70: 194–202.

Isticato, R., Lanzilli, M., Petrillo, C. et al. (2020). Bacillus subtilis builds structurally and functionally different spores in response to the temperature of growth. *Environmental Microbiology* 22 (1): 170–182.

Ivanova, A., Kizilova, A., Kravchenko, I. et al. (2013). Carbohydrates oxidizing acidophilic thermotolerant bacterial consortium from sulfur soils. *Mikrobiologiia* 82 (4): 464–472.

Jacquet, P., Daudé, D., Bzdrenga, J. et al. (2016). Current and emerging strategies for organophosphate decontamination: special focus on hyperstable enzymes. *Environmental Science and Pollution Research* 23 (9): 8200–8218.

Jameson, E., Rowe, O., Hallberg, K., and Johnson, D. (2010). Sulfidogenesis and selective precipitation of metals at low pH mediated by *Acidithiobacillus* spp. and acidophilic sulfate-reducing bacteria. *Hydrometallurgy* 104 (3-4): 488–493.

Jamieson, A.J., Fujii, T., Mayor, D.J. et al. (2010). Hadal trenches: the ecology of the deepest places on Earth. *Trends in Ecology & Evolution* 25 (3): 190–197.

Janto, B., Ahmed, A., Ito, M. et al. (2011). Genome of alkaliphilic *Bacillus pseudofirmus* OF4 reveals adaptations that support the ability to grow in an external pH range from 7.5 to 11.4. *Environmental Microbiology* 13 (12): 3289–3309.

Jardine, J., Stoychev, S., Mavumengwana, V., and Ubomba-Jaswa, E. (2018). Screening of potential bioremediation enzymes from hot spring bacteria using conventional plate assays and liquid chromatography-tandem mass spectrometry (Lc-Ms/Ms). *Journal of Environmental Management* 223: 787–796.

Javaid, A., Latif, S., Imran, M. et al. (2022). Nanohybrids-assisted photocatalytic removal of pharmaceutical pollutants to abate their toxicological effects—a review. *Chemosphere* 291, 133056.

Jeong, S.-W. and Choi, Y.J. (2020). Extremophilic microorganisms for the treatment of toxic pollutants in the environment. *Molecules* 25 (21): 4916.

Jeong, S.-W., Jung, J.-H., Kim, M.-K. et al. (2016). The three catalases in *Deinococcus radiodurans*: only two show catalase activity. *Biochemical and Biophysical Research Communications* 469 (3): 443–448.

Jiang, Y., Ma, H., Zong, L. et al. (2016). The screening and capability of heat resistant oil-degrading bacteria. *Zhongguo Weishengtaxixue Zazhi/Chinese Journal of Microecology* 28 (6): 674–677.

Johnson, D. (2008). Biodiversity and interactions of acidophiles: key to understanding and optimizing microbial processing of ores and concentrates. *Transactions of Nonferrous Metals Society of China* 18 (6): 1367–1373.

Joshi, S. and Satyanarayana, T. (2013). Biotechnology of cold-active proteases. *Biology* 2 (2): 755–783.

Joshi, S. and Satyanarayana, T. (2013). Characteristics and applications of a recombinant alkaline serine protease from a novel bacterium *Bacillus lehensis*. *Bioresource Technology* 131: 76–85.

Kambura, A.K., Mwirichia, R.K., Kasili, R.W. et al. (2016). Bacteria and Archaea diversity within the hot springs of Lake Magadi and Little Magadi in Kenya. *BMC Microbiology* 16 (1): 1–12.

Kanaly, R.A. and Harayama, S. (2010). Advances in the field of high-molecular-weight polycyclic aromatic hydrocarbon biodegradation by bacteria. *Microbial Biotechnology* 3 (2): 136–164.

Kato, C. and Nogi, Y. (2001). Correlation between phylogenetic structure and function: examples from deep-sea Shewanella. *FEMS Microbiology Ecology* 35 (3): 223–230.

Kato, C., Inoue, A., and Horikoshi, K. (1996). Isolating and characterizing deep-sea marine microorganisms. *Trends in Biotechnology* 14 (1): 6–12.

Kausar, F., Bagheri, A.R., Rasheed, T. et al. (2022). Nanomaterials for removal of heavy metals from wastewater. *Nano-Biosorbents for Decontamination of Water, Air, and Soil Pollution* 135–161.

Khan, A., Malik, S., Ali, N. et al. (2023). Metal-organic frameworks for reduction of heavy metals. In: *Emerging Techniques for Treatment of Toxic Metals from Wastewater* (ed. A. Ahmad, R. Kumar, and M. Jawaid), 457–474. Elsevier.

Khemili-Talbi, S., Kebbouche-Gana, S., Akmoussi-Toumi, S. et al. (2015). Isolation of an extremely halophilic arhaeon *Natrialba* sp. C21 able to degrade aromatic compounds and to produce stable biosurfactant at high salinity. *Extremophiles* 19 (6): 1109–1120.

Kiadehi, M.S.H., Amoozegar, M.A., Asad, S., and Siroosi, M. (2018). Exploring the potential of halophilic archaea for the decolorization of azo dyes. *Water Science and Technology* 77 (6): 1602–1611.

Kitjanukit, S., Sasaki, K., and Okibe, N. (2019). Production of highly catalytic, archaeal Pd (0) bionanoparticles using Sulfolobus tokodaii. *Extremophiles* 23: 549–556.

Kolenc, R., Inniss, W., Glick, B. et al. (1988). Transfer and expression of mesophilic plasmid-mediated degradative capacity in a psychrotrophic bacterium. *Applied and Environmental Microbiology* 54 (3): 638–641.

Konings, W.N., Albers, S.-V., Koning, S., and Driessen, A.J. (2002). The cell membrane plays a crucial role in survival of bacteria and archaea in extreme environments. *Antonie Van Leeuwenhoek* 81 (1): 61–72.

Krulwich, T.A. (1995). Alkaliphiles: 'basic' molecular problems of pH tolerance and bioenergetics. *Molecular Microbiology* 15 (3): 403–410.

Kuldeep, C., Singhal, H., Khare, A.K. et al. (2011). Randomized comparison of topical betamethasone valerate foam, intralesional triamcinolone acetonide and tacrolimus ointment in management of localized *Alopecia areata*. *International Journal of Trichology* 3 (1): 20.

Kumar, S., Dangi, A.K., Shukla, P. et al. (2019). Thermozymes: adaptive strategies and tools for their biotechnological applications. *Bioresource Technology* 278: 372–382.

Kumar, A., Singh, A.K., Bilal, M. et al. (2022). Paper and pulp mill wastewater: characterization, microbial-mediated degradation, and challenges. In: *Nanotechnology in Paper and Wood Engineering* (ed. R. Bhat, A. Kumar, T.A. Nguyen, and S. Swati), 371–387. Elsevier.

Kumari, A., Kishor, N., and Guptasarma, P. (2018). Characterization of a mildly alkalophilic and thermostable recombinant *Thermus thermophilus* laccase with applications in decolourization of dyes. *Biotechnology Letters* 40 (2): 285–295.

Lefebvre, O., Vasudevan, N., Torrijos, M. et al. (2005). Halophilic biological treatment of tannery soak liquor in a sequencing batch reactor. *Water Research* 39 (8): 1471–1480.

Li, J., Li, Q., Ma, X. et al. (2016). Biosynthesis of gold nanoparticles by the extreme bacterium *Deinococcus radiodurans* and an evaluation of their antibacterial properties. *International Journal of Nanomedicine* 11: 5931.

Liaquat, H., Imran, M., Latif, S. et al. (2022). Citric acid-capped NiWO4/Bi2S3 and rGO-doped NiWO4/Bi2S3 nanoarchitectures for photocatalytic decontamination of emerging pollutants from the aqueous environment. *Environmental Research* 212: 113276.

Liu, X.H., Zheng, H., Zhong, L. et al. (2011). Anisotropic swelling and fracture of silicon nanowires during lithiation. *Nano Letters* 11 (8): 3312–3318.

Liu, W., Sutton, N.B., Rijnaarts, H.H., and Langenhoff, A.A. (2016). Pharmaceutical removal from water with iron-or manganese-based technologies: a review. *Critical Reviews in Environmental Science and Technology* 46 (19-20): 1584–1621.

Liu, Z., Kim, M.C., Wang, L. et al. (2017). *Deinococcus taklimakanensis* sp. nov., isolated from desert soil. *International Journal of Systematic and Evolutionary Microbiology* 67 (11): 4311–4316.

Liu, L., Bilal, M., Duan, X., and Iqbal, H.M. (2019). Mitigation of environmental pollution by genetically engineered bacteria – current challenges and future perspectives. *Science of the Total Environment* 667: 444–454.

Lofthus, S., Almås, I.K., Evans, P. et al. (2018). Biodegradation in seawater of PAH and alkylphenols from produced water of a North Sea platform. *Chemosphere* 206: 465–473.

Lončar, N., Colpa, D.I., and Fraaije, M.W. (2016). Exploring the biocatalytic potential of a DyP-type peroxidase by profiling the substrate acceptance of *Thermobifida fusca* DyP peroxidase. *Tetrahedron* 72 (46): 7276–7281.

Lusk, B.G. (2019). Thermophiles; or, the modern Prometheus: the importance of extreme microorganisms for understanding and applying extracellular electron transfer. *Frontiers in Microbiology* 10: 818.

Mainka, T., Weirathmüller, D., Herwig, C., and Pflügl, S. (2021). Potential applications of halophilic microorganisms for biological treatment of industrial process brines contaminated with aromatics. *Journal of Industrial Microbiology and Biotechnology* 48 (1-2): kuab 015.

Mao, X. and Guo, J. (2018). Functional protein research progression in extremophiles. *Chinese Bulletin of Life Sciences* 30 (01): 107–112.

Margesin, R. and Feller, G. (2010). Biotechnological applications of psychrophiles. *Environmental Technology* 31 (8–9): 835–844.

Master, E.R., Agar, N.Y., Gómez-Gil, L. et al. (2008). Biphenyl dioxygenase from an arctic isolate is not cold adapted. *Applied and Environmental Microbiology* 74 (12): 3908–3911.

Matsuno, T., Goto, T., Ogami, S. et al. (2018). Formation of proton motive force under low-aeration alkaline conditions in alkaliphilic bacteria. *Frontiers in Microbiology* 9: 2331.

Maynard, R.L. and Chilcott, R.P. (2009). Toxicology of chemical warfare agents. *General, Applied and Systems Toxicology*.

McMullan, G., Meehan, C., Conneely, A. et al. (2001). Microbial decolourisation and degradation of textile dyes. *Applied Microbiology and Biotechnology* 56 (1): 81–87.

Merino, N., Aronson, H.S., Bojanova, D.P. et al. (2019). Living at the extremes: extremophiles and the limits of life in a planetary context. *Frontiers in Microbiology* 10: 780.

Michels, M. and Bakker, E.P. (1985). Generation of a large, protonophore-sensitive proton motive force and pH difference in the acidophilic bacteria *Thermoplasma acidophilum* and *Bacillus acidocaldarius*. *Journal of Bacteriology* 161 (1): 231–237.

Milatovic, D., Gupta, R.C., Zaja-Milatovic, S., and Aschner, M. (2009). Excitotoxicity, oxidative stress, and neuronal injury. In: *Handbook of Toxicology of Chemical Warfare Agents* (ed. R.C. Gupta), 633–651. Elsevier.

Miyake, R., Kawamoto, J., Wei, Y.-L. et al. (2007). Construction of a low-temperature protein expression system using a cold-adapted bacterium, *Shewanella* sp. strain Ac10, as the host. *Applied and Environmental Microbiology* 73 (15): 4849–4856.

Mohamed, Y.M., Ghazy, M.A., Sayed, A. et al. (2013). Isolation and characterization of a heavy metal-resistant, thermophilic esterase from a Red Sea Brine Pool. *Scientific Reports* 3 (1): 1–8.

Mole, D.R., Blancher, C., Copley, R.R. et al. (2009). Genome-wide association of hypoxia-inducible factor (HIF)-1α and HIF-2α DNA binding with expression profiling of hypoxia-inducible transcripts. *Journal of Biological Chemistry* 284 (25): 16767–16775.

Morita, R.Y., Horikoshi, K., and Grant, W.D. (1999). Extremes of biodiversity extremophiles: microbial life in extreme environments. *Current Science* 89: 78–90.

Moussa, T.A. and Khalil, N.M. (2022). Extremozymes from extremophilic microorganisms as sources of bioremediation. In: *Microbial Extremozymes* (ed. M. Kuddus), 135–146. Elsevier.

Muddemann, T., Haupt, D., Sievers, M., and Kunz, U. (2019). Electrochemical reactors for wastewater treatment. *ChemBioEng Reviews* 6 (5): 142–156.

Mushtaq, S., Yun, S.-J., Yang, J.E. et al. (2017). Efficient and selective removal of radioactive iodine anions using engineered nanocomposite membranes. *Environmental Science: Nano* 4 (11): 2157–2163.

Okibe, N., Maki, M., Nakayama, D., and Sasaki, K. (2016). Microbial recovery of vanadium by the acidophilic bacterium, *Acidocella aromatica*. *Biotechnology Letters* 38 (9): 1475–1481.

Okumura, T., Kawagucci, S., Saito, Y. et al. (2016). Hydrogen and carbon isotope systematics in hydrogenotrophic methanogenesis under H2-limited and H2-enriched conditions: implications for the origin of methane and its isotopic diagnosis. *Progress in Earth and Planetary Science* 3 (1): 1–19.

Oren, A., Gurevich, P., Azachi, M., and Henis, Y. (1992). Microbial degradation of pollutants at high salt concentrations. *Biodegradation* 3 (2): 387–398.

Özdemir, S., Kilinc, E., Poli, A. et al. (2012). Cd, Cu, Ni, Mn and Zn resistance and bioaccumulation by thermophilic bacteria, *Geobacillus toebii* subsp. decanicus and *Geobacillus thermoleovorans* subsp. *stromboliensis*. *World Journal of Microbiology and Biotechnology* 28 (1): 155–163.

Özdemir, S., Kılınç, E., Poli, A., and Nicolaus, B. (2013). Biosorption of heavy metals ($Cd2^+$, $Cu2^+$, $Co2^+$, and $Mn2^+$) by thermophilic bacteria, *Geobacillus thermantarcticus* and *Anoxybacillus amylolyticus*: equilibrium and kinetic studies. *Bioremediation Journal* 17 (2): 86–96.

Panak, P., & Nitsche, H. (2001). Interaction of aerobic soil bacteria and plutonium (VI). *Radiochimica Acta*,. 89 (8), 499–504. doi:https://doi.org/10.1524/ract.2001.89.8.499

Parales, R.E. and Resnick, S.M. (2004). Aromatic hydrocarbon dioxygenases. In: *Biodegradation and Bioremediation* (ed. A. Singh and O.P. Ward), 175–195. Springer.

Park, M., Song, J., Nam, G.G. et al. (2018). *Deinococcus lacus* sp. nov., a gamma radiation-resistant bacterium isolated from an artificial freshwater pond. *International Journal of Systematic and Evolutionary Microbiology* 68 (4): 1372–1377.

Paudyn, K., Rutter, A., Rowe, R.K., and Poland, J.S. (2008). Remediation of hydrocarbon contaminated soils in the Canadian Arctic by landfarming. *Cold Regions Science and Technology* 53 (1): 102–114.

Pereira, L., Mehboob, F., Stams, A.J. et al. (2015). Metallic nanoparticles: microbial synthesis and unique properties for biotechnological applications, bioavailability and biotransformation. *Critical Reviews in Biotechnology* 35 (1): 114–128.

Pereira, S.B., Sousa, A., Santos, M. et al. (2019). Strategies to obtain designer polymers based on cyanobacterial extracellular polymeric substances (EPS). *International Journal of Molecular Sciences* 20 (22): 5693.

Poirier, L., Brun, L., Jacquet, P. et al. (2017). Enzymatic degradation of organophosphorus insecticides decreases toxicity in planarians and enhances survival. *Scientific Reports* 7 (1): 1–11.

Prakash, D., Gabani, P., Chandel, A.K. et al. (2013). Bioremediation: a genuine technology to remediate radionuclides from the environment. *Microbial Biotechnology* 6 (4): 349–360.

Preiss, L., Hicks, D.B., Suzuki, S. et al. (2015). Alkaliphilic bacteria with impact on industrial applications, concepts of early life forms, and bioenergetics of ATP synthesis. *Frontiers in Bioengineering and Biotechnology* 3: 75.

Rafeeq, H., Qamar, S.A., Nguyen, T.A. et al. (2022). Microbial degradation of environmental pollutants. In: *Biodegradation and Biodeterioration at the Nanoscale* (ed. H.M.N. Iqbal, M. Bilal, T.A. Nguyen, and G. Yasin), 509–528. Elsevier.

Rafiq, A., Khan, S.A., Akbar, A. et al. (2018). Isolation and identification of antibiotic producing microorganisms from soil. *International Journal of Pharmaceutical Sciences and Research* 9 (3): 1002–1011.

Rajpurohit, Y.S., Bihani, S.C., Waldor, M.K., and Misra, H.S. (2016). Phosphorylation of *Deinococcus radiodurans* RecA regulates its activity and may contribute to radioresistance. *Journal of Biological Chemistry* 291 (32): 16672–16685.

Rambabu, K., Banat, F., Pham, Q.M. et al. (2020). Biological remediation of acid mine drainage: review of past trends and current outlook. *Environmental Science and Ecotechnology* 2: 100024.

Rasuk, M.C., Ferrer, G., Moreno, J. et al. (2016). The diversity of microbial extremophiles. *Molecular Diversity of Environmental Prokaryotes* 1: 87–126.

Rizwan, K., & Bilal, M. (2022) Environmental sensing and detection of toxic heavy metals by metal organic frameworks-based electrochemical sensors. In Chowdhary, P., Kumar, V., Kumar, S., Vishvas, H., (Eds) *Environmental Management Technologies* (pp. 215–227). CRC Press.

Rizwan, K., Bilal, M., Slimani, Y. et al. (2022). Hydrogen-based sono-hybrid catalytic degradation and mitigation of industrially-originated dye-based pollutants. *International Journal of Hydrogen Energy* 48: 6597–6612.

Röling, W., Ortega-Lucach, S., Larter, S., and Head, I. (2006). Acidophilic microbial communities associated with a natural, biodegraded hydrocarbon seepage. *Journal of Applied Microbiology* 101 (2): 290–299.

Rothschild, L.J. and Mancinelli, R.L. (2001). Life in extreme environments. *Nature* 409 (6823): 1092–1101.

Saavedra, A., Aguirre, P., and Gentina, J.C. (2020). Biooxidation of iron by *Acidithiobacillus ferrooxidans* in the presence of D-galactose: understanding its influence on the production of EPS and cell tolerance to high concentrations of iron. *Frontiers in Microbiology* 11: 759.

Sadegh, H., Ali, G.A., Gupta, V.K. et al. (2017). The role of nanomaterials as effective adsorbents and their applications in wastewater treatment. *Journal of Nanostructure in Chemistry* 7 (1): 1–14.

Saeed, M.U., Hussain, N., Sumrin, A. et al. (2021). Microbial bioremediation strategies with wastewater treatment potentialities – a review. *Science of the Total Environment* 818: 151754.

Salvador-Castell, M., Tourte, M., and Oger, P.M. (2019). In search for the membrane regulators of Archaea. *International Journal of Molecular Sciences* 20 (18): 4434.

Salwan, R. and Sharma, V. (2020). Molecular and biotechnological aspects of secondary metabolites in actinobacteria. *Microbiological Research* 231: 126374.

Santos, C.A., Vieira, A.M., Fernandes, H.L. et al. (2001). Optimisation of the biological treatment of hypersaline wastewater from *Dunaliella salina* carotenogenesis. *Journal of Chemical Technology & Biotechnology: International Research in Process, Environmental & Clean Technology* 76 (11): 1147–1153.

Sarkar, S., Banerjee, A., Halder, U. et al. (2017). Degradation of synthetic azo dyes of textile industry: a sustainable approach using microbial enzymes. *Water Conservation Science and Engineering* 2 (4): 121–131.

Sarkar, S., Ponce, N.T., Banerjee, A. et al. (2020). Green polymeric nanomaterials for the photocatalytic degradation of dyes: a review. *Environmental Chemistry Letters* 18 (5): 1569–1580.

Sarker, A., Kim, J.E., Islam, A.R.M. et al. (2021). Heavy metals contamination and associated health risks in food webs– a review focuses on food safety and environmental sustainability in Bangladesh. *Environmental Science and Pollution Research* 29 (3): 3230–3245.

Schelert, J., Dixit, V., Hoang, V. et al. (2004). Occurrence and characterization of mercury resistance in the hyperthermophilic archaeon *Sulfolobus solfataricus* by use of gene disruption. *Journal of Bacteriology* 186 (2): 427–437.

Serour, E. and Antranikian, G. (2002). Novel thermoactive glucoamylases from the thermoacidophilic Archaea *Thermoplasma acidophilum*, *Picrophilus torridus* and *Picrophilus oshimae*. *Antonie Van Leeuwenhoek* 81 (1): 73–83.

Sghaier, H., Ghedira, K., Benkahla, A., and Barkallah, I. (2008). Basal DNA repair machinery is subject to positive selection in ionizing-radiation-resistant bacteria. *BMC Genomics* 9 (1): 1–7.

Sharma, A. and Satyanarayana, T. (2012). Production of acid-stable and high-maltose-forming α-amylase of *Bacillus acidicola* by solid-state fermentation and immobilized cells and its applicability in baking. *Applied Biochemistry and Biotechnology* 168 (5): 1025–1034.

Sharma, A., Parashar, D., and Satyanarayana, T. (2016). Acidophilic microbes: biology and applications. In: *Biotechnology of Extremophiles*, 215–241. Springer.

Siani, L., Papa, R., Di Donato, A., and Sannia, G. (2006). Recombinant expression of Toluene o-Xylene Monooxygenase (ToMO) from *Pseudomonas stutzeri* OX1 in the marine Antarctic bacterium *Pseudoalteromonas haloplanktis* TAC125. *Journal of Biotechnology* 126 (3): 334–341.

Singh, S.K., Singh, S.K., Tripathi, V.R. et al. (2011). Comparative one-factor-at-a-time, response surface (statistical) and bench-scale bioreactor level optimization of thermoalkaline protease production from a psychrotrophic *Pseudomonas putida* SKG-1 isolate. *Microbial Cell Factories* 10 (1): 1–13.

Sinha, A., Pant, K.K., and Khare, S.K. (2012). Studies on mercury bioremediation by alginate immobilized mercury tolerant *Bacillus cereus* cells. *International Biodeterioration & Biodegradation* 71: 1–8.

Slonczewski, J.L., Fujisawa, M., Dopson, M., and Krulwich, T.A. (2009). Cytoplasmic pH measurement and homeostasis in bacteria and archaea. *Advances in Microbial Physiology* 55: 1–317.

Sorokin, D.Y., Berben, T., Melton, E.D. et al. (2014). Microbial diversity and biogeochemical cycling in soda lakes. *Extremophiles* 18 (5): 791–809.

Srinivasan, S., Lim, S., Lim, J.-H. et al. (2017). *Deinococcus rubrus* sp. Nov., a bacterium isolated from Antarctic coastal sea water. *Journal of Microbiology and Biotechnology* 27 (3): 535–541.

Stetter, K.O. (1998). Hyperthermophiles and their possible role as ancestors of modern life. *The Molecular Origins of Life* 315–335.

Stetter, K.O., Fiala, G., Huber, R. et al. (1990). Hyperthermophilic microorganisms. *FEMS Microbiology Reviews* 75: 117–124.

Tanaka, M., Earl, A.M., Howell, H.A. et al. (2004). Analysis of *Deinococcus radiodurans*'s transcriptional response to ionizing radiation and desiccation reveals novel proteins that contribute to extreme radioresistance. *Genetics* 168 (1): 21–33.

Tanner, K., Molina-Menor, E., Latorre-Pérez, A. et al. (2020). Extremophilic microbial communities on photovoltaic panel surfaces: a two-year study. *Microbial Biotechnology* 13 (6): 1819–1830.

Thomassin-Lacroix, E., Eriksson, M., Reimer, K., and Mohn, W. (2002). Biostimulation and bioaugmentation for on-site treatment of weathered diesel fuel in Arctic soil. *Applied Microbiology and Biotechnology* 59 (4): 551–556.

Tin, T., Fleming, Z.L., Hughes, K.A. et al. (2009). Impacts of local human activities on the Antarctic environment. *Antarctic Science* 21 (1): 3–33.

Tkavc, R., Matrosova, V.Y., Grichenko, O.E. et al. (2018). Prospects for fungal bioremediation of acidic radioactive waste sites: characterization and genome sequence of *Rhodotorula taiwanensis* MD1149. *Frontiers in Microbiology* 8: 2528.

Vieille, C. and Zeikus, J.G. (1996). Thermozymes: identifying molecular determinants of protein structural and functional stability. *Trends in Biotechnology* 14 (6): 183–190.

Waigi, M.G., Sun, K., and Gao, Y. (2017). Sphingomonads in microbe-assisted phytoremediation: tackling soil pollution. *Trends in Biotechnology* 35 (9): 883–899.

Watling, H.R., Shiers, D.W., and Collinson, D.M. (2015). Extremophiles in mineral sulphide heaps: some bacterial responses to variable temperature, acidity and solution composition. *Microorganisms* 3 (3): 364–390.

Weber, M.H. and Marahiel, M.A. (2003). Bacterial cold shock responses. *Science Progress* 86 (1–2): 9–75.

Wernick, D.G., Pontrelli, S.P., Pollock, A.W., and Liao, J.C. (2016). Sustainable biorefining in wastewater by engineered extreme alkaliphile *Bacillus marmarensis*. *Scientific Reports* 6 (1): 1–10.

Wiegel, J. and Canganella, F. (2011). Extremophiles: from abyssal to terrestrial ecosystems and possibly. *Naturwissenschaften* 98: 253–279.

Xia, W., Dong, H., Zheng, C. et al. (2015). Hydrocarbon degradation by a newly isolated thermophilic *Anoxybacillus* sp. with bioemulsifier production and new alkB genes. *RSC Advances* 5 (124): 102367–102377.

Xin, M. and Zhou, P. (1998). Advance of research for microbial life in low temperature environments. *Wei Sheng wu xue bao=Acta Microbiologica Sinica* 38 (5): 400–403.

Yadav, A.N. (2021). Beneficial plant-microbe interactions for agricultural sustainability. *Journal of Applied Biology and Biotechnology* 9 (1): i–v.

Yang, H.-C. and Rosen, B.P. (2016). New mechanisms of bacterial arsenic resistance. *Biomedical Journal* 39 (1): 5–13.

Yang, D., Hassan, Q.U., Chen, Q.W. et al. (2022). Development of novel K0. 8Ni0. 4Ti1. 6O4 nano bamboo leaves, microstructural characterization, double absorption, and photocatalytic removal of organic pollutant. *Environmental Research* 211: 113118.

Yemashova, N.A., Murygina, V.P., Zhukov, D.V. et al. (2007). Biodeterioration of crude oil and oil derived products: a review. *Reviews in Environmental Science and Bio/Technology* 6 (4): 315–337.

Yergeau, E., Arbour, M., Brousseau, R. et al. (2009). Microarray and real-time PCR analyses of the responses of high-arctic soil bacteria to hydrocarbon pollution and bioremediation treatments. *Applied and Environmental Microbiology* 75 (19): 6258–6267.

Zahradka, K., Slade, D., Bailone, A. et al. (2006). Reassembly of shattered chromosomes in *Deinococcus radiodurans*. *Nature* 443 (7111): 569–573.

Zeng, X., Xiao, X., Wang, P., and Wang, R. (2004). Screening and characterization of psychrotrophic, lipolytic bacteria from deep-sea sediments. *Journal of Microbiology and Biotechnology* 14 (5): 952–958.

Zhang, C., Wu, D., and Ren, H. (2020). Bioremediation of oil contaminated soil using agricultural wastes via microbial consortium. *Scientific Reports* 10 (1): 1–8.

Zhang, S., Khan, A., Ali, N., et al. (2022). Designing, characterization, and evaluation of chitosan-zinc selenide nanoparticles-based photocatalytic materials for visible-light-induced degradation of tartrazine and sunset yellow dye pollutants. *Environmental Research*, 213, 113722.

Zhao, Q., Bilal, M., Yue, S. et al. (2017a). Identification of biphenyl 2,3-dioxygenase and its catabolic role for phenazine degradation in *Sphingobium yanoikuyae* B1. *Journal of Environmental Management* 204: 494–501.

Zhao, Q., Yue, S., Bilal, M. et al. (2017b). Comparative genomic analysis of 26 *Sphingomonas* and *Sphingobium* strains: dissemination of bioremediation capabilities, biodegradation potential and horizontal gene transfer. *Science of the Total Environment* 609: 1238–1247.

Zhao, B., Al Rasheed, H., Ali, I., and Hu, S. (2021). Efficient enzymatic saccharification of alkaline and ionic liquid-pretreated bamboo by highly active extremozymes produced by the co-culture of two halophilic fungi. *Bioresource Technology* 319: 124115.

Zhuang, W.-Q., Fitts, J.P., Ajo-Franklin, C.M. et al. (2015). Recovery of critical metals using biometallurgy. *Current Opinion in Biotechnology* 33: 327–335.

# 7

# Bacterial/Fungal Inoculants: Application as Bio Stimulants

*V. Mamtha, Swati, K. Sowmiya, and Haralakal Keerthi Kumari*

*Department of Agricultural Microbiology, University of Agricultural Sciences, Bangalore, India*

## 7.1 Introduction

Biostimulants are substances that encourage and control plant development on an organic basis. They nourish plants and increase crop growth and yield in an environmentally friendly manner. Kauffman et al. defined the term "biostimulant" as substances/material's apart from chemical fertilizer's that encourage the growth of crops, which, when supplemented in low quantities. These materials are biological and organic in nature, which consist of hormones that are found in plants, amino acids, molecules of polysaccharides, compounds with humus, and plants that can uptake readily. The plant may receive nutrients from these organic materials like biostimulants in addition to being stimulated metabolically and having biotic and abiotic stressors reduced.

There are several different ways that biostimulants work to increase crop yields, such as by boosting essential soil-enzyme activities, stimulating soil microbial activity, or producing phytohormones. Thirty nations attended the first global congress regarding employing these biological stimulants in agricultural practices, which happened in Strasburg, France, in November 2012. This congress's primary goal was to bring together experts from academia, business, and regulatory organizations who are studying the characteristics of biostimulants. As a result, these biological compounds' usage started to become profitable (Kumari et al. 2022b).

Microbial inoculants are preparations of potentially advantageous microflora that have a positive impact on the biome of soil in an environmentally sustainable way (Babalola and Glick 2012; Puranik et al. 2023; Singh et al. 2022a). The diversity of agriculturally significant microorganisms found in soil impart an advantageous impact on soil properties and crops by supplying nutrients and also shielding the crop arising different natural enemies and illnesses.

The two main categories of microbial inoculants are biopesticides and biofertilizers, respectively, while those that act as biofertilizers are classified as biostimulants (Bashan 1998; Calvo et al. 2014; Singh et al. 2022b). They are also known as bioinoculants, and whenever it is sprayed to plants/crops, seeds, or soil, they can increase the formation of roots, biomass and plant matter, provide nutrients and also improve the potential to absorb nutrients and minerals. Bioinoculants include lifeforms in general, particularly microbes, and they encourage the growth of plants through an array of strategies (Kloepper et al. 1989).

According to Vessey (2003) and Dodd and Ruiz-Lozano (2012), the majority of microorganisms that operate as biostimulants are Arbuscular mycorrhizal fungus (AMF) and Free-living bacteria. Free-living bacteria isolated from the plant's root system called plant growth promoting bacteria (PGPB) and plant growth promoting rhizobacteria (PGPR) can operate as biofertilizers while inciting plant growth (Gond et al. 2021; Gupta et al. 2020; Kloepper et al. 1989; Singh et al. 2020). Dalmastri et al. (1999) and Remans et al. (2008) found that there exist numerous aspects, for instance, the variety of plants, which contribute to the growth of microbial inoculants as biofertilizers.

Compatibility with various soil types, chemical fertilizers and environmental factors (Calvo et al. 2014). Exudates from the root system are plant extracellular secretions, which posses major contribution to the efficacy of inoculants containing microbial cells and function as a substrate for the production of compounds with biological activity (Khalid et al. 2004).

*Microbes Based Approaches for the Management of Hazardous Contaminants*, First Edition. Edited by Ajay Kumar, Livleen Shukla, Joginder Singh, and Luiz Fernando R. Ferreira.

Two categories of microbes employed as inoculants (bioinoculants) can be distinguished; these categories are addressed below:

1) **Plant growth promoting bacteria (PGPB):** Rhizospheric, rhizoplanic, phylospheric, and free-living bacteria that are beneficial roles are called PGPB plays an essential and crucial part in the developmental processes in plants (Kumar et al. 2016; Shridhar 2012). They are members of various genera like *Pseudomonas* sp., *Bacillus* sp., *Azotobacter* sp., *Azospirillum* sp., *Enterobacter* sp., etc. Bashan and de Bashan put forward numerous advantages of the growth and development of plants, among which are enhanced nutrient uptake and reduced abiotic and biotic stresses (Alavi et al. 2013; Lucy et al. 2004) tolerance, enhanced growth and production, in addition to plant pathogen control (Alam et al. 2011; Lucy et al. 2004). They come from multiple distinct genera, notably, *Azospirillum*, *Pseudomonas*, *Azotobacter*, and *Bacillus* species. Boosted nutrition in plants and reduced biotic and abiotic stresses are just a few of the favorable consequences of plant development that Bashan and de Bashan proposed.

   They are versatile and can be administered either directly to the plant and seed or mixed with a carrier material, such as peat, sawdust, compost, compost, or vermiculite, that creates an environment that, is conducive to their growth (Smith 1992; Van Veen et al. 1997). They provide plants nutrients while encouraging plant growth by means of an array of processes and the processes encompass:

   i) Biological nitrogen fixation (BNF)
   ii) Production of iron-chelating compounds
   iii) Phytohormones production
   iv) Solubilization of inorganic Phosphorous (P)

This is thoroughly examined in the subsequent sections

### 7.1.1 Biological Nitrogen Fixation (BNF)

All organisms that exist, including plants, require nitrogen for enhancement and growth. It's a crucial component of proteins, amino acids, nucleic acids, and energy currency (Guanosine-5′-triphosphate (GTP), Adenosine diphosphate (ADP), and Adenosine triphosphate (ATP). As a result, it is referred to as a cell's building block. Nitrogen makes up approximately 78% of the atmosphere; plants and animals are unable to absorb it as a result of its complicated structure, in particular the triple connection that exists between two nitrogen atoms. It exists in the form of gas (Ahemad and Kibret 2014). By using the nitrogenase enzyme system, PGPB can transform nitrogen from its gaseous state into a form that can be utilized, i.e., ammonia and make it available to plants. Microbial agents that have the ability to fix nitrogen mainly comprise a variety of genera, such as *Azotobacter* spp. (Mrkovacki and Milic 2001), *Bacillus polymyxa* (Omar et al. 1996), *Gluconoacebacter diazotrophicus* (Boddey et al. 2001), and *Burkholderia* spp (Divan Baldani et al. 2000).

### 7.1.2 Production of an Iron Chelating Compound

Iron(Fe) is the 4th most commonly encountered micronutrient in the biosphere. Fe is mainly noticed as ferric ions or $Fe^{+3}$,which are not soluble and difficult for plants and microbes to access in an oxygen-rich environment (Gupta et al. 2015). In calcareous soil, the alkaline conditions render iron less soluble and thus making it unavailable for plants (Masalha et al. 2000). Siderophores, an assembly of low-molecular-weight substances/compounds that chelate iron, which is secreted by microbes, particularly PGPB (Ahemad and Kibret 2014). The aforementioned compounds make iron available to plants and contribute with the delivery of iron into bacterial cells. Siderophores additionally serve as a substance that controls biological activity, and they cause inadequate levels of iron in places close to plant roots by prohibiting diseases that affect plants from flourishing therein (Gupta et al. 2015).

When compared to sterile soil, nonsterile soils that are calcareous significantly enhance the uptake of Fe by many different plants, including sunflowers and maize. This emerges primarily as a consequence of soil microbes' function in facilitating plants uptake of iron as a result of soil microbes' role in assisting plants' uptake of iron (Masalha et al. 2000). Rice which was inoculated with the strain of *Pseudomonas*, is found to produce siderophores and this mainly contributed to an enormous increase in iron content in the rice crop. The greater concentrations of iron additionally enhanced the nutritional content of the grains of rice. Therefore, siderophore synthesis is an essential characteristic of PGPB and boosts the capacity of plants to absorb iron (Sharma et al. 2013).

### 7.1.3 Phytohormone Production

A variety of plant hormones or plant growth regulators are also known to be synthesized by microbial inoculants, including PGPB, and these have an impact on the morphology of root systems in plants and their development (Bhattacharyya and Jha 2012; Dodd et al. 2010;.Idris et al. 2007.). Auxins, abscisic acid, gibberellins, cytokinins, and ethylene make up this group of plant hormones (Hayat et al. 2010). These hormones are capable of regulating the growing length of roots, the emergence of hairs in the root system and the initial development of new roots, along with other physiological activities (Calvo et al. 2014).

Additionally, they have extensively stated that microbial inoculants produce the naturally occurring auxin, Indole-3-acetic acid (IAA). Indole-3-acetic acid (IAA) features a favorable effect on a variety of plant processes, including the emergence of roots, the growth and development of tissues from the vascular system, the expressions of a variety of genes present in plants and the modulation of climatic response (Calvo et al. 2014).

Additionally, cytokinins are crucial for plants, as they delay the aging of leaves and encourage the division of mitotic cells in shoots and roots (De Pascale et al. 2017).

Gibberellin hormones influence the growth and development of flowers along with fruits, the germination of seeds, and the dormancy of vegetative organs (Ali et al. 2009). Furthermore, abscisic acid also plays a vital part in the manner that helps plants to establish and respond to environmental challenges such as elevated salt levels and drought (Miyakawa et al. 2013).

Aside from its well known function as a ripening. Hormones and ethylene have been implicated in a number of other processes, including seed germination, cell expansion and the aging of flowers and leaves.

Due to the fact that it is synthesized during both abiotic and biotic stresses, ethylene is also referred to as a stress hormone (Spaepen et al. 2009).

Since the generation of elevated levels of ethylene concentrations inevitably limits down growth of plants, an inhibiting impact on the development of roots has also been noticed. In order to overcome negative impacts, PGPBs produces an essential enzyme called L-aminocyclopropane-1-carboxylatedeaminase (ACC deaminase), which regulates the biosynthesis of ethylene and catalyzes the conversion of the intermediate precursor of ethylene, ACC (L-amino cyclopropane-L-carboxylic acid.), into -ketobutyrate and ammonia (Calvo et al. 2014).

Subsequently, PGPB generated a number of low molecular weight volatile organic compounds (VOCs), such as alcohols, ketones, aldehydes, and hydrocarbons, which typically possess a high vapor pressure and are released into the environment (Ortíz-Castro et al. 2009). Microbial volatile organic compounds (mVOCs) are a collective term for these substances (Kai et al. 2009). In addition, PGPB changes the configuration of roots and gives plants nourishment (Kumari et al. 2022b) (Figure 7.1).

### 7.1.4 Solubilization of Phosphate (P)

In terms of plant nutrition, the second-most significant nutrient is phosphorus after nitrogen. Only 01 mg/kg of phosphorus is present in the forms of dihydrogen phosphate ($H_2O_4P^-$) and hydrogen phosphate ($HO_4P^{-2}$ (Halpern et al. 2015). Considering agricultural soils, where the sum of the phosphorus concentrations typically ranges between 400 and 1200 mg/kg. Most of the organic and insoluble inorganic forms of P found in soil are in these two categories (Halpern et al. 2015).

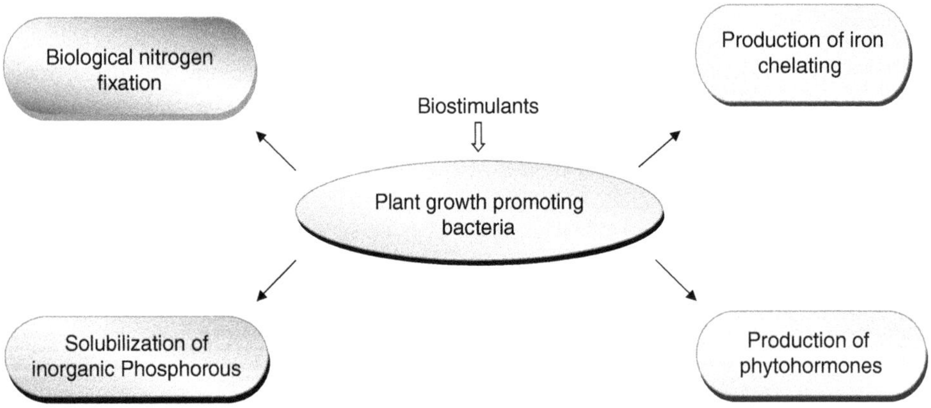


**Figure 7.1** Role of biostimulants as plant growth promoting bacteria.

Approximately between 20% and 50% of the amount of P is found in the form of inorganic compounds in soil (Richardson 2001) and is usually accessible in the form of $PO_4-$ ions, which are only partially soluble because of positively charged components in soil and are adhered to them, which are also precipitated with certain metals, that is Aluminum (Al), Iron (Fe), and Calcium (Ca). Insoluble organic P, which is present in the soil, is also additionally available in the form of inositol phosphates, phosphate esters and uncharacterized large organic molecules and contains 50–80% of total P (Richardson 2001).

In agricultural systems, the inadequate accessibility of P in soils is a serious problem (Calvo et al. 2014). During the process of solubilizing P, PGPB can improve the nutritional requirements of plants (Halpern et al. 2015). Inorganic and organic P, which is insoluble, is solubilized by bacteria by employing an array of strategies. The two most common methods for bacteria to solubilize P are they can be generated by producing organic acids or developing the enzymes phosphatases and phytases (Goldstein 1995; Rodríguez et al. 2006). By virtue of the existence of hydroxyl and carboxyl groups, which chelate the cations attached to phosphates and convert them into more soluble forms, insoluble inorganic P is converted into a more soluble form as the outcome of the creation of organic acids.

By generating organic acids, the pH of the soil surrounding it is likewise lowered. The release of phosphate ions enhances the solubility of P and decreases the pH (Rodríguez and Fraga 1999). According to Richardson et al. (2009), PGPB is also recognized for solubilizing organic P via the production of phosphatase and phytase enzymes, transforming it into soluble forms that plants may readily absorb from the soil.

The major rhizospheric bacteria, as well as other microbes that can solubilize P, originate from the genera, *Micrococcus*, *Burkholderia*, *Erwinia*, *Agrobacterium*, *Bacillus*, *Pseudomonas*, *Rhizobium*, *Achromobacter*, *Streptomyces*, etc. Most phosphate-solubilizing bacteria may not function well in soil scenarios despite the fact that they perform well in lab settings (Richardson 2001). In order to accomplish and characterize field-compatible phosphate solubilizing highly effective bacteria, intensive field studies are now being conducted.

## 7.2 Arbuscular Mycorrhizal Fungi (AMF)

There are two additional methods fungi, which are present in the soil, are connected to plant roots by means of parasitism and mutualistic symbiosis. While both organisms coexist and forge positive interactions in mutualistic symbiosis, parasitism interactions result in one partner's gains to the detriment of the other (Kashyap et al. 2023; Kumari et al. 2022a; Remy et al. 1994). The term "mycorrhizal fungi" refers to an advantageous and eclectic group of fungi that form symbiotic connections with over 90% of plant species (Du Jardin 2015).

Different groups of mycorrhizal fungi exist, but arbuscular mycorrhizal fungi are a widespread type of endomycorrhiza and are frequently associated with horticultural and crop plants. Arbuscular mycorrhizal fungus (phylum Glomeromycota) were the first to develop and were connected to a variety of plant species early in the history of terrestrial plants (Bonfante and Genre 2010; Schüßler et al. 2001).

Arbuscules constitute a unique kind of branched structure that typically develops when a Glomeromycota species of fungus penetrates the root cortical cells of plants (Behie and Bidochka 2014; Remy et al. 1994). These mycorrhizal fungi, which have been demonstrated to deliver improved nutrients (macro and micronutrients), water intake and also to assist plants in withstanding biotic and abiotic stress, are now of significant importance in the application for environmentally friendly agriculture.

According to a current study, there exists not only an interrelationship among plants and fungi developed by the hyphal network, but additionally there is also a connecting individual species of plants within a community and contributing in signaling among interplant species (Johnson and Gilbert 2015; Simard et al. 2012).

In a variety of ways, AMF contributes a significant part in promoting the growth of plants.

i) AMF enhances water uptake and boosts the surface area of the plant root system, which further enables water uptake by roots
ii) It makes nutrients available, especially phosphorus(P), when there is a nutrient shortage
iii) Renovations at the root architecture
iv) Associated with physiological and metabolic mechanisms in plants, which impact the manner in which they react to stress caused by oxidative stress and
v) Stress induces the production of abscisic acid (ABA) hormones that regulate plants.

In multiple species of plants, notably wheat, onions, soybean, lettuce and corn, Begum, Auge, and Brundertt and discovered that a symbiotic relationship with mycorrhizal had some alleviating advantages on the physiological effects of drought. This develops as a result of more effective endurance to drought spurred due to increased development of roots. When AMF invades plants, good water efficiency and rapid growth and development are maintained (Augé 2001).

Additionally, the water potential of plants may also be affected by the changes in the structure of soil by the production of glomalin, which is a soil-binding material such as an insoluble glue-like substance in nature by the hyphae of AMF. Taking advantage of AMF in order to protect the roots of plants from the damaging effects of toxic metals (Leyval et al. 1997).

The administration of co-inoculation with AMF and PGPR enhances a plant's capacity to cope with drought. As an instance, when the lettuce plants had been co-inoculated alongside AMF *Bacillus* spp., *Glomus intraradices*, *Glomus mosseae* and (PGPR), more accurate growth of the plant, conductance of stomatal cells, and the effectiveness of water consumption as well as more substantial rates of photosynthetic activity have been observed. In contrast to the use of merely a single organism, co-inoculation with *Bacillus* spp. and AMF achieved more favorable outcomes. This developed as the outcome of PGPR, or *Bacillus* spp., facilitating the proliferation and colonization of AMF (Vivas et al. 2003). However, there exist certain constraints regarding the usage of AMFs as biostimulants that might be related to them being biotrophic in nature. They are not easy to propagate on a considerable amount, and scientists had no luck in unraveling the variables that influence host particular characteristics along with extra assistance on dynamics of the population of the mycorrhizal fungus in farming systems.

The capability of some fungi to establish colonies in root systems and to deliver vital nutrients to their hosts, such as *Sebacinales* and *Trichoderma* spp. which are not mycorrhizal species, has additionally been observed. However, the mechanisms beneath these capabilities are still not completely understood. These fungal organisms can, however, be employed as biological inoculants to increase the nutritional quality of plants. *Trichoderma* species are widely recognized for their effective biocontrol and biopesticide properties; however, they also have certain stimulating impacts on plants, that lead to improved absorption of nutrients, morphogenesis, and organ development, as well as higher tolerance to non-living factors. In light of their impacts on plants, these fungal endophytes are often considered to be simultaneously biostimulants and biopesticides (Pathak et al. 2022; Verma et al. 2021). Consequently, inoculants made up of microbial species, which include both fungal organisms and beneficial microorganisms, constitute a promising means of promoting environmentally friendly agriculture. In conjunction with improving the nutritional status of crops, they additionally assist plants in dealing with a variety of environmental challenges. They boost the system that supports agriculture without incurring any adverse consequences (Figure 7.2).

### 7.2.1 Microbial Inoculants as Pathogens or Parasites

Mycorrhizal fungi and rhizobia are two examples of frequent microbial taxa employed as inoculants. However, their advantages are typically context-dependent. Arbuscular mycorrhizal fungi (AMF), for instance, are able to function parasitically whenever phosphorus (P) is readily accessible, absorbing photosynthate from the plants they colonize and not contributing any nutrients in exchange. As a result, administering AMF as inoculants into regulated systems may actually slow down the development of plants. A particular instance was the application of commercially available mycorrhizal inoculums on coastal sage scrub rehabilitation sites, which significantly affected plant height. If native fungi that have adapted to the local plants are displaced by ruderal species, several detrimental impacts may result.

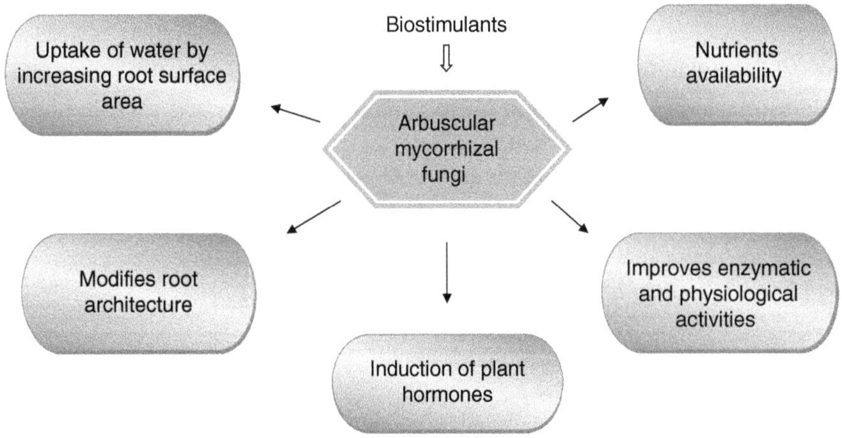


**Figure 7.2** Role of biostimulants as arbuscular mycorrhizal fungi (AMF).

Furthermore, microorganisms that are used as biofertilizers or bioprotectants for plants that promote plant growth have the potential to develop harmful lifestyles. Extensive numbers of fungi, for instance, are closely connected to diseases and can convert from mutually beneficial organisms to microorganisms that cause death, especially when the host environment changes. As an illustration, under a tropical palm tree with more intense light, the helpful endophyte of the fungus *Diplodiamutila* exhibited symptoms of an illness (Jack et al. 2021).

### 7.2.2 Other than Bacterial/Fungal Inoculants Algal Extracts also Play Important Role

Seaweed extracts (SWE) from algae have been used as biostimulants as plant growth-promoting factors. They are emerging commercial formulations as well as an approach that enhances the resistance to humidity, drought and heat. Extracts of algae are designed to encourage adaptation to stress through a wide range of routes (Figure 7.3). A total of 10% of marine production can be made up of the green, red, and brown macroalgae that are referred to as seaweeds. Since ancient times, macroalgae have been utilized as organic fertilizers, and they are still in use today.

Nearly every formulation is rendered from *Ascophyllum nodosum*, the brown algae that is produced and marketed by over 47 businesses for use in agriculture. Even while the impacts of seaweed extracts on boosting growth have been identified in an abundance of different species, little is truly understood about the processes behind these effects. It can be challenging to pinpoint precisely which elements of these chemicals are crucial because of their complexity and variability.

The composition of commercial SWE formulations, which are frequently proprietary, is heavily influenced by the extraction process. Indeed, a desirable initial step to more accurately hypothesize and/or show a cause-and-effect link of their effective method would be, to characterize the precise composition of the majority of popular algal-based commercial products.

Some of the more widely used techniques include mechanical disruption, pulverization, acid or alkali extractions, and so on. The majority of industrial products are made from the red macroalgae, *Lithothamnium calcareum* and the brown macroalgae *Durvillaea potatorum* and *Ascophyllum nodosum*. SWEs along with cold tolerance have recently been acknowledged to be significant considerations. The capability of SWEs to boost resistance to cold stress has become the subject of recent research. Particularly, extracts with elevated levels of zinc and manganese had the potential to elevate the resistance to cold in maize when a variety of extracts were evaluated to see if they could do so through improved ROS responses.

Providing plants with micronutrients that function as co-factors in anti-oxidative enzymes is likely what causes the protective effects in this situation. According to these findings, SWEs rich in micronutrients can be used to reduce the nutrient shortage stress brought on by cold and increase resistance to oxidative stress. Nutrient seed priming's value has been shown in previous research with maize seedlings under root chilling stress that received micronutrient supplements (Van Oosten et al. 2017).

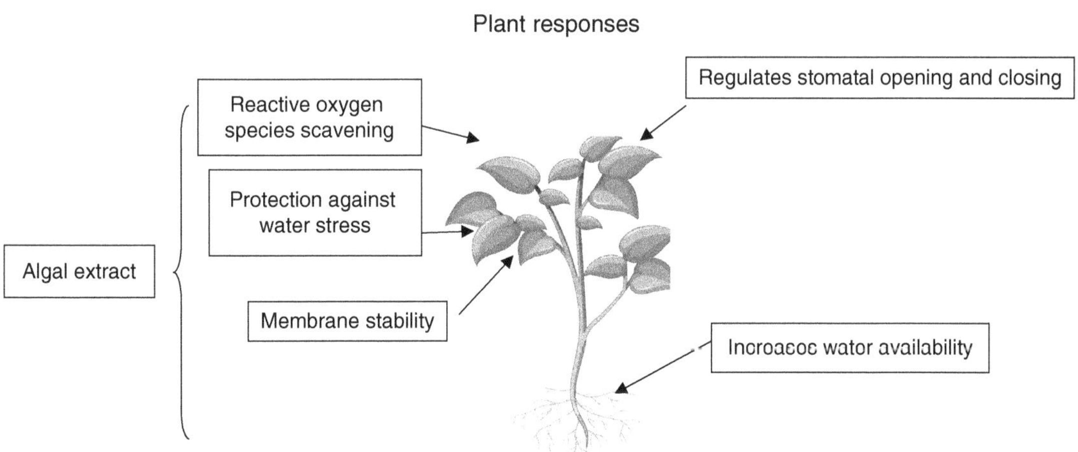


**Figure 7.3** Algal biostimulants influence plant growth.

### 7.2.3 Disruption of Ecosystem Services

Numerous ecosystem services, including the frequency of disease (regulation), the availability of clean water (provisioning), and cultural practices involving religion, are all impacted by microbes, either directly or indirectly. Although they can even emerge in severely disturbed agricultural systems, microbes-mediated ecosystem services are often thought to arise in pristine ecosystems. Many cases of potato scab (*Streptomyces scabies*) being suppressed as a result of prolonged potato monoculture have been documented. Antagonistic and competitive interactions between *Streptomyces* strains's present in the soil that leads to the disease-suppression of potato scab. Continuous potato cropping raises the number of saprotrophic *Streptomyces*, which creates an environment in the potato rhizosphere that is extremely competitive. As a result, *Streptomyces* strains engage in an arms race that intensifies, producing novel antibiotics in an effort to get the upper hand. This, in turn, suppresses disease in plants.

Potato disease resistance is however an instance of how infection-suppressive soils may be found in other farming practices, including barley, cotton and wheat. In addition, they can also spread widely within a system; for instance, the Pacific Northwest is home to over 800,000 ha of wheat that is affected by disease suppression to take all diseases. Given that it takes many years for disease-suppressive communities to emerge, it is risky to introduce microbial inoculum to geotechnical soils without first knowing local rhizosphere interactions. Monitoring the soil environment with this question in mind would be necessary to find evidence for it; however, there is evidence that administering plant growth-promoting rhizobacterial (PGPR) inoculants has the potential to modify the rhizosphere's bacterial population structure (Jack et al. 2021).

### 7.2.4 World Market for PGPR-Based Biostimulants

Biostimulants are becoming increasingly important in practices for environmentally friendly farming. Excessive and inappropriate use of products that are derived from chemicals on crops has led to instances of soil pollution and environmental hazards, which have emerged as a significant issue in industry in recent years.

The Dimensions of the international marketplace for biostimulants were evaluated at USD 1.74 billion in 2016 and is anticipated to grow at a CAGR of 10.2% from 2017 to 2025. Over the course of the projection period, the market is anticipated to be driven by a growing emphasis on productivity improvement as well as rapid soil degradation. The size of the worldwide biostimulants market was predicted to be USD 2.30 billion in 2019 and USD 2.53 billion in 2020.

Dunhamtrimmer.com (2018) projects that the worldwide marketplace for biostimulants is anticipated to grow at a yearly average acceleration in growth rate of 10.2% between 2017 and 2025 to reach USD 4.14 billion. Despite the fact that not every one of the biostimulants has biological roots, a plant's most important ally is bacteria in all environments. Additionally, a mere of 25% of biostimulants' commercial offerings have their basis on microorganisms, according to the literature that is currently available. Despite the fact that certain formulations include fungal connections; the majority of conceptualization is centered on PGPR (Hamid et al. 2021).

## 7.3 Conclusion

In farming and environmental management, bacterial and fungal inoculants have emerged as promising tools with their potential application as biostimulants featuring several advantages. Microorganisms that serve as biostimulants are Arbuscular mycorrhizal fungus (AMF) and Free-living bacteria; they promote plant growth, enhance nutrient uptake and defend against pathogens. Additionally, they play an important role in reducing the need for chemical fertilizers, soil health and promoting sustainable farming practices. Besides Bacterial/fungal inoculants, seaweed extracts (SWE) are a new commercial formulation that enhances plant resilience to acclimatization to stress and facilitates plant development. As our understanding of these biostimulants continues to grow, their role in strengthening agricultural productivity and environmental sustainability will likely become even more significant. Therefore, these microbial inoculants as biostimulants represent a valuable avenue for the future of agriculture and ecosystem management.

## References

Ahemad, M. and Kibret, M. (2014). Mechanisms and applications of plant growth promoting rhizobacteria: current perspective. *Journal of King Saud University, Science* 26 (1): 1–20.

Alam, M., Khaliq, A., Sattar, A. et al. (2011). Synergistic effect of arbuscular mycorrhizal fungi and *Bacillus subtilis* on the biomass and essential oil yield of rose-scented geranium (*Pelargonium graveolens*). *Archives of Agronomy and Soil Science* 57 (8): 889–898.

Alavi, P., Starcher, M.R., Zachow, C. et al. (2013). Root-microbe systems: the effect and mode of interaction of stress protecting agent (SPA) *Stenotrophomonas rhizophila* DSM14405T. *Frontiers in Plant Science* 4: 141.

Ali, B., Sabri, A.N., Ljung, K. et al. (2009). Auxin production by plant associated bacteria: impact on endogenous IAA content and growth of *Triticum aestivum* L. *Letters in Applied Microbiology* 48 (5): 542–547.

Augé, R.M. (2001). Water relations, drought and vesicular-arbuscular mycorrhizal symbiosis. *Mycorrhiza* 11 (1): 3–42.

Babalola, O.O. and Glick, B.R. (2012). The use of microbial inoculants in African agriculture: current practice and future prospects. *Journal of Food, Agriculture and Environment* 10 (3–4): 540–549.

Bashan, Y. (1998). Proposal for the division of plant growth-promoting rhizobacteria into two classifications: biocontrol-PGPB (plant-growth-promoting bacteria) and PGPB. *Soil Biology and Biochemistry* 30: 1225–1228.

Behie, S.W. and Bidochka, M.J. (2014). Nutrient transfer in plant–fungal symbioses. *Trends in Plant Science* 19 (11): 734–740.

Bhattacharyya, P.N. and Jha, D.K. (2012). Plant growth-promoting rhizobacteria (PGPR): emergence in agriculture. *World Journal of Microbiology and Biotechnology* 28: 1327–1350.

Boddey, R.M., Polidoro, J.C., Resende, A.S. et al. (2001). Use of the15N natural abundance technique for the quantification of the contribution of $N_2$ fixation to sugar cane and other grasses. *Functional Plant Biology* 28 (9): 889–895.

Bonfante, P. and Genre, A. (2010). Interactions in mycorrhizal symbiosis. *Nature Communications* 1: 1–11.

Calvo, P., Nelson, L., and Kloepper, J.W. (2014). Agricultural uses of plant biostimulants. *Plant and Soil* 383: 3–41.

Dalmastri, C., Chiarini, L., Cantale, C. et al. (1999). Soil type and maize cultivar affect the genetic diversity of maize root–associated *Burkholderia cepacia* populations. *Microbial Ecology* 38: 273–284.

De Pascale, S., Rouphael, Y., and Colla, G. (2017). Plant biostimulants: innovative tool for enhancing plant nutrition in organic farming. *European Journal of Horticultural Science* 82 (6): 277–285.

Divan Baldani, V.L., Baldani, J.I., and Döbereiner, J. (2000). Inoculation of rice plants with the endophytic diazotrophs *Herbaspirillum seropedicae* and *Burkholderia* spp. *Biology and Fertility of Soils* 30: 485–491.

Dodd, I.C. and Ruiz-Lozano, J.M. (2012). Microbial enhancement of crop resource use efficiency. *Current Opinion in Biotechnology* 23 (2): 236–242.

Dodd, I.C., Zinovkina, N.Y., Safronova, V.I. et al. (2010). Rhizobacterial mediation of plant hormone status. *Annals of Applied Biology* 157 (3): 361–379.

Du Jardin, P. (2015). Plant biostimulants: definition, concept, main categories and regulation. *Scientia Horticulturae* 196: 3–14.

Dunhamtrimmer.com. 2018. Available online: http://dunhamtrimmer.com/products-services (accessed on 11 January 2021).

Goldstein, A.H. (1995). Recent progress in understanding the molecular genetics and biochemistry of calcium phosphate solubilization by gram negative bacteria. *Biological Agriculture & Horticulture* 12 (2): 185–193.

Gond, D.P., Jha, S.S., Kumar, A. et al. (2021). Plant growth promoting bacteria and its role in green remediation. In: *Sustainable Environmental Clean-up* (ed. V.K. Mishra and A. Kumar), 149–163. Cambridge, MA: Elsevier https://doi.org/10.1016/B978-0-12-823828-8.00007-4.

Gupta, G., Parihar, S.S., Ahirwar, N.K. et al. (2015). Plant growth promoting rhizobacteria (PGPR): current and future prospects for development of sustainable agriculture. *Journal of Microbial and Biochemical Technology* 7 (2): 096–102.

Gupta, A., Singh, S.K., Singh, M.K. et al. (2020). Plant growth–promoting rhizobacteria and their functional role in salinity stress management. In: *Abatement Environ Pollutants* (ed. P. Singh, A. Kumar, and A. Borthakur), 151–160. Cambridge, MA: Elsevier https://doi.org/10.1016/B978-0-12-818095-2.00007-2.

Halpern, M., Bar-Tal, A., Ofek, M. et al. (2015). The use of biostimulants for enhancing nutrient uptake. *Advances in Agronomy* 130: 141–174.

Hamid, B., Zaman, M., Farooq, S. et al. (2021). Bacterial plant biostimulants: a sustainable way towards improving growth, productivity, and health of crops. *Sustainability* 13 (2856): 1–24. https://doi.org/10.3390/su13052856.

Hayat, R., Ali, S., Amara, U. et al. (2010). Soil beneficial bacteria and their role in plant growth promotion: a review. *Annals of Microbiology* 60: 579–598.

Idris, E.E., Iglesias, D.J., Talon, M. et al. (2007). Tryptophan-dependent production of indole-3-acetic acid (IAA) affects level of plant growth promotion by *Bacillus amyloliquefaciens* FZB42. *Molecular Plant-Microbe Interactions* 20 (6): 619–626.

Jack, C.N., Petipas, R.H., Cheeke, T.E. et al. (2021). Microbial inoculants: silver bullet or microbial Jurassic Park? *Trends in Microbiology* 29 (4): 299–308.

Johnson, D. and Gilbert, L. (2015). Interplant signalling through hyphal networks. *New Phytologist* 205 (4): 1448–1453.

Kai, M., Haustein, M., Molina, F. et al. (2009). Bacterial volatiles and their action potential. *Applied Microbiology and Biotechnology* 81: 1001–1012.

Kashyap, N., Singh, S.K., Yadav, N. et al. (2023). Biocontrol screening of endophytes: applications and limitations. *Plants* 12 (13): 2480. https://doi.org/10.3390/plants12132480.

Khalid, A., Arshad, M., and Zahir, Z.A. (2004). Screening plant growth-promoting rhizobacteria for improving growth and yield of wheat. *Journal of Applied Microbiology* 96 (3): 473–480.

Kloepper, J.W., Lifshitz, R., and Zablotowicz, R.M. (1989). Free-living bacterial inocula for enhancing crop productivity. *Trends in Biotechnology* 7 (2): 39–44.

Kumar, A., Singh, M., Singh, P.P. et al. (2016). Isolation of plant growth promoting rhizobacteria and their impact on growth and curcumin content in Curcuma longa L. *Biocatalysis and Agricultural Biotechnology* 8: 1–7.

Kumari, M., Qureshi, K.A., Jaremko, M. et al. (2022a). Deciphering the role of endophytic microbiome in postharvest diseases management of fruits: opportunity areas in commercial up-scale production. *Frontiers in Plant Science* 13: 4390. https://doi.org/10.3389/fpls.2022.1026575.

Kumari, M., Swarupa, P., Kesari, K.K. et al. (2022b). Microbial inoculants as plant biostimulants: a review on risk status. *Life* 13 (1): 12.

Leyval, C., Turnau, K., and Haselwandter, K. (1997). Effect of heavy metal pollution on mycorrhizal colonization and function: physiological, ecological and applied aspects. *Mycorrhiza* 7: 139–153.

Lucy, M., Reed, E., and Glick, B.R. (2004). Applications of free living plant growth-promoting rhizobacteria. *Antonie Van Leeuwenhoek* 86: 1–25.

Masalha, J., Kosegarten, H., Elmaci, Ö. et al. (2000). The central role of microbial activity for iron acquisition in maize and sunflower. *Biology and Fertility of Soils* 30: 433–439.

Miyakawa, T., Fujita, Y., Yamaguchi-Shinozaki, K. et al. (2013). Structure and function of abscisic acid receptors. *Trends in Plant Science* 18 (5): 259–266.

Mrkovacki, N. and Milic, V. (2001). Use of *Azotobacter chroococcum* as potentially useful in agricultural application. *Annals of Microbiology* 51 (2): 145–158.

Omar, M.N.A., Hamouda, A.M., and Mahrous, N.M. (1996). Evaluating the efficiency of inoculating some diazotrophs on yield and protein content of 3 wheat cultivars under graded levels of nitrogen fertilization. *Annals of Agricultural Science* 41: 579–590.

Ortíz-Castro, R., Contreras-Cornejo, H.A., Macías-Rodríguez, L. et al. (2009). The role of microbial signals in plant growth and development. *Plant Signaling & Behavior* 4 (8): 701–712.

Pathak, P., Rai, V.K., Can, H. et al. (2022). Plant-endophyte interaction during biotic stress management. *Plants* 11 (17): 2203. https://doi.org/10.3390/plants11172203.

Puranik, S., Shukla, L., Kundu, A. et al. (2023). Exploring potent fungal isolates from sanitary landfill soil for in vitro degradation of dibutyl phthalate. *Journal of Fungi* 9 (1): 125. https://doi.org/10.3390/jof9010125.

Remans, R., Beebe, S., Blair, M. et al. (2008). Physiological and genetic analysis of root responsiveness to auxin-producing plant growth-promoting bacteria in common bean (*Phaseolus vulgaris* L.). *Plant and Soil* 302: 149–161.

Remy, W., Taylor, T.N., Hass, H. et al. (1994). Four hundred-million-year-old vesicular arbuscular mycorrhizae. *Proceedings of the National Academy of Sciences* 91 (25): 11841–11843.

Richardson, A.E. (2001). Prospects for using soil microorganisms to improve the acquisition of phosphorus by plants. *Functional Plant Biology* 28 (9): 897–906.

Richardson, A.E., Barea, J.M., McNeill, A.M. et al. (2009). Acquisition of phosphorus and nitrogen in the rhizosphere and plant growth promotion by microorganisms. *Plant Soil* 321: 305–339.

Rodríguez, H. and Fraga, R. (1999). Phosphate solubilizing bacteria and their role in plant growth promotion. *Biotechnology Advances* 17 (4–5): 319–339.

Rodríguez, H., Fraga, R., Gonzalez, T. et al. (2006). Genetics of phosphate solubilization and its potential applications for improving plant growth-promoting bacteria. *Plant and Soil* 287: 15–21.

Schüßler, A., Schwarzott, D., and Walker, C. (2001). A new fungal phylum, the Glomeromycota: phylogeny and evolution. *Mycological Research* 105 (12): 1413–1421.

Sharma, A., Shankhdhar, D., and Shankhdhar, S.C. (2013). Enhancing grain iron content of rice by the application of plant growth promoting rhizobacteria. *Plant, Soil and Environment* 59 (2): 89–94.

Shridhar, B.S. (2012). Nitrogen fixing microorganisms. *International Journal of Microbiology* 3 (1): 46–52.

Simard, S.W., Beiler, K.J., Bingham, M.A. et al. (2012). Mycorrhizal networks: mechanisms, ecology and modelling. *Fungal Biology Reviews* 26 (1): 39–60.

Singh, D., Singh, S.K., Singh, V.K. et al. (2020). Plant growth–promoting bacteria and their role in environmental management. In: *Abatement Environ Pollutants* (ed. P. Singh, A. Kumar, and A. Borthakur), 161–175. Cambridge, MA: Elsevier https://doi.org/10.1016/B978-0-12-818095-2.00008-4.

Singh, M., Qureshi, K.A., Jaremko, M. et al. (2022a). Bioprospects of endophytic bacteria in plant growth promotion and Ag-nanoparticle biosynthesis. *Plants* 11 (14): 1787.

Singh, R., Pandey, K.D., Singh, M. et al. (2022b). Isolation and characterization of endophytes bacterial strains of *Momordica charantia* L. and their possible approach in stress management. *Microorganisms* 10 (2): 290.

Smith, R.S. (1992). Legume inoculant formulation and application. *Canadian Journal of Microbiology* 38 (6): 485–492.

Spaepen, S., Vanderleyden, J., and Okon, Y. (2009). Plant growth-promoting actions of rhizobacteria. *Advances in Botanical Research* 51: 283–320.

Van Oosten, M.J., Pepe, O., De Pascale, S. et al. (2017). The role of biostimulants and bioeffectors as alleviators of abiotic stress in crop plants. *Chemical and Biological Technologies in Agriculture* 4: 1–12.

Van Veen, J.A., van Overbeek, L.S., and van Elsas, J.D. (1997). Fate and activity of microorganisms introduced into soil. *Microbiology and Molecular Biology Reviews* 61 (2): 121–135.

Verma, H., Kumar, D., Kumar, V. et al. (2021). The potential application of endophytes in management of stress from drought and salinity in crop plants. *Microorganisms* 9 (8): 1729. https://doi.org/10.3390/microorganisms9081729.

Vessey, J.K. (2003). Plant growth promoting rhizobacteria as biofertilizers. *Plant and Soil* 255: 571–586.

Vivas, A., Marulanda, A., Ruiz-Lozano, J.M. et al. (2003). Influence of a *Bacillus* sp. on physiological activities of two arbuscular mycorrhizal fungi and on plant responses to PEG-induced drought stress. *Mycorrhiza* 13: 249–256.

# 8

# Microbial Inoculants and Their Potential Application in Bioremediation: Emphasis on Agrochemicals

*Shriniketan Puranik[1], Kallinkal Sobha Sruthy[1], Menpadi Manoj[2], Konaghatta Vijayakumar Vikram[1], Praveen Karijadar[2], Sandeep Kumar Singh[1], and Livleen Shukla[1]*

[1] Division of Microbiology, ICAR-Indian Agricultural Research Institute, New Delhi, India
[2] Department of Agricultural Microbiology, University of Agricultural Sciences, GKVK, Bengaluru, India

## 8.1 Introduction

Although modern-day chemicals are beneficial, their unregulated over-exploitation results in adverse effects. Human activities, especially pertaining to industries and farming, release large quantities of synthetic compounds into the environment (Cribb 2021). Natural ecosystems like air, water, and soil are greatly polluted by these chemicals, causing complications in biotic entities of nature, thereby posing a threat to their lives. The human contribution in pollution is so strong that the contaminants are ubiquitous in an array of environments, even including extreme ones like deep ocean, mountain tops, North and South poles, uninhabited niches, deserts, etc. (Luo et al. 2019; Rochman and Hoellein 2020; Xu et al. 2020; Ji et al. 2021). Further, pollutant accumulation in various matrices also enters the food chain through plants, thereby impairing normal human functioning by mutation (Ali et al. 2019; Hu et al. 2020a). They have now been found in plants, animals, and certain human body fluids like blood, urine, milk, tissues, including those of infants (Gruber 2018; Kumari and Kaur 2020; Caballero-Casero et al. 2021; Gbadamosi et al. 2021; Hu et al. 2021; Mathiesen et al. 2021; Bertram et al. 2022; Porta et al. 2023). Some of these compounds are recalcitrant in nature and take several days to degrade. Thus, their accumulation in the environmental matrices is a menace. The pollutant property greatly depends on its chemistry. Based on their chemical properties, they can be classified as organic and inorganic compounds. While organic contaminants are mainly water-insoluble pesticides, phenols, hydrocarbons, heavy metals like chromium, lead, arsenic, mercury, etc., comprise the inorganic pollutants. Most of these compounds are toxic, even at minute concentrations, depending on the matrices that they are present in. While some aggregate with other compounds to form more toxic molecules, others breakdown, thereby forming intermediates with increased toxicity (Zeng et al. 2021). Thus, it is necessary for their complete elimination or conversion to less/nontoxic forms in the matrices, a process called remediation.

Several processes are advocated for the remediation of contaminated matrices. Physiological and chemical processes like filtration (Rodriguez-Narvaez et al. 2017), evaporation (Sandy and DiSante 2010), sorption (Thue et al. 2018), photocatalysis (Fast et al. 2017) are hypothesized by various studies, and they are considered costly and laborious. Thus, finding cheaper, alternative methods is important. Bioremediation is an effective method for degradation of contaminants using microbial sources. It is a relatively effective process because of the wide plethora of microbes that can be explored for decontamination of several poisoned matrices, because of their wide metabolic capacity (Bhadouria et al. 2020; Giri et al. 2021; Gond et al. 2021). US Environmental Protection Agency (USEPA) also supports bioremediation as exo-friendly tactic for cleansing of nature and promotion of sustainability (Bharagava et al. 2020). Extensive studies have reported utilization of efficient microorganisms for degradation of various xenobiotic compounds present in different environmental matrices (Kour et al. 2022; Gupta et al. 2018). Pre-examination of polluted-matrix conditions, chemistry of pollutants, extent of pollution, and environmental conditions govern the method of remediation as well as the microbe to be employed. Profiling of polluted sites is a prerequisite too. Furthermore, water-insoluble contaminants need special additives like surfactants externally or biosurfactant-producing strains that increase their solubility by reducing the surface tension,

*Microbes Based Approaches for the Management of Hazardous Contaminants*, First Edition. Edited by Ajay Kumar, Livleen Shukla, Joginder Singh, and Luiz Fernando R. Ferreira.

thus increasing their biodegradability. Considering these notions, the subsequent sections focus on different chemical compounds that contaminate environmental matrices, various techniques in bioremediation, microbial inoculants involved and novel enzymes that are involved in the process. Since agriculture is one of the most important sources of pollution due to overexploitation of certain pesticides, insecticides, etc., the chapter emphasizes on microbial inoculants used in the remediation of some agriculturally important chemicals (called "agrochemicals") that pollute the soil and water bodies.

## 8.2 Pollution of Different Matrices by Agrochemicals

The production of anthropogenic chemicals rose from 1 to 400 million tons annually between 1930 and 2000 (Ozkara and Akyil 2018). The growing population and industrial and agricultural advancements have considerably contributed to increased production and pollution of the environment due to the uncontrolled utility of hazardous, cancer-causing, and threat-inducing chemicals. Pesticides and related chemicals, crude oil and its by-products, polymers, plasticizers, and wastes persist in the environment for a long time, where they are absorbed by the atmosphere, stored, and biomagnified to amounts that may be hazardous to species at higher levels in the food chain (Osman 2018). Their occurrence is constantly posing grave hazards to living things through extensive dispersion in different environmental matrices such as soil, water, and air is a concern for human health and aquatic ecosystems, as they are highly persistent and potent of bioaccumulation. This section emphasizes on contamination of environmental entities surrounding the site of pollution.

### 8.2.1 Soil

The term "soil pollution" often describes the entry and persistence of harmful chemicals in soil in amounts greater than the ecological limit, thus altering soil fertility, hampering agricultural and animal production, and groundwater contamination, thereby negatively impacting human health. Soil contamination is primarily caused by two factors: natural processes and human interventions. The former includes natural disasters like earthquakes, tsunamis, volcanoes, etc. (Candeias et al. 2022), while the latter includes agrochemicals and heavy metals (Singh and Singh 2020). Current-day farming greatly depends on several agricultural chemicals, which aid in the improvement of economical crop production in order to fulfill the consumption demand of fast-increasing world population (Koli et al. 2019). Agricultural chemicals. Also called "agrochemicals" are a group of chemical compounds utilized during farming and comprise of fertilizers, pesticides, insecticides and others that aid in plant resistance improvement. They are among the infamous pollutants since they are used in farming works worldwide. Chemicals like 2,4-D, synthetic pyrethroids, neonicotinoids (NEOs) (acetamiprid, thiamethoxam, clothianidin, thiacloprid, imidacloprid), organophosphorus (chlorpyrifos), etc., are globally used by farmers. According to Aktar et al. (2009), Indian production of technical-grade pesticides in 1998 was 102,240 metric tons, that increased from 5000 metric tons in 1958. The FAO estimates that between 2015 and 2030, global fertilizer usage would increase by 1.0% per year, from 138 million tons in 1997/98 to 182 million tons (Bruinsma 2003).

The rate at which some agrochemicals break down in the soil (or their "half-life") is used to determine how persistent they are. The chemicals with longer half-life remain in water or soil for longer duration, making them more likely to leach through soil (Das and Mukherjee 2012). The pesticide classes with the highest environmental persistence are organochlorines (OCs) and have been classified as exceedingly persistent, bioaccumulative, and toxic (PBT) substances, including aldrin, dieldrin, chlordane, p-dichlorodiphenyltrichloroethane (DDT), mirex, and toxaphene. A study showed that OC load was found to be up to 49.74 μg/kg, majorly constituted by DDT and endosulfan in Kenyan soils (Sun et al. 2016). Similarly, Kafaei et al. (2020) reported DDT concentration in soil Shabankare and Dalaki areas of Iran to be 4.37 and 11.99 ng/g, respectively. They also observed that chlordane concentrations in both areas were 1.62 and 2.04 ng/g, respectively. However, a study in India showed relatively lower levels of DDT (1–66 ng/kg), endosulfan (2–39 ng/kg) and chlordane (0.2–19 ng/kg) in comparison to former studies (Kumari et al. 2008). Atrazine is the most often used OC herbicide and is used in agriculture all over the globe for the control of both pre- and post-emergent weeds. It can stay in the environment for a prolonged period because of its high-water solubility and persistence (Barchanska et al. 2017). Atrazine levels in greenhouses were found to be substantially higher than those in the neighboring open-field soils (Dou et al. 2020). Further, agricultural soils surrounding the Yangtze River Delta of China contained an atrazine load of 1.0–113 μg/kg (Sun et al. 2017). After the ban on several OCs, organophosphate (OP) pesticides have emerged with extensive use in agriculture, which are basically chief insect-managing chemicals (Soltaninejad and Shadnia 2014; Costa 2018). According to the Codex Committee on Pesticide Residues (38th Session) Fortaleza, Brazil (3–8 April 2006), the average Acute Reference

Dose (ARfD) for OPs is in the range of 0.01–0.3 (mg/kg body weight/day). OPs such as quinalphos (0.001–0.010 μg/g), malathion (0.002–0.008 μg/g) and chlorpyriphos (0.002–0.172 μg/g) were found in soils (Kumari et al. 2008).

Pyrethroids are another class of insecticides that are broad-spectrum and efficient in pest control. Their frequent usage causes in their residue to build-up in farm-related matrices, and especially surface water and soil (DeMars et al. 2021). The residuary concentration of cypermethrin, fenvalerate, and deltamethrin of Indian agricultural soils (average: 16.85, 6.75, and 5.5 μg/g, respectively) were much greater than South-east China (7.59, 2.78, and 3.05, respectively) (Kumari et al. 2008). A big-scale study at farm conditions in South-east China from 2015 to 2017 on the dispersal and partition of pyrethroids revealed that 644 surface soil samples showed pyrethroid residues along with 630 surface water samples. They further stated that cyfluthrin and bifenthrin were found in higher quantities (0.1 and 12.14 ng/g, respectively) (Tang et al. 2021).

In recent research, Zhang et al. (2020a) reported NEO residues and distribution in the soil, water, and sediments in the city and village regions of South China. Their findings revealed that each sample included evidence of at least one NEO. The summed levels of five target NEOs such as Imidacloprid, acetamiprid, clothianidin thiamethoxam and thiacloprid, ranged between 0.003–229 ng/g dw soil, 7.94–636 ng/l water, and 0.017–31.3 ng/g dw sediments. Imidacloprid predominated in the agricultural soils, where residual NEO levels were highest, while acetamiprid prevailed in the other five functional zones. Thus, in general, contamination of soil by agrochemicals and related xenobiotic entities largely affects soil microbiome (Walder et al. 2022), plant health (Parween et al. 2016), and surrounding water bodies (Riza et al. 2021).

### 8.2.2 Water

Industrial, commercial, and agricultural waste disposals all contribute significantly to wastewater pollution. Such waters are barely treated (10%), and a large proportion of untreated water is let into natural water systems (Yadav et al. 2015). Pollutants like organic compounds are found in water bodies more often (Teodosiu et al. 2018). The significant pollutants that have an impact on water contamination are agrochemicals. As discussed earlier, these contaminants are harmful to people and can result in serious health issues like cancer, reproductive disorders, and changes at the chromosomal level as a result of oxidative stress and mutation in DNA, further leading to diseases like Alzheimer's and Parkinson's (Sabarwal et al. 2018).

A number of researchers have reported pesticide traces in sub-ground level and drinking water across the globe and India (Jaipieam et al. 2009; Kaushik et al. 2012; Mekonen et al. 2016; Wang et al. 2022; Ganaie et al. 2023). HCHs and DDTs were found in ponds, hand pumps, wells, and other water sources in Bhopal, India. Well water samples in Bhopal included 5794 g/l of DDT and 4640 g/l of HCHs (Bouwer 1989). Further, drinking water in Ahmedabad, India, contained total DDT concentrations of 10.90–314 ng/l and total HCH concentrations of 23.90 and 2488.70 ng/l (Jani et al. 1991). Ray (1992) found a significant amount of pesticide remains in water samples at ground level around the Ganga River, obtained from residential irrigation wells, household wells, and canals in Aligarh, India.

Of many pesticides in water bodies, one of the most commonly used hazardous pesticides is OPs, like chlorpyrifos (Huang et al. 2020). According to John and Shaike (2015), the half-life of chlorpyrifos in soil can range from a few days to four years. Chlorpyrifos concentrations that often exceed water quality standards have been found (Corbin et al. 2009; Ensminger et al. 2011). It has been reported that, of 1575 samples from the San Joaquin valley area, 6.5% contained more than 0.04 g/l chlorpyrifos, a level higher than the recommended limit (Zhang et al. 2012a). Further, Ccanccapa et al. (2016) observed that chlorpyrifos was one among the most frequent contaminants of total water samples (95%) from the Ebro River basin. They also observed that other OP, like diazonin was also in highest frequency (95%). They also noted that a fungicide belonging to the carbamate group, carbendazim was present in 70% water samples. Other carbamates like carbaryl and carbofuran are also prime contaminants, found in pond and canal water due to agriculture in concentrations: 0.095–0.714 and 0.546–4.820 μg/l, respectively (Bhuiyan et al. 2021).

Another class of insecticides, NEO, have also been detected at high concentrations in water bodies (Sanchez-Bayo and Goka 2006). They have a lethal dosage (LC50) that can range from 7 to 719 mg/l in aquatic species (Ding et al. 2004; Sanchez-Bayo et al. 2016). In comparison to urban regions, rural areas had greater levels of NEO contamination in their soil, water, and sediments. Of the 29 water samples, 28 had more than 35 ng/l NEO concentration. Furthermore, 45% of the water samples had a load of more than 200 ng/l, especially in rural areas. According to ecological standards, surface water NEO concentrations should be fewer than 35 ng/l (chronic) or 200 ng/l (acute) in order to protect aquatic life (Morrissey et al. 2015). Worldwide, six NEOs are ubiquitous. Imidacloprid and thiamethoxam have been found in the highest concentrations (320 and 225 g/l, respectively), while others were present in an average load ranging between 0.08 g/l (dinotefuran) and 0.73 g/l (imidacloprid) (Sanchez-Bayo et al. 2016). Also, Zhang et al. (2020b) found thiamethoxam and acetamiprid in sediment and water samples of South China.

Apart from insecticides and fungicides, herbicides are prime contaminants in water bodies. The herbicide that is most commonly found in surface aquatic systems is atrazine. Given that it is a frequent herbicide sprayed on crops like maize and soybeans, atrazine has a huge surface area. It degrades slowly in water and barely in neutral or slightly alkaline solutions. According to Peng et al. (2018), the highest levels of atrazine in China were reported to be 1726 ng/l. The complexity and diversity of emerging organic pollutants from over 21 sites across United States were examined by Battaglin et al. (2016). They found that levels of atrazine were as high as 183 ng/l with a frequency of 18% of water samples studied. The same group also stated that glyphosate concentrations were more than ~8.9 times that of atrazine (1630 ng/l). However, the highest concentration was that of Clomazone (~15.7 times, 2880 ng/l). Likewise, glyphosate concentration in the Pinios River in Greece was 1.16 g/l (Tsaboula et al. 2016). Another herbicide, diuron, is a phenyl urea weed-killer that is used to combat annual weeds and leaves, among other pesticides (Li et al. 2021). Due to its excessive use, the diuron is characterized in the surface and groundwater worldwide. In a study conducted over five years, the authors examined 14 different locations (135 water samples). They are long with dimethoate and propanil, diuron had one of the highest quantities in the studied water samples. With a half-life of 43–2180 days in the aquatic medium, it is quite persistent (Moon et al. 2019). La Mula micro-catchment in Costa Rica was found to have the greatest concentration of diuron in rivers, with a value of 22 770 ng/l (Carazo-Rojas et al. 2018). This was a very high concentration, as per Huovinen et al. (2015), who demonstrated that a concentration of 429 ng/l impacted human development (through cyto-toxicological tests using human cells). Furthermore, Mohammed et al. (2018) observed that very low concentrations (0.429 ng/l) of diuron can cause fetotoxicity through metabolism in placenta. For such reasons, these pollutants should be completely eliminated from natural aquatic systems and ensured good quality potable water required for maintenance of health (Mekonen et al. 2016).

### 8.2.3 Air

Agrochemical contamination may be monitored using a variety of media, and it is evident that airborne dispersal is a key method for spreading harmful chemicals in the environment (Gibbs et al. 2017). Using samplers, Chakraborty et al. (2010) examined the levels of OCs in the air in major cities, including Agra, Bangalore, Chennai, Goa, Mumbai, New Delhi, and Kolkata. They determined the concentration of HCHs detected in Indian towns, which varied from 890 to 17,000 pg/$m^3$, the highest ever recorded worldwide. DDT levels varied from 250 to 6110 pg/$m^3$, chlordane from 290 to 5260 pg/$m^3$, endosulfan from 240 to 4650 pg/$m^3$, and hexachlorobenzene from 120 to 2890 pg/$m^3$. In addition to this, Devi et al. (2011) observed that the ambient air at high altitudes in Manipur, India, has considerable amounts of OCs with the DDT profile.

By employing an air sampler (polyurethane PUF and polyester PEF filter matrices) in 15 districts of eastern Austria with various land uses for up to eight months, Zaller et al. (2022) reported the concentration of agrochemicals in the atmosphere. Gas chromatography/mass spectrometry (GC–MS) was used to identify 566 compounds in the samples. In total, 67 chemicals were observed, including 24 herbicides, 30 fungicides and 13 insecticides. Folpet (412–465 ng/sample) and Prosulfocarb (725–1218 ng/sample) were the herbicides with the highest concentrations, including glyphosate. Chlorpyrifos-ethyl (110 ± 98 ng/sample was the most exploited insecticide, exposed in 93% of the test samples. Predominant OCs such as dichlorodiphenyldichloroethylene (DDE), DDT, heptachlor, and mirex are found during the rainy season in Tamil Nadu (Srimurali et al. 2015). Msibi et al. (2021) reported that the indoor air of certain households in Eswatini (South Africa) had considerable amounts of following pesticides: ametryn (0.75 μg/$m^3$), atrazine (0.32 μg/$m^3$), pendimethalin (0.57 μg/$m^3$), and glyphosate (0.004 μg/$m^3$). Apart from indoor, outdoors in Germany were found to be contaminated with agrochemicals like metolachlor, chlorothalonil, terbuthylazine, pendimethalin, dimethenamid, prothioconazole-desthio, flufenacet, prosulfocarb, aclonifen, tebuconazole, hexachlorobenzene, chlorflurenol (Kruse-Plaß et al. 2021). They also reported the unexpected presence of around 6 and 13 chemicals in the mountain tops of Bavarian Forest and national park "Harz," respectively. Further, every sample analyzed contained glyphosate. These molecules not only affect human health but also undesirably influence agriculture by polluting soil, creating stress in plants and altering the population of agriculturally important insects, like bees. They cause acute intoxication to bees especially by guttation droplets contaminated of seed-coating insecticides (Tapparo et al. 2011) and direct contact of bees to the emitted dust during treated-seed sowing done by drilling machine (30 μg/$m^3$) (Krupke et al. 2012).

Numerous investigations on the examination and analyses of persistent organic pesticides have shown that these chemicals have polluted every environmental region in India; their residues are discovered above maximum residue limit values in the air, water, and soil across the country. Despite India's current low usage of pesticides, the bulk of the research indicates that there are substantial residual levels of pesticides in the air, water, and soil of India. This might be as a consequence of the extensive usage of several chemical pesticides prior to the majority of them being outlawed.

Another factor contributing to residue levels in the environment of India is the application of pesticides at the improper dosage, which results in an excess and off-target quantity.

Considering the mounting evidence that these systemic and non-systemic pesticides have negative effects on some non-target creatures as well as mankind (Table 8.1; Figure 8.1). The entire public, especially end users, has to be made aware of and educated on pesticide management best practices. The general public is exposed to possible health concerns as a result of these practices. Integrated pest management and integrated farming systems, two more cost-effective and ecologically beneficial methods, should be used and promoted. The user group should also be provided with information on the proper use of pesticides. Farmers might receive guidance from plant health centers. This will not reduce the levels of contaminants in different matrices but will help against further accumulation.

## 8.3 Different Strategies Employed in Bioremediation

Bioremediation is a broad term that basically is a green technology that employs microorganisms to breakdown toxic pollutants into less toxic/non-toxic molecules, thereby reducing chemical load in matrices. It is an ecofriendly, slow, but effective operation for bringing down pollutant concentrations in soil, air, water, etc. It implies mediation of microorganisms either directly or indirectly, wherein they metabolize these harmful compounds by assimilation or bioconversion (Rehman et al. 2020). It can occur in both absence or presence of molecular oxygen and is governed by anaerobic and aerobic microorganisms, respectively. Furthermore, considering the site of bioremediation, type of pollutant and matrix polluted, there are two types of bioremediation: *in situ* and *ex situ* (Sales da Silva et al. 2020).

### 8.3.1 *In Situ* Biodegradation Strategies

Different methods help in combating various chemicals in a wide array of matrices. *In situ* bioremediation is carried out at the point of pollution, without the transfer of the polluted matrix (Banerjee et al. 2016). It largely depends on the native microbiome for degradation of pollutants. It is comparatively cost-effective and needs certain additions like oxygen, nutrients, and moisture in order to enhance the population and activity of innate microflora. There are many types of *in situ* bioremediation techniques, like biopiling, biosparging, bioventing, biostimulation, bioaugmentation, etc., that work on similar phenomenon but with varied mechanization.

Biopiling is a preliminary remediation method in which the contaminated soil is placed as a heap on a workable area, mixed with nutrients and pumped with air for better degradation. It is an aerobic process and uses autochthonous microbes to consume the pollutants, thereby reducing their concentration. Preliminary biosparging is a process that involves localized air injection under pressure at a saturated zone, along with certain nutrients needed for growth of native microbes. This also increases the volatilization and upward movement of pollutants. In fact, biosparging has proved to be an excellent strategy for degradation of oil leakages and organic pollutants like dioxin in ground water (Macaulay and Rees, 2014; Bell et al. 2022). Another method similar to biosparging is bioventing. However, unlike biosparging, it mainly concentrates on unsaturated areas of contamination. It involves the slow release of air into contaminated sites under relatively reduced pressure, thus preventing release of contaminants into atmosphere through volatilization (Frutos et al. 2010). It is also affected by reduced temperature and moisture content (Azubuike et al. 2016). A combined method of biosparging and bioventing can be effective against petroleum hydrocarbon contamination in ground water. Likewise, other methods like biostimulation and bioaugmentation are also closely related processes that are used as remediation strategies. Biostimulation is a process in which innate microbial proliferation is accelerated due to the supplementation of essential nutrients like nitrogen, phosphorus, etc. In addition, certain growth limiting factors, electron donors, or acceptors can also be added to stimulate their growth (Kanissery and Sims 2011). This usually happens via injection wells (Zeneli et al. 2019). Biostimulation is one of the key techniques for bioremediation of oil spills and agrochemicals (Kanissery and Sims 2011; Aldas-Vargas et al. 2021; Sun et al. 2023). Contrastingly, bioaugmentation is a method that involves supplementation of previously screened or genetically altered autochthonous microorganisms with potential of degradation. This increases the population of more efficient microbial strains that have the upper hand in the degradation of selected pollutants.

The main advantage is that the microbes selected and added are well-adapted to the existing conditions and pollutant concentrations. Gao et al. (2015) reported the remediation of dichlorodiphenyltrichloroethane (DDT) in contaminated soil by *Escherichia coli* carrying a catabolic pDOD plasmid from *Sphingomonas* sp. through bioaugmentation. Dueholm et al. (2015) added autochthonous *Pseudomonas monteilii* SB3078 for remediation of benzene, toluene, ethylbenzene

**Table 8.1** Matrices polluted and potential health hazards posed by various chemical pollutants.

| Sl. No. | Compound | Chemical nature | Matrix polluted | Possible health hazard | References |
|---|---|---|---|---|---|
| 1. | As, Cd, Cr, Cu, Pb, Hg, Se, U and Zn | Metals and non-metals (elements) | Water and Soil | Acute contact leads to vomiting, nausea, gastro intestinal irregularities, dermatitis, degeneration of neurons like in Alzheimer's, Parkinson's diseases, dystrophy in muscles, liver and kidney damage | Crawford and Bhattacharya (1987) and Kampa and Castanas (2008) |
| 2. | Polycyclic aromatic contaminants (Naphthalene and Biphenyl) | Multiple benzene rings (>3) comprising of C and H | Soil, Air and Water | Diabetes, cancer, oxidative stress, deprived fetal development, heart-related diseases | Hiller et al. (2011) |
| 3. | DDD, dichlorodiphenyl-dichloroethene (DDE), Dichlorodiphenyltrichloroethane (DDT) and their derivatives | Organochlorine | Soil, Water and Air | Allergies and rashes, the sensitivity of skin, deleterious effects on the neural system, damage to lungs, problems related to the reproduction system, malfunctioning of endocrine and immune systems, birth impairments, non-Hodgkin's lymphoma, Cancer in breasts | Longnecke (1997), Turusov et al. (2002), and Mansour (2004) |
| 4. | Atrazine | Organochlorine | Water and Soil | Endocrine disruptor, immune system function impairment, leukemia, brain and testicular cancers | Mills (1998), Hooghe et al. (2000), and Rohr (2021) |
| 5. | Aldrin and dieldrin | Organochlorine | Soil, Water and Air | Irritation, headaches, giddiness, vomiting, chronic convulsion, nervous system impairment, reduce immunity, high child death rate, kidney damage and involuntary muscle movement | Belta et al. (2006) |
| 6. | Endosulfan | Organochlorine | Soil, Water and Air | Central nervous system, hepatic toxicity in liver, hypotension, metabolic disturbances, tremors, convulsions, breathing disorder, congestive heart failure. Liver, kidney and myocardial poisoning | Li et al. (2009) and Moon and Chun (2009) |
| 7. | BHC and its derivatives | Organochlorine | Soil, Water and Air | Neural disorder, β-BHC impairs thyroid hormonal equilibrium and brain development, carcinogenic, photosensitivity, perpetual hair loss | World Health Organization (1990), Chopra et al. (2011), and Debnath and Khan (2017) |
| 8. | Endrin, Endrin-ketone and Endrin-aldehyde | Organochlorine | | Ingestion of larger quantities causes convulsions and leads to death, headaches, giddiness, perplexity, nausea, nervousness and vomiting | Lango-Reynoso et al. (2013) |
| 9. | Chlordane | Organochlorine | Soil, Water and Air | Reproductive and immune toxic effects in wildlife migraines, respiratory infections, diabetes, anxiety and depression | ATSDR: Agency for Toxic Substances and Disease Registry 2000 and Fosu-Mensah et al. (2016) |

*(Continued)*

**Table 8.1** (Continued)

| Sl. No. | Compound | Chemical nature | Matrix polluted | Possible health hazard | References |
|---|---|---|---|---|---|
| 10. | Chlorpyrifos | Organophosphate | Soil, Water and Air | Inhibition of central brain enzyme and acetyl choline esterase activity, highly toxic to prenatal and infant neural-endocrine development | Grandjean and Landrigan (2014) and Burke et al. (2017) |
| 11. | Profenofos | Organophosphate | Soil and Water | Genotoxic effects | Hammam and Abd el Mottaleb (2007) and Kushwaha et al. (2016) |
| 12. | Quinalphos | Organophosphate | Soil and Water | | Eskenazi et al. (1999) and Parven et al. (2021) |
| 13. | Phorate | Organophosphate | Soil and Water | Inhibition of the activity of acetylcholine esterase leads to acetylcholine accumulation and associated neural-poisoning, prostate cancer, pulmonary poisoning | Fulton and Key (2001), Oruç and Usta (2007), Saquib et al. (2011) |
| 14. | Dimethoate | Organophosphate | Water and Soil | Mice study showed linkage with mammalian neuromuscular transmission blockage, chromosomal aberrations, immunotoxicological and genotoxic effects | Rupa et al. (1991) and Farag (2006) |
| 15. | Dichlorvos | Organophosphate | Soil, Air and Water | Inhibiting neural acetyl cholin esterase with disruption of nerve function. Nausea, tearing, diarrhea, vomiting, bronchial secretion increase, fasciculation, weakness and flaccidity in muscles | ATSDR: Agency for Toxic Substances and Disease Registry (1997) and Wang et al. (2004) |
| 16. | Parathion | Organophosphate | Water and Soil | Endocrine and neural-disruptors, impairment of cholinesterase enzymes, decreased insulin production, interruption in regular cellular assimilation of carbohydrates, proteins and lipids | Karami-Mohajeri and Abdollahi (2011) |
| 17. | Malathion | Organophosphate | Air, Water and Soil | Respirational issues, giddiness, fetal toxicity due to high exposure, anti-cholinesterase activity cause respiratory paralysis | Chen et al. (2006) |
| 18. | Profenofos | Organophosphate | Air, Surface Water, and Soil | Inhibiting neural acetylcholine esterase with disruption of nerve function, carcinogenic, and reproductive toxicity, Multiorgan dysfunction | Kushwaha et al. (2016) |
| 19. | Monocrotophos | Organophosphate | Soil, Water | Potent carcinogen. Damage to DNA, programmed cell death, and heart issues. It might lead to chronic problems like eye inflammation, myosis, vision blurring, giddiness, nausea, dyspnea, convulsions, excessive saliva production, cramps in the abdominal area, diarrhea and vomiting | Sundarmurugasan et al. (2016) |

**Table 8.1** (Continued)

| Sl. No. | Compound | Chemical nature | Matrix polluted | Possible health hazard | References |
|---|---|---|---|---|---|
| 20. | Carbofuran, aldicarb | Carbamate | Water and Soil | Anti-choline esterase activity, reproductive hormone impairment activity, genotoxic and cytotoxic irregularities | Gupta (2006) and WHO (2009) |
| 21. | Carbaryl and methomyl | Carbamate | | The short-term exposure causes twitching of muscles, headache, nausea, giddiness, memory loss, feebleness, reduced heartbeat, trembling, diarrhea, perspiration, salivation, lacrimation, and pupil constriction<br>Long-term exposure causes delayed neural toxicity, followed by extreme tingling and burning. Further might lead to paralysis. Hepatic, renal immune, bone marrow damages | Pope et al. (2005) and Senarathne et al. (2022) |
| 22. | Pyrethroids | Neonicotenoid | Soil, Water | Potent neuro poisons, endocrine disruptors, and paralysis | Costa (2008) and Mnif et al. (2011) |
| 23. | 2,4-D | Auxin | Water and Soil | Potent carcinogen. Also causes DNA damage, diabetes, hypertension, strokes, autism, renal failure, Alzheimer's and Parkinson's diseases | Swanson et al. (2014) |
| 24. | 2,4-D | Auxin | Air, Water and Soil | Cause irritation, itching, burning sensation of skin, shortness of breath and burning sensation in the upper lung, while its ingestion results in vomiting, diarrhea, headache, skeletal muscle injury, irritation, hypertension | Ganguli et al. (2014) |
| 25. | Plasticizers: Di-(2-ethylhexyl)-phthalate (DEHP) and Bisphenol A (BPA) | Phthalates | Water and Soil | Endocrine-disrupting chemicals, carcinogenic, mutagenic, immunotoxic, neurotoxic and hepatotoxic | Xing et al. (2010) and Schug et al. (2016) |
| 26. | Polychlorinated biphenyls (PCBs) | Organochlorine compounds | Water and Air | PCBs impair reproduction, immune function and carcinogenicity in humans | Desforges et al. (2018) |
| 27. | Nitrapyrin | Organochlorine, [2-chloro-6-(trichloromethyl) pyridine] Nitrification inhibitor | Water and Air | Irritation in eyes, dermal irritation and classified as "likely to be a human carcinogen" | Zaller et al. (2023) |

and xylene in contaminated wastewater effluent. However, they also reported that these native strains fail to degrade because of reduced survivability and activity *in situ*. Needless to say, both biostimulation and bioaugmentation have been employed for cleaning up atrazine-contaminated soils (Lima et al. 2009). Thiabendazole, a widely used fungicide, was also eliminated from contaminated soils via bioaugmentation of a bacterial consortium (Papadopoulou et al. 2018). Thus, it is clear that *in situ* mechanisms are promising, provided they are used tactically considering the nature of the pollutant and matrix.

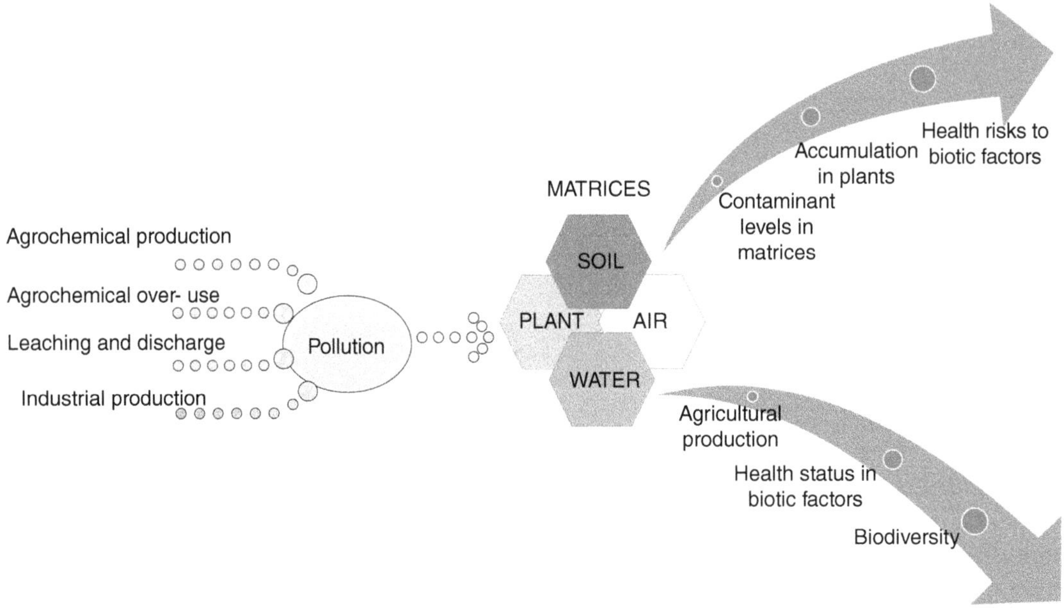


**Figure 8.1** Causes and impact of contamination of environmental matrices.

### 8.3.2 *Ex Situ* Biodegradation Strategies

*Ex situ* bioremediation involves excavation and conveyance of polluted matrices, mainly soil, to a newer, monitored area for treatment. Like *in situ*, *the* volume of a matrix, type of pollutant, depth of pollution, cost of treatment and geography are considered (Sharma 2020). This includes techniques like bioreactors, landfarming, composting, etc. (Figure 8.2) Bioreactors are large stainless-steel vessels that support bioreactions or a bioactive system. It has a monitoring and sampling system that can be altered to provide congenial conditions for remediation process. Sludges, wastewater, and sometimes soils are also decontaminated in bioreactors. The polluted matrix is mixed with potent microbial strains (allochthonous) and provided with optimum pH, temperature, aeration, and moisture contents. After a due course of time, the samples are analyzed for the amount of degradation, thereby understanding efficacy and period of the process (Dzionek et al. 2016). Bioreactors are effectively used in breakdown of agrochemicals in the soil and related matrices. Whittington et al. (2020) studied in microbial consortium-mediated accelerated breakdown of ametoctradin. Abdi et al. (2020) reported employment of woodchip bioreactors for agrochemical degradation and observed a significant reduction in quantities of selected chemicals (bifenthrin: 76%; chlorpyrifos: 63%; Oxyfluorfen: 31%). Air-pulse fluidized bioreactors were shown to achieve 20% and 64.7% degradation of acetamiprid and imidacloprid (4 mg/l) due to inoculation of *Trametes versicolor* (Hu et al. 2022).

Land farming is another effective strategy of pulverizing the contaminated soil for better aeration and contact with microorganisms, so as to improve the physical and chemical properties of soil, which further assists in remediation (Pavel and Gavrilescu 2008). It has a very low footprint and cost, thereby being popular. Generally, the polluted area is arranged in layers of 50 cm in thickness and is mixed with agricultural equipment (Cristorean et al. 2016). Landfarming is also used in bioremediation of certain pollutants. Rubinos et al. (2007) mitigated hexachlorocyclohexane-isomers pollution using this technique. They reported >80% of decrease in isomer load. de Quadros et al. (2016) observed increased microbial activity and microbial structural alterations in oil sludge-amended land farm soil. Another efficient technique, composting, is an aerobic process, which involves degradation of complex polymers of pollutants into simpler molecules, mediated by microorganisms. It involves three stages and is dominated by different groups of microorganisms at each stage. The initial phase is mesophilic stage, in which mesophiles like *Pseudomonas, Flavobacterium, Enterobacter, Alternaria, Mucor,* and *Aspergillus* govern the degradation process. The next stage is the thermophilic stage, dominated by thermophiles like *Thermus, Streptomyces, Thermoactinomycetes, Chaetomium, Humicola,* and *Absidia*. The final stage is again a mesophilic one (maturation phase), wherein microbes like *Bacillus, Flavobacterium, Pseudomonas, Cellulomonas, Aspergillus, Bipolaris, Streptomyces,* etc. are in higher loads (Srivastava et al. 2016). There are reports on remediation of pollutants via composting. Ali et al. (2014) observed high degradation rates of aldrin (85.76%), endosulfan α (84.95%), endosulfan β (83.20%) and

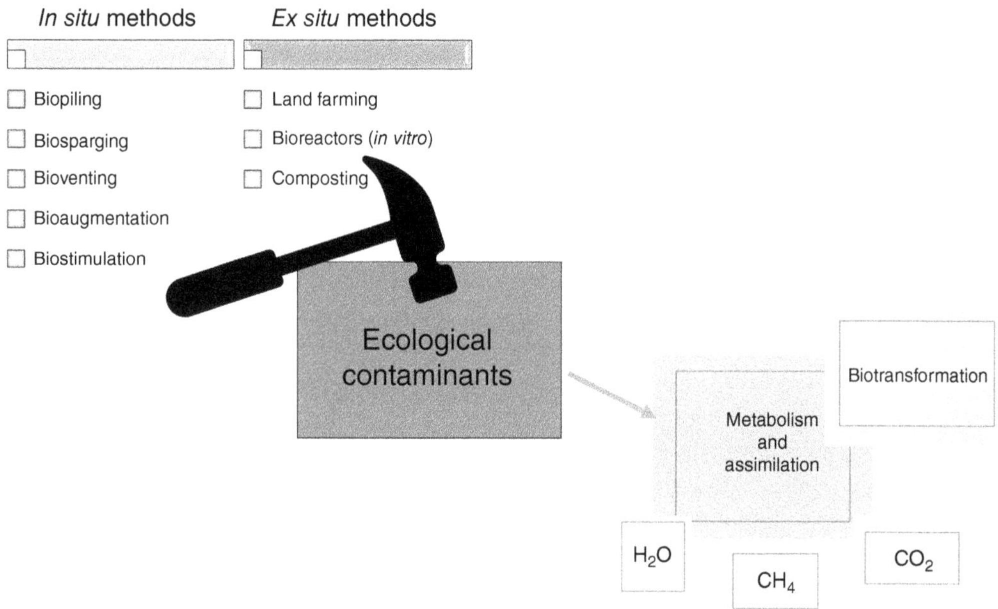


**Figure 8.2** Different strategies of bioremediation and possible outcomes of the process.

lindane (81.36%) with reduced half-lives (25.54, 18.43, 18.43 and 27.43 days, respectively) under rotary drum composting conditions. Ali et al. (2016) also studied the effect of windrow and drum composting on degradation of aldrin and endosulfan. They concluded that both the methods promised more than 65% degradation at same time period, and that breakdown of agrochemicals in rotary drum composting was higher (>85%) as compared to window (>66%). This was attributed to the higher aeration in the former than in the latter. Sayara and Sánchez (2020) reported an enhanced degradation rate of polycyclic aromatic hydrocarbons through composting.

Although *ex situ* method is a good alternative, it has certain limitations. It is a faster, uniform, well-monitored, but costly and laborious process. It needs a large workspace and is difficult to manage. Some of them release volatile organic compounds as waste. This might also expose or spread the contaminants during excavation and transportation (Azubuike et al. 2016). Thus, it is crucial to select the right technique for the right pollutants and conditions (Figure 8.2).

## 8.4 Microbe-Mediated Bioremediation and Recent Advances

It is evident that effective remediation of contaminants, especially those related to agriculture, is possible by microbial intervention (Kour et al. 2022). Amidst wide plethora of microorganisms, many members belonging to Prokaryota and Eukarya aid in degradation. While some use them as C-source, others unintentionally degrade the contaminants by their enzyme production. *Ex situ* bioremediation is found to be relatively rapid and more effective in comparison with *in situ,* since the former can be more closely monitored as compared to latter. The present section explores a different class of microorganisms like bacteria, fungi and microalgae that were observed to be effective in *ex situ* degradation of pollutants, mainly agrochemicals, under *in vitro* or large-scale conditions. By utilizing these natural processes, they promise solutions for addressing environmental pollution and restoring ecosystems.

### 8.4.1 Bacterial Bioremediation

The success of bacterial biodegradation is influenced by many factors, including the selection of appropriate bacterial strains, understanding the specific pollutants present in the surroundings, optimizing environmental conditions, and ensuring the availability of necessary nutrients for bacterial activity and growth. One of the key advantages of bacterial bioremediation is its potential to remediate contaminated sites without causing additional environmental damage. Traditional remediation methods often involve excavation and disposal of contaminated materials, which can be costly, disruptive, and may lead to the generation of secondary waste. In contrast, bacterial bioremediation offers an *in situ* approach, minimizing the need for extensive site excavation and reducing the overall environmental impact. Moreover, bacterial bioremediation could be harnessed in a variety of matrices, involving soil, subsurface water, sediments, and even marine environments. Bacterial strains with diverse metabolic capabilities have been isolated and characterized, allowing for targeted approaches to address specific contaminants in different matrices.

Several studies showed bacteria-mediated pollutant degradation. Gallego et al. (2001) developed a simple laboratory methodology for *in situ* diesel fuel-polluted soil bioremediation. Biostimulation with inorganic nitrogen and phosphorus, along with the addition of activated sludge, significantly enhanced biodegradation, achieving extents higher than 90% after 45 days in a bioreactor setup. *Acinetobacter* bacteria were found to be predominant in both laboratory and field samples, with their presence linked to the degradation of hydrocarbons. The study also observed the production of a biosurfactant-like substance by *Acinetobacter* colonies, which was evident at the field site, suggesting the feasibility of *in situ* breakdown after physically removing the spilled diesel fuel. Chlorinated herbicides such as s-triazines and polychlorinated biphenyls (PCBs) are persistent organic pollutants that pose environmental and health risks. Seeger et al. (2010) identified potent bacteria that degraded s-triazines and PCBs through hydrolytic reactions and reductive dehalogenation, respectively.

Divya and Deepak Kumar (2011) have highlighted the potential of plant-associated bacteria, such as rhizospheric and endophytic bacteria, in degrading toxic compounds in soil and improving phytoremediation (Kashyap et al. 2023; Kumari et al. 2022; Pathak et al. 2022). Introducing pollutant-degrading bacteria to plant seeds and genetically engineering endophytic and rhizospheric bacteria are considered promising approaches for efficient and environmentally friendly remediation of contaminated sites using transgenic plants (Singh et al. 2022a,b; Verma et al. 2021). Islas-Garcia et al. (2015) conducted a study in Tepeaca, Mexico, to assess the feasibility of bioremediation for agricultural soil contaminated with hydrocarbons and OCs. The soil exhibited concentrations of OCs such as aldrin, heptachlor, endosulfan I, endosulfan II, 4,4′-DDT, 4,4′-DDE, *trans*-chlordane, and endrin aldehyde. Hydrocarbonoclastic as well as pesticideclastic fungi and bacteria were detected in the soil with tolerances to diesel and OCs, respectively. The presence of these tolerant microorganisms suggested the potential for stimulating native microorganisms to degrade both hydrocarbons and pesticides, making bioremediation feasible for the contaminated soil.

Singh et al. (2013), in a study focused on the biotreatment of wastewater from a petroleum refinery, assessed the degradation potential of four bacteria: *Bacillus subtilis*, *Corynebacterium propinquum*, *Alcaligenes odorans,* and *P. aeruginosa*. They were found to have the ability to metabolize crude oil and phenol in the wastewater. After inoculating the bacterial consortium, it was observed that the levels of oil and phenol in the wastewater were reduced by 70% and 85%, respectively, through bioremediation. Guermouche Mrassi et al. (2015) focused on isolating and characterizing bacteria from polluted sites in a refinery in Arzew, Algeria capable of degrading hydrocarbons. They showed good metabolic activity toward high molecular weight hydrocarbon compounds like pristane and benzo[a]pyrene. Some strains demonstrated the capability of metabolizing various tested hydrocarbons, including octadecane, squalene, phenanthrene, and pyrene, while others particularly broke down specific substrates. Notably, a strain designated as *Pseudomonas* sp. (LGM2) exhibited the ability to metabolize linear and branched alkanes, in addition to high and low molecular weight polyaromatic hydrocarbons (PAHs). The presence of the *alkB* gene needed for the degradation of alkane was found in LGM2. These findings highlighted the practical significance of the isolated bacterial strains in bioremediation efforts targeting oil-contaminated environments. Bhakta et al. (2014) isolated cadmium (Cd) and arsenic (As) resistant strains from coastal sediments. Four selected isolates, namely *A. brisouii* RCd3, *P. abietaniphila* RCd6, *Exiguobacterium aestuarii* RAs7, and *P. rifietoensis* RAs10, demonstrated enhanced proliferation rates in Cd- or As-augmented culture media. These isolates exhibited significant resistance to Cd (with minimum inhibitory concentrations ranging from 60 to 400 mg/l) and As (with concentrations ranging from 400 to 2200 mg/l), highlighting their remarkable metal(loid) resistance capabilities. The selected isolates also displayed higher efficiency in removing Cd (0.032–0.268 µg Cd/h/mg) compared to As (0.0003–0.0172 µg As/h/mg) from water. Dey et al. (2016) studied the development of innovative strategies for As elimination from aqueous systems. The bacteria, identified as *Bacillus* sp. and *Aneurinibacillus aneurinilyticus*, demonstrated high tolerance to As concentrations, with one strain tolerating up to 4500 ppm of arsenate and the other up to 550 ppm of arsenite. These isolates, named SW2 and SW4, respectively, exhibited significant arsenic removal capabilities, removing over 50% of both arsenite and arsenate from culture media. Additionally, they were capable of oxidizing arsenite to less toxic arsenate. They have reported that these arsenic-resistant bacteria could serve as a novel bioremediation pathway for addressing arsenic contamination. Kang et al. (2016) studied the investigated the positively-interactive influence of bacterial co-cultures (*Viridibacillus arenosi*, *Sporosarcina soli* and two strains of *Enterobacter cloacae*) on the biodegradation of heavy metal-mixture [Pb, Cd and Cu] from the polluted site. After 48 hours, the bacterial mixtures achieved significant remediation rates of Pb (98.3%), Cd (85.4%)

and Cu (5.6%). In comparison to individual strains, the co-culture exhibited greater efficiency and resistance in heavy metal bioremediation. These findings highlighted the usefulness of bacterial mixtures for the remediation of heavy metal contamination in the environment. According to Lopez-Casaperalta et al. (2022), two Hg-resistant bacteria, *Kocuria* sp. and *Zhihengliuella* sp., exhibited significant Hg removal capacities of approximately 28.4% and 33.3% within 45 days of incubation.

While agrochemicals are of increasing menace, some studies have shown their bioremediation strategies. According to Topp (2001), a study evaluated the degradation and mineralization of the herbicide atrazine in loam soil using three atrazine-degrading bacteria: *Pseudomonas* sp. strain ADP, *Pseudaminobacter* sp., and *Nocardioides* sp. These bacteria dechlorinate the herbicide hydrolytically and exhibit comparable specific activities for atrazine degradation in pure culture. Zhang et al. (2012b) also reported two bacteria *Rhodobacter sphaeroides* W16 and *Acinetobacter lwoffii,* with atrazine-metabolizing ability, coupled with reduced oxidative stress mechanism. John et al. (2018) performed *in situ* bioremediation and revealed that *Klebsiella* sp. effectively degraded chlorpyrifos into nontoxic byproducts, resulting in increased microbial growth and soil dehydrogenase activity compared to soil treated with chlorpyrifos alone. Birolli et al. (2019) investigated the biodegradation of lambda-cyhalothrin (LC), a chiral pyrethroid insecticide widely used in agriculture, by bacteria from the Brazilian savannah. The bacterial strains showed varying efficiencies between 56.9% and 96.3%. Mehta et al. (2021) conducted a study in the Himachal region of India to isolate malathion and chlorpyrifos-degrading bacteria from agricultural soil samples. They identified *Kocuria assamensis* (PDM-15) as a potential candidate for bioremediation of up to 71.3% (chlorpyrifos) and 85% (malathion). Erguven and Demirci (2021) conducted a study in Adana province, Turkey, to investigate the bioremediation potential of *O. thiophenivorans* and *S. melonis* for imidacloprid, a widely used selective insecticide in Mediterranean climate agriculture. The results showed that *O. thiophenivorans* and *S. melonis* effectively bioremediated the imidacloprid insecticide with an efficiency of over 90%. Hassan and Alghuthaymi (2022) employed the single-strand-conformation polymorphism (SSCP) technique based on 16S rRNA genes to assess the bacterial diversity in soil samples contaminated with PCBs. The SSCP analysis revealed an increasing bacterial community diversity in the PCB-contaminated soil. From contaminated sites, 23 isolates were obtained, belonging to the genera *Bacillus*, *Brevibacillus*, *Burkholderia*, *Pandoraea*, *Pseudomonas,* and *Rhodococcus*. These isolates exhibited the capacity to metabolize PCBs as a sole C-source and harbored the enzyme 2,3-dihydroxybiphenyl dioxygenase (DHBDO), which can serve as a molecular marker for detecting PCB-metabolizing bacteria in polluted sites.

In conclusion, bacteria have demonstrated significant potential in the biodegradation of organic pollutants, particularly pesticides. Overall, the use of bacteria in bioremediation offers a promising and environmentally friendly solution for reducing the persistent and toxic effects of agrochemicals in soil and aquatic systems. Continued research and development pertaining to this area can contribute to effectively cleaning pesticide-contaminated sites, ultimately protecting ecosystems and human health.

### 8.4.2 Fungal Bioremediation

Fungi are very useful organisms due to their ability and adaptability, especially in pollutant degradation. They have gained substantial attention due to their versatile enzymatic systems capable of breaking down a broad variety of agrochemicals and structurally diverse pollutants. Degradation mechanisms varied depending on the pollutant, involving hydrolyzation, dichlorination, and oxidation, as inferred from identified metabolites (Hu et al. 2020b). Extracellular and oxidative enzymes, like laccase (Lac), lignin peroxidase (LiP), versatile peroxidase (VP), and manganese peroxidase (MnP), position them as prominent members for pollutant biodegradation endeavors (Zhuo and Fan 2021; Puranik et al. 2023). Several fungi are reported of pollutant metabolism. Fungal degradation of diverse pesticides, such as lindane, endosulfan, DDT, atrazine, etc., has also been reported (Maqbool et al. 2016; Kumar et al. 2019; Bokade et al. 2021; Bose et al. 2021). Certain fungi, like *Trametes*, *Ganoderma*, *Aspergillus*, *Fusarium*, and more, are recognized as pesticide degraders, spanning various fungal groups. These findings underscore fungi's potential for pesticide remediation and the wide range of species involved in this degradation process.

Fungi like *Trichoderma*, *Aspergillus*, and *Penicillium* are also involved in remediation. A study spotlighted seven ocean-dwelling fungi, encompassing *Aspergillus* spp. and *Penicillium* spp., to evaluate their efficacy in removing methyl parathion (MP), an insecticide. Among these, *Aspergillus sydowii* CBMAI 935 emerged as the most proficient strain, achieving complete MP removal within three weeks of incubation. Metabolite analyses unveiled a degradation pathway involving the creation of the hazardous intermediate methyl paraoxon. Subsequently, this intermediate underwent full degradation into *p*-nitrophenol, resulting in a striking 120-fold reduction in toxicity levels (Alvarenga et al. 2014). Furthermore, the same strain exhibited new phosphoesterases, phosphodiesterase and methyltransferases capable of degrading chlorpyrifos, methyl parathion, and profenofos (Soares et al. 2021).

Notably, *P. miczynskii* exhibited remarkable efficacy, achieving a 90% degradation of Dieldrin within a span of 14 days. Orji et al. (2022) conducted a study focusing on the breakdown of glyphosate soil contamination through fungal cultures. Among them, *A. niger*, *A. terreus*, *A. tamari*, *Fusarium oxysporium*, and *A. flavus*, *A. terreus* and *F. oxysporium* exhibited the highest glyphosate degradation capacity, individually reducing residual glyphosate to 21.17 and 17.22 μg/ml, respectively. When combined, their bi-culture treatment achieved 91.41% degradation with a residual glyphosate value of 8.87 μg/ml. Puranik et al. (2023) degraded dibutyl phthalate (DBP), an endocrine disruptor leached from plastic surfaces in agricultural soils, under *in vitro* conditions using *A. flavus* isolated from sanitary landfill soil. They observed that the fungus almost completely mineralized (99.34%) in 15 days after incubation by the production of esterase enzyme. During the degradation, benzyl-butyl phthalate, dimethyl-phthalate, di-iso-butyl-phthalate and phthalic acid were found as intermediates through GC-MS analysis. However, none of these intermediates were detected at the end of the degradation process.

White rot fungi (WRF) are among the prominent pollutant-degrading class of fungi (Barr and Aust 1994; Tortella et al. 2005; Marco-Urrea and Reddy 2012). Zhuo and Fan (2021) highlighted significant environmental challenges posed by organic pollutants and emphasized the crucial role of WRF in addressing these issues. Their versatility has led to successful applications of WRF and their ligninolytic enzymes in mitigating various pollutants such as PAHs, pharmaceutical compounds, endocrine disruptors, pesticides and synthetic dyes. *Trametes versicolor* showcased its efficacy in degrading three different water-insoluble insecticides: cypermethrin, and chlorpyriphos, dicofol, representing synthetic pyrethroid and OCs, respectively. Over 14 days, at 25°C, significant removal rates were observed: 93% for cypermethrin, 95% for chlorpyriphos and 88% for dicofol. The degradation process encompassed initial rapid adsorption onto fungal pellets followed by subsequent biodegradation. The same organism also exhibited the capability to eliminate herbicides, bentazon and diuron from agricultural wastewater (Beltran-Flores et al. 2021).

Furthermore, degradation of diuron has also been exhibited by WRF, *Ganoderma lucidum*, by concurrent laccase production. Apart from extracellular ligninolytic enzymes that aid in handling hydrophobic pollutants, intracellular enzymes like cytochrome P450 play a crucial role in pesticide degradation. For example, cytochrome P450 facilitated the intracellular metabolism of fipronil, a phenylpyrazole insecticide with limited aqueous solubility. This was affirmed by the formation of hydroxylated and glycosylated transformation intermediates. This enzyme further exhibited the ability to degrade water-soluble insecticides like imidacloprid and acetamiprid (Hu et al. 2022). *Phanerochaete chrysosporium*, which exhibits approximately 150 genes related to P450 monooxygenase in its genome, employs two cytochrome P450 isozymes for breaking down four NEOs: acetamiprid, imidacloprid, clothianidin, and thiacloprid (Mori et al. 2021).

Studies have demonstrated that co-culture systems of various fungi, as opposed to individual trains, enhance the degradation efficiency of certain pesticides. For instance, a consortium comprising five fungi (three from the *Pleurotus* genus, one from *Coriolopsis* genus, and one unknown) demonstrated improved degradation of methomyl and diazinon compared to individual strains (Nyakundi et al. 2011). A concurrent advantageous effect was observed for dual culture involving fungi (*Mortierella* LEJ702) and bacteria (*Variovorax* SRS16 and *A. globiformis* D47) in the degradation of diuron. *Mortierella* spp. LEJ702, when as a consortium with *Aminobacter* spp., also displayed enhanced remediation of the herbicide 2,6-dichlorobenzamide (Ellegaard-Jensen et al. 2014). In certain instances, the combination of fungi and specific nutrients has demonstrated higher efficacy in pesticide removal compared to the use of fungi or minerals alone. For instance, some utilized a combination of *P. chrysosporium* and the boron silicate mineral tourmaline to remediate OCs- and PAHs-contaminated soils (Wang et al. 2014). This synergistic approach led to a remarkable 44% degradation of OCs in soils, surpassing the singular treatments with the fungus (34%) or tourmaline (26%). The mineral tourmaline was suggested to contribute by enhancing soil enzymatic activities and fostering biodiversity among active soil microorganisms.

### 8.4.3 Microalgae and Diatom-Based Bioremediation

Apart from bacteria and fungi, algal groups like microalgae and diatoms are also known for pollutant degradation. Jin et al. (2012) observed that green alga *Chlamydomonas reinhardtii* exhibited remarkable proficiency in both accumulating and degrading the herbicide prometryne. This efficient uptake and catabolism of prometryne resulted in its quick removal from the surrounding environment. These findings suggest that *Chlamydomonas reinhardtii* possesses an inherent tolerance mechanism, making it a potent member for bioremediation in prometryne-polluted aquatic ecosystems. Kurade et al. (2016) reported that *Chlorella vulgaris* exhibited remarkable efficiency in removing diazinon (0–100 ppm) from aqueous solutions. Among four microalga species tested, *C. vulgaris* demonstrated maximum biodegradation ability, successfully removing 94% diazinon at a concentration of 20 mg/l. However, the growth of *C. vulgaris* was considerably hindered when diazinon concentrations exceeded 40 mg/l, leading to over-increasing growth inhibition (30%) after 12 days of incubation.

Interestingly, the study found that diazinon by-products generated during microalgal metabolism appeared to be less toxic, as they enhanced microalgal growth during the log phase. Biochemical properties of *C. vulgaris*, including carotenoid, chlorophyll, and antioxidant enzymes, were altered due to diazinon levels. GC–MS analysis revealed the formation of 2-isopropyl-6-methyl-4-pyrimidinol, which exhibited lower toxicity levels. The study highlighted the high tolerance of *C. vulgaris* to diazinon and its potential application in removing traces of diazinon from contaminated wastewater, showcasing the promising role of algae in detoxification. Hussein et al. (2017) noted that *C. vulgaris* demonstrated significant potential for the bioremoval of a mixture of pesticides at a certain concentration (0.1 mg/ml each). The research encompassed both short- and long-duration studies, utilizing both free and immobilized cells. In short-duration experiments, live cells achieved removal percentages ranging from 86% to 89%, while lyophilized algal cells exhibited higher removal rates (96–99%). During long-duration study, live growing cells showed good removal rates (87% and 96.5%). The primary mechanism behind this pesticide elimination from the aquatic phase was attributed to adsorption on algal cells, evident from the rapid removal observed in the short duration of the experiments. These findings underscore the potential of microalgae, particularly *C. vulgaris*, in the bioremediation of water contaminated with pesticides. *Chlorella sorokiniana* was involved in the degradation of malathion (100 mg/l), wherein, it could degrade more than 70% of the pollutant, with the aid of the carboxylesterase enzyme (Nanda et al. 2019).

Wang et al. (2020) investigated the response of diatom, *Nitzschia palea,* to OPs: acephate and trichlorfon. They found that trichlorfon significantly inhibited the growth of *N. palea*, leading to a substantial decrease in cell density after seven days of exposure. In contrast, while acephate had less impact on diatom growth, it induced significant abnormal morphological changes in diatom valves at a higher concentration (50 mg/l) after the same exposure period. Interestingly, both pesticides had no adverse effects on the chlorophyll content of *N. palea*.

Furthermore, the study revealed that acephate and trichlorfon triggered an increase in the activities of superoxide dismutase and catalase, which are biocatalysts known to play a defensive role against environmental stresses. Additionally, malondialdehyde and extracellular polysaccharide content enhanced in response to the OPs. Importantly, *N. palea* demonstrated a good acephate removal ability (24.44–32.75%) from its environment. These biochemical responses in *N. palea* suggest that the organism makes adaptive alterations against oxidative stress stimulated by acephate and trichlorfon.

In another study by Nasiri et al. (2023), the capability of microalgae, *Dunaliella salina* in the removal of four OPs from subsurface water (wells) was explored. The research showed the growth rate of *D. salina* declined with an increase in levels of pesticides. Notably, at specific pesticide concentrations (5 ppm), the microalga exhibited the ability to grow and remove 70.25% diazinon, 47.65% ethion, 53.89% profenofos, and 41.84% chlorpyrifos. Kinetic studies indicated biodegradation rates of 0.0235 /d and 0.1769 /d, with half-lives of 2.76 and 29.49 days for the observed chemicals. Similarly, *Scenedesmus* sp. was involved in the degradation of NEO, thiamethoxam. The microalgae completely metabolized the pesticide up to concentrations of 5–40 mg/l, and exhibited 97.5% at 60 mg/l, at the end of 12 days after inoculation (Quan et al. 2023). Thus, these findings underscore the remarkable resilience of microalga and diatoms to environmental stress, their capacity to thrive in pesticide-contaminated water, and their efficacy in pesticide removal.

## 8.5 Novel Enzymes or Genes Involved in Bioremediation of Pollutants

Enzymes are protein that acts as bio-catalysts aiding in speeding up the transformation of reactants into products by lowering the activation energy of the reaction. It is believed that enzyme-based bioremediation is one of the best strategies of pollutant degradation that is currently being researched for various reasons. Enzymes are the most efficient tools of bioremediation that advance all the chemical changes on the pollutants. The specificity of enzymes is typically versatile enough to affect several chemicals with related forms. Enzymes can also be engineered to be more efficient and stable under particular conditions or with certain substrates (Mousavi et al. 2021). Enzymes work well in the presence of microbial antagonists or predators as well as low levels of pollutant concentrations. Moreover, they are more mobile than microbes and act on a specific substrate because of their smaller size (Rao et al. 2010). The use of enzymes is advantageous in various ways, such as it offers a more rapid process, easier to manipulate, more effective, no nutrient requirement and reduces by-product formation when compared to the microbial process. Alternatively, genetically modified organisms (GMOs) can be harnessed in the production process to overcome wild strain adversities, such as stress conditions (Saravanan et al. 2021; Bittencourt et al. 2023). Even though enzymatic degradation produces better outcomes than other bioremediation techniques, choosing the right enzyme-producing microbial counterparts is crucial, and involves higher costs in enzymatic production and purification on a large scale. Hence, immobilization strategies are employed to increase the process efficacy and to reduce costs pertaining to production and downstream processing. Immobilized enzymes are capable of working on a broader spectrum of pH and temperature, further improving enzyme stability (Ba et al. 2013; Antecka et al. 2018; Bilal and Asgher 2016). Additionally, genetic engineering approaches are adopted to produce large amounts of purified enzymes to degrade pollutants in an industrial scale (Huy et al. 2021).

The common representatives of the microbe-derived enzymes used for remediation techniques include cytochrome P450, laccases, dehydrogenases, hydrolases, dehalogenases, proteases, lipases, esterases, oxygenases and peroxidases that possess the potential to degrade pollutants including polymers, halogenated compounds, hydrocarbons, dyes, detergents, agrochemicals. Such compounds are degraded or detoxified by mechanisms like oxidation, reduction, elimination and ring opening (Sharma et al. 2018; Bhandari et al. 2021). Several strains of fungi, bacteria, archaea and algae produce such kinds of enzymes for bioremediation of pollutants. For example, fungal laccases have been used to oxidize an array of environmental contaminants, including pesticides, dyes, pharmaceuticals, and inorganic substances that originate from different industries. Tanaka et al. (2001) used laccases from *Trametes* sp. for the degradation of bidphenol A, phenolic endocrine-disrupting chemicals, nonylphenols, octylphenol, and ethynylestradiol. Wang et al. (2018) used immobilized laccases for the removal of PAHs, pyrene and benzo[α]pyrene in polluted soil. The immobilized bacterial laccases from *B. cereus* and *P. parafulva* isolated from the Western Ghats of India were used by Khaled et al. (2022) for the decolorization of three azo dyes T-blue, yellow GR and orange 3R. The enzymes of *P. parafulva* showed better removal of dyes with an efficiency of 89.21–91.69%. Furthermore, the enzyme cytochrome P450 also has greater potential in the biotransformation of toxic chemicals in our ecosystem. It is a superfamily of heme enzymes found across all three domains of life (Bhandari et al. 2021). The degradation of fipronil by cytochrome P450 of white rot fungi *Trametes versicolor* was demonstrated by Wolfand et al. (2016). The heterologous expression of cytochrome P450 from the novel bacterial strain *Microbacterium keratanolyticum* ZY dechlorinated dichloromethane with an efficiency of 90.6% (Hu et al. 2021). The amidase enzyme encoded by *ramA* gene derived from *Enterobacter* sp. Z1 was used by Zhang et al. (2020b) to degrade triazophos, methamidophos, and carbofuran in wastewater simultaneously with efficiencies of 98.3%, 100%, and 98.7%, respectively, using a batch bioreactor. Esterases are another class of enzymes that are involved in bioremediation. Puranik et al. (2023) reported the involvement of esterases in the degradation of phthalate esters like dibutyl phthalate by *A. flavus*. Further, in *E. coli*, the phthalate esters were also degraded by a thermostable esterase enzyme EstS1 from *Sulfobacillus acidophilus* DSM10332, which was expressed in the host *E. coli* cells heterologously (Zhang et al. 2014).

Recent advances in enzyme technology, such as enzyme engineering, are employed to improve the enzyme characteristics such as optimum pH, optimum temperature, efficiency, substrate specificity and stereoselectivity. It involves several techniques such as DNA shuffling, random priming, staggered extension process and site-directed mutagenesis (Bhatt et al. 2021). Kumar et al. (2012) studied engineered CYP102A1, which is a cytochrome P450 monooxygenase for the oxidation of hydrocarbons. The enhanced activity toward PAHs, PCBs, and linear alkanes were obtained. An enhanced degradation of 1,2,3-trichloropropane was obtained by trichloropropane by an engineered haloalkane dehalogenase with the constitutive *dhlA* promoter *P. putida* MC4 (Samin et al. 2014). A laccase enzyme LacH5 from *Bacillus atrophaeus,* which has been engineered for fusion expression with *Vitreoscilla* hemoglobin and a replacement from a strong promoter, showed improved catalytic activity toward the oxidation of 2,2′-azinobis(3-ethylbenzothiazolin-6-sulfnic acid) ammonium salt (ABTS) which is a polycyclic aromatic hydrocarbon (Wang et al. 2022). The wild-type lignin peroxidase enzyme from *P. chrysosporium* subjected to saturation mutagenesis at the catalytic Trp171 environment shows 13-folds of higher catalytic activity toward decolorization of azo dyes (Ilić Đurđić et al. 2021).

Over 99% of microorganisms in terrestrial or aqueous ecosystems are uncultured, which means that most of the promising pollutants-degrading candidates have remained uncharacterized. Advanced technologies such as next-generation sequencing and omics have enabled numerous insights into the functional potential of bacteria and fungi to catabolize various pollutants. In recent years, through the functional metagenomics approach, several genes and assimilation pathways involved in contaminant degradation have been deciphered. Functional metagenomics is a highly effective strategy for accelerating the discovery of novel biocatalysts from the enormous diversity of uncultured microorganisms since it involves the allocation of functions to proteins encoded by all of the genomes of a microbial community without the need for isolation and cultivation (Ufarté et al. 2015). Extraction of DNA and screening of metagenomes are the two major steps in metagenomic approach. Several important pollutant-degrading enzymes have been identified using this method. Suenaga et al. (2007) through a metagenomics approach, retrieved novel estradiol dioxygenases (EDOs) that are capable of degrading aromatic compounds from the metagenomic DNA isolated from sludge. Jiao et al. (2013) identified cold-active phthalate esters hydrolase that degrades phthalate esters by screening the metagenomics library obtained in biofilms formed in wastewater treatment plants. A novel gene encoding LC-cutinase, which is a homolog of the cutinase enzyme, was identified from fosmid library of a leaf-branch compost metagenome which had ability to degrade poly (ε-caprolactone) and polytheneterephthalate (PET) with specific PET degrading activity of 12 mg/h/mg of enzyme at pH 8 and 50°C (Sulaiman et al. 2012). Some of the novel pollutants, including pesticide-degrading enzymes/genes identified by molecular and metagenomics approaches are shown in (Table 8.2).

**Table 8.2** List of novel enzymes/genes with special alterations/abilities involved in bioremediation.

| Sl. No. | Novel enzyme/gene | Organism of origin | Substrate specificity | Optimum conditions for degradation | Special characters or remarks | References |
|---|---|---|---|---|---|---|
| 1. | *pytY* (esterase) | *Ochrobactrum anthropi* YZ-1 | Pyrethroid | pH 7.5 and temperature of 35°C | Ability to degrade a broad spectrum of pyrethroids. Stable even at a wider spectrum of pH and temperature | Ruan et al. (2013) |
| 2. | Ring hydroxylating dioxygenases | Unknown | Polycyclic Aromatic Hydrocarbons (PAH) | – | Retrieved from prokaryotic metagenomic DNA in polluted soil. The enzymes exhibited higher activity when supplemented with electron carriers from a Sphingomonad | Chemerys et al. (2014) |
| 3. | Laccase | *T. versicolor* | Bisphenol A and sulphamethoxazole | – | The enzyme is expressed on the cell surface of *Saccharomyces cerevisiae* to improve activity and renewability | Chen et al. (2016) |
| 4. | CE_Ubrb (carboxyl-ester hydrolase) | Unknown | Carbamates | – | Retrieved from the metagenomics DNA of Bovine rumen microbiome | Ufarté et al. (2017) |
| 5. | PETase | *Ideonella sakaiensis* 201-F6 | Polyethylene terephthalate (PET) and polyethylene-2,5-furandicarboxylate (PEF) | – | The binding cleft in the active site is narrowed down by mutating two active-site residues to conserved amino acids in cutinases | Austin et al. (2018) |
| 6. | *cpd* (chlorpyriphos hydrolase) | *Paracoccus* sp. TRP | Chlorpyriphos | – | Ser155 in the catalytic site is the key amino acid for chlorpyriphos degradation | Fan et al. (2018) |
| 7. | PETase | *Ideonella sakaiensis* 201-F6 | PET | – | Protein engineering of the enzyme was performed by replacing Arg280 present at a distance of 23 Å from the catalytic site to Ala to enhance PET degradation | Joo et al. (2018) |
| 8. | Est3385 (Esterase) | *Rhodopseudomonas palustris* PSB-S | Pyrethroid | pH 6.0 and temperature 35°C | Enzymes can degrade several pyrethroids. It was found to This enzyme could detoxify fenpropathrin with a notable $K_m$ (0.734 ± 0.013 mmol/l) and $V_{max}$ (0.918 ± 0.025 U/μg) | Luo et al. (2018) |

(*Continued*)

**Table 8.2** (Continued)

| Sl. No. | Novel enzyme/gene | Organism of origin | Substrate specificity | Optimum conditions for degradation | Special characters or remarks | References |
|---|---|---|---|---|---|---|
| 9. | PETase | *I. sakaiensis* 201-F6 | PET | – | Mutation of six key residues in substrate-binding groove for creation of space in active site, enhancing the hydrophobic nature of the active site, enhancing the binding ability of amino acids and PET in the active site | Ma et al. (2018) |
| 10. | (AzoRed2) Azoreductase | *Streptomyces* sp. S27 | Azo dyes | pH 7 and 55°C | To overcome the requirement of the co-enzyme by azoreductase while degrading Azo dyes, glucose 1-dehydrogenase (BsGDH) from *B. subtilis* was introduced as a regenerating enzyme for construction of cellular recycling system | Dong et al. (2019) |
| 11. | IifC_IDO3 (Indole oxygenase) | *Burkholderia* sp. IDO3 | Indole | Temperature 25–35°C; pH 4.0–9.0 | It belongs to a novel group of indole oxygenases | Ma et al. (2019) |
| 12. | Dihydroxybiphenyl 1,2-dioxygenase (BphC-SD3) and catechol 2,3-dioxygenase (C23O-RW1) | Unknown | Catecholic compounds | – | Isolated from the metagenomics DNA of sludge and river water samples by applying a functional metagenomics approach | Sidhu et al. (2019) |
| 13. | CarEW (Esterase) | *Bacillus* sp. *K91* | Diisobutyl Phthalate (DiBP) | Temperature of 45°C and pH 7.5 | It is expressed on the cell surface of *E. coli* for utilization as a whole-cell biocatalyst in DiBP cleavage | Ding et al. (2020) |
| 14. | *cehA* (hydrolase) | *Sphingobium* sp. CFD-1 | Carbofuran | Temperature of 40°C and pH 7.0 | The site-directed mutation proved that His313, His315, His453, and His495 had crucial parts in the CehA-mediated hydrolysis of carbofuran | Jiang et al. (2020) |
| 15. | *amdS* (EC hydrolase) | *Aspergillus oryzae* | Ethyl carbamate | pH 9.6 and temperature of 40°C | The gene was expressed in the yeast receptor, which exhibited better degradation of ethyl carbamate | Masaki et al. (2020) |
| 16. | ($EthABD_{B2}$) Cytochrome P450 family | *Rhodococcus* sp. *B2* | Chloracetanilide herbicides | Temperature of 30°C and pH 7.5 | The recombinant enzyme $EthABD_{B2}$ has a dual function of O-dealkylation and N-dealkoxymethylation toward chloroacetanilide herbicides | Liu et al. (2021) |
| 17. | $pLac_{Gy}$ (Laccase) | *Geobacillus yumthangensis* | Aromatic organic pollutants | pH 5 and temperature of 60°C | The enzyme is stable at wide spectrum of temperatures and pH. It can catalyze the cleavage of organic dyes and toxic pollutants like guaiacol, bisphenol A and phenol with a redox mediator | Sharma and Leung (2021) |

**Table 8.2** (Continued)

| Sl. No. | Novel enzyme/gene | Organism of origin | Substrate specificity | Optimum conditions for degradation | Special characters or remarks | References |
|---|---|---|---|---|---|---|
| 18. | CYP5208A3 (Cytochrome p450 monooxygenase) | *Cunninghamella elegans* | Pyrethroids | – | Co-expression with cytochrome P450 reductase of *C. elegans* in heterologous system (*Pichia pastoris X-33*) gave biotransformation of pyrethroids | Khan and Murphy (2022) |
| 19. | Amidase (TccS) and phenol oxidase ($PH_{IND}$) | *Rhodococcus rhodochrous* BX2 | Triclocarban (TCC) | – | These enzymes together are accountable for the degradation of TCC and the intermediate 4-chloroaniline (4-CA) | Li et al. (2022) |

## 8.6 Conclusion

Pollutants, especially agrochemicals, are finding their way into environmental matrices due to their production and abuse. These not only hamper the soil, water and air, but hinder plant growth. They enter plant systems and often end up being consumed by humans and animals, thus causing health risks. However, scientists have found numerous ways of detoxifying them, among which, microbe-mediated process is the most eco-friendly, cost-effective, and efficient one. Microorganisms either consume these chemicals due to their wide metabolic capability or break them down, thereby transforming them into less toxic products. Thus, there is a need to harness and explore these microorganisms for fast and potent bioremediation of pollutants, especially agrochemicals. The use of novel techniques like metagenomics and genetic engineering can help in improving the process of biodegradation, thus yielding sustainability.

## References

Abdi, D.E., Owen, J.S. Jr., Brindley, J.C. et al. (2020). Nutrient and pesticide remediation using a two-stage bioreactor-adsorptive system under two hydraulic retention times. *Water Research* 170: 115311.

Aktar, M.W., Sengupta, D., and Chowdhury, A. (2009). Impact of pesticides use in agriculture: their benefits and hazards. *Interdisciplinary Toxicology* 2 (1): 1.

Aldas-Vargas, A., van der Vooren, T., Rijnaarts, H.H., and Sutton, N.B. (2021). Biostimulation is a valuable tool to assess pesticide biodegradation capacity of groundwater microorganisms. *Chemosphere* 280: 130793.

Ali, M., Kazmi, A.A., and Ahmed, N. (2014). Study on effects of temperature, moisture and pH in degradation and degradation kinetics of aldrin, endosulfan, lindane pesticides during full-scale continuous rotary drum composting. *Chemosphere* 102: 68–75.

Ali, M., Gani, K.M., Kazmi, A.A., and Ahmed, N. (2016). Degradation of aldrin and endosulfan in rotary drum and windrow composting. *Journal of Environmental Science and Health, Part B* 51 (5): 278–286.

Ali, H., Khan, E., and Ilahi, I. (2019). Environmental chemistry and ecotoxicology of hazardous heavy metals: environmental persistence, toxicity, and bioaccumulation. *Journal of Chemistry* 2019.

Alvarenga, N., Birolli, W.G., Seleghim, M.H., and Porto, A.L. (2014). Biodegradation of methyl parathion by whole cells of marine-derived fungi *Aspergillus sydowii* and *Penicillium decaturense*. *Chemosphere* 117: 47–52.

Antecka, K., Zdarta, J., Siwińska-Stefańska, K. et al. (2018). Synergistic degradation of dye wastewaters using binary or ternary oxide systems with immobilized laccase. *Catalysts* 8 (9): 402. https://doi.org/10.3390/catal8090402.

ATSDR: Agency for Toxic Substance and Disease Registry (1997). *Toxicological Profiles of Dichlorvos*, 1–178. Atlanta, GA: U.S. Department of Health and Human Services.

ATSDR: Agency for Toxic Substance and Disease Registry (2000). *Prepared by Clement International Corp., Under Contract*, 205, 88-0608.

Austin, H.P., Allen, M.D., Donohoe, B.S. et al. (2018). Characterization and engineering of a plastic-degrading aromatic polyesterase. *Proceedings of the National Academy of Sciences of the United States of America* 115 (19): E4350–E4357. https://doi.org/10.1073/pnas.1718804115.

Azubuike, C.C., Chikere, C.B., and Okpokwasili, G.C. (2016). Bioremediation techniques–classification based on site of application: principles, advantages, limitations and prospects. *World Journal of Microbiology and Biotechnology* 32: 1–18.

Ba, S., Arsenault, A., Hassani, T. et al. (2013). Laccase immobilization and insolubilization: from fundamentals to applications for the elimination of emerging contaminants in wastewater treatment. *Critical Reviews in Biotechnology* 33 (4): 404–418. https://doi.org/10.3109/07388551.2012.725390.

Banerjee, P., Mukhopadhayay, A., and Das, P. (2016). Advances in bioremediation for removal of toxic dye from different streams of wastewater. In: *Toxicity and Waste Management Using Bioremediation* (ed. A.K. Rathoure and V.K. Dhatwalia), 266–278. IGI Global.

Barchanska, H., Sajdak, M., Szczypka, K. et al. (2017). Atrazine, triketone herbicides, and their degradation products in sediment, soil and surface water samples in Poland. *Environmental Science and Pollution Research* 24: 644–658.

Barr, D.P. and Aust, S.D. (1994). Pollutant degradation by white rot fungi. *Reviews of Environmental Contamination and Toxicology: Continuation of Residue Reviews* 49–72.

Battaglin, W.A., Smalling, K.L., Anderson, C. et al. (2016). Potential interactions among disease, pesticides, water quality and adjacent land cover in amphibian habitats in the United States. *Science of the Total Environment* 566: 320–332.

Bell, C.H., Wong, J., Parsons, K. et al. (2022). First full-scale in situ propane biosparging for co-metabolic bioremediation of 1,4-dioxane. *Groundwater Monitoring and Remediation* 42 (4): 54–66.

Belta, G.D., Likata, P., Bruzzese, A. et al. (2006). Level and congener pattern of PCBs and OCPs residues in blue-fin tuna (*Thunnus thynnus*) from the straits of Messina (Sicily, Italy). *Environment International* 32: 705–710.

Beltran-Flores, E., Sarra, M., and Blanquez, P. (2021). Pesticide bioremediation by *Trametes versicolor*: application in a fixed-bed reactor, sorption contribution and bioregeneration. *Science of the Total Environment* 794: 148386.

Bertram, M.G., Martin, J.M., McCallum, E.S. et al. (2022). Frontiers in quantifying wildlife behavioural responses to chemical pollution. *Biological Reviews* 97 (4): 1346–1364.

Bhadouria, R., Das, S., Kumar, A. et al. (2020). Mycoremediation of agrochemicals. In: *Agrochemicals Detection, Treatment and Remediation* (ed. M. Narasimha Vara Prasad), 593–620. Oxford: Butterworth-heinemann https://doi.org/10.1016/B978-0-08-103017-2.00022-2.

Bhakta, J.N., Munekage, Y., Ohnishi, K. et al. (2014). Isolation and characterization of cadmium-and arsenic-absorbing bacteria for bioremediation. *Water, Air, & Soil Pollution* 225: 1–10.

Bhandari, S., Poudel, D.K., Marahatha, R. et al. (2021). Microbial enzymes used in bioremediation. *Journal of Chemistry* 2021: 1–17. https://doi.org/10.1155/2021/8849512.

Bharagava, R.N., Saxena, G., and Mulla, S.I. (2020). Introduction to industrial wastes containing organic and inorganic pollutants and bioremediation approaches for environmental management. In: *Bioremediation of Industrial Waste for Environmental Safety: Volume I: Industrial Waste and Its Management* (ed. G. Saxena and R. Bharagava), 1–18. Singapore: Springer.

Bhatt, P., Gangola, S., Bhandari, G. et al. (2021). New insights into the degradation of synthetic pollutants in contaminated environments. *Chemosphere* 268: 128827. https://doi.org/10.1016/j.chemosphere.2020.128827.

Bhuiyan, M.A.H., Rahman, M.H., Uddin, M.A. et al. (2021). Contamination of pond and canal water by residues of organophosphorus and carbamate pesticides in Feni district, Bangladesh. *Environmental Sustainability* 4: 191–197.

Bilal, M. and Asgher, M. (2016). Enhanced catalytic potentiality of *Ganoderma lucidum* IBL-05 manganese peroxidase immobilized on sol-gel matrix. *Journal of Molecular Catalysis B: Enzymatic* 128: 82–93. https://doi.org/10.1016/j.molcatb.2016.03.013.

Birolli, W.G., Arai, M.S., Nitschke, M., and Porto, A.L. (2019). The pyrethroid (±)-lambda-cyhalothrin enantioselective biodegradation by a bacterial consortium. *Pesticide Biochemistry and Physiology* 156: 129–137.

Bittencourt, G.A., de Souza Vandenberghe, L.P., Martínez-Burgos, W.J. et al. (2023). Emerging contaminants bioremediation by enzyme and nanozyme-based processes – a review. *Iscience* 26: 106785. https://doi.org/10.1016/j.isci.2023.106785.

Bokade, P., Purohit, H.J., and Bajaj, A. (2021). Myco-remediation of chlorinated pesticides: insights into fungal metabolic system. *Indian Journal of Microbiology* 61: 237–249.

Bose, S., Kumar, P.S., Vo, D.V.N. et al. (2021). Microbial degradation of recalcitrant pesticides: a review. *Environmental Chemistry Letters* 19: 3209–3228.

Bouwer, H. (1989). Agriculture and groundwater quality. *Civil Engineering* 59 (7): 60.

Bruinsma, J. (ed.) (2003). *World Agriculture: Towards 2015/2030: An FAO Perspective*. Earthscan.

Burke, R.D., Todd, S.W., Lumsden, E. et al. (2017). Developmental neurotoxicity of the organophosphorus insecticide chlorpyrifos: from clinical findings to preclinical models and potential mechanisms. *Journal of Neurochemistry* 142: 162–177.

Caballero-Casero, N., Castro, G., Bastiaensen, M. et al. (2021). Identification of chemicals of emerging concern in urine of Flemish adolescents using a new suspect screening workflow for LC-QTOF-MS. *Chemosphere* 280: 130683.

Candeias, C., Avila, P.F., Sequeira, C. et al. (2022). Potentially toxic elements dynamics in the soil rhizospheric-plant system in the active volcano of Fogo (Cape Verde) and interactions with human health. *Catena* 209: 105843.

Carazo-Rojas, E., Pérez-Rojas, G., Pérez-Villanueva, M. et al. (2018). Pesticide monitoring and ecotoxicological risk assessment in surface water bodies and sediments of a tropical agro-ecosystem. *Environmental Pollution* 241: 800–809.

Ccanccapa, A., Masiá, A., Navarro-Ortega, A. et al. (2016). Pesticides in the Ebro River basin: occurrence and risk assessment. *Environmental Pollution* 211: 414–424.

Chakraborty, P., Zhang, G., Li, J. et al. (2010). Selected organochlorine pesticides in the atmosphere of major Indian cities: levels, regional versus local variations, and sources. *Environmental Science & Technology* 44 (21): 8038–8043.

Chemerys, A., Pelletier, E., Cruaud, C. et al. (2014). Characterization of novel polycyclic aromatic hydrocarbon dioxygenases from the bacterial metagenomic DNA of a contaminated soil. *Applied and Environmental Microbiology* 80 (21): 6591–6600. https://doi.org/10.1128/AEM.01883-14.

Chen, X.Y., Shao, J.Z., Xiang, L.X., and Liu, X.M. (2006). Involvement of apoptosis in malathion-induced cytotoxicity in a grass carp (Ctenopharyngodon idellus) cell line. *Comparative Biochemistry and Physiology Part C: Toxicology & Pharmacology* 142 (1–2): 36–45.

Chen, Y., Stemple, B., Kumar, M., and Wei, N. (2016). Cell surface display fungal laccase as a renewable biocatalyst for degradation of persistent micropollutants bisphenol A and sulfamethoxazole. *Environmental Science & Technology* 50 (16): 8799–8808. https://doi.org/10.1021/acs.est.6b01641.

Chopra, A.K., Sharma, M.K., and Chamoli, S. (2011). Bioaccumulation of organochlorine pesticides in aquatic system – an overview. *Environmental Monitoring and Assessment* 173: 905–916.

Corbin, M., Daiss, R., Wolf, J. K., Hurley, P., Spatz, D., & Branch, E. R. A. (2009). *Risks of Chlorpyrifos Use to Federally Threatened and Endangered California Red-Legged Frog (Rana Aurora Draytonii) California Tiger Salamander (Ambystoma Californiense)* USEPA Office of Pesticide Program. Available online via: http://www. epa. gov/espp/litstatus/effects/redleg-frog/chlorpyrifos/analysis. pdf

Costa, L.G. (2008). Toxic effects of pesticides. *Casarett and Doull's Toxicology: The Basic Science of Poisons* 8: 883–930.

Costa, L.G. (2018). Organophosphorus compounds at 80: some old and new issues. *Toxicological Sciences* 162 (1): 24–35.

Crawford, A.J. and Bhattacharya, S.K. (1987). Excessive intracellular zinc accumulation in cardiac and skeletal muscles of dystrophic hamsters. *Experimental Neurology* 95 (2): 265–276.

Cribb, J. (2021). *Earth Detox: How and Why We Must Clean Up Our Planet*. Cambridge University Press.

Cristorean, C., Micle, V., and Sur, I.M. (2016). A critical analysis of ex-situ bioremediation technologies of hydrocarbon polluted soils. *EcoTerra (Journal of Environmental Research and Protection)* 13: 17–29.

Das, S.K. and Mukherjee, I. (2012). Effect of moisture and organic manure on persistence of flubendiamide in soil. *Bulletin of Environmental Contamination and Toxicology* 88: 515–520.

Debnath, M. and Khan, M.S. (2017). Health concerns of pesticides. In: *Pesticide Residue in Foods* (ed. M. Khan and M. Rahman), 103–118. Cham: Springer.

DeMars, C., Wang, R., Grieneisen, M.L. et al. (2021). Assessment of pyrethroid contamination and potential mitigation strategies in California central coast surface waters. *Journal of Environmental Management* 278: 111507.

Desforges, J.P., Hall, A., McConnell, B. et al. (2018). Predicting global killer whale population collapse from PCB pollution. *Science* 361 (6409): 1373–1376.

Devi, N.L., Qi, S., Chakraborty, P. et al. (2011). Passive air sampling of organochlorine pesticides in a northeastern state of India, Manipur. *Journal of Environmental Sciences* 23 (5): 808–815.

Dey, U., Chatterjee, S., and Mondal, N.K. (2016). Isolation and characterization of arsenic-resistant bacteria and possible application in bioremediation. *Biotechnology Reports* 10: 1–7.

Ding, Z., Yang, Y., Jin, H. et al. (2004). Acute toxicity and bio-concentration factor of three pesticides on *Brachydanio rerio*. *Ying Yong Sheng Tai Xue Bao = The Journal of Applied Ecology* 15 (5): 888–890.

Ding, J., Zhou, Y., Wang, C. et al. (2020). Development of a whole-cell biocatalyst for diisobutyl phthalate degradation by functional display of a carboxylesterase on the surface of *Escherichia coli*. *Microbial Cell Factories* 19: 1–11. https://doi.org/10.1186/s12934-020-01373-6.

Divya, B. and Deepak Kumar, M. (2011). Plant-microbe interaction with enhanced bioremediation. *Research Journal of Biotechnology* 6: 4.

Dong, H., Guo, T., Zhang, W. et al. (2019). Biochemical characterization of a novel azoreductase from *Streptomyces* sp.: application in eco-friendly decolorization of azo dye wastewater. *International Journal of Biological Macromolecules* 140: 1037–1046. https://doi.org/10.1016/j.ijbiomac.2019.08.196.

Dou, R., Sun, J., Deng, F. et al. (2020). Contamination of pyrethroids and atrazine in greenhouse and open-field agricultural soils in China. *Science of the Total Environment* 701: 134916.

Dueholm, M.S., Marques, I.G., Karst, S.M. et al. (2015). Survival and activity of individual bioaugmentation strains. *Bioresource Technology* 186: 192–199.

Dzionek, A., Wojcieszyńska, D., and Guzik, U. (2016). Natural carriers in bioremediation: a review. *Electronic Journal of Biotechnology* 23: 28–36.

Ellegaard-Jensen, L., Knudsen, B.E., Johansen, A. et al. (2014). Fungal–bacterial consortia increase diuron degradation in water-unsaturated systems. *Science of the Total Environment* 466: 699–705.

Ensminger, M., Bergin, R., Spurlock, F., and Goh, K.S. (2011). Pesticide concentrations in water and sediment and associated invertebrate toxicity in Del Puerto and Orestimba Creeks, California, 2007–2008. *Environmental Monitoring and Assessment* 175: 573–587.

Erguven, G.O. and Demirci, U. (2021). Using *Ochrobactrum thiophenivorans* and *Sphingomonas melonis* for bioremediation of imidacloprid. *Environmental Technology & Innovation* 21: 101236.

Eskenazi, B., Bradman, A., and Castorina, R. (1999). Exposures of children to organophosphate pesticides and their potential adverse health effects. *Environmental Health Perspectives* 107 (suppl 3): 409–419.

Fan, S., Li, K., Yan, Y. et al. (2018). A novel chlorpyrifos hydrolase CPD from *Paracoccus* sp. TRP: molecular cloning, characterization and catalytic mechanism. *Electronic Journal of Biotechnology* 31: 10–16. https://doi.org/10.1016/j.ejbt.2017.10.009.

Farag, A.T., Karkour, T.A.Z., and El Okazy, A. (2006). Developmental toxicity of orally administered technical dimethoate in rats. *Birth Defects Research Part B: Developmental and Reproductive Toxicology* 77 (1): 40–46.

Fast, S.A., Gude, V.G., Truax, D.D. et al. (2017). A critical evaluation of advanced oxidation processes for emerging contaminants removal. *Environmental Processes* 4: 283–302.

Fosu-Mensah, B.Y., Okoffo, E.D., Darko, G., and Gordon, C. (2016). Assessment of organochlorine pesticide residues in soils and drinking water sources from cocoa farms in Ghana. *Springerplus* 5: 1–13.

Frutos, F.J.G., Escolano, O., García, S. et al. (2010). Bioventing remediation and ecotoxicity evaluation of phenanthrene-contaminated soil. *Journal of Hazardous Materials* 183 (1–3): 806–813.

Fulton, M.H. and Key, P.B. (2001). Acetylcholinesterase inhibition in estuarine fish and invertebrates as an indicator of organophosphorus insecticide exposure and effects. *Environmental Toxicology and Chemistry: An International Journal* 20 (1): 37–45.

Gallego, J.L., Loredo, J., Llamas, J.F. et al. (2001). Bioremediation of diesel-contaminated soils: evaluation of potential in situ techniques by study of bacterial degradation. *Biodegradation* 12: 325–335.

Ganaie, M.I., Jan, I., Mayer, A.N. et al. (2023). Health risk assessment of pesticide residues in drinking water of upper Jhelum region in Kashmir Valley-India by GC-MS/MS. *International Journal of Analytical Chemistry* 2023: 1–16.

Ganguli, A., Choudhury, D., and Chakrabarti, G. (2014). 2,4-Dichlorophenoxyacetic acid induced toxicity in lung cells by disruption of the tubulin-microtubule network. *Toxicology Research* 3 (2): 118–130.

Gao, C., Jin, X., Ren, J. et al. (2015). Bioaugmentation of DDT-contaminated soil by dissemination of the catabolic plasmid pDOD. *Journal of Environmental Sciences* 27: 42–50.

Gbadamosi, M.R., Abdallah, M.A.E., and Harrad, S. (2021). A critical review of human exposure to organophosphate esters with a focus on dietary intake. *Science of the Total Environment* 771: 144752.

Gibbs, J.L., Yost, M.G., Negrete, M., and Fenske, R.A. (2017). Passive sampling for indoor and outdoor exposures to chlorpyrifos, azinphos-methyl, and oxygen analogs in a rural agricultural community. *Environmental Health Perspectives* 125 (3): 333–341.

Giri, D.D., Singh, S.K., Giri, A. et al. (2021). Bioremediation potential of methylotrophic bacteria. In: *Microbe Mediated Remediation of Environmental Contaminants* (ed. A. Kumar, V.K. Singh, P. Singh, and V.K. Mishra), 199–207. Cambridge, MA: Woodhead Publishing https://doi.org/10.1016/B978-0-12-821199-1.00017-1.

Gond, D.P., Jha, S.S., Kumar, A., and Singh, S.K. (2021). Plant growth promoting bacteria and its role in green remediation. In: *Sustainable Environmental Clean-up* (ed. V.K. Mishra and A. Kumar), 149–163. Cambridge, MA: Elsevier https://doi.org/10.1016/B978-0-12-823828-8.00007-4.

Grandjean, P. and Landrigan, P.J. (2014). Neurobehavioural effects of developmental toxicity. *The Lancet Neurology* 13 (3): 330–338.

Gruber, K. (2018). Cleaning up pollutants to protect future health. *Nature*, 555(7695), S20–S22. doi:https://doi.org/10.1038/d41586-018-02481-5

Guermouche Mrassi, A., Bensalah, F., Gury, J., and Duran, R. (2015). Isolation and characterization of different bacterial strains for bioremediation of n-alkanes and polycyclic aromatic hydrocarbons. *Environmental Science and Pollution Research* 22: 15332–15346.

Gupta, R. (2006). *Toxicology of Organophosphate and Carbamate Compounds*. San Diego, Estados Unidos: Academic Press.

Gupta, A., Singh, M., Pandey, K.D., and Kumar, A. (2018). Biodegradation of xenobiotics using cyanobacteria. *Journal of Scientific Research* 62: 35–40.

Hammam, F.M. and Abd el Mottaleb, E.M. (2007). Studies of the genotoxic and histopathological effects of the organophosphorous insecticide 'Profenofos' on white rats. *The Egyptian Journal of Hospital Medicine* 29 (1): 685–706.

Hassan, H.A. and Alghuthaymi, M.A. (2022). Biotechnology methods for succession of bacterial communities in polychlorinated biphenyls (PCBs) contaminated soils and isolation novel PCBs-degrading bacteria. *Scientific Reports* 12 (1): 19223.

Hiller, E., Zemanová, L., Sirotiak, M., and Jurkovič, L.U. (2011). Concentrations, distributions, and sources of polychlorinated biphenyls and polycyclic aromatic hydrocarbons in bed sediments of the water reservoirs in Slovakia. *Environmental Monitoring and Assessment* 173: 883–897.

Hooghe, R.J., Devos, S., and Hooghe-Peters, E.L. (2000). Effects of selected herbicides on cytokine production in vitro. *Life Sciences* 66 (26): 2519–2525.

Hu, B., Shao, S., Ni, H. et al. (2020a). Current status, spatial features, health risks, and potential driving factors of soil heavy metal pollution in China at province level. *Environmental Pollution* 266: 114961.

Hu, K., Peris, A., Torán, J. et al. (2020b). Exploring the degradation capability of *Trametes versicolor* on selected hydrophobic pesticides through setting sights simultaneously on culture broth and biological matrix. *Chemosphere* 250: 126293.

Hu, J., Zhang, Y., Wu, Y. et al. (2021). Heterologous expression of bacterial cytochrome P450 from *Microbacterium keratanolyticum* ZY and its application in dichloromethane dechlorination. *Environmental Pollution* 287: 117597. https://doi.org/10.1016/j.envpol.2021.117597.

Hu, K., Barbieri, M.V., López-García, E. et al. (2022). Fungal degradation of selected medium to highly polar pesticides by *Trametes versicolor*: kinetics, biodegradation pathways, and ecotoxicity of treated waters. *Analytical and Bioanalytical Chemistry* 414: 439–449.

Huang, X., Cui, H., and Duan, W. (2020). Ecotoxicity of chlorpyrifos to aquatic organisms: a review. *Ecotoxicology and Environmental Safety* 200: 110731.

Huovinen, M., Loikkanen, J., Naarala, J., and Vähäkangas, K. (2015). Toxicity of diuron in human cancer cells. *Toxicology In Vitro* 29 (7): 1577–1586.

Hussein, M.H., Abdullah, A.M., Badr El-Din, N.I., and Mishaqa, E.S.I. (2017). Biosorption potential of the microchlorophyte *Chlorella vulgaris* for some pesticides. *Journal of Fertilizers & Pesticides* 8 (1): 1000177.

Huy, N.D., Le, N.T.M., Chew, K.W. et al. (2021). Characterization of a recombinant laccase from *Fusarium oxysporum* HUIB02 for biochemical application on dyes removal. *Biochemical Engineering Journal* 168: 107958. https://doi.org/10.1016/j. bej.2021.107958.

Ilić Đurđić, K., Ostafe, R., Prodanović, O., Đurđević Đelmaš, A., Popović, N., Fischer, R., ... Prodanović, R. (2021). Improved degradation of azo dyes by lignin peroxidase following mutagenesis at two sites near the catalytic pocket and the application of peroxidase-coated yeast cell walls. *Frontiers of Environmental Science & Engineering*, 15, 1–10. DOI:https://doi.org/10.1007/s11783-020-1311-4.

Islas-Garcia, A., Vega-Loyo, L., Aguilar-Lopez, R. et al. (2015). Evaluation of hydrocarbons and organochlorine pesticides and their tolerant microorganisms from an agricultural soil to define its bioremediation feasibility. *Journal of Environmental Science and Health, Part B* 50 (2): 99–108.

Jaipieam, S., Visuthismajarn, P., Sutheravut, P. et al. (2009). Organophosphate pesticide residues in drinking water from artesian wells and health risk assessment of agricultural communities, Thailand. *Human and ecological risk assessment* 15 (6): 1304–1316.

Jani, J.P., Raiyani, C.V., Mistry, J.S. et al. (1991). Residues of organochlorine pesticides and polycyclic aromatic hydrocarbons in drinking water of Ahmedabad City, India. *Bulletin of Environmental Contamination and Toxicology* 47: 381–385.

Ji, X., Abakumov, E., Chigray, S. et al. (2021). Response of carbon and microbial properties to risk elements pollution in arctic soils. *Journal of Hazardous Materials* 408: 124430.

Jiang, W., Gao, Q., Zhang, L. et al. (2020). Identification of the key amino acid sites of the carbofuran hydrolase CehA from a newly isolated carbofuran-degrading strain *Sphingbium* sp. CFD-1. *Ecotoxicology and Environmental Safety* 189: 109938. https://doi.org/10.1016/j.ecoenv.2019.109938.

Jiao, Y., Chen, X., Wang, X. et al. (2013). Identification and characterization of a cold-active phthalate esters hydrolase by screening a metagenomic library derived from biofilms of a wastewater treatment plant. *PLoS One* 8 (10): e75977.

Jin, Z.P., Luo, K., Zhang, S. et al. (2012). Bioaccumulation and catabolism of prometryne in green algae. *Chemosphere* 87 (3): 278–284.

John, E.M. and Shaike, J.M. (2015). Chlorpyrifos: pollution and remediation. *Environmental Chemistry Letters* 13: 269–291.

John, E.M., Varghese, E.M., Krishnasree, N., and Jisha, M.S. (2018). In situ bioremediation of chlorpyrifos by *Klebsiella* sp. isolated from pesticide contaminated agricultural soil. *International Journal of Current Microbiology and Applied Sciences* 7 (3): 1418–1429.

Joo, S., Cho, I.J., Seo, H. et al. (2018). Structural insight into molecular mechanism of poly (ethylene terephthalate) degradation. *Nature Communications* 9 (1): 382.

Kafaei, R., Arfaeinia, H., Savari, A. et al. (2020). Organochlorine pesticides contamination in agricultural soils of southern Iran. *Chemosphere* 240: 124983.

Kampa, M. and Castanas, E. (2008). Human health effects of air pollution. *Environmental Pollution* 151 (2): 362–367.

Kang, C.H., Kwon, Y.J., and So, J.S. (2016). Bioremediation of heavy metals by using bacterial mixtures. *Ecological Engineering* 89: 64–69.

Kanissery, R.G. and Sims, G.K. (2011). Biostimulation for the enhanced degradation of herbicides in soil. *Applied and Environmental Soil Science* 2011.

Karami-Mohajeri, S. and Abdollahi, M. (2011). Toxic influence of organophosphate, carbamate, and organochlorine pesticides on cellular metabolism of lipids, proteins, and carbohydrates: a systematic review. *Human & Experimental Toxicology* 30 (9): 1119–1140.

Kashyap, N., Singh, S.K., Yadav, N. et al. (2023). Biocontrol screening of endophytes: applications and limitations. *Plants* 12 (13): 2480. https://doi.org/10.3390/plants12132480.

Kaushik, C.P., Sharma, H.R. and Kaushik, A. (2012). Organochlorine pesticide residues in drinking water in the rural areas of Haryana, India. *Environmental monitoring and assessment* 184: 103–112.

Khaled, J.M., Alyahya, S.A., Govindan, R. et al. (2022). Laccase producing bacteria influenced the high decolorization of textile azo dyes with advanced study. *Environmental Research* 207: 112211. https://doi.org/10.1016/j.envres.2021.112211.

Khan, M.F. and Murphy, C.D. (2022). Cytochrome P450 5208A3 is a promiscuous xenobiotic biotransforming enzyme in *Cunninghamella elegans*. *Enzyme and Microbial Technology* 161: 110102. https://doi.org/10.1016/j.enzmictec.2022.110102.

Koli, P., Bhardwaj, N.R., and Mahawer, S.K. (2019). Agrochemicals: harmful and beneficial effects of climate changing scenarios. In: *Climate Change and Agricultural Ecosystems* (ed. K.K. Choudhary, A. Kumar, and A.K. Singh), 65–94. Woodhead Publishing.

Kour, D., Khan, S.S., Kour, H. et al. (2022). Microbe-mediated bioremediation: current research and future challenges. *Journal of Applied Biology and Biotechnology* 10 (2): 6–24.

Krupke, C.H., Hunt, G.J., Eitzer, B.D. et al. (2012). Multiple routes of pesticide exposure for honey bees living near agricultural fields. *PLoS One* 7 (1): e29268.

Kruse-Plaß, M., Hofmann, F., Wosniok, W. et al. (2021). Pesticides and pesticide-related products in ambient air in Germany. *Environmental Sciences Europe* 33: 1–21.

Kumar, S., Jin, M., & Weemhoff, J. L. (2012). Cytochrome P450-mediated phytoremediation using transgenic plants: a need for engineered cytochrome P450 enzymes. *Journal of Petroleum & Environmental Biotechnology*, 3(5). doi: https://doi.org/10.4172/2157-7463.1000127.

Kumar, S.S., Ghosh, P., Malyan, S.K. et al. (2019). A comprehensive review on enzymatic degradation of the organophosphate pesticide malathion in the environment. *Journal of Environmental Science and Health, Part C* 37 (4): 288–329.

Kumari, A. and Kaur, R. (2020). A review on morpho-physiological traits of plants under phthalates stress and insights into their uptake and translocation. *Plant Growth Regulation* 91 (3): 327–347.

Kumari, B., Madan, V.K., and Kathpal, T.S. (2008). Status of insecticide contamination of soil and water in Haryana, India. *Environmental Monitoring and Assessment* 136: 239–244.

Kumari, M., Qureshi, K.A., Jaremko, M. et al. (2022). Deciphering the role of endophytic microbiome in postharvest diseases management of fruits: opportunity areas in commercial up-scale production. *Frontiers in Plant Science* 13: 4390. https://doi.org/10.3389/fpls.2022.1026575.

Kurade, M.B., Kim, J.R., Govindwar, S.P., and Jeon, B.H. (2016). Insights into microalgae mediated biodegradation of diazinon by *Chlorella vulgaris*: microalgal tolerance to xenobiotic pollutants and metabolism. *Algal Research* 20: 126–134.

Kushwaha, M., Verma, S., and Chatterjee, S. (2016). Profenofos, an acetylcholinesterase-inhibiting organophosphorus pesticide: a short review of its usage, toxicity, and biodegradation. *Journal of Environmental Quality* 45 (5): 1478–1489.

Lango-Reynoso, F., Castañeda-Chávez, M.D.R., Landeros-Sánchez, C. et al. (2013). Cd, Cu, Hg and Pb, and organochlorines pesticides in commercially important benthic organisms coastal lagoons SW Gulf of Mexico. *Agricultural Sciences* 1 (1): 63–79.

Li, W., Dai, Y., Xue, B. et al. (2009). Biodegradation and detoxification of endosulfan in aqueous medium and soil by *Achromobacter xylosoxidans* strain CS5. *Journal of Hazardous Materials* 167 (1–3): 209–216.

Li, J., Zhang, W., Lin, Z. et al. (2021). Emerging strategies for the bioremediation of the phenylurea herbicide diuron. *Frontiers in Microbiology* 12: 686509.

Li, C., Sun, Y., Sun, G. et al. (2022). An amidase and a novel phenol hydroxylase catalyze the degradation of the antibacterial agent triclocarban by *Rhodococcus rhodochrous*. *Journal of Hazardous Materials* 430: 128444. https://doi.org/10.1016/j.jhazmat.2022.128444.

Lima, D., Viana, P., André, S. et al. (2009). Evaluating a bioremediation tool for atrazine contaminated soils in open soil microcosms: the effectiveness of bioaugmentation and biostimulation approaches. *Chemosphere* 74 (2): 187–192.

Liu, H.M., Yuan, M., Liu, A.M. et al. (2021). A bifunctional enzyme belonging to cytochrome P450 family involved in the *O*-dealkylation and *N*-dealkoxymethylation toward chloroacetanilide herbicides in *Rhodococcus* sp. B2. *Microbial Cell Factories* 20: 1–13. https://doi.org/10.1186/s12934-021-01544-z.

Longnecker, M.P., Rogan, W.J., and Lucier, G. (1997). The human health effects of DDT (dichlorodiphenyltrichloroethane) and PCBS (polychlorinated biphenyls) and an overview of organochlorines in public health. *Annual Review of Public Health* 18 (1): 211–244.

Lopez-Casaperalta, P., Febres-Molina, C., Aguilar-Pineda, J.A. et al. (2022). Peruvian native bacterial strains as potential bioremediation agents in Hg-polluted soils by artisanal mining activities in Southern Peru. *Sustainability* 14 (16): 10272.

Luo, X., Zhang, D., Zhou, X. et al. (2018). Cloning and characterization of a pyrethroid pesticide decomposing esterase gene, *Est3385*, from *Rhodopseudomonas palustris* PSB-S. *Scientific Reports* 8 (1): 7384.

Luo, X., Bing, H., Luo, Z. et al. (2019). Impacts of atmospheric particulate matter pollution on environmental biogeochemistry of trace metals in soil-plant system: a review. *Environmental Pollution* 255: 113138.

Ma, Y., Yao, M., Li, B. et al. (2018). Enhanced poly (ethylene terephthalate) hydrolase activity by protein engineering. *Engineering* 4 (6): 888–893.

Ma, Q., Liu, Z., Yang, B. et al. (2019). Characterization and functional gene analysis of a newly isolated indole-degrading bacterium *Burkholderia* sp. IDO3. *Journal of Hazardous Materials* 367: 144–151. https://doi.org/10.1016/j.jhazmat.2018.12.068.

Macaulay, B.M. and Rees, D. (2014). Bioremediation of oil spills: a review of challenges for research advancement. *Annals of Environmental Science* 8: 9–37.

Mansour, S.A. (2004). Pesticide exposure – Egyptian scene. *Toxicology* 198 (1–3): 91–115.

Maqbool, Z., Hussain, S., Imran, M. et al. (2016). Perspectives of using fungi as bioresource for bioremediation of pesticides in the environment: a critical review. *Environmental Science and Pollution Research* 23: 16904–16925.

Marco-Urrea, E. and Reddy, C.A. (2012). Degradation of chloro-organic pollutants by white rot fungi. In: *Microbial Degradation of Xenobiotics* (ed. S. Singh), 31–66. Berlin, Heidelberg: Springer.

Masaki, K., Fujihara, K., Kakizono, D. et al. (2020). *Aspergillus oryzae* acetamidase catalyzes degradation of ethyl carbamate. *Journal of Bioscience and Bioengineering* 130 (6): 577–581. https://doi.org/10.1016/j.jbiosc.2020.07.015.

Mathiesen, L., Buerki-Thurnherr, T., Pastuschek, J. et al. (2021). Fetal exposure to environmental chemicals; insights from placental perfusion studies. *Placenta* 106: 58–66.

Mehta, A., Bhardwaj, K.K., Shaiza, M., and Gupta, R. (2021). Isolation, characterization and identification of pesticide degrading bacteria from contaminated soil for bioremediation. *Biologia Futura* 72: 317–323.

Mekonen, S., Argaw, R., Simanesew, A. et al. (2016). Pesticide residues in drinking water and associated risk to consumers in Ethiopia. *Chemosphere* 162: 252–260.

Mills, P.K. (1998). Correlation analysis of pesticide use data and cancer incidence rates in California counties. *Archives of Environmental Health: An International Journal* 53 (6): 410–413.

Mnif, W., Hassine, A.I.H., Bouaziz, A. et al. (2011). Effect of endocrine disruptor pesticides: a review. *International Journal of Environmental Research and Public Health* 8 (6): 2265–2303.

Mohammed, A.M., Karttunen, V., Huuskonen, P. et al. (2018). Transplacental transfer and metabolism of diuron in human placenta. *Toxicology Letters* 295: 307–313.

Moon, J.M. and Chun, B.J. (2009). Acute endosulfan poisoning: a retrospective study. *Human & Experimental Toxicology* 28 (5): 309–316.

Moon, Y.S., Kim, M., Hong, C.P. et al. (2019). Overlapping and unique toxic effects of three alternative antifouling biocides (Diuron, Irgarol 1051®, Sea-Nine 211®) on non-target marine fish. *Ecotoxicology and Environmental Safety* 180: 23–32.

Mori, T., Ohno, H., Ichinose, H. et al. (2021). White-rot fungus *Phanerochaete chrysosporium* metabolizes chloropyridinyl-type neonicotinoid insecticides by an *N*-dealkylation reaction catalyzed by two cytochrome P450s. *Journal of Hazardous Materials* 402: 123831.

Morrissey, C.A., Mineau, P., Devries, J.H. et al. (2015). Neonicotinoid contamination of global surface waters and associated risk to aquatic invertebrates: a review. *Environment International* 74: 291–303.

Mousavi, S.M., Hashemi, S.A., Iman Moezzi, S.M. et al. (2021). Recent advances in enzymes for the bioremediation of pollutants. *Biochemistry Research International* 2021: https://doi.org/10.1155/2021/5599204.

Msibi, S.S., Chen, C.Y., Chang, C.P. et al. (2021). Indoor air exposure to multiple agricultural pesticides potentially posing the highest risk to young children. *Aerosol and Air Quality Research* 21 (9): 210062.

Nanda, M., Kumar, V., Fatima, N. et al. (2019). Detoxification mechanism of organophosphorus pesticide via carboxylestrase pathway that triggers de novo TAG biosynthesis in oleaginous microalgae. *Aquatic Toxicology* 209: 49–55.

Nasiri, M., Ahmadzadeh, H., and Amiri, A.H. (2023). Biodegradation and metabolic fate of organophosphorus pesticides in well water using *Dunaliella salina*. *International Journal of Environmental Science and Technology* 20 (1): 981–992.

Nyakundi, W.O., Magoma, G., Ochora, J., and Nyende, A.B. (2011). Biodegradation of diazinon and methomyl pesticides by white rot fungi from selected horticultural farms in rift valley and Central Kenya. *Journal of Applied Technology in Environmental Sanitation* 1: 107–124.

Orji, M., Agu, K., Ikele, M. et al. (2022). Bioremediation of glyphosate polluted soil using fungal species. *International Journal of Trend in Scientific Research and Development* 6 (7): 726–732.

Oruç, E.Ö. and Usta, D. (2007). Evaluation of oxidative stress responses and neurotoxicity potential of diazinon in different tissues of *Cyprinus carpio*. *Environmental Toxicology and Pharmacology* 23 (1): 48–55.

Osman, K.T. (2018). Polluted soils. In: *Management of Soil Problems* (ed. Osman), 333–408.

Ozkara, A. and Akyil, D. (2018). Environmental pollution and the effects of the pollutants on the ecosystem. *Türk Bilimsel Derlemeler Dergisi* 11 (2): 11–17.

Papadopoulou, E.S., Genitsaris, S., Omirou, M. et al. (2018). Bioaugmentation of thiabendazole-contaminated soils from a wastewater disposal site: factors driving the efficacy of this strategy and the diversity of the indigenous soil bacterial community. *Environmental Pollution* 233: 16–25.

Parven, A., Khan, M.S.I., Prodhan, M.D.H. et al. (2021). Human health risk assessment through quantitative screening of insecticide residues in two green beans to ensure food safety. *Journal of Food Composition and Analysis* 103: 104121.

Parween, T., Jan, S., Mahmooduzzafar, S. et al. (2016). Selective effect of pesticides on plant—A review. *Critical Reviews in Food Science and Nutrition* 56 (1): 160–179.

Pathak, P., Rai, V.K., Can, H. et al. (2022). Plant-endophyte interaction during biotic stress management. *Plants* 11 (17): 2203. https://doi.org/10.3390/plants11172203.

Pavel, L.V. and Gavrilescu, M. (2008). Overview of ex situ decontamination techniques for soil cleanup. *Environmental Engineering & Management Journal (EEMJ)* 7 (6): 815–834.

Peng, J., Lu, X., Jiang, X. et al. (2018). Degradation of atrazine by persulfate activation with copper sulfide (CuS): kinetics study, degradation pathways and mechanism. *Chemical Engineering Journal* 354: 740–752.

Pope, C., Karanth, S., and Liu, J. (2005). Pharmacology and toxicology of cholinesterase inhibitors: uses and misuses of a common mechanism of action. *Environmental Toxicology and Pharmacology* 19 (3): 433–446.

Porta, M., Pumarega, J., Gasull, M. et al. (2023). Individual blood concentrations of persistent organic pollutants and chemical elements, and COVID-19: a prospective cohort study in Barcelona. *Environmental Research* 223: 115419.

Puranik, S., Shukla, L., Kundu, A. et al. (2023). Exploring potent fungal isolates from sanitary landfill soil for in vitro degradation of dibutyl phthalate. *Journal of Fungi* 9 (1): 125.

de Quadros, P.D., Cerqueira, V.S., Cazarolli, J.C. et al. (2016). Oily sludge stimulates microbial activity and changes microbial structure in a landfarming soil. *International Biodeterioration & Biodegradation* 115: 90–101.

Quan, L., Cheng, Y., Wang, J. et al. (2023). Efficient removal of thiamethoxam by freshwater microalgae *Scenedesmus* sp. TXH: removal mechanism, metabolic degradation and application. *Journal of Environmental Management* 332: 117388.

Rao, M. A., Scelza, R., Scotti, R., & Gianfreda, L. (2010). Role of enzymes in the remediation of polluted environments. *Journal of Soil Science and Plant Nutrition*, 10(3), 333–353. DOI: https://doi.org/10.4067/S0718-95162010000100008.

Ray, P.K. (1992). *Measurement of Ganga River Quality-Heavy Metals and Pesticides. Project report.* Lucknow: Industrial Toxicology Research Centre.

Rehman, K., Ahmad, R., and Qadri, H. (ed.) (2020). *Bioremediation and Biotechnology: Sustainable Approaches to Pollution Degradation.* Springer.

Riza, M., Ehsan, M.N., and Hoque, S. (2021). Portrayal of textile based pollutants and its impact on soil, plants and fisheries. *Nature Environment and Pollution Technology* 20 (3): 1269–1275.

Rochman, C.M. and Hoellein, T. (2020). The global odyssey of plastic pollution. *Science* 368 (6496): 1184–1185.

Rodriguez-Narvaez, O.M., Peralta-Hernandez, J.M., Goonetilleke, A., and Bandala, E.R. (2017). Treatment technologies for emerging contaminants in water: a review. *Chemical Engineering Journal* 323: 361–380.

Rohr, J.R. (2021). The atrazine saga and its importance to the future of toxicology, science, and environmental and human health. *Environmental Toxicology and Chemistry* 40 (6): 1544–1558.

Ruan, Z., Zhai, Y., Song, J. et al. (2013). Molecular cloning and characterization of a newly isolated pyrethroid-degrading esterase gene from a genomic library of *Ochrobactrum anthropi* YZ-1. *PLoS One* 8 (10): e77329. https://doi.org/10.1371/journal.pone.0077329.

Rubinos, D.A., Villasuso, R., Muniategui, S. et al. (2007). Using the landfarming technique to remediate soils contaminated with hexachlorocyclohexane isomers. *Water, Air, and Soil Pollution* 181: 385–399.

Rupa, D.S., Reddy, P.P., Sreemannarayana, K. et al. (1991). Frequency of sister chromatid exchange in peripheral lymphocytes of male pesticide applicators. *Environmental and Molecular Mutagenesis* 18 (2): 136–138.

Sabarwal, A., Kumar, K., and Singh, R.P. (2018). Hazardous effects of chemical pesticides on human health–cancer and other associated disorders. *Environmental Toxicology and Pharmacology* 63: 103–114.

Sales da Silva, I.G., Gomes de Almeida, F.C., Padilha da Rocha e Silva, N.M. et al. (2020). Soil bioremediation: overview of technologies and trends. *Energies* 13 (18): 4664.

Samin, G., Pavlova, M., Arif, M.I. et al. (2014). A *Pseudomonas putida* strain genetically engineered for 1,2,3-trichloropropane bioremediation. *Applied and Environmental Microbiology* 80 (17): 5467–5476. https://doi.org/10.1128/AEM.01620-14.

Sanchez-Bayo, F. and Goka, K. (2006). Influence of light in acute toxicity bioassays of imidacloprid and zinc pyrithione to zooplankton crustaceans. *Aquatic Toxicology* 78 (3): 262–271.

Sanchez-Bayo, F., Goka, K., and Hayasaka, D. (2016). Contamination of the aquatic environment with neonicotinoids and its implication for ecosystems. *Frontiers in Environmental Science* 4: 71.

Sandy T, DiSante C (2010) *Review of Available Technologies for Removal of Selenium From Water*. Prepared for North American Metals Council http://www.namc.org/docs/00062756.pdf. Accessed 20 Aug 2023

Saquib, Q., Al-Khedhairy, A.A., Siddiqui, M.A. et al. (2011). Preferential binding of insecticide phorate with sub-domain IIA of human serum albumin induces protein damage and its toxicological significance. *Food and Chemical Toxicology* 49 (8): 1787–1795.

Saravanan, A., Kumar, P.S., Vo, D.V.N. et al. (2021). A review on catalytic-enzyme degradation of toxic environmental pollutants: microbial enzymes. *Journal of Hazardous Materials* 419: 126451.

Sayara, T. and Sánchez, A. (2020). Bioremediation of PAH-contaminated soils: process enhancement through composting/compost. *Applied Sciences* 10 (11): 3684.

Schug, T.T., Johnson, A.F., Birnbaum, L.S. et al. (2016). Minireview: endocrine disruptors: past lessons and future directions. *Molecular Endocrinology* 30 (8): 833–847.

Seeger, M., Hernandez, M., Mendez, V. et al. (2010). Bacterial degradation and bioremediation of chlorinated herbicides and biphenyls. *Journal of Soil Science and Plant Nutrition* 10 (3): 320–332.

Senarathne, R., Hettiaratchi, U., Athiththan, L. et al. (2022). Selected liver markers in predicting the severity of organophosphate and carbamate poisoning. *Journal of Environmental and Public Health* 2022.

Sharma, I. (2020). Bioremediation techniques for polluted environment: concept, advantages, limitations, and prospects. In: *Trace Metals in the Environment-New Approaches and Recent Advances* (ed. M.A. Murillo-Tovar, H. Saldarriaga-Noreña, and A. Saeid), 221–236. IntechOpen.

Sharma, N. and Leung, I.K. (2021). Novel thermophilic bacterial laccase for the degradation of aromatic organic pollutants. *Frontiers in Chemistry* 9: 711345. https://doi.org/10.3389/fchem.2021.711345.

Sharma, B., Dangi, A.K., and Shukla, P. (2018). Contemporary enzyme based technologies for bioremediation: a review. *Journal of Environmental Management* 210: 10–22. https://doi.org/10.1016/j.jenvman.2017.12.075.

Sidhu, C., Solanki, V., Pinnaka, A.K., and Thakur, K.G. (2019). Structure elucidation and biochemical characterization of environmentally relevant novel extradiol dioxygenases discovered by a functional metagenomics approach. *Msystems* 4 (6): e00316–e00319. https://doi.org/10.1128/msystems.00316-19.

Singh, S.P. and Singh, M.K. (2020). Soil pollution and human health. *Plant Responses to Soil Pollution* 205–220.

Singh, A., Kumar, V., and Srivastava, J.N. (2013). Assessment of bioremediation of oil and phenol contents in refinery waste water via bacterial consortium. *Journal of Petroleum & Environmental Biotechnology* 4 (3): 1–4.

Singh, M., Qureshi, K.A., Jaremko, M. et al. (2022a). Bioprospects of endophytic bacteria in plant growth promotion and Ag-nanoparticle biosynthesis. *Plants* 11 (14): 1787.

Singh, R., Pandey, K.D., Singh, M. et al. (2022b). Isolation and characterization of endophytes bacterial strains of *Momordica charantia* L. and their possible approach in stress management. *Microorganisms* 10 (2): 290.

Soares, P.R.S., Birolli, W.G., Ferreira, I.M., and Porto, A.L.M. (2021). Biodegradation pathway of the organophosphate pesticides chlorpyrifos, methyl parathion and profenofos by the marine-derived fungus *Aspergillus sydowii* CBMAI 935 and its potential for methylation reactions of phenolic compounds. *Marine Pollution Bulletin* 166: 112185.

Soltaninejad, K. and Shadnia, S. (2014). History of the use and epidemiology of organophosphorus poisoning. In: *Basic and Clinical Toxicology of Organophosphorus Compounds* (ed. M. Balali-Mood and M. Abdollahi), 25–43. London: Springer.

Srimurali, S., Govindaraj, S., Krishna Kumar, S., and Babu Rajendran, R. (2015). Distribution of organochlorine pesticides in atmospheric air of Tamilnadu, southern India. *International Journal of Environmental Science and Technology* 12: 1957–1964.

Srivastava, A.K., Srivastava, M., Kashyap, P.L., and Srivastava, A.K. (2016). Prospects of biocomposting in organic farming and environment management. In: *Environmental Biotechnology: A New Approach*, 147–178. New Delhi: Daya Publishing House.

Suenaga, H., Ohnuki, T., and Miyazaki, K. (2007). Functional screening of a metagenomic library for genes involved in microbial degradation of aromatic compounds. *Environmental Microbiology* 9 (9): 2289–2297. https://doi.org/10.1111/j.1462-2920.2007.01342.x.

Sulaiman, S., Yamato, S., Kanaya, E., Kim, J. J., Koga, Y., Takano, K., & Kanaya, S. (2012). Isolation of a novel cutinase homolog with polyethylene terephthalate-degrading activity from leaf-branch compost by using a metagenomic approach. *Applied and Environmental Microbiology*, 78(5), 1556–1562. DOI: https://doi.org/10.1128/AEM.06725-11.

Sun, H., Qi, Y., Zhang, D. et al. (2016). Concentrations, distribution, sources and risk assessment of organohalogenated contaminants in soils from Kenya, Eastern Africa. *Environmental Pollution* 209: 177–185.

Sun, J.T., Pan, L.L., Zhan, Y. et al. (2017). Atrazine contamination in agricultural soils from the Yangtze River Delta of China and associated health risks. *Environmental Geochemistry and Health* 39: 369–378.

Sun, X., Fu, H., Bao, M. et al. (2023). Preparation of slow-release microencapsulated fertilizer-biostimulation remediation of marine oil spill pollution. *Journal of Environmental Chemical Engineering* 11 (2): 109283.

Sundarmurugasan, R., Gumpu, M.B., Ramachandra, B.L. et al. (2016). Simultaneous detection of monocrotophos and dichlorvos in orange samples using acetylcholinesterase–zinc oxide modified platinum electrode with linear regression calibration. *Sensors and Actuators B: Chemical* 230: 306–313.

Swanson, N.L., Leu, A., Abrahamson, J., and Wallet, B. (2014). Genetically engineered crops, glyphosate and the deterioration of health in the United States of America. *Journal of Organic Systems* 9 (2): 6–37.

Tanaka, T., Tonosaki, T., Nose, M. et al. (2001). Treatment of model soils contaminated with phenolic endocrine-disrupting chemicals with laccase from *Trametes* sp. in a rotating reactor. *Journal of Bioscience and Bioengineering* 92 (4): 312–316. http://doi.org/10.1016/S1389-1723(01)80232-2.

Tang, T., Wu, R., Zhang, L. et al. (2021). Distribution and partitioning of pyrethroid insecticides in agricultural lands: critical influencing factors. *Environment International* 156: 106736.

Tapparo, A., Giorio, C., Marzaro, M. et al. (2011). Rapid analysis of neonicotinoid insecticides in guttation drops of corn seedlings obtained from coated seeds. *Journal of Environmental Monitoring* 13 (6): 1564–1568.

Teodosiu, C., Gilca, A.F., Barjoveanu, G., and Fiore, S. (2018). Emerging pollutants removal through advanced drinking water treatment: a review on processes and environmental performances assessment. *Journal of Cleaner Production* 197: 1210–1221.

Thue, P.S., Sophia, A.C., Lima, E.C. et al. (2018). Synthesis and characterization of a novel organic-inorganic hybrid clay adsorbent for the removal of acid red 1 and acid green 25 from aqueous solutions. *Journal of Cleaner Production* 171: 30–44.

Topp, E. (2001). A comparison of three atrazine-degrading bacteria for soil bioremediation. *Biology and Fertility of Soils* 33: 529–534.

Tortella, G.R., Diez, M.C., and Durán, N. (2005). Fungal diversity and use in decomposition of environmental pollutants. *Critical Reviews in Microbiology* 31 (4): 197–212.

Tsaboula, A., Papadakis, E.N., Vryzas, Z. et al. (2016). Environmental and human risk hierarchy of pesticides: a prioritization method, based on monitoring, hazard assessment and environmental fate. *Environment International* 91: 78–93.

Turusov, V., Rakitsky, V., and Tomatis, L. (2002). Dichlorodiphenyltrichloroethane (DDT): ubiquity, persistence, and risks. *Environmental Health Perspectives* 110 (2): 125–128.

Ufarté, L., Laville, É., Duquesne, S., and Potocki-Veronese, G. (2015). Metagenomics for the discovery of pollutant degrading enzymes. *Biotechnology Advances* 33 (8): 1845–1854. http://dx.doi.org/10.1016/j.biotechadv.2015.10.009.

Ufarté, L., Laville, E., Duquesne, S. et al. (2017). Discovery of carbamate degrading enzymes by functional metagenomics. *PLoS One* 12 (12): e0189201. https://doi.org/10.1371/journal.pone.0189201.

Verma, H., Kumar, D., Kumar, V. et al. (2021). The potential application of endophytes in management of stress from drought and salinity in crop plants. *Microorganisms* 9 (8): 1729. https://doi.org/10.3390/microorganisms9081729.

Walder, F., Schmid, M.W., Riedo, J. et al. (2022). Soil microbiome signatures are associated with pesticide residues in arable landscapes. *Soil Biology and Biochemistry* 174: 108830.

Wang, J.J., Cheng, W.X., Ding, W., and Zhao, Z.M. (2004). The effect of the insecticide dichlorvos on esterase activity extracted from the psocids, *Liposcelis bostrychophila* and *L. entomophila*. *Journal of Insect Science* 4 (1): 23.

Wang, C., Yu, L., Zhang, Z. et al. (2014). Tourmaline combined with *Phanerochaete chrysosporium* to remediate agricultural soil contaminated with PAHs and OCPs. *Journal of Hazardous Materials* 264: 439–448.

Wang, X., Sun, S.Y., Ni, Z.J. et al. (2018). Degradation of polycyclic aromatic hydrocarbons in contaminated soil by immobilized laccase. *Journal of the Serbian Chemical Society* 83 (5): 549–559. https://doi.org/10.2298/JSC171004022W.

Wang, Y., Mu, W., Sun, X. et al. (2020). Physiological response and removal ability of freshwater diatom *Nitzschia palea* to two organophosphorus pesticides. *Chemistry and Ecology* 36 (9): 881–902.

Wang, L., Tan, Y., Sun, S., Zhou, L., Wu, G., Shao, Y., ... Xin, Z. (2022). Improving degradation of polycyclic aromatic hydrocarbons by *Bacillus atrophaeus* laccase fused with *Vitreoscilla* hemoglobin and a novel strong promoter replacement. *Biology*, 11(8), 1129. doi: https://doi.org/10.3390/biology11081129.

Whittington, H.D., Singh, M., Ta, C. et al. (2020). Accelerated biodegradation of the agrochemical ametoctradin by soil-derived microbial consortia. *Frontiers in Microbiology* 11: 1898.

WHO (2009). The WHO recommended classification of pesticide by hazard. In: *World Health 1020 Organization*. Geneva.

Wolfand, J.M., LeFevre, G.H., and Luthy, R.G. (2016). Metabolization and degradation kinetics of the urban-use pesticide fipronil by white rot fungus *Trametes versicolor*. *Environmental Science: Processes & Impacts* 18 (10): 1256–1265.

World Health Organization (1990). *Public Health Impact of Pesticides Used in Agriculture*. World Health Organization.

Xing, L., Xu, Y., Xiao, Y. et al. (2010). Embryotoxic and teratogenic effects of the combination of bisphenol A and genistein on in vitro cultured postimplantation rat embryos. *Toxicological Sciences* 115 (2): 577–588.

Xu, Z., Mi, W., Mi, N. et al. (2020). Characteristics and sources of heavy metal pollution in desert steppe soil related to transportation and industrial activities. *Environmental Science and Pollution Research* 27: 38835–38848.

Yadav, I.C., Devi, N.L., Syed, J.H. et al. (2015). Current status of persistent organic pesticides residues in air, water, and soil, and their possible effect on neighboring countries: a comprehensive review of India. *Science of the Total Environment* 511: 123–137.

Zaller, J.G., Kruse-Plaß, M., Schlechtriemen, U. et al. (2022). Pesticides in ambient air, influenced by surrounding land use and weather, pose a potential threat to biodiversity and humans. *Science of the Total Environment* 838: 156012.

Zaller, J.G., Kruse-Plaß, M., Schlechtriemen, U. et al. (2023). Unexpected air pollutants with potential human health hazards: nitrification inhibitors, biocides, and persistent organic substances. *Science of the Total Environment* 862: 160643.

Zeneli, A., Kastanaki, E., Simantiraki, F., and Gidarakos, E. (2019). Monitoring the biodegradation of TPH and PAHs in refinery solid waste by biostimulation and bioaugmentation. *Journal of Environmental Chemical Engineering* 7 (3): 103054.

Zeng, J., Li, Y., Dai, Y. et al. (2021). Effects of polycyclic aromatic hydrocarbon structure on PAH mineralization and toxicity to soil microorganisms after oxidative bioremediation by laccase. *Environmental Pollution* 287: 117581.

Zhang, X., Starner, K., and Spurlock, F. (2012a). Analysis of chlorpyrifos agricultural use in regions of frequent surface water detections in California, USA. *Bulletin of Environmental Contamination and Toxicology* 89: 978–984.

Zhang, Y., Meng, D., Wang, Z. et al. (2012b). Oxidative stress response in atrazine-degrading bacteria exposed to atrazine. *Journal of Hazardous Materials* 229: 434–438.

Zhang, X.Y., Fan, X., Qiu, Y.J. et al. (2014). Newly identified thermostable esterase from *Sulfobacillus acidophilus*: properties and performance in phthalate ester degradation. *Applied and Environmental Microbiology* 80 (22): 6870–6878.

Zhang, C., Yi, X., Chen, C. et al. (2020a). Contamination of neonicotinoid insecticides in soil-water-sediment systems of the urban and rural areas in a rapidly developing region: Guangzhou, South China. *Environment International* 139: 105719.

Zhang, Y., Xu, Z., Chen, Z., and Wang, G. (2020b). Simultaneous degradation of triazophos, methamidophos and carbofuran pesticides in wastewater using an *Enterobacter* bacterial bioreactor and analysis of toxicity and biosafety. *Chemosphere* 261: 128054. https://doi.org/10.1016/j.chemosphere.2020.128054.

Zhuo, R. and Fan, F. (2021). A comprehensive insight into the application of white rot fungi and their lignocellulolytic enzymes in the removal of organic pollutants. *Science of the Total Environment* 778: 146132.

# 9

# Porous Nanomaterials for Enzyme Immobilization and Bioremediation Applications

*Nazim Hussain[1], Areej Shahbaz[1], Hafiza Ayesha Malik[1], Farhana Ehsan[1], José Cleiton Sousa dos Santos[2], and Aldona Balčiūnaitė[3]*

[1] *Center for Applied Molecular Biology (CAMB), University of the Punjab, Lahore, Pakistan*
[2] *Instituto de Engenharias e Desenvolvimento Sustentável, Universidade da Integração Internacional da Lusofonia Afro-Brasileira, Campus das Auroras, Redenção, Brazil*
[3] *Department of Catalysis, Center for Physical Sciences and Technology, Vilnius, Lithuania*

## 9.1 Introduction

Synthetic chemical compounds have grown increasingly important in business in recent years. Because of their low biodegradability, these chemicals are unsuitable for widespread usage (Ning et al. 2007). Consequently, the treatment and remediation of these substances, including endocrine disrupters, pesticides, hormones, medicines, and synthetic textile dye pollution, has and will continue to acquire substantial attention. In other words, these chemicals may create problems with the neurological, hormonal, and reproductive systems, putting an individual's health at risk (Bilal et al. 2018a; Hernandez-Vargas et al. 2018). Due to certain synthetic substances' genotoxic or cytotoxic properties, exposure to them may cause diseases, including obesity and diabetes, cardiovascular and reproductive disorders, and even cancer (Tiwari et al. 2012). Discharging dye-containing effluents from municipal wastewater treatment systems in paper or industrial textile companies, agricultural activities, or residential areas contributes substantially to environmental degradation. As a result, enzyme-aided biodegradation of hazardous compounds has been widely accepted, despite the availability of chemical and physical treatment (Chatha et al. 2017; Bilal et al. 2019).

Bioremediation was thought to be the least damaging to the environment and the most successful at cleaning up ecosystems of anthropogenically introduced toxins. Many microorganisms and enzymes released by microbes have been found and characterized as having the potential to be employed in bioremediation (Dvořák et al. 2017; Wu et al. 2017). Even if they can completely degrade the pollutant they are targeting; in some instances, they are either ineffective inside the natural environment or incapable of doing so. Due to advances in understanding how enzymes break down contaminants, there has been a considerable surge in interest in using engineered enzymes for this purpose. Among the many applications for enzymes that break down pollutants in industry and agriculture today are biosensors and the synthesis of drugs (Han et al. 2017; Takano et al. 2020).

Enzyme enzymes are highly efficient and environmentally benign biocatalysts and exhibit strong regio-, chemo-, and stereoselectivity characteristics (Bilal et al. 2018b; Khan et al. 2018). By reducing the number of synthesis steps, toxic solvents, and the activation energy needed, enzymes provide a more environmentally friendly, cost-effective, and environmentally sensitive alternative to catalytic procedures (Cowan and Fernandez-Lafuente, 2011) (Figure 9.1). Immobilized monooxygenases and oxygenase equivalents for environmental protection have no known references in the academic literature. Toxic compounds may be broken down by oxidation–reduction catalysis, made possible by oxidoreductases. The compounds include phenol, Ethylene Dichloride, natural and synthetic colors, pesticides/herbicides/fungicides, Pharmaceutical and Personal Care Products, and pharmaceuticals (Kalyani et al. 2016; Bilal et al. 2017; Kalsoom et al. 2022; Ayodeji et al. 2022; Riaz et al. 2022; Gan et al. 2022).

*Microbes Based Approaches for the Management of Hazardous Contaminants*, First Edition. Edited by Ajay Kumar, Livleen Shukla, Joginder Singh, and Luiz Fernando R. Ferreira.

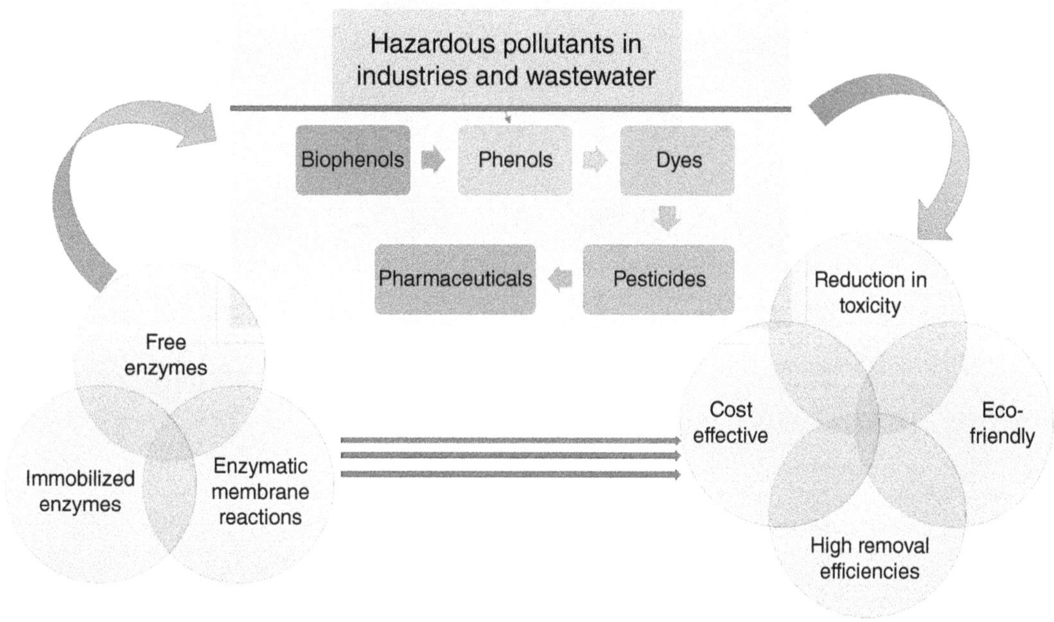


**Figure 9.1** Immobilized enzymes for environment protection.

In this review, oxidoreductases and peroxidases have received particular attention because they serve as biocatalyst support for porous nanomaterials, which may potentially be used as potential biocatalysts for reducing environmental pollution.

## 9.2 Enzyme Immobilization

Enzymes have long been recognized as biocatalytic agents due to their high chemo/regio/stereoselectivity levels. Enzymes, also known as biocatalysts, may perform a range of functions. Enzymes also minimize the toxic solvents, synthesis stages, and activation energy required for the catalytic process (Cowan and Fernandez-Lafuente 2011). The most frequent way of immobilizing enzymes is to bind enzyme molecules to an insoluble solid carrier in the reaction's environment. This is considered the most prevalent approach (Zhao 2010). The end product is an immobilized biocatalyst in which the enzyme has changed from its original homogeneous form into a heterogeneous form. When the enzyme is immobilized, it can be used or recycled as much as possible continuously. In these processes, the substrate is constantly added to the reaction, which makes the biocatalytic process more productive (amount of substrate molecules converted per amount of enzyme). Extremes of pH, temperature, and pressure often do less damage to immobilized enzymes than they would to free enzymes (Zhang et al. 2015; Al-Maqdi et al. 2021a; Tan et al. 2021; Morsi et al. 2020; Rafeeq et al. 2022). Compared to the free enzyme, the immobilized enzyme is also more stable during storage and can be used more than once (Iyer and Ananthanarayan 2008). Because this procedure consistently delivers the desired results, immobilization is an easy and practical way to enhance the properties of enzymes (Adeel et al. 2018). Five basic immobilization methods may be used: covalent binding, adsorption, encapsulation, entrapment, and crosslinking (Figure 9.2). However, it is important to emphasize that no technique can be applied universally to any enzyme in particular (Krajewska et al. 1997). It is important to consider the enzyme type and the biocatalytic process when determining the best support material and immobilization technique. Toxic substances may be effectively degraded, whereas enzymes like laccases and tyrosinases are preserved from denaturation. They can be reused if the proper support is provided for their usage in environmental applications (Ba et al. 2013). There are several materials to select from as a support material for enzyme immobilization. However, several conditions must be met before the material may be used as a carrier. Enzymes and support materials need specific physicochemical properties to form strong and long-lasting bonds. These include, but are not limited to, the existence of a chemical moiety(s), a wide surface area, and good sorption properties (Mateo et al. 2005). Decomposing support materials must be biocompatible and nontoxic. The hydrophilicity of the carrier impacts the parameters of the biocatalytic system

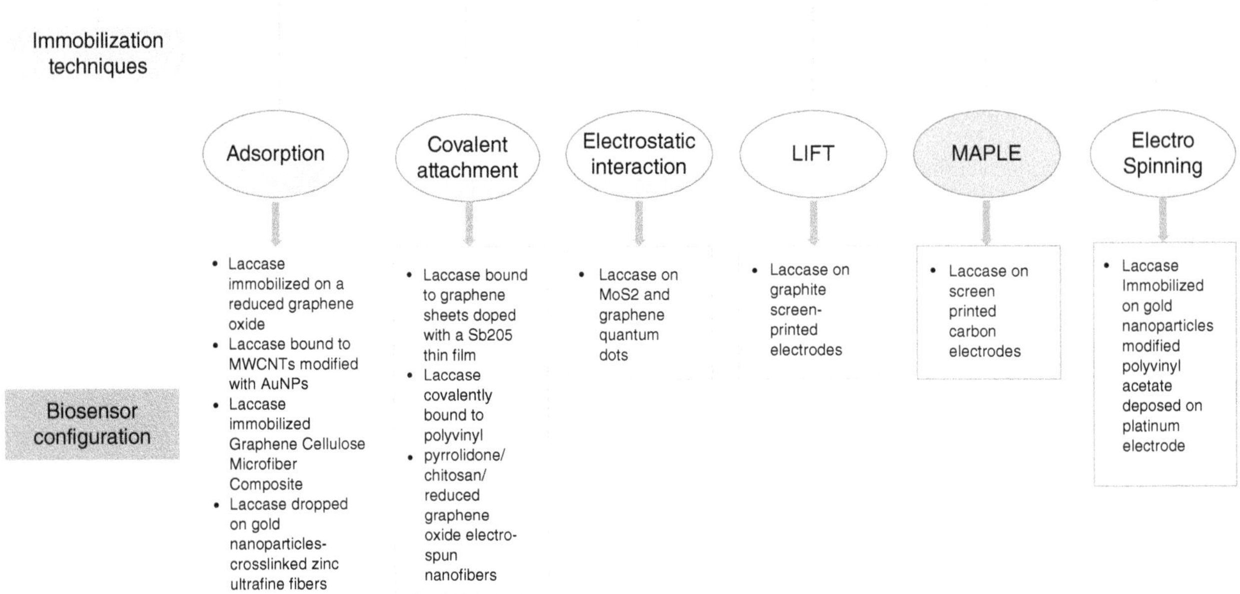


**Figure 9.2** Immobilization techniques (Shahbaz et al. 2021) / with permission of Springer Nature.

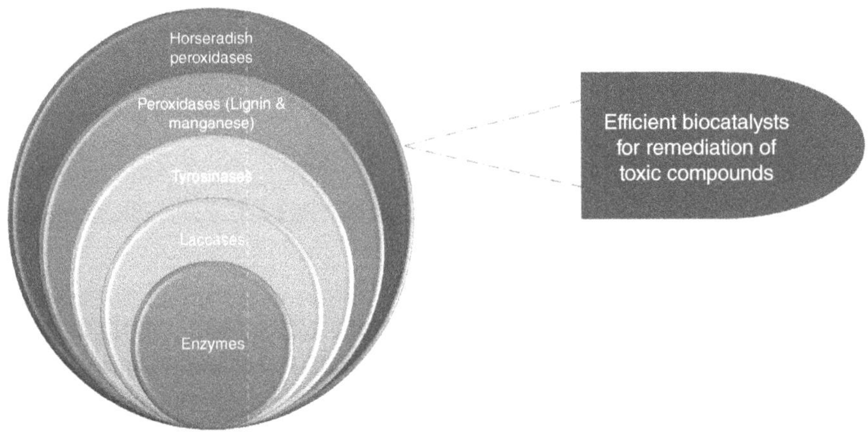


**Figure 9.3** Major enzymes used for immobilization.

and, ultimately, immobilization success (Barbosa et al. 2013). It should be emphasized that organic, inorganic, hybrid, or composite materials have been studied widely to attach a biocatalyst for use in environmental applications (Mateo et al. 2007). The enzyme type, biocatalytic conversion type, process nature, and needed form and shape of biocatalytic beads determine how effectively the bioremediation process will work (Rao et al. 2014).

Different reactor designs may be employed because of immobilization, which leads to better control over the process. Furthermore, the immobilized enzymes show a higher degree of selectivity and specificity (Basso and Serban 2019). There is a possibility that enzymes may become inactive or denatured during the immobilization procedure. In the case of multiple interactions between the enzyme and the matrix, distortion is too responsible for this problem. According to certain theories, secondary structural modifications may occur when proteins are immobilized on solid substrates (Homaei et al. 2013). These factors can result in proteins having β-sheet structure instead of an alpha-helical structure. On the other hand, the biocatalyst's altered properties may be due to distortions (Fujiwara et al. 2012).

Oxidoreductases (EC 1), such as laccases, tyrosinases, manganese peroxidases, horseradish peroxidases, and phenoloxidases, are used in the majority of environmental applications (Bilal et al. 2017). Many potentially toxic chemicals, including phenolics, bisphenols, dyes, and pharmaceuticals, are being investigated for their possible application in "green catalyst" bioremediation processes (Ba et al. 2014). Even though biocatalysts are efficient in harmful detoxifying compounds, each type of catalyst has its unique structure, cofactors, and mechanisms of action. The following section will describe some examples of enzymes often used in environmental protection (Figure 9.3).

## 9.3 Model Enzymes With Multifunctional Attributes

### 9.3.1 Laccases

Laccases are enzymes with minimal substrate specificity that may catalyze a wide range of reactions. Laccases may execute a variety of responses, the most common of which is one-electron oxidation of monophenolics, diphenolics, polyphenols, aromatic amines, N-heterocycles, phenothiazines, and many other chemicals. Reactive phenoxy radicals are produced during these oxidation reactions, which convert oxygen to water. This approach does not need hydrogen peroxide since it eliminates its need (Bronikowski et al. 2017). Laccases, in reality, are part of a larger enzyme family that requires several copper ions to operate. They each contain four copper ions and have distinct characteristics. Laccases catalyze oxidation and reduction reactions, including copper atoms of type 1, 2, or 3. As a consequence (Bakka et al. 2016), the blue color of the laccase is due to the presence of three copper atoms: one type 1 (T1), one type 2 (T2), and two type 3 (T3) (Zdarta et al. 2018a). A His-Cys-His triad transfers electrons from substrate molecules to the T2/T3 group through type-1 copper atoms, where the bioconversion process begins in particular terms. Because of this, the oxidation process may start while molecular oxygen is reduced to water due to the accumulation of electrons (Giardina et al. 2010). In addition, it is important to note that laccases are used in various applications, including food, textiles, fuel, and medicine (Mogharabi and Faramarzi, 2014; Noreen et al. 2022; Aslam et al. 2022; Sarker et al. 2022; Bhatt et al. 2021; Perveen et al. 2022). Because they can degrade a wide spectrum of phenolic chemicals, they are utilized to bioremediate soils and water (Barrios-Estrada et al. 2018).

### 9.3.2 Tyrosinases

Tyrosinases are quite common in nature. They are in various taxa, including plants, fungi, bacteria, insects, and mammals (Selinheimo et al. 2007). They play a crucial role in creating melanin, the pigment responsible for human skin and hair color. The process fundamentally differs when tyrosinases rather than laccases oxidize phenol and its derivatives (Hussain et al. 2021). Even though tyrosinases need oxygen as a cofactor, they create quinones as a byproduct rather than free radicals and water molecules (Land et al. 2003). Tyrosinase catalytic activity is a two-step sequential reaction of monophenol hydroxylation to ortho-diphenols and their subsequent oxidation to ortho-quinones. This is tyrosinase catalysis. Water molecules are released, and macromolecular substances are produced during catalytic oxidation (Shuster and Fishman 2009). As tyrosinases may be found in various molecular configurations, an extensive range of intermediates can be generated during the catalysis of phenolic substances (Sánchez-Ferrer et al. 1995).

### 9.3.3 Peroxidases, i.e., Lignin and Manganese

Lignin peroxidases extract phenolic chemicals from water and soil. It may oxidize aromatic phenolic and non-phenolic chemicals and xenobiotics with a 1.4 V redox potential. LiP has limited substrate specificity and an oxidation–reduction potential. It oxidizes phenolic and non-phenolic aromatic compounds (Valli et al. 1990). Molecular weights typically range between 40 and 50 kDa for manganese peroxidase, a protein in various forms (Hofrichter, 2002). MnPs have increased catalytic capacities in the presence of calcium cations and reach their maximum activity at Mn(II) concentrations greater than 100 M (Glenn and Gold, 1985). Manganese peroxidases prefer temperatures between 30°C and 40°C. However, this might vary depending on the source. There are, of course, a few notable exceptions to this rule. These enzymes perform at their highest catalytic efficiency at pH values of about 4, which is mildly acidic (Yehia and Rodriguez-Couto, 2017).

### 9.3.4 Horseradish Peroxidases

In hydrogen peroxide, horseradish peroxidase oxidizes phenolic acids, aromatic phenols such as pyrogallol, and their derivatives (i.e., different bisphenols), nonaromatic amines such as 4-amino antipyrine, indoles, and other chemicals to yield two molecules of water. Aromatic and inorganic acids, as well as their decomposition products, are all subject to oxidation. When oxidation reactions produce polymeric compounds, the second product of the catalytic process may be a precursor to the formation of these polymeric compounds (Bilal et al. 2018 a, b, c). Environmental remediation, removing hazardous substances from drinking water, such as dyes, and industrial applications are the most major practical uses of heat-resistant proteins (HRPs) (Bilal et al. 2018 a, b, c).

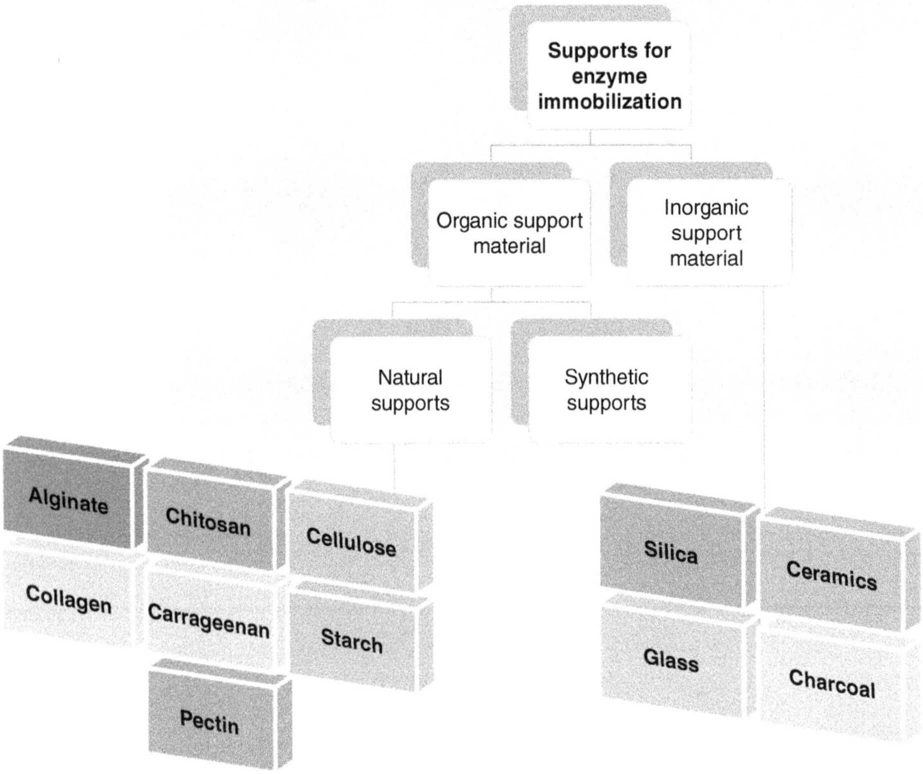


**Figure 9.4** Categorization of support matrices for enzyme immobilization.

## 9.4 Supports for Enzyme Immobilization

Immobilized enzymes may be compared to heterogeneous catalysts in several ways. Compared to unbound enzymes, supported catalysts are significantly easier to deal with. It is very easy to extract them from the reaction solution, and once separated, they may either be retained or recycled. Protein aggregation may be reduced by the immobilization of protease enzymes, which reduces the effects of autolysis. Heat and chemical inactivation may be more resistant to enzymes that are immobilized. It should also be noted that immobilized enzymes should be able to catalyze chemical reactions in organic solvents, even if the native enzyme cannot dissolve in them. A whole new product will be created as a consequence. These distributions may be compared to those seen in aqueous media. It is possible to see an increase in activity after being immobilized in certain instances (Kreiner et al. 2001). Because reactants and products cannot move freely, immobilization nearly invariably results in decreased activity. It is thus important to understand how to manage the support's chemistry, porosity, and surface area. Customized supports are expected to have higher production costs than normal high-area silicas, but better performance may be able to pay for these higher costs. Enzymes may be immobilized on both organic and inorganic substrates. Glass, silica gel, alumina, clay minerals, and the oxides of transition metals are examples of inorganic carriers (e.g., NiO, $TiO_2$, ZrOa). Polysaccharides (naturally occurring polymers) and synthetic polymers that have been functionalized (such as nylon and polystyrene) are employed as organic carriers (Tischer and Wedekind, 1999).

Separated into two categories by the materials they include, support matrices may be classified as either "inorganic" or "organic" (organic supports are further subdivided into natural and synthetic supports) (Figure 9.4).

## 9.5 Inorganic Materials as Support Matrices

Materials such as silica gel, metal oxides, zirconia, and other silica-based compounds are preferable because of their resistance to heat and mechanical stress. On the other hand, bacteria and fungi cannot grow on inorganic supports because they do not act as a substrate for their growth. The combination of stiffness and porosity is one of the characteristics that sets inorganic supports apart from their organic counterparts. Another advantage is that they maintain a constant ratio of pore diameter to volume, allowing them to offer consistent support in terms of their volume and shape (Hartmann and Kostrov, 2013) (Figure 9.5).

Chemically inert silica carriers must be modified and activated before they can be utilized. Therefore, they are often treated with aminoalkyl triethoxysilanes to add an amino group and then started using a variety of ways to immobilize enzymes. The thermal stability improved when penicillin G amidase was immobilized on amino-activated silica gel (Burteau et al. 1989). Similar to the immobilization of the enzymes on an activated silica substrate, it was discovered that immobilized enzymes derived from horseradish and lignin peroxidase (HRP) could eliminate chlorolignins from eucalyptus kraft effluent (Dezott et al. 1995). The immobilization of beta-amylase on silica nanoparticles has also improved the cleaning effectiveness of detergents (Figure 9.6) (Soleimani et al. 2011).

Ceramics are frequently used in enzyme immobilization because of their porous nature. Candida Antarctica lipase enzymes can be immobilized on a ceramic membrane to decrease the influence of feedback inhibition on hydrolytic and synthesis processes. Ceramic foams have been proven to improve specific surface areas while decreasing diffusion rates (Magnan et al. 2004). It is a distinct kind of inorganic support system that is often employed. Because of its naturally high viscosity, this material may be used to immobilize a broad spectrum of enzymes, including amylase. It was revealed during this procedure that the phthaloyl chloride-containing amino group functionalized glass beads produced a long-lasting and reliable product. When put on glass pH electrodes and employed in this way, urease could operated as a biosensor for detecting urea levels in blood samples at concentrations as low as 52 mg/mL (Sahney et al. 2005).

Figure 9.5 Categories of inorganic material as supports.

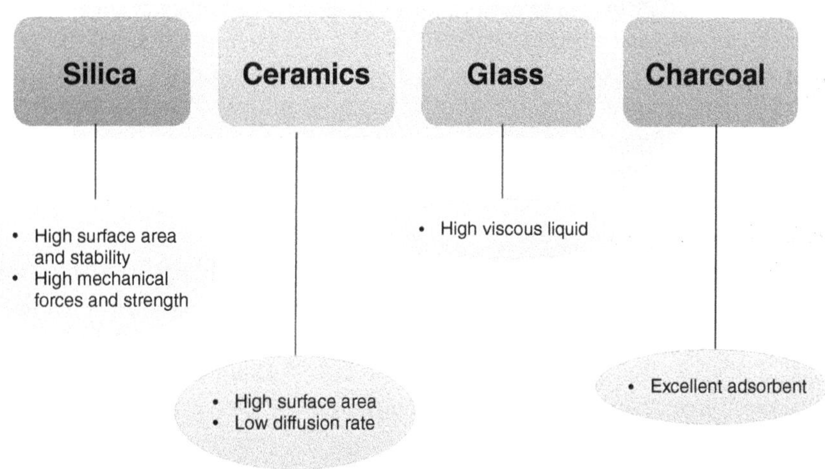


Figure 9.6 Categories of organic material as supports.

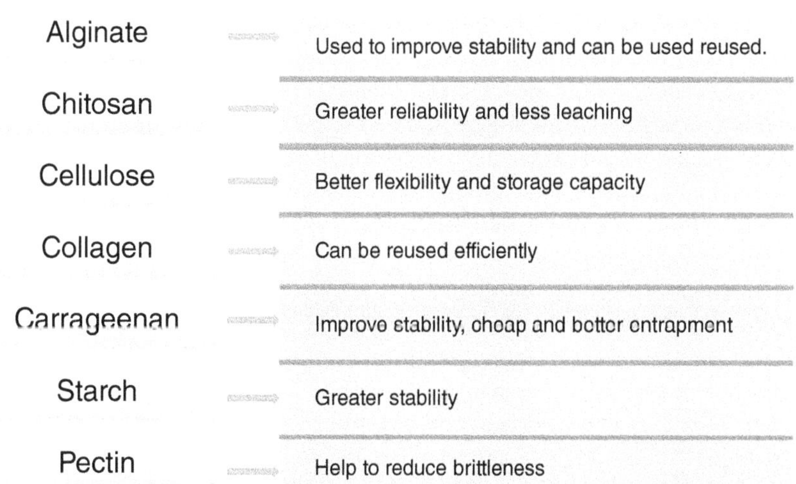

Charcoal has been shown to have 90% catalytic activity when used to immobilize enzymes such as amyloglucosidase without crosslinking agents for starch hydrolysis. Adsorbent charcoal has a high adsorption capacity and produces less particulate matter than other adsorbents (Kibarer and Akovali, 1996).

## 9.6 Organic Materials as Support Matrices

Collagen, chitosan, carrageenan, alginate, cellulose, starch, and agarose are the natural polymers that are exploited for enzyme immobilization the most frequently. Due to their chemical structure, these polymers can form inert gels and bind to proteins and enzymes in both a reversible and an irreversible manner. When crosslinked with bifunctional reagents, they demonstrate thermal solid and mechanical resistance and are widely accessible in large numbers at a low cost. In addition, they may be obtained without much difficulty. Because they possess all of these qualities, they make for a great support system (Figure 9.5). This sulfated polysaccharide is found in the cell walls of brown algae. Immobilizing enzymes in an alginate salt, such as xanthan–alginate beads, polyacrylamide gel, or calcium–alginate beads, may increase activity and reusability. Increased enzyme stability can be attributed to alginate's ability to form crosslinks with divalent ions and glutaraldehyde (FloresMaltos et al. 2011). A polysaccharide has been employed as a support matrix (Vaillant 2000; Kapoor and Kuhad, 2007). It is a good contender for usage as a carrier system due to its numerous benefits. Examples include mechanical stability, stiffness, hydrophilicity, active and reactive groups interacting with enzymes, and protein binding.

It is the carrier molecule used most often for the immobilization of enzymes. It has a lesser potential for binding, is affordable, and is readily accessible in both fibrous and globular forms in commercial markets (Namdeo and Bajpai 2009). It was found that by immobilizing-amylase on cellulose dialdehyde-coated magnetite nanoparticles and performing starch degradation in this manner, a novel starch degrading system could be created. This finding was made public. Immobilizing the enzyme onto an ionic liquid-cellulose film activated with glutaraldehyde produced similar results. As a consequence, the enzyme's tolerance, as well as its practicality, improved. Collagen's proteinaceous composition, great water-holding ability, and good porosity make it an ideal structural support system. Collagen, on the other hand, is highly porous. The collagen side chains and the enzyme establish a covalent bond during immobilization. This anchors the enzyme on the support via the covalent link. Glutaraldehyde was previously used as a crosslinking agent to immobilize tannase (Katwa et al. 1981). Chen et al. (2011) reported that catalase was immobilized using $Fe^{3+}$-collagen fibers, and its activity was demonstrated to be mainly retained even after it had been reused 26 times.

Red algae have been reported to contain this linear sulfated polysaccharide. Because of their high protein-holding capacity, robust gelling capabilities, and pseudoplastic character, carrageenans are an excellent support system. This nature causes them to become less viscous when subjected to shear stress and assists them in regaining their viscosity after being disturbed. Carrageenans are helpful to support systems not only because of their high protein-holding capacity but also because of their pseudoplastic nature and high protein-holding capacity. According to one piece of research, using coextrusion with carrageenans as the support matrix resulted in an encapsulation efficacy of 42% for biodiesel synthesis (Jegannathan et al. 2010 research). It has been discovered that galactosidase and lactic acid can be transported using an inexpensive and stable carrier (Girigowda and Mulimani 2006; Rao et al. 2008).

Sirisha et al. reported that the enzyme bitter gourd peroxidase was immobilized using a calcium alginate-starch hybrid platform. Because of the intrinsic carbohydrate moieties, the encapsulated enzyme was also demonstrated to be more stable in the presence of urea. Surface-immobilized enzymes, on the other hand, increased their activity significantly (Matto and Husain, 2009).

It is a polysaccharide that may be found in plant cell walls. It increases intercellular connections similarly to plant tissues. Because it holds water, it is an excellent gelling agent. Ceniceros (2003) used 0.2–0.7% glycerol to immobilize papain on pectin. Pectin-chitin and pectin-calcium alginate increased the enzyme's heat and denaturant resistance. The immobilized enzyme's catalytic capabilities are improved Satar et al. (2008).

## 9.7 Synthetic Polymers as Support Matrices

Synthetic polymers' ion-exchange resins cannot be dissolved in their natural environment. These polymers effectively immobilize enzymes due to their porous structure. Microbes cannot infect them. Examples include polystyrene, polyvinyl chloride (PVC), 1-Phenyl-2-propylaminopentane , Diethylaminoethyl, cellulose, UV-activated PE Glycerol, polyacrylate polyamide polypropylene and PVC. Enzymes were initially immobilized in polystyrene, a synthetic polymer.

## 9.8 Nanomaterials as Supports for Enzyme Immobilization

Chitin and chitosan are abundant and widely accessible. Chitosan and magnetic nanoparticles have been suggested as hybrid support for enzyme immobilization. Moreover, enzymes were immobilized on octyl-agarose hybrid nano support to improve their thermal stability and activity levels (Sirisha et al. 2016). Synthetic polymers and hybrid nano-supports, including synthetic polymers, can immobilize microbial-resistant enzymes (Yang et al. 2012). Researchers have recently been interested in nanocomposites and hybrid supports because they blend the best qualities of organic and inorganic materials (Al-Maqdi et al. 2021b). Mixing organic and inorganic elements can result in organic–inorganic hybrids: (i) Polyaniline and polyacrylonitrile; epoxy-activated acrylate copolymers; polyethyleneimine; and polyaniline and polyacrylonitrile, among others (Xue et al. 2005; Rajdeo et al. 2016); (ii) two biopolymers that contain (chitosan and alginate, chitin and lignin, cellulose, and dextran) (De Paula et al. 2015; Pervez et al. 2017); and (iii) the mixing of a synthetic polymer with a biopolymer (for example, polyacrylic acid and cellulose, or polyvinyl alcohol and chitosan) (Badgujar and Bhanage, 2017; Riccardi et al. 2017). Nanocomposites and hybrid supports are popular because they blend organic and inorganic compounds. Organic–inorganic hybrids occur by mixing inorganic compounds. Epoxy-activated acrylate copolymers, polyaniline/acrylonitrile, polyethyleneimine (Yang et al. 2012). One other benefit of hybrid nanomaterials is that they are suitable for various enzymes that correspond to all types of catalytic reactions. It is also possible to create enzyme-friendly and reusable support materials by combining high-temperature resistant and mechanically stable synthetic polymers with biopolymers, predominantly composed of polysaccharides. When these two ingredients are combined, one advantage is the ability to control the size and shape of the emerging creatures while they are still developing. Mineral-organic hybrids are created when carbon-based polymers like chitosan or polyvinyl chloride are mixed with inorganic substances like zinc or titanium oxide. This creates a new organic–inorganic compound (Li et al. 2013; Zdarta et al. 2018b). This class of hybrid nanomaterials has a lot of potential as enzyme immobilization supports due to their exceptional mechanical and thermal durability and strong affinity for biological molecules. This is partly due to the strong affinity of these hybrid nanomaterials for biomolecular components. Their chemical inertness and high enzyme binding affinity are due to their organic polymer component, not their inorganic component, which is responsible for their high stability and chemical inertness.

Increasing the biocatalyst's resistance to extreme environmental conditions like high temperatures and pH values is one of the key purposes for immobilizing enzymes. Still, it also enables them to be recycled and reused, which lowers the total cost of the process. Due to the significant influence, it has had in recent years on enhancing enzyme stability and refining immobilization methods, the application of nanotechnology in biological systems has increased significantly over the past several years. It has been demonstrated that decreasing the support size increases the immobilized enzyme's productivity. As a result of their low cost and large specific surface area, nanoparticles are good substrates for enzyme immobilization. They have a lot of surface area to bind to, which increases the number of enzymes per mass unit (Ansari and Husain 2012). The ability to recycle and reuse enzymes immobilized on nano supports has been demonstrated, as have enhanced stability and resilience to unfavorable situations. Study results showed this to be true. Numerous studies have been conducted on organic nanomaterials, such as chitin, cellulose nanoparticles, or starch nanofibers (Bilal and Iqbal 2019). Macroscopic comparison of nanofibers and their conventional counterparts indicated a higher enzyme-binding capacity and uniform dispersion in nanofibers (Shojaei et al. 2017; Bilal and Iqbal 2019). Aside from their high porosity, lower resistance to mass transfer, and the ability to prevent or reduce protein denaturation or enzyme dysfunctions, they also feature a wide range of other relevant properties. Nanofibers have been suggested as a viable support material for immobilizing a range of enzymes in many applications because of the abovementioned features. Covalent attachment and surface adsorption are commonly used to immobilize enzymes on nanofiber surfaces (Datta et al. 2013). Lysozyme was immobilized utilizing bacterial cellulose nanofibers as a support material in recent work by Bayazidi and colleagues. According to studies, immobilized enzyme activity was 12% lower than soluble free enzyme. Reduced substrate and product diffusion, loss of molecular flexibility, or modification of the enzyme's active site may impair immobilized enzyme activity. These variables may explain the occurrence (Bayazidi et al. 2018).

The treatment of wastewater is of the highest importance since the number of companies and the number of effluents they release into the environment are increasing. Effluents from the textile, paper, and leather industries (Rosales-Hernández et al. 2007) are particularly high in dyes, which, even in low quantities, can cause cancer (Akhtar et al. 2005). Enzyme-based bioremediation has emerged as a viable alternative to conventional techniques and microbiological treatment, which both have substantial drawbacks regarding the degradation and removal of phenolic contaminants (Chen et al. 2011). In this regard, enzymes like peroxidase, laccase, and azoreductase can significantly impact. However, these enzymes are rendered ineffective by the harsh environmental conditions. Immobilized enzyme-based water treatment is the most effective method for removing contaminants from water. Calcium alginate beads, for example, can protect HRP against degradation at high temperatures. An immobilized form of laccase can break down various dyes, including anthracinoid dye, lancet blue, and others (Wu et al. 2005).

## 9.9 Porous Nanomaterials as Supports for Enzyme Immobilization

Since their larger surface areas allow for higher protein loading levels, porous supports have attracted much attention in the scientific community. Micro, Meso, and macroporous materials are all defined by the IUPAC, the International Union of Pure and Applied Chemistry. Protein molecules, ranging from 2 to 10 nm, are too big to be successfully accepted by microporous materials with pore sizes of 0.2 nm. This is because pore sizes are moving closer and closer to those of molecular scales. Macroporous materials, defined as those with pore sizes greater than 50 nm, have been widely used in enzyme immobilization. Popular macroporous supports include commercially available resins like Eupergits and Sepabeads. Each has between 42 and 46 functional groups that can be covalently linked to reactive surface moieties, and the amount of functional groups varies (Bayne et al. 2013).

## 9.10 Advantages of Enzyme Immobilization

First, it is essential to quickly review the factors contributing to the widespread acceptance of immobilized enzymes as the form of choice when using biocatalysts (Figure 9.7).

### 9.10.1 Stabilization

The stability of folded enzymes is between 5 and 20 Kcal/mol compared to the stability of unfolded enzymes (Gupta 1993). This is the basis for the widely believed yet valid opinion that enzymes are more sensitive than chemical catalysts. Before recombinant DNA was invented, scientists utilized chemical alteration and stabilizer addition methods (Himmel and Georgiou, 1991). The only methods available for additional stability were immobilization on solid supports and using these methods. In contrast to the widespread belief, immobilization does not necessarily result in a significant improvement in stability (Tischer and Kasche 1999). It should also be mentioned that immobilization is capable of achieving both storage stability and operating stability at the same time (Bickerstaff 1997).

### 9.10.2 Recovery and Reusability

For economic reasons, it is better to use the enzyme several times in batch reactor cycles or for a more extended period if the reactor is operated continuously (with the appropriate design). This characteristic is renowned for its operational stability, which prioritizes reusability. Recovery after use, not to be confused with reusability, refers to the ease with which immobilized enzymes may be removed or separated from the reaction components to be reused. To ensure optimal reusability, the suitable technique preceding reuse or cleaning-in-place activities must be precisely established. This may be performed either before or following the surgeries. Centrifugation and membrane filtration are two techniques that readily

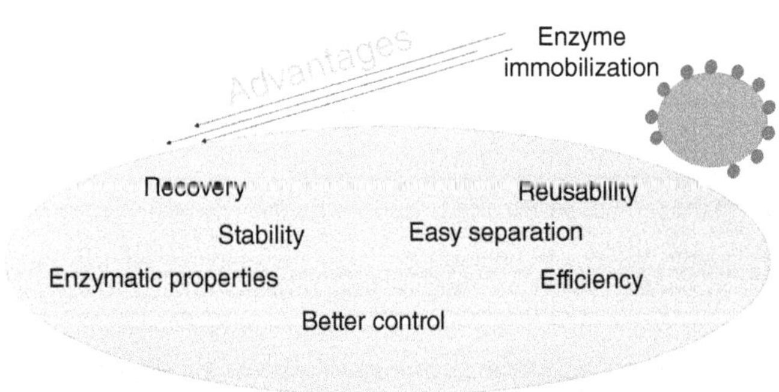


Figure 9.7 Advantages of enzyme immobilization.

spring to mind when solid supports are applied. It is feasible to develop stimuli-sensitive immobilized enzymes using intelligent carriers. These enzymes can behave as homogeneous catalysts but can also readily dissociate from the reaction mixture, much like heterogeneous catalysts (Raghava et al. 2006). Enable magnetic carriers and apply magnetic fields; enzymes immobilized on solid supports may be easily separated (Safarik and Safarikova 2009). In this respect, magnetic nanocarriers have shown themselves to be an alternative that is both truly interesting and extremely helpful.

### 9.10.3 Flexibility

After immobilizing enzymes, one can use batch, packed bed, or fluidized bed column reactors. Micron-sized particles, such as agarose, dextran, or cellulose (or their composite materials), have been the most often used carriers for immobilization. Early on, it was observed that the beads' inside surfaces had a far larger surface area than the beads' external surfaces. Due to the nanoscale dimensions of enzymes and proteins, the beads could hold more enzymes and proteins than the beads' holes. Adsorption or covalent coupling was used to immobilize the molecule in this case (Guisan 2006).

## 9.11 Metal–Organic Frameworks as Supports for Enzyme Immobilization

Metal–organic frameworks, also known as MOFs, are structures of metal nodes and organic ligands coupled by coordination bonds. These structures have the potential to be used in a wide variety of different fields (Ding et al. 2019; Abánades Lázaro et al. 2020). MOFs enable enzymes by participating in interactions such as hydrogen bonding, van der Waals forces, covalence, and coordinative bonding, all of which are possible because of the metal nodes and ligands attached to MOFs (Majewski et al. 2017; Xv et al. 2022; Du et al. 2022). In addition, MOFs have a very high surface area and a large pore volume and are simple to modify, making them a suitable supporting matrix for immobilizing enzymes (Doonan et al. 2017; Gkaniatsou et al. 2017). Finally, MOFs have an ordered crystal structure and great tunability, which can provide uniform loading and reduced leaching, as well as an enzyme's microenvironment, which protects enzymes from hostile thermal and chemical environments. This is significant because enzymes play a crucial role in the functioning of biological systems (Drout et al. 2019). Since their crystalline structure, pore volume, and physical and chemical features may be easily adjusted, MOFs have emerged as an all-encompassing platform for enzyme immobilization and other hybrid materials.

Enzyme immobilization and production of other mixed materials are possible using MOFs. Since enzyme-MOF composites may be reused and recycled, this feature of the materials helps to keep their price lower. Magnetic moieties were also incorporated into magnetic MOFs to aid magnetic separation from reaction mixtures, resulting in the creation of a magnetic MOF that is more user-friendly for recycling (Huo et al. 2015; Falcaro et al. 2016). High temperatures, pH, chemical solvents, and trypsin digestion affect free enzymes. When the enzymes lose their enzymatic function due to this, it is because of the external stimuli that these enzymes are exposed to. It is possible to give enzymes immobilized with microenvironments, reducing the loss of enzyme structure when exposed to denaturing conditions. Immobilized MOF enzymes can improve catalytic activity by utilizing a well-designed matrix or altering the enzymes. A larger concentration of substrates near enzymes, for example, might boost enzymatic activity, provided that the substrate and product adsorption and desorption on support surfaces are properly balanced. Equitable distribution of substrates and products would result in this (Zhang et al. 2015). For nanoreactors, this local enrichment of substrate, often referred to as the positive partition effect, allows MOFs to be utilized, which ostentatiously improves apparent enzyme activity. The positive partition effect is another name for this phenomenon.

Computational organic synthesis (COS) has emerged as an important step toward developing custom catalysts that may be used in various applications. COFs are ideal hosts for enzymes because of their simplicity of modification and the ability to enhance some of the features that work against them in large-scale applications that are theoretically viable. Low stability, operating range, recyclability, and inhibition byproducts and byproducts are only a few of these qualities (Sun et al. 2019). Compared to other micro/nanoparticles, COFs have better process throughput because of their higher porosity. Because of the high porosity, a higher volume of biocatalyst may be immobilized per unit support mass (Kandambeth et al. 2015). Covalent attachment, physical adsorption, covalent attachment via a linker molecule, and direct covalent attachment between enzymes and COFs are all options for integrating the two (Oliveira et al. 2019).

## 9.12 Bioremediation Applications of Enzyme Immobilized Porous Nanomaterials

Because of their small size in comparison to other materials, nanoparticles are perfect for wastewater remediation. Various applications can benefit from their distinct chemical, physical, and biological properties. Various nanomaterials have made wastewater treatment possible, including carbon-based compounds such as nanotubes (nanocomposite materials) and metallic materials (metal oxides). Photocatalytic degradation, adsorption, nanoparticle filtration, and monitoring of various toxins and pollutants are all part of wastewater management (Palit and Hussain, 2020).

Adsorbents made of nanoparticles are becoming increasingly popular for use in removing hazardous substances from wastewater produced by industrial processes. It is possible to employ nano-adsorbents to remove both organic and inorganic contaminants from the environment. The most prevalent materials used to categorize nanoparticles are carbon, metal, and metal oxide nanoparticles. Carbon nanotubes (also known as CNTs), activated carbon, graphene, and fullerene are all examples of different types of carbon nanoparticles. These nanoparticles may be categorized into four distinct groups (Kumari et al. 2019). Nano-magnets made of activated carbon altered to extract fluoride ions from wastewater have also been employed. The removal of the heavy metal was accomplished by the use of magnetic separation and sorption techniques (Nogueira et al. 2019). In addition, nano-adsorbents comprised of metals and metal oxides are a significant contributor to removing contaminants from wastewater. According to the findings of several studies, the adsorption capacity of magnetic nanoparticles may be improved by covering them with various supports (Najafpoor et al. 2020).

When combining enzymes with nanotechnology, it is critical to minimize the harmful environmental impact of nanomaterials. Because of steric hindrances, the cell contacts between enzyme molecules and nanomaterials are diminished, reducing the combined system's surface energy (Dwevedi 2018). Consequently, nanoparticles are more flexible and effective in bioremediation and green energy generation. Catalysis is a unique process since it uses enzymes, which are excellent for the environment. Because they resist unfolding, are less vulnerable to diffusional constraints, can be used repeatedly, and have better kinetic characteristics, enzymes immobilized on nanomaterials are exceptionally stable (Ding et al. 2015). Nanomaterials have a large surface area, making loading enzymes more accessible and increasing their immobilization effectiveness. When using magnetic nanoparticles as an immobilizing matrix, it is much easier to remove immobilized enzymes from the reaction mixture than when using alternative matrices. Another approach to enhancing enzyme stability involves using nanomaterials to immobilize multimeric enzymes like oxidoreductases. For example, a rise in alpha-sheet structure and a reduction in the beta-helical structure are seen when enzymes are immobilized on solid substrates. Nanomaterial-immobilized enzymes do not exhibit these structural alterations (Secundo, 2013).

Several studies have demonstrated that integrating these two technologies results in a quality product. Darwesh et al. (2019) showed the influence on wastewater bioremediation of an immobilized peroxidase enzyme. They discovered that magnetic iron oxide nanoparticles treated with glutaraldehyde could bind pH- and temperature-resistant enzymes. Both the green and the red azo dyes were eliminated from the system in just four hours by the immobilized peroxidase enzyme, independently of one another. When both dyes were provided at the same time, it took six hours to completely remove the dyes from the sample in experiments that were conducted on a laboratory scale (Darwesh et al. 2019). Laccase is an enzyme that is frequently utilized in the process of treating industrial effluents. Several different composites of magnetic nanoparticles have been tried to immobilize laccase for biodegradation.

## 9.13 Future Directions

Nanotechnology's favorable impacts include a large surface area, many applications, the capacity to withstand challenging circumstances, quick and efficient manipulations of materials, greater contact, and many more advantages that have attracted the curiosity of researchers. Integrating nanotechnology with microorganisms and enzymes has resulted in a more ecologically friendly solution for treating industrial effluents (Mandeep et al. 2020). Utilizing microorganisms as a risk mitigation strategy is one way to reduce the dangers posed by nanoparticles produced by chemical synthesis. The residues that are left are either biocompatible or may be easily separated by utilizing processes such as filtering and precipitation that are quite straightforward. Commercializing these nanotechnological features will be the more difficult barrier to overcome. Only 1% of these nanotechnological characteristics have been commercially available (Dwevedi 2018). Therefore, the widespread use of simple and effective nanotechnology processes assisted by microbes will serve as a stepping stone for the industrial sector to progress. It is harnessing the potential of nanotechnology for usage in enterprises that produce goods sustainably and cost-effectively has to be nurtured with the help of academics and government funding.

## 9.14 Conclusion

The present worldwide environmental problem is restricting and eliminating the widespread presence of toxic chemicals and their derivatives, such as dye pollution, phenolic compounds, and medications, in the environment or aquatic systems. Many people are worried about the existence of these compounds and their derivatives in the environment. Biocatalysts classed as "oxidoreductases" include enzymes. These enzymes may catalyze the oxidation and removal of various chemical pollutants and micropollutants without using cofactors other than hydrogen peroxide and molecular oxygen, both of which are abundant. Enzymes belong to the biocatalyst family. Enzyme-assisted biodegradation can be very efficient and cost-effective in moderate environmental conditions. To deal with the bothersome impurities, using high-tech equipment, a lot of energy, or toxic chemicals is unnecessary. Diagrammatically depicts a step-by-step procedure for developing, producing, upgrading, and deploying a unique catalyst of interest. Because of their increased performance, significant progress has been achieved in designing and constructing various insoluble biocatalytic structures using immobilization procedures for their continued use in the biotechnological and environmental sectors.

## References


Abánades Lázaro, I., Wells, C.J.R., and Forgan, R.S. (2020). Multivariate modulation of the Zr MOF UiO-66 for defect-controlled combination anticancer drug delivery. *Angewandte Chemie* 132 (13): 5249–5255.

Adeel, M., Bilal, M., Resheed, T. et al. (2018). Graphene and graphene oxide: functionalization and nano-bio-catalytic system for enzyme immobilization and biotechnological perspective. *International Journal of Biological Macromolecules* 120:1430–1440.

Akhtar, S., Khan, A.A., and Husain, Q. (2005). Partially purified bitter gourd (*Momordica charantia*) peroxidase catalyzed decolorization of textile and other industrially important dyes. *Bioresource Technology* 96 (16): 1804–1811.

Al-Maqdi, K.A., Bilal, M., Alzamly, A. et al. (2021a). Enzyme-loaded flower-shaped nanomaterials: a versatile platform with biosensing, biocatalytic, and environmental promise. *Nanomaterials* 11 (6): 1460.

Al-Maqdi, K.A., Elmerhi, N., Athamneh, K. et al. (2021b). Challenges and recent advances in enzyme-mediated wastewater remediation – a review. *Nanomaterials* 11 (11): 3124.

Ansari, S.A. and Husain, Q. (2012). Potential applications of enzymes immobilized on/in nano materials: a review. *Biotechnology Advances* 30 (3): 512–523.

Aslam, S., Ali, A., Asgher, M. et al. (2022). Fabrication and catalytic characterization of laccase-loaded calcium-alginate beads for enhanced degradation of dye-contaminated aqueous solutions. *Catalysis Letters* 152 (6): 1729–1741.

Ayodeji, F.D., Shava, B., Iqbal, H. et al. (2022). Biocatalytic versatilities and biotechnological prospects of laccase for a sustainable industry. *Catalysis Letters* 1–25.

Ba, S., Arsenault, A., Hassani, T. et al. (2013). Laccase immobilization and insolubilization: from fundamentals to applications for the elimination of emerging contaminants in wastewater treatment. *Critical Reviews in Biotechnology* 33 (4): 404–418.

Ba, S., Haroune, L., Cruz-Morató, C. et al. (2014). Synthesis and characterization of combined cross-linked laccase and tyrosinase aggregates transforming acetaminophen as a model phenolic compound in wastewaters. *Science of the Total Environment* 487: 748–755.

Badgujar, K.C. and Bhanage, B.M. (2017). Investigation of deactivation thermodynamics of lipase immobilized on polymeric carrier. *Bioprocess and Biosystems Engineering* 40 (5): 741–757.

Bakka, A., Mamouni, R., Saffaj, N. et al. (2016). The treated eggshells as a new biosorbent for elimination of carbaryl pesticide from aqueous solutions: kinetics, thermodynamics and isotherms. *Scientific Study & Research. Chemistry & Chemical Engineering, Biotechnology, Food Industry* 17 (3): 271.

Barbosa, O., Torres, R., Ortiz, C. et al. (2013). Heterofunctional supports in enzyme immobilization: from traditional immobilization protocols to opportunities in tuning enzyme properties. *Biomacromolecules* 14 (8): 2433–2462.

Barrios-Estrada, C., de Jesús Rostro-Alanis, M., Parra, A.L. et al. (2018). Potentialities of active membranes with immobilized laccase for bisphenol A degradation. *International Journal of Biological Macromolecules* 108: 837–844.

Basso, A. and Serban, S. (2019). Industrial applications of immobilized enzymes – a review. *Molecular Catalysis* 479: 110607.

Bayazidi, P., Almasi, H., and Asl, A.K. (2018). Immobilization of lysozyme on bacterial cellulose nanofibers: characteristics, antimicrobial activity and morphological properties. *International Journal of Biological Macromolecules* 107: 2544–2551.

Bayne, L., Ulijn, R.V., and Halling, P.J. (2013). Effect of pore size on the performance of immobilised enzymes. *Chemical Society Reviews* 42 (23): 9000–9010.

Bhatt, P., Tiwari, M., Parmarick, P. et al. (2021). Insights into the catalytic mechanism of ligninolytic peroxidase and laccase in lignin degradation. *Bioremediation Journal* 1–11.

Bickerstaff, G.F. (1997). *Immobilization of Enzymes and Cells*, 1–11. Springer.

Bilal, M. and Iqbal, H.M. (2019). Chemical, physical, and biological coordination: an interplay between materials and enzymes as potential platforms for immobilization. *Coordination Chemistry Reviews* 388: 1–23.

Bilal, M., Asgher, M., Parra-Saldivar, R. et al. (2017). Immobilized ligninolytic enzymes: an innovative and environmental responsive technology to tackle dye-based industrial pollutants – a review. *Science of the Total Environment* 576: 646–659.

Bilal, M., Rasheed, T., Iqbal, H.M.N. et al. (2018a). Horseradish peroxidase immobilization by copolymerization into cross-linked polyacrylamide gel and its dye degradation and detoxification potential. *International Journal of Biological Macromolecules* 113: 983–990.

Bilal, M., Rasheed, T., Sosa-Hernández, J.E. et al. (2018b). Biosorption: an interplay between marine algae and potentially toxic elements – a review. *Marine Drugs* 16 (2): 65.

Bilal, M., Rasheed, T., Zhao, Y. et al. (2018c). "Smart" chemistry and its application in peroxidase immobilization using different support materials. *International Journal of Biological Macromolecules* 119: 278–290.

Bilal, M., Rasheed, T., Zhao, Y. et al. (2019). Agarose-chitosan hydrogel-immobilized horseradish peroxidase with sustainable bio-catalytic and dye degradation properties. *International Journal of Biological Macromolecules* 124: 742–749.

Bronikowski, A., Hagedoorn, P.-L., Koschorreck, K. et al. (2017). Expression of a new laccase from *Moniliophthora roreri* at high levels in *Pichia pastoris* and its potential application in micropollutant degradation. *AMB Express* 7 (1): 1–13.

Burteau, N., Burton, S., and Crichton, R.R. (1989). Stabilisation and immobilisation of penicillin amidase. *FEBS Letters* 258 (2): 185–189.

Ceniceros, H.D. (2003). The effects of surfactants on the formation and evolution of capillary waves. *Physics of Fluids* 15 (1): 245–256.

Chatha, S.A.S., Asgher, M., and Iqbal, H.M.N. (2017). Enzyme-based solutions for textile processing and dye contaminant biodegradation – a review. *Environmental Science and Pollution Research* 24 (16): 14005–14018.

Chen, G.-C., Kuan, I.C., Hong, J.-R. et al. (2011). Activity enhancement and stabilization of lipase from *Pseudomonas cepacia* in polyallylamine-mediated biomimetic silica. *Biotechnology Letters* 33 (3): 525–529.

Cowan, D.A. and Fernandez-Lafuente, R. (2011). Enhancing the functional properties of thermophilic enzymes by chemical modification and immobilization. *Enzyme and Microbial Technology* 49 (4): 326–346.

Darwesh, O.M., Matter, I.A., and Eida, M.F. (2019). Development of peroxidase enzyme immobilized magnetic nanoparticles for bioremediation of textile wastewater dye. *Journal of Environmental Chemical Engineering* 7 (1): 102805.

Datta, S., Christena, L.R., and Rajaram, Y.R.S. (2013). Enzyme immobilization: an overview on techniques and support materials. 3 *Biotech* 3 (1): 1–9.

De Paula, R.C.M., Feitosa, J.P.A., and Paula, H.C.B. (2015). Polysaccharide based copolymers as supramolecular systems in biomedical applications. *Current Drug Targets* 16: 1591–1605.

Dezott, M., Innocentini-Mei, L.H., and Durán, N. (1995). Silica immobilized enzyme catalyzed removal of chlorolignins from eucalyptus kraft effluent. *Journal of biotechnology* 43 (3): 161–167.

Ding, S., Cargill, A.A., Medintz, I.L. et al. (2015). Increasing the activity of immobilized enzymes with nanoparticle conjugation. *Current Opinion in Biotechnology* 34: 242–250.

Ding, M., Cai, X., and Jiang, H.-L. (2019). Improving MOF stability: approaches and applications. *Chemical Science* 10 (44): 10209–10230.

Doonan, C., Riccò, R., Liang, K. et al. (2017). Metal–organic frameworks at the biointerface: synthetic strategies and applications. *Accounts of Chemical Research* 50 (6): 1423–1432.

Drout, R.J., Robison, L., and Farha, O.K. (2019). Catalytic applications of enzymes encapsulated in metal–organic frameworks. *Coordination Chemistry Reviews* 381: 151–160.

Du, Y., Jia, X., Zhong, L. et al. (2022). Metal-organic frameworks with different dimensionalities: an ideal host platform for enzyme@ MOF composites. *Coordination Chemistry Reviews* 454: 214327.

Dvořák, P., Nikel, P.I., Damborskýc, J. et al. (2017). Bioremediation 3.0: engineering pollutant-removing bacteria in the times of systemic biology. *Biotechnology Advances* 35 (7): 845–866.

Dwevedi, A. (2018). *Solutions to Environmental Problems Involving Nanotechnology and Enzyme Technology*. Academic Press.

Falcaro, P., Ricco, R., Yazdi, A. et al. (2016). Application of metal and metal oxide nanoparticles@ MOFs. *Coordination Chemistry Reviews* 307: 237–254.

Flores-Maltos, A., Rodríguez-Durán, L.V., Renovato, J. et al. (2011). Catalytical properties of free and immobilized Aspergillus niger tannase. *Enzyme research* 2011 (768183): 1–6.

Fujiwara, K., Toda, H., and Ikeguchi, M. (2012). Dependence of α-helical and β-sheet amino acid propensities on the overall protein fold type. *BMC Structural Biology* 12 (1): 1–15.

Gan, J., Ashraf, S.S., Bilal, M., and Iqbal, H.M. (2022). Biodegradation of environmental pollutants using catalase-based biocatalytic systems. *Environmental Research* 113914.

Giardina, P., Faraco, V., Pezzella, C. et al. (2010). Laccases: a never-ending story. *Cellular and Molecular Life Sciences* 67 (3): 369–385.

Girigowda, K. and Mulimani, V.H. (2006). Hydrolysis of galacto-oligosaccharides in soymilk by κ-carrageenan-entrapped α-galactosidase from Aspergillus oryzae. *World Journal of Microbiology and Biotechnology* 22: 437–442.

Gkaniatsou, E., Sicard, C., Ricoux, R. et al. (2017). Metal–organic frameworks: a novel host platform for enzymatic catalysis and detection. *Materials Horizons* 4 (1): 55–63.

Glenn, J.K. and Gold, M.H. (1985). Purification and characterization of an extracellular Mn(II)-dependent peroxidase from the lignin-degrading basidiomycete, *Phanerochaete chrysosporium*. *Archives of Biochemistry and Biophysics* 242 (2): 329–341.

Guisan, J.M. (2006). *Immobilization of Enzymes and Cells*. Springer.

Gupta, M.N. (1993). *Thermostability of Enzymes*. Narosa Publishing House.

Han, H., Vila-Aiub, M.M., Jalaludin, A. et al. (2017). A double EPSPS gene mutation endowing glyphosate resistance shows a remarkably high resistance cost. *Plant, Cell & Environment* 40 (12): 3031–3042.

Hartmann, M. and Kostrov, X. (2013). Immobilization of enzymes on porous silicas–benefits and challenges. *Chemical Society Reviews* 42 (15): 6277–6289.

Hernandez-Vargas, G., Sosa-Hernández, J.E., Saldarriaga-Hernandez, S. et al. (2018). Electrochemical biosensors: a solution to pollution detection with reference to environmental contaminants. *Biosensors* 8 (2): 29.

Himmel, M.E. and Georgiou, G. (1991). *Biocatalyst Design for Stability and Specificity*. Washington, DC: American Chemical Society.

Hofrichter, M. (2002). Lignin conversion by manganese peroxidase (MnP). *Enzyme and Microbial Technology* 30 (4): 454–466.

Homaei, A.A., Sariri, R., Vianello, F. et al. (2013). Enzyme immobilization: an update. *Journal of Chemical Biology* 6 (4): 185–205.

Huo, J., Aguilera-Sigalat, J., El-Hankari, S. et al. (2015). Magnetic MOF microreactors for recyclable size-selective biocatalysis. *Chemical Science* 6 (3): 1938–1943.

Hussain, A., Rafeeq, H., Qasim, M. et al. (2021). Engineered tyrosinases with broadened bio-catalysis scope: immobilization using nanocarriers and applications. *3 Biotech* 11 (8): 1–18.

Iyer, P.V. and Ananthanarayan, L. (2008). Enzyme stability and stabilization – aqueous and non-aqueous environment. *Process Biochemistry* 43 (10): 1019–1032.

Jegannathan, K.R., Jun-Yee, L., Chan, E.S., and Ravindra, P. (2010). Production of biodiesel from palm oil using liquid core lipase encapsulated in κ-carrageenan. *Fuel* 89 (9): 2272–2277.

Kalsoom, U., Khalid, N., Ibrahim, A. et al. (2022). Biocatalytic degradation of reactive blue 221 and direct blue 297 dyes by horseradish peroxidase immobilized onto iron oxide nanoparticles with improved kinetic and thermodynamic characteristics. *Chemosphere* 312: 137095.

Kalyani, D., Munk, L., Mikkelsen, J.D. et al. (2016). Molecular and biochemical characterization of a new thermostable bacterial laccase from *Meiothermus ruber* DSM 1279. *RSC Advances* 6 (5): 3910–3918.

Kandambeth, S., Venkatesh, V., Shinde, D.B. et al. (2015). Self-templated chemically stable hollow spherical covalent organic framework. *Nature Communications* 6 (1): 1–10.

Kapoor, M. and Kuhad, R.C. (2007). Immobilization of xylanase from Bacillus pumilus strain MK001 and its application in production of xylo-oligosaccharides. *Applied biochemistry and biotechnology* 142: 125–138.

Katwa, L.C., Ramakrishna, M., and Rao, M.R. (1981). Spectrophotometric assay of immobilized tannase. *Journal of Biosciences* 3: 135–142.

Khan, A.Z., Bilal, M., Rasheed, T. et al. (2018). Advancements in biocatalysis: from computational to metabolic engineering. *Chinese Journal of Catalysis* 39 (12): 1861–1868.

Kibarer, G.D. and Akovali, G. (1996). Optimization studies on the features of an activated charcoal-supported urease system. *Biomaterials* 17 (15): 1473–1479.

Krajewska, B., Zaborska, W., and Leszko, M. (1997). Inhibition of chitosan-immobilized urease by boric acid as determined by integration methods. *Journal of Molecular Catalysis B: Enzymatic* 3 (5): 231–238.

Kreiner, M., Moore, B.D., and Parker, M.C. (2001). Enzyme-coated micro-crystals: a 1-step method for high activity biocatalyst preparation. *Chemical Communications* (12): 1096–1097.

Kumari, P., Alam, M., and Siddiqi, W.A. (2019). Usage of nanoparticles as adsorbents for waste water treatment: an emerging trend. *Sustainable Materials and Technologies* 22: e00128.

Land, E.J., Ramsden, C.A., and Riley, P.A. (2003). Tyrosinase autoactivation and the chemistry of ortho-quinone amines. *Accounts of Chemical Research* 36 (5): 300–308.

Li, J., Wu, H., Liang, Y. et al. (2013). Facile fabrication of organic–inorganic hybrid beads by aminated alginate enabled gelation and biomimetic mineralization. *Journal of Biomaterials Science, Polymer Edition* 24 (2): 119–134.

Mandeep, Liu, H., Luo, J. et al. (2020). Effluents detoxification from pulp and paper industry using microbial engineering and advanced oxidation techniques. *Journal of Hazardous Materials* 398: 122998.

Magnan, E., Catarino, I., Paolucci-Jeanjean, D. et al. (2004). Immobilization of lipase on a ceramic membrane: activity and stability. *Journal of Membrane Science* 241 (1): 161–166.

Majewski, M.B., Howarth, A.J., Li, P. et al. (2017). Enzyme encapsulation in metal–organic frameworks for applications in catalysis. *CrystEngComm* 19 (29): 4082–4091.

Mateo, C., Abian, O., Bernedo, M. et al. (2005). Some special features of glyoxyl supports to immobilize proteins. *Enzyme and Microbial Technology* 37 (4): 456–462.

Mateo, C., Grazu, V., Palomo, J.M. et al. (2007). Immobilization of enzymes on heterofunctional epoxy supports. *Nature Protocols* 2 (5): 1022–1033.

Matto, M. and Husain, Q. (2009). Calcium alginate–starch hybrid support for both surface immobilization and entrapment of bitter gourd (Momordica charantia) peroxidase. *Journal of Molecular Catalysis B: Enzymatic* 57 (1–4), 164–170.

Mogharabi, M. and Faramarzi, M.A. (2014). Laccase and laccase-mediated systems in the synthesis of organic compounds. *Advanced Synthesis & Catalysis* 356 (5): 897–927.

Morsi, R., Bilal, M., Iqbal, H.M., and Ashraf, S.S. (2020). Laccases and peroxidases: the smart, greener and futuristic biocatalytic tools to mitigate recalcitrant emerging pollutants. *Science of the Total Environment* 714: 136572.

Najafpoor, A., Norouzian-Ostad, R., Alidadi, H. et al. (2020). Effect of magnetic nanoparticles and silver-loaded magnetic nanoparticles on advanced wastewater treatment and disinfection. *Journal of Molecular Liquids* 303: 112640.

Namdeo, M. and Bajpai, S.K. (2009). Immobilization of α-amylase onto cellulose-coated magnetite (CCM) nanoparticles and preliminary starch degradation study. *Journal of Molecular Catalysis B: Enzymatic* 59 (1–3): 134–139.

Ning, B., Graham, N., Zhang, Y. et al. (2007). Degradation of endocrine disrupting chemicals by ozone/AOPs. *Ozone: Science and Engineering* 29 (3): 153–176.

Nogueira, H.P., Toma, S.H., Silveira, A.T. et al. (2019). Efficient Cr(VI) removal from wastewater by activated carbon superparamagnetic composites. *Microchemical Journal* 149: 104025.

Noreen, S., Asgher, M., Qamar, S.A. et al. (2022). Poly(vinyl alcohol)-alginate immobilized *Trametes versicolor* IBL-04 laccase as eco-friendly biocatalyst for dyes degradation. *Catalysis Letters* 152 (6): 1869–1879.

Oliveira, F.L., de Souza, S.P., Bassut, J. et al. (2019). Enzyme-decorated covalent organic frameworks as nanoporous platforms for heterogeneous biocatalysis. *Chemistry - A European Journal* 25 (69): 15863–15870.

Palit, S. and Hussain, C.M. (2020). Functionalization of nanomaterials for industrial applications: recent and future perspectives. In: *Handbook of Functionalized Nanomaterials for Industrial Applications*, 3–14. Elsevier.

Perveen, S., Noreen, S., Shahid, S. et al. (2022). Carrier-free cross-linked laccase crystals for biocatalytic degradation of textile industrial effluents. *Applied Biochemistry and Biotechnology* 1–15.

Pervez, S., Aman, A., and Qader, S.A.U.I. (2017). Role of two polysaccharide matrices on activity, stability and recycling efficiency of immobilized fungal amyloglucosidase of GH15 family. *International Journal of Biological Macromolecules* 96: 70–77.

Rafeeq, H., Qamar, S.A., Shah, S.Z.H. et al. (2022). Biodegradation of environmental pollutants using horseradish peroxidase. In: *Biodegradation and Biodeterioration at the Nanoscale*, 603–633. Elsevier.

Raghava, S., Mondal, K., Gupta, M.N. et al. (2006). Preparation and properties of thermoresponsive bioconjugates of trypsin. *Artificial Cells, Blood Substitutes, and Biotechnology* 34 (3): 323–336.

Rajdeo, K., Harini, T., Lavanya, K. et al. (2016). Immobilization of pectinase on reusable polymer support for clarification of apple juice. *Food and Bioproducts Processing* 99: 12–19.

Rao, C.S., Prakasham, R.S., Rao, A.B. and Yadav, J.S. (2008). Functionalized alginate as immobilization matrix in enantioselective L (+) lactic acid production by Lactobacillus delbrucekii. *Applied biochemistry and biotechnology* 149: 219–228.

Rao, M., Scelza, R., Acevedo, F. et al. (2014). Enzymes as useful tools for environmental purposes. *Chemosphere* 107: 145–162.

Riaz, R., Ashraf, M., Hussain, N. et al. (2022). Redesigning robust biocatalysts by engineering enzyme microenvironment and enzyme immobilization. *Catalysis Letters* 1–15.

Riccardi, C.M., Kasi, R.M., and Kumar, C.V. (2017). Nanoarmoring of enzymes by interlocking in cellulose fibers with poly (acrylic acid). *Methods in Enzymology* 590: 475–500. Elsevier.

Rosales-Hernández, M., Kispert, L., Torres-Ramírez, E. et al. (2007). Electron paramagnetic resonance analyses of biotransformation reactions with cytochrome P-450 immobilized on mesoporous molecular sieves. *Biotechnology Letters* 29 (6): 919–924.

Safarik, I. and Safarikova, M. (2009). Magnetic nano-and microparticles in biotechnology. *Chemical Papers* 63 (5): 497–505.

Sahney, R., Puri, B.K., and Anand, S. (2005). Enzyme coated glass pH-electrode: Its fabrication and applications in the determination of urea in blood samples. *Analytica Chimica Acta* 542 (2): 157–161.

Sánchez-Ferrer, Á., Rodríguez-López, J.N., García-Cánovas, F. et al. (1995). Tyrosinase: a comprehensive review of its mechanism. *Biochimica et Biophysica Acta (BBA) – Protein Structure and Molecular Enzymology* 1247 (1): 1–11.

Sarker, A., Islam, T., Bilal, M., and Kim, J.E. (2022). A pilot study for enhanced transformation of a metabolite 3,5-dichloroaniline derived from dicarboximide fungicides through immobilized laccase mediator system. *Environmental Science and Pollution Research* 1–16.

Satar, R., Matto, M. and Husain, Q. (2008). Studies on calcium alginate-pectin gel entrapped concanavalin A-bitter gourd (Momordica charantia) peroxidase complex. *Journal of Scientific & Industrial Research* 67 (8): 609–615.

Secundo, F. (2013). Conformational changes of enzymes upon immobilisation. *Chemical Society Reviews* 42 (15): 6250–6261.

Selinheimo, E., NiEidhin, D., Steffensen, C. et al. (2007). Comparison of the characteristics of fungal and plant tyrosinases. *Journal of Biotechnology* 130 (4): 471–480.

Shahbaz, A., Hussain, N., Intisar, A. et al. (2021). Immobilized enzymes-based biosensing cues for strengthening biocatalysis and biorecognition. *Catalysis Letters* 152: 2637–2649.

Shojaei, F., Homaei, A., Taherizadeh, M.R. et al. (2017). Characterization of biosynthesized chitosan nanoparticles from *Penaeus vannamei* for the immobilization of *P. vannamei* protease: an eco-friendly nanobiocatalyst. *International Journal of Food Properties* 20 (supp. 2): 1413–1423.

Shuster, V. and Fishman, A. (2009). Isolation, cloning and characterization of a tyrosinase with improved activity in organic solvents from *Bacillus megaterium*. *Microbial Physiology* 17 (4): 188–200.

Sirisha, V.L., Jain, A., and Jain, A. (2016). Enzyme immobilization: an overview on methods, support material, and applications of immobilized enzymes. *Advances in Food and Nutrition Research* 79: 179–211.

Soleimani, R., Heytens, E., Darzynkiewicz, Z., and Oktay, K. (2011). Mechanisms of chemotherapy-induced human ovarian aging: double strand DNA breaks and microvascular compromise. *Aging (Albany NY)* 3 (8): 782.

Sun, Q., Pan, Y., Wang, X. et al. (2019). Mapping out the degree of freedom of hosted enzymes in confined spatial environments. *Chem* 5 (12): 3184–3195.

Takano, H.K., Fernandes, V.N., Adegas, F.S. et al. (2020). A novel TIPT double mutation in EPSPS conferring glyphosate resistance in tetraploid *Bidens subalternans*. *Pest Management Science* 76 (1): 95–102.

Tan, Z., Bilal, M., Raza, A. et al. (2021). Expanding the biocatalytic scope of enzyme-loaded polymeric hydrogels. *Gels* 7 (4): 194.

Tischer, W. and Kasche, V. (1999). Immobilized enzymes: crystals or carriers? *Trends in Biotechnology* 17 (8): 326–335.

Tischer, W. and Wedekind, F. (1999). *Immobilized enzymes: methods and applications*. In: *Biocatalysis From Discovery to Application*, 95–126. Springer.

Tiwari, D., Kamble, J., Chilgunde, S. et al. (2012). Clastogenic and mutagenic effects of bisphenol A: an endocrine disruptor. *Mutation Research, Genetic Toxicology and Environmental Mutagenesis* 743 (1–2): 83–90.

Valli, K., Wariishi, H., and Gold, M.H. (1990). Oxidation of monomethoxylated aromatic compounds by lignin peroxidase: role of veratryl alcohol in lignin biodegradation. *Biochemistry* 29 (37): 8535–8539.

Vaillant, G.E. (2000). Adaptive mental mechanisms: their role in a positive psychology. *American Psychologist* 55 (1): 89.

Wu, L., Yuan, X., and Sheng, J. (2005). Immobilization of cellulase in nanofibrous PVA membranes by electrospinning. *Journal of Membrane Science* 250 (1–2): 167–173.

Wu, M., Li, W., Dick, W.A. et al. (2017). Bioremediation of hydrocarbon degradation in a petroleum-contaminated soil and microbial population and activity determination. *Chemosphere* 169: 124–130.

Xue, H., Shen, Z., and Li, C. (2005). Improved selectivity and stability of glucose biosensor based on in situ electropolymerized polyaniline–polyacrylonitrile composite film. *Biosensors and Bioelectronics* 20 (11): 2330–2334.

Xv, J., Zhang, Z., Pang, S. et al. (2022). Accelerated $CO_2$ capture using immobilized carbonic anhydrase on polyethyleneimine/dopamine co-deposited MOFs. *Biochemical Engineering Journal* 189: 108719.

Yang, X.-Y., Tian, G., Jiang, N. et al. (2012). Immobilization technology: a sustainable solution for biofuel cell design. *Energy & Environmental Science* 5 (2): 5540–5563.

Yehia, R. and Rodriguez-Couto, S. (2017). Discoloration of the azo dye Congo Red by manganese-dependent peroxidase from *Pleurotus sajor caju*. *Applied Biochemistry and Microbiology* 53 (2): 222–229.

Zdarta, J., Meyer, A.S., Jesionowski, T. et al. (2018a). Developments in support materials for immobilization of oxidoreductases: a comprehensive review. *Advances in Colloid and Interface Science* 258: 1–20.

Zdarta, J., Meyer, A.S., Jesionowski, T. et al. (2018b). A general overview of support materials for enzyme immobilization: characteristics, properties, practical utility. *Catalysts* 8 (2): 92.

Zhang, Y., Ge, J., and Liu, Z. (2015). Enhanced activity of immobilized or chemically modified enzymes. *ACS Catalysis* 5 (8): 4503–4513.

Zhao, H. (2010). Methods for stabilizing and activating enzymes in ionic liquids – a review. *Journal of Chemical Technology & Biotechnology* 85 (7): 891–907.

# 10

# Effects of Microbial Inoculants on Soil Nutrients and Microorganisms

*D. Vijaysri[1], Konderu Niteesh Varma[1], Haralkal Keerthi Kumari[2], D. Sai Srinivas[3], S.T.M. Aravindharajan[1], Dilbag Singh[1], Livleen Shukla[1], T. Kavya[1], and Sandeep Kumar Singh[1]*

[1] *Division of Microbiology, ICAR-Indian Agricultural Research Institute, New Delhi, India*
[2] *Department of Agricultural Microbiology, University of Agricultural Sciences, Bengaluru, India*
[3] *Department of Microbiology, School of Life Sciences, JSS Academy of Higher Education and Research, Mysore, India*

## 10.1 Introduction

Microorganisms have a significant role in agricultural systems. A healthy soil acts as a home to a variety of microorganisms, which includes bacteria, fungi, and protozoa. Microorganisms perform several functions in soil, which include nitrogen fixation, phosphorous solubilization, preservation of soil structure, regulation of bio-geochemical cycles, suppression of pathogens and toxins and thus ultimately help in sustainable agriculture. The constant diminishment of soil nutrient supply capacity is attributed to various factors, including soil degradation, erosions, salt deposition, presence of undesired metals and elements, water scarcity or surplus, and an imbalanced nutrient delivery system (Sammauria et al. 2020). The rising need for food production in the face of diminishing land resources poses a significant obstacle to the long-term viability of agriculture. Efficient utilization of determinate resources is crucial for attaining sustainable food production (Owen et al. 2015). The continuous utilization of fertilizers, herbicides, and pesticides has resulted in reduced agricultural productivity, decreased fertility of the soil, adverse economic outcomes, food poisoning, soil degradation, loss of biodiversity, and significant environmental hazards. Nevertheless, the escalating consequences on the environment by many of agricultural practices have progressively influenced the soil's quality. Consequently, it is imperative to enhance soil productivity in a manner that safeguards its ability to serve as a healthy medium (Trivedi et al. 2012).

Soil microorganisms are an important part of the soil, which plays a significant role in the integrated nutrient management and soil biodiversity system. The growth and development of plants are significantly influenced by their pivotal role. Microbial inoculants refer to living organisms that are intentionally introduced into a certain environment with specific objectives, such as the control of pests or the enhancement of plant growth. These organisms encompass a range of microorganisms, including bacteria, fungi, and other microbial entities (Shahwar et al. 2023). The majority of microorganisms utilized in the microbial inoculants inhabit or have the potential to reside in soil, where they fulfill diverse roles and activities within the soil ecosystem. Microbial inoculants have been utilized for over a century, but their significance has significantly increased in the past three decades, as seen by the availability of numerous commercial products in the market made up of microbial inoculants (Babalola and Glick 2012). Although inconsistent reports have been found on the beneficial effects of using microbial inoculants for enhanced plant growth and health, a significant trend is observed with the use of microbial inoculants to meet the needs of sustainable agriculture. Microbial inoculants have a significant role in various biological and chemical processes within ecosystems, including the control of diseases and the cycling of nutrients (Kumari et al. 2022; Verma et al. 2022). Consequently, they contribute to the enhancement of nutrient availability. In addition to the production of chemicals that stimulate plant growth, these organisms play an important role in fixing atmospheric nitrogen and mobilizing/solubilizing soil nutrients. The application of microbial inoculants has been shown to enhance biodiversity, hence establishing favorable conditions for the proliferation of beneficial microorganisms. In addition,

*Microbes Based Approaches for the Management of Hazardous Contaminants*, First Edition. Edited by Ajay Kumar, Livleen Shukla, Joginder Singh, and Luiz Fernando R. Ferreira.

they enhance the physical characteristics of soil, including the enhancement of soil particle structure and aggregation, reduction of soil compaction, and increase in pore spaces and water infiltration (Alori et al. 2017).

In order to increase the productivity of seeds, plants, and soil, microbial inoculants are administered either alone or in combination. To achieve agricultural sustainability, they are specialized formulations that make use of the functional characteristics of the microbes in a variety of soils and cropping patterns. These kinds of microorganisms have been referred to by a variety of terms, includes biostimulant, bio-inoculant, and bio-fertilizer. They are environmentally friendly, extremely effective, and may also be used as biopesticides that have no harmful impacts on plant products (Ansari et al. 2015; Halpern et al. 2015; Shahwar et al. 2023; Singh et al. 2013).

However, deliberately releasing microorganisms into a new environment raises serious concerns because of the possible effects on ecosystem functioning and biodiversity. The process of inoculation involves the introduction of high number of viable and effective microorganisms into the host rhizosphere, resulting in a temporary disturbance in the stability of soil microbial populations. Changes in microbial compositions could negatively impact the next crops and cause unfavorable changes in the microbial makeup if important native species are lost (Trabelsi and Mhamdi, 2013). This emphasizes the possible evolutionary and ecological implications of inoculation of microbial inoculants to soil native microbial communities, even though the microbial community of soil may be able to reorganize and revert to its natural form (resilience) following the disturbances caused by inoculation. However, when more sustainable farming practices are implemented, the use of microbial inoculants increases, and sometimes, the effects may deviate from the anticipated purposes (Verma et al. 2019). Therefore, further awareness is required to determine the exact impact that these introductions – whether successful or not – have impacts on the local microbial communities.

## 10.2 Microbial Inoculants and Soil Nutrients

The three main categories of microbial inoculants are plant growth-promoting rhizobacteria (PGPR), nitrogen-fixing rhizobia, and arbuscular mycorrhiza fungi (that are typically not regarded as PGPR). It is well recognized that these groups have the ability to increase nutrient availability and nutrient uptake as well as maintain plant health in order to stimulate plant development (Yadav and Verma 2014). Microbial inoculants are not themselves, but they are microorganisms that might enhance the availability of nutrients in soil through their biophysical as well as biochemical activities (Alori et al. 2017). Another name for microbial inoculants used in soil nutrient management is biofertilizers. Biofertilizers are low-cost, highly effective, and environmentally beneficial agricultural inputs (Pathak and Kumar, 2016). Using biofertilizers for different crops benefits the soils in a real way, whether directly or indirectly. Biofertilizers are more effective even in smaller quantities, have more targeted activity, and are safer with reduced environmental damage (Suyal et al. 2016). They are the specialized formulations that make use of the functional characteristics of the microorganisms in a variety of cropping patterns and soil systems to achieve sustainable agriculture. PGPR contains many well-known microbes, such as *Rhizobium*, *Bacillus*, *Azospirillum*, *Burkholderia*, *Azotobacter*, *Klebsiella*, *Pseudomonas*, and *Enterobacter*. Some of these organisms also include endophytic forms. Plant growth can be directly or indirectly stimulated by PGPRs (Gond et al. 2021; Gupta et al. 2020; Kumar et al. 2016). PGPRs directly help by different mechanisms, which include the release of nutrients, the synthesis of phytohormones, and the stimulation of induced systemic resistance, for example, nitrogen fixers, viz., Rhizobia group, *Azospirillum*, *Pseudomonas*, Phosphate (P) solubilizing bacteria (PSB), etc. Suyal et al. 2014; Singh et al. 2020, 2022).

## 10.3 Influence of Microbial Inoculants on Soil Nutrient Quality

### 10.3.1 Nitrogen

Nitrogen is one of the main elements on the Earth involved in the production of proteins, nucleic acids, chlorophyll, secondary metabolites, phytohormones, and co-enzymes (Suyal et al. 2016). The two main sources of nitrogen for agricultural soils are mineral fertilizers and nitrogen fixation, which is a microbiological process. Biological nitrogen fixation is the initial stage of cycling nitrogen from the atmosphere into the biosphere. It is a microbiological process that transforms atmospheric nitrogen into a usable form to plants. It provides an alluring and effective way to decrease external inputs and enhance plant internal resources. The limitation of available nitrogen in the rhizospheric soils makes nitrogen fixation a very desired phenomenon in the soil by diazotrophs (Vijaysri et al. 2023). The biological nitrogen fixation process is the main reason for 65% of the nitrogen used in agriculture today, and it will remain crucial for crop production systems that are sustainable in the future (Matiru and Dakora 2004). The inoculation of nitrogen-fixing PGPRs on plant yield results in disease control, growth-promoting activities, and nitrogen retention (Shahwar et al. 2023). Microbes, mainly bacteria, are the main reasons for N cycling. Bacteria are exclusively known organisms that fix nitrogen from the atmosphere, either symbiotically or asymbiotically, because of the presence of an essential enzyme called nitrogenase, which precisely converts atmospheric nitrogen to ammonia (Wagner 2011).

#### 10.3.1.1 Symbiotic Nitrogen Fixation

Symbiotic nitrogen fixation (SNF) is a mutualistic association between plants and microbes. Legume and nonlegume plants are involved in the symbiotic process of fixing nitrogen in the soil. Several eubacteria, blue-green algae and actinomycetes are some of the microorganisms which are involved in nitrogen fixation (Pandey et al. 2023; Singh et al. 2023). The bacterial group called as rhizobia, which includes Rhizobium, Allorhizobium, Sinorhizobium, Mesorhizobium, Bradyrhizobium, and Azorhizobium, are mainly responsible for the legume N fixation. They establish close symbiotic relationships with different legumes by growing inside their root nodules during the vegetative stage and reacting chemotactically to flavonoid molecules. Frankia and Actinobacteria, on the other hand, fix nitrogen in nonlegumes. Besides bacteria, there are some small fern which is also known for its nitrogen fixation symbiotically. For instance, cyanobacteria (Anabaena) and the tiny, free-floating aquatic fern Azolla symbiotically fix atmospheric nitrogen. Azolla mainly helps in rice farming by adding nitrogen and biomass to the soil. Actinorhizal plants can develop root nodules through the action of actinomycetes, such as *Frankia*. Some genera, including *Allocasuarina*, *Casuarina*, *Myrica*, *Coriaria*, and *Eleagnus* are nodulated by *Frankia*. These plants are monocots with a promising future in the reclamation of agricultural land (Shahwar et al. 2023). It is obvious that legumes with rhizobial inoculation have the capacity to increase the N status of soil. According to reports, BNF generates 200 million tons of nitrogen annually worldwide (Peoples et al. 2009); symbiotic nitrogen fixation accounts for 80% of this amount, with free-living nitrogen fixation providing the remaining portion (Adesemoye et al. 2009). Legumes are used in crop sequence in the majority of crop rotation systems as they may fix nitrogen and improve the soil's N status in the following crop. This method of crop rotation has been used extensively in the management of soil nutrients to minimize the use of artificial fertilizers. Thus, sustainable agricultural production will ultimately profit from the appropriate and effective application of microbial inoculants (Alori et al. 2017).

#### 10.3.1.2 Nonsymbiotic Nitrogen Fixation

N can be fixed into the soil by a class of free-living bacteria. These are usually known as PGPR, plant growth-promoting microorganisms, or plant growth-promoting bacteria (Calvo et al. 2014). Additionally, nitrogen fixers that promote the growth of non-leguminous plants like rice, wheat, sugarcane, etc., carry out nonsymbiotic nitrogen fixation. *Gluconacetobacter diazotrophicus*, *Pseudomonas*, *Burkholderia*, *Azoarcus*, *Enterobacter*, and *Azospirillum* are a few more rhizospheric bacteria that function as nonsymbiotic nitrogen fixers. They absorb carbon exudates from plants, and they fix nitrogen while they are free-living. *Azospirillum*, *Acetobacter diazotrophicus* and *Herbaspirillum* are associated with the roots of the Gramineae family (Pathak and Kumar 2016). Many of these organisms, either alone or in combination with other microorganisms, are employed as microbial inoculants to enhance crop growth. When compared to rhizobial inoculants, the utilization of these nonsymbiotic nitrogen-fixing bacteria has not been very effective as a single inoculant to enhance soil N content for crop use (Alori et al. 2017). Though many PGPR have extremely variable abilities to raise the soil's N content, some instances of a noticeable rise in soil N have been noted, particularly when more than one organisms are used in inoculants (Rawat et al. 2013).

### 10.3.2 Phosphorous

Phosphorus is one of the important macronutrients required for the growth and development of plants. It is also necessary for the functioning of essential enzymes, which regulate the metabolic pathways. The major part of the phosphorus in soil is insoluble and not readily available to plants (Vijaysri et al. 2023). Approximately, 40% of crop output worldwide on arable land is restricted by low soil available P. To add on the problem of deficiency of available phosphorus, added P fertilizers get fixed due to the exchange of complexes within the soil, and thus limits the accessibility of P to plants. A variety of microbes have been responsible for the biogeochemical cycling of organic and inorganic P in the rhizosphere of plants. It is expected that P-solubilizing microorganisms as inoculants will grow rapidly on the commercial market in the future (Rafi et al. 2019). Owing to the inherent P shortage in a lot of agricultural soils, these kinds of microorganisms are often proposed as possible P biofertilizers. Researchers are interested in using numerous PGPRs as microbial inoculants because of their ability to solubilize P. They were reported to release organic phosphorus up to 30–50 kg $P_2O_5$/ha into soil solution by the action of enzymes (Pindi and Satyanarayana 2012).

Two approaches can be used to analyze the importance of microbial inoculants in enhancing P availability in soil for plant growth: first, the mobilization of P and helps the plant roots for uptake, and second, the P solubilization from the mineral rock thereby, increases the P availability in soil solution (Owen et al. 2015). It has been reported that certain bacterial and fungal genera can solubilize and make insoluble P in soil accessible to plant roots are designated as phosphate solubilizing and phosphate mobilizing microorganisms. P-solubilizing bacteria (PSB) are known as heterotrophic bacteria and are crucial in the process of converting unavailable forms to available forms through various mechanisms. These mechanisms include the release of organic acids (such as gluconic acid, lactic acid, acetic acid, oxalic acid, and malonic acid) that complex with the cations which are bound to phosphate and transform them into soluble form, and the production of enzymes (phosphatases) which dephosphorylate the phytates, which is one of the dominant organic P in the soil (60% of organic P) and discharge P and makes it available to plants (Bakhshandeh et al. 2017). According to several reports, the important PSBs include *Bacillus*, *Pseudomonas*, *Burkholderia*, *Beijerinckia*, *Enterobacter*, *Flavobacterium*, *Erwinia*, *Microbacterium*, *Arthrobacter*, *Flavobacterium*, *Azotobacter*, *Rhizobium*, *Rhodococcus*, *Serratia*, etc.

Phosphate-mobilizing microorganisms are fungi that can use their hyphal structures to mobilize or solubilize immobile forms of phosphorous, enabling plants to absorb it. Arbuscular mycorrhizal fungi are the important researched P mobilizing fungus. As a major factor in P nutrition to plants and its growth in the natural environment, the arbuscular mycorrhiza is widely distributed across the plant kingdom. AMFs, or arbuscular mycorrhizal fungi, belong to the phylum Glomeromycota and have the ability to form mutualistic symbioses with a variety of land plants. By expanding the surface area of the plant root system, mycorrhizae improve nutrient absorption from soil, particularly in soils deficient in phosphorus (Sammauria et al. 2020). Additionally, they explore the soil more thoroughly and provide the host roots with more N, P, K, Zn, Cu, S, Fe, Ca, Mg, and Mn. By releasing organic acids and phosphatase enzymes, certain other fungi, such *Penicillium* and *Aspergillus* species, are also responsible to solubilize inorganic phosphate and mineralize organic phosphate. Hence, the utilization of these organisms, whether in the form of bacterial or fungal inoculants, offers a distinct benefit compared to the administration of P fertilizers, which tend to form complexes in the soil upon application. Thus, the continuous P supply to crops can be very beneficial for soil management techniques that include the application of microbial inoculants.

### 10.3.3 Potassium

Potassium (K) is the third most important macronutrient for plants. Over 90% of the potassium in the soil is found in the form of silicate minerals and insoluble rocks. Soluble potassium concentrations in the soil are normally quite low. It plays a specific role in regulating stomatal opening and closure, synthesis of proteins, food uptake, improving product quality, and providing resistance under adverse environmental conditions (Santosh et al. 2022). It has been demonstrated that microorganisms that solubilize K in soil and the rhizosphere of plants are one of the main components in the K cycles (Liu et al. 2012). They secrete organic acids that can solubilize potassium rock. The organic acid production either directly or indirectly enhances K dissolution by creating complexes with reaction products in solution or through ligand- or proton-mediated mechanisms. This is one of the ways that KSB provides plants access to K (Etesami et al. 2017). It has been found that PGPR such as *Bacillus edaphicus*, *Acidothiobacillus ferrooxidans*, *Paenibacillus* sp., *Burkholderia*, *Bacillus mucilaginosus*, and *Pseudomonas* are also able to solubilize potassium from minerals which contains unavailable potassium in soils into an available form. It has also been reported that AM fungal inoculants also can enhance K absorption. The production of organic anions, including oxalate, citrate and malate, as well as proton $H^+$ or $CO_2$, by arbuscular mycorrhizal fungal inoculants increases the K solubility (Meena et al. 2014). Therefore, the role of PGPR and Arbuscular mycorrhizal fungi (AMF) in the solubilization and mobilization of potassium (K) in the soil is crucial, as it ultimately leads to increased availability of K to plants, thereby contributing to sustainable improvements in plant nutrition.

### 10.3.4 Zinc

The widespread utilization of chemical fertilizers in agricultural practices has significantly depleted agricultural land and resulted in elevated soil pH levels, hence disrupting the accessibility of vital micronutrients to plants. Zinc plays a significant role in the metabolic activities of crop plants, hence contributing significantly to their developmental processes (Kumar et al. 2019). It serves as a chief component of many enzymes and it acts as a precursor of indole-3-acetic acid (IAA) for the synthesis of tryptophan. At high pH, Zn becomes unavailable in soil and in turn to plants. Zinc deficiency is a widespread problem, mainly in rice cropland of Asia and in soil belonging to orders aridisols, mollisols, alfisols, and vertisols throughout the world. In most of the agricultural soils, zinc deficiency is more commonly caused by low soluble zinc than by low total zinc concentration. Application of zinc in the inorganic form typically satisfies plant requirements partially. Since, after a few days of application, 96.99% of zinc is converted into various insoluble forms depending on different soil types and physicochemical processes (Saravanan et al. 2004). Microbes that solubilize complex zinc in soil are a potential substitute that could meet plants' needs for zinc. *Pseudomonas*, *Bacillus*, *Acinetobacter*, *Burkholderia*, etc., are reported to solubilize Zn compounds. Zinc solubilizing microbial inoculants help to release the unavailable form of Zn and increase Zn uptake, and lead to the fortification of grains with Zn (Barbagelata and Mallarino 2013).

## 10.4 Impact of Microbial Inoculants on Natural Soil Microbial Communities

The utilization of inoculants is seen as highly advantageous due to its potential to significantly decrease the reliance on chemical fertilizers and pesticides. Moreover, there has been a notable rise in the commercialization of inoculants for various crops (Trabelsi and Mhamdi 2013). The awareness on the survival of inoculated bacterial and fungal strains in natural environments, as well as the impact of their introduction on the existing native microbiota, has been of great interest since the practical use of selected microorganisms, either natural or genetically modified microorganisms, has been developed. The soil microbial community is characterized by its complexity and dynamic nature, exhibiting variations in composition across different compartments and layers, which represents a real challenge in soil ecology. Inoculation will be advantageous when the inoculant can multiply in the soil and colonize the roots despite competition from the native soil microbiota (Lugtenberg and Kamilova 2009). The large number of viable and effective microorganisms are added to the soil through seed inoculation and soil inoculation in order to quickly colonize the host rhizosphere. This can significantly affect the stability of native soil microbial population (Babalola 2014). If significant native species disappear, it could negatively impact the microbial composition, thus affecting the subsequent crops. The selection of inoculants is based on their beneficial characteristics that can influence plant development and overall health. However, the possible impacts on the microbial population are often not assessed frequently. In order to effectively attain the benefits of this practice, it is imperative to introduce a substantial concentration of inoculant into the surrounding environment that can create alterations to the local biological structure (Lupwayi et al. 2000; Litchman 2010). These alterations may arise due to either direct competition or the synergetic or antagonistic interactions between the new microorganisms and the indigenous microorganisms. The complexity and diversity of soil microbial communities often render them resistant to the colonization of introduced microbes (Mallon et al. 2015). Since it is believed that inoculants and plants would actively collaborate in the zone of plant roots, these interactions may help in the establishment of relationships that will improve environmental adaption and thereby alter, in some way, the surroundings. These modifications may occur within the taxonomic classification or within the functional capacities of the soil microbial population.

## 10.5 Microbial Inoculants: Mechanisms Involved in Affecting the Resident Microbial Community

Although many studies reported that the inoculation of microbial inoculants frequently influences the native soil microbial population, but the mechanisms behind the influence of inoculating microbial inoculants on resident soil microbial populations remain poorly understood. Some of the research reports suggested potential mechanisms of inoculants involved in affecting the soil microbial population; however, none of the studies thoroughly addresses the relative significance of the several potential processes. Here, a few mechanisms are elucidated by which inoculation influences the alterations in resident microbial communities.

### 10.5.1 Competition

The first mechanism refers to the competition for resources. Numerous studies have demonstrated that the fate of invading species is determined by the quantity of (limited) resources that are left unused by resident microbes and the rate at which invaders and native species consume resources. This indicates that the successful establishment of inoculant in their new environment increases with the large numbers of vacant habitats (those that are not occupied by the resident group) (Yang et al. 2017). The inoculation of microbial inoculants may outperform some taxa after establishment and use available resources to proliferate. This might be relevant if the invaders have unique characteristics that enable them to use resources more effectively than their competitors. For example, they may encourage acidification of soil or have more access to iron in the soil as a result of siderophore synthesis (Zhang et al. 2010). Once the inoculants establish themselves, their numbers may inhibit resident populations' functionally equivalent species, or those that compete for the same resources, and promote the emergence of new taxa that are not functionally related to the inoculants.

### 10.5.2 Antagonism

Microbial inoculants have the ability to exert an impact directly on the multiplication and functioning of resident communities, mostly by the process of antibiosis, which involves the release of chemical compounds that kills or inhibits neighboring resident bacteria. This mechanism can be seen in several microbial inoculants that are used for agricultural purposes, especially those that are used to control diseases. For example, specific strains of *Enterobacter*, *Lysobacter*, *Burkholderia*, *Pseudomonas*, *Burkholderia*, and *Bacillus* are primarily responsible for the synthesis of antibiotics (Dukare et al. 2019). These chemical may also affect microbial taxa that are not their intended targets, even when they are directed against certain pathogens. For example, the genetic diversity of several species of rhizobia in the rhizosphere of sugar beet was reduced by 2,4-diacetylphloroglucinol (Phl), an antibiotic that was produced by *Pseudomonas fluorescens* F113Rif (Walsh et al. 2003).

### 10.5.3 Synergism

The introduction of microorganisms can also have an impact on the existing microbial communities through a phenomenon known as synergism. Synergism refers to a cooperative interaction among microorganisms, ranging from providing marginal assistance to establishing a state of absolute mutual dependency. The expansion of resident microbial communities is stimulated by the introduction of microbial inoculants, which produce secondary metabolites, including vitamins, precursors, and specific amino acids (Schink 2002). Furthermore, some bacteria can rely heavily on one another, and neither species can function to its full potential when the other species is absent (Kato et al. 2012).

### 10.5.4 Indirect Effect Through Root Exudation

The modification of the soil microbial population that occurs by the influence of plant root exudates is an indirect process. Numerous microbial inoculants, such as PGPR, have been observed to exert a significant impact on the development of the plant root system by means of phytohormone synthesis and the generation of several other bioactive compounds. These chemicals have the ability to stimulate lateral root branching and alter the functionality of the roots (Vacheron et al. 2013). Specifically, the introduction of PGPR was found to enhance the rates of root exudation, leading to potential modifications in the microbial communities inside the rhizosphere. For example, Florio et al. (2017) found that the *Azospirillum lipoferum* CRT1, a PGPR, which is able to encourage root exudation, increases the abundance of denitrifiers in soils where the availability of carbon limits denitrifiers. Beyond the quantity of exudates, studies also found that microbial inoculants have the ability to alter the root exudate composition in addition to their amount, particularly with regard to amino acids and various groups of flavonoids (Matilla et al. 2010). These exudates also contain a variety of chemical substances that are best metabolized by particular microorganisms (Dardanelli et al. 2010).

## 10.6 Effect of Monoinoculation Versus Coinoculation

Most of the inoculants commonly depend on the use of a particular strain, which could potentially contribute to the observed variations in field outcomes. One of the methods to overcome this problem is by adding many beneficial microbial species or strains to the same microbial mixture. Without the need for genetic engineering, coinoculation would integrate many mechanisms, improve performance of the plant, and increase the reliability and efficacy of beneficial impacts on crops (Janisiewicz 1996; Marimuthu et al. 2002). Coinoculation has the capacity to withstand more variable conditions than monocultures and can outcompete single microbes at multiple tasks (Brenner, You, and Arnold 2008). In order to better compete with resident soil microorganisms, more diverse communities are expected to inhabit a larger resource habitat in soil when applied and designed deliberately. This includes combining different microbial strains and/or species. However, choosing the members of a consortium to maximize the performance presents an important challenge. As of now, the majority of artificial microbial communities used in agriculture have been created and assessed in highly controlled environments with an emphasis on the plant microbiome (Bano et al. 2021). In order to facilitate the future deployment of successful synthetic microbial consortia, further investigation is required to study the intricate interactions that exist between plants, microbiota, and the environment.

## 10.7 Conclusion

The exponential growth of the world population has necessitated a significant escalation in agricultural production in order to reach the increased food demand. The constant utilization of fertilizers, insecticides, and herbicides has resulted in diminished agricultural output, decreased fertility of the soil, soil degradation, loss of biodiversity, unfavorable economic outcomes, instances of food poisoning, and significant environmental hazards. Therefore, it is crucial to use environmentally sustainable approaches for nutrient management. Soil microbes have great potential for agriculture. Numerous microorganisms have been used in place of traditional inorganic fertilizers to increase crop yield by various mechanisms. However, microbial inoculation has the potential to drastically alter the number and composition of taxonomic groups. Intentional release of organisms into a new environment may have a potential influence on nontarget effects on biodiversity and ecosystem functioning. Hence, it is imperative to acquire a better understanding of the underlying mechanisms that drive alterations in the composition and the soil microbiome functioning following invasion. Therefore, to accomplish the objective of sustainable agriculture, the implementation of appropriate microbial inoculant technology with a better understanding on its effect on soil microbial community could be adopted to ensure safe, enhanced productivity and long-term agricultural sustainability.

## References

Adesemoye, A.O., Torbert, H.A., and Kloepper, J.W. (2009). Plant growth-promoting rhizobacteria allow reduced application rates of chemical fertilizers. *Microbial Ecology* 58: 921–929.

Alori, E.T., Dare, M.O., and Babalola, O.O. (2017). Microbial inoculants for soil quality and plant health. In: *Sustainable Agriculture Reviews* (ed. E. Lichtfouse), 281–307. Springer.

Ansari, M.F., Tipre, D.R., and Dave, S.R. (2015). Efficiency evaluation of commercial liquid biofertilizers for growth of *Cicer aeritinum* (chickpea) in pot and field study. *Biocatalysis and Agricultural Biotechnology* 4 (1): 17–24.

Babalola, O.O. (2014). Does nature make provision for backups in the modification of bacterial community structures? *Biotechnology and Genetic Engineering Reviews* 30 (1): 31–48.

Babalola, O.O. and Glick, B.R. (2012). Indigenous African agriculture and plant associated microbes: current practice and future transgenic prospects. *Scientific Research and Essays* 7 (28): 2431–2439.

Bakhshandeh, E., Pirdashti, H., and Lendeh, K.S. (2017). Phosphate and potassium-solubilizing bacteria effect on the growth of rice. *Ecological Engineering* 103: 164–169.

Bano, S., Wu, X., and Zhang, X. (2021). Towards sustainable agriculture: rhizosphere microbiome engineering. *Applied Microbiology and Biotechnology* 105: 7141–7160.

Barbagelata, P.A. and Mallarino, A.P. (2013). Field correlation of potassium soil test methods based on dried and field-moist soil samples for corn and soybean. *Soil Science Society of America Journal* 77 (1): 318–327.

Brenner, K., You, L., and Arnold, F.H. (2008). Engineering microbial consortia: a new frontier in synthetic biology. *Trends in Biotechnology* 26 (9): 483–489.

Calvo, P., Nelson, L., and Kloepper, J.W. (2014). Agricultural uses of plant biostimulants. *Plant and Soil* 383: 3–41.

Dardanelli, M.S., Manyani, H., González-Barroso, S. et al. (2010). Effect of the presence of the plant growth promoting rhizobacterium (PGPR) *Chryseobacterium balustinum* Aur9 and salt stress in the pattern of flavonoids exuded by soybean roots. *Plant and Soil* 328: 483–493.

Dukare, A.S., Paul, S., Nambi, V.E. et al. (2019). Exploitation of microbial antagonists for the control of postharvest diseases of fruits: a review. *Critical Reviews in Food Science and Nutrition* 59 (9): 1498–1513.

Etesami, H., Emami, S., and Alikhani, H.A. (2017). Potassium solubilizing bacteria (KSB): mechanisms, promotion of plant growth, and future prospects: a review. *Journal of Soil Science and Plant Nutrition* 17 (4): 897–911.

Florio, A., Pommier, T., Gervaix, J. et al. (2017). Soil C and N statuses determine the effect of maize inoculation by plant growth-promoting rhizobacteria on nitrifying and denitrifying communities. *Scientific Reports* 7 (1): 8411.

Gond, D.P., Jha, S.S., Kumar, A. et al. (2021). Plant growth promoting bacteria and its role in green remediation. In: *Sustainable Environmental Clean-Up* (ed. V.K. Mishra and A. Kumar), 149–163. Cambridge, MA: Elsevier https://doi.org/10.1016/B978-0-12-823828-8.00007-4.

Gupta, A., Singh, S.K., Singh, M.K. et al. (2020). Plant growth–promoting rhizobacteria and their functional role in salinity stress management. In: *Abatement of Environmental Pollutants* (ed. P. Singh, A. Kumar, and A. Borthakur), 151–160. Cambridge, MA: Elsevier https://doi.org/10.1016/B978-0-12-818095-2.00007-2.

Halpern, M., Bar-Tal, A., Ofek, M. et al. (2015). The use of biostimulants for enhancing nutrient uptake. *Advances in Agronomy* 130: 141–174.

Janisiewicz, W. (1996). Ecological diversity, niche overlap, and coexistence of antagonists used in developing mixtures for biocontrol of postharvest diseases of apples. *Phytopathology* 86 (5): 473–479.

Kato, S., Hashimoto, K., and Watanabe, K. (2012). Methanogenesis facilitated by electric syntrophy via (semi) conductive iron-oxide minerals. *Environmental Microbiology* 14 (7): 1646–1654.

Kumar, A., Singh, M., Singh, P.P. et al. (2016). Isolation of plant growth promoting rhizobacteria and their impact on growth and curcumin content in *Curcuma longa* L. *Biocatalysis and Agricultural Biotechnology* 8: 1–7.

Kumar, A., Dewangan, S., Lawate, P. et al. (2019). Zinc-solubilizing bacteria: a boon for sustainable agriculture. In: *Plant Growth Promoting Rhizobacteria for Sustainable Stress Management. 1: Rhizobacteria in Abiotic Stress Management*, 139–155. Springer Nature Singapore Pte Ltd.

Kumari, M., Qureshi, K.A., Jaremko, M. et al. (2022). Deciphering the role of endophytic microbiome in postharvest diseases management of fruits: opportunity areas in commercial up-scale production. *Frontiers in Plant Science* 13: 4390. https://doi.org/10.3389/fpls.2022.1026575.

Litchman, E. (2010). Invisible invaders: non-pathogenic invasive microbes in aquatic and terrestrial ecosystems. *Ecology Letters* 13 (12): 1560–1572.

Liu, D., Lian, B., and Dong, H. (2012). Isolation of *Paenibacillus* sp. and assessment of its potential for enhancing mineral weathering. *Geomicrobiology Journal* 29 (5): 413–421.

Lugtenberg, B. and Kamilova, F. (2009). Plant-growth-promoting rhizobacteria. *Annual Review of Microbiology* 63: 541–556.

Lupwayi, N.Z., Olsen, P.E., Sande, E.S. et al. (2000). Inoculant quality and its evaluation. *Field Crops Research* 65 (2–3): 259–270.

Mallon, C.A., Van Elsas, J.D., and Salles, J.F. (2015). Microbial invasions: the process, patterns, and mechanisms. *Trends in Microbiology* 23 (11): 719–729.

Marimuthu, S., Subbian, P., Ramamoorthy, V. et al. (2002). Synergistic effect of combined application of *Azospirillum* and *Pseudomonas fluorescens* with inorganic fertilizers on root rot incidence and yield of cotton. *Journal of Plant Diseases and Protection* 109 (6): 569–577.

Matilla, M.A., Ramos, J.L., Bakker, P.A. et al. (2010). *Pseudomonas putida* KT2440 causes induced systemic resistance and changes in *Arabidopsis* root exudation. *Environmental Microbiology Reports* 2 (3): 381–388.

Matiru, V.N. and Dakora, F.D. (2004). Potential use of rhizobial bacteria as promoters of plant growth for increased yield in landraces of African cereal crops. *African Journal of Biotechnology* 3 (1): 1–7.

Meena, V.S., Maurya, B.R., and Verma, J.P. (2014). Does a rhizospheric microorganism enhance K+ availability in agricultural soils? *Microbiological Research* 169 (5-6): 337–347.

Owen, D., Williams, A.P., Griffith, G.W. et al. (2015). Use of commercial bio-inoculants to increase agricultural production through improved phosphorus acquisition. *Applied Soil Ecology* 86: 41–54.

Pandey, K.D., Singh, S.K., Shukla, L. et al. (2023). Mechanistic approaches and factors regulating microalgae mediated heavy metal remediation from the aquatic ecosystem. In: *Advances in Chemical Pollution, Environmental Management and Protection*, vol. 9 (ed. L.F.R. Ferreira, A. Kumar, and M. Bilal), 285–299. Cambridge, MA: Elsevier, Academic Press https://doi.org/10.1016/bs.apmp.2022.10.011.

Pathak, D.V. and Kumar, M. (2016). Microbial inoculants as biofertilizers and biopesticides. In: *Microbial Inoculants in Sustainable Agricultural Productivity. 1: Research Perspectives* (ed. D.P. Singh, H.B. Singh, and R. Prabha), 197–209. Springer.

Peoples, M.B., Brockwell, J., Herridge, D.F. et al. (2009). The contributions of nitrogen-fixing crop legumes to the productivity of agricultural systems. *Symbiosis* 48: 1–17.

Pindi, P.K. and Satyanarayana, S.D.V. (2012). Liquid microbial consortium – a potential tool for sustainable soil health. *Journal of Biofertilizers & Biopesticides* 3 (4): 124.

Rafi, M.M., Krishnaveni, M.S., and Charyulu, P.B.B.N. (2019). Phosphate-solubilizing microorganisms and their emerging role in sustainable agriculture. In: *Recent Developments in Applied Microbiology and Biochemistry* (ed. V. Buddolla), 223–233. Elsevier.

Rawat, A.K., Rao, D.L.N., and Sahu, R.K. (2013). Effect of soybean inoculation with *Bradyrhizobium* and wheat inoculation with *Azotobacter* on their productivity and N turnover in a vertisol. *Archives of Agronomy and Soil Science* 59 (11): 1559–1571.

Sammauria, R., Kumawat, S., Kumawat, P. et al. (2020). Microbial inoculants: potential tool for sustainability of agricultural production systems. *Archives of Microbiology* 202 (4): 677–693.

Santosh, S., Velmourougane, K., Idapuganti, R.G. et al. (2022). Potassium solubilizing potential of native bacterial isolates from cotton rhizosphere of rainfed vertisols. *National Academy Science Letters* 45 (3): 209–212.

Saravanan, V.S., Subramoniam, S.R., and Raj, S.A. (2004). Assessing *in vitro* solubilization potential of different zinc solubilizing bacterial (ZSB) isolates. *Brazilian Journal of Microbiology* 35: 121–125.

Schink, B. (2002). Synergistic interactions in the microbial world. *Antonie Van Leeuwenhoek* 81: 257–261.

Shahwar, D., Mushtaq, Z., Mushtaq, H. et al. (2023). Role of microbial inoculants as bio fertilizers for improving crop productivity: a review. *Heliyon* 9 (6): 16134.

Singh, R., Soni, S.K., Patel, R.P. et al. (2013). Technology for improving essential oil yield of *Ocimum basilicum* L. (sweet basil) by application of bioinoculant colonized seeds under organic field conditions. *Industrial Crops and Products* 45: 335–342.

Singh, D., Singh, S.K., Singh, V.K. et al. (2020). Plant growth–promoting bacteria and their role in environmental management. In: *Abatement of Environmental Pollutants* (ed. P. Singh, A. Kumar, and A. Borthakur), 161–175. Cambridge, MA: Elsevier https://doi.org/10.1016/B978-0-12-818095-2.00008-4.

Singh, M., Qureshi, K.A., Jaremko, M. et al. (2022). Bioprospects of endophytic bacteria in plant growth promotion and Ag-nanoparticle biosynthesis. *Plants* 11 (14): 1787.

Singh, S.K., Shukla, L., Singh, R.P. et al. (2023). Microalgae mediated wastewater treatment and its production for biofuels and bioproducts. In: *Advances in Chemical Pollution, Environmental Management and Protection*, vol. 9 (ed. L.F.R. Ferreira, A. Kumar, and M. Bilal), 153–165. Cambridge, MA: Elsevier, Academic Press https://doi.org/10.1016/bs.apmp.2022.10.007.

Suyal, D.C., Shukla, A., and Goel, R. (2014). Growth promotory potential of the cold adapted diazotroph *Pseudomonas migulae* S10724 against native green gram (*Vigna radiata* (L.) Wilczek). *Biotech* 4: 665–668.

Suyal, D.C., Soni, R., Sai, S. et al. (2016). Microbial inoculants as biofertilizer. In: *Microbial Inoculants in Sustainable Agricultural Productivity: 1. Research Perspectives* (ed. D.P. Singh, H.B. Singh, and R. Prabha), 311–318. Springer.

Trabelsi, D. and Mhamdi, R. (2013). Microbial inoculants and their impact on soil microbial communities: a review. *BioMed Research International* 2013: 1–11.

Trivedi, P., Pandey, A., and Palni, L.M.S. (2012). Bacterial inoculants for field applications under mountain ecosystem: present initiatives and future prospects. In: *Bacteria in Agrobiology: Plant Probiotics* (ed. D.K. Maheshwari), 15–44. Springer.

Vacheron, J., Desbrosses, G., Bouffaud, M.L. et al. (2013). Plant growth-promoting rhizobacteria and root system functioning. *Frontiers in Plant Science* 4: 356.

Verma, M., Mishra, J., and Arora, N.K. (2019). Plant growth-promoting rhizobacteria: diversity and applications. In: *Environmental Biotechnology: For Sustainable Future* (ed. R. Sobti, A.K. Arora, and R. Kothari), 129–173. Springer.

Verma, H., Kant, C., Singh, S.K. et al. (2022). Biocontrol potential of microbial consortia: approaches in food security and disease management. In: *Microbial Biocontrol: Sustainable Agriculture and Phytopathogen Management*, vol. 1 (ed. A. Kumar), 187–203. Cham: Springer International Publishing https://doi.org/10.1007/978-3-030-87512-1_7.

Vijaysri, D., Priya, M., Singh, D. et al. (2023). Plant growth promoting rhizobacteria: an efficient tool for sustainable agriculture. In: *Multidisciplinary Approach: Enhanced Agriculture Production in a Sustainable Way*, 1–10. Ujjain Road, Indore (M.P.): Department of Plant Protection, Shri Institute of Agriculture Gram Baroli.

Wagner, S.C. (2011). Biological nitrogen fixation. *Nature Education Knowledge* 3 (10): 15.

Walsh, U.F., Moënne-Loccoz, Y., Tichy, H.V. et al. (2003). Residual impact of the biocontrol inoculant *Pseudomonas fluorescens* F113 on the resident population of rhizobia nodulating a red clover rotation crop. *Microbial Ecology* 45: 145–155.

Yadav, J. and Verma, J.P. (2014). Effect of seed inoculation with indigenous Rhizobium and plant growth promoting rhizobacteria on nutrients uptake and yields of chickpea (*Cicer arietinum* L.). *European Journal of Soil Biology* 63: 70–77.

Yang, T., Wei, Z., Friman, V.P. et al. (2017). Resource availability modulates biodiversity-invasion relationships by altering competitive interactions. *Environmental Microbiology* 19 (8): 2984–2991.

Zhang, N.N., Sun, Y.M., Li, L. et al. (2010). Effects of intercropping and Rhizobium inoculation on yield and rhizosphere bacterial community of faba bean (*Vicia faba* L.). *Biology and Fertility of Soils* 46: 625–639.

# 11

# Bacterial Treatment of Industrial Wastewaters: Applications and Challenges

*Christina Saran[1], Anuradha Devi[1], Ganesh Dattatraya Saratale[2], Rijuta Ganesh Saratale[3], Luiz Fernando R. Ferreira[4], Sikandar I. Mulla[5], and Ram Naresh Bharagava[1]*

[1] *Laboratory of Bioremediation and Metagenomics Research (LBMR), Department of Environmental Microbiology (DEM), Babasaheb Bhimrao Ambedkar University (A Central University), Lucknow, India*
[2] *Department of Food Science and Biotechnology, Dongguk University-Seoul, Goyang-si, Republic of Korea*
[3] *Department of Food Science and Biotechnology, Research Institute of Biotechnology and Medical Converged Science, Dongguk University-Seoul, Goyang-si, Republic of Korea*
[4] *Post Graduate Program in Genomic Sciences and Biotechnology, Genomic Sciences and Biotechnology Lab, Catholic University of Brasília, Brasília, Brazil*
[5] *Department of Biochemistry, School of Applied Sciences, REVA University, Bangalore, India*

## 11.1 Introduction

Our depleting natural resources (water, air, and land) are being stressed out by the rapidly growing human population and the desire for industrial growth for economic progress. Since the rise of industry and technology, environmental management and maintenance have also been lacking. Therefore, the past few decades have seen an increase in environmental pollution on a global scale as an outcome of populace growth, fast industrial development, growth in human-caused activities, and unregulated and hazardous agricultural practices, which have led to the buildup of harmful waste in the environment, such as agricultural, metallic, domestic, plastic, rubber, and industrial waste. In addition to heavy metals, pesticides, aliphatic, petroleum oil, aromatic, nitroaromatic compounds, polycyclic aromatic hydrocarbons, chlorinated hydrocarbons like polychlorinated biphenyls (PCBs) and organophosphorus compounds, perchloroethylene, organic solvents (phenolic compounds), and phthalates, this hazardous waste may also contain phthalates and organic solvents (phenolic compounds) (Tyagi and Kumar 2021).

Water is crucial to daily living and several industrial growth processes. The need for recycling and reuse of wastewater for various uses is brought on by rising water demands caused by pollution. In some developing nations, the sole supply of water utilized to irrigate crop plants is industrial effluent. Since freshwater supplies are finite, it is necessary to consider and identify ways to recycle or reuse wastewater. The use and growth of the economy are related to the water quality criteria. There are several methods to pollute water. A significant cause of water contamination is industrial effluent. Industrial discharge in metropolitan areas is the main cause of poor water quality (Garg 2022).

Environmentalists have employed bioremediation, a biotechnological technique that uses biologically mediated changes in pollutants to reconstruct the recalcined environment and control pollutants through detoxification and mineralization, the most often among the many biological techniques. It is a process that makes use of green plants, microbes, or their enzymes to clean up contaminated areas and return them to their original condition. It also helps to enhance soil and water health, cleaning up water sources and the quality of the air and water in all parts of the world. Microbial bioremediation uses both aerobic and anaerobic bacteria as part of a natural process to remove contaminants from the environment (García-Depraect et al. 2021). Microbes need organic carbon and a source of energy even if they have distinct dietary demands (Haripriyan et al. 2022).

Halogenated petroleum hydrocarbons, metals, pesticides, polyaromatic hydrocarbons, nitroaromatic compounds, and industrial solvents are just a few of the chemical contaminants that have had their concentration and toxicity reduced via the bioremediation method. Co-metabolism, biotransformation kinetics, biotreatment, and biogeochemical modeling are

*Microbes Based Approaches for the Management of Hazardous Contaminants*, First Edition. Edited by Ajay Kumar, Livleen Shukla, Joginder Singh, and Luiz Fernando R. Ferreira.

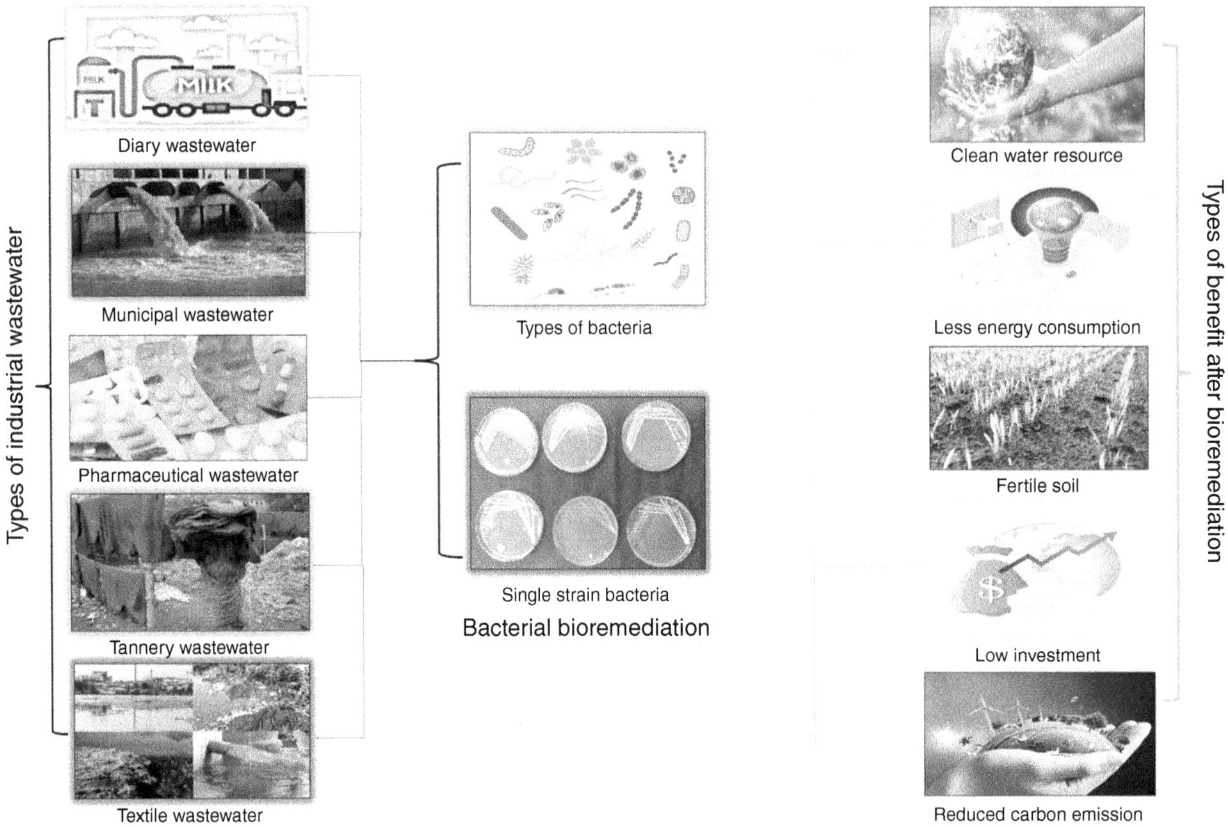


**Figure 11.1** Bioremediation of industrial wastewater using bacterial strains.

core study topics in the subject of bioremediation. Subjected application areas of research include biogeochemical evaluation techniques, ecofriendly attenuation, fate modeling, and co-metabolic approaches.

The chapter describes the properties of wastewater, several conventional and modern wastewater treatment technologies, and modern strategies for water purification for reusing and reclamation. This chapter's main goal is to draw attention to the subtle aspects of bioremediation, including its principles, the different bioremediation techniques, such as *in-situ* and *ex-situ* remediation, as shown in Figure 11.1, the factors that affect bioremediation, their benefits and drawbacks, their futures, and potential solutions.

## 11.2 Composition and Nature of Various Industrial Wastewater

Due to its excessive water usage and the adverse effects of industrial wastewater, industrial wastewater is a big issue of discussion in the environmental movement if released into the environment. Additionally, industrial water is any water utilized in industry for manufacturing and processing activities. For instance, water may be used to prepare rinses, acid treatments, or active baths, among many other things. In all industries, commercial items are produced using water, either directly or as a byproduct of cleaning operations. Water that has been utilized is regarded as wastewater (Garg 2022).

### 11.2.1 Types and Sources of Wastewater on the Basis of Wastewater Production

Depending on the source of a particular industrial wastewater, contaminants in industrial water vary widely (Lipps et al. 2017).

- **Alkaline contamination**: The metal or chemical industries (such as those that employ tanning chemicals in textile mills) are the sources of alkaline kinds of industrial wastewater.
- **Cooling and radioactive contaminants**: In nuclear power premises, clean water from neighboring towns is frequently used to assure safety. Since no water is released into the environment throughout this procedure, the cooling water is entirely evaporated. Wastewater from uranium mining operations and medical institutions can include radioactive pollution (X-ray).

- **Organic pollution**: As an example, the paper industry (this water frequently also has a high percentage of suspended particulates) and the meat processing sector are sources of industrial wastewater that are contaminated with organic materials.
- **Oil or grease contamination**: It arises from industries that process animal products, such as meat factories or dairies, as well as from metal processing (like chipping) or refinery activities.
- **Acid or contamination by heavy metals**: Some possible sources for such types of contaminants in industrial wastewater include the electroplating business, bleaching and pickling factories, and coal processing industries.
- **Toxic contaminants**: As in the production of pesticides, explosives, and in the textile industry (dyeing, tanning).
- **Contamination by cleaning agents**: Laundries, ordinary homes, or the surfactant industry are a few examples of sources of industrial wastewater that have been contaminated by cleaning compounds.

### 11.2.2 Characteristics of Industrial Wastewater

Many pollutants, fertilizers, and organic materials are present in the wastewater produced by many businesses. When wastewater is released into the ecosystem or any aquatic body, the toxins it contains, which may be nonbiodegradable, biodegradable, or inactive in the atmosphere, can create pollution or health issues. Wastewater features are categorized into physical, chemical, and biological categories (Lipps et al. 2017).

#### 11.2.2.1 Physical Characteristics of Wastewater

The physical properties constitute color, Odor, turbidity, temperature, and suspended solids.

- **Color**: The color of the wastewater effluent indicates its quality. The color of the wastewater changes over time; the older the wastewater, the darker the coloration. Decomposition in anaerobic instances causes the change in color. The development of sulfides during decomposition is usually the cause of the black coloration.
- **Odor**: The public is affected by the wastewater's odor, which is a serious problem. The main cause of the odor is the anaerobic breakdown of organic substances in wastewater. The majority of sulfur products are unpleasant. Different types of industrial wastewater have distinctive smells; for instance, wastewater from the organic compost, beverage, and most of the food sectors contains ammonia, sulfur compounds, and alcohols, products of anaerobic decomposition, which have an unpleasant odor.
- **Turbidity**: Turbidity in wastewater is caused by suspended solids or colloidal particles. The procedure involves the use of coagulation and filtration techniques to remove turbidity. The blocking or reduced penetration of light into the water body may hinder the activities of benthic organisms and photosynthesis. By blocking their gills with wastewater sediments, certain fish populations are also wiped out by turbidity.
- **Temperature**: Wastewater has a higher temperature than normal drinking water because of industrial activities and the presence of various contaminants in it. The temperature of the wastewater changes from season to season and with different countries, ranging from 7°C to 18°C in colder winter months to 13–24°C in warmer summer months. Before treating wastewater, temperature must be measured since several biological treatment processes depend on temperature.
- **Solids**: Solids (organic and inorganic) material in effluent that is soluble or insoluble. Equally volatile (organic materials, including carbohydrates, proteins, and lipids) and fixed (nonvolatile) solids can be found in wastewater. The wastewater's soluble and suspended particles are included in the total solids. The evaporation and sedimentation processes can assist in removing the suspended particles. According to Zainith et al. (2016), wastewater typically contains 40–60% total solids.

#### 11.2.2.2 Chemical Characteristics of Wastewater

The inorganic and organic nature of wastewater is often determined by its chemical features, such as its biochemical oxygen demand (BOD) and chemical oxygen demand (COD). The usage of raw materials, the creation of a product, and industrial processes all have an impact on the chemical composition of wastewater (Lipps et al. 2017).

- **Inorganic chemicals**: It contains both organic and inorganic phosphorus, as well as organic nitrogen, nitrates, and nitrites. Phosphorus and nitrogen are important nutrients needed for plant development. In general, effluent from the pesticide and fertilizer industries contains significant amounts of these nutrients. During aquatic environments, they encourage plant development, but in larger quantities, they pollute and contaminate the water. Heavy metals such as copper (Cu), cobalt (Co), zinc (Zn), and iron (Fe) are examples of trace elements. Arsenic (As), cadmium (Cd), mercury (Hg), chromium (Cr), and other heavy metals create hazardous consequences when they are present in significant amounts (Devi et al. 2023).

It is crucial to identify and deal with these harmful compounds. While gases like oxygen, methane, carbon dioxide, and sulfide are necessary for the system to function, their excess leads to environmental imbalance and pollution because hydrogen sulfide, which is produced during anaerobic processes, is an odorous and harmful gas that also corrodes long sewage pipes. Measurement of dissolved oxygen is necessary prior to treatment because it impacts the aerobic biological therapy processes.

- **Organic chemical**: Total organic carbon (TOC), BOD, and COD are used to determine in wastewater. The organic compound may occasionally be biodegradable and cause harm. Therefore, prior to disposal, such compounds must be identified and treated. In addition to oils, phenols, and insecticides (Devi et al. 2022), harmful organic chemicals also include them. For the purpose of determining the concentration of priority pollutants, some particular organic pollutants are monitored. Complex calcium is contaminated by the polyphosphates found in soaps, detergents, and other products. These contaminants are well suited for use as coolants and insulating fluids in transformers and capacitors due to their great thermal, biological, and chemical stability.
- **Volatile organic compounds (VOC)**: Most frequently, chemicals like benzene, xylenes, trichloroethane, and dichloromethane absorb from underground storage tanks, poorly disposed solvents, and landfills in built-in industrial and commercial districts. Common soil contaminants in industrial regions include VOC. The transport mechanism across cell walls is impeded when heavy metals, including cadmium, mercury, lead, and copper ions, interact with cell membranes. Enzymes are disrupted by cadmium, mercury, and lead's affinity for sulfur. Because it is often employed in metal plating, cadmium may leak into water through mining wastes and industrial effluent. Red blood cell destruction, renal failure, and high blood pressure are all consequences of this water poisoning to humans. Leaded petrol and inorganic lead discharge from mining sources are the main causes of air and land pollution. Humans, with its contamination, have severely dysfunctional renal, reproductive, neurological, and brain systems. Appliances and machinery used in both the home and industry are harmed by the hardness of water brought on by the presence of calcium and magnesium.

### 11.2.3 Biological Characteristics of Wastewater

Water-borne diseases for public well-being concerns and biological procedures like bioaccumulation or eutrophication from nutrient inputs are two large sub-categories of biological characteristics. Examples of the organisms that cause waterborne infections include viruses, bacteria, protozoans, and helminths. Hepatitis, giardiasis, jaundice, and other conditions are a few instances of waterborne illnesses. In the United States and certain developing nations in the 1800s, cholera, the most severe bacterial waterborne disease, caused substantial mortality and conditions (Englande et al. 2015).

Excessive growth of bacteria and a deterioration in the quality of water resources are results of nutrient pollution of water, such as nitrogen or phosphorus. Lowering the oxygen content of water bodies also has an impact on the development of aquatic animals. Additionally, contaminants in wastewater might be classified as hazardous, conventional, or nonconventional. The following are examples of conventional pollutants: pH, coliforms, BOD, suspended solid particles, oil and grease, etc. Priority pollutants are among the toxic contaminants. COD, TOC, and nutrients are nonconventional pollutants. Wastewater also contains nanoscale particles, such as sub-colloidal elements containing viruses and organic debris from nature. Fe, Mg, and Pb oxidation compounds are a few examples of nanoscale particles (Garg 2022).

## 11.3 Role of Bacteria in Biodegradation of Specific Pollutant Found in Wastewater

Environmental contamination has recently emerged as a sensitive topic on the international stage. Increased industrialization, potentially hazardous farming practices, urbanization, population increase, and trash discharge all contributed to significant environmental contamination. Domestic solid waste, sewage, industrial effluent, and municipal solid waste are sources of pollutants that contaminate agricultural soil, groundwater, and rivers with inorganic chemicals ($NH_3$, $N_2$, P, $NO_3$, $NO_2$) as well as organic chemicals as and heavy metals (Cd, Pb, Hg, Cu, As, Cr, etc.) (Tegene 2020).

According to research from industrial everyday operations, 310 kg of harmful elements is released into the water, air, and land every second, totaling nearly 10 million tons annually (Tegene 2020). According to another estimate, humans, animals, and crops globally produce about 38 billion metric tons of the organic waste per year (Maji et al. 2020). Municipal Solid Waste (MSW) generation is also predicted to be between 1.7 and 1.9 billion metric tons globally. Causes of waste production include households (76%), institutions, companies, and hotels (18%), and street sweeping (6%). There are both biodegradable and nonbiodegradable wastes in these municipal solid wastes. According to Zambrano-Monserrate et al. (2021), inappropriate textile dye effluent discarding in aqueous environments reduces sunlight penetration, which in turn lowers photosynthetic activity, water quality, dissolved oxygen (DO), and exhibits acute lethal effects on aquatic fauna and flora, resulting in severe environmental complications across the globe.

Since bacteria use wastewater as a feed source and produce a variety of metabolic products from it, the features and environmental circumstances of the wastewater are susceptible to change. Most changes in wastewater properties and their surroundings are caused by heterotrophic and autotrophic bacteria, depending on the carbon source they are utilizing. Different biochemical processes can occur when bacteria play a role in wastewater treatment which is summarized in Table 11.1. Heavy metals and other nonbiodegradable hazardous organic pollutants might interfere with microbial bioremediation processes (Igiri et al. 2018), which is summarized in Table 11.2.

Bacterial involvement in the procedure of biodegradation can take the form of bio-piling, venting, attenuation, and bioaugmentation. The native microorganisms added to the biological wastewater treatment methods include wastewater

**Table 11.1** Bacterial strains remediated different wastewater.

| Source of wastewater | Chemical | Bacterial strains | References |
|---|---|---|---|
| Sea | Sodium ($Na^+$) | *Rhodobacter sphaeroides, Rhodovulum* sps. | Sasaki et al. (2017) |
| Saline | Calcium ($Ca^{2+}$) | *Bacillus licheniformis SRB2* | Zhao et al. (2019) |
| Hard water | | *Sporosarcina pasteurii, Bacillus megaterium* | Chaparro-Acuña et al. (2018) |
| Hard water | Magnesium ($Mg^{2+}$) | *Bacillus licheniformis SBR2* | Zhao et al. (2019) |
| Industrial | Aluminium ($Al^{3+}$) | *Vibrio alginolyticus* | Purwanti et al. (2019) |
| Drinking | Iron ($Fe^{3+}$) | *Rhodobacter capsulatus, Pelobacter carbinolicus, Geobacter sulfurreducens, Gallionella capsiferriformas* strain ES-2 | Gnanaprakasam et al. (2017) |
| Groundwater | Nitrogen ammonical ($NH^{4+}$) | *Nitrosomonadales* convert $NH^{4+}$ to $NO^{2-}$, *Nitrospirales convert* $NO^{2-}$ to $NO^{3-}$ | Maharjan et al. (2020) |
| Urban | | *Chlorella vulgaris, Bacillus cereus, Pseudomonas putida* | Gómez-Guzmán et al. (2017) |
| Sewage | Chloride ($Cl^-$) | *Escherichia coli* | Owoseni et al. (2017) |
| Contaminated soil | Fluoride ($F^-$) | *Bacillus flexus PN4* | Sakthi Thesai et al. (2020) |
| Groundwater | | *Providencia vermicola KX926492* | Mukherjee et al. (2017) |
| Drinking | Sulfate ($SO_4^{2-}$) | *Salmonella typhimurium, Clostridium pasteurianum* | Gnanaprakasam et al. (2017) |
| Manufacturing | | *Desulfovibrio sps, Desulfotomaculum* sps | Piacenza et al. (2018) |
| Inorganic | Sulfite ($SO_3^{2-}$) | *Chromatium vinosum* (as hydrogen sulfite degrades) | Syed et al. (2006) |
| Drinking | Nitrate ($NO^{3-}$) | *Massilia brevitalea, Psychrobacter glacincola, Arthrobacter defluvi, Pseudomonas antarctivca, Rhodobacter* sps. | Gnanaprakasam et al. (2017) |
| Urban | | *Chlorella vulgaris, Bacillus cereus, Pseudomonas putida* | Gómez-Guzmán et al. (2017) |
| Drinking | Nitrite ($NO_2^-$) | *Nitrospora* sps., *Bradyrhizobium, Nitrospira moscoviensis* | Gnanaprakasam et al. (2017) |
| Synthetic | Phosphate ($PO_4^{3-}$) | *Pseudomonas* sp. *JPSB12, Enterobacter* sp. *TPSB20, Flavobacterium* sp. *TPSB23* | Paul and Sinha (2015) |
| Urban | | *Chlorella vulgaris, Bacillus cereus, Pseudomonas putida* | Guerra-Renterla et al. (2019) |
| Sludge | | *Accumulibacter* | Zou et al. (2014) |
| Urban | Silica ($Sio_2$) | *Rhodococcus* sp. *BH4* | Lee et al. (2020) |
| Textile | Potassium ($K^+$) | *Sapindus mukorossi* | Jassal et al. (2015) |

**Table 11.2** Bacterial strains remediated heavy metals from various industrial wastewater.

| Industrial wastewater | Chemical | Bacterial strains | Reference |
|---|---|---|---|
| Drinking | Arsenic | *Pseudomonas chengduensis As11, Bacillus flexus As12* | Jebelli et al. (2018) |
| Groundwater | | *Pseudomonas putida strain WB, Geobacter lovleyi, Bacillus selenatarsenatis, Hydrogenophaga* sp. *strain CL3, Sinorhizobium, Arthrobacter aurescens, Stenotrophomonas* sp. *strain MM7* | Gnanaprakasam et al. (2017) |
| Industrial | | *Klebsiella pneumonia, Enterobacter* sps. | Abbas et al. (2014a) |
| Industrial | | *Corynebacterium glutamicum* | Mateos et al. (2006) |
| Drinking | Boron | *Escherichia coli, Enterococcus faecium* | Heim et al. (2015) |
| Municipal | | *Candida tropicalis, Rhodotorula mucilaginosa, Micrococcus luteus, Bacillus thuringiensis, B. cereus, B. megaterium, B. pumilus, Pseudomonas aeruginosa, Aspergillus versicolor* | Laçin et al. (2015) |
| Industrial | | *Lysinibacillus* sp. *21019, B. horneckial DSM23495, Microbacterium* sp. *CRRI-B* | Raja and Omine (2013) |
| Active sludge | | *Variovarox, Shewanella, Mycobacterium, Rhodococcus, B. simplex* | Miwa and Fujiwara (2009) |
| Industrial | Cadmium | *Lactobacillus plantarum CCFM8610* | Zhai et al. (2017) |
| Industrial | | *Pseudomonas* sp. *M3* | Abbas et al. (2014b) |
| Industrial | | *Pseudomonas aeruginosa strain KUCd1* | Sinha and Mukherjee (2009) |
| Industrial | Lead | *Pseudomonas aeruginosa ATCC27853* | Babiker et al. (2020) |
| Sludge | | *Cupriavidus metallidurans CH3* | Jarosławiecka and Piotrowska-Seget (2014) |
| Battery waste | | *Providencia alcalifaciens strain 2EA* | Naik et al. (2013) |
| Industrial | Cadmium + lead | *Bifidobacterium longum46, Lactobacillus fermentum ME3, Bifidobacterium lactis Bb12* | Halttunen et al. (2007) |
| Industrial | Chromium | *Lactobacillus plantarum MF042018* | Ameen et al. (2020) |
| Tannery | | *Klebsiella pneumoniae strain MS 1.5, Mangrovibacter yixingensis strain MS2.4* | Sanjay et al. (2020) |
| Tannery | | *Lactobacillus rhamnosus MTCC 1408, L. casei MTCC1423* | Mishra et al. (2012) |
| Industrial | | *Bacillus coagulans, Desulfomacculum reducens, E. coli, Pseudomonad, P. ambigua G-1, P. putida, Enterobacter cloacae, E. coli ATCC33456, Alcaligens eutrophus AE104, P. fluorescens, B. mycoids, Shewanella oneidensis strain MR-1* | Singh (2008) |
| Industrial | Copper | *Enterococcus faecium* | Yilmaz et al. (2010) |
| Manufacturing | | *Geobacter metallireducens, Geobacter sulfurreducens* | Fang and Achal (2019) |
| Mining industry | Cyanide | *Pseudomonas pseudoalcaligenes CECT5344* | Luque-Almagro et al. (2016) |
| Cassava industry | | *Bacillus pumilus* | Kandasamy et al. (2015) |
| Metal refinery | Manganese | *Pseudomonas putida strain MnB1, Pseudomonas* sp. *strain SK3* | Kitjanukit et al. (2017) |
| Groundwater | Selenium | *Lysinibacillus* sps, *Azospirillum* sps, *Burkholderia fungorum, Bacillus cereus, Bacillus safensis JG-B5T, Alishewanella* sp. *WH16-1, Stenotrophomonas maltophilia SeITE02* | Sinharoy and Lens (2020) |
| Manufacturing | | *Aeromonas* sp.*VS6, Citrobactor freundii KS8, Pseudomonas flurescens K27, Enterobacter cloacae SLS1a-1, R. spharoids, R. rubrum S1* | Piacenza et al. (2018) |
| Refinery industry | | *Pseudomonas stutzeri NT-1* | Kuroda et al. (2011) |

**Table 11.2** (Continued)

| Industrial wastewater | Chemical | Bacterial strains | Reference |
|---|---|---|---|
| Industrial | Mercury | *Pseudomonas aeruginosa ATCC27853* | Babiker et al. (2020) |
| Industrial | | *Vibrio fluvialis* | Saranya et al. (2017) |
| Petrochemical | | *Pseudomonas* sp. *B50A* | Giovanella et al. (2016) |
| Mining industry | | *Cupriavidus metallidurans strain MSR3* | Rojas et al. (2011) |
| Industrial | | *Pseudomonas putida* spi3 | Von canstein et al. (1999) |
| Sewage | Zinc | *Bacillus megaterium EMCC 1013, Rhizobium rhizogenes EMCC1743, Rhizobium leguminosarum EMCC1130, Azotobacter vinelandii, Nocardiopsis dassenvillei* | El-Barbary and El-Badry (2019) |
| Industrial | | *Bacterium VMSDCM* | Mishra et al. (2014) |
| Mine soils | Antimony | *Sinorhizobium* sp. *GW3* | Li et al. (2019) |
| Mine soils | | *Cupriavidus, Moraxella* sp. *S2* | Li et al. (2018) |
| Manufacturing | Tellurium | *E. coli, Lactococcus lactis, R. capsulatus, R. rubrum G9, R. capsulatus, P. fluorescens K27, D. gigas, P. aeruginosa ML4262, Stearothermophilus, Mycobacteruim tuberculosis, B. beveridgei, B. selenitireduceus, S. barnesii, Shewanella frigidimarins ER-Te-48, Bacillus* sp. *GT-83* | Piacenza et al. (2018) |
| Power plants | | *Pseudoalteromonas* sp. *strain EPR3* | Bonificio and Clarke (2014) |
| Industrial | | *Stenotrophomonas maltophilia TI-1, Ochrobacterum anthropi TI-2, Ochrobacterum anthropi TI-2* | Kagami et al. (2012) |

heterotrophic and natural autotrophic bacteria. By employing wastewater as an electron supplier for respiration, heterotrophic bacteria biodegrade easily biodegradable organic compounds in wastewater (Aragaw 2021).

Under aerobic, anaerobic, or anoxic conditions, respiration may use DO and other nutrients, such as nitrate ($NO_3^-$) or sulfate ($SO_4^{2-}$), present in the wastewater as an electron acceptor. Depending on the effluent source, the major products of respiration are carbon dioxide ($CO_2$), nitrogen gas ($N_2$), sulfide ($S^{2-}$), and biomass, with varied levels of quantity and quality. According to Fang and Achal (2019), heterotrophic bacteria also reduce by utilizing organic compounds as carbon sources during fermentation activities that take place in an aerobic environment. These processes result in the production of carbon dioxide and biomass. Alkalinity is used as a carbon source by autotrophic nitrifiers as they oxidize ammonia to nitrate or nitrite for energy. Sulfide, a byproduct of anaerobic respiration utilizing sulfate, produces offensive odors and is harmful to human health, making it uninteresting for use in wastewater treatment operations. Additionally, the pipeline systems' sulfide oxidation processes produce sulfuric acid, which is corrosive to the pipeline (Gómez-Guzmán et al. 2017).

In their bacterial cells, bacteria have a number of defense mechanisms that make them resistant to hazardous contaminants. Examples include the establishment of an extracellular sequestration, active movement of pollutants (efflux), extracellular barrier, intracellular sequestration, and the reduction of large molecules into small ones and their oxidation states (Okoduwa et al. 2017).

This chapter describes various bacterial strain that biodegrades a specific pollutant found in wastewater as an aspect of natural bioremediation (Aragaw 2021).

## 11.4 Different Approaches and Mechanism of Bacterial Bioremediation in Industrial Wastewater

Since they have the capacity to grow under-regulated settings and can withstand harsh environmental conditions, bacteria are essential biosorbents for the cleanup of polluted ecosystems and are highly beneficial in bioremediation. In contaminated settings, they show strong biosorbent activity against heavy metals. As opposed to "enzymes, glycoproteins, lipopolysaccharides, lipoproteins, and phospholipids found in Gram-negative bacteria cell walls, peptidoglycan layers containing alanine, glutamic acid, and teichoic acid found in Gram-positive bacteria serve as active sites for binding heavy metal ions" (Pal et al. 2020) resulting in the treatment of heavy metals as of a contaminated environment (Mishra et al. 2012).

The removal of Cr can be effectively accomplished by using bacteria, i.e., *Bacillus cereus*, *Pseudomonas putida* and *Bacillus subtilis* (Gómez-Guzmán et al. 2017), whereas *Pseudomonas veronii*, *Kocuria flava*, and *Sporosarcina ginsengisoli* have the to be expected to remove As, Cu, Cd, and Zn from the polluted environment (Pal et al. 2020). For the bioremediation of heavy metals, bacterial consortiums have been proven to be more effective than monoculture strains. We conclude the application, concepts, benefits, and limitations in this chapter. *In-situ* and *ex-situ* bioremediation are the two categories into which bioremediation belongs, which is summarized in Figure 11.2.

### 11.4.1 *In-Situ* Bioremediation

Although it includes treating pollutants at actual polluted sites without any excavation or disruption, *in-situ* bioremediation is regarded as the most suitable method. It may be classified as intrinsic *in-situ* bioremediation (without any enhancement practices) or enhanced *in-situ* bioremediation (with enhancement practices such as bioaugmentation, biostimulation, bioventing, and biosparging). While the *in-situ* technique is more economically viable than *ex-situ* bioremediation technique because no extra costs for excavation or conveyance are required (only the cost of designing and installing the equipment used to enhance bacterial activity), it is less controllable and less effective. According to Tyagi and Kumar (2021), this technique is utilized to effectively remediate areas that have been polluted with dyes, chlorine-based solvents, toxic metals, and hydrocarbons.

#### 11.4.1.1 Intrinsic *In-Situ* Bioremediation

One of the well-known *in-situ* bioremediation techniques, this process is known as natural attenuation. With this technique, polluted places are remedied passively and without human intervention. This method treats wastewater and biodegradable contaminants using both (aerobic and anaerobic) microbial processes. Inspite of the lack of any peripheral influences in this process, constant monitoring is necessary to develop a successful and environmentally friendly process; for this reason, it is also known as monitored natural attenuation (Azubuike et al. 2016).

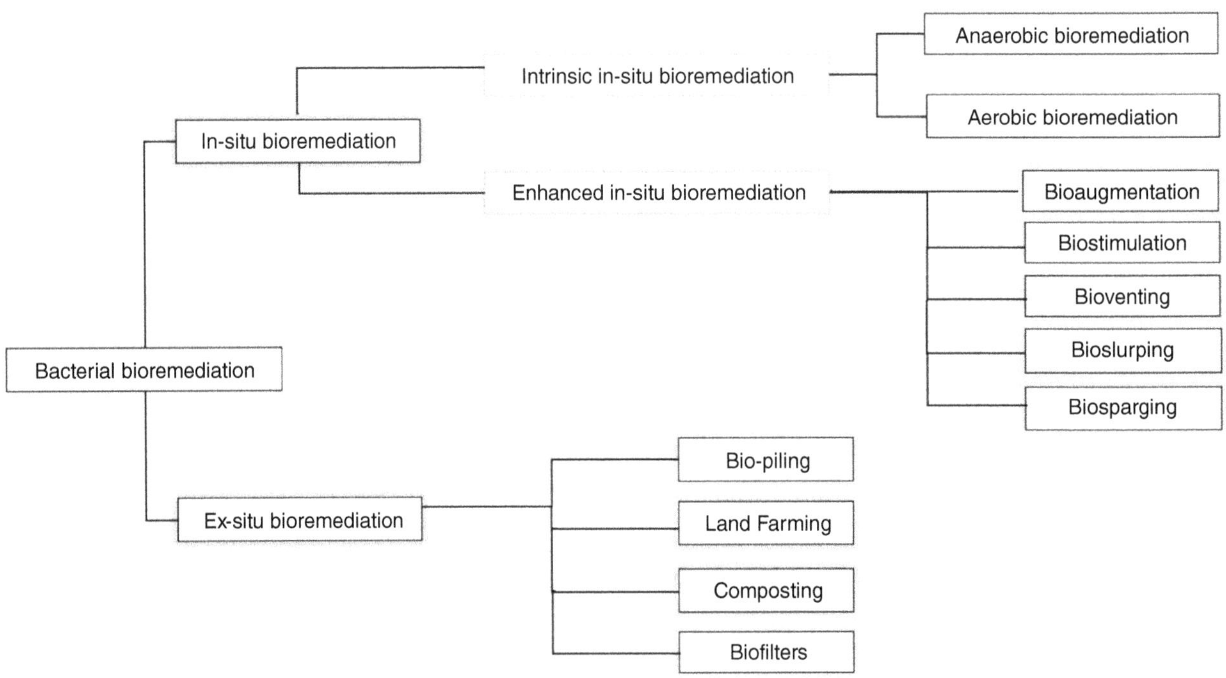


**Figure 11.2** Different types of bioremediation approaches.

***Anaerobic Bioremediation Approach for Wastewater Treatment*** In the absence of oxygen, bacteria decompose waste during anaerobic wastewater treatment procedures. These procedures frequently take place in an enclosed, airtight bioreactor that is loaded with sludge. Anaerobic bacteria and other helpful microorganisms are present in the sludge. The microorganisms in the bioreactor break down the organic material in the wastewater (Li et al. 2019).

The many types of anaerobic wastewater treatment equipment are described here:

- **Lagoons**: Large ponds are anaerobic lagoons, they are often employed by plants in the first stages of the industrial, agricultural, and municipal wastewater treatment processes. Water settles down into various strata in the lagoons, covering the sludge to keep oxygen from contacting it throughout microbial decomposition (Rana and Gupta 2022).
- **Sludge blanket reactors**: In anaerobic sludge blanket reactors, wastewater is passed through a sludge layer that floats like a blanket. Bacteria (anaerobic) break down the organic material as the effluent travels through the sludge, and the pieces that collect at the bottom of the tank can be recycled by the facility for use in other procedures (Rana and Gupta 2022).
- **Filter reactors**: An anaerobic filter reactor kinds use of a tank with a filter media connected. A biofilm is created when microbes attach to the filter media. The biofilm and associated bacteria are efficient in reducing organic materials in wastewater after they have developed (Rana and Gupta 2022).

***Aerobic Bioremediation Approach for Wastewater Treatment*** After primary treatment eliminates most of the grit and substantial solid particles from the wastewater, aerobic treatment often takes place as part of secondary treatment. Equipment like surface aerators or diffusers are frequently used in aerobic wastewater treatment to mix air into the water column. When using surface aerators, the mechanical agitation on the wastewater's surface mixes in oxygen. The air bubbles that rise from the tank's bottom with diffusers aid in the transport of oxygen. Kandasamy et al. (2015), the existence of oxygen in water facilitates good oxygen-feeding bacteria, protozoa, and other microorganisms to treat the waste by decomposing organic materials.

There are different types of aerobic wastewater processes.

- **Activated sludge**: Aerators or diffusers are used in an aeration tank during the activated sludge process. Large bacteria-filled into pieces known as flocs are created as the organic content in waste decomposes. The flocs then sink to the tank's bottom, where it is simple to remove them. The resultant activated sludge can subsequently be reintroduced into earlier stages of treatment by wastewater treatment facilities so that bacteria may help with waste breakdown.
- **Fixed-bed bioreactors**: The thousands of microscopic components known as media are packed into small chambers by fixed-bed bioreactors (FBBRs). The media elements remain stationary and the large amount of surface area they have traps bacteria, which stick to the surfaces and create a biofilm. The bacteria then aid in the degradation of waste. Additionally, several of these chambers permit anoxic denitrification zones.
- **Moving-bed bioreactors**: Thousands of microscopic media carriers are suspended across the wastewater in a large tank in moving-bed bioreactors (MBBRs). The suspended media carriers are free to float and offer microorganisms a favorable habitat for growth. The media carriers aid in the digestion of organic waste material because they offer a large surface area for helpful microorganisms to develop.
- **Membrane bioreactors**: Membrane bioreactors (MBRs) utilize cutting-edge technology for treating wastewater that has a high level of suspended particles. MBRs operate by fusing membrane filtration with activated sludge processes. They successfully separate and recycle the suspended solids to treat the waste.

#### 11.4.1.2 Enhanced *In-Situ* Bioremediation

The improvement of contaminated locations by excavating or adding nutrients, air, and bacterial strains to promote bacterial growth for the purpose of biodegrading contaminants essentially is meant by this form of *in-situ* bioremediation. The improved *in-situ* bioremediation approach includes methods including bioaugmentation, biostimulation, bioslurping, biosparging, and bioventing.

***Bioaugmentation*** In order to speed up the procedure of decomposition, a group of native bacteria, bacteria from other places, or bacteria created through genetic engineering are added to the polluted site. This technique is mostly employed to remediate ground polluted with aromatic and chlorinated hydrocarbons as well as municipal effluent. It has been shown that nonindigenous bacteria may compete with native ones for the expansion of their populations, necessitating adequate management of the land treatment unit (Tyagi and Kumar 2021).

***Biostimulation*** Using this *in-situ* bioremediation technique, the growth of the local microbial population is encouraged in polluted soil or groundwater by adding the nutrients required for the efficient removal of contaminants. This approach is effective for cleaning up metal and hydrocarbon-contaminated locations (Haripriyan et al. 2022).

***Bioventing*** The bioventing bioremediation technology stimulates the activity of bacteria for the degradation procedure by supplying air (oxygen) and nutrients in a precise manner through water sources to the polluted soil. This approach involves adding barely any oxygen to the polluted soil at low airflow rates, which is adequate for efficient microbial biodegradation while reducing pollutant volatilization and release to the environment. The bioventing approach may be used to remove contaminants from the area known as the vadose zone, which is deep below the surface. According to Höhener and Ponsin (2014), this method is mostly utilized to biodegrade absorbed fuel residues, low-molecular-weight hydrocarbons, spilt light petroleum, and volatile contaminants from soil.

***Bioslurping*** In order to eliminate pollutants from groundwater and soil though promoting microbial biodegradation, the approach of bioslurping combines a number of events, including vacuum-enhanced pumping, bioventing, and soil vapor extraction. The limited permeability of the soil, which lowers the oxygen transfer rate and further lowers microbial activity, is the technique's deficit. These methods are often used to remove organic pollutants that are volatile and semi-volatile from soil and liquid.

***Biosparging*** In contrast to the bioventing process, the biosparging process introduces air into the subsurface of the soil (i.e., saturated zone), which might trigger the upward movement of volatile organic contaminants to the surface zone for the biodegradation procedure (Kumar et al. 2018). This process is aided by two main factors, namely "soil permeability and pollutant biodegradability, and it also enhances pollutant removal from the polluted site by causing microbial activity, For the treatment of aquifers contaminated with hydrocarbons (benzene, xylene, toluene, and ethylbenzene) and petroleum products, the biosparging technique has been widely used" (Tyagi and Kumar 2021).

## 11.4.2 *Ex-Situ* Bioremediation

The *ex-situ* bioremediation approach entails the elimination of pollutants using various bioremediation procedures after the pollutants have been excavated and transported from the original contaminated site to another location. Based on the types of pollutants, the depth and level of pollution, the expenses of treatment, and the physical and geographical characteristics of the polluted site, the *ex-situ* bioremediation techniques are additionally separated into various categories, such as bio piling, land farming, composting, biofilters, and bioreactors (Tyagi and Kumar 2021).

### 11.4.2.1 Bio-Piling

Land farming and composting techniques are combined in the bio-piling or heap approach. This method includes laboratory testing of the polluted sample to determine its potential for degradation, mechanical separation of the soil sample to homogenize the soil, piling of the excavated polluted soil, additional nutrient amendment to the soil, and forced aeration to promote microbial degradation. Additionally, potent microorganisms might be introduced to the bio-pile for efficient pollution removal. The entire set-up of this system consists of a leachate collecting system, an irrigation system buried beneath the soil, a treatment bed and nutrient system, and an aeration system (Azubuike et al. 2016). The bio-piling method is increasingly used for bioremediation due to its cost-effectiveness, and the temperature, pH, and nutrient conditions can be controlled for bioremediation (Whelan et al. 2015). It is primarily used to treat soil contaminated with low-molecular-weight pollutants and petroleum hydrocarbons, such as PAHs and phenols, and may work well in the remediation of an extremely contaminated environment, such as in cold regions.

### 11.4.2.2 Land Farming

To enable the aerobic biodegradation of contaminants via indigenous bacteria, contaminated soil is frequently excavated, distributed, and tilled over the permanent soil layer support above the ground. According to Kumar et al. (2018), land farming can only be used to remediate soil that is 10–35 cm thick at the surface. Agricultural ash can be used to keep the pH balanced. Land farming is typically thought of as an *ex-situ* bioremediation technique, although there are a few instances when it may be considered an *in-situ* technique. This approach is mostly utilized to address locations that have been polluted with PCBs, polycyclic aromatic hydrocarbons, and aliphatic hydrocarbons. As minimal equipment, money, and maintenance are needed for this method to be effective, it is regarded as the easiest and most affordable choice. As a result, it is receiving the most scrutiny as a substitute dumping method.

#### 11.4.2.3 Composting

An *ex-situ* bioremediation technique that dates back thousands of years, composting is still used today on a wide range of scales, from little garden sites to massive industrial operations. The 4Rs, which include reductions, reuse, recycling, and recovery techniques, are fundamental and accepted concepts in integrated solid waste management (ISWM) (Lin et al. 2022).

Microorganisms that are naturally present in soil seek to break down organic matter into compost. Under natural circumstances, the majority of the first mechanical breakdown of organic materials into smaller particles is carried out by earthworms, nematodes, and soil insects, i.e., ants, sow bugs, mites, springtails, and beetles. At high temperatures, which are between 55°C and 65°C, bacteria break down the waste. Microbes produce heat during the degradation process, raising the temperature, which increases the solubility of waste and boosts metabolic activity in composts. When organic materials are combined and heaped high, compost can form either aerobically or anaerobically.

The quickest way to make completed compost is by aerobic composting, which is also the most effective kind of decomposition. Composters use chopping or grinding to break down big particles under carefully regulated circumstances. The composting process begins when soil microorganisms colonize the organic material and favorable physical conditions are created. The process of creating finished compost has no set duration. The length of time depends on the feedstocks, the composting technique, and management. It might take three months or two years to complete. Furthermore, to give some control over the carbon–nitrogen ratio, adding plant-based elements to a compost combination can also be important for providing structure and channels for aeration (Tegene 2020). Bulking and air access are also important.

#### 11.4.2.4 Biofilters

The biofilter technique is mostly employed to eliminate gaseous contaminants. This method uses microbe-encrusted columns to remove gaseous contaminants from the air.

### 11.4.3 Bioreactors

In this method, the pollutant is typically degraded under controlled circumstances in a reactor or vessel. Liquid and polluted solid materials (soil, sludge, and sediment) are treated using engineered inhibition mechanisms, such as slurry or aqueous reactors. The bioreactor cantered bioremediation method is the most effective because of elements like regulated bioaugmentation, nutritional supplementation, mass transfer, bioavailability of contaminants, and optimized reaction conditions. For the remediation of (soil/water) polluted with volatile organic pollutants, including xylene, ethylbenzene, toluene, and benzene, the bioreactor technique is utilized. As a bioreactor may be used to regulate, manage, and alter the process parameters necessary for the biological reactions utilized in the biodegradation of pollutants, it is much more advantageous to use one for the bioremediation process than other approaches (Azubuike et al. 2016).

#### 11.4.3.1 Microbial Fuel Cells

Microbial fuel cell (MFC) technology is a brand-new type of renewable technology that eliminates organic and inorganic pollutants in addition to producing bioelectricity. There are aspects in an MFC type of bioreactor, including anode and cathode chambers, electrodes, a salt bridge, and an external circuit wire (Saran et al. 2022).

The production of bioelectricity in MFCs is caused by the progress of a bio-potential as a result of metabolic activity (bacterial) in the anodic chamber, i.e., reduction reactions, which produce both $e^-$ (electrons) and $H^+$ (protons) and $e^-$ acceptor conditions at the cathode (which are separated by a membrane) (Fang and Achal 2019). The electrochemically active microorganisms in the anodic chamber can give electrons to an anode, which are freed by oxidizing organic/inorganic waste (such as fuel), generating a basis of energy. Anode-respiring bacteria are electrochemically active bacteria and electricigens, as are other microorganisms that may provide $e^-$. Exoelectrophs are the name given to microbes that can take $e^-$. Through a proton exchange membrane (PEM) or salt bridge, $H^+$ generated by the electrochemically active bacteria in the anode permeates into the cathodic compartment. Due to its availability and high reduction potential, oxygen is typically utilized as the oxidant in the cathodic compartment. MFCs have a cathodic chamber that handles metal oxidants like copper (Cu), cadmium (Cd), and chromium (Cr). The majority of MFC configurations are often used as anaerobic devices. In order to be categorized as an MFC, it must be capable of having its fuel source, which is oxidized at the substrate-anode interface (e.g., wastewater), supplied either irregularly or constantly. The reason for this is that anaerobic conditions are necessary for bacterial species that are currently regarded as the "gold standard" in terms of their ability to transport electrons (Slate et al. 2019).

## 11.5 Factors Influencing Bacterial Degradation Efficiency

Due to their metabolic and ability to acclimatize towards unfavorable conditions, bacteria may destroy a wide range of organic contaminants. As a result, bacteria play a significant role in site restoration. However, a variety of parameters, such as the chemical nature and quantity of pollutants, their accessibility to bacteria, and the physicochemical properties of the environment, affect how effective they are. Therefore, variables that affect the rate at which bacteria degrade pollutants are either connected to the bacteria and their nutritional needs (biological factors) or related to the environment factors.

### 11.5.1 Biological Factors

The capacity for the metabolism of bacteria is a biotic factor. Direct suppression of enzyme activity and the methods by which degrading microorganisms proliferate are two examples of biotic variables that have an impact on the microbial degradation of organic molecules. This suppression may happen, for instance, if bacteria compete to limit carbon sources or engage in hostile interactions. This inhibition may take place, for instance, in situations where microbes compete for scarce carbon sources, interact negatively, or are preyed upon (Moënne-Loccoz et al. 2015).

The concentration of the contamination and the amount of catalyst present are frequently correlated with the rate of contaminant breakdown. The number of organisms that can metabolize the contaminant in this context, as well as the number of enzymes generated by each cell, are represented by the amount of catalyst. Additionally, the particular enzymes involved, their affinity for the contamination, and the availability of the contaminant all have a significant role in how much of a contaminant is metabolized. Furthermore, for unlimited microbial development to take place, adequate amounts of nutrients and oxygen must be present in a practical form and in the correct ratios. pH, temperature, and moisture are other variables that affect the pace of biodegradation by regulating the rates of enzyme-catalyzed processes (Koh and Khor 2022).

The metabolic activity of biological enzymes engaged in the degradation procedure varies depending on the temperature and has an optimal range. In fact, every 10°C drop in temperature results in an approximately 50% reduction in the rate of biodegradation. A pH of 6.5–8.5 is typically ideal for biodegradation in the majority of aquatic and terrestrial systems, while biodegradation can occur under a wide variety of pH conditions. Moisture affects the kind and quantity of soluble materials that are accessible, as well as the osmotic pressure and pH of terrestrial and aquatic systems, which all affect the pace of pollutant metabolism (Ghosh et al. 2020).

#### 11.5.1.1 Bacterial Community Structure and Diversity

The dynamics of the bacterial population must be better understood in order to design bioremediation methods or degradation strategies. Pollution-contaminated environments alter the variety, richness, and evenness of microbial communities, which has an impact on the metabolic activity of bacteria that break down contaminants. According to several research, when pollutants are added to the soil environment, bacterial communities change. Several bacterial species may experience immediate and large declines, while other bacterial species may become considerably more abundant (Omrani et al. 2018). According to the structure and diversity of the hydrocarbon-degrading bacterial community, bacterial consortiums, whether native or introduced, are more effective than single strains at completely degrading hydrocarbon pollutants in soil (Purwanti et al. 2019).

This is because mixed bacterial isolates (polyculture consortium) are effective at metabolizing highly concentrated and a wide variety of pollutants, but a single bacterial isolate (monoculture) can only use an inadequate number of substrates (Yap et al. 2021). This is due to the fact that bacterial consortia have a wide range of degradative enzymes, can break down a variety of materials, are more tolerant of physical and chemical conditions (such as temperature, pH, and salinity), synthesize surface-active biomolecules (such as biosurfactants) at a faster rate, and work in concert to speed up the process of bioremediation.

The opposed action of some other bacteria in bacterial consortia may reduce the effectiveness of bacterial remediation. Therefore, it is necessary to restrict the actions of these hostile bacteria in order to support the ongoing growth of bacterial consortia with a high potential for viability. Numerous studies have demonstrated the important role that microbial communities, including algae, bacteria, fungi, and yeasts role in the biodegradation of contaminants. As a result, biodegradation may become more rapid, effective, and thorough when utilizing a group of prospective bacteria having broadly catabolic enzymes and genes (Kebede et al. 2021).

#### 11.5.1.2 Biofilm Formation

For sustenance, bacteria gather in the form of flocks, mats, or biofilms. Up to 90% of these collections are made up of microbial cells and extracellular matrix. An adverse environment can cause microbial communities to respond to stimuli by creating extracellular polymers that mount them and aid in their adhesion to the substrate (Kebede et al. 2021). This is crucial in order to alter how they respond to their environment's redox status, catabolic gene exchange, gene transcription, and/or horizontal gene transfer, as well as how they respond to their growth (Saran et al. 2022).

However, biofilm can be created by a single or a group of microbial species, *Pseudomonas* sp., *Bacillus* sp., and other bacterial species are used to assess their biodegradation capacity. Bacterial communities that form biofilm have the capacity to increase the bio-accessibility of the carbon sources from the various contaminated wastewater sites through cooperative effects, environmental adaptation, and the acquisition of metabolic modifications to survive in an environment with a constrained range of nutrients (Giovanella et al. 2016). Thus, the biodegradation rates of the contaminants will be accelerated or slowed down depending on whether biofilm development occurs. The rate of biodegradation will increase as more biofilms are formed.

#### 11.5.1.3 Redox Potential of the Bacteria

Energy is needed by bacteria for reproduction, cell maintenance, and biological processes. According to physiologically, redox reactions in bacterial cells provide this energy (Oktyabrskii and Smirnova 2012). Therefore, bacteria use contaminants as a source of energy by catalyzing the transport of $e^-$ during aerobic or anaerobic bioremediation from $e^-$ donors to $e^-$ acceptors. This behavior depends on redox potentials, which are crucial for driving the oxidation of pollutants originating from hydrocarbons for bacterial respiration and promoting their development by lowering the redox potential.

Additionally, to serve as the ultimate electron acceptor in aerobic settings, oxygen is crucial for the enzymatic activation of the substrate (aromatic hydrocarbons as mono or dioxygenases). As an effect of the high Gibbs free energy change of these compounds' oxidation with various $e^-$ acceptors, aromatic compounds are suitable $e^-$ donors for bacterial growth. Furthermore, under anaerobic settings, compounds with high reduction potential, such as chlorate and perchlorate, offer excellent e-acceptors for microbial metabolism during bioremediation procedures (Liebensteiner et al. 2014).

### 11.5.2 Environmental Factors

Since traditional physicochemical approaches for remediation appear technically and financially difficult, bioremediation of wastewater is a viable alternative. When it comes to pollutant elimination, bioremediation is the best option for wastewater toxins since it is nonaggressive and potentially economical. However, there are several drawbacks to using bioremediation approaches, including the contaminant characteristics, limited field competence of microbial communities, lower bioavailability of pollutants, and a predominance of environmental factors in bacterial development. Therefore, considering and understanding, particularly the environmental elements that might impact bacterial degradation, is of major study interest at this time. The availability of nutrients, aeration and oxygen, temperature, and pH are a few examples of environmental elements that have an impact on bacterial degradation (Aragaw 2021).

#### 11.5.2.1 Nutrient Availability

The microbial/bacterial community depends on nutrient availability in the polluted environment since bacteria need resources for their metabolism and proliferation. In the context of nature, inorganic nutrients operate as limiting variables to control how much bacteria can degrade (Kebede et al. 2021). This suggests that the distribution and degradation of pollutants in polluted areas are influenced by the occurrence of macronutrients (nitrogen, phosphate, and potassium). Organic matter in the wastewater provides nutrients and improves some aspects to encourage the development and activity of bacteria. Additionally, a high concentration of organic nutrients causes toxicity because the biodegradation of hydrocarbons produces hazardous intermediates.

Due to the quick microbial metabolic activities that occur when there is a significant buildup of organic carbon in polluted sites, other inorganic nutrients like nitrogen, phosphate, and potassium (N, P, and K), as well as trace amounts of calcium, sulfur, magnesium, iron, and manganese are rapidly depleted. This restricts biodegradation and biotransformation's pace and scope. Studies advised adjusting the C/N/P (carbon, nitrogen, and phosphorus) ratio to 100:10:1 to 100:20:1 for significant hydrocarbon biodegradation to minimize this constraint (Du et al. 2020).

Urea, phosphate, NPK fertilizers, ammonium, and phosphate salts must be added in order to apply this. Although certain bacteria may thrive or adapt well to the high levels of carbon, nitrogen, potassium, and phosphorus found in hydrocarbon-contaminated locations, biostimulation, the addition of additional nutrients or fertilizers, has a substantial impact on the local microbial population.

For instance, excessive concentrations of hydrocarbon pollutants may also change the NPK ratio, ultimately leading to a lack of oxygen. In general, an inadequate supply (high or low) and/or a lack of mineral nutrients prevent soil bacteria that use hydrocarbons from growing. As a result, successful hydrocarbon biodegradation requires the addition of the right amount of nutrients (N and P) at polluted areas (Pandolfo et al. 2022). Therefore, the hydrocarbon biodegradation depends on soil environment modifications, and in that manner, the maximum degradation rate may be increased with the ideal degree of nutrient input.

#### 11.5.2.2 Aeration/Oxygen Availability

In research, hydrocarbon-degrading bacteria (oleophilic) were found to be capable of both aerobic and facultative anaerobic respiration. The final e-acceptor and chemical reactant (the enzymatic oxygenase reactions) for a terminal, subterminal, and biterminal oxidation and ring cleavage of aromatic hydrocarbon contaminants, as well as the limiting feature for aerobic bioremediation, is the presence of molecular oxygen in the soil. Additionally, the research found that without seeing the total biomass of potential bacteria that can break down hydrocarbons, oxygen (3.1 mg/ml) needed to break down hydrocarbon (1 mg/ml) contaminants, and oxygen (10–40%) is needed for effective hydrocarbon biodegradation (Aragaw 2021).

Compared to anaerobic metabolism, aerobic catabolism produces a more rapid rate of biodegradation. According to reports, isolated bacterial strains could degrade 20–25% of the overall amount of oil under aerobic conditions in just 10 days, but under anaerobic conditions, similar strains could only degrade 15–18% of the total petroleum present in 50 days, and in anoxic soil regions, there was little to no hydrocarbon pollutant degradation.

The low oxygen levels in hydrocarbon-contaminated locations, however, restrict the aerobic degradation pathways and the low removal effectiveness that results. Bacterial oxygen consumption rate, oxygen concentration, physical and chemical characteristics of contaminated soils (type and porosity), and the presence of usable substrates or accrual of organic matter are the factors causing slow biodegradation and free oxygen reduction in the soil. As a result, successful biodegradation depends on the incidence of oxygen in the soil (Kebede et al. 2021).

#### 11.5.2.3 Effect of Temperature

Temperature influences both the physical condition of the pollutants found in the wastewater (solubility, viscosity, ionization, etc.) and the bacteria that consume them (growth rate, metabolism, etc.). As a result, raising the reaction temperature causes a proportional increase in pollutant solubility, improves diffusion, and reduces viscosity; conversely, lowering the reaction temperature causes a decrease in biodegradation capability. According to certain research, the temperature has an impact on the generation of biocatalysts (enzymes), which are in charge of bioremediation processes, as well as the physicochemical characteristics of the contaminant, which have an impact on bacterial effectiveness (Aragaw 2021).

#### 11.5.2.4 Effect of pH

Depending on the pH range at which a particular bacterial species may live, bioremediation can take place. On acidic, neutral, and basic conditions, respectively, acidophile, neutralophilic, and alkaliphile bacteria may biodegrade materials most effectively. The life cycles of microbial communities in wastewater frequently expose neutralophilic bacteria, in particular, to variable pH fluctuation stressors, and they are crucial to system stability. Because the ideal pH is critical to controlling microbial biomass and enzyme activity and promoting pollutant biodegradation, reaction temperature plays a crucial role in bioremediation. In order to maximize the effects of bioremediation, it is crucial to modify the pH in polluted areas to an ideal range between 5 and 8 (Kebede et al. 2021).

#### 11.5.2.5 Effect of Salinity

For halophile or halotolerant species in their ecosystem, the deposition and buildup of salt can have an adverse effect on microbial hydrocarbon clastic activity. High salt concentrations can lead to an oxygen deficiency in bacteria by increasing osmotic pressure and impairing bacterial metabolism, solubility, and ion transportation systems (Kanekar and Kanekar 2022).

This may have occurred as a result of the discharge of salty solution into water or soil as well as the introduction of several soluble fertilizers (such as nitrates or ammonium salts). Halophilic bacteria may live and break down the contaminants in such hypersaline settings that are polluted with petroleum chemicals. The varied hydrocarbon catabolism of those prospective bacteria is caused by genes encoding hydrocarbon clastic enzymes (Arif et al. 2020).

A high level of soil salt reduced the pace and extent of hydrocarbon biodegradation, according to certain research, which showed that considerable hydrocarbon degradation was identified with 0.1–2 M NaCl but up to 0.4 M. The kind of possible bacteria that might degrade hydrocarbons and hydrocarbon-contaminated soil conditions are what determine the optimum salt content (Varjani 2016).

## 11.6 Conclusion and Future Prospects

The most prevalent and significant idea for managing water resources is wastewater treatment technology. Wastewater treatment for specific industries is possible using a variety of treatment facilities and methods. In recent years, scientists and researchers in the sector have also become aware of the many difficulties associated with wastewater treatment and reuse. The treatment facilities put in place in industries depend on the quality, volume, and amount of wastewater produced. Specific wastewater treatment methods reduce potential threats to the available freshwater resources. In addition to treatment, wastewater may be disinfected and recovered for a variety of uses.

For the treatment of heavy metal wastewater, chlorinated compound, hydrocarbon petroleum, plastic waste, volatile compound pollutants, organic and inorganic municipal solid waste, lignocellulose organic agricultural-based waste, industry effluent waste, nuclear waste, different dye and agrochemical waste, *ex-situ* and *in-situ* bioremediation methods such as bio-piling, composting, land farming, bioventing, biosparging, biostimulation, and bioaugmentation are used. A diverse class of enzymes is produced by microorganisms that fall under six classes. During the bioremediation process, oxidoreductases, transferases, hydrolases, lyases, isomerases, and ligases (synthetases) are employed for mineralization and degradation. Utilizing the potential of microorganisms and developing waste management strategies require an understanding of their mechanism, mode of action, and involvement in the bioremediation procedure.

Numerous studies have demonstrated that bacteria may degrade contaminants using a variety of enzymes. When biotic and abiotic circumstances allow, the process might begin in polluted areas. To promote the best possible bacterial growth and metabolism, environmental conditions must be optimized. Furthermore, a viable approach to removing the obstacle to microbial absorption of polluted sites appears to be the development of innovative biosurfactants that increase the bioavailability of wastewater pollutants to bacteria.

Future possibilities for wastewater treatment technologies include the development of innovative, secure, economical, and dependable procedures. The two main necessary areas for the research are industrial performance and quality assurance, as well as health and environmental concerns. The researcher can also discuss the capability for epidemiological research, government response infrastructure, and the evaluation of the impact of pollutants or other components in reclaimed water on human health.

## References

Abbas, S.Z., Rafatullah, M., Ismail, N., and Lalung, J. (2014a). Isolation, identification, and characterization of cadmium resistant *Pseudomonas* sp. M3 from industrial wastewater. *Journal of Waste Management* 2014.

Abbas, S.Z., Riaz, M., Ramzan, N. et al. (2014b). Isolation and characterization of arsenic resistant bacteria from wastewater. *Brazilian Journal of Microbiology* 45 (4): 1309–1315.

Ameen, F.A., Hamdan, A.M., and El-Naggar, M.Y. (2020). Assessment of the heavy metal bioremediation efficiency of the novel marine lactic acid bacterium, *Lactobacillus plantarum* MF042018. *Scientific Reports* 10 (1): 1–11.

Aragaw, T.A. (2021). Functions of various bacteria for specific pollutants degradation and their application in wastewater treatment: a review. *International Journal of Environmental Science and Technology* 18: 2063–2076.

Arif, Y., Singh, P., Siddiqui, H. et al. (2020). Salinity induced physiological and biochemical changes in plants: an omic approach towards salt stress tolerance. *Plant Physiology and Biochemistry* 156: 64–77.

Azubuike, C.C., Chikere, C.B., and Okpokwasili, G.C. (2016). Bioremediation techniques – classification based on site of application: principles, advantages, limitations and prospects. *World Journal of Microbiology and Biotechnology* 32: 1–18.

Babiker, R.A., Elsharief, U.A., Salim, J.B. et al. (2020). Bacterial removal of lead and mercury elements from water using *Pseudomonas aeruginosa* in vitro. *Applied Microbiol Open Access* 6: 169.

Bonificio, W.D. and Clarke, D.R. (2014). Bacterial recovery and recycling of tellurium from tellurium-containing compounds by *Pseudoalteromonas* sp. EPR 3. *Journal of Applied Microbiology* 117 (5): 1293–1304.

Chaparro-Acuña, S.P., Becerra-Jiménez, M.L., Martínez-Zambrano, J.J., and Rojas-Sarmiento, H.A. (2018). Soil bacteria that precipitate calcium carbonate: mechanism and applications of the process. *Acta Agronómica* 67 (2): 277–288.

Devi, A., Ferreira, L.F.R., Saratale, G.D. et al. (2022). Microbe-assisted phytoremediation of environmental contaminants. In: *Advances in Microbe-Assisted Phytoremediation of Polluted Sites* (ed. K. Bauddh and Y. Ma), 3–26. Elsevier.

Devi, A., Verma, M., Saratale, G.D. et al. (2023). Microalgae: a green eco-friendly agents for bioremediation of tannery wastewater with simultaneous production of value-added products. *Chemosphere* 139192.

Du, E., Terrer, C., Pellegrini, A.F. et al. (2020). Global patterns of terrestrial nitrogen and phosphorus limitation. *Nature Geoscience* 13 (3): 221–226.

El-barbary, T. and El-Badry, M. (2019). Adsorption of zinc ion by different bacterial species. *Journal of Environmental Science and Technology* 7 (2): 559–567.

Englande, A.J. Jr., Krenkel, P., and Shamas, J. (2015). Wastewater treatment & water reclamation. In: *Reference Module in Earth Systems and Environmental Sciences*. Elsevier http://www.10.1016/B978-0-12-409548-9.09508-7.

Fang, C. and Achal, V. (2019). The potential of microbial fuel cells for remediation of heavy metals from soil and water – review of application. *Microorganisms* 7 (12): 697.

García-Depraect, O., Lebrero, R., Rodriguez-Vega, S. et al. (2021). Biodegradation of bioplastics under aerobic and anaerobic aqueous conditions: kinetics, carbon fate and particle size effect. *Bioresource Technology* 344: 126265.

Garg, S. (2022). Industrial wastewater: characteristics, treatment techniques and reclamation of water. In: *Advanced Industrial Wastewater Treatment and Reclamation of Water: Comparative Study of Water Pollution Index during Pre-Industrial, Industrial Period and Prospect of Wastewater Treatment for Water Resource Conservation*, 1–23. Cham: Springer.

Ghosh, S., Sharma, I., Nath, S., and Webster, T.J. (2020). Bioremediation – the natural solution. *Microbial Ecology of Wastewater Treatment Plants* 11–40.

Giovanella, P., Cabral, L., Bento, F.M. et al. (2016). Mercury (II) removal by resistant bacterial isolates and mercuric (II) reductase activity in a new strain of *Pseudomonas* sp. B50A. *New Biotechnology* 33 (1): 216–223.

Gnanaprakasam, E.T., Lloyd, J.R., Boothman, C. et al. (2017). Microbial community structure and arsenic biogeochemistry in two arsenic-impacted aquifers in Bangladesh. *MBio* 8 (6).

Gómez-Guzmán, A., Jiménez-Magaña, S., Guerra-Rentería, A.S. et al. (2017). Evaluation of nutrients removal ($NO^3$-N, $NH^3$-N and $PO^4$-P) with *Chlorella vulgaris*, *Pseudomonas putida*, *Bacillus cereus* and a consortium of these microorganisms in the treatment of wastewater effluents. *Water Science and Technology* 76 (1): 49–56.

Guerra-Renteria, A.S., García-Ramírez, M.A., Gómez-Hermosillo, C. et al. (2019). Metabolic pathway analysis of nitrogen and phosphorus uptake by the consortium between *C. vulgaris* and *P. aeruginosa*. *International Journal of Molecular Sciences* 20 (8): 1978.

Halttunen, T., Salminen, S., and Tahvonen, R. (2007). Rapid removal of lead and cadmium from water by specific lactic acid bacteria. *International Journal of Food Microbiology* 114 (1): 30–35.

Haripriyan, U., Gopinath, K.P., Arun, J., and Govarthanan, M. (2022). Bioremediation of organic pollutants: a mini review on current and critical strategies for wastewater treatment. *Archives of Microbiology* 204 (5): 286.

Heim, C., de Vivanco, M.U., Rajab, M. et al. (2015). Rapid inactivation of waterborne bacteria using boron-doped diamond electrodes. *International journal of Environmental Science and Technology* 12 (10): 3061–3070.

Höhener, P. and Ponsin, V. (2014). In situ vadose zone bioremediation. *Current Opinion in Biotechnology* 27: 1–7.

Igiri, B.E., Okoduwa, S.I., Idoko, G.O. et al. (2018). Toxicity and bioremediation of heavy metals contaminated ecosystem from tannery wastewater: a review. *Journal of Toxicology* 2018.

Jarosławiecka, A. and Piotrowska-Seget, Z. (2014). Lead resistance in micro-organisms. *Microbiology* 160 (1): 12–25.

Jassal, V., Shanker, U., Kaitha, B.S., and Shankarb, S. (2015). Green synthesis of potassium zinc hexacyano-ferrate nanocubes and their potential application in photocatalytic degradation of organic dyes. *RSC Advances* 5 (33): 26141–26149.

Jebelli, M.A., Maleki, A., Amoozegar, M.A. et al. (2018). Isolation and identification of the native population bacteria for bioremediation of high levels of arsenic from water resources. *Journal of Environmental Management* 212: 39–45.

Kagami, T., Fudemoto, A., Fujimoto, N. et al. (2012). Isolation and characterization of bacteria capable of reducing tellurium oxyanions to insoluble elemental tellurium for tellurium recovery from wastewater. *Waste and Biomass Valorization* 3 (4): 409–418.

Kandasamy, S., Dananjeyan, B., Krishnamurthy, K., and Benckiser, G. (2015). Aerobic cyanide degradation by bacterial isolates from cassava factory wastewater. *Brazilian Journal of Microbiology* 46 (3): 659–666.

Kanekar, P.P. and Kanekar, S.P. (2022). Halophilic and halotolerant microorganisms. In: *Diversity and Biotechnology of Extremophilic Microorganisms from India*, 13–69. Singapore: Naveen Kumar Arora, Springer Nature Singapore.

Kebede, G., Tafese, T., Abda, E.M. et al. (2021). Factors influencing the bacterial bioremediation of hydrocarbon contaminants in the soil: mechanisms and impacts. *Journal of Chemistry* 2021: 1–17.

Kitjanukit, S., Takamatsu, K., Takeda, K. et al. (2017). Manganese removal from metal refinery wastewater using Mn (II)-oxidizing bacteria. In: *Solid State Phenomena*, vol. 262, 673–676. Trans Tech Publications Ltd.

Koh, L.M. and Khor, S.M. (2022). Biodegradation process: basics, factors affecting, and industrial applications. In: *Handbook of Biodegradable Materials* (ed. G.A.M. Ali and A.S.H. Makhlouf), 1–39. Cham: Springer International Publishing.

Kumar, V., Shahi, S.K., and Singh, S. (2018). Bioremediation: an eco-sustainable approach for restoration of contaminated sites. *Microbial Bioprospecting for Sustainable Development* 115–136.

Kuroda, M., Notaguchi, E., Sato, A. et al. (2011). Characterization of *Pseudomonas stutzeri* NT-I capable of removing soluble selenium from the aqueous phase under aerobic conditions. *Journal of Bioscience and Bioengineering* 112 (3): 259–264.

Laçin, B., Ertit Taştan, B., and Dönmez, G. (2015). Detection of boron removal capacities of different microorganisms in wastewater and effective removal process. *Water Science and Technology* 72 (10): 1832–1839.

Lee, K., Choo, K.H., Ng, H.Y., and Lee, C.H. (2020). Preparation of a mesoporous silica quorum quenching medium for wastewater treatment using a membrane bioreactor. *Biofouling* 36: 369–377.

Li, J., Yu, H., Wu, X. et al. (2018). Novel hyper antimony-oxidizing bacteria isolated from contaminated mine soils in China. *Geomicrobiology Journal* 35 (8): 713–720.

Li, J., Zhang, Y., Zheng, S. et al. (2019). Anaerobic bacterial immobilization and removal of toxic Sb(III) coupled with Fe(II)/Sb(III) oxidation and denitrification. *Frontiers in Microbiology* 10: 360.

Liebensteiner, M.G., Tsesmetzis, N., Stams, A.J., and Lomans, B.P. (2014). Microbial redox processes in deep subsurface environments and the potential application of (per)chlorate in oil reservoirs. *Frontiers in Microbiology* 5: 91086.

Lin, C., Cheruiyot, N.K., Bui, T., and Ngo, H.H. (2022). Composting and its application in bioremediation of organic contaminants. *Bioengineered* 13 (1): 1073–1089.

Lipps, W.C., Baxter, T.E., Braun-Howland, E.B. et al. (ed.) (2017). *Standard Methods for the Examination of Water and Wastewater*. American Public Health Association.

Luque-Almagro, V.M., Moreno-Vivián, C., and Roldán, M.D. (2016). Biodegradation of cyanide wastes from mining and jewellery industries. *Current Opinion in Biotechnology* 38: 9–13.

Maharjan, A.K., Kamei, T., Amatya, I.M. et al. (2020). Ammonium-nitrogen ($NH^{4+}$-N) removal from groundwater by a dropping nitrification reactor: characterization of $NH^{4+}$-N transformation and bacterial community in the reactor. *Water* 12 (2): 599.

Maji, S., Dwivedi, D. H., Singh, N., et al. (2020). Agricultural waste: its impact on environment and management approaches. Emerging Eco-Friendly Green Technologies for Wastewater Treatment, 329–351, Springer.

Mateos, L.M., Ordóñez, E., Letek, M., and Gil, J.A. (2006). *Corynebacterium glutamicum* as a model bacterium for the bioremediation of arsenic. *International Microbiology* 9 (3): 207–215.

Mishra, R., Sinha, V., Kannan, A., and Upreti, R.K. (2012). Reduction of chromium-VI by chromium resistant lactobacilli: a prospective bacterium for bioremediation. *Toxicology International* 19 (1): 25.

Mishra, V., Balomajumder, C., and Agarwal, V.K. (2014). Biological removal of heavy metal zinc from industrial effluent by zinc sequestering bacterium VMSDCM. *Clean Technologies and Environmental Policy* 16 (3): 555–568.

Miwa, H. and Fujiwara, T. (2009). Isolation and identification of boron-accumulating bacteria from contaminated soils and active sludge. *Soil Science and Plant Nutrition* 55 (5): 643–646.

Moënne-Loccoz, Y., Mavingui, P., Combes, C. et al. (2015). Microorganisms and biotic interactions. In: *Environmental Microbiology: Fundamentals and Applications: Microbial Ecology*, 395–444. Springer Nature.

Mukherjee, S., Sahu, P., and Halder, G. (2017). Microbial remediation of fluoride-contaminated water via a novel bacterium *Providencia vermicola* (KX926492). *Journal of Environmental Management* 204: 413–423.

Naik, M.M., Khanolkar, D., and Dubey, S.K. (2013). Lead-resistant *Providencia alcalifaciens* strain 2 EA bioprecipitates $Pb^{+2}$ as lead phosphate. *Letters in Applied Microbiology* 56 (2): 99–104.

Okoduwa, S.I.R., Igiri, B., Udeh, C.B. et al. (2017). Tannery effluent treatment by yeast species isolates from watermelon. *Toxics* 5 (1): 6.

Oktyabrskii, O.N. and Smirnova, G.V. (2012). Redox potential changes in bacterial cultures under stress conditions. *Microbiology* 81: 131–142.

Omrani, R., Spini, G., Puglisi, E., and Saidane, D. (2018). Modulation of microbial consortia enriched from different polluted environments during petroleum biodegradation. *Biodegradation* 29: 187–209.

Owoseni, M.C., Olaniran, A.O., and Okoh, A.I. (2017). Chlorine tolerance and inactivation of *Escherichia coli* recovered from wastewater treatment plants in the Eastern Cape, South Africa. *Applied Sciences* 7 (8): 810.

Pal, A.K., Singh, J., Soni, R. et al. (2020). The role of microorganism in bioremediation for sustainable environment management. In: *Bioremediation of Pollutants* (ed. V.C. Pandey and V. Singh), 227–249. Elsevier.

Pandolfo, E., Barra Caracciolo, A., and Rolando, L. (2022). Recent advances in bacterial degradation of hydrocarbons. *Water* 15 (2): 375.

Paul, D. and Sinha, S.N. (2015). Biological removal of phosphate using phosphate solubilizing bacterial consortium from synthetic wastewater: a laboratory scale. *Environment Asia* 8 (1): 1–8.

Piacenza, E., Presentato, A., Zonaro, E. et al. (2018). Microbial-based bioremediation of selenium and tellurium compounds. In: *Biosorption* (ed. J. Derco), 117–147.

Purwanti, I.F., Kurniawan, S.B., and Simanjuntak, D.Y. (2019). Removal of aluminium in contaminated soil using locally isolated *Vibrio alginolyticus*. *Journal of Ecological Engineering* 20 (3): 135–140.

Raja, C.E. and Omine, K. (2013). Characterization of boron tolerant bacteria isolated from a fly ash dumping site for bacterial boron remediation. *Environmental Geochemistry and Health* 35 (4): 431–438.

Rana, N. and Gupta, P. (2022). Microbes: an eco-friendly tool in wastewater treatment. In: *Synergistic Approaches for Bioremediation of Environmental Pollutants: Recent Advances and Challenges*, 161–183. Academic Press.

Rojas, L.A., Yáñez, C., González, M. et al. (2011). Characterization of the metabolically modified heavy metal-resistant *Cupriavidus metallidurans* strain MSR33 generated for mercury bioremediation. *PLoS ONE* 6 (3): e17555.

Sakthi Thesai, A., Rajakumar, S., and Ayyasamy, P.M. (2020). Removal of fluoride in aqueous medium under the optimum conditions through intracellular accumulation in *Bacillus flexus* (PN4). *Environmental Technology* 41 (9): 1185–1198.

Sanjay, M.S., Sudarsanam, D., Raj, G.A., and Baskar, K. (2020). Isolation and identification of chromium reducing bacteria from tannery effluent. *Journal of King Saud University, Science* 32 (1): 265–271.

Saran, C., Purchase, D., Saratale, G.D. et al. (2022). Microbial fuel cell: a green eco-friendly agent for tannery wastewater treatment and simultaneous bioelectricity/power generation. *Chemosphere* 312: 137072.

Saranya, K., Sundaramanickam, A., Shekhar, S. et al. (2017). Bioremediation of mercury by *Vibrio fluvialis* screened from industrial effluents. *BioMed Research International* 2017: 1–6.

Sasaki, K., Hosokawa, Y., Takeno, K., and Sasaki, K. (2017). Removal of sodium from seawater medium using photosynthetic bacteria. *Journal of Agricultural Chemistry and Environment* 6 (3): 133–143.

Singh, A.L. (2008). Removal of chromium from waste water with the help of microbes: a review. *Journal of Science and Technology* 3 (3).

Sinha, S. and Mukherjee, S.K. (2009). *Pseudomonas aeruginosa* KUCd1, a possible candidate for cadmium bioremediation. *Brazilian Journal of Microbiology* 40 (3): 655–662.

Sinharoy, A. and Lens, P.N. (2020). Biological removal of selenate and selenite from wastewater: options for selenium recovery as nanoparticles. *Current Pollution Reports* 6: 321–351.

Slate, A.J., Whitehead, K.A., Brownson, D.A., and Banks, C.E. (2019). Microbial fuel cells: an overview of current technology. *Renewable and Sustainable Energy Reviews* 101: 60–81.

Syed, M., Soreanu, G., Falletta, P., and Béland, M. (2006). Removal of hydrogen sulfide from gas streams using biological processes a review. *Canadian Biosystems Engineering* 48: 2.

Tegene, B.G. (2020). Mode of action, mechanism, and role of microbes in bioremediation service for environmental pollution management. *Journal of Biotechnology & Bioinformatics Research* 2 (3): 1–18.

Tyagi, B. and Kumar, N. (2021). Bioremediation: principles and applications in environmental management. In: *Bioremediation for Environmental Sustainability* (ed. G. Saxena, V. Kumar, and M.P. Shah), 3–28. Elsevier.

Varjani, S.J. (2016). Microbial degradation of petroleum hydrocarbons. *Bioresource Technology* 223: 277–286.

Von Canstein, H., Li, Y., Timmis, K.N. et al. (1999). Removal of mercury from chloralkali electrolysis wastewater by a mercury-resistant *Pseudomonas putida* strain. *Applied and Environmental Microbiology* 65 (12): 5279–5284.

Whelan, M.J., Coulon, F., Hince, G. et al. (2015). Fate and transport of petroleum hydrocarbons in engineered biopiles in polar regions. *Chemosphere* 131: 232–240.

Yap, H.S., Zakaria, N.N., Zulkharnain, A. et al. (2021). Bibliometric analysis of hydrocarbon bioremediation in cold regions and a review on enhanced soil bioremediation. *Biology* 10 (5): 354.

Yilmaz, M., Tay, T., Kivanc, M., and Turk, H. (2010). Removal of copper (II) ions from aqueous solution by a lactic acid bacterium. *Brazilian Journal of Chemical Engineering* 27 (2): 309–314.

Zainith, S., Sandhya, S., Saxena, G., and Bharagava, R.N. (2016). Microbes an ecofriendly tool for the treatment of industrial waste waters. *Microbes and Environmental Management* 2016: 75–100.

Zambrano-Monserrate, M.A., Ruano, M.A., and Ormeño-Candelario, V. (2021). Determinants of municipal solid waste: a global analysis by countries' income level. *Environmental Science and Pollution Research* 28: 62421–62430.

Zhai, Q., Xiao, Y., Zhao, J. et al. (2017). Identification of key proteins and pathways in cadmium tolerance of *Lactobacillus plantarum* strains by proteomic analysis. *Scientific Reports* 7 (1): 1–17.

Zhao, Y., Yan, H., Zhou, J. et al. (2019). Bio-precipitation of calcium and magnesium ions through extracellular and intracellular process induced by *Bacillus licheniformis* SRB2. *Minerals* 9 (9): 526.

Zou, H., Lu, X., and Abualhail, S. (2014). Characterization of denitrifying phosphorus removal microorganisms in a novel two-sludge process by combining chemical with microbial analysis. *Journal of Chemistry* 2014: 1–7.

# 12

## Sustainable Algal Industrial Wastewater Treatment: Applications and Challenges

*Anuradha Devi*[1], *Christina Saran*[1], *Ganesh Dattatraya Saratale*[2], *Rijuta Ganesh Saratale*[3], *Luiz Fernando R. Ferreira*[4], *Sikandar I. Mulla*[5], *and Ram Naresh Bharagava*[1]

[1] *Laboratory of Bioremediation and Metagenomics Research (LBMR), Department of Environmental Microbiology (DEM), Babasaheb Bhimrao Ambedkar University (A Central University), Lucknow, India*
[2] *Department of Food Science and Biotechnology, Dongguk University-Seoul, Goyang-si, Republic of Korea*
[3] *Department of Food Science and Biotechnology, Research Institute of Biotechnology and Medical Converged Science, Dongguk University-Seoul, Goyang-si, Republic of Korea*
[4] *Post Graduate Program in Genomic Sciences and Biotechnology, Genomic Sciences and Biotechnology Lab, Catholic University of Brasília, Brasília, Brazil*
[5] *Department of Biochemistry, School of Applied Sciences, REVA University, Bangalore, India*

## 12.1 Introduction

The hasty growth of the anthropological population has encouraged industrial development, which plays a crucial role in the economic progress of nations. The surge in wastewater in evolving nations is closely tied to the expansion of people (Sathinathan et al. 2023). The wastewater originates from various industrial sources, including pharmaceuticals, food, electronics, beverage, agrochemicals, gas industries, petrochemicals, and oil, which carry significant organic and inorganic loads, oil, grease and heavy metals (Kumar et al. 2020). Industrial wastewater (IWW) is characterized as high in chemical oxygen demand (COD), high biological oxygen demand (BOD), loads of phosphorus (P), and nitrogen (N) present in wastewater, causing eutrophication and generating unpleasant emissions to the air as well as solid wastes (Kanu and Achi 2011; Kumar et al. 2020). Pollutant-loaded wastewater also promotes the microbial load outburst, which deteriorates the drinking water quality, contributing the major health-related problems. Heavy metals (HMs) are a significant part of IWW (tannery, textile, packaging, etc.); direct contact, inhalation, and ingestion, even at very minimum concentrations, cause noteworthy intimidation to human health and elevated the risk of inducing cancers (Hafizur et al. 2017). The emerging pollutants or synthetic organic compounds include various categories, including flame retardants, pharmaceutical compounds (PCs), per- and polyfluoroalkyl substances (PFASs), artificial sweeteners, illicit drugs, and pesticides have recently been documented because of increasing knowledge of their harmful effects and developments in analytical procedures (i.e., lower detection limits (Sutherland and Ralph 2019). The harmful effects on the environment and human health are still being researched. Hence, the disposal/discharge of wastewater produced from various industries directly to the freshwater resources imparts environmental damage and health-related issues (Singh et al. 2021). To prevent environmental pollution and ensure the safety of water supplies, a substantial volume of IWW must undergo treatment. The wastewater treatment (WWT) method primarily depends on the effluent composition. Due to the occurrence of various pollutants, high organic matter content, and poorly biodegradable components, it is highly challengeable (Awuchi et al. 2020; Singh et al. 2021). Conventional WWT methods consist of different chemical and physical treatments. The most effective methods for WWT involve chemical processes aimed at modifying water characteristics. This includes adjusting pH, reducing turbidity (removing suspended solids), and eliminating dissolved materials to enhance water quality. Common chemical treatment approaches comprise chloramination, chlorination, coagulation/flocculation, ozonation, and ultraviolet light

*Microbes Based Approaches for the Management of Hazardous Contaminants*, First Edition. Edited by Ajay Kumar, Livleen Shukla, Joginder Singh, and Luiz Fernando R. Ferreira.

(Singh et al. 2021). However, these procedures, while more active in secondary WWT, are associated with significant drawbacks such as high costs, limitations in dewatering, and substantial maintenance efforts (Singh et al. 2021).

In divergence, biological approaches rely on the metabolic actions of microorganisms to break down and change pollutants in wastewater into biomass and associated gases ($CO_2$, $N_2$, $CH_4$, and $SO_2$). This helps reduce the values of COD and BOD in the effluents and improves water quality (Singh et al. 2021). Biological treatments include the use of various microorganisms, with microalgae, fungi, yeast, and bacteria (Kapoor et al. 2021). While biological WWT may seem straightforward, it operates at the node of biochemistry and biology, making it complex and not yet fully understood. These methods are generally cost-effective under aerobic or anaerobic conditions, as they require a significant land area and time for adequate treatment and may not completely remove certain pollutants. Also, intricate chemical pollutants are resistant to degradation by conventional methods (Devi et al. 2023).

Conventional treatment methods have several drawbacks, including high energy consumption, extensive land requirements, and substantial operational and maintenance costs.

Algae that are too tiny to be seen with the naked eye are referred to as "microalgae". This covers both bacterial cyanobacteria and eukaryotic microalgae. Chlorophyta (green algae), Chrysophyta (golden algae), Bacillariophyta (diatoms), and Rhodophyta (red algae) are some of the major taxonomic orders of algae. Primarily, Cyanophyceae was grouped with eukaryotic algae, but their characteristics (morphological) align more with bacteria, leading to their renaming as cyanobacteria (Chapman 2013). Taxonomic orders may vary among different authors, and microalgae members have diverse evolutionary backgrounds.

Microalgae provide a substitute method to biological treatment, as it can effectively eliminate nutrients and change them into biomass (Udaiyappan et al. 2017). Certain microalgae species can also uptake other pollutants like nitrogen compounds, heavy metals, and harmful chemicals. Microalgae have been previously explored for the treatment of various wastewaters, such as rubber mill wastewater, palm oil mill effluent (POME), sago starch wastewater, textile wastewater, tannery wastewater, and wastewater containing heavy metals (Priya et al. 2022; Udaiyappan et al. 2017). The resilience of microalgae allows them to prosper even in harsh surroundings, which proves advantageous in the treatment of these wastewater. Several microalgal species, which include *Scenedesmus, Chlamydomonas, Phormidium, Botryococcus, Chlorella*, and *Limnospira*, have demonstrated excellent capabilities in bioremediating nutrients, emerging pollutants, heavy metals and pathogens from wastewater (Priya et al. 2022). Some microalgae species, i.e., *Scenedesmus, Chlorella, Oscillatoria, Euglena, Ankistrodesmus,* and *Chlamydomonas* have grown competently in wastewater and exhibit remarkable tolerance to wastewater toxins (Abdelfattah et al. 2023; Hashmi et al. 2023). Microalgae-based procedures have gained substantial consideration for treating industrial, agro-industrial, municipal, and livestock wastewater. Microalgae could help mitigate the risk of eutrophication by eliminating phosphorus (P) and nitrogen (N) components, making them multifunctional options for biological treatment. They can transform undesirable organic and inorganic substances into valuable biomass (Udaiyappan et al. 2017). A source of food and a feedstock for several goods, microalgae collected from certain treatment ponds play various roles in diverse industries.

Addressing global water shortages is expected to be one of the utmost challenges of the 21$^{st}$ century; microalgae-based WWT has the potential to recover nutrients though generating clean water and can effectively treat agricultural, industrial, and municipal wastewater. Microalgae strains could endure the extreme environments condition of modern industrial (Abdelfattah et al. 2023). The shift toward a circular bioeconomy, with a focus on resource diversification, has led to the reorganization of traditional WWT processes into a low-carbon, integrated biorefinery model capable of accommodating multiple waste streams (Abdelfattah et al. 2023). The growing interest in microalgal WWT can be attributed to changes in how WWT is perceived. Microalgae-based WWT is now a serious competitor to conventional methods, as it has partly overcome major challenges related to nutrient assimilation and microalgae population. This chapter aims to compile recent advances and emerging knowledge in microalgae-based WWT. An overview of this chapter is shown in Figure 12.1.

## 12.2 Characteristics and Composition of Industrial Wastewater (IWW)

Compared to freshwater and marine water and with media like BG11, Chu13 resources are generally used for cultivating microalgal culture. Synthetic wastewater exhibits substantial disparities in its chemical composition and physical characteristics (Chakraborty et al. 2023). In developing nations, the majority of wastewater is generated by domestic, municipal, agricultural, and industrial activities, often without undergoing adequate treatment. Industries, in particular, produce

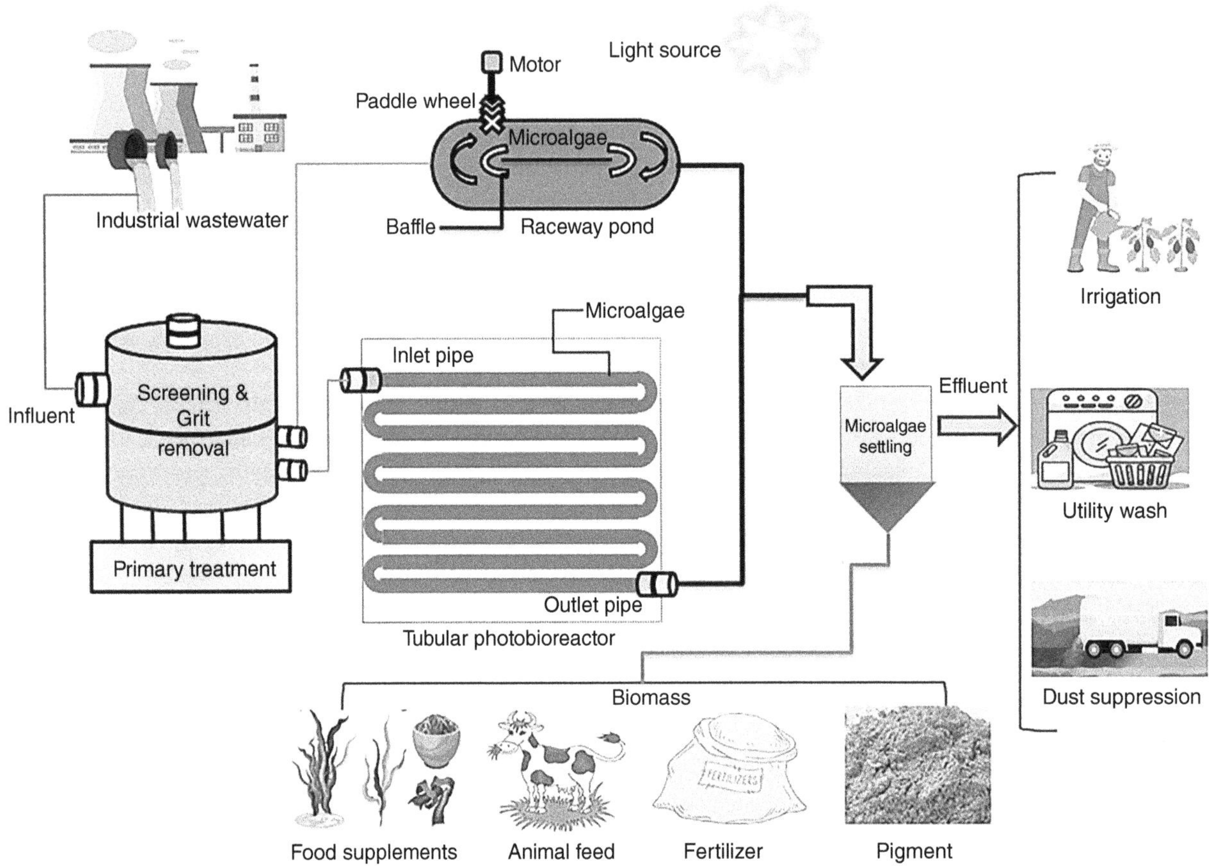


**Figure 12.1** Treatment of industrial wastewater and production biomass for value-added products (VAPs) using microalgae.

substantial volumes of wastewater, which can contain high concentrations of heavy metals and sometimes nanomaterials (Bathi et al. 2021). The specific composition of these industrial wastes depends on the raw materials and processes employed within each industry.

For instance, wastewater from the leather industry typically contains substances like keratin proteins, hair, fats, coloring dyes, and dissolved and suspended solids (Devi et al. 2023). Paper industry waste is characterized by components such as cellulose, lignin, various coloring agents, acids, alkalis, bleaching agents, ammonia, nitrates, benzene, and mercury. The dairy, distillery, and wine industries produce their own sets of waste materials, including casein, lactose, carbohydrates, organic acids, and alcohol. Overall, IWW, originating from manufacturing units and food processing facilities, combines with domestic wastewater, rainwater, and groundwater seepage as it enters the municipal sewage network, forming what is known as municipal wastewater. It is a complex mixture comprising organic matter, several nutrients like nitrogen, phosphorus, and potassium (which can be beneficial for plants), suspended solids, volatile compounds, suspended and colloidal impurities, pathogens like viruses, bacteria, and helminth eggs, biodegradable organics, stable organics, dissolved inorganic substances such as sodium, calcium, magnesium, chloride, and boron, heavy metals like cadmium, lead, nickel, zinc, arsenic, and mercury, and residual chlorine (Wijaya and Soedjono 2018). Numerous studies have demonstrated that using municipal wastewater to grow microalgae can boost biomass output and aid in the removal of contaminants, such as nutrients like nitrate and phosphate (Ahmed et al. 2022). Municipal wastewater contains significant quantities of ammonia ($NH_3$), phosphate ($PO_4^{3-}$), and other essential elements that promote the growth of algal biomass. Additionally, trace metals such as manganese, copper, zinc, and iron aid in the photosynthesis process. Therefore, large-scale microalgae cultivation using conventional municipal wastewater has the potential to remove dissolved nitrogen, phosphorus, and other nutrients, all while producing algal biomass suitable for biofuel production (Ahmed et al. 2022).

Wastewater generated by various industrial sectors typically contains micropollutants, nutrients, i.e., sulfur, phosphorus, nitrogen, and copper, carbon-based pollutants like antibiotics, aromatic hydrocarbons, phenolic compounds, biocides, and surfactants, as well as heavy metals, which include nickel, cadmium, mercury, chromium, lead, copper, and zinc. Heavy metals are especially common in wastewater and pose a risk to aquatic ecosystems and organisms. Any WWT method that leverages microalgae's ability to eliminate pollutants offers the additional use of being ecologically friendly, making it commercially viable (Tkaczyk et al. 2020). Textile wastewater is notably characterized by its pH, intense salinity, color, temperature, total phosphorus (TP), total nitrogen (TN), total dissolved solids (TDS), BOD, COD, and trace heavy metals, for example, arsenic (As), copper (Cu), chromium (Cr), and zinc (Zn) (Rajasimman et al. 2017). The concentrations of nutrients in textile wastewater can vary depending on their source (Kligerman and Bouwer 2015). Agricultural wastewaters, especially those stemming from animal manure, are rich in nutrients. They can also be contaminated by pesticides (organophosphates, carbamates, pyrethroids, dichlorodiphenyltrichloroethane (DDT), fertilizers used to enhance productivity, and metals like selenium, lead, copper, and pathogens, all of which significantly reduce water quality (Mateo-Sagasta et al. 2017).

Animal farms often generate wastewater with high levels of nitrogen and phosphorus. Aside from manure, agricultural runoff may also contain nitrate and phosphate components, pesticides, fungicides, and herbicides resulting from various agricultural activities (Mateo-Sagasta et al. 2017). IWW originates from a range of industries like tannery petroleum, chemical textile, pharmaceutical, pulp and paper factories, and distilleries. The main determinants of IWW features are ammoniacal nitrogen ($NH_4^+$-N), BOD, COD, heavy metals, pH, turbidity, color, and other elements. Furthermore, it includes phenols, volatile organic compounds, aromatics, dyes, fluorinated or chlorinated chemicals, and other poisonous elements (Tkaczyk et al. 2020).

## 12.3 Perks of Microalgae in Wastewater Treatment (WWT)

Microalgae are photosynthetic organisms that are capable of utilizing wastewater as a nutrient source (organic and inorganic contaminants) for their growth and metabolism. The main objective of using wastewater is to replace growth medium (BG11, Zarrouk's medium, etc.) for reduction in expenditure and, at the same time, remediation of wastewater also took place (Chernova et al. 2020). The harvested microalgal biomass generated from wastewater could be used for several highly value-added products. The scarcity of freshwater, 450 billion $m^3$ annually is consumed in industries and domestics (Changming et al. 2001). From this, about 70% of water is used for domestic use only, which is about 315 billion $m^3$ annually (Mekonnen et al. 2015). Hypothetically, if that 70% water is meant for microalgae cultivation, it will produce 23.5 billion tons of oil. Also, the composition of microalgae contains ~50% protein, ~23% lipid, and ~27% of carbohydrate content, which can be further used for various industries like food, nutraceuticals, transport sectors, cosmetics, pharmaceuticals, and textile. (Bhagea et al. 2019).

The advancing in technology, such as high-rate algal ponds (HRAP) and photobioreactors, are being meant for the cultivation of microalgae at large scale and using wastewater as a nutrient growth medium. This technology simultaneously treats wastewater while cultivating microalgae that can be further harvested to obtain desired value-added products (Song et al. 2022). Microalgal WWT follows mechanisms like biosorption, bioaccumulation, and biodegradation (Abdelfattah et al. 2023). While treating wastewater microalgae assimilate nutrients like nitrogen, sulfur, carbon, and phosphates into their cells, which help in development. Ammonium, nitrates, and nitrites are essential nutrient sources for microalgae, while phosphates are also a significant nutrient existing in wastewater. Phosphates have a crucial role as mediates in carbohydrate metabolism (Abdelfattah et al. 2023). Microalgae take up the phosphates and nitrogen source, converting them into ammonium and adenosine diphosphate (ADP) to regulate metabolic activity within the cell. As microalgae grow through photosynthesis, they release oxygen, which is subsequently used up by bacteria, thereby reducing input energy in wastewater systems. Additionally, microalgae also capture $CO_2$ generated by bacterial respiration, effectively sequestering carbon dioxide through the action of the enzyme Carbonic anhydrase (Gonçalves et al. 2017). Past research on Chlorophyta species has been well explored for WWT by utilizing its nutrients. The research is mostly confined to lab scale in a batch process or chemostat mode, where chemostat system is twofold efficient in biomass production. These lab-scale researches are potential explanatory of feasibility and need of microalgal WWT. The high BOD content of wastewater enhances microalgal biomass production (Gonçalves et al. 2017).

However, the presence of other microbes challenges the microalgal biomass productivity. Microalgae being a photosynthetically active organism, heterotrophic mode had also been practiced, which depicts the slow growth of microalgae in devoid of light energy (Chen and Wang 2021; Abinandan and Shanthakumar 2015). Compared to traditional biological treatment methods, the utilization of microalgae in untreated wastewater has several benefits like microalgae functions as flocculants, which improves the sedimentation rate of particles in the wastewater (Javed et al. 2019). Microalgae are highly adaptable to different wastewater conditions, making them well-suited for IWW treatment. Furthermore, industrial gases released during the WWT process, when supplied to microalgae, doubled the benefit by facilitating both $CO_2$ fixation and WWT simultaneously (Kumar and Jujjavarapu 2023).

The core application of cultivating microalgae is biomass production, and it also facilitates avenues for producing value-added products (VAPs) based on the bimolecular composition of microalgal biomass. Microalgae are screened based on their bimolecular makeup for efficient biomass use in WWT and significant potential for commercial applications. The microalgal cultivation of microalgae using nutrient-rich IWW serves as a cost-efficient and readily available medium, as it holds both macronutrients and micronutrients (Kurniawan et al. 2022). For example, *Chlorella pyrenoidosa*, exhibited the uppermost growth at 75% concentration of textile WW, leading to the removal of nitrate (82%), phosphate, (87%), and BOD (63%), respectively (Kumar et al. 2022). Similarly, *Scenedesmus quadricauda* was used to remove antibiotic tetracycline in dairy effluent. In the same WW, the microalga *Tetraselmis suecica* showed less efficient tetracycline removal (Goswami et al. 2021).

## 12.4 Cultivation System for IWW Treatment

Large-scale cultivation of microalgae is a need for sustainable biomass production with cost-effective generation of VAPs. For microalgal cultivation, it is crucial to select the right wastewater type and cultivation system for successful cultivation, WWT and production of biomass. For the biomass application and effective sustainable biofuel production with other feedstocks, a substantial quantity of microalgae biomass is needed (Khan et al. 2022).

However, very limited number of systems are suitable for using wastewater in diluted form as a culture medium. For ideal microalgae cultivation system necessitates precise characteristics, including low installation and operational costs, adequate light supply, and maintenance costs (Song et al. 2022). Cultivation systems can be categorized into two main types: indoor and outdoor cultivation systems, and they can further be categorized into closed and open systems. Both open and closed systems offer various options for large-scale microalgae cultivation, depending on the purpose of cultivation and the intended use of microalgal biomass. Each of these techniques has its own set of pros and cons in terms of cost, biomass production, energy efficiency, and space requirements (Bhatia et al. 2022).

### 12.4.1 Open Cultivation Systems

Open cultivation systems have significant drawbacks as they are susceptible to contamination in both abiotic and biotic ways. Also, adverse weather conditions can hinder microalgae growth, making it not suitable for microalgal cultivation (Bhatia et al. 2022). Microbial contamination from other microalgae is also very communal in open pond systems, and rotifers, in particular, can harm algae cultures of genera such as *Nannochloropsis*, *Tetraselmis*, *Chlorella*, and *Scenedesmus* (Devi et al. 2023). Maintaining parameters such as irradiance, supply evaporation, and temperature is challenging in an open environment. Apart from this, some microalgae species are commercially grown in open ponds due to their lower manufacturing costs. In open cultivation systems, there are two main types: non-stirred and stirred ponds (Kurniawan et al. 2022).

Non-stirred ponds are intermediate to large-scale cultivation systems having shallow depths of $\geq 1.5$ m on average. As open systems, these are exposed to fluctuating abiotic factors, including contamination, biotic interactions, grazing, and competition. However, the contamination risk of instinctive algae and algae grazers confines the amount of microalgae that can be produced in non-stirred ponds (Purushanahalli Shivagangaiah et al. 2021).

On the contrary, stirred ponds are HRAPs like circular ponds or raceway ponds that are usually installed type. Raceway ponds are extensively applied in large-scale microalgal biomass production and accounts ~95% of total microalgae production (Dasan et al. 2019). Microalgal cultivation (in raceway ponds) suitable for some extremophilic microalgae. Rather than being fully exposed to the environment, the extreme culture environment helps exclude certain contaminants. HRAPs are the economically feasible option for cultivating microalgae, which can be further processed into biofuels with a nominal environmental deteriorating effect (Kohlheb et al. 2020). HRAPs also act as a sink for greenhouse gases (GHGs) (Kurniawan et al. 2022). Mostly, algal biomass in open raceway ponds uses synthetic/artificial media, which elevates the production cost, whereas wastewater can be replaced as the medium for producing biomass in HRAPs with some additional advantage of phycoremediation. The productivity of HRAPs can be comparable to photobioreactors (PBRs), with a maximum dry biomass of 20 g/m/d under ideal conditions. However, still the presence of pathogens and zooplankton grazers lowers productivity than predictable (Ali et al. 2022). Cultivation of

*C. pyrenoidosa* in an open raceway pond using domestic wastewater, achieving an average biomass production (1.71 g/l) while reducing TN (95%), COD (78%), and TP (81%) (Dahmani et al. 2016).

### 12.4.2 Closed Cultivation System for IWW Treatment

Tubular and flat-panel reactors are commonly employed for microalgae closed cultivation systems (Jerney and Spilling 2020). Closed cultivation systems have significant attention due minimum risk of contamination, making them particularly beneficial for applications in the food and pharmaceutical industries. Moreover, closed systems offer higher growth rates by optimized irradiance supply and effective water evaporation mitigation compared to open systems (Ozkan et al. 2012).

Tubular PBRs are cylindrical in design, available in both orientations (vertical and horizontal), maximizing surface area to capture sunlight. They are the best choices for large-scale outdoor microalgae cultivation (Sathinathan et al. 2023). Typically, these cylinders range from 10 to 60 mm and extend to 100 m. The flow velocity of the growth medium within these tubes typically falls between 0.2 and 0.5 m/s (Sathinathan et al. 2023). Various pilot-scale experiments have been conducted using tubular PBRs to cultivate various algae, including *Tetraselmis, Haematococcus, Chlorella, Porphyridium, Dunaliella, Limnospira*, and *Phaeodactylum* (Ali et al. 2022).

PBRs (Flat-panel) are energy-efficient and give the best ratio (surface area: volume) with short light paths compared to tubular PBRs (Sathinathan et al. 2023). These reactors usually have cube-like shapes and are constructed from glass or polycarbonate (transparent materials). However, flat-panel systems require more installation area associated to tubular systems, resulting in maintenance costs per unit area and higher operational and installation. Despite that sometimes considered the utmost promising kind for microalgae cultivation systems with drawbacks of biofouling on surfaces and challenging cleaning processes (Sathinathan et al. 2023).

### 12.4.3 Hybrid Cultivation System for IWW Treatment Systems

Hybrid microalgal cultivation systems are a combination of closed and open cultivation systems, having various advantages and challenges. One of their key benefits is the potential for reducing capital expenditure (CAPEX) and operating expenditure (OPEX) (Salama et al. 2017). These systems are least prone to contamination and achieve higher productivity through the optimization of resources and environmental conditions. The lower risk of contamination makes it feasible for applications in food, pharmaceuticals, and other industries with strict quality requirements. Additionally, they are operated, easy to maintain, and monitored with a capacity to effectively absorb carbon dioxide ($CO_2$) from the environment, contributing to environmental sustainability and enhancing their overall efficiency (Mehariya et al. 2021). However, hybrid systems also have certain drawbacks, including the need for extensive infrastructure and increased initial costs. Furthermore, their complex design and operation can pose challenges that require careful planning and management to fully harness their potential (Abinandan & Shanthakumar, 2015).

## 12.5 Algal Nutrient Uptake Mechanisms

Mechanisms for remediation and affecting variables Utilising microalgae to remediation the environment uses a variety of methods, including biosorption, bioaccumulation, and biodegradation (Abdelfattah et al. 2023).

### 12.5.1 Biosorption

Biosorption is energy-deprived process with the use of biological materials to bind and concentrate pollutants from water. It is a mass transfer process where substances are removed after the liquid phase and adhere toward solid surface material. Biosorption encompasses chemical, physical, and metabolic-independent processes, each maintained by diverse mechanisms such as absorption, ion exchange, precipitation, electrostatic interaction, surface complexation, and adsorption (Singh and Saini 2020). The key components of this phenomenon are a biosorbent (a solid phase sorbent) and a target sorbate present in the dispersed or dissolved form in water. The biomaterial intricate can be either living or nonliving microbes or even their constituents. The attraction between biosorbent and target sorbate is the driving force for adhesion, while the total capacity dictates the number of sorbate molecules adsorbed (Abdelfattah et al. 2023). This process continues until reaches equilibrium is established within the adsorbed material on the biosorbent and its residual concentration in the liquid phase. Additionally, the biosorbent's affinity for a particular sorbate determines its dispersal between the phases (solid and liquid) (Singh and Saini 2020). The first discovery of biosorption in several microalgae was observed in the early 1970s when heavy metals and radionuclides released from a nuclear reactor were concentrated within these microorganisms (Abdelfattah et al. 2023). The microalgae cell wall contributes a central role in biosorption, and its chemical composition mainly influences the process and the underlying mechanism (Spain et al. 2021). Also, microalgae possess surface pores and surface charge, which aids in biosorption. Various chemical groups existing in the microalgal cell wall, including hydroxyl, sulfate, and carboxyl, serve as binding sites and function as active ion exchangers, facilitating the complexation of metal ions and adsorption of organic constituents from wastewater (Saravanan et al. 2022).

Additionally, molecules, i.e., lipids, proteins, and nucleic acids, might be present on the cell surface, with their many efficient groups (aminic, carboxylic, imidazolic, thiolic, thioesteric, and oxygen in peptide links) accomplished of binding to metal cations (Abdelfattah et al. 2023). The cell wall typically consists of a fibril matrix that imparts mechanical strength and an amorphous fraction that provides flexibility. Both these fractions, along with intercellular spaces on a cell wall, enhance the biosorption process (Abdelfattah et al. 2023). Moreover, the surface roughness and porosity of microalgae could affect the adsorption of heavy metals (Marella et al. 2020). Smaller cell sizes result in a greater surface area for adsorption per unit biomass, resulting to an increased number of active sites for biosorption. However, higher biomass does not certainly equate to greater biosorption properties, as the presence of more microalgae surfaces can lead to active site repulsion, reducing the removal efficiency of metal cations (Dönmez and Aksu 2002).

Heavy metals, i.e., boron (B), cobalt (Co), copper (Cu), iron (Fe), molybdenum (Mo), manganese (Mn), zinc (Zn), and trace elements are vital for enzymatic processes and metabolism. Although some of these elements are advantageous at trace levels, others, i.e., arsenic (As), cadmium (Cd), chromium (Cr), mercury (Hg), and lead (Pb), are usually toxic to microalgae. Interestingly, trace levels of toxic metal ions can enhance the growth and metabolism of microalgae due to a phenomenon known as hormesis (Alloway 2013).

Additionally, certain contaminants in water can form compounds with active binding sites on the cell surface, supporting flocculation and lowering levels of total dissolved and suspended particles (Kumar et al. 2015).

The biosorption of heavy metal ions by microalgae is a two-step mechanism. The first step is a metabolism-independent procedure characterized by the quick and reversible binding of the adsorbate to active sites on the surface of microalgae. This is followed by a second and slower stage involving positive intracellular diffusion, primarily driven by the metabolic motion on microalgae. Microalgae are capable to capture anionic and cationic species of heavy metals due to the abundance of deprotonated carboxyl and sulfate groups, as well as monomeric alcohols on their surface, which facilitates biosorption (Abdelfattah et al. 2023). The extracellular polymeric substances (EPS) derivatives from microalgae could accelerate metal biosorption, with their effectiveness depending on solution properties, metal species, EPS structure, and operational conditions (Xiao and Zheng 2016). The interaction between microalgae and heavy metals is complex and extremely dependent on factors, i.e., pH, temperature, and competition among metal ions (Abdelfattah et al. 2023). The pH level typically ranges from (3.0–6.5), with values less than 3.0 causing competition between dissociated $H^+$ ions and active sites on the cell surface, while values above 6.5 can lead to the precipitation of metal hydroxides, rendering the biosorbents ineffective and subsequently in the accretion of pollutant sludge (Abdelfattah et al. 2023). Temperature variations have been observed to influence heavy metal adsorption differently in various studies, with either positive or negative effects, making it a factor to consider in WWT applications (Abdelfattah et al. 2023).

### 12.5.2 Bioaccumulation

Bioaccumulation, unlike biosorption, is an energy-mediated metabolic process that involves the utilization of various substrates within the cell. It demands energy and proceeds at a slower pace compared to biosorption. Living cells use bioaccumulation to detoxify waste by absorbing substances and either accumulating or metabolizing them. This process plays a crucial role in removing both organic pollutants and inorganic contaminants. Contaminants such as sulfates, phosphates, nitrates, heavy metals, and pesticides are eliminated by transporting them into microalgal cells (Abdelfattah et al. 2023). Mechanisms followed by microalgae to remediate wastewater can be done by bioaccumulation or biosorption, which are significantly different. Quantifying the biosorbed or bioaccumulated contaminants is very challenging due to their interchangeable properties (Gondi et al. 2022). Microalgae can endure contaminants even at very small concentrations because of their adaptability toward the surrounding environment and substantial tolerance to variety of pollutants released from industrial sectors (Abdelfattah et al. 2023). The bioconcentration factor is a mode of calculating bioaccumulation, which is defined by the concentration of a particular contaminate present in the medium to the concentration absorbed. But it is notable that it is not clear that all the bioadsorbed contaminants could be bioaccumulated (Abdelfattah et al. 2023).

For example, microalgae *Spirogyra* could able to absorb radioactive phosphorus at rates of 850 000 times higher than surrounding water (Krumholz et al. 1954). Also, *Cladophora* sp. is able to accumulate up to 50–400 ng $g^{-1}$ of triclocarban and triclosan (lipophilic organic compounds) used as antimicrobial compounds surpassing concentrations found in environmental not more than 200 ng $Kg^{-1}$ (Abdelfattah et al. 2023).

Contaminants move from a high concentration (external) gradient to a low concentration (internal) without requiring energy (Abdelfattah et al. 2023). Due to the hydrophobic nature of the cell membrane, lipid-soluble materials and low-molecular-weight nonpolar can enter the cell membrane through the process of passive diffusion. In disparity, polar compounds, molecules (molecular weight) are unable to diffuse through the cell wall to cannot diffuse through the cell membrane (Vo et al. 2019).

Changes in permeability of cell membrane ether due to contaminants or environmental stress lead to depolarization or hyperpolarization of cell membrane. Exposure of *Anabaena* CPB4337 to perfluorinated alkyl acid compounds altered its sensitivity toward herbicides, making some herbicides more or less harmful (Rodea-Palomares et al. 2015). Also, *Chlorella vulgaris enhances* bioaccumulation levofloxacin by adding 1% (w/v) NaCl from 34 to 101 mg $g^{-1}$ (Rodea-Palomares et al. 2015). Passive diffusion involves transporter proteins that facilitate the entry of polar molecules into the cell. However, active transport along the cell wall requires energy to move contaminations against the concentration gradient (Vo et al. 2019).

Apart from this, physicochemical factors (external or internal) like temperature, pH, contact time, and pollutant concentration affect bioaccumulation significantly (Abdelfattah et al. 2023). For example, variation in temperature influences the bioaccumulation of antibiotics in the planktonic food web (Tang et al. 2020). There is a biphasic relationship between bioaccumulation and temperature, varying with the specific pollutant. *Scenedesmus obliquus* and *Chlamydomonas mexicana* improved of carbamazepine bioaccumulation with increase in carbamazepine concentration and cultivation time (Kurade et al. 2016). However, some challenges like accumulated substances stimulate the production of reactive oxygen species (ROS), including free radicals (superoxide radical ($O_2$), alkoxy radicals, hydroxyl radical (-OH), singlet oxygen ($^1O_2$), perhydroxyl radical ($HO_2$), and non-radicals like hydrogen peroxide ($H_2O_2$), which can profoundly affect cellular metabolism, potentially causing oxidative cellular dysfunction, damage to biomolecules, and eventually cell death (Abdelfattah et al. 2023). To improve the efficiency of microalgal bioremediation, optimizing physicochemical environment is essential due to the rate of microalgal capacity dependence on these parameters in the bioaccumulation process. Additionally, screening of microalgal species that are tolerant to elevated pollutant concentrations is an effective way to achieve higher bioaccumulation rates and capacities (Gonçalves et al. 2017).

### 12.5.3 Biodegradation

Biodegradation is the most effective method for eliminating pollutants from effluents. In this process, complex compounds break down into simpler and less toxic states (Abdelfattah et al. 2023). Unlike bioaccumulation and biosorption, which need microorganisms acting like biological filters to separate pollutants from the wastewater, bioremediation involves the breakdown of contaminants using biotransformation or complete mineralization to $H_2O$ and $CO_2$ which includes an array of enzymatic reactions resulting in various metabolic intermediates (Abdelfattah et al. 2023).

The basic mechanisms of biodegradation are categorized into two groups first one is metabolic degradation, where contaminants act as carbon sources and electron donors/acceptors for microalgae, and the second is cometabolism, where contaminants act as both electron donors and carbon sources for non-living matter (Song et al. 2022). The biodegradation process through microalgae can occur extracellularly, intracellularly, or through a mixture of both processes with extracellular breakdown leading to intracellular decomposition of breakdown intermediates (Song et al. 2022). The Extracellular degradation depends on the EPS primarily excreted by microalgae, which remain limited to outside of the cell, permitting microalgae to mineralize pollutants in their dissolved state and serve as an external digestive system. EPS also works in enhancing pollutant bioavailability and acts as a biosurfactant facilitating bioaccumulation in microalgae. The precise mechanisms governing the interaction between EPS and pollutants emitted utilized by microalgae are still unknown, necessitating further research (Xiao and Zheng 2016).

The biodegradation of organic pollutants by microalgae involves three stages of complex enzymatic processes. Stage I entails cytochrome P450 enzymes, including monooxygenase, hydroxylase, carboxylase, and decarboxylase, which increase the hydrophilicity of pollutants by adding or exposing hydroxyl groups through oxidation–reduction reactions or hydrolysis (Abdelfattah et al. 2023). In Stage II, enzymes such as glucosyltransferases and glutathione-S-transferases promote the conjugation of pollutants with glutathione, safeguarding cells against oxidative damage (Abdelfattah et al. 2023). Stage III is the detoxification of various enzymes like glutamyl-tRNA reductase, dehydrogenase, mono(di)oxygenase, carboxylase, laccases, pyrophosphatase, transferase, dehydratase and hydrolases, which biotransform molecules into less/nontoxic intermediates (Abdelfattah et al. 2023).

Cometabolism technique degraded contaminants by cofactor or an enzyme produced through microbial metabolism of another compound. To enhance the microalgae-based biodegradation of recalcitrant pollutants (ROPs), an effective approach of introducing additional organic substrates or nutrient substances to start cometabolic system (Tran et al. 2013). For example, the introduction of sodium acetate ($CH_3COONa$) showed a significant boost in removal efficiency from 13% to 56% of ciprofloxacin by *Ceratozamia Mexicana* (Ghaffar et al. 2023). This improvement in removal efficiency, along with the control of transformation products and pathways, was basically associated with the concentration of acetate. Remarkably, the concentration and the type of nutrient substance had the most significant impact among various factors (Abdelfattah et al. 2023).

The impact of carbon sources on cometabolism of pollutants using microalgae has been extensively investigated. While the findings confirmed that sugar-based carbon sources led to the highest EPS production, optimal removal efficiency and enzyme concentrations, the presence of certain organic substrates might reduce the removal efficiency of micropollutants, possibly due to catabolite repression (Xiao and Zheng 2016). Therefore, it is crucial to evaluate the effects of various carbon sources to select the most suitable source and ensure its suitability for using in the pollutants biodegradation in wastewater. In the meantime, this innovation can potentially enhance biomass recovery and generate revenue across various sectors.

In general, it has been observed that the main mechanisms (biosorption, bioaccumulation, and biodegradation) can work in tandem in microalgal WWT. Biodegraded particles could be adsorbed onto the cell membrane of the microalgae, and these particles into the cells through active processes. Subsequently, enzymatic reactions facilitate precipitation, biotransformation, and bioaccumulation within the cells (Abdelfattah et al. 2023).

## 12.6 Bioremediation of Industrial Effluents

The prime causes of water contamination could be attributed to the Industrial Revolution and technological advancements, which have headed to the indirect or direct release of toxic pollutants into natural water bodies. Various treatment methods, including membrane technology, activated sludge, electrochemical treatments, and advanced oxidation, have been developed to address these contaminations (Priya et al. 2022). However, mostly these treatment techniques had their own drawbacks like high operational and maintenance costs, incomplete removal of heavy metals (HMs), and the generation of toxic sludge or waste by-products. Microalgae-based WWT offers a promising, environment-friendly, sludge-free, and sustainable solution. Microalgae could metabolize hazardous substances to non/less toxic compounds, making it an inexpensive process that simultaneously purifies wastewater and allows for biomass harvesting, which can be further utilized for various VAPs production. This section focuses on the application of microalgae in IWW treatment for toxic contaminants, including HMs, pharmaceuticals, and dyes (Abdelfattah et al. 2023). Various contaminant removals by using microalgae are shown in Table 12.1.

### 12.6.1 Removal of Heavy Metals

HMs are a group of metals with high atomic densities and are nonbiodegradable. They have a significant pollution issue globally due to their toxicity, nonbiodegradability and tendency to accumulate in the environment. HMs can be categorized into three groups: radionuclides, precious metals, and toxic metals. These contaminants enter aquatic systems by agricultural runoff and industrial discharges (Priya et al. 2022). Conventional methods for removing HMs from wastewater are often costly and generate secondary waste. Thus, there is a need for eco-friendly and economically viable solutions. Microalgae offer a promising approach for HM removal due to their unique cell wall properties, which contain functional groups like amino, hydroxyl, carboxyl, and sulfate that can bind with HMs. Various species of freshwater microalgae, including *C. vulgaris, Chlorella miniata, Sphaeroplea,* and *C. reinhardtii,* have demonstrated successful removal of toxic HMs from wastewater. The removal efficiency depends on the properties of the microalgae, the type of HMs, and effluent characeristics (Abdelfattah et al. 2023). *Anabaena sphaerica* demonstrated its biosorption capability in reducing cadmium and lead concentrations, highlighting its potential in HM remediation (Priya et al. 2022).

**Table 12.1** Various microalgae used in different wastewater with removal efficiency of targeted pollutant.

| Wastewater type | Microalgae | Pollutant | Removal efficiency | References |
|---|---|---|---|---|
| Urban WW | *Chlorella* sp. and *Scenedesmus* sp. | Carbamazepine | 20% | Matamoros et al. (2016) |
| Textile effluent | *Oscillatoria* sp. | Blue dye, red dye | 76.48% and 62.63% | Brahmbhatt and Jasrai (2016) |
| Textile effluent | *Spirogyra* sp. | Blue dye, red dye | 78.29% and 64.21% | |
| Textile wastewater | *Cosmarium* sp. | Malachite green (conc. $10\,\text{mg}\,\text{l}^{-1}$) | 87.1% | Daneshvar et al. (2005) |
| Textile wastewater | *Spirogyra* sp. | Synazol | 85% | Khalaf (2008) |
| Electroplating and galvanizing industrial effluents | *Chlorella vulgaris* (Non-living) | Chromium (conc. $50\,\text{mg}\,\text{l}^{-1}$) | – | Sibi (2016) |
| Piggery wastewater | *Chlamydomonas mexicana* | Total nitrogen and phosphate | 62% and 28% | Abou-Shanab et al. (2013) |
| Dairy wastewater | *Chlorella* sp. | Total nitrogen and phosphate | 75.7–82.5%, 62.5–74.7% | Wang et al. (2010) |
| Mariculture wastewater | *Chlorella vulgaris* | Ammonical nitrogen, phosphate, COD | 70.70%, 90.00%, 98.10% | Gao et al. (2021) |
| Urine, anaerobically treated black water and synthetic urine | *Chlorella sorokiniana* | Ibuprofen, paracetamol metoprolol, carbamazepine, trimethoprim | 100%, 100%, 100%, 30%, and 40% | de Wilt et al. (2016) |

Furthermore, *Pseudochlorococcum typicum* was found to be efficient in biosorbing mercury and lead from wastewater, emphasizing the role of living microalgae in HM removal (Dhar et al. 2021). Also, *Maugeotia genuflexa* exhibited biosorption capabilities in reducing arsenic concentrations in contaminated water (Hadiani et al. 2019). These studies collectively demonstrate the potential of microalgae-based remediation techniques to effectively remove HMs from wastewater.

### 12.6.2 Removal of Pharmaceuticals

Pharmaceutical compounds (PCs) are extensively used worldwide, and their residues find their way into aquatic environments through domestic sources, agricultural and industrial. Even at low concentrations, PCs can negatively impact microbial communities and ecosystems. Microalgae-based PC bioremediation is gaining interest due to its solar irradiation usage, low cost, and environmentally friendly nature. Microalgae can surpass some limitations related with bacteria and fungi when it on the process of pollutant breakdown. Various species of microalgae have been found to remove and absorb PCs from wastewater (Kumar et al. 2023). The mechanisms involved by microalgae in eliminating PCs include biosorption, bioaccumulation, and biodegradation, with their effectiveness depending on the properties (physicochemical) of the aimed PCs. Overall, microalgae offer a promising solution for the removal of PCs from wastewater, contributing to both environmental clean-up and sustainable biomass production. The bioaccumulation of carbamazepine, trimethoprim, florfenicol and sulfamethoxazole has been reported by passive diffusion in microalgae cells (Kumar et al. 2023).

### 12.6.3 Removal of Dyes

Dyes are widely used in many industries contributing to aesthetic but environmental problems when discharged into ecosystems. Phycoremediation technology employed microalgae to dye contaminated wastewater to clean it and has attained prominence. Microalgae possess cell walls with functional groups that help in dye removal. Several factors like pH, initial dye concentration, contact time, adsorbent, and dosage temperature influence the efficacy of microalgal dye removal (Kapoor et al. 2021). Traditional methods for dye WWT involve chemical, physicochemical, and biological approaches, but they have limitations such as ineffective dye removal and the generation of sludge. Bioremediation, particularly phycoremediation, offers an alternative that is environmentally friendly and cost-effective (Kapoor et al. 2021).

## 12.7 Recovery of Valuable Nutrients

The cultivation of microalgae and harvesting biomass using wastewater offers a sustainable solution with multiple potential applications, including biofuel production, fertilizers/biofertilizers, and animal feed (Song et al. 2022). The expenses associated with microalgal production are typically by the initial investment in WWT plant infrastructure (Song et al. 2022). This approach not only proves cost-effective but also reduces the environmental impact by utilizing wastewater instead of freshwater (Ali et al. 2022). Several research studies have indicated that microalgal biomass can serve as a valuable resource for various purposes, such as biodiesel production, biofuel production through lipid extraction, energy generation, animal feed, and the extraction of bioactive compounds (Ali et al. 2022).

### 12.7.1 Fertilizers and Biofertilizers

Microalgal biomass could act as a valuable resource for enhancing the mineral content of nutrient-deficient soils and improving their water-retaining capacity (Song et al. 2022). In the realm of agro-industries, this approach exhibits the potential to reduce conventional fertilizer expenses and disadvantages by recovering essential nutrients back into the soil. However, if wastewater is used directly and deprived of appropriate treatment, it may have adverse effects on soil health. For instance, the use of untreated sago wastewater for irrigation as a fertilizer has been observed to result in less growth, yield, and overall soil health (Song et al. 2022).

Biofertilizers, comprised of microalgae, play a pivotal role in providing nitrogen to plants, essentially functioning as miniature nitrogen-fixing factories within the soil. These biofertilizers commonly include nitrogen-fixing cyanobacteria like *Anabaena* sp., *Nostoc* sp., *Scytonema* sp., *Tolypothrix* sp., *Aulosira* sp. and *Plectonema* sp. (Oinam et al. 2015). Furthermore, macroalgae have gained attention as plant growth enhancers recently. Extracts and substances derived from macroalgae have been employed to stimulate processes such as stem or leaf growth, germination, and flowering in plants.

### 12.7.2 Animal Feed

Animal health is profoundly affected by the quality of their diet, as it plays a crucial role in their well-being. Research has explored the potential of microalgae as an animal feed source due to their rich composition of carbohydrates, proteins, and essential nutrients, which are vital for promoting animal growth (Song et al. 2022).

Moreover, microalgae stand out for their higher protein content compared to traditional land-based plants. For instance, Pulz and Gross (2004) found that *Isochrysis galbana*, when cultivated in open ponds, yields more fatty acids appropriate for animal nutrition. Similarly, Phang et al. (2000) discovered that *Spirulina* when cultivated in an HRAP using sago wastewater, can serve as high-end animal feed, particularly in the aquaculture sector. Given that fish contribute to a significant portion of global animal protein consumption, incorporating microalgae into their diet has the potential to enhance the quality of fish protein (Hua et al. 2019).

### 12.7.3 Biofuel

The global demand for energy continues to rise; a substantial 88% of this demand is currently met by fossil fuels. Unfortunately, the usage of fossil fuels contributes to greenhouse gas emissions (Moravvej et al. 2019). On the other hand, biofuels derived from microalgae act as promising candidates for reducing these emissions. Biofuels encompass various types, such as biohydrogen, biodiesel, bioethanol, and biomethane derived from microalgal biomass. The increased interest in biofuels is due to the concerns over the exhaustion of fossil fuel resources due to over-extraction and the associated raised levels of $CO_2$ emissions (Song et al. 2022). Microalgae are particularly excellent for biofuel production because of their remarkable attributes, like highly efficient photosynthesis, rapid growth rates, and the ability to thrive in non-arable land, saline water, and wastewater environments (Ali et al. 2022).

Fatty acid methyl esters and biodiesel are generally derived from oils from biomass, primarily from plant oils. However, first-generation biofuels are outstretched due to competing with global food resources (Moravvej et al. 2019). Rapeseed, palm, and soybean are oil-producing plants, but these crops demand vast expanses of land for cultivation (Moravvej et al. 2019). The second-generation biofuels use non-food crops such as *Jatropha* plant, but its slow growth rate and high demand for cultivation land make it economically unviable (Moravvej et al. 2019).

Microalgae come under the third generation of biofuels due to their elevated lipid-producing capabilities, their noncompetitive food-related issues and their small land requirement makes them win over the other two generations of biofuels (Wahlen et al. 2011). They have the potential to store a substantial amount of lipids that could be converted into biodiesel by process of transesterification. Certain microalgae species can accumulate as much as 20–50% (w/w) triglycerides of total dry weight (Wahlen et al. 2011). Algal biofuel is are potential replacement of petrol and diesel, although further advancements are needed in the efficient extraction of oil from microalgae.

Biogas production by microalga as substrate, particularly biomethane and biohydrogen, is considered as best energy-efficient method (Schenk et al. 2008).

In biomethane and biohydrogen production, microalgae plays a very important role as a source of carbon for methanogens having hydrogenase enzymes. Also, some microalgae have the ability to liberate hydrogen directly. Like *Chlamydomonas reinhardtii*, under sulfur-deprived conditions can switch to hydrogen production from oxygen production (Torzillo et al. 2015). However, further exploration is required to identify other strains with similar hydrogen-producing capabilities. In ethanol production, microalgal biomass could be served as a carbon source for ethanol-producing yeast or bacteria. Hence, microalgae themselves did not produce ethanol but instead used as a crucial carbon source in the ethanol production process (Torzillo et al. 2015).

### 12.7.4 Bioactive Compounds

Microalgae are recognized for their ability to produce secondary metabolites that can be applied in pharmaceuticals and food supplements, including compounds like astaxanthin, omega-3 fatty acids, sterols, enzymes, proteins, pigments and vitamins (Skjånes et al. 2013). Primary metabolites are essential components for growth and building proteins, carbohydrates and lipids. On the contrary, secondary metabolites, often referred to as bioactive compounds, have different functions, like defense tactics. These metabolites can be harnessed and operated to develop dietary supplements or valuable drugs (Park et al. 2022).

Antioxidants are the valuable compounds produced by microalgae (Banskota et al. 2019). Due to the phototrophic nature of microalgae it is exposed to radical stresses and elevated oxygen levels. Consequently, they have evolved highly efficient protection against oxidative stressors and radicals. This system allows microalgae to eliminate the accumulated free radicals and reactive oxygen species (ROS), thereby safeguarding their cells from damage. However, these bioactive compounds represent a small fraction of the overall microalgal biomass (Banskota et al. 2019).

## 12.8 Future Directions and Research Frontiers

Microalgal WWT is still facing challenges such as contamination by other microorganisms, improper downstream processing, low biomass productivity, difficult harvesting requirements, complex monitoring and operation systems, and high demand for specific nutrients.

Most of the research is still on a pilot scale. Industrial-scale cultivation will expose microalgae to varied seasonal ecological and environmental conditions and variations on the composition or concentration of wastewater. The variations and spontaneous conditions may reduce the biomass yield and treatment efficiency. To tackle this problem, resilient microalgal consortia with greater adaption capabilities need to be used to challenge environmental conditions.

Also, some primary challenges faced by industrial-scale microalgal WWT plants are the inhibition of microalgal development because of turbidity and the presence of hazardous pollutants in the wastewater. To encounter this, it is recommended to implement influent control measures and, if necessary, pretreatment processes prior to microalgae introduction (Abinandan and Shanthakumar 2015). Additionally, various ranges of nutrients (contaminants of wastewater) and their varying concentrations in wastewater demand multiple pretreatment stages to attain an optimal balanced nutrient for optimum growth of microalgae. In such scenarios, researchers should consider using optimized blends of different wastewater sources to create a well-balanced nutrient medium for cultivating microalgae. One of the noteworthy hurdles in operating large-scale microalgal WWT plants, particularly those employing microalgae-based processes, is the complexity associated is regulating the entire operation. Maintaining factors like temperature, pH levels, BOD, and DO is very critical and time-consuming. To address this challenge, innovative technologies like online monitoring systems and remote control capabilities need development. Evaluating the economic viability of microalgal WWT and comparing it to conventional methods is very important. Due to the absence of a large area of research, the absence of fundamental guidelines for designing and operating microalgae-based WWT systems are area for huge development and research.

In the above-mentioned studies, researchers utilized microalgae to reduce common pollutants like nutrients and heavy metals. However, it is imperative to expand the use of microalgae for the elimination of additional contaminants, such as dyes, other types of heavy metals, and emerging pollutants (personal care products and antibiotics). Moreover, numerous investigations are concentrated on employing microalgae to extract excess nutrients from secondary treated wastewater. Consequently, further research is required to integrate microalgae as a biological treatment process to gain a comprehensive understanding of the interactions between the existing other microbes in wastewater and microalgae.

## 12.9 Conclusion

Various type of IWW (effluents of textile tannery food industries, pharmaceutical, printing, and electroplating industries etc.) possesses a variety of contaminants as well as nutrient sources. A number of promising research are going on to utilize wastewater as a nutrient source for microalgal cultivation. Being an autotrophic and huge biomass producer, microalgae can be a game changer in terms of wastewater remediation and biomass production, which can serve as a raw material for VAPs generation (bioactive compounds, biofuel, biofertilizers, nutraceuticals etc.). Based on the past research microalgal-based treatment is outperformed as compared to the conventional techniques with no sludge generation, minimizing secondary pollutants, carbon capture, and HMs removal. However, certain challenges in cultivation, contamination, and bio-refining need to be still researched to make it economically available.

## References

Abdelfattah, A., Ali, S.S., Ramadan, H. et al. (2023). Microalgae-based wastewater treatment: mechanisms, challenges, recent advances, and future prospects. *Environmental Science and Ecotechnology* 13: 100205.

Abinandan, S. and Shanthakumar, S. (2015). Challenges and opportunities in application of microalgae (Chlorophyta) for wastewater treatment: a review. *Renewable and Sustainable Energy Reviews* 52: 123–132.

Abou-Shanab, R.A., Ji, M.K., Kim, H.C. et al. (2013). Microalgal species growing on piggery wastewater as a valuable candidate for nutrient removal and biodiesel production. *Journal of Environmental Management* 115: 257–264.

Ahmed, S.F., Mofijur, M., Parisa, T.A. et al. (2022). Progress and challenges of contaminate removal from wastewater using microalgae biomass. *Chemosphere* 286: 131656.

Ali, S.S., El-Sheekh, M., Manni, A. et al. (2022). Microalgae-mediated wastewater treatment for biofuels production: a comprehensive review. *Microbiological Research* 265: 127187.

Alloway, B.J. (2013). Heavy metals and metalloids as micronutrients for plants and animals. In: *Heavy Metals in Soils: Trace Metals and Metalloids in Soils and Their Bioavailability* (ed. B.J. Alloway), 195–209. Dordrecht: Springer.

Awuchi, C.G., Hannington, T., Awuchi, C.G. et al. (2020). Industrial waste management, treatment, and health issues: wastewater, solid, and electronic wastes. *European Academic Research* 8 (2): 1081–1119.

Banskota, A.H., Sperker, S., Stefanova, R. et al. (2019). Antioxidant properties and lipid composition of selected microalgae. *Journal of Applied Phycology* 31: 309–318.

Bathi, J.R., Moazeni, F., Upadhyayula, V.K. et al. (2021). Behavior of engineered nanoparticles in aquatic environmental samples: current status and challenges. *Science of the Total Environment* 793: 148560.

Bhagea, R., Bhoyroo, V., and Puchooa, D. (2019). Microalgae: the next best alternative to fossil fuels after biomass. A review. *Microbiology Research* 10 (1): 7936.

Bhatia, S.K., Ahuja, V., Chandel, N. et al. (2022). Advances in algal biomass pretreatment and its valorisation into biochemical and bioenergy by the microbial processes. *Bioresource Technology* 358: 127437.

Brahmbhatt, N.H. and Jasrai, R.T. (2016). The role of algae in bioremediation of textile effluent. *Methods* 21: 28.

Chakraborty, B., Gayen, K., and Bhowmick, T.K. (2023). Transition from synthetic to alternative media for microalgae cultivation: a critical review. *Science of the Total Environment* 897: 165412.

Changming, L., Jingjie, Y., and Kendy, E. (2001). Groundwater exploitation and its impact on the environment in the North China Plain. *Water International* 26 (2): 265–272.

Chapman, R.L. (2013). Algae: the world's most important "plants"—an introduction. *Mitigation and Adaptation Strategies for Global Change* 18: 5–12.

Chen, H. and Wang, Q. (2021). Regulatory mechanisms of lipid biosynthesis in microalgae. *Biological Reviews* 96 (5): 2373–2391.

Chernova, N., Kiseleva, S., Vlaskin, M., and Grigorenko, A. (2020). Recycling of aqueous phase from hydrothermal liquefaction and municipal wastewater by microalgae. In: *IOP Conference Series: Materials Science and Engineering*. 877 (1), 012045.

Dahmani, S., Zerrouki, D., Ramanna, L. et al. (2016). Cultivation of *Chlorella pyrenoidosa* in outdoor open raceway pond using domestic wastewater as medium in arid desert region. *Bioresource Technology* 219: 749–752.

Daneshvar, N., Ayazloo, M., Khataee, A.R., and Pourhassan, M. (2005). Biodegradation of the textile dye malachite green by *Microalgae Cosmarium* sp. *International Journal for Science and high Technology and Environmental Sciences* 1: 1–7.

Dasan, Y.K., Lam, M.K., Yusup, S. et al. (2019). Life cycle evaluation of microalgae biofuels production: effect of cultivation system on energy, carbon emission and cost balance analysis. *Science of the Total Environment* 688: 112–128.

Devi, A., Verma, M., Saratale, G.D. et al. (2023). Microalgae: a green eco-friendly agents for bioremediation of tannery wastewater with simultaneous production of value-added products. *Chemosphere* 336: 139192.

Dhar, S., Das, A., and Bhattacharjee, P. (2021). Heavy metal removal using microbial bioremediation techniques. In: *New Trends in Removal of Heavy Metals from Industrial Wastewater* (ed. M.P. Shah, S.R. Couto, and V. Kumar), 649–673. Elsevier.

Dönmez, G. and Aksu, Z. (2002). Removal of chromium (VI) from saline wastewaters by *Dunaliella* species. *Process Biochemistry* 38 (5): 751–762.

Gao, Y., Guo, L., Liao, Q. et al. (2021). Mariculture wastewater treatment with Bacterial-Algal Coupling System (BACS): effect of light intensity on microalgal biomass production and nutrient removal. *Environmental Research* 201: 111578.

Ghaffar, I., Hussain, A., Hasan, A., and Deepanraj, B. (2023). Microalgal-induced remediation of wastewaters loaded with organic and inorganic pollutants: an overview. *Chemosphere* 320: 137921.

Gonçalves, A.L., Pires, J.C., and Simões, M. (2017). A review on the use of microalgal consortia for wastewater treatment. *Algal Research* 24: 403–415.

Gondi, R., Kavitha, S., Kannah, R.Y. et al. (2022). Algal-based system for removal of emerging pollutants from wastewater: a review. *Bioresource Technology* 344: 126245.

Goswami, R.K., Agrawal, K., Mehariya, S., and Verma, P. (2021). Current perspective on wastewater treatment using photobioreactor for *Tetraselmis* sp.: an emerging and foreseeable sustainable approach. *Environmental Science and Pollution Research* 29 (41): 61905–61937.

Hadiani, M.R., Khosravi-Darani, K., and Rahimifard, N. (2019). Optimization of As (III) and As (V) removal by *Saccharomyces cerevisiae* biomass for biosorption of critical levels in the food and water resources. *Journal of Environmental Chemical Engineering* 7 (2): 102949.

Hafizur, R.M., Nuralam, H.M., and Rumainul, I.M. (2017). Investigation of physicochemical parameter, heavy metal in Turag river water and adjacent industrial effluent in Bangladesh. *Journal of Science, Technology and Environment Informatics* 5 (1): 347–360.

Hashmi, Z., Bilad, M.R., Fahrurrozi et al. (2023). Recent progress in microalgae-based technologies for industrial wastewater treatment. *Fermentation* 9 (3): 311.

Hua, K., Cobcroft, J.M., Cole, A. et al. (2019). The future of aquatic protein: implications for protein sources in aquaculture diets. *One Earth* 1 (3): 316–329.

Javed, F., Aslam, M., Rashid, N. et al. (2019). Microalgae-based biofuels, resource recovery and wastewater treatment: a pathway towards sustainable biorefinery. *Fuel* 255: 115826.

Jerney, J. and Spilling, K. (2020). Large scale cultivation of microalgae: open and closed systems. In: *Biofuels from Algae: Methods and Protocols* (ed. K. Spilling), 1–8. New York, NY: Springer, Humana.

Kanu, I. and Achi, O.K. (2011). Industrial effluents and their impact on water quality of receiving rivers in Nigeria. *Journal of Applied Technology in Environmental Sanitation* 1 (1): 75–86.

Kapoor, R.T., Danish, M., Singh, R.S. et al. (2021). Exploiting microbial biomass in treating azo dyes contaminated wastewater: mechanism of degradation and factors affecting microbial efficiency. *Journal of Water Process Engineering* 43: 102255.

Khalaf, M.A. (2008). Biosorption of reactive dye from textile wastewater by non-viable biomass of *Aspergillus niger* and *Spirogyra* sp. *Bioresource Technology* 99 (14): 6631–6634.

Khan, S., Naushad, M., Iqbal, J. et al. (2022). Challenges and perspectives on innovative technologies for biofuel production and sustainable environmental management. *Fuel* 325: 124845.

Kligerman, D.C. and Bouwer, E.J. (2015). Prospects for biodiesel production from algae-based wastewater treatment in Brazil: a review. *Renewable and Sustainable Energy Reviews* 52: 1834–1846.

Kohlheb, N., van Afferden, M., Lara, E. et al. (2020). Assessing the life-cycle sustainability of algae and bacteria-based wastewater treatment systems: high-rate algae pond and sequencing batch reactor. *Journal of Environmental Management* 264: 110459.

Krumholz, L. A., Authority, T. V., & Norris, T. (1954). 33'W. L. Minckley 111111P W. L. Minckley.

Kumar, K.S., Dahms, H.U., Won, E.J. et al. (2015). Microalgae–a promising tool for heavy metal remediation. *Ecotoxicology and Environmental Safety* 113: 329–352.

Kumar, N., Banerjee, C., Chang, J.S., and Shukla, P. (2022). Valorization of wastewater through microalgae as a prospect for generation of biofuel and high-value products. *Journal of Cleaner Production* 362: 132114.

Kumar, T. and Jujjavarapu, S.E. (2023). Carbon dioxide sequestration and wastewater treatment via an innovative self-sustaining algal microbial fuel cell. *Journal of Cleaner Production* 415: 137836.

Kumar, V., Kumar, P., Singh, J., and Kumar, P. (2020). Current status of water pollution by integrated industrial hubs (IIHs) in India. *Environmental Degradation: Causes and Remediation Strategies* 1 (1): 104.

Kumar, V., Lakkaboyana, S.K., Sharma, N. et al. (2023). A critical assessment of technical advances in pharmaceutical removal from wastewater – a critical review. *Case Studies in Chemical and Environmental Engineering* 100363.

Kurade, M.B., Kim, J.R., Govindwar, S.P., and Jeon, B.H. (2016). Insights into microalgae mediated biodegradation of diazinon by *Chlorella vulgaris*: microalgal tolerance to xenobiotic pollutants and metabolism. *Algal Research* 20: 126–134.

Kurniawan, S.B., Ahmad, A., Imron, M.F. et al. (2022). Potential of microalgae cultivation using nutrient-rich wastewater and harvesting performance by biocoagulants/bioflocculants: mechanism, multi-conversion of biomass into valuable products, and future challenges. *Journal of Cleaner Production* 365: 132806.

Marella, T.K., Saxena, A., and Tiwari, A. (2020). Diatom mediated heavy metal remediation: a review. *Bioresource Technology* 305: 123068.

Matamoros, V., Uggetti, E., García, J., and Bayona, J.M. (2016). Assessment of the mechanisms involved in the removal of emerging contaminants by microalgae from wastewater: a laboratory scale study. *Journal of Hazardous Materials* 301: 197–205.

Mateo-Sagasta, J., Zadeh, S.M., Turral, H., and Burke, J. (2017). Water pollution from agriculture: a global review. *Executive summary. Food and agriculture organization of the united nations* 1: 4–33.

Mehariya, S., Goswami, R.K., Verma, P. et al. (2021). Integrated approach for wastewater treatment and biofuel production in microalgae biorefineries. *Energies* 14 (8): 2282.

Mekonnen, M.M., Pahlow, M., Aldaya, M.M. et al. (2015). Sustainability, efficiency and equitability of water consumption and pollution in Latin America and the Caribbean. *Sustainability* 7 (2): 2086–2112.

Moravvej, Z., Makarem, M.A., and Rahimpour, M.R. (2019). The fourth generation of biofuel. In: *Second and Third Generation of Feedstocks* (ed. A. Basile and F. Dalena), 557–597. Elsevier.

Oinam, G., Indira, W., Singh, O.A. et al. (2015). Identification of diazotrophic nostocalean cyanobacteria of north eastern region of India and evaluation for nitrogenase activity and extracellular ammonium excretion. *Journal of Applied Biology and Biotechnology* 3 (6): 001–008.

Ozkan, A., Kinney, K., Katz, L., and Berberoglu, H. (2012). Reduction of water and energy requirement of algae cultivation using an algae biofilm photobioreactor. *Bioresource Technology* 114: 542–548.

Park, Y.H., Han, S.I., Oh, B. et al. (2022). Microalgal secondary metabolite productions as a component of biorefinery: a review. *Bioresource Technology* 344: 126206.

Priya, A.K., Jalil, A.A., Vadivel, S. et al. (2022). Heavy metal remediation from wastewater using microalgae: recent advances and future trends. *Chemosphere* 305: 135375.

Pulz, O. and Gross, W. (2004). Valuable products from biotechnology of microalgae. *Applied Microbiology and Biotechnology* 65: 635–648.

Purushanahalli Shivagangaiah, C., Sanyal, D., Dasgupta, S., and Banik, A. (2021). Phycoremediation and photosynthetic toxicity assessment of lead by two freshwater microalgae *Scenedesmus acutus* and *Chlorella pyrenoidosa*. *Physiologia Plantarum* 173 (1): 246–258.

Rajasimman, M., Babu, S.V., and Rajamohan, N. (2017). Biodegradation of textile dyeing industry wastewater using modified anaerobic sequential batch reactor–start-up, parameter optimization and performance analysis. *Journal of the Taiwan Institute of Chemical Engineers* 72: 171–181.

Rodea-Palomares, I., Makowski, M., Gonzalo, S. et al. (2015). Effect of PFOA/PFOS pre-exposure on the toxicity of the herbicides 2, 4-D, Atrazine, Diuron and Paraquat to a model aquatic photosynthetic microorganism. *Chemosphere* 139: 65–72.

Salama, E.S., Kurade, M.B., Abou-Shanab, R.A. et al. (2017). Recent progress in microalgal biomass production coupled with wastewater treatment for biofuel generation. *Renewable and Sustainable Energy Reviews* 79: 1189–1211.

Saravanan, A., Kumar, P.S., Hemavathy, R.V. et al. (2022). A comprehensive review on sources, analysis and toxicity of environmental pollutants and its removal methods from water environment. *Science of the Total Environment* 812: 152456.

Sathinathan, P., Parab, H.M., Yusoff, R. et al. (2023). Photobioreactor design and parameters essential for algal cultivation using industrial wastewater: a review. *Renewable and Sustainable Energy Reviews* 173: 113096.

Schenk, P.M., Thomas-Hall, S.R., Stephens, E. et al. (2008). Second generation biofuels: high-efficiency microalgae for biodiesel production. *Bioenergy Research* 1: 20–43.

Sibi, G. (2016). Biosorption of chromium from electroplating and galvanizing industrial effluents under extreme conditions using *Chlorella vulgaris*. *Green Energy & Environment* 1 (2): 172–177.

Singh, A. and Saini, P. (2020). Biosorption: a novel biotechnological application for removal of hazardous pollutants. In: *Innovations in Food Technology: Current Perspectives and Future Goals* (ed. P. Mishra, R.R. Mishra, and C.O. Adetunji), 341–360. Singapore: Springer.

Singh, S., Sharma, A., and Malviya, R. (2021). Industrial wastewater: health concern and treatment strategies. *The Open Biology Journal* 9 (1): 1–10.

Skjånes, K., Rebours, C., and Lindblad, P. (2013). Potential for green microalgae to produce hydrogen, pharmaceuticals and other high value products in a combined process. *Critical Reviews in Biotechnology* 33 (2): 172–215.

Song, Y., Wang, L., Qiang, X. et al. (2022). The promising way to treat wastewater by microalgae: approaches, mechanisms, applications and challenges. *Journal of Water Process Engineering* 49: 103012.

Spain, O., Plöhn, M., and Funk, C. (2021). The cell wall of green microalgae and its role in heavy metal removal. *Physiologia Plantarum* 173 (2): 526–535.

Sutherland, D.L. and Ralph, P.J. (2019). Microalgal bioremediation of emerging contaminants-opportunities and challenges. *Water Research* 164: 114921.

Tang, J., Wang, S., Tai, Y. et al. (2020). Evaluation of factors influencing annual occurrence, bioaccumulation, and biomagnification of antibiotics in planktonic food webs of a large subtropical river in South China. *Water Research* 170: 115302.

Tkaczyk, A., Mitrowska, K., and Posyniak, A. (2020). Synthetic organic dyes as contaminants of the aquatic environment and their implications for ecosystems: a review. *Science of the Total Environment* 717: 137222.

Torzillo, G., Scoma, A., Faraloni, C., and Giannelli, L. (2015). Advances in the biotechnology of hydrogen production with the microalga *Chlamydomonas reinhardtii*. *Critical Reviews in Biotechnology* 35 (4): 485–496.

Tran, N.H., Urase, T., Ngo, H.H. et al. (2013). Insight into metabolic and cometabolic activities of autotrophic and heterotrophic microorganisms in the biodegradation of emerging trace organic contaminants. *Bioresource Technology* 146: 721–731.

Udaiyappan, A.F.M., Hasan, H.A., Takriff, M.S., and Abdullah, S.R.S. (2017). A review of the potentials, challenges and current status of microalgae biomass applications in industrial wastewater treatment. *Journal of Water Process Engineering* 20: 8–21.

Vo, H.N.P., Ngo, H.H., Guo, W. et al. (2019). Identification of the pollutants' removal and mechanism by microalgae in saline wastewater. *Bioresource Technology* 275: 44–52.

Wahlen, B.D., Willis, R.M., and Seefeldt, L.C. (2011). Biodiesel production by simultaneous extraction and conversion of total lipids from microalgae, cyanobacteria, and wild mixed-cultures. *Bioresource Technology* 102 (3): 2724–2730.

Wang, L., Li, Y., Chen, P. et al. (2010). Anaerobic digested dairy manure as a nutrient supplement for cultivation of oil-rich green microalgae *Chlorella* sp. *Bioresource Technology* 101 (8): 2623–2628.

Wijaya, I.M.W. and Soedjono, E.S. (2018). Domestic wastewater in Indonesia: challenge in the future related to nitrogen content. *Geomate Journal* 15 (47): 32–41.

de Wilt, A., Butkovskyi, A., Tuantet, K. et al. (2016). Micropollutant removal in an algal treatment system fed with source separated wastewater streams. *Journal of Hazardous Materials* 304: 84–92.

Xiao, R. and Zheng, Y. (2016). Overview of microalgal extracellular polymeric substances (EPS) and their applications. *Biotechnology Advances* 34 (7): 1225–1244.

# 13

# Immobilization of Microbial Inoculants for Improving Soil Nutrient Bioavailability

*Swati, V. Mamtha, and Haralakal Keerthi Kumari*

*Department of Agricultural Microbiology, University of Agricultural Sciences, Bangalore, India*

## 13.1 Introduction

Inoculated microorganisms often fail to locate/settle in soil because of plant and microbe interactions, microorganisms load, variety of soil, competition and carnivores, etc. (Young et al. 2006). For a bio-fertilizer to maintain a high microbial count during several processes like manufacturing, transportation, storage time, and soil application, the initial microbial population needs to be large. Innovative delivery methods for microorganisms that can be used to fertilize soil are still lacking. A suitable carrier and additive in the formulation should help stabilize and protect the microbial cells while delivering them to the target zone.

In order to generate biofertilizer with maximum microbial load and prolonged life span of bacterial cells, an attempt was made to examine the effectiveness of utilizing immobilization technique. A high amount of microbial load must be maintained at the fixed site of supplementation in the rhizosphere through the inert and constant liberation of the immobilized microorganisms. Immobilization is also an effective method for extending life span of biofertilizers, which is a constraint especially in Gram-negative microorganisms that deprive protective structures (Pereg and McMillan 2015).

The development of microbial cells improves plant growth, suppresses plant diseases, and lessens the requirement of inorganic/synthetic fertilizers and pesticides (Gond et al. 2021; Gupta et al. 2020; Singh et al. 2020). Advantageous microorganisms aid in the release of inaccessible nutrients from soil and compost making nutrients available for plant assimilation. These microorganisms also release plant hormones which aid in plant root development and reduce the disease spreading by producing organic acids, antibiotics, enzymes, etc. The development of bio-fertilizer heavily relies on the targeted isolation of microorganisms with superior features.

Microorganisms play a crucial part in phosphorous bioavailability to crops through solubilizing and mineralizing unavailable forms of phosphorous (Mundra et al. 2011; Pereg and McMillan 2015). In another way, this helps in sequestration of iron from soil solution by siderophore production, which has strong affinity toward iron (Rashid et al. 2016). Sequestration enhances plant nutrition and also prevents the emergence of fungi (Souza et al. 2015). Additionally, indole acetic acid (IAA), which can be secreted by microorganisms, has been shown to have antifungal properties and enhance plant roots growth and development (Arruda et al. 2013; Jha et al. 2009). In the meantime, nitrogen-fixing microbes use a sophisticated nitrogenase system to transform free environmental nitrogen into ammonia, a form that plants may use (Ahemad and Kibret 2014).

Immobilization of microbial cells is a technique that involves the attachment or confinement of microorganisms to a solid support matrix. Microorganisms, enzymes, cellular organelles, and animal and plant cells can be immobilized. Confinement of microorganisms is also referred to as the action of confining intact cells onto specific regions in a device or material without the loss of requisite biological function. This can be generally performed by following various methods such as physical adsorption, encapsulation, entrapment, and covalent bonding.

In this process, viable microbial cells are physically isolated or localized to a specific area of space in order to restrict their ability to move freely, display hydrodynamic properties that are distinct from the surrounding environment, and maintain their catalytic activities for repeated and continuous use.

*Microbes Based Approaches for the Management of Hazardous Contaminants*, First Edition. Edited by Ajay Kumar, Livleen Shukla, Joginder Singh, and Luiz Fernando R. Ferreira.

The key point of immobilization is to safeguard newly introduced soil microbes and to enable their gradual and prolonged release. The physiological function of the microbes will directly relate to the deterioration level of the encapsulating material used. Since Chibata's group revealed the effective use of continuous fermentation of L-aspartic acid in the beginning of the 1970s. Numerous research teams have tried a variety of microbial applications with immobilized cells.

## 13.2 History of Immobilization

The immobilization concept already exists in environment, as shown by the fact that microalgal samples obtained along the Gulf shores were covered by biological films of bacteria that use oil to break down hydrocarbons in saltwater. This is proof that immobilization is a natural phenomenon that occurs all over the world.

These biological films are superficially inclined microbial communities that comprise many layers of cells embedded in moist matrices. They can be extended superficially or inside themself structures *viz.*, body, tooth, grains, glass, a water pipe or conduit, etc. This natural happening motivated humans to use them for their work (Bayat et al. 2015). In 1969 for the initial period, enzyme immobilization was implemented for regular making of L-amino acids from acyl DL-amino acids. By immobilizing aminoacylase enzymes since the late 1970s, immobilization strategies are used frequently in many laboratories.

## 13.3 Support Material Selection

The carrier selection during immobilizing microbial cells is very crucial. The following requirements must be met by carriers

- Nontoxic, nonpolluting, nonbiodegradable
- It should have maximum amount of cells holding ability
- Extremely high chemical, biological, and physical persistent
- Extended life span
- appropriate function groupings
- Reasonably low cost
- Ideal separation from liquid media to the core of carrier
- Simple removal of carriers and cells from medium
- It should be regenerative and handy

## 13.4 Support Materials Used for Immobilization of Microbes

The characteristics of the support material, microbial cell surface, and ambient circumstances are all factors that influence how effectively microbial cells are immobilized. The process of immobilization can be impacted by the choice of the best support material, and the features of the support microbial inoculant preparation are determined by considering characteristics of each of the bacteria along with the support material. As a result, the relationship of each of them results in immobilized microbial inoculants having particular mechanical, chemical, metabolic, and kinetic properties. Suitable support material characteristics are listed as being hydrophilic nature, inertness to enzymes, biological compatibility, protection from microbial contamination, resilience to contraction, and ease of availability at a reasonable price.

The carriers are categorized into two types: organic polymers and inorganic materials such as zeolite, clay, anthracite, porous glass, activated charcoal, and ceramics. Inorganic carriers have been used to immobilize bacteria since carriers are thermostable and resistant to microbial destruction (Lu and Toy 2009). Organic polymeric carriers, which include both natural and manmade polymeric carriers, are more prevalent than inorganic carriers. Alginate, carrageenan, agar, agarose, several synthetic (acrylamide, polyurethane, polyvinyl resins), and organic polymer compounds of algal polymers, including chitosan, an amino polysaccharide produced by chitin, are all used in experiments (Manger 2013).

Among all, alginate and carrageenan are the most often used polymers;however, they are more fragile in liquid waste than artificial polymers. Alginates that are polymeric comprised of various ratios and blends of mannuronic and guluronic acids are convenient to operate, harmless to people, to the atmosphere, and confined microbes, permissible to use by people, accessible in large numbers, and reasonably priced. From a physiological stand point, alginate has many benefits, including the fact that the gel is translucent and porous and that confined cells do not undergo significant changes in their physical and chemical state throughout the confining process. The entrapped microbe in the encapsulation can be readily disrupted during the procedure, hence this substance cannot be maintained in aqueous solution for an prolonged duration (Buque et al. 2002).

The amino functional groups in chitosan have the ability to be reactive, nontoxic, and also cheap, which can increase the carrier's affinity with the microbes. Yet, the structural integrity of the carrier will decline due to renewed ability during usage. As carriers for microbial encapsulation, other naturally derived gels including agar, collagen, and agarose can also be employed. Certain natural polymers have been more susceptible to environmental microbial breakdown. Natural polymers, on the other hand, are less dangerous to create and have a higher diffusivity. Synthetic polymeric supports outperform natural carriers in terms of mechanical performance and are not readily biodegradable.

### 13.4.1 Modifications of the Support Material for Immobilization of Microbial Inoculants

Modification of the support material surface physical and chemical properties will be useful in promoting immobilization. One of the physical methods is to increase roughness of the surface, it provides protection to the cells against solvent shearing, and it also increases surface area for immobilization process. Change in chemical properties to generate interactions like ionic and covalent bonds between support and cells (microorganisms). So, increased cations and hydrophobicity of the support surface are favorable for the confined process (Zhu 2007). It was demonstrated that adhesion of *Pseudomonas putida* to hydrophobic and positively charged glass beads was substantially higher than normal beads.

### 13.4.2 Types of Immobilization

Compared to free-floating cell systems, immobilized cells are much more resilient to change in atmospheric situations and low susceptible to toxins at the major stage. Common methods of immobilization are adsorption upon surfaces, covalent attachment to carriers, cross-linking, entrapment, and encapsulation. It should be emphasized that the bulk of work on the application of living, immobilized microbes is currently carried out on a small scale in the lab.

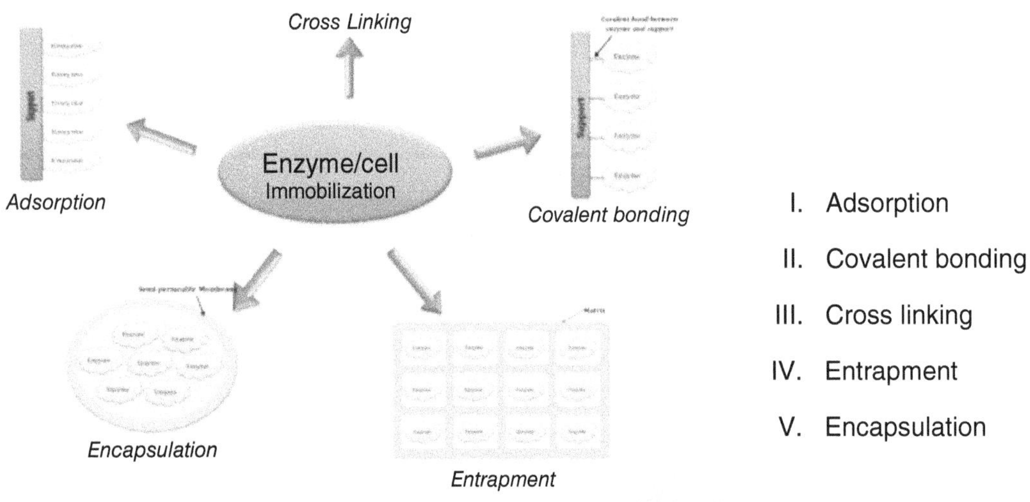


Enzyme/cell immobilization methods

#### 13.4.2.1 Adsorption

Adsorption involves physically attaching the microorganisms or cells on the exterior of a neutral support. Adsorption is due to electrostatic forces, hydrophobic interaction, and hydrogen bonds between atoms and ions of carriers and microbes (Zhu 2007). The carrier materials used are organic and inorganic compounds (e.g., inorganic:alumina, silica gel, calcium phosphate gel, glass; organic:starch, carboxymethyl cellulose, DEAE-cellulose, DEAE-Sephadex). Absorbed compounds will be quickly eliminated by small modifications in pH, ionic strength, or temperature. Adsorption can be done at appropriate pH and ionic strength. This method is less damaging to protein compared to chemical methods. Loading rate is 1 g enzyme/1 g matrix.

#### 13.4.2.2 Types of Adsorption

Because of random adhesion of various substances or materials on the sensors interface might contaminate the signal and produce storage stability issues for adsorbed cells or enzyme, in addition to these issues (Nguyen and Kim 2017) the following three subcategories of immobilization by adsorption are frequently used: physical adsorption, electrostatic binding and hydrophobic adsorption.

***Physical Adsorption*** Enzymatic biosensors have been developed using this immobilization technique frequently. In this process, the support material has to be placed in a solution containing enzyme and incubated for a predetermined amount of time (hours). Hydrogen bonds, Van der Waals forces, and hydrophobic interactions are only a few examples of the weak, nonspecific forces that help enzymes adhere to the supporting matrix.

***Electrostatic Binding*** When employing electrostatic forces for enzyme immobilization, it is essential to take into account both the pH of the reaction solution and the enzyme's isoelectric point. Enzyme molecules exhibit positive or negative charges on their surfaces, facilitating their binding to surfaces carrying opposite charges through robust ionic and polar interactions.

***HydrophobicAdsorption*** Another strategy hydrophobic interaction between enzyme/cells and the carrier/support is another type of immobilization. Increase in entropy during immobilization causes displacement of water atoms from the surface of the support layer, and this interaction causes enzyme molecules/cells to surface. The hydrophobic property of the enzyme/cells and adsorbent significantly depends on the effectiveness of the interaction.

#### 13.4.2.3 Covalent Bonding

In this method, immobilization occurs by forming covalent bond among an inorganic matrix and cell (microorganism) with the action of cross-linking agent (binding agent). It is usually used in enzyme immobilization and is rarely used in whole-cell immobilization because it leads to loss of cell viability and cell damage by the toxicity of coupling agents this is the main disadvantage of this method (Zhu 2007). Enzyme loading is low (0.02 g/gm matrix).

***Methods of Covalent Bonding*** Following is a quick explanation of the methods (Nguyen and Kim 2017) most frequently used in confining enzymes/cells on enhanced membranes (via triggering of carboxylic group and amino group).

1.1 **Cyanogen bromide activation**: Cyanogen bromide activated by the support material containing glycol groups (cellulose, sepharose, and sephadex), which is then binds to the enzyme and immobilizes it.
1.2 **Diazotation:** Bonding between two amine groups (amino benzyl cellulose, amino derivatives of polystyrene, aminosilanzed porous glass).
1.3 **Peptide bond formation:** Bond formation between carboxyl group of enzymes and amino groups of the support matrix.
1.4 **Polyfunctionalagents:** Bond formation between two amino groups (between support material and enzyme) by reagents like glutaraldehyde.

***Cross-Linking*** When covalent bonds are used to create intermolecular cross-links between the enzymes, an irreversible process known as enzyme immobilization by cross-linking is attained. In order to join enzyme molecules various operative testing agents, these serve as a linker during the procedure. Agents like Glutaraldehyde and Diazonium salt are frequently employed and this approach is inexpensive and quick.

***Entrapment*** An enzyme, cell, or microorganism is actually imprisoned inside a matrix. Bonds involved could be covalent or non-covalent. Surface material/matrices employed water-soluble polymer, polyacrylamide gel, cellulose triacetate, calcium alginate, and gelatin.To stop the loss of bacteria, cells, and enzymes, the matrix's pore size is modified. Gel is loaded with 1 g of enzyme. After entrapment of enzymes/cells inside matrix substrate and product diffusion creates difficulties.

Immobilization of cells by entrapment method using polyurithryl foam was also done. Here support material polyurithryl foam is a very porous material in which cells get entrapped easily and they can grow in the cubes by utilizing the medium imbibed by the foam.

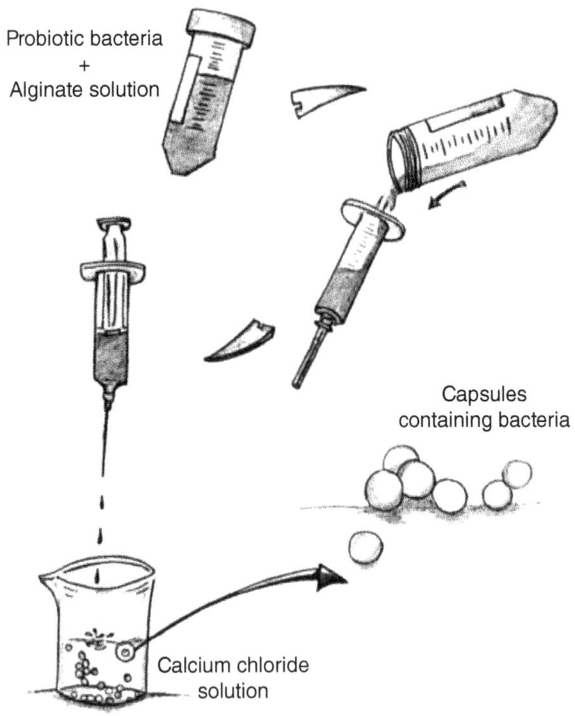


Using this method an aqueous solution of sodium alginate that contains the required microbial agent is dispersed on a solution that contains divalent cations, like calcium chloride. After the droplets solidify and the sodium alginate's charge-negative polymer chain interacts with the $Ca^{2+}$ cations, a hydrogel is then produced. Depending on the needle size, this method produces homogeneous beads with sizes ranging from a few micrometers to several millimeters.

***Encapsulation*** Encapsulation of microbial inoculants is achieved by adding a biological agent into the matrix, which is subjected to a mechanical process and stabilized by a chemical or physical–chemical process. This phrase describes the creation of spherical particles, wherein a liquid or suspension exists inside a semipermeable barrier. The membrane may be polymeric, lipoidal, lipoprotein based or non-ionic in nature. It contains three methods, physical, chemical, and physicochemical method. It has no impact on enzyme activity, it possesses a high enzyme loading efficiency but it utilizes a carrier with restricted pore size.

### 13.4.3 Mode of Action of Immobilized Microorganisms in Improving Soil Bioavailability

The bioavailability of soil is increased by microorganisms in a variety of ways. For example, biological processes such as nitrogen, sulfur, or phosphorous fixation, the production of siderophore to increase iron bioavailability helps to incorporate or increase nutrients in the soil, as well as their bioavailability to plants. By digesting plant and animal waste, microorganisms' extracellular secretions increase the amount of biological matter in soil and can improve the stability of the soil aggregates.

### 13.4.4 Characteristics of Immobilized Microorganisms

They have increased stability, improved productivity, and enhanced selectivity of the biocatalyst;make separation simple and the biocatalyst can be recycled; and reduce the overall cost of the process.

**Advantages of immobilization:**
- Tolerate higher substrate concentrations, resulting in higher yields.
- By single step different enzymes can be introduced.
- Microbes are long-lasting and stable.

- It is the most effective method to maintain microbial inoculant native conformation.
- Same microbial cells can be received.
- Inoculum preparation time is reduced.
- Ensure high level of purity and yield can be achieved.
- Separation/purification of product became easy.
- Can be used when heat-labeled enzyme/extracellular is our target.
- It improves stability and efficiency of the fermentation.
- It gives an immense amount of biomass.
- By creating optimal microenvironmental circumstances.
- Enhancing the genetic stability.
- It offers defense against compressive injury.

**Disadvantages:**
- Microbial inoculant concentrations will be lower.
- Undesirable enzymes are also synthesized
- Some enzymes may interfere with the product.
- Possibility of leakage of cells from carrier material.

## 13.5 Conclusion

The demand for agricultural products has grown in tandem with the world population's expansion. Food security must be maintained for both people and animals while also protecting and enhancing ecosystems. Increased agrarian output through the overuse of agrochemicals had a significant negative impact on ecosystems, human health, and the environment as a whole. Soil and water quality have declined, and the environment has generally changed. The employment of advantageous microorganisms, however, is one of the alternatives to agrochemicals. Their application favorably influences crop production, increases tolerance to water and salt stress, and increases resilience to phytopathogens, potentially displacing the overuse of artificial fertilizers and pesticides. However, using microorganisms in the soil comes with a number of difficulties, including preserving cell viability and microbial resistance to diverse environmental factors.

## References

Ahemad, M. and Kibret, M. (2014). Mechanisms and applications of plant growth promoting rhizobacteria: current perspective. *Journal of King Saudi University-Science* 26 (1): 1–20.

Arruda, L., Beneduzi, A., Martins, A. et al. (2013). Screening of rhizobacteria isolated from maize (*Zea mays* L.) in Rio Grande do Sul State (South Brazil) and analysis of their potential to improve plant growth. *Applied Soil Ecology* 63: 15–22.

Bayat, Z., Hassanshahian, M., and Cappello, S. (2015). Immobilization of microbes for bioremediation of crude oil polluted environments: a mini review. *The Open Microbiology Journal* 9: 48.

Buque, E.M., Chin-Joe, I., Straathof, A.J. et al. (2002). Immobilization affects the rate and enantioselectivity of 3-oxo ester reduction by baker's yeast. *Enzyme and Microbial Technology* 31 (5): 656–664.

Gond, D.P., Jha, S.S., Kumar, A. et al. (2021). Plant growth promoting bacteria and its role in green remediation. In: *Sustainable Environmental Clean-Up* (ed. V.K. Mishra and A. Kumar), 149–163. Cambridge, MA: Elsevier https://doi.org/10.1016/B978-0-12-823828-8.00007-4.

Gupta, A., Singh, S.K., Singh, M.K. et al. (2020). Plant growth-promoting rhizobacteria and their functional role in salinity stress management. In: *Abatement of Environmental Pollutants* (ed. P. Singh, A. Kumar, and A. Borthakur), 151–160. Cambridge, MA: Elsevier https://doi.org/10.1016/B978-0-12-818095-2.00007-2.

Jha, B.K., Gandhi Pragash, M., Cletus, J. et al. (2009). Simultaneous phosphate solubilization potential and antifungal activity of new fluorescent pseudomonad strains, *Pseudomonas aeruginosa*, *P. plecoglossicida* and *P. mosselii*. *World Journal of Microbiology and Biotechnology* 25: 573–581.

Lu, J. and Toy, P.H. (2009). Organic polymer supports for synthesis and for reagent and catalyst immobilization. *Chemical Reviews* 109 (2): 815–838.

Magner, E. (2013). Immobilisation of enzymes on mesoporous silicate materials. *Chemical Society Reviews* 42 (15): 6213–6222.

Mundra, S., Arora, R., and Stobdan, T. (2011). Solubilization of insoluble inorganic phosphates by a novel temperature-, pH-, and salt-tolerant yeast, *Rhodotorula* sp. PS4, isolated from seabuckthorn rhizosphere, growing in cold desert of Ladakh, India. *World Journal of Microbiology and Biotechnology* 27: 2387–2396.

Nguyen, H.H. and Kim, M. (2017). An overview of techniques in enzyme immobilization. *Applied Science and Convergence Technology* 26 (6): 157–163.

Pereg, L. and McMillan, M. (2015). Scoping the potential uses of beneficial microorganisms for increasing productivity in cotton cropping systems. *Soil Biology and Biochemistry* 80: 349–358.

Rashid, M.I., Mujawar, L.H., Shahzad, T. et al. (2016). Bacteria and fungi can contribute to nutrients bioavailability and aggregate formation in degraded soils. *Microbiological Research* 183: 26–41.

Singh, D., Singh, S.K., Singh, V.K. et al. (2020). Plant growth–promoting bacteria and their role in environmental management. In: *Abatementof Environmental Pollutants* (ed. P. Singh, A. Kumar, and A. Borthakur), 161–175. Cambridge, MA: Elsevier https://doi.org/10.1016/B978-0-12-818095-2.00008-4.

Souza, R.D., Ambrosini, A., and Passaglia, L.M. (2015). Plant growth-promoting bacteria as inoculants in agricultural soils. *Genetics and Molecular Biology* 38: 401–419.

Young, C.C., Rekha, P.D., Lai, W.A. et al. (2006). Encapsulation of plant growth-promoting bacteria in alginate beads enriched with humic acid. *Biotechnology and Bioengineering* 95 (1): 76–83.

Zhu, Y. (2007). Immobilized cell fermentation for production of chemicals and fuels. In: *Bioprocessing for Value-Added Products from Renewable Resources*, 373–396. Elsevier Science.

# 14

# Insight Into the Factors Inhibiting the Anammox Process in Wastewater

*Surbhi Sinha, Anamika Singh, and Rachana Singh*

*Amity Institute of Biotechnology, Amity University Uttar Pradesh, Noida, India*

## 14.1 Introduction

The anaerobic ammonium oxidation (anammox) process has been considered a promising method for treating high-nitrogen wastewater since it produces little sludge and does not require the addition of exogenous carbon sources or aeration (Zekker et al. 2021). Anammox was initially observed in a denitrifying fluidized bed reactor in 1995 (Xu et al. 2018). Autotrophic anammox bacteria convert ammonium in anoxic circumstances into dinitrogen gas while using nitrite as an electron acceptor (Shaw et al. 2020). Anammox, a more sustainable biological nitrogen-removal alternative to standard nitrification–denitrification technology, has a greater nitrogen removal rate (NRR), reduced operational costs, and fewer space requirements (Ali and Okabe 2015). Anammox has been used successfully in the lab, pilot scale, and/or full scale to treat ammonium-rich wastewater, such as sludge-digestion liquid, landfill leachate, coke-oven wastewater, swine wastewater, and pharmaceutical wastewater, among others. Vineyard et al. (2020) reported that the anammox process is projected to cost US$ 4.09/kg N removed, which is significantly less expensive than the conventional nitrification/denitrification method. Studies and implementations of high throughput anammox processes have been reported repeatedly (Tang et al. 2011; Wang et al. 2022). Numerous studies have demonstrated the presence of anammox bacteria in a wide range of man-made and natural environments with oxygen-depleted zones and readily available $NO_x^-$ ($NO^-$ and $NO_2^-$), including wastewater treatment facilities, marine, freshwater, terrestrial, and even sea ice ecosystems (Zhu et al. 2015; Kuliński et al. 2022). However, the application and industrialization of the Anammox process have been impeded by the growth kinetics of the anammox bacteria and the prevalence of inhibitory factors in nitrogen-rich wastewater. Anammox bacteria are extremely difficult to cultivate due to their slow growth rate (their doubling period at 30–40°C is typically 10–14 days), low cell yield (0.11 g VSS $g^{-1}$ $NH_4^+$—N), and susceptibility to fluctuating environments (Cho et al. 2019). Even though the rate of growth can be increased by optimizing the operational circumstances, i.e., the doubling time can be lowered to 4.8 or even 1.8 days, cultivation remains a challenge. Moreover, several influencing elements, such as temperature, pH, and the presence of heavy metals, antibiotics, and nanoparticles (NPs), pose threats to the stable operation of the anammox process. For instance, research demonstrated that the high concentrations of ammonium ($NH^{4+}$) and nitrite ($NO^{2-}$) impacted the growth and metabolism of anammox bacteria (Lin et al. 2020). Trace elements are necessary for anaerobic bacteria; however, too much metal inhibits the activity of bacteria and worsens the ability of anammox process to remove nitrogen (Gutwiński et al. 2021). Antibiotics, which are typical pollutants, are said to slow down the activity of anammox. Also, antibiotics cause antibiotic-resistant bacteria (ARB) and antibiotic-resistant genes (ARGs) to grow and spread in the environment (Zhang et al. 2019). Some substances, like microplastics and NPs, may potentially have a negative impact on anaerobic bacteria (Tang et al. 2021). The heterogeneity of industrial waste makes it more challenging to commence the anammox process, makes the technique more susceptible to suppression, and hinders the recuperation of Anammox performance from inhibition.

Although several reviews on this subject have been published, most of them focus primarily on conventional inhibitors and their impact on the anammox process. This chapter provides a thorough assessment of the most recent research on anammox inhibition, covering the inhibitory effect of developing contaminants, as well as the recovery process.

*Microbes Based Approaches for the Management of Hazardous Contaminants*, First Edition. Edited by Ajay Kumar, Livleen Shukla, Joginder Singh, and Luiz Fernando R. Ferreira.

The chapter also covers the knowledge gap on anammox inhibition mechanisms and provides a foundation for nitrogenous wastewater treatment research.

## 14.2 Substrate Inhibition

### 14.2.1 Ammonium Inhibition

In most modern wastewater treatment plants (WWTPs), ammonium is one of the main forms of nitrogen (Van Hulle et al. 2010). Nitrogen-containing wastewater, such as sludge digestion liquid, landfill leachate, monosodium glutamate wastewater, nitrogen fertilizer production wastewater, coking wastewater, and other industrial waste, is often rich in ammonium. It has been established that concentrations of ammonium as high as 1 g N $l^{-1}$ do not impede the Anammox process (Jin et al. 2012). High ammonia levels, however, have been shown to inhibit anammox activities in some research (Fernández et al. 2012).

### 14.2.2 Nitrite Inhibition

Numerous anammox bacteria are significantly inhibited by the presence of high nitrite concentrations (Zhou et al. 2011; Cao et al. 2021). One of the substrates for anammox is nitrite, and because nitrite is biotoxic, anammox will be suppressed if the nitrite concentration rises over a particular threshold value. Nitrite inhibition threshold values were lower than ammonium inhibition threshold values. In other words, nitrite inhibition was more likely to cause the Anammox process to fail as compared to ammonium inhibition (Isaka et al. 2007; Dapena-Mora et al. 2007). The stability of the anammox mechanism is largely dependent on the nitrite concentration, which has been shown in numerous studies to cause severe inhibition in certain experimental circumstances (Lackner et al. 2014; Ma et al. 2016). These investigations, however, disagreed on the nitrite inhibition threshold value, with reported threshold concentrations ranging from 5 to 280 mg N $l^{-1}$ under various experimental settings and operating modes (Van Hulle et al. 2010). It has also been mentioned that in continuous feeding mode and with a small quantity of nitrite, anammox activity is not inhibited (Zhang et al. 2018).

## 14.3 Heavy Metals Inhibition

Heavy metals are difficult to biodegrade and can accumulate in organisms, resulting in biological accumulation toxicity. Heavy metal ions are frequently found in nitrogen-rich wastewater, particularly landfill leachate. A better understanding of heavy metal inhibition of Anammox can aid in nitrogen removal from wastewater. Van de Graaf et al. (1995) reported that 1 mmol $l^{-1}$ $HgCl_2$ completely inhibited the activity of Anammox. Limited studies on heavy-metal inhibition of anammox have been reported and thus more research needs to be done in this area. Table 14.1 summarizes some of the relevant research corresponding to the anammox process.

**Table 14.1** Effect of heavy metals on anammox process.

| Inhibitor | Operation mode | Concentration (mg $l^{-1}$) | Effect | References |
|---|---|---|---|---|
| $Cu^{+2}$ | Batch Study | 19.3 | $IC_{50}$ | Val del Río et al. (2017) |
| $Cu^{+2}$ | – | 16.3 | – | Zhang et al. (2016) |
| $Cr^{+2}$ | Batch Study | 26.9 | $IC_{50}$ | Val del Río et al. (2017) |
| $Pb^{+2}$ | Batch Study | 45.6 | $IC_{50}$ | Val del Río et al. (2017) |
| $Cd^{+2}$ | Batch Study | 174.6 | $IC_{50}$ | Val del Río et al. (2017) |
| $Cd^{+2}$ | Batch Study | 11.16 ± 0.42 | $IC_{50}$ | Zhen Bi et al. (2014) |
| Cr(VI) | Batch Study | 296.5 | $IC_{50}$ | Jiang et al. (2018) |
| $Zn^{+2}$ | – | 20.0 | – | Zhang et al. (2016) |
| $Zn^{+2}$ | –. | 1100 | $IC_{50}$ | Lotti et al. (2012) |
| Hg | Batch Study | 11.52 ± 0.49 | $IC_{50}$ | Zhen Bi et al. (2014) |

## 14.4 Organic Matter Inhibition

Anammox bacteria are chemoautotrophic microorganisms that only consume $CO_2$ as their sole carbon source. The concentration of bicarbonate in the anammox process influent is thus essential for the growth of anammox bacteria. Generally, the growth or activity of anammox bacteria can be promoted by the right addition of inorganic carbon. The common consensus is that anammox bacteria are negatively impacted by organic materials (organic carbon). However, several types of wastewater rich in nitrogen, such as landfill leachate, digested animal waste wastewater, and monosodium glutamate wastewater, typically contain a lot of organic matter. Understanding organic matter inhibition in anammox would aid research and application of the anammox process in the treatment of both nitrogen- and organic-matter-containing wastewater.

### 14.4.1 Nontoxic Organic Matter Inhibition

Anammox is affected by nontoxic organic matter in a concentration-dependent manner. Anammox activity was inhibited by high levels of organic matter, whereas low levels of organic matter had little to no effect and even stimulated biological responses (Yang et al. 2019). In general, there are two postulated pathways for anammox organic matter inhibition. First, heterotrophic bacteria in anammox systems compete with autotrophic anammox bacteria. Heterotrophic bacteria outcompete autotrophic anammox bacteria when organic matter concentrations are high. This results in the loss of anammox bacteria and a consequent decrease in the latter's ability to remove nitrogen from the environment. The second mechanism proposes that when organic matter concentrations are high, anammox bacteria continue to dominate the system, but they switch to utilize organic matter as a substrate instead of ammonium and nitrite (Jin et al. 2012). In a nutshell, the anammox bacteria exhibit either a diverse substrate or a diverse metabolic profile. Due to the decreased anammox activity, a decline in nitrogen removal efficiency was noted.

According to some studies, anammox bacteria can co-occur with other bacteria, such as the denitrifying bacteria nitrosobacteria, and this coexistence can play an important role in the treatment of nitrogen- and COD-containing wastewater (Lan et al. 2011). The low availability of easily biodegradable organic carbon is thought to limit heterotrophic denitrifying growth; thus, denitrifiers are unable to dominate and cannot outcompete anammox in environments such as digested liquor and landfill leachate. Dissolved oxygen (DO), pH, and temperature controls are also important for keeping anammox in balance with other processes that help remove nitrogen and organic carbon from the same system at the same time (Lan et al. 2011).

### 14.4.2 Toxic Organic Matter Inhibition

The anammox process is inhibited by toxic organic materials by poisoning of anammox bacteria or inactivation of enzymes, and the suppression is typically irreversible. Some typical classes of harmful organic material that can impede anammox include alcohol, aldehydes, phenol, and antibiotics. These are addressed in Sections 14.4.2.1–14.4.2.3.

#### 14.4.2.1 Alcohol and Aldehydes

Alcohol's ability to inhibit anammox action has previously been studied (Oshiki et al. 2011; Tang et al. 2009; Isaka et al. 2008). Methanol and ethanol are the two most hazardous alcohols for anammox activity. During the inhibition mechanism, an enzyme called methanol dehydrogenase turns methanol into aldehyde groups, and ethanol turns into acetyl-coenzyme before being oxidized to acetaldehyde. The addition of methanol inhibits the coupling mechanism and prevents the formation of nitrite in the bioreactor, hence reducing denitrification. Anammox showed no activity at a concentration of 3–4 mmol methanol, and the suppression was irreversible (You et al. 2020). In a separate investigation, methanol was found to be the most toxic inhibitor. At a concentration of just as 16 mg $l^{-1}$, methanol stopped anammox activity completely and irreversibly (Güven et al. 2005). The anammox activity, on the other hand, was shown to be reduced by 71% in a batch test when there was 160 mg $l^{-1}$ of methanol present (Isaka et al. 2008). In the anammox process, the difference in methanol resistance could be due to the different anammox species.

#### 14.4.2.2 Phenols

While phenolic compounds are rarely found in municipal wastewater, they are frequently found in the industrial wastewater from pulp mills, textile and petrochemical industries, refineries, coal-gasification plants, coking plants, pharmaceuticals, fertilizers, degreasing and paint-stripping operations, and fiberboard manufacturing (Zhao et al. 2018; Dosta et al. 2011).

Toh et al. (2002) successfully grew and acclimated anammox *Candidatus Brocadia* from municipal treatment plant sludge using synthetic wastewater containing phenol. When the concentration of phenol was gradually increased from 50 to 550 mg $l^{-1}$, the removal of ammonium nitrogen was initially inhibited but then gradually recovered. According to these findings, phenol harmed anammox metabolism, but acclimatization increased anammox activity.

#### 14.4.2.3 Antibiotics

Wastewater contains high concentrations of antibiotics, which inhibit or kill anammox bacteria and denitrifying bacteria owing to their antibacterial properties. Anammox inhibition is brought on by antibiotics such as polymyxin B, penicillin G-Na, chloramphenicol, and kanamycin sulfate (Fu et al. 2021). Few studies have been done on how antibiotics affect anammox, and the ones that have been done focus on only three types of antibiotics: Tetracycline, b-lactams, and chloramphenicol. It was reported that chloramphenicol (200 mg $l^{-1}$) or ampicillin (800 mg $l^{-1}$) reduced more than 90% of the anammox activity. For the first three days following exposure, chloramphenicol could reduce anammox activity by 98%; however, due to acclimatization, anammox activity marginally increased after three days (Van de Graaf et al. 1995). Like this, in a batch test utilizing 25 ml serum vials with either chloramphenicol at 250–1000 mg $l^{-1}$ or tetracycline hydrochloride at 100–1000 mg $l^{-1}$, anammox activity was reduced by 20–80% (Fernandez et al. 2009). However, the activity of anammox was not significantly inhibited by low concentrations of antibiotics such as allylthiourea and chloramphenicol (Dapena-Mora et al. 2007). A minimal study has been done on the impacts of other types of antibiotics, and there is currently no literature on how specific Anammox species are affected by antibiotic inhibition. There is a need for additional investigation.

## 14.5 Salinity Inhibition

Oil refineries, ship decks, tanning industries, and fish canneries are the primary sources of saline wastewater (Giustinianovich et al. 2018; Xiao and Roberts 2010). Because there are not enough freshwater resources, greater amounts of seawater are being used for municipal and industrial water supplies. This results in the discharge of additional wastewater that contains high levels of salt (Huang et al. 2018). As the cell loses free water, salinity stress at high concentrations alters the internal and exterior osmotic pressures of the cell membrane. Additionally, metabolic activity of anaerobic bacteria is impacted by salt stress, which also damages the structure of the enzyme. High salinity causes high osmotic pressure, which causes microorganisms to die (Jin et al. 2012). As a result, the osmotic pressure present in highly concentrated waste may severely inhibit the bacteria. Although salinity inhibits anaerobic biological treatment processes, the anammox reaction is still a promising process for treating wastewater with high salt concentrations (Xiao and Roberts 2010). To alleviate the inhibitory effect of salinity, two solutions were implemented by anaerobic bacteria: (i) pumping inorganic ions into cells to balance the osmotic pressure of the extracellular space (salt-in strategy), and (ii) synthesizing or concentrating the organic solvent (compatible solute strategy) (Li et al. 2018). Since the anammox pathway is present in all anoxic marine environments, some researchers have enriched anammox bacteria from marine sediments or acclimated freshwater anammox bacteria to higher salt concentrations (Lu et al. 2019; Gonzalez-Silva et al. 2017).

When using the anammox method to remediate high-salt wastewater, it is important to keep in mind that salinity could be a factor that slows down the process. The salt content in certain concentrations (3–15 g NaCl $l^{-1}$) was reported to promote the establishment of anammox granular sludge and increase bacteria retention in the reactor, but high salinity inhibited anammox (Dapena-Mora et al. 2010). However, acclimatization improved anammox performance in high salinity conditions.

## 14.6 Microplastic Inhibition

Microplastics are described as plastic particles that are less than 5 mm in size and are the result of the breakdown of larger plastic fragments (Mason et al. 2016). When present in large quantities, microplastics alter the composition of the microbes and thus the efficiency with which they remove nitrogen during the anammox process (Tang et al. 2021). Microplastics altered the structural elements of EPS and prevented anammox bacteria from aggregating. Nanoscale microplastics adhered to the cell membrane and had an immediate inhibitory effect on critical enzymes, metabolic intermediates, and metabolic end products. Nanoscale microplastics harm DNA and material exchange routes and can also kill cells (Zhang and Chen 2020). Microplastics are carriers for contaminants such as antibiotics, heavy metals, and persistent organic contaminants. The desorption of these contaminants from the surfaces of microplastics had an impact on anammox bacteria (Zhang and Chen 2020).

## 14.7 Nanoparticle (NPs) Inhibition

NPs are widely employed in cosmetics, coatings, electronics, energy materials, and catalytic materials owing to the demands of industry and daily life (Xu et al. 2020). The principal NPs are carbon nanotubes (CNTs), nanoscale zero-valent iron (nZVI), graphene oxide NPs, CuO NPs, ZnO NPs, $Fe_3O_4$ NPs, and Ag NPs. It is unavoidable that some of these NPs will be released into the environment and sewage treatment facilities. However, not all NPs are detrimental to the anammox process. Depending on concentration, NPs can affect anammox process positively or negatively. When present in low quantities, NPs enhanced both the efficiency with which the anammox process removed nitrogen and the richness of the microbiome (Elreedy et al. 2021; Liu et al. 2021; Yan et al. 2019). NPs suppress anammox process at high concentrations. The anammox process was suppressed by ZnO NPs, and the inhibitory impact was enhanced with rising ZnO NP concentration (Vale et al. 2016; Yamindago et al. 2018). Metals like $Cu^{2+}$ and $Zn^{2+}$ were adsorbed by Ag NPs. The adsorbed metals are then desorbed into the environment, increasing the concentration of metals in the aqueous environment and their negative impact on the anammox process (Liu et al. 2018; Tao et al. 2016). Similarly, GO induces oxidative stress and absorbs metals and organic contaminants (De Marchi et al. 2018).

## 14.8 Control Strategies

Reduced inhibition of the anammox process could improve nitrogen elimination efficacy. The following methods can be used to reduce anammox activity inhibition.

### 14.8.1 Temperature

The optimal temperature range for the anammox process is between 30°C and 40°C. At temperatures above 45°C and below 30°C, anammox activity declines irreversibly. However, it has been suggested that anammox activity might be detected at temperatures as low as 20°C and that the anammox microorganisms themselves could acclimate to even colder environments. These reported results suggested that the anammox method may be acclimatized and then used to treat diverse nitrogen-rich wastewater at various temperatures (Table 14.2). However, at the optimal temperature range, anammox's ability to remove nitrogen was still unsatisfactory even after acclimation (Ma et al. 2019). For example, nitrogen remediation rates were lower at 6.3–25°C. However, the fluctuations in temperature that occur throughout the anammox

**Table 14.2** Effect of temperature on Anammox performance.

| Temperature | Anammox bacteria | Reactor | Nitrogen removal rate | References |
|---|---|---|---|---|
| 18–35°C | *C. Jettenia* | Continuous Anammox Bio-carrier Reactor (CABR) | 0.1–0.3 kg N $m^{-3}$ $d^{-1}$ | Le et al. (2022) |
| 15°C | *Kuenenia stuttgartiensis* | Sequencing batch reactor | 0.05 g N $l^{-1}$ $d^{-1}$ | Dosta et al. (2008) |
| 24°C | *Candidatus Brocadia* | Sequencing batch reactor | 0.05–0.16 g N $l^{-1}$ $d^{-1}$ | Del Rio et al. (2018) |
| 13–18°C | *Candidatus Kuenenia* | Up-flow anaerobic sludge blanket reactor | 5.16–5.40 kg N $m^{-3}$ $d^{-1}$ | He et al. (2018) |
| 30°C | *Candidatus Brocadia* | Membrane bioreactor | 0.33 g N $l^{-1}$ $d^{-1}$ | Hoekstra et al. (2018) |
| 40°C | *Candidatus Brocadia* | Sequencing batch reactor | 1.3 g N $l^{-1}$ $d^{-1}$ | Sobotka et al. (2016) |
| 30°C | Ca. *Brocadia* | Sequencing batch reactor | 79.4 ± 7.83% | Banach-Wiśniewska et al. (2021) |

process influence the concentrations of unionized ammonia and unionized nitrous acid. Therefore, temperature control is crucial for regulating anammox inhibition.

Anammox bacteria perform and thrive best at 30–40°C, preventing other negative effects including inhibitory effects. Due to the presence of free ammonia, excessive nitrite concentrations, and other hazardous elements, a few studies have suggested that temperatures above 35°C can limit the growth of some types of substrates. According to certain studies, temperatures between 40°C and 45°C can be harmful to anammox bacteria because they cause bacterial cell lysis due to the liberation of cytochrome c following a temperature shock, which lowers anammox activity. There has recently been an increase in interest in using the anammox technique at higher latitudes with low temperatures between 10°C and 20°C. However, at lower temperatures (below 20°C), the bacterial growth rate decreases due to the reduced enzymatic activity, decreased cytoplasm density, and reduced rate of mass transfer. Additionally, anammox systems running at lower temperatures might result in an increase in ladderane fatty acid content, which in turn limits membrane permeability and passive protons passage out of the anammoxosome, inhibiting ATP synthesis and so slowing metabolism. Consequently, temperature variability in anammox processes has an impact on the effectiveness of those processes.

### 14.8.2 pH Control

The pH shift in the culture can have a significant impact on the anammox process owing to the deposition of toxic chemicals that are responsible for the suppression of anammox activity. To sustain a high NRR over time, it is vital to maintain a steady pH. The pH range of 7–8 is ideal for the anammox process and appears to be the perfect range for high anammox system performance (Table 14.3). The pH range of 7–8 reduces the inhibition of the anammox process due to the presence of non-ionized nitrous acid and ammonia. Due to low pH (less than 7), free ammonia concentration falls, whereas unionized nitrous acid concentration increases. At high pH, free ammonia concentration increases and free nitrous acid concentration declines. To prevent unionized ammonia and nitrous acid inhibition, the pH is often adjusted to a neutral value.

### 14.8.3 Nitrogen Loading Rate and Substrate Concentration

Controlling the substrate concentration is a crucial strategy for avoiding anammox bacteria inhibition. When present in concentrations lower than the inhibition threshold, substrates do not affect the activity of anammox. Controlling the substrate concentration at a lower level is the most straightforward way to prevent anammox inhibition but doing so too early in the starting process can lead to substrate limitation, as the substrate is critical for anammox biomass growth (Tsushima et al. 2007). Additionally, because the anammox process needs influents with low substrate concentrations, it cannot be used to treat wastewater with increased nitrogen concentrations, which is typical of many different types of wastewater.

**Table 14.3** Effect of pH on Anammox performance.

| pH | Anammox bacteria | Reactor | Nitrogen removal rate | References |
|---|---|---|---|---|
| 6.5–8.5 | *Candidatus Brocadia* | Sequencing batch reactor<br>Biofilter | 1.1 g $N^{-1}$ $d^{-1}$ | Yin et al. (2016) |
| 7.5 | *Candidatus_Kuenenia* | – | More than 85% | Ma et al. (2022) |
| 7.50–8.50 | *Brocadia anammoxidans*, *Brocadia* sp., *Candidatus Kuenenia*, and *Brocadia fulgida* | Sequencing batch reactor | 0.25 g N $l^{-1}$ $d^{-1}$ | Isanta et al. (2015) |
| 7.3 | *Candidatus Brocadia fulgida* | Membrane bioreactor | 0.4 g N $l^{-1}$ $d^{-1}$ | Lotti et al. (2014) |
| 7–8 | *Kuenenia stuttgartiensis* | Sequencing batch reactor | 0.05 g N $l^{-1}$ $d^{-1}$ | Dosta et al. (2008) |

As a result, this issue needs to be fixed by employing the technique of effluent recirculation during the treatment of nitrogen-rich wastewater that has a low substrate concentration.

With other appropriate operating conditions, effluent recirculation helps to enhance the NRR. Based on the substrate concentrations, Chen et al. (2011) used various effluent recirculation ratios. Using this method, the NRR reached up to $34.5\,kg\,N\,m^{-3}\,d^{-1}$ with a doubling time of 4.3–7.4 days for the anammox bacteria. An increased NRR of $26\,kg\,N\,m^{-3}\,d^{-1}$ was reached at a higher influent ammonium-to-nitrite ratio. The influent ammonium-to-nitrite ratio was a key factor in determining nitrogen removal efficiency.

Moreover, it was observed that an increased NRR of $26\,kg\,N\,m^{-3}\,d^{-1}$ was reached at a higher influent ammonium-to-nitrite ratio. Thus, the influent ammonium-to-nitrite ratio was also considered a key factor in determining nitrogen removal efficiency (Tsushima et al. 2007). A higher nitrogen loading rate (NLR) makes the anammox process susceptible to variations in operational conditions and reduces its resilience to adverse conditions. Particularly, overloads frequently cause system instability (Trigo et al. 2006). To be safe, an NLR level lower than the maximum tolerated value should be maintained to prevent the Anammox from becoming unstable. Jaroszynski et al. (2011) found that the pH did not affect the NRR when reactors were running at low or medium loads. However, when the NLR was high, changing the pH from 7.5 to 8.1 reduced the NRR by 61%.

### 14.8.4 Dissolved Oxygen (DO) Control

Anammox process relies heavily on DO as an important operational parameter. Anammox activity has been studied at extremely low oxygen concentrations, even within the detection limit of $0.01\,mg\,l^{-1}$, to address the susceptibility of anammox bacteria to oxygen levels (Dalsgaard et al. 2012). Anoxic circumstances promote anammox metabolism, especially enzyme synthesis, and stimulation of the anammox reaction. Wang et al. (2016) reported that increasing the DO concentration from 2 to $8\,mg\,O_2\,l^{-1}$ in a bioreactor caused a decrease in the NRR. Similarly, Liu et al. (2008) claimed that an increase in the DO concentration above $0.4\,mg\,O_2\,l^{-1}$ led to a reduced NRR and high nitrate concentrations. However, it was also suggested that DO at a modest level (2% air saturation) reversibly hindered the anammox process (Jin et al. 2012). Egli et al. (2001) discovered that reversible suppression of the anammox process occurred at low oxygen concentrations (1% air saturation), whereas irreversible inhibition occurred at greater oxygen concentrations (>18% air saturation). Therefore, DO must be closely regulated in anammox systems to prevent oxygen inhibition.

### 14.8.5 Oxidation–Reduction Potential (ORP) Control

ORP shows the concentration of DO, the organic substrate, the activity of organisms, and some harmful byproducts in reactor systems under certain operational conditions (Kim and Hao 2001). It is reported that ORP is used to monitor and control the nitrification and denitrification process online (Lackner and Horn 2012). When compared to pH or DO levels, the ORP was thought to be a more reliable measure for process control. According to Lackner and Horn (2012), the ORP was a useful control parameter in a nitritation-anammox system.

### 14.8.6 Sludge/Biofilm Acclimatization

Anammox bacteria can be found in a variety of hazardous settings, including coke-oven wastewater sludge, high-organic load, and methanol methanogenesis sludge (Toh et al. 2002). By acclimating to sludge, anammox bacteria become resistant to inhibitory factors. For example, anammox bacteria are adapted to phenolic wastewater, antibiotic pharmaceutical wastewater, and high-salinity wastewater (Jin et al. 2011). In addition to this, the incorporation of auxiliary agents during the acclimatization phase was beneficial to the sludge culture and enhanced inhibition relief. Hydrazine or the anammox intermediate hydroxylamine, for example, can help to relieve nitrite inhibition (Strous et al. 1999b).

There are many types of anammox biomass in reactor systems. These include biofilm, granular sludge, and free cell or flocculent sludge. Granule and biofilm systems provide an advantage for the growth of anammox bacteria since anammox bacteria tend to form in aggregates. The granule and biofilm methods are frequently employed to increase the number of anammox bacteria. The inner layers of the anammox bacteria are less exposed to inhibitors like DO and hence the granule and biofilm systems can effectively support biomass (You et al. 2020).

Biomass retention is important to anammox's strong NRR. Biomass is simple to remove in Anammox systems that use free cells or flocculent sludge. The membrane bioreactor may hold flocculent anammox sludge better (Trigo et al. 2006). Therefore, the choice of reactor configuration is equally essential for sludge/biofilm acclimation.

### 14.8.7 Addition of Fresh Anammox Sludge

An anammox bacterial concentration of $10^{10}$–$10^{11}$ cells $ml^{-1}$ is required for the bacteria to multiply (Strous et al. 1999b). The NRR was shown to be directly related to the number of anammox bacteria, as reported by Isaka et al. (2006). The addition of fresh anammox sludge to the anammox reactor raised sludge concentration and decreased unit-sludge loading rate in a short period and may have introduced growth stimulants, reducing Anammox inhibition (Tang et al. 2011).

## 14.9 Conclusion and Prospects

The anammox method is a cost-effective and efficient biological wastewater treatment system. However, anammox bacteria are susceptible to environmental contaminants and require strict survival conditions. Inhibition of anammox process is a significant challenge to the wide-scale implementation of biological nitrogen eradication. Substrates such as nitrite and ammonia, organic substances including hazardous and nontoxic chemical substances, salinity, heavy metals, phosphate, and sulfur dioxide all limit the anammox activity at concentrations over the inhibition thresholds. The anammox species, exposure dosage, and all operating parameters have an inhibitory impact on these anammox inhibitors. At moderate doses, some inhibitors (such as NaCl and a few species of organic materials) unexpectedly promote the nitrogen removal of anammox.

However, proper procedures can control anammox inhibitions. The prevention and relief of anammox inhibition can be achieved by adjusting the substrate content and loading rate, the pH, the DO, and ORP, sludge acclimation, adding auxiliary agents, or adding fresh anammox sludge to inhibited anammox systems. Variations in anammox sludge (or dominating species) and the intricacy of real wastewater present obstacles to the anammox method for removing nitrogen. The following areas of anammox inhibition need more research: Anammox's ability to remove nitrogen from specific wastewater must be studied. Research into the specific inhibition issues posed by the many different types of wastewater would make it possible to develop a more tailored strategy for the relief of anammox inhibition. It is necessary to study how different Anammox species respond to inhibitors. More research is needed to better understand the factors that hinder Anammox, such as the ammonium-to-nitrite ratio, organic matter (both innocuous and harmful organic matter), heavy metals, salts, phosphate, and sulfide. Research on composite inhibition is desired. The effects of various inhibitors on the Anammox system are frequently interdependent. As there are few reports on anammox's composite inhibition, which results from the combined effects of numerous inhibitory factors, it is recommended that more studies be done on this topic. It is necessary to research inhibitory mechanisms using molecular ecology technologies. Further research that integrates the novel pure bacteria cultural methods with technology for molecular ecology will help in identifying anammox molecular ecology processes and enhance our knowledge of anammox inhibition.

## References

Ali, M. and Okabe, S. (2015). Anammox-based technologies for nitrogen removal: advances in process start-up and remaining issues. *Chemosphere* 141: 144–153.

Banach-Wiśniewska, A., Ćwiertniewicz-Wojciechowska, M., and Ziembińska-Buczyńska, A. (2021). Effect of temperature shifts and anammox biomass immobilization on sequencing batch reactor performance and bacterial genes abundance. *International Journal of Environmental Science and Technology* 18 (7): 1719–1730.

Cao, S., Du, R., and Zhou, Y. (2021). Coupling anammox with heterotrophic denitrification for enhanced nitrogen removal: a review. *Critical Reviews in Environmental Science and Technology* 51 (19): 2260–2293.

Chen, T., Zheng, P., Shen, L. et al. (2011). Kinetic characteristics and microbial community of Anammox-EGSB reactor. *Journal of Hazardous Materials* 190 (1–3): 28–35.

Cho, S., Kambey, C., and Nguyen, V.K. (2019). Performance of anammox processes for wastewater treatment: a critical review on effects of operational conditions and environmental stresses. *Water* 12 (1): 20.

Dalsgaard, T., Thamdrup, B., Farías, L., and Revsbech, N.P. (2012). Anammox and denitrification in the oxygen minimum zone of the eastern South Pacific. *Limnology and Oceanography* 57 (5): 1331–1346.

Dapena-Mora, A., Fernandez, I., Campos, J.L. et al. (2007). Evaluation of activity and inhibition effects on Anammox process by batch tests based on the nitrogen gas production. *Enzyme and Microbial Technology* 40 (4): 859–865.

Dapena-Mora, A., Vázquez-Padín, J.R., Campos, J.L. et al. (2010). Monitoring the stability of an Anammox reactor under high salinity conditions. *Biochemical Engineering Journal* 51 (3): 167–171.

De Marchi, L., Pretti, C., Gabriel, B. et al. (2018). An overview of graphene materials: properties, applications and toxicity on aquatic environments. *Science of the Total Environment* 631: 1440–1456.

Del Rio, A.V., Pichel, A., Fernandez-Gonzalez, N. et al. (2018). Performance and microbial features of the partial nitritation-anammox process treating fish canning wastewater with variable salt concentrations. *Journal of Environmental Management* 208: 112–121.

Dosta, J., Fernández, I., Vázquez-Padín, J.R. et al. (2008). Short-and long-term effects of temperature on the Anammox process. *Journal of Hazardous Materials* 154 (1–3): 688–693.

Dosta, J., Nieto, J.M., Vila, J. et al. (2011). Phenol removal from hypersaline wastewaters in a membrane biological reactor (MBR): operation and microbiological characterisation. *Bioresource Technology* 102 (5): 4013–4020.

Egli, K., Fanger, U., Alvarez, P.J. et al. (2001). Enrichment and characterization of an anammox bacterium from a rotating biological contactor treating ammonium-rich leachate. *Archives of Microbiology* 175 (3): 198–207.

Elreedy, A., Ismail, S., Ali, M. et al. (2021). Unraveling the capability of graphene nanosheets and $\gamma$-$Fe_2O_3$ nanoparticles to stimulate anammox granular sludge. *Journal of Environmental Management* 277: 111495.

Fernandez, I., Mosquera-Corral, A., Campos, J.L., and Mendez, R. (2009). Operation of an Anammox SBR in the presence of two broad-spectrum antibiotics. *Process Biochemistry* 44 (4): 494–498.

Fernández, I., Dosta, J., Fajardo, C. et al. (2012). Short-and long-term effects of ammonium and nitrite on the Anammox process. *Journal of Environmental Management* 95: S170–S174.

Fu, J., Zhang, Q., Huang, B. et al. (2021). A review on anammox process for the treatment of antibiotic-containing wastewater: linking effects with corresponding mechanisms. *Frontiers of Environmental Science & Engineering* 15 (1): 1–15.

Giustinianovich, E.A., Campos, J.L., Roeckel, M.D. et al. (2018). Influence of biomass acclimation on the performance of a partial nitritation-anammox reactor treating industrial saline effluents. *Chemosphere* 194: 131–138.

Gonzalez-Silva, B.M., Rønning, A.J., Andreassen, I.K. et al. (2017). Changes in the microbial community of an anammox consortium during adaptation to marine conditions revealed by 454 pyrosequencing. *Applied Microbiology and Biotechnology* 101 (12): 5149–5162.

Gutwiński, P., Cema, G., Ziembińska-Buczyńska, A. et al. (2021). Long-term effect of heavy metals Cr(III), Zn(II), Cd(II), Cu(II), Ni(II), Pb(II) on the anammox process performance. *Journal of Water Process Engineering* 39: 101668.

Güven, D., Dapena, A., Kartal, B. et al. (2005). Propionate oxidation by and methanol inhibition of anaerobic ammonium-oxidizing bacteria. *Applied and Environmental Microbiology* 71 (2): 1066–1071.

He, S., Chen, Y., Qin, M. et al. (2018). Effects of temperature on anammox performance and community structure. *Bioresource Technology* 260: 186–195.

Hoekstra, M., de Weerd, F.A., Kleerebezem, R., and van Loosdrecht, M.C. (2018). Deterioration of the anammox process at decreasing temperatures and long SRTs. *Environmental Technology* 39 (5): 658–668.

Huang, Z., Wang, Y., Jiang, L. et al. (2018). Mechanism and performance of a self-flocculating marine bacterium in saline wastewater treatment. *Chemical Engineering Journal* 334: 732–740.

Isaka, K., Sumino, T., Yoshie, S., and Tsuneda, S. (2006). Growth characteristic of anaerobic ammonium-oxidizing bacteria in an anaerobic biological filtrated reactor. *Applied Microbiology and Biotechnology* 70 (1): 47–52.

Isaka, K., Sumino, T., and Tsuneda, S. (2007). High nitrogen removal performance at moderately low temperature utilizing anaerobic ammonium oxidation reactions. *Journal of Bioscience and Bioengineering* 103 (5): 486–490.

Isaka, K., Suwa, Y., Kimura, Y. et al. (2008). Anaerobic ammonium oxidation (anammox) irreversibly inhibited by methanol. *Applied Microbiology and Biotechnology* 81 (2): 379–385.

Isanta, E., Bezerra, T., Fernández, I. et al. (2015). Microbial community shifts on an anammox reactor after a temperature shock using 454-pyrosequencing analysis. *Bioresource Technology* 181: 207–213.

Jaroszynski, L.W., Cicek, N., Sparling, R., and Oleszkiewicz, J.A. (2011). Importance of the operating pH in maintaining the stability of anoxic ammonium oxidation (anammox) activity in moving bed biofilm reactors. *Bioresource Technology* 102 (14): 7051–7056.

Jin, R.C., Ma, C., Mahmood, Q. et al. (2011). Anammox in a UASB reactor treating saline wastewater. *Process Safety and Environmental Protection* 89 (5): 342–348.

Jin, R.C., Yang, G.F., Yu, J.J., and Zheng, P. (2012). The inhibition of the Anammox process: a review. *Chemical Engineering Journal* 197: 67–79.

Kim, H. and Hao, O.J. (2001). pH and oxidation–reduction potential control strategy for optimization of nitrogen removal in an alternating aerobic–anoxic system. *Water Environment Research* 73 (1): 95–102.

Kuliński, K., Rehder, G., Asmala, E. et al. (2022). Biogeochemical functioning of the Baltic Sea. *Earth System Dynamics* 13 (1): 633–685.

Lackner, S. and Horn, H. (2012). Evaluating operation strategies and process stability of a single stage nitritation–anammox SBR by use of the oxidation–reduction potential (ORP). *Bioresource Technology* 107: 70–77.

Lackner, S., Gilbert, E.M., Vlaeminck, S.E. et al. (2014). Full-scale partial nitritation/anammox experiences – an application survey. *Water Research* 55: 292–303.

Lan, C.J., Kumar, M., Wang, C.C., and Lin, J.G. (2011). Development of simultaneous partial nitrification, anammox and denitrification (SNAD) process in a sequential batch reactor. *Bioresource Technology* 102 (9): 5514–5519.

Le, L.T., Jeon, J., Dang, B.T. et al. (2022). Influence of temperature on anammox reaction and microbial diversity in a bio-carriers reactor under mainstream conditions. *Environmental Technology & Innovation* 25: 102178.

Li, J., Li, J., Gao, R. et al. (2018). A critical review of one-stage anammox processes for treating industrial wastewater: optimization strategies based on key functional microorganisms. *Bioresource Technology* 265: 498–505.

Lin, L., Pratt, S., Rattier, M., and Ye, L. (2020). Individual and combined effect of salinity and nitrite on freshwater Anammox bacteria (FAB). *Water Research* 169: 114931.

Liu, S., Yang, F., Xue, Y. et al. (2008). Evaluation of oxygen adaptation and identification of functional bacteria composition for anammox consortium in non-woven biological rotating contactor. *Bioresource Technology* 99 (17): 8273–8279.

Liu, W., Zeng, Z., Chen, A. et al. (2018). Toxicity effects of silver nanoparticles on the freshwater bivalve *Corbicula fluminea*. *Journal of Environmental Chemical Engineering* 6 (4): 4236–4244.

Liu, Y., Li, Y., Pan, B. et al. (2021). Application of low dosage of copper oxide and zinc oxide nanoparticles boosts bacterial and fungal communities in soil. *Science of the Total Environment* 757: 143807.

Lotti, T., Kleerebezem, R., Lubello, C., and Van Loosdrecht, M.C.M. (2014). Physiological and kinetic characterization of a suspended cell anammox culture. *Water Research* 60: 1–14.

Lu, H., Li, Y., Shan, X. et al. (2019). A holistic analysis of ANAMMOX process in response to salinity: from adaptation to collapse. *Separation and Purification Technology* 215: 342–350.

Ma, B., Wang, S., Cao, S. et al. (2016). Biological nitrogen removal from sewage via anammox: recent advances. *Bioresource Technology* 200: 981–990.

Ma, H., Zhang, Y., Xue, Y. et al. (2019). Relationship of heme c, nitrogen loading capacity and temperature in anammox reactor. *Science of the total environment* 659: 568–577.

Ma, B., Zhang, X., Wei, D. et al. (2022). Simultaneous nitrogen and sulfur removal using Anammox coupled sulfide-denitrification process: impact of pH. *Journal of Water Process Engineering* 49: 103176.

Mason, S.A., Garneau, D., Sutton, R. et al. (2016). Microplastic pollution is widely detected in US municipal wastewater treatment plant effluent. *Environmental Pollution* 218: 1045–1054.

Oshiki, M., Shimokawa, M., Fujii, N. et al. (2011). Physiological characteristics of the anaerobic ammonium-oxidizing bacterium 'Candidatus Brocadia sinica'. *Microbiology* 157 (6): 1706–1713.

Shaw, D.R., Ali, M., Katuri, K.P. et al. (2020). Extracellular electron transfer-dependent anaerobic oxidation of ammonium by anammox bacteria. *Nature Communications* 11 (1): 1–12.

Sobotka, D., Czerwionka, K., and Makinia, J. (2016). Influence of temperature on the activity of anammox granular biomass. *Water Science and Technology* 73 (10): 2518–2525.

Strous, M., Fuerst, J.A., Kramer, E.H. et al. (1999a). Missing lithotroph identified as new planctomycete. *Nature* 400 (6743): 446–449.

Strous, M., Kuenen, J.G., and Jetten, M.S. (1999b). Key physiology of anaerobic ammonium oxidation. *Applied and Environmental Microbiology* 65 (7): 3248–3250.

Tang, C.J., Zheng, P., Mahmood, Q., and Chen, J.W. (2009). Start-up and inhibition analysis of the Anammox process seeded with anaerobic granular sludge. *Journal of Industrial Microbiology and Biotechnology* 36 (8): 1093.

Tang, C.J., Zheng, P., Wang, C.H. et al. (2011). Performance of high-loaded ANAMMOX UASB reactors containing granular sludge. *Water Research* 45 (1): 135–144.

Tang, L., Su, C., Chen, Y. et al. (2021). Influence of biodegradable polybutylene succinate and non-biodegradable polyvinyl chloride microplastics on anammox sludge: performance evaluation, suppression effect and metagenomic analysis. *Journal of Hazardous Materials* 401: 123337.

Tao, W., Chen, G., Zeng, G. et al. (2016). Influence of silver nanoparticles on heavy metals of pore water in contaminated river sediments. *Chemosphere* 162: 117–124.

Toh, S.K., Webb, R.I., and Ashbolt, N.J. (2002). Enrichment of autotrophic anaerobic ammonium-oxidizing consortia from various wastewaters. *Microbial Ecology* 154–167.

Trigo, C., Campos, J.L., Garrido, J.M., and Mendez, R. (2006). Start-up of the Anammox process in a membrane bioreactor. *Journal of Biotechnology* 126 (4): 475–487.

Tsushima, I., Ogasawara, Y., Kindaichi, T. et al. (2007). Development of high-rate anaerobic ammonium-oxidizing (anammox) biofilm reactors. *Water Research* 41 (8): 1623–1634.

Vale, G., Mehennaoui, K., Cambier, S. et al. (2016). Manufactured nanoparticles in the aquatic environment-biochemical responses on freshwater organisms: a critical overview. *Aquatic Toxicology* 170: 162–174.

Van de Graaf, A.A., Mulder, A., de Bruijn, P.E.T.E.R. et al. (1995). Anaerobic oxidation of ammonium is a biologically mediated process. *Applied and Environmental Microbiology* 61 (4): 1246–1251.

Van Hulle, S.W., Vandeweyer, H.J., Meesschaert, B.D. et al. (2010). Engineering aspects and practical application of autotrophic nitrogen removal from nitrogen rich streams. *Chemical Engineering Journal* 162 (1): 1–20.

Vineyard, D., Hicks, A., Karthikeyan, K.G., and Barak, P. (2020). Economic analysis of electrodialysis, denitrification, and anammox for nitrogen removal in municipal wastewater treatment. *Journal of Cleaner Production* 262: 121145.

Wang, D., Wang, Q., Laloo, A. et al. (2016). Achieving stable nitritation for mainstream deammonification by combining free nitrous acid-based sludge treatment and oxygen limitation. *Scientific Reports* 6 (1): 1–10.

Wang, L., Gu, W., Liu, Y. et al. (2022). Challenges, solutions and prospects of mainstream anammox-based process for municipal wastewater treatment. *Science of the Total Environment* 820: 153351.

Xiao, Y. and Roberts, D.J. (2010). A review of anaerobic treatment of saline wastewater. *Environmental Technology* 31 (8–9): 1025–1043.

Xu, X., Liu, G., Wang, Y. et al. (2018). Analysis of key microbial community during the start-up of anaerobic ammonium oxidation process with paddy soil as inoculated sludge. *Journal of Environmental Sciences* 64: 317–327.

Xu, J.J., Cheng, Y.F., and Jin, R.C. (2020). Long-term effects of $Fe_3O_4$ NPs on the granule-based anaerobic ammonium oxidation process: performance, sludge characteristics and microbial community. *Journal of Hazardous Materials* 398: 122965.

Yamindago, A., Lee, N., Woo, S. et al. (2018). Acute toxic effects of zinc oxide nanoparticles on *Hydra magnipapillata*. *Aquatic Toxicology* 205: 130–139.

Yan, Y., Wang, Y., Wang, W. et al. (2019). Comparison of short-term dosing ferrous ion and nanoscale zero-valent iron for rapid recovery of anammox activity from dissolved oxygen inhibition. *Water Research* 153: 284–294.

Yang, J., Li, J., Zheng, Z. et al. (2019). Effect of organic matters on anammox coupled denitrification system: when nitrite was sufficient. *Royal Society Open Science* 6 (11): 190771.

Yin, Z., dos Santos, C.E.D., Vilaplana, J.G. et al. (2016). Importance of the combined effects of dissolved oxygen and pH on optimization of nitrogen removal in anammox-enriched granular sludge. *Process Biochemistry* 51 (9): 1274–1282.

You, Q.G., Wang, J.H., Qi, G.X. et al. (2020). Anammox and partial denitrification coupling: a review. *RSC Advances* 10 (21): 12554–12572.

Zekker, I., Artemchuk, O., Rikmann, E. et al. (2021). Start-up of anammox SBR from non-specific inoculum and process acceleration methods by hydrazine. *Water* 13 (3): 350.

Zhang, Z. and Chen, Y. (2020). Effects of microplastics on wastewater and sewage sludge treatment and their removal: a review. *Chemical Engineering Journal* 382: 122955.

Zhang, W., Wang, D., and Jin, Y. (2018). Effects of inorganic carbon on the nitrous oxide emissions and microbial diversity of an anaerobic ammonia oxidation reactor. *Bioresource Technology* 250: 124–130.

Zhang, Q.Q., Bai, Y.H., Wu, J. et al. (2019). Microbial community evolution and fate of antibiotic resistance genes in anammox process under oxytetracycline and sulfamethoxazole stresses. *Bioresource Technology* 293: 122096.

Zhao, L., Wu, Q., and Ma, A. (2018). Biodegradation of phenolic contaminants: current status and perspectives. *IOP Conference Series: Earth and Environmental Science* 111 (1): 012024. IOP Publishing.

Zhou, Y., Oehmen, A., Lim, M. et al. (2011). The role of nitrite and free nitrous acid (FNA) in wastewater treatment plants. *Water Research* 45 (15): 4672–4682.

Zhu, G., Xia, C., Shanyun, W. et al. (2015). Occurrence, activity and contribution of anammox in some freshwater extreme environments. *Environmental Microbiology Reports* 7 (6): 961–969.

# 15

# Chitinolytic Microbes for Pest Management in Organic Agriculture: Challenges and Strategies

*Vikram Poria[1], Sandeep Kumar[1], Babett Greff[2], Pawan Kumar[1], Prakriti Jhilta[1], Balkar Singh[3], and Surender Singh[1]*

[1] *Department of Microbiology, Central University of Haryana, Mahendergarh, Haryana, India*
[2] *Department of Food Science, Faculty of Agricultural and Food Sciences, Széchenyi István University, Mosonmagyaróvár, Hungary*
[3] *Department of Botany, Arya PG College, Panipat, Haryana, India*

## 15.1 Introduction

During the last few decades, organic agricultural practices have gained popularity due to the growing demand for organic food products and increased awareness about the harmful effects of pesticides. Organic practices solely rely on organic inputs derived from botanicals, microorganisms, and organic material for nutrients and pest control (Ferdous et al. 2021; Singh et al. 2010). According to the Food and Agriculture Organization of the United States (FAO), nearly 72.3 million hectares of agricultural land were under organic cultivation worldwide in 2021, with a market size of approximately 106 billion Euros (Willer et al. 2021). Nonetheless, different pests (e.g., pathogenic bacteria, fungi, pests, and weeds) pose a major threat to organic agriculture, resulting in 20–40% yield loss every year (Khan et al. 2021). Overall, an estimated average of 30% of rice, 22.5% of maize, 21.5% of wheat, 21.4% of soybean, and 17.2% of potato yields are lost due to diseases and pests (Kashyap and Kumar 2021).

Chemical pesticides are widely used to control phytopathogens and pests. In 2020, a total of 3.1 million tons of agrochemicals were used globally (FAO 2022). Although the average pesticides used per unit of cropland is around 2.63 kg/ha (Zhu and Wang 2021; Santos et al. 2021), the overuse of these products leads to environmental degradation, contamination of surface and groundwater, pathogen resistance, and an increased cost of crop production (Nicolopoulou-Stamati et al. 2016; Pandit et al. 2022). As a result, there is a rising consumer demand for food produced without synthetic chemicals across the world (Fruscella et al. 2021).

Pest management in organic cultivation presents unique challenges compared to conventional agriculture due to the restricted use of synthetic chemical pesticides. Biological control using natural enemies, such as predators and parasitoids, is a key component of organic pest management. However, the effectiveness of biological control can be limited by factors such as weather, habitat availability, and the presence of nontarget organisms. In addition, the use of biological control agents may require specialized knowledge and skills and may not be cost-effective in some situations (Hamadttu Abdel Farag 2019; Headrick 2021).

The use of chitinolytic microorganisms as biocontrol agents is a promising approach to enhance and complement existing pest management strategies in organic agriculture. Chitinase is an enzyme that degrades chitin, which is a structural component of the cell walls of many pests and pathogens, including insects, fungi, and nematodes. Chitinase disrupts the integrity of the cell walls, leading to the death of the insect or pathogen (Liu et al. 2019). Various groups of organisms, including bacteria, actinomycetes, fungi, or even plants, may be used to control different pests by employing chitinases. Many BCAs containing organisms or the purified form of chitinase are available commercially (Poria et al. 2021a).

In this chapter, we discuss the importance of chitinolytic microorganisms in organic agriculture and their potential role as biocontrol agents for pest management. Additionally, we will address the challenges associated with their use and propose strategies for their sustainable integration into organic farming systems.

*Microbes Based Approaches for the Management of Hazardous Contaminants*, First Edition. Edited by Ajay Kumar, Livleen Shukla, Joginder Singh, and Luiz Fernando R. Ferreira.

## 15.2 Alternatives to Agrochemicals in Organic Agriculture for Pest Management

Approximately 40% of the annual food production is lost owing to insect infestation; consequently, effective pest management employing a wide variety of pesticides is essential to combat pests and boost crop yield. In 2019, nearly two million tons of pesticides were used around the world, with 47.5% of herbicides, 29.5% of insecticides, 17.5% of fungicides, and 5.5% of other agents (Sharma et al. 2019). The excessive use of agrochemicals may damage the environment and negatively influence diverse life forms by posing health hazards (respiratory problems, skin irritation, and neurological disorders), environmental pollution (contaminating soil, water, and air, leading to environmental pollution), killing of non-target organisms, such as pollinators and natural enemies of pests, biomagnification, development of resistance in pests, soil degradation, and economic burden (Rajmohan et al. 2020; Mahmood et al. 2016). Some of these agrochemicals, like Bensulide, Chloropicrin, Dichlobenil, Dicrotophos, EPTC, Norflurozon, Oxytetracycline, Paraquet, Phorate, Terbufos, and Tribufos, are banned for use in the European Union, China, and some other countries (Donley 2019).

Several pests like western corn root worm, cotton ball worm, Colorado potato beetle, *Pseudomonas aeruginosa*, *Plasmopara viticola*, *Zymoseptoria* spp., *Aspergillus fumigatus*, *Botrytis cinerea*, *Rhynchosporium commune,* and weeds like common ragweed, waterhemp, and *Palmer amaranth* have been reported to become resistant to chemical pesticides due to their overuse in agricultural practices (Sharma et al. 2019; Hawkins et al. 2019). Thus, the use of organic or inorganic amendments, natural chemicals, and microorganisms or microbial enzymes can be more sustainable and effective in controlling pests and pathogens. Beneficial bacteria and fungi, as well as organic or inorganic amendments, can assist in increasing soil fertility and plant growth while stifling the growth and activity of undesirable microbes (Yatoo et al. 2021; Singh et al. 2020). Botanicals and microbial enzymes can manage diseases and pests efficiently since they are frequently more targeted than chemical pesticides and less harmful to species other than their intended target (Kumar et al. 2021).

### 15.2.1 Climate-Friendly Practices

Climate-friendly practices are based on the strategic deployment of knowledge for making pest habitat inhospitable to pests, as well as good agricultural practices (GAPs) (i.e., integrated polyculture design, monitoring and management of surrounding environments, crop rotations, use of companion crops, trap crops, soil quality enhancement, timely seeding, weeding, harvesting, sanitation, mechanical control, genetic resistance, choice of resistant varieties, and plant diversity), to achieve a significant reduction in pest numbers to allow natural enemies or biological control agents to act effectively (Walgenbach 2018; Singh et al. 2013; Prasanna et al. 2014).

For instance, total sugar and crude protein of a crop have a positive correlation with fruit borer infestation, while phenols, tannins, and flavonoids increase insect resistance in plants (Haldhar et al. 2018). This knowledge can be utilized to manage pest populations by using plants with high levels of phenols, tannins, and flavonoids as natural insect repellents.

Another such approach is the use of resistant varieties. There are several wild varieties of crops that are resistant to different pests. In a study, Liu et al. (2020a) screened the resistant varieties of *Ipomoea batatas* and reported higher expression of chitinase of the GH19 family, which inhibited the growth of *Ceratocystis fimbriata*, *Fusarium solani*, *Ustilaginoidea virens*, and *Verticillium dahlia*. Increasing vegetation diversity in agroecosystems can reduce pests and diseases through resource dilution and stimuli-deterrent diversion, disruption of the spatial cycle and temporal cycle, allelopathic effects, general and specific soil suppressiveness, crop physiological resistance, conservation of natural enemies and facilitation of their action against aerial pests, and direct as well as indirect architectural/physical effects (Ratnadass et al. 2012).

Although climate-friendly practices make the environment less attractive to pests and make it more difficult for them to survive, disperse, grow, and reproduce, they are not a foolproof method of pest control.

### 15.2.2 Use of Microbial Enzymes

Beneficial microbes, such as bacteria and fungi, can produce a variety of compounds with antimicrobial properties that can control pathogens. These compounds include lytic enzymes, which can break down the cell wall of pathogenic microorganisms, and antibiotics, which can inhibit the growth of these microorganisms (Köhl et al. 2019). For example, *Rhizobium* and *Azospirillum* can colonize plant roots and produce enzymes such as cellulases, proteases, and chitinases that can degrade plant cell walls and inhibit the growth of pathogenic fungi. Similarly, fungi such as *Trichoderma* and mycorrhizal

**Table 15.1** Significance of lytic enzymes in pest management in organic agriculture.

| Enzyme | Pest | Crop | References |
|---|---|---|---|
| β-1,3-Glucanase | *Botrytis cinerea, Alternaria panax* | Ginseng | Wang et al. (2021b) |
| β-1,3-Glucanase | *F. oxysporum* | Spinach | Ueki et al. (2020) |
| β-1,3-Glucanase | *Magnaporthe oryzae* | Rice | Wang et al. (2021c) |
| β-1,3-Glucanase | *Botrytis cinerea* | Strawberry | Wang et al. (2021a) |
| β-1,3-Glucanase | *Penicillium* spp. | Citrus | Rodrigues et al. (2020) |
| Chitinase, protease, and lipase | *Fusarium oxysporum, Rhizoctonia solani* | Chickpea | Kumar et al. (2022a) |
| Chitinase, protease, and lipase | *Uromyces appendiculatus* | Common bean | Abo-Elyousr et al. (2021) |
| Protease and lipase | *Bemisia tabaci* | Eggplant | Islam et al. (2019) |
| Protease and lipase | *Meloidogyne javanica* | Tomato | Kavari et al. (2019) |
| Lipase | *Spodoptera litura* | Castor, tobacco | Supakdamrongkul et al. (2010) |
| Lipase | *Xylella fastidiosa* | Grapes | Zhang et al. (2015) |
| Lipase | *Pieris brassicae* | Cabbage | Dhawan and Joshi (2017) |
| Chitinase, protease, and pectinase | *Fusarium solani* | Common bean | Mohamadpoor et al. (2022) |
| Chitinase | *Phytophthora* spp. | Cucumber | Liu et al. (2022a) |
| Chitinase | *Fusarium, Rhizoctonia* | Tomato | Malik et al. (2022) |
| Chitinase | *Verticillium dahliae* | Olive | Reghmit et al. (2021) |
| Chitinase | *Fusarium oxysporum* | Chili | Mohiddin et al. (2021) |
| Chitinase | *Fusarium* | Onion | Moka et al. (2021) |
| Chitinase | *Leptinotarsa decemlineata* | Potato | Aktas et al. (2022) |

fungi can produce a range of lytic enzymes and antibiotics that can protect plants from fungal pathogens (Lahlali et al. 2022). The production of these lytic enzymes is the most common biocontrol strategy employed by the majority of antagonistic agents, as shown in Table 15.1.

While the use of microbial enzymes for pest management in organic agriculture is an eco-friendly approach, it may not be as effective as chemical pesticides. Microbial enzymes may have a specific activity toward targeted pests and may not control other pathogens. Their efficacy is influenced by various environmental factors, and their production, formulation, and storage pose significant challenges (Mishra et al. 2020). Therefore, it is crucial to consider the limitations and challenges associated with the use of microbial enzymes and evaluate them carefully before utilizing them as part of an integrated pest management strategy in organic farming.

### 15.2.3 Use of Biocontrol and Plant Growth-Promoting Microorganisms

In recent years, microorganisms and their bioactive products have gained attention as a possible alternative to chemicals to manage pests. Several genera of bacteria, actinomycetes, and fungi can control phytopathogens. Among these microbes, *Pseudomonas, Bacillus, Streptomyces, Agrobacterium, Serratia, Trichoderma, Aspergillus, Metarrhizium, Beauvaria, Globisporangium, Hyphochytrium*, and nonpathogenic *Fusarium* species have been previously utilized as BCAs and have great potential in sustainable pest management (Pal and Gardener 2006; Kshetri et al. 2018).

In addition to their antagonistic activities, BCAs may also possess certain plant growth-promoting (PGP) properties (Kumar et al. 2017; Gond et al. 2021; Gupta et al. 2020). PGP microorganisms may have direct and indirect effect on pests by secreting lytic enzymes (chitinase, cellulase, protease, pectinase, β-1,3-glucanase, and lipase, etc.), antibiotics, antifungal metabolites, volatile organic compounds, and siderophores, competing for space and nutrients, or by improving plant resistance by triggering induced systemic resistance (Ahmad and Rasool 2021; Carmona-Hernandez et al. 2019; Fiodor et al. 2021; Poria et al. 2021b). Different strategies for pest management are also shown in Table 15.2.

**Table 15.2** Effective strategies for pest control in organic crop production.

| Strategy | Mechanism | Pathogen | Crop | References |
|---|---|---|---|---|
| *Climate-friendly practices* | Fallow-based crop rotation reduced the seed-borne disease incidence in wheat | *Fusarium* | Wheat | Keler et al. (2020) |
| *A. Crop rotation* | Rotating *Allium fistulosum*, *Brassica juncea*, and *Triticum aestivum* reformed the soil microbial structure | *Fusarium* | Apple | Wang et al. (2022) |
| | Crop rotation changed the microbial communities in the soil and decreased the incidence of disease | *Fusarium* | Banana | Fan et al. (2022) |
| *B. Soil quality enhancement* | Soil fumigation using ammonium bicarbonate, and application of organic fertilizer combined with biocontrol agent *Bacillus amyloliquefaciens* suppressed the pathogen and reformed the bacterial and fungal microbial community in soil | *Fusarium oxysporum* | Watermelon | Xue et al. (2019) |
| | Biochar addition improved physiochemical properties and increased bacterial richness and diversity in soil | *Ralstonia solanacearum* | Tobacco | Zhang et al. (2017) |
| | Various composts suppressed pathogens by reshaping the soil microbial communities | *Fusarium oxysporum* | Tomato, melon, and basil | De Corato et al. (2019) |
| *C. Resistant varieties* | *Cuminum cyminum*, viz., var. GC-4 (Resistant) secreted two PR proteins, viz. β-1,3-glucanase and chitinase | *Fusarium oxysporum* | *C. cyminum* | Bhardwaj et al. (2021) |
| | Soybean-resistant variety has shown early induction of bioactive jasmonate and increased phenylpropanoid pathway intermediates | *Sclerotinia sclerotiorum* | Soybean | Ranjan et al. (2019) |
| | In *Gossypium barbadense*, the increased expression landscape of lncRNAs (long noncoding RNAs) involved in plant responses to infection | *Verticillium dahliae* | *G. barbadense* | Zhang et al. (2018) |
| *Use of purified/crude microbial enzymes* | Purified β-glucanase from *Aspergillus niger* degraded the fungal cell wall | *Fusarium oxysporum* and *Penicillium digitatum* | *In vitro* | El-Shora et al. (2021) |
| | Purified chitinases from *Bacillus cereus* degraded the chitin present in the fungal cell wall | *Fusarium verticillioides* | *In vitro* | Morales-Ruiz et al. (2021) |
| | Cellulase enzyme extract from co-culture of *Talaromyces* sp., *Aspergillus* sp., and *Penicillium* sp. hydrolyzed the cellulose present in the cell wall | *Phytophthora* sp. | Tomato | Jayasekara et al. (2021) |
| *Use of Biocontrol agents (BCAs)* | *Paenibacillus polymyxa* inhibited the mycelial growth of pathogens by secreting hydrolytic enzymes | *Phytophthora tropicalis* | Cacao | Chávez-Ramírez et al. (2021) |
| | *Bacillus siamensis* suppressed the pathogen by increasing the activity of defense-related enzymes | *Alternaria alternata* | Tobacco | Xie et al. (2021) |
| | *Pantoea vagans* showed antifungal activity by inhibiting the biosynthesis of ergosterol through production of metabolites | *Penicillium expansum* | Citrus | Bi et al. (2021) |

(Continued)

**Table 15.2** (Continued)

| Strategy | Mechanism | Pathogen | Crop | References |
|---|---|---|---|---|
| *Use of Plant growth-promoting microbes* | *Bacillus velezensis* CE 100 secretes enzymes that breakdown the β-glucan and protein components of phytopathogenic oomycetes cell walls (β-1,3-glucanase and protease). *B. velezensis* also showed PGP activity by producing IAA | *Phytophthora* spp. | *Chamaecyparis obtusa* Endlicher | Moon et al. (2021) |
| | *Penicillium aurantiogriseum* secretes cell wall-degrading enzymes, i.e., cellulase, β-1,3-glucanase, and protease. *Penicillium aurantiogriseum* also showed phosphate solubilizing activity as well as IAA and siderophores secretion | *Fusarium oxysporum* | *Sesamum indicum* L. | Zhao et al. (2021) |
| | *Bacillus subtilis* showed different plant growth-promoting traits and reduced pathogen growth by producing different metabolites (lipopeptides, polyketides, and dipeptides) | *Fusarium* spp. | Potato | Ben Khedher et al. (2021) |

## 15.3 Pest Management in Organic Agriculture Using Chitinolytic Microbial Agents

Since chitin is an important component of the integral structure of several pests (e.g., peritrophic matrix of insects, cell wall of fungi, eggshell of nematodes), microbial chitinases are a promising option for plant protection (Berini et al. 2018, 2019). Chitinolytic microorganisms use several mechanisms to inhibit pests' growth, and these mechanisms include cell wall lysis of fungi, cuticle degradation, parasitism, secretion of antifungal and antibacterial metabolites, and activation of systemic responses (Poria et al. 2021a; Kumar et al. 2022b).

Chitin-degrading microorganisms like *Bacillus*, *Streptomyces*, *Nocardiopsis*, *Brevibacillus*, *Paenibacillus*, *Serratia*, and *Pseudomonas* are already reported for their pest control activity (Veliz et al. 2017). Jabeen et al. (2018) demonstrated that a chitinolytic bacteria (*Stenotrophomonas maltophilia*) was highly effective against two termite species, *Coptotermes heimi* (53% mortality after 24 hours) and *Heterotermes indicola* (57% mortality after 24 hours), that are responsible for great economic losses in Asia. In a similar study, Dukare et al. (2020) isolated chitin-degrading bacteria from rhizosphere of pigeon pea, which could inhibit soil-borne *Fusarium udum* and lower the disease incidence to 14.62%. These studies highlight the potential of using chitinolytic microorganisms as a sustainable and environmentally friendly approach to pest management in agriculture.

Due to changing environmental conditions, crops face different abiotic stresses, and these stresses are also known to aggravate the pathogenic effects of different pathogens, so commercially available bio-products tend to be less efficient. Therefore, research on the isolation of abiotic stress-tolerant chitinase-producing microbes is picking up pace. Philip et al. (2020) isolated *Paenibacillus elgii* from marine soil, which inhibited the growth of *A. niger*, *T. rubrum*, *M. gypseum*, *C. albicans*, and *S. cerevisiae*. Few of the limitations with the use of these microbes are the longer period of time it takes to affect phytopathogens and their specificity, as most of the microbial agents are specific only to a group of disease-causing organisms (Canan, 2013). Moreover, certain pathogens are known to secrete suppressor molecules that downregulate or block the translation of hydrolytic enzymes in the BCAs (Yarullina et al. 2016; Jadhav et al. 2017). One viable alternative is the use of purified enzyme, which overcomes these difficulties since they act immediately after application, are active against a wide range of pathogens, and their expression cannot be controlled by the pathogens.

In the past, researchers have demonstrated the efficacy of purified chitinases. For example, a purified chitinase from Asian corn borer possessed antifungal activity against *F. graminearum*, *B. cinerea*, *R. solani*, *P. capsici*, and *C. gloeosporioides* (Liu et al. 2020b). Similarly, a recombinant enzyme isolated from *Chromobacterium violaceum* and expressed in *E. coli* was effective against *F. oxysporum* and *F. guttiforme* by inhibiting conidia germination (Sousa et al. 2019). Farag et al. (2016)

isolated, characterized, and purified chitinase from *Aspergillus terreus* that showed antifungal activity against *A. niger*, *R. solani*, *A. oryzae*, *F. solani*, *C. albicans*, and *F. oxysporum*. This enzyme also inhibited the growth of some bacteria, including *S. aureus*, *S. typhi*, and *P. aeruginosa*. Chitinase from *P. elgii* HOA73 could inhibit the spore germination of *Cladosporium* and *Botrytis cinerea* (Kim et al. 2017). Similarly, chitinase from *Paenibacillus ehimensis* MA2012 was effective against *Colletotrichum gloeosporioides* (Seo et al. 2016).

Chitinases from microorganisms adapted to harsh environmental conditions may provide extra benefits in stressful conditions: Chitinase from *Pseudomonas* adapted to cold environments was utilized against *V. dahlia*, *F. oxysporum* f. sp. *cucumerinum*, *A. niger*, and *P. macrosclerotiorum* (Liu et al. 2019). Chitinase from halotolerant bacteria *Bacillus pumilus* MCB-7 has shown both antifungal and insecticidal activity. This enzyme inhibited the egg hatching of *Scirpophaga incertulas* and prevented the growth of *A. flavus*, *A. niger*, *A. fumigates*, *C. hydrophila*, and *F. oxysporum* (Rishad et al. 2017). Chitinase is also reported for its role in regulation of pathogenesis-related (PR) genes and triggers defense signaling in plants. The inoculation of *Phytophthora capsici* along with chitin induced the ChiIV3 gene in *Capsicum annuum,* which encodes chitinase protein containing a chitin-binding domain and targets the plasma membrane of plants. ChiIV3 acts as a positive regulator of plant cell death and triggers defense signaling and upregulation of PR genes against *Phytophthora capsici* (Liu et al. 2017) (Table 15.3).

**Table 15.3** Efficacy of chitinolytic microbes against different pests and diseases.

| Chitinase source | Plant pathogen | Experiment type | Effects | References |
|---|---|---|---|---|
| *Chitinophaga ginsengisoli* | *Fusarium oxysporum*<br>*Meloidogyne incognita* | *In vitro* | 10% decrease in disease severity<br>~85% mortality (1 mg/ml) | Sharma et al. (2020) |
| *Xenorhabdus nematophila* | *Helicoverpa armigera* | *In vitro* | 11.3% mortality (2.5 μg/g of diet)<br>72% mortality (20 μg/g of diet)<br>100% mortality (40 μg/g of diet) on 6th day | Mahmood et al. (2020) |
| *Xenorhabdus indica* | *Helicoverpa armigera* | *In vitro* | 67.9% mortality (16.56 μg/g of diet) within one week | Kumar et al. (2013) |
| *Ipomoea batatas* | *Ceratocystis fimbriata*, *Fusarium solani*, *Ustilaginoidea virens*, *Verticillium dahliae* | *In vitro* | 55–80% antifungal index against different fungal pathogens at enzyme concentration (0.45 U) | Liu et al. (2020a) |
| *Corallococcus* sp. EGB | *Magnaporthe oryzae* | *In vitro* | Germination of conidia and formation of appressorium are inhibited<br>Inhibition is dose-dependent with a minimum concentration of 0.08 mg/ml | Li et al. (2019) |
| *Chromobacterium violaceum* | *Fusarium oxysporum*, *Fusarium guttiforme* | *In vitro* | The conidia germination of both species is inhibited by the enzyme at a concentration of 1 mg/ml | Sousa et al. (2019) |
| *Bacillus pumilus* RST25 | *Fusarium solani*<br>*Aspergillus niger* | *In vitro* | Inhibitory zone of 13 mm<br>Inhibitory zone of 11 mm (crude chitinase) | Gurav et al. (2017) |
| *Glycine max* | *Callosobruchus maculatus* | *In vitro* | Larval survival decreased by about 77% | Silva et al. (2018) |
| *B. pumilus* MCB-7 | *Scirpophaga incertulas* | *In vitro* | 28% larval mortality when co-cultured with 10 ml of bacterial culture ($1 \times 10^8$ cfu/ml) | Rishad et al. (2017) |
| *Pseudomonas fluorescens* | *Helopeltis theivora* | *In vitro* | 100% tea mosquito bug mortality with chitinase (0.048 U/ml) enzyme | Suganthi et al. (2017) |
| *Metarhizium anisopliae* | *Bemisia tabaci* | *In vivo* | 18.37% mortality after six days on transgenic crop | Anwar et al. (2019) |
| *Beauveria bassiana* | *Myzus persicae* | *In vivo* | When treated with three-day-old cultures in the pH 7 and 8 media, the highest mortality (over 70%) is seen | Cheong et al. (2020) |

## 15.4 Challenges Associated With the Use of Chitinolytic Microorganisms

Organic agriculture has unique challenges when it comes to pest control. As intended, organic agriculture does not allow synthetic inputs; thus, the focus shifts toward the prevention of crop from pest infestation rather than control. Chitinolytic microorganisms have shown great potential as biocontrol agents and are an important strategy for controlling insects and diseases in a sustainable and eco-friendly manner, as they offer several advantages over traditional chemical pesticides, such as non-toxicity and minimal impact on nontarget organisms (Veliz et al. 2017). But there are several challenges associated with the use of these biocontrol agents (Fig. 15.1).

### 15.4.1 Challenges Associated With the Formulations and Delivery Systems of Chitinolytic Microorganisms

Major challenge associated with the formulations are the stability and storage of chitinolytic microorganisms. Several factors, like temperature, humidity, etc., affects the storage and viability of the microbes. Developing suitable formulations that can maintain desired population during storage and enable efficient distribution to the target place is one of the major challenges (Gopalakrishnan et al. 2016; Mejri et al. 2013).

Additionally, the delivery mechanism needs to be designed to ensure effective and precise application of the biocontrol agent. To maximize the efficiency while minimizing the possibility of adverse effects on nontarget organisms or the environment, factors like application rate, timing, and mode of application must be taken into account (Wong et al. 2019; Adetunji et al. 2018).

Low levels of pests in the crop may not cause significant losses in yield; however, if not monitored, infestation levels rise suddenly and pose a substantial decrease in yield. Models derived from available data on pest incidence in organic agriculture must be utilized for the management of infection-causing elements and focus on targeted therapy for growth management (Kashyap and Kumar 2021; Headrick 2021; Isman 2019). The establishment time for microbial agents is very long and is dependent on several factors; therefore, early pest infestation detection techniques and timely application of formulated chitinolytic microorganism is mandatory while using these agents to control pests (Bale et al. 2007).

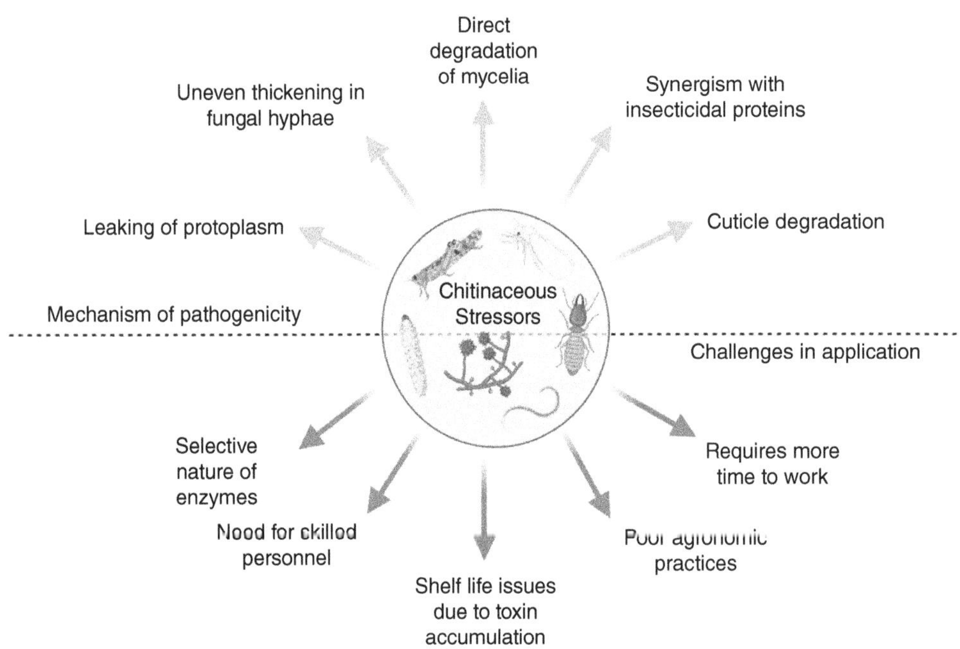


**Figure 15.1** Illustration of possible mechanism of pathogenicity of chitinases and challenges in application of enzyme/agents at field level.

### 15.4.2 Challenges Associated With Integration With Other Pest Management Strategies

Integrating chitinolytic microorganisms with other pest management strategies can present several challenges (Figure 15.1). One challenge is the potential for interactions between different control measures that can result in unintended consequences. For example, the use of pesticides alongside chitinolytic microorganisms can negatively affect the viability and activity of the microorganisms (Patibanda and Ranganathswamy 2018; Adams and Wong 1991).

Another challenge is determining the optimal timing and sequence of different control measures to achieve the desired level of pest control while minimizing negative impacts on nontarget organisms and the environment (Tang et al. 2010). The effectiveness of chitinolytic microorganisms can also be affected by factors such as crop rotation, tillage practices, and the use of cover crops, which may impact the populations and diversity of microorganisms in the soil (Gajda et al. 2018). Availability of appropriate technical expertise and infrastructure to support production, distribution, and application of chitinolytic microorganisms are crucial factors playing significant role in the success of integrating these agents with other pest management strategies (Warrior et al. 2002). Further, to promote this strategy of integrated pest management, public–private partnerships, education and training programs, and investment in research and development are required.

One of the main factors limiting the development of biological control products and their integration with IPM is their availability and dependability. Chitinolytic agents currently in use have a relatively narrow spectrum of activity and are preventative in nature, whereas chemical pesticides can be used even after infection has occurred, thus defining the priority of the end user (Jacobsen and Backman 1993). On the other hand, the integration of chitinolytic microorganisms with other strategies is also affected by different regulatory frameworks, which pose challenges, particularly in the context of international trade and harmonization of standards (Desai et al. 2016; Ehlers 2011). For instance, when supplied interstate, microbial pest control products are regulated by environmental and toxicological safety criteria issued by various governments as they are regarded as exotic or new to environments (Jacobsen and Backman 1993). Additionally, the present government programs and subsidies in developing economies have generally been targeted at medium- to large-scale farmland owners (Goodell 1984; Rodríguez and Niemeyer 2005). These challenges increase the cost as well as complexity of implementing integrated pest management strategies. Yelitza et al. (2016) have summarized the IPM adoption challenges in developing countries.

### 15.4.3 Environmental Challenges

Environmental factors such as temperature, relative humidity, and soil pH can significantly affect the efficacy of chitinolytic microorganisms as biocontrol agents. These factors play a crucial role in the performance of any microorganism in the environment. If these conditions differ greatly from the optimum conditions required for the chitinolytic microorganism to function effectively, it can result in stressful conditions for the microbe, leading to its death or failure to exhibit the desired activity (He et al. 2021; Smolińska and Kowalska 2018). De Curtis et al. (2012) conducted a study to demonstrate the influence of two environmental factors, relative humidity and temperature, on the biocontrol activity of *Metschnikowia pulcherrima* and *Aureobasidium pullulans* against fungal pathogen *Aspergillus carbonarius* and reported that the environmental factors significantly influence the biocontrol activity of biocontrol agents. Higher humidity influences the colonization rate of biocontrol agents only up to a certain range of temperature (at 20°C and 25°C).

### 15.4.4 Other Challenges

In addition, there are significant practical challenges, such as the selection of an appropriate control agent, quality of available products, availability of skilled manpower, and timely deployment of the agent (Isman 2019; Michaud 2018). Cell wall-degrading enzymes such as cellulase and pectinase are produced by several plant pathogens, as these enzymes help in degrading plant cell walls. Several chitinolytic microorganisms are also reported to produce these enzymes along with chitinases, which can lead to some detrimental effects on plant growth. The production of these enzymes is directly related to the pathogenicity of the pest (Hassan et al. 2020; Hashem and Ali 2004). Therefore, by producing cellulases and pectinases, chitinolytic microbes can indirectly increase the pathogenicity of pests. These microorganisms must first be registered for use as biopesticides, which implies that they must clear a number of legal requirements in order to be used as a pest management option. Numerous studies have already reported that the lengthy and expensive regulatory and registration procedures are a major barrier to the availability of commercial biopesticides (Lahlali et al. 2022; Saeed et al. 2021; Rani et al. 2023).

Chitinolytic microorganisms target the chitinous cell walls of few pests (mainly insects and few fungi), and therefore they are target-specific. While these pests can cause significant damage to crops, there are many other pests, such as nematodes, bacteria, and viruses, that are not affected by chitinolytic microorganisms (Kramer and Muthukrishnan, 1997). Furthermore, even among the pests that are susceptible to chitinolytic microorganisms, there may be variations in susceptibility between different strains or species because the chitin in the cell walls or exoskeletons of insects vary in structure (Casadidio et al. 2019; Elieh-Ali-Komi and Hamblin, 2016) and composition, or they may have developed a mechanism to resist the actions of chitinase enzymes (Poria et al. 2021a). However, these microbes use several other techniques, such as secondary metabolite production, induction of plant defense mechanism, and competition for resources, but a single method for pest management is unlikely to be effective against all pests (Döring et al. 2012). Thus, it requires multiple strategies to reduce pest levels. Low enzyme titers and high purification costs during product development are also the main challenges in using enzymes as a pest control strategy. Another difficulty is maintaining enzyme stability and activity for a long time to allow storage and commercialization. Enzymes are more susceptible to degradation and have a shorter shelf life when used in aqueous solutions (Mishra et al. 2020).

## 15.5 Strategies for Sustainable Use of Chitinolytic Microorganisms in Organic Agriculture

Several strategies can be used to overcome the challenges associated with using chitinolytic microorganisms for pest management. Good cultural practices include the selection of cultivars with resistance to common pests, weeds, and diseases. Since chemical inputs and genetic engineering are prohibited in organic agriculture, breeding programs can have a positive effect on pest control by producing resistant cultivars, but only against specific pests (Headrick 2021; Döring et al. 2012). These practices can be integrated with the use of chitinolytic microorganisms to lower down the chances of pest attack.

### 15.5.1 Development of Novel Formulations and Delivery Systems

Nanotechnology applications in agriculture have received much interest during the last decade (Mittal et al. 2020). Nanotools and materials can help turn contemporary agriculture into precision agriculture with minimal inputs, either acting as protectants or carriers and delivering active ingredients in a controlled and targeted manner (Mittal et al. 2020; Zehra et al. 2021). These can be used as carriers to deliver chitinolytic microorganisms or chitinase enzymes. Nanoparticles have already been reported for their antimicrobial/pesticidal activity by several researchers (Akther and Hemalatha 2019; Shantamma et al. 2020; Dhiman et al. 2021), and by using these as carriers or protectants, the efficacy of chitinolytic microorganisms can be increased.

Encapsulation is another technique that involves the use of a protective coating to improve the shelf life and stability of microorganisms, and under adverse environmental conditions, it can also improve survival. Several materials have been used for encapsulation, including alginate, chitosan, and proteins (Bonaterra et al. 2012). Various microorganisms have been encapsulated through extrusion encapsulation using alginates, as well as through emulsification using carrageenan or pectin with a solidifying agent (Rojas-Sánchez et al. 2022). Similarly, a potential method for protecting chitinolytic microorganisms during delivery in the field is through the utilization of biofilm formation techniques (Pandin et al. 2017). This approach creates a protective barrier, limiting chemical transport in the matrix and promoting intercellular signaling and metabolite production by microorganisms, potentially increasing their efficacy against pests (Muhammad et al. 2020; Rieusset et al. 2020). To create a biofilm, the microorganisms can be sprayed onto a surface coated with a biofilm-promoting agent (Kinniment and Wimpenny 1990). Such surfaces include molecules, nanoparticles, and polymers, depending upon method of application. In a similar study, Zhang et al. (2019) investigated the formation of *S. oneidensis* biofilms using a conductive polymer, poly(3-[3′-N,N,N-triethyloamino-1′-propyloxy]-4-methyl-2,5-thiophene hydrochloride), and found that the bacteria exhibited enhanced survival and metabolic activity inside the matrix. However, synthetic biofilm-promoting substances can hinder the biocontrol process depending on their surface chemistry, leading to reduced efficacy (Muhammad et al. 2020; Cegelski et al. 2009).

### 15.5.2 Integration With Other Pest Management Strategies

Chitinolytic microorganisms can be incorporated in various ways with conventional pest control methods. One strategy is to develop a consortium by combining chitinolytic microorganisms with other biocontrol agents, including bacteria, entomopathogenic fungi, nematodes, etc. The most efficient combination should be determined through laboratory and field tests of compatibility and synergistic biocontrol actions (Someya et al. 2001; Masri et al. 2022). The chitinase-producing microbes with biocontrol potential against different pests can be combined together to provide protection against a wider range of plant pests. This can be due to the fact that various chitinase-producing microorganisms may produce various types or concentrations of chitinase, which might target different kinds or stages of pest development. Combining several strains of chitinase-producing bacteria may further improve the efficacy of biocontrol by creating a more diverse and robust microbial population that is more capable of adapting to changing environmental conditions (Hongxiang et al. 2004).

Chitinolytic microbes can also be used in combination with different climate-friendly practices like crop rotation, soil quality enhancement methods, and with different pest-resistant varieties. Different amendments like organic manure, vermicompost, biochar, pressmud, and chitin-based treatments can also be used in combination with chitinolytic microbes, which can provide a suitable environment for these microbes to grow by acting as a suitable carrier and by providing nutrition to the microbes to proliferate (Sharp 2013; de Araújo et al. 2010; Bonanomi et al. 2020). These organic amendments are reported to have a positive impact on soil microflora, by reducing the pathogen load and promoting the growth of beneficial microorganisms in soil; therefore, they are commonly used to improve soil fertility and enhance plant growth. The effectiveness of this strategy, however, may vary depending on the type and quality of the organic amendment and the particular chitinolytic bacteria employed (Bonanomi et al. 2020; Su et al. 2022). Using chitinolytic microorganisms along with resistant plant cultivars can offer a multifaceted strategy for pest control in organic agriculture (Herrera-Estrella and Chet 1999).

### 15.5.3 Importance of Education and Training

Education and training play a crucial role in overcoming the challenges associated with the use of chitinolytic microorganisms in pest management. Farmers and other stakeholders need to be educated on the benefits, limitations, and proper use of these biocontrol agents. This includes information on the appropriate methods of application, compatibility with other pest management strategies, and optimal environmental conditions for their effectiveness (Bale et al. 2007). Training programs can be organized to educate farmers and extension workers on the use of chitinolytic microorganisms. These programs can include hands-on training on the application of these agents, monitoring and evaluation of their efficacy, and troubleshooting potential challenges. By providing adequate education and training, stakeholders can effectively incorporate chitinolytic microorganisms into their pest management strategies, leading to increased effectiveness and sustainability in organic agriculture (Liu et al. 2022b; Marian and Shimizu 2019).

## 15.6 Conclusion and Prospects

Eco-friendly substitutes of chemical pesticides are required for pest management in sustainable agriculture. Because of their selectivity, safety, and capacity to break down the chitin found in pests, the employment of chitinolytic bacteria has become a promising option. Chitinolytic microorganisms are compatible with organic farming practices and have no negative effect on nontarget organisms or the environment. However, there are a number of difficulties with regard to the environmental conditions, formulations, integration, and practicality that must be addressed for the effective use of chitinolytic microorganisms in pest management. In addition to develop new formulations and integrating them with existing pest management strategies, farmers also need to be trained and educated about use and benefits of these formulations. The combinations of chitinolytic microorganisms with other biocontrol agents can be explored. The employment of chitinolytic bacteria in organic agriculture has the potential to develop into a workable and sustainable pest management technique, fostering both environmental and economic sustainability. The development of regulatory framework and research on cost-effective production and formulation techniques for chitinolytic microorganisms could enhance their availability and affordability for farmers. Hence, further research is needed to explore the full potential of chitinolytic microorganisms in pest management and to develop effective strategies for their use.

## Acknowledgments

V.P. acknowledges the DBT-JRF fellowship (award letter no. DBTHRDPMU/JRF/BET-22/I/2022-23/45) provided by the Department of Biotechnology, Government of India.

## References


Abo-Elyousr, K.A.M., Abdel-Rahim, I.R., Almasoudi, N.M., and Alghamdi, S.A. (2021). Native endophytic *Pseudomonas putida* as a biocontrol agent against common bean rust caused by *Uromyces appendiculatus*. *Journal of Fungi* 7 (9): 745. https://doi.org/10.3390/jof7090745.

Adams, P.B. and Wong, J.A.L. (1991). The effect of chemical pesticides on the infection of sclerotia of *Sclerotinia minor* by the biocontrol agent *Sporidesmium sclerotivorum*. *Phytopathology* 81: 1340–1343. https://doi.org/10.1094/PHYTO-81-1340.

Adetunji, C.O., Oloke, J.K., and Osemwegie, O.O. (2018). Environmental fate and effects of granular pesta formulation from strains of *Pseudomonas aeruginosa* C1501 and *Lasiodiplodia pseudotheobromae* C1136 on soil activity and weeds. *Chemosphere* 195: 98–107. https://doi.org/10.1016/j.chemosphere.2017.12.056.

Ahmad, T. and Rasool, A. (2021). Hydrolytic enzymes and integrated pest management. In: *Microbes for Sustainable Insect Pest Management: Hydrolytic Enzyme & Secondary Metabolite – Volume 2* (ed. M.A. Khan and W. Ahmad), 59–74. Cham: Springer International Publishing https://doi.org/10.1007/978-3-030-67231-7_3.

Aktas, C., Ruzgar, D., Gurkok, S., and Gormez, A. (2022). Purification and characterization of *Stenotrophomonas maltophilia* chitinase with antifungal and insecticidal properties. *Preparative Biochemistry & Biotechnology* 53: 797–806. https://doi.org/10.1080/10826068.2022.2142942.

Akther, T. and Hemalatha, S. (2019). Mycosilver nanoparticles: synthesis, characterization and its efficacy against plant pathogenic fungi. *BioNanoScience* 9 (2): 296–301. https://doi.org/10.1007/s12668-019-0607-y.

Anwar, W., Javed, M.A., Shahid, A.A. et al. (2019). Chitinase genes from *Metarhizium anisopliae* for the control of whitefly in cotton. *Royal Society Open Science* 6 (8): 190412. https://doi.org/10.1098/rsos.190412.

Bale, J.S., van Lenteren, J.C., and Bigler, F. (2007). Biological control and sustainable food production. *Philosophical Transactions of the Royal Society, B: Biological Sciences* 363 (1492): 761–776. https://doi.org/10.1098/rstb.2007.2182.

Ben Khedher, S., Mejdoub-Trabelsi, B., and Tounsi, S. (2021). Biological potential of *Bacillus subtilis* V26 for the control of *Fusarium* wilt and tuber dry rot on potato caused by *Fusarium* species and the promotion of plant growth. *Biological Control* 152: 104444. https://doi.org/10.1016/j.biocontrol.2020.104444.

Berini, F., Katz, C., Gruzdev, N. et al. (2018). Microbial and viral chitinases: attractive biopesticides for integrated pest management. *Biotechnology Advances* 36 (3): 818–838. https://doi.org/10.1016/j.biotechadv.2018.01.002.

Berini, F., Casartelli, M., Montali, A. et al. (2019). Metagenome-sourced microbial chitinases as potential insecticide proteins. *Frontiers in Microbiology* 10: 1358. https://doi.org/10.3389/fmicb.2019.01358.

Bhardwaj, G., Ravi, I., Kumar, S., and Sharma, V. (2021). Induction of β-1,3-glucanase and chitinase activity, cloning and their characterization in the defense response of *Cuminum cyminum* plant against the fungal pathogen *Fusarium oxysporum*. *Archives of Phytopathology and Plant Protection* 54 (19–20): 1617–1632. https://doi.org/10.1080/03235408.2021.1932016.

Bi, W., Wang, R., Yang, Y. et al. (2021). *Pantoea vagans* strain BWL1 controls blue mold in mandarin fruit by inhibiting ergosterol biosynthesis in *Penicillium expansum*. *Biological Control* 161: 104639. https://doi.org/10.1016/j.biocontrol.2021.104639.

Bonanomi, G., Alioto, D., Minutolo, M. et al. (2020). Organic amendments modulate soil microbiota and reduce virus disease incidence in the TSWV-tomato pathosystem. *Pathogens* 9 (5): 379. https://doi.org/10.3390/pathogens9050379.

Bonaterra, A., Badosa, E., Cabrefiga, J. et al. (2012). Prospects and limitations of microbial pesticides for control of bacterial and fungal pomefruit tree diseases. *Trees* 26 (1): 215–226. https://doi.org/10.1007/s00468-011-0626-y.

Canan, U. (2013). Microorganisms in biological pest control – a review (bacterial toxin application and effect of environmental factors). In: *Current Progress in Biological Research* (ed. S.-O. Marina), 287–317. Rijeka, Chapter 13: IntechOpen https://doi.org/10.5772/55786.

Carmona-Hernandez, S., Reyes-Pérez, J.J., Chiquito-Contreras, R.G. et al. (2019). Biocontrol of postharvest fruit fungal diseases by bacterial antagonists: a review. *Agronomy* 9 (3): 121. https://doi.org/10.3390/agronomy9030121.

Casadidio, C., Peregrina, D.V., Gigliobianco, M.R. et al. (2019). Chitin and chitosans: characteristics, eco-friendly processes, and applications in cosmetic science. *Marine Drugs* 17 (6): 369. https://doi.org/10.3390/md17060369.

Cegelski, L., Pinkner, J.S., Hammer, N.D. et al. (2009). Small-molecule inhibitors target *Escherichia coli* amyloid biogenesis and biofilm formation. *Nature Chemical Biology* 5 (12): 913–919. https://doi.org/10.1038/nchembio.242.

Chávez-Ramírez, B., Rodríguez-Velázquez, N.D., Mondragón-Talonia, C.M. et al. (2021). *Paenibacillus polymyxa* NMA1017 as a potential biocontrol agent of *Phytophthora tropicalis*, causal agent of cacao black pod rot in Chiapas, Mexico. *Antonie Van Leeuwenhoek* 114 (1): 55–68. https://doi.org/10.1007/s10482-020-01498-z.

Cheong, P.C.H., Glare, T.R., Rostás, M. et al. (2020). Lack of involvement of chitinase in direct toxicity of *Beauveria bassiana* cultures to the aphid *Myzus persicae*. *Journal of Invertebrate Pathology* 169: 107276. https://doi.org/10.1016/j.jip.2019.107276.

de Araújo, A.S.F., de Melo, W.J., and Singh, R.P. (2010). Municipal solid waste compost amendment in agricultural soil: changes in soil microbial biomass. *Reviews in Environmental Science and Bio/Technology* 9 (1): 41–49. https://doi.org/10.1007/s11157-009-9179-6.

De Corato, U., Patruno, L., Avella, N. et al. (2019). Composts from green sources show an increased suppressiveness to soilborne plant pathogenic fungi: relationships between physicochemical properties, disease suppression, and the microbiome. *Crop Protection* 124: 104870. https://doi.org/10.1016/j.cropro.2019.104870.

De Curtis, F., de Felice, D.V., Ianiri, G. et al. (2012). Environmental factors affect the activity of biocontrol agents against ochratoxigenic *Aspergillus carbonarius* on wine grape. *International Journal of Food Microbiology* 159 (1): 17–24. https://doi.org/10.1016/j.ijfoodmicro.2012.07.023.

Desai, S., Kumar, G.P., Amalraj, E.L.D. et al. (2016). Challenges in regulation and registration of biopesticides: an overview. In: *Microbial Inoculants in Sustainable Agricultural Productivity*, Functional Applications, vol. 2 (ed. D.P. Singh, H.B. Singh, and R. Prabha), 301–308. New Delhi: Springer India https://doi.org/10.1007/978-81-322-2644-4_19.

Dhawan, M. and Joshi, N. (2017). Enzymatic comparison and mortality of *Beauveria bassiana* against cabbage caterpillar *Pieris brassicae* LINN. *Brazilian Journal of Microbiology* 48 (3): 522–529. https://doi.org/10.1016/j.bjm.2016.08.004.

Dhiman, S., Singh, S., Varma, A., and Goel, A. (2021). Phytofabricated zinc oxide nanoparticles as a nanofungicide for management of *Alternaria* blight of *Brassica*. *Biometals* 34 (6): 1275–1293. https://doi.org/10.1007/s10534-021-00342-9.

Donley, N. (2019). The USA lags behind other agricultural nations in banning harmful pesticides. *Environmental Health* 18 (1): 44. https://doi.org/10.1186/s12940-019-0488-0.

Döring, T.F., Pautasso, M., Wolfe, M.S., and Finckh, M.R. (2012). *Pest and disease management in organic farming: implications and inspirations for plant breeding*. In: *Organic Crop Breeding* (ed. E.T. Lammerts van Bueren and J.R. Myers), 39–59. Wiley https://doi.org/10.1002/9781119945932.ch3.

Dukare, A., Paul, S., and Arambam, A. (2020). Isolation and efficacy of native chitinolytic rhizobacteria for biocontrol activities against *Fusarium* wilt and plant growth promotion in pigeon pea (*Cajanus cajan* L.). *Egyptian Journal of Biological Pest Control* 30 (1): 56. https://doi.org/10.1186/s41938-020-00256-7.

Ehlers, R.-U. (2011). Regulation of biological control agents and the EU policy support action REBECA. In: *Regulation of Biological Control Agents* (ed. R.-U. Ehlers), 3–23. The Netherlands, Dordrecht: Springer https://doi.org/10.1007/978-90-481-3664-3_1.

Elieh-Ali-Komi, D. and Hamblin, M.R. (2016). Chitin and chitosan: production and application of versatile biomedical nanomaterials. *International Journal of Advanced Research* 4 (3): 411–427.

El-Shora, H.M., El-Sharkawy, R.M., Khateb, A.M., and Darwish, D.B. (2021). Production and immobilization of β-glucanase from *Aspergillus niger* with its applications in bioethanol production and biocontrol of phytopathogenic fungi. *Scientific Reports* 11 (1): 21000. https://doi.org/10.1038/s41598-021-00237-2.

Fan, P., Lai, C., Yang, J. et al. (2022). Crop rotation suppresses soil-borne *Fusarium* wilt of banana and alters microbial communities. *Archives of Agronomy and Soil Science* 68 (4): 447–459. https://doi.org/10.1080/03650340.2020.1839058.

FAO (2022). *FAO Statistical Database*. FAO- Food and Agriculture Organization of the United Nations https://www.fao.org/faostat/en/#data. Accessed 13-11-2022.

Farag, A.M., Abd-Elnabey, H.M., Ibrahim, H.A.H., and El-Shenawy, M. (2016). Purification, characterization and antimicrobial activity of chitinase from marine-derived *Aspergillus terreus*. *The Egyptian Journal of Aquatic Research* 42 (2): 185–192. https://doi.org/10.1016/j.ejar.2016.04.004.

Ferdous, Z., Zulfiqar, F., Datta, A. et al. (2021). Potential and challenges of organic agriculture in Bangladesh: a review. *Journal of Crop Improvement* 35 (3): 403–426. https://doi.org/10.1080/15427528.2020.1824951.

Fiodor, A., Singh, S., and Pranaw, K. (2021). The contrivance of plant growth promoting microbes to mitigate climate change impact in agriculture. *Microorganisms* 9 (9): 1841. https://doi.org/10.3390/microorganisms9091841.

Fruscella, L., Kotzen, B., and Milliken, S. (2021). Organic aquaponics in the European Union: towards sustainable farming practices in the framework of the new EU regulation. *Reviews in Aquaculture* 13 (3): 1661–1682. https://doi.org/10.1111/raq.12539.

Gajda, A., Czyż, E., Dexter, A. et al. (2018). Effects of different soil management practices on soil properties and microbial diversity. *International Agrophysics* 32: 81–91. http://dx.doi.org/10.1515/intag-2016-0089.

Gond, D.P., Jha, S.S., Kumar, A., and Singh, S.K. (2021). Plant growth promoting bacteria and its role in green remediation. In: *Sustainable Environmental Clean-up* (ed. V.K. Mishra and A. Kumar), 149–163. Cambridge, MA: Elsevier https://doi.org/10.1016/B978-0-12-823828-8.00007-4.

Goodell, G. (1984). Challenges to international pest management research and extension in the third world: do we really want IPM to work? *Bulletin of the Entomological Society of America* 30 (3): 18–26. https://doi.org/10.1093/besa/30.3.18.

Gopalakrishnan, S., Sathya, A., Vijayabharathi, R., and Srinivas, V. (2016). Formulations of plant growth-promoting microbes for field applications. In: *Microbial Inoculants in Sustainable Agricultural Productivity*, Functional Applications, vol. 2 (ed. D.P. Singh, H.B. Singh, and R. Prabha), 239–251. New Delhi: Springer India https://doi.org/10.1007/978-81-322-2644-4_15.

Gupta, A., Singh, S.K., Singh, M.K. et al. (2020). Plant growth–promoting rhizobacteria and their functional role in salinity stress management. In: *Abatement of Environmental Pollutants* (ed. P. Singh, A. Kumar, and A. Borthakur), 151–160. Cambridge, MA: Elsevier https://doi.org/10.1016/B978-0-12-818095-2.00007-2.

Gurav, R., Tang, J., and Jadhav, J. (2017). Novel chitinase producer *Bacillus pumilus* RST25 isolated from the shellfish processing industry revealed antifungal potential against phyto-pathogens. *International Biodeterioration & Biodegradation* 125: 228–234. https://doi.org/10.1016/j.ibiod.2017.09.015.

Haldhar, S.M., Samadia, D.K., Bhargava, R. et al. (2018). Host plant accessions determine bottom-up effect of snapmelon (*Cucumis melo* var. *momordica*) against melon fly (*Bactrocera cucurbitae* [Coquillett]). *Breeding Science* 68 (5): 499–507 https://doi.org/10.1270/jsbbs.17065.

Hamadttu Abdel Farag, E.-S. (2019). Insect pest management in organic farming system. In: *Multifunctionality and Impacts of Organic and Conventional Agriculture* (ed. M. Jan, M. Kassio Ferreira, B. Jaroslav, et al.). Rijeka: IntechOpen Chapter 8. https://doi.org/10.5772/intechopen.84483.

Hashem, M. and Ali, E. (2004). *Epicoccum Nigrum* as biocontrol agent of *Pythium* damping-off and root-rot of cotton seedlings. *Archives of Phytopathology and Plant Protection* 37 (4): 283–297. https://doi.org/10.1080/03235400310001612955.

Hassan, A.G., Amal, M.I.E., Abdel-Aal, H.M. et al. (2020). Formulations as biocontrol and role of pectinase and cellulose in pathogenicity. *Journal of Agriculture, Food and Environment* 1 (1): 1–5. https://doi.org/10.47440/JAFE.2020.1101.

Hawkins, N.J., Bass, C., Dixon, A., and Neve, P. (2019). The evolutionary origins of pesticide resistance. *Biological Reviews* 94 (1): 135–155. https://doi.org/10.1111/brv.12440.

He, D.-C., He, M.-H., Amalin, D.M. et al. (2021). Biological control of plant diseases: an evolutionary and eco-economic consideration. *Pathogens* 10 (10): 1311. https://doi.org/10.3390/pathogens10101311.

Headrick, D. (2021). The future of organic insect pest management: be a better entomologist or pay for someone who is. *Insects* 12 (2): 140. https://doi.org/10.3390/insects12020140.

Herrera-Estrella, A. and Chet, I. (1999). Chitinases in biological control. *EXS* 87: 171–184. https://doi.org/10.1007/978-3-0348-8757-1_12.

Hongxiang, Y., Jie, Z., Ruotong, H. et al. (2004). Isolation and identification of a chitinase-producing bacterium and its synergistic effect on locust biocontrol. *Plant Protection* 30 (2): 37–41.

Islam, T., Gupta, D.R., Surovy, M.Z. et al. (2019). Identification and application of a fungal biocontrol agent *Cladosporium cladosporioides* against *Bemisia tabaci*. *Biotechnology & Biotechnological Equipment* 33 (1): 1698–1705. https://doi.org/10.1080/13102818.2019.1695541.

Isman, M.B. (2019). Challenges of pest management in the twenty first century: new tools and strategies to combat old and new foes alike. *Frontiers in Agronomy* 1: 1–4. https://doi.org/10.3389/fagro.2019.00002.

Jabeen, F., Hussain, A., Manzoor, M. et al. (2018). Potential of bacterial chitinolytic, *Stenotrophomonas maltophilia*, in biological control of termites. *Egyptian Journal Of Biological Pest Control* 28 (1): 86. https://doi.org/10.1186/s41938-018-0092-6.

Jacobsen, B.J. and Backman, A. (1993). Biological and cultural plant disease controls: alternatives and supplements to chemicals in IPM systems. *Plant Disease* 77 (3): 311–315.

Jadhav, H.P., Shaikh, S.S., and Sayyed, R.Z. (2017). Role of hydrolytic enzymes of Rhizoflora in biocontrol of fungal phytopathogens: an overview. In: *Rhizotrophs: Plant Growth Promotion to Bioremediation* (ed. S. Mehnaz), 183–203. Singapore: Springer Singapore https://doi.org/10.1007/978-981-10-4862-3_9.

Jayasekara, S., Karunarathna, K., Kumara, K., and Ratnayake, R. (2021). Suppression of damping-off disease causative agents using cellulase enzyme extracts of locally isolated cellulolytic fungi and their co-cultures. *Ceylon Journal of Science* 50 (4): 487–496. http://doi.org/10.4038/cjs.v50i4.7947.

Kashyap, B. and Kumar, R. (2021). Sensing methodologies in agriculture for monitoring biotic stress in plants due to pathogens and pests. *Inventions* 6 (2): 29. https://doi.org/10.3390/inventions6020029.

Kavari, M., Mahdikhani-Moghadam, E., and Rouhani, H. (2019). Evaluation of some *Trichoderma* spp. protease and lipase activity and its effect on biocontrol of tomato root-knot nematode (*Meloidogyne javanica*). *Journal of Plant Protections* 33 (4): 363–374. https://doi.org/10.22067/jpp.v33i4.35150.

Keler, V., Khizhnyak, S., and Eskova, E. (2020). Effects of crop rotation, pesticides and fertilizers on wheat seed contamination with seed-borne *Fusarium* pathogens. In: *IOP Conference Series: Earth and Environmental Science*, vol. 7, 072005. IOP Publishing https://doi.org/10.1088/1755-1315/548/7/072005.

Khan, M., Khan, A.U., Hasan, M.A. et al. (2021). Agro-nanotechnology as an emerging field: a novel sustainable approach for improving plant growth by reducing biotic stress. *Applied Sciences* 11 (5): 2282. https://doi.org/10.3390/app11052282.

Kim, Y.H., Park, S.K., Hur, J.Y., and Kim, Y.C. (2017). Purification and characterization of a major extracellular chitinase from a biocontrol bacterium, *Paenibacillus elgii* HOA73. *The Plant Pathology Journal* 33 (3): 318–328. https://doi.org/10.5423/PPJ.FT.01.2017.0022.

Kinniment, S. and Wimpenny, J.W.T. (1990). Biofilms and biocides. *International Biodeterioration* 26 (2): 181–194. https://doi.org/10.1016/0265-3036(90)90058-F.

Köhl, J., Kolnaar, R., and Ravensberg, W.J. (2019). Mode of action of microbial biological control agents against plant diseases: relevance beyond efficacy. *Frontiers in Plant Science* 10: 845. https://doi.org/10.3389/fpls.2019.00845.

Kramer, K.J. and Muthukrishnan, S. (1997). Insect chitinases: molecular biology and potential use as biopesticides. *Insect Biochemistry and Molecular Biology* 27 (11): 887–900. https://doi.org/10.1016/S0965-1748(97)00078-7.

Kshetri, P., Roy, S.S., Sharma, S.K. et al. (2018). Feather degrading, phytostimulating, and biocontrol potential of native actinobacteria from North Eastern Indian Himalayan region. *Journal of Basic Microbiology* 58 (9): 730–738. https://doi.org/10.1002/jobm.201800169.

Kumar, P., Surender, S., Debjani, D. et al. (2013). Extracellular novel metalloprotease from *Xenorhabdus indica* and its potential as an insecticidal agent. *Journal of Microbiology and Biotechnology* 23 (11): 1536–1543. https://doi.org/10.4014/jmb.1306.06062.

Kumar, A., Verma, H., Singh, V.K. et al. (2017). Role of *Pseudomonas* sp. in sustainable agriculture and disease management. In: *Agriculturally Important Microbes for Sustainable Agriculture: Volume 2: Applications in Crop Production and Protection* (ed. V.S. Meena, P.K. Mishra, J.K. Bisht, and A. Pattanayak), 195–215. Singapore, Singapore: Springer https://doi.org/10.1007/978-981-10-5343-6_7.

Kumar, J., Ramlal, A., Mallick, D., and Mishra, V. (2021). An overview of some biopesticides and their importance in plant protection for commercial acceptance. *Plants* 10 (6): https://doi.org/10.3390/plants10061185.

Kumar, A., Kumar, S.P.J., Chintagunta, A.D. et al. (2022a). Biocontrol potential of *Pseudomonas stutzeri* endophyte from *Withania somnifera* (Ashwagandha) seed extract against pathogenic *Fusarium oxysporum* and *Rhizoctonia solani*. *Archives of Phytopathology and Plant Protection* 55 (1): 1–18. https://doi.org/10.1080/03235408.2021.1983384.

Kumar, P., Singh, S., Pranaw, K. et al. (2022b). Bioinoculants as mitigators of multiple stresses: a ray of hope for agriculture in the darkness of climate change. *Heliyon* 8 (11): e11269. https://doi.org/10.1016/j.heliyon.2022.e11269.

Lahlali, R., Ezrari, S., Radouane, N. et al. (2022). Biological control of plant pathogens: a global perspective. *Microorganisms* 10 (3): 596. https://doi.org/10.3390/microorganisms10030596.

Li, Z., Xia, C., Wang, Y. et al. (2019). Identification of an endo-chitinase from *Corallococcus* sp. EGB and evaluation of its antifungal properties. *International Journal of Biological Macromolecules* 132: 1235–1243. https://doi.org/10.1016/j.ijbiomac.2019.04.056.

Liu, Z., Shi, L., Yang, S. et al. (2017). Functional and promoter analysis of ChiIV3, a chitinase of Pepper Plant, in response to *Phytophthora capsici* infection. *International Journal of Molecular Sciences* 18 (8): 1661. https://doi.org/10.3390/ijms18081661.

Liu, K., Ding, H., Yu, Y., and Chen, B. (2019). A cold-adapted chitinase-producing bacterium from Antarctica and its potential in biocontrol of plant pathogenic fungi. *Marine Drugs* 17 (12): 695. https://doi.org/10.3390/md17120695.

Liu, M., Gong, Y., Sun, H. et al. (2020a). Characterization of a novel chitinase from sweet potato and its fungicidal effect against *Ceratocystis fimbriata*. *Journal of Agricultural and Food Chemistry* 68 (29): 7591–7600. https://doi.org/10.1021/acs.jafc.0c01813.

Liu, T., Guo, X., Bu, Y. et al. (2020b). Structural and biochemical insights into an insect gut-specific chitinase with antifungal activity. *Insect Biochemistry and Molecular Biology* 119: 103326. https://doi.org/10.1016/j.ibmb.2020.103326.

Liu, Y., He, P., He, P. et al. (2022a). Potential biocontrol efficiency of *Trichoderma* species against oomycete pathogens. *Frontiers in Microbiology* 13: 1–11. https://doi.org/10.3389/fmicb.2022.974024.

Liu, Y., Shi, R., Peng, Y. et al. (2022b). Impacts of technology training provided by agricultural cooperatives on Farmers' adoption of biopesticides in China. *Agriculture* 12 (3): –316. https://doi.org/10.3390/agriculture12030316.

Mahmood, I., Imadi, S.R., Shazadi, K. et al. (2016). Effects of pesticides on environment. In: *Plant, Soil and Microbes: Volume 1: Implications in Crop Science* (ed. K.R. Hakeem, M.S. Akhtar, and S.N.A. Abdullah), 253–269. Cham: Springer International Publishing https://doi.org/10.1007/978-3-319-27455-3_13.

Mahmood, S., Kumar, M., Kumari, P. et al. (2020). Novel insecticidal chitinase from the insect pathogen *Xenorhabdus nematophila*. *International Journal of Biological Macromolecules* 159: 394–401. https://doi.org/10.1016/j.ijbiomac.2020.05.078.

Malik, M.S., Haider, S., Rehman, A. et al. (2022). Biological control of fungal pathogens of tomato (*Lycopersicon esculentum*) by chitinolytic bacterial strains. *Journal of Basic Microbiology* 62 (1): 48–62. https://doi.org/10.1002/jobm.202100512.

Marian, M. and Shimizu, M. (2019). Improving performance of microbial biocontrol agents against plant diseases. *Journal of General Plant Pathology* 85 (5): 329–336. https://doi.org/10.1007/s10327-019-00866-6.

Masri, M., Sukmawaty, E., and Awalia Amir, A. (2022). Anti-fungal activity of chitinolytic bacteria *Lysinibacillus fusiformis* and *Brevibacillus reuszeri* against the fungal pathogens *Rhizoctonia solani* and *Fusarium oxysporum*. *Microbiology Indonesia* 15 (4): 3. https://doi.org/10.5454/mi.15.4.3.

Mejri, D., Gamalero, E., and Souissi, T. (2013). Formulation development of the deleterious rhizobacterium *Pseudomonas trivialis* X33d for biocontrol of brome (*Bromus diandrus*) in durum wheat. *Journal of Applied Microbiology* 114 (1): 219–228. https://doi.org/10.1111/jam.12036.

Michaud, J.P. (2018). Problems inherent to augmentation of natural enemies in open agriculture. *Neotropical Entomology* 47 (2): 161–170. https://doi.org/10.1007/s13744-018-0589-4.

Mishra, P., Mishra, J., Dwivedi, S.K., and Arora, N.K. (2020). Microbial enzymes in biocontrol of phytopathogens. In: *Microbial Enzymes: Roles and Applications in Industries* (ed. N.K. Arora, J. Mishra, and V. Mishra), 259–285. Singapore, Singapore: Springer https://doi.org/10.1007/978-981-15-1710-5_10.

Mittal, D., Kaur, G., Singh, P. et al. (2020). Nanoparticle-based sustainable agriculture and food science: recent advances and future outlook. *Frontiers in Nanotechnology* 2: 1–38. https://doi.org/10.3389/fnano.2020.579954.

Mohamadpoor, M., Amini, J., Ashengroph, M., and Azizi, A. (2022). Evaluation of biocontrol potential of *Achromobacter xylosoxidans* strain CTA8689 against common bean root rot. *Physiological and Molecular Plant Pathology* 117: 101769. https://doi.org/10.1016/j.pmpp.2021.101769.

Mohiddin, F.A., Padder, S.A., Bhat, A.H. et al. (2021). Phylogeny and optimization of *Trichoderma harzianum* for chitinase production: evaluation of their antifungal behaviour against the prominent soil borne phyto-pathogens of temperate India. *Microorganisms* 9 (9): 1962. https://doi.org/10.3390/microorganisms9091962.

Moka, S., Singh, N., and Buttar, D.S. (2021). Identification of potential native chitinase-producing *Trichoderma* spp. and its efficacy against damping-off in onion. *European Journal of Plant Pathology* 161 (2): 289–300. https://doi.org/10.1007/s10658-021-02321-9.

Moon, J.-H., Won, S.-J., Maung, C.E. et al. (2021). *Bacillus velezensis* CE 100 inhibits root rot diseases (*Phytophthora* spp.) and promotes growth of Japanese cypress (*Chamaecyparis obtusa* endlicher) seedlings. *Microorganisms* 9 (4): 821. https://doi.org/10.3390/microorganisms9040821.

Morales-Ruiz, E., Priego-Rivera, R., Figueroa-López, A.M. et al. (2021). Biochemical characterization of two chitinases from *Bacillus cereus sensu lato* B25 with antifungal activity against *Fusarium verticillioides* P03. *FEMS Microbiology Letters* 368 (2): fnaa218. https://doi.org/10.1093/femsle/fnaa218.

Muhammad, M.H., Idris, A.L., Fan, X. et al. (2020). Beyond risk: bacterial biofilms and their regulating approaches. *Frontiers in Microbiology* 11: 928. https://doi.org/10.3389/fmicb.2020.00928.

Nicolopoulou-Stamati, P., Maipas, S., Kotampasi, C. et al. (2016). Chemical pesticides and human health: the urgent need for a new concept in agriculture. *Frontiers in Public Health* 4: 148. https://doi.org/10.3389/fpubh.2016.00148.

Pal, K.K. and Gardener, B.M. (2006). Biological control of plant pathogens. *The Plant Health Instructor.* 1–25. https://doi.org/10.1094/PHI-A-2006-1117-02.

Pandin, C., Le Coq, D., Canette, A. et al. (2017). Should the biofilm mode of life be taken into consideration for microbial biocontrol agents? *Microbial Biotechnology* 10 (4): 719–734. https://doi.org/10.1111/1751-7915.12693.

Pandit, M.A., Kumar, J., Gulati, S. et al. (2022). Major biological control strategies for plant pathogens. *Pathogens* 11 (2): 273. https://doi.org/10.3390/pathogens11020273.

Patibanda, A.K. and Ranganathswamy, M. (2018). Effect of agrichemicals on biocontrol agents of plant disease control. In: *Microorganisms for Green Revolution: Volume 2: Microbes for Sustainable Agro-Ecosystem* (ed. D.G. Panpatte, Y.K. Jhala, H.N. Shelat, and R.V. Vyas), 1–21. Singapore: Springer https://doi.org/10.1007/978-981-10-7146-1_1.

Philip, N.V., Koteshwara, A., Kiran, G.A. et al. (2020). Statistical optimization for coproduction of chitinase and beta 1,4-endoglucanase by chitinolytic *Paenibacillus elgii* PB1 having antifungal activity. *Applied Biochemistry and Biotechnology* 191 (1): 135–150. https://doi.org/10.1007/s12010-020-03235-8.

Poria, V., Rana, A., Kumari, A. et al. (2021a). Current perspectives on chitinolytic enzymes and their agro-industrial applications. *Biology* 10 (12): 1319. https://doi.org/10.3390/biology10121319.

Poria, V., Singh, S., Nain, L. et al. (2021b). Rhizospheric microbial communities: occurrence, distribution, and functions. In: *Microbial Metatranscriptomics Belowground* (ed. M. Nath, D. Bhatt, P. Bhargava, and D.K. Choudhary), 239–271. Singapore, Singapore: Springer https://doi.org/10.1007/978-981-15-9758-9_12.

Prasanna, R., Singh, S., and Nain, L. (2014). Microbe mediated integrated nutrient management strategies for rice-wheat cropping system. *Journal of Bioresources for Sustainable Plant Nutrient Management.*

Rajmohan, K.S., Chandrasekaran, R., and Varjani, S. (2020). A review on occurrence of pesticides in environment and current technologies for their remediation and management. *Indian Journal of Microbiology* 60 (2): 125–138. https://doi.org/10.1007/s12088-019-00841-x.

Rani, A., Rana, A., Dhaka, R.K. et al. (2023). Bacterial volatile organic compounds as biopesticides, growth promoters and plant-defense elicitors: current understanding and future scope. *Biotechnology Advances* 63: 108078. https://doi.org/10.1016/j.biotechadv.2022.108078.

Ranjan, A., Westrick, N.M., Jain, S. et al. (2019). Resistance against *Sclerotinia sclerotiorum* in soybean involves a reprogramming of the phenylpropanoid pathway and up-regulation of antifungal activity targeting ergosterol biosynthesis. *Plant Biotechnology Journal* 17 (8): 1567–1581. https://doi.org/10.1111/pbi.13082.

Ratnadass, A., Fernandes, P., Avelino, J., and Habib, R. (2012). Plant species diversity for sustainable management of crop pests and diseases in agroecosystems: a review. *Agronomy for Sustainable Development* 32 (1): 273–303. https://doi.org/10.1007/s13593-011-0022-4.

Reghmit, A., Benzina-tihar, F., López Escudero, F.J. et al. (2021). *Trichoderma* spp. isolates from the rhizosphere of healthy olive trees in northern Algeria and their biocontrol potentials against the olive wilt pathogen, *Verticillium dahliae. Organic Agriculture* 11 (4): 639–657. https://doi.org/10.1007/s13165-021-00371-1.

Rieusset, L., Rey, M., Muller, D. et al. (2020). Secondary metabolites from plant-associated *Pseudomonas* are overproduced in biofilm. *Microbial Biotechnology* 13 (5): 1562–1580. https://doi.org/10.1111/1751-7915.13598.

Rishad, K.S., Rebello, S., Shabanamol, P.S., and Jisha, M.S. (2017). Biocontrol potential of halotolerant bacterial chitinase from high yielding novel *Bacillus Pumilus* MCB-7 autochthonous to mangrove ecosystem. *Pesticide Biochemistry and Physiology* 137: 36–41. https://doi.org/10.1016/j.pestbp.2016.09.005.

Rodrigues, P.L., da Silva, J.L., de Alfaia, J.P. et al. (2020). Biocontrol potential of yeasts in citrus postharvest by production of β-1,3-glucanase enzyme and killer activity: a review. *Citrus Research and Technology* 41: e1056. https://doi.org/10.4322/crt.20020.

Rodríguez, L.C. and Niemeyer, H.M. (2005). Integrated pest management, semiochemicals and microbial pest-control agents in Latin American agriculture. *Crop Protection* 24 (7): 615–623. https://doi.org/10.1016/j.cropro.2004.11.006.

Rojas-Sánchez, B., Guzmán-Guzmán, P., Morales-Cedeño, L.R. et al. (2022). Bioencapsulation of microbial inoculants: mechanisms, formulation types and application techniques. *Applied Biosciences* 1 (2): 198–220. https://doi.org/10.3390/applbiosci1020013.

Saeed, Q., Xiukang, W., Haider, F.U. et al. (2021). Rhizosphere bacteria in plant growth promotion, biocontrol, and bioremediation of contaminated sites: a comprehensive review of effects and mechanisms. *International Journal of Molecular Sciences* 22 (19): 10529. https://doi.org/10.3390/ijms221910529.

Santos, M.S., Rodrigues, T.F., Nogueira, M.A., and Hungria, M. (2021). The challenge of combining high yields with environmentally friendly bioproducts: a review on the compatibility of pesticides with microbial inoculants. *Agronomy* 11 (5): –870. https://doi.org/10.3390/agronomy11050870.

Seo, D.-J., Lee, Y.-S., Kim, K.-Y., and Jung, W.-J. (2016). Antifungal activity of chitinase obtained from *Paenibacillus ehimensis* MA2012 against conidial of *Collectotrichum gloeosporioides* in vitro. *Microbial Pathogenesis* 96: 10–14. https://doi.org/10.1016/j.micpath.2016.04.016.

Shantamma, Rangaswamy, K.T., Prakash, N.B., and Achari, R. (2020). Silver nanoparticle: synthesized and antimicrobial activity on target plant pathogens. *Current Journal of Applied Science and Technology* 39 (48): 335–344. https://doi.org/10.9734/cjast/2020/v39i4831239.

Sharma, A., Kumar, V., Shahzad, B. et al. (2019). Worldwide pesticide usage and its impacts on ecosystem. *SN Applied Sciences* 1 (11): 1446. https://doi.org/10.1007/s42452-019-1485-1.

Sharma, S., Kumar, S., Khajuria, A. et al. (2020). Biocontrol potential of chitinases produced by newly isolated *Chitinophaga* sp. S167. *World Journal of Microbiology and Biotechnology* 36 (6): 90. https://doi.org/10.1007/s11274-020-02864-9.

Sharp, R.G. (2013). A review of the applications of chitin and its derivatives in agriculture to modify plant-microbial interactions and improve crop yields. *Agronomy* 3 (4): 757–793. https://doi.org/10.3390/agronomy3040757.

Silva, N.C.M., Conceição, J.G., Ventury, K.E. et al. (2018). Soybean seed coat chitinase as a defense protein against the stored product pest *Callosobruchus maculatus*. *Pest Management Science* 74 (6): 1449–1456. https://doi.org/10.1002/ps.4832.

Singh, B., Singh, S., Singh, S. et al. (2010). Ethnobotanical significance and antimicrobial activity of *Blumea lacera* (Roxb.) DC. *International Journal of Pharmaceutical & Biological Archives* 1 (3): 314–316.

Singh, R.J., Ahlawat, I.P.S., and Singh, S. (2013). Effects of transgenic Bt cotton on soil fertility and biology under field conditions in subtropical inceptisol. *Environmental Monitoring and Assessment* 185 (1): 485–495. https://doi.org/10.1007/s10661-012-2569-1.

Singh, D., Singh, S.K., Singh, V.K. et al. (2020). Plant growth–promoting bacteria and their role in environmental management. In: *Abatement of Environmental Pollutants* (ed. P. Singh, A. Kumar, and A. Borthakur), 161–175. Cambridge, MA: Elsevier https://doi.org/10.1016/B978-0-12-818095-2.00008-4.

Smolińska, U. and Kowalska, B. (2018). Biological control of the soil-borne fungal pathogen *Sclerotinia sclerotiorum* – a review. *Journal of Plant Pathology* 100 (1): 1–12. https://doi.org/10.1007/s42161-018-0023-0.

Someya, N., Nakajima, M., Hirayae, K. et al. (2001). Synergistic antifungal activity of chitinolytic enzymes and prodigiosin produced by biocontrol bacterium, *Serratia marcescens* strain B2 against gray mold pathogen, *Botrytis cinerea*. *Journal of General Plant Pathology* 67 (4): 312–317. https://doi.org/10.1007/PL00013038.

Sousa, A.J.S., Silva, C.F.B., Sousa, J.S. et al. (2019). A thermostable chitinase from the antagonistic *Chromobacterium violaceum* that inhibits the development of phytopathogenic fungi. *Enzyme and Microbial Technology* 126: 50–61. https://doi.org/10.1016/j.enzmictec.2019.03.009.

Su, J.-Y., Liu, C.-H., Tampus, K. et al. (2022). Organic amendment types influence soil properties, the soil bacterial microbiome, and tomato growth. *Agronomy* 12 (5): 1236. https://doi.org/10.3390/agronomy12051236.

Suganthi, M., Senthilkumar, P., Arvinth, S., and Chandrashekara, K.N. (2017). Chitinase from *Pseudomonas fluorescens* and its insecticidal activity against *Helopeltis theivora*. *The Journal of General and Applied Microbiology advpub.* 63: 222–227. https://doi.org/10.2323/jgam.2016.11.001.

Supakdamrongkul, P., Bhumiratana, A., and Wiwat, C. (2010). Characterization of an extracellular lipase from the biocontrol fungus, *Nomuraea rileyi* MJ, and its toxicity toward *Spodoptera litura*. *Journal of Invertebrate Pathology* 105 (3): 228–235. https://doi.org/10.1016/j.jip.2010.06.011.

Tang, S., Tang, G., and Cheke, R.A. (2010). Optimum timing for integrated pest management: modelling rates of pesticide application and natural enemy releases. *Journal of Theoretical Biology* 264 (2): 623–638. https://doi.org/10.1016/j.jtbi.2010.02.034.

Ueki, A., Takehara, T., Ishioka, G. et al. (2020). β-1,3-Glucanase production as an anti-fungal enzyme by phylogenetically different strains of the genus *Clostridium* isolated from anoxic soil that underwent biological disinfestation. *Applied Microbiology and Biotechnology* 104 (12): 5563–5578. https://doi.org/10.1007/s00253-020-10626-8.

Veliz, E.A., Pilar, M.-H., and Ann, M.H. (2017). Chitinase-producing bacteria and their role in biocontrol. *AIMS Microbiology* 3 (3): 689–705. https://doi.org/10.3934/microbiol.2017.3.689.

Walgenbach, J.F. (2018). Chapter 16—Integrated pest management strategies for field-grown tomatoes. In: *Sustainable Management of Arthropod Pests of Tomato* (ed. W. Wakil, G.E. Brust, and T.M. Perring), 323–339. San Diego: Academic Press https://doi.org/10.1016/B978-0-12-802441-6.00016-4.

Wang, F., Xiao, J., Zhang, Y. et al. (2021a). Biocontrol ability and action mechanism of *Bacillus halotolerans* against *Botrytis cinerea* causing grey mould in postharvest strawberry fruit. *Postharvest Biology and Technology* 174: 111456. https://doi.org/10.1016/j.postharvbio.2020.111456.

Wang, R., Long, Z., Liang, X. et al. (2021b). The role of a β-1,3-1,4-glucanase derived from *Bacillus amyloliquefaciens* FS6 in the protection of ginseng against *Botrytis cinerea* and *Alternaria panax*. *Biological Control* 164: 104765. https://doi.org/10.1016/j.biocontrol.2021.104765.

Wang, Y., Li, D., Dong, C. et al. (2021c). Heterologous expression and characterization of a novel glycoside hydrolase family 55 β-1,3-glucanase, AcGluA, from *Archangium* sp. strain AC19. *Applied Microbiology and Biotechnology* 105 (18): 6793–6803. https://doi.org/10.1007/s00253-021-11513-6.

Wang, H., Sheng, Y., Jiang, W. et al. (2022). The effects of crop rotation combinations on the soil quality of old apple orchard. *Horticultural Plant Journal* 8 (1): 1–10. https://doi.org/10.1016/j.hpj.2021.04.010.

Warrior, P., Konduru, K., and Vasudevan, P. (2002). Formulation of biological control agents for pest and disease management. In: *Biological Control of Crop Diseases*, 1e (ed. S.S. Gnanamanickam). Boca Raton: CRC press http://dx.doi.org/10.1201/9780203910955.ch18.

Willer, H., Meier, C., Schlatter, B. et al. (2021). The World of Organic Agriculture Statistics and Emerging Trends 2021, 1–351. Research Institute of Organic Agriculture FiBL, Frick, and IFOAM-Organics International, Bonn, Germany.

Wong, C.K.F., Saidi, N.B., Vadamalai, G. et al. (2019). Effect of bioformulations on the biocontrol efficacy, microbial viability and storage stability of a consortium of biocontrol agents against *Fusarium* wilt of banana. *Journal of Applied Microbiology* 127 (2): 544–555. https://doi.org/10.1111/jam.14310.

Xie, Z., Li, M., Wang, D. et al. (2021). Biocontrol efficacy of *Bacillus siamensis* LZ88 against brown spot disease of tobacco caused by *Alternaria alternata*. *Biological Control* 154: 104508. https://doi.org/10.1016/j.biocontrol.2020.104508.

Xue, C., Shen, Z., Hao, Y. et al. (2019). Fumigation coupled with bio-organic fertilizer for the suppression of watermelon *Fusarium* wilt disease re-shapes the soil microbiome. *Applied Soil Ecology* 140: 49–56. https://doi.org/10.1016/j.apsoil.2019.04.007.

Yarullina, L.G., Akhatova, A.R., and Kasimova, R.I. (2016). Hydrolytic enzymes and their proteinaceous inhibitors in regulation of plant–pathogen interactions. *Russian Journal of Plant Physiology* 63 (2): 193–203. https://doi.org/10.1134/S1021443716020151.

Yatoo, A.M., Ali, M.N., Baba, Z.A., and Hassan, B. (2021). Sustainable management of diseases and pests in crops by vermicompost and vermicompost tea. A review. *Agronomy for Sustainable Development* 41 (1): 7. https://doi.org/10.1007/s13593-020-00657-w.

Yelitza, C., Carlos, V., Natália, C., and Javier, F. (2016). Implementation and adoption of integrated pest management approaches in Latin America: challenges and potential. In: *Integrated Pest Management (IPM)* (ed. G. Harsimran Kaur and G. Gaurav), 1–18. Rijeka, Chapter 1: IntechOpen https://doi.org/10.5772/64098.

Zehra, A., Rai, A., Singh, S.K. et al. (2021). Chapter 10 – An overview of nanotechnology in plant disease management, food safety, and sustainable agriculture. In: *Food Security and Plant Disease Management* (ed. A. Kumar and S. Droby), 193–219. Woodhead Publishing https://doi.org/10.1016/B978-0-12-821843-3.00009-X.

Zhang, S., Chakrabarty, P.K., Fleites, L.A. et al. (2015). Three new Pierce's disease pathogenicity effectors identified using *Xylella fastidiosa* biocontrol strain EB92-1. *PLoS One* 10 (7): e0133796. https://doi.org/10.1371/journal.pone.0133796.

Zhang, C., Lin, Y., Tian, X. et al. (2017). Tobacco bacterial wilt suppression with biochar soil addition associates to improved soil physiochemical properties and increased rhizosphere bacteria abundance. *Applied Soil Ecology* 112: 90–96. https://doi.org/10.1016/j.apsoil.2016.12.005.

Zhang, L., Wang, M., Li, N. et al. (2018). Long noncoding RNAs involve in resistance to *Verticillium dahliae*, a fungal disease in cotton. *Plant Biotechnology Journal* 16 (6): 1172–1185. https://doi.org/10.1111/pbi.12861.

Zhang, P., Zhou, X., Qi, R. et al. (2019). Conductive polymer–exoelectrogen hybrid bioelectrode with improved biofilm formation and extracellular electron transport. *Advanced Electronic Materials* 5 (8): 1900320. https://doi.org/10.1002/aelm.201900320.

Zhao, X., Ni, Y., Zhao, H. et al. (2021). Plant growth-promoting ability and control efficacy of *Penicillium aurantiogriseum* 44M-3 against sesame *Fusarium* wilt disease. *Biocontrol Science and Technology* 31 (12): 1314–1329. https://doi.org/10.1080/09583157.2021.1946011.

Zhu, W. and Wang, R. (2021). Impact of farm size on intensity of pesticide use: evidence from China. *Science of the Total Environment* 753: 141696. https://doi.org/10.1016/j.scitotenv.2020.141696.

# 16

# Microbial Bioremediation of Metals and Radionuclides: Approaches and Advancements

*Sobia Riaz*[1], *Muhammad Sohail*[2], *and Rashba Sahar*[3]

[1] *Institute of Soil & Environmental Sciences, University of Agriculture Faisalabad, Faisalabad, Pakistan*
[2] *Department of Forestry and Range Management, University of Agriculture Faisalabad, Faisalabad, Pakistan*
[3] *Department of Plant Pathology, University of Agriculture, Faisalabad, Pakistan*

## 16.1 Introduction

Pollutants of all kinds are being discharged into the environment directly as a consequence of industrialization. Pollutant transport, dispersion, properties, transformation, and mineralization are all processes in which microorganisms participate. They have been utilized effectively to clean up a diverse array of hazardous substances in the natural surroundings, from organic chemicals to metals (Isakovski et al., 2020). Microbial metabolism can totally destroy organic substances, allowing them to be eliminated from the environment. However, heavy metals (HMs) exhibit persistence in the natural environment because they are not prone to biodegradation (Dhankhar and Hooda 2011). Therefore, microbial activity can only reduce the toxicity, bioavailability, or mobility of HMs; it cannot entirely breakdown or eliminate them. Metals in the environment can be eliminated more easily after being absorbed and accumulated by microorganisms. Any element having an atomic number exceeding 20 and metallic characteristics (such as ligand specificity, conductivity, ductility, and stability as cations) are considered HMs. Copper (Cu), cadmium (Cd), lead (Pb), chromium (Cr), mercury (Hg), and nickel (Ni) are most often occurring ions of HMs in pollution. Because metals often constitute common elements found in soil hence plants require them as micronutrients. On the other hand, from the time when the industrial revolution began, the number of toxic contaminants in the biosphere has significantly increased. Human activities that are causing irreversible damage to the environment include metal extraction and smelting, gas emissions, fertilizer production, electroplating, energy and fuel generation, sewage treatment, pesticide application, and the creation of urban waste (Zubair et al. 2016).

This shows that microorganisms are a powerful tool in the fight against global pollution. Microorganisms are found everywhere in the biosphere because they have remarkable metabolic capabilities and can thrive in a broad variety of environments. Microbes' adaptability in the food web can be used for biodegradation of contaminants. Bioremediation is the word used to describe such processes. Through the procedure, energy and biomass production are obtained, which is maintained based on the capacity of particular microorganisms to modify, convert, and utilize harmful contaminants (Tang et al. 2007). Rather than solely collecting and storing the pollution, bioremediation refers to a microbiological systematically structured procedural operation that breaks down or converts the contaminants into less harmful forms or harmless elemental and/or compound states. In the bioremediation process, live organisms are called bioremediators. Bioremediation is typically carried out by bacteria, archaea, and fungi (Strong and Burgess 2008). The biotechnological process of bioremediation uses microorganisms to remedy environmental issues and eliminate dangerous pollutants from the environment through biodegradation (Bhadouria et al. 2020; Giri et al. 2021; Gond et al. 2021; Gupta et al. 2020). There is a growing tendency to use the phrases bioremediation and biodegradation interchangeably. Because of their effectiveness relative to conventional remediation methods, microorganisms play a pivotal role in the remediation of contaminated soil, water, and sediments. For bioremediation to work, bacteria must use their enzymes to break down the pollutants and produce safer byproducts (Callaghan 2013; Kour et al. 2021). Since bioremediation is merely effective in places where microbes can thrive, adjusting environmental factors to speed up the degradation process is a common part of the process

*Microbes Based Approaches for the Management of Hazardous Contaminants*, First Edition. Edited by Ajay Kumar, Livleen Shukla, Joginder Singh, and Luiz Fernando R. Ferreira.

(Mishra et al. 2021; Pandey et al. 2023; Singh et al. 2021a). Using *in silico* methods to identify important components and modify microbial and plant systems to meet our needs through genetic manipulation of key candidate genes has enormous potential to speed up the method of bioremediation.

## 16.2 Sources and Effects of Heavy Metals

Table 16.1 details the hazardous consequences of some of the most common contaminants that have entered environmental effects brought about by human urbanization and industrialization. Major contributors to soil and water contamination are pollutants resulting from a variety of industrial operations. HMs of varying sorts and quantities are released both during and as a result of industrial production. Antimony, chromium, and mercury have all been linked to wastewater from dye factories (Methneni et al. 2021). Al, Cu, Zn, Ni, Pb, and As are just some of the pollutants released into the environment via fertilizers application, stimulants, herbicides, and pesticides in agriculture (Ayilara et al. 2020; Prabagar et al. 2021). Pollutants sourcing agri-food effluents and sewage water dumped into canals and rivers can also damage ecosystems if not properly treated (Siric et al. 2022a; AL-Huqail et al. 2022).

As a result of damage to pipelines, transportation leaks, and/or unintentional spills, crude oil is also a significant environmental pollutant (Ogunlaja et al. 2019). Toxic metals and chemicals are often emitted throughout the mining process in the surrounding natural environment (Liu et al. 2020). Cyanide and sulfuric acid are just two examples of environmentally hazardous compounds used in mining (Ayangbenro and Babalola 2018). Zn, Cu, and Cd are also released from other industrial wastes present in the upper soil layers (Jafari et al. 2019). Plastics including Hg, Cd, Pb, Mg, Fe, Cu, Cr, Ag, and Sb (Zhou et al. 2019), as well as Cr and Pb derived from pharmaceutical effluents (Kumari and Tripathi 2020), all contribute to water pollution. Cu, As, Hg, Cr, Pb, Ni, Cd, and Zi are also released into the environment as a result of the coal industry (Sun et al. 2019).

Entire HMs are extremely dangerous to both aquatic and terrestrial ecosystems. Lead causes liver and kidney dysfunction, cardiovascular disease, reproductive and immune system dysfunction, and developmental delays in humans (Ayangbenro and Babalola 2020; Zwolak et al. 2019). According to multiple studies (Ayangbenro and Babalola 2020;

**Table 16.1** Impacts of environmental contaminants on wildlife and human health.

| Pollutants | Sources | Organism affected | Effect on the organism | References |
|---|---|---|---|---|
| Mercury | Mining and industrial production | Human | Injuries to the human central nervous system (CNS), hepatotoxicity, and renal dysfunction | Zhang et al. (2020) |
| Aluminum | Weathering, mining, and industrial activities | Plants wall | Inhibition of cell division, cell membrane disruption, and disturbance of calcium homeostasis | Rehman et al. (2021) |
| Pesticides (such as deltamethrin, fenthion, spinosyn) and heavy metals (aluminum, copper, and zinc) | Agricultural and mining activities | Animals (bats) | DNA damage and morphology hepatocytes | de Souza et al. (2020) |
| Cadmium | Agricultural amendments | Plants | Chlorosis, stunted growth, Modifications in water balance | Rehman et al. (2021) |
| Chromium and lead | Industry and mining | Plants | Declined growth, reduced photosynthesis, and root growth | Zeng et al. (2012) |
| Lead | Industrial activities | Humans | Pulmonary impairment and hepatic injury, neurological impairment and cardiovascular complications | Balali-Mood et al. (2021) |
| Arsenic | Industrial activity | Humans | CNS injury, skin and hair infection, cardiovascular dysfunction and liver damage | Balali-Mood et al. (2021) |
| Chloride | Industrial activities | Animals (rats) | Kidney destruction and central nervous system injury | Aragao et al. (2018) |

Fashola et al. 2020; Zwolak et al. 2019), cadmium exposure can lead to cancer, skeletal diseases, neurotoxic and nephrotoxic complexity, and reproductive system malfunction. Soil and water contamination is a common result of inappropriate disposal of wastes containing HMs (Fashola et al. 2016). They are biomagnified and give rise to long-term illnesses in both people and animals, and they can kill fish and other aquatic life if disposed of in waterways. Therefore, physical, chemical, or biological cleanup of these contaminants is required. Both physical and chemical approaches have been tried and tested over the years, but they have significant limitations; chemical bioremediation requires a specialist and expensive equipment, while physical bioremediation is prohibitively costly (Mahmood et al. 2021).

## 16.3 Biotic and Abiotic Factors Affecting Microbial Bioremediation

The effectiveness of microorganisms in bioremediating HMs depends on various factors, such as the overall metal ion concentration, redox potential, chemical speciation of the metals, microbial competition, pH, temperature, soil composition, oxygen availability, moisture content, soil type, and the solubility of HMs in water (Medfu Tarekegn et al. 2020; Singh et al. 2019). HMs create unbound ionic species at acidic pH, which allows for a greater abundance of protons to saturate the binding site. Between 6.5 and 8.5 is the ideal pH range for bioremediation (Kharangate-Lad and D'Souza 2021) because the pH of a particular environment influences the configuration of pollutants and determines whether a microorganism can survive there. Microbes battle over carbon, a scarce fuel source and a commodity in the microbial food chain. Therefore, the rate of biodegradation would be influenced by the innate capacity of microorganisms that participate well to digest the HM pollution. To maximize the rate of biodegradation in environments where nitrogen (N) and phosphorus (P) are scarce, striking a balance between the three elements is necessary. Incorporating them into a system has been shown to boost microbial activity (Bala et al. 2022). Bioremediation rates and outcomes are dependent on the microbes used; while one organism strain may bioremediate a specific HM in the lab, this may not be the case in the field, where the presence of microbial consortium is required (Patel et al. 2022). Different microorganisms have different bioremediation capacities depending on their molecular makeup, the genes and enzymes they induce, the metabolites they produce, how well they grow, and how long they live (Kebede et al. 2021). Rates of microbial biodegradation are also affected by factors such as the ionization of chemical moieties in the cell wall, the microbial cell wall's structure, and the location of the sorption site (Mahmoud 2021).

The availability, pH, and osmotic pressure of HMs in water, along with their solubility in it, are all affected by the moisture content within the environment (Medfu Tarekegn et al. 2020). The rate of microbial biodegradation is negligible at high moisture content. This may occur when aerobic microbes are killed off by the presence of an anaerobic environment. Furthermore, bacteria cannot survive at low moisture levels; thus, an optimal moisture level is essential for an effective microbial biodegradation procedure. The rate of microbial breakdown of HMs is poor in frigid places where only psychrophiles may survive. While high temperature is beneficial to the majority of bioremediation processes. Each compound degrades at its specific temperature (Ren et al. 2018; Bala et al. 2022; Sharma et al. 2022). Because the subzero water freezes the microbial transport pathways, metabolic activity is reduced. At increasing temperatures, both the availability of HMs and their microbial biodegradation speed increase (Mahmoud 2021).

Several factors influence the rate of microorganism-mediated degradation of metals or contaminants, including their chemical structure, bioavailability, concentration, toxicity, and stability, as noted by Kebede et al. (2021). Microorganisms tend to more effectively remediate HMs with simple chemical structures and low concentrations, compared to those with complex structures and high temperatures. For instance, compounds with four or more rings, highly condensed cycloalkane compounds, present more significant challenges for degradation compared to substances like anthracene, naphthalene, phenanthrene, and unbranched alkanes (typically ranging from C10 to C25 in chain length), as described by Kebede et al. (2021). Consequently, *n*-alkanes degrade more readily than asphaltenes, low molecular weight aromatics, branched alkanes, and high molecular weight hydrocarbons, as stated by Imam et al. (2019). Biodegradation processes encompass both aerobic and anaerobic pathways, with the survival of nutrient-degrading organisms greatly influenced by their specific characteristics, as emphasized by Jacob et al. (2018). For instance, aerobically active organisms only create the enzyme oxygenase when oxygen is present. The rate of microbial biodegradation is impacted by several factors in the soil medium including its location, capacity to retain moisture, texture, and particle size (Alvarez et al. 2020). The top 10 cm of soil has the greatest concentration and variety of microorganisms. This is in contrast to what occurs in silt soils, where both oxygen and organic matter get depleted (Ndeddy Aka and Babalola 2017). Hydrocarbon retention is more surface-oriented in fine-particle soils like clayey soils, making soil nutrients and oxygen unavailable to plants. Since well-drained soil allows for greater oxygen availability and houses a greater diversity of soil microbes, it is ideal for promoting microbial biodegradation (Huang et al. 2019). Hydrocarbonoclastic activity of halotolerant and halophilic bacteria is altered by salinity's presence, and soil microbes are put under stress as a result. When exposed to higher levels of salt, bacteria experience greater osmotic pressure. Imron et al. (2020) and Kebede et al. (2021) stated that the solubility of the HMs, the transportation system, and the bacterial metabolic processes are all directly harmed by this.

## 16.4 Approaches for Bioremediation of Heavy Metals Through Microbial Processes: An Introduction

Bioremediation is a technique for cleaning polluted areas with naturally occurring organisms, such as microorganisms, plants, or their enzymes.. Mineralization is a more technical term describing either a single species or a consortium of microorganisms entirely transforming an organic contaminant into inorganic constituents this idea includes both partial and complete biodegradation, alteration, and decontamination of toxins by microorganisms. When microorganisms assist in converting a pollutant without releasing carbon or energy, this process is called co-metabolism. In order to restore a site's soil, deposits, and groundwater to their natural state, bioremediation typically employs the utilization of living organisms (Giri et al. 2017a, b; Kumar et al. 2011, 2020). Various methods employed for the bioremediation of different pollutants are listed in Figure 16.1.

Biodegradation, often known as trash recycling or the breakdown of organic materials into usable nutrients for other organisms, was first coined by ecologist and environmental scientist Alexander (1994). Bioremediation is performed by a wide variety of organisms, including bacteria, fungi, insects, worms, and plants. These organisms obtain nutrients from the pollutant and convert xenobiotics into molecules that are safer for the environment (Vidali 2001; Debbarma et al. 2017; Bhatt et al. 2020, 2021a, b). Bioremediation is crucial in restoring clean water supplies. Industrial effluent-induced water body contamination reduces the quantity of clean water that is readily available for use in agriculture and by humans. There are several types of bioremediation based on physical, biological, and chemical remediation techniques as listed in Figure 16.1. Bioremediation can be carried out either *ex-situ* or *in-situ* methods, with the choice depending on a range of factors. These factors encompass but are not restricted to cost considerations, site-specific features, and the type as well as the concentration of pollutants present.

Enhancing the degradation process can be achieved by providing native microorganisms with additional nutrients, carbon sources, or electron donors (referred to as biostimulation or bio-restoration). Alternatively, an enriched culture of microorganisms with specific capabilities to accelerate the biodegradation of the target contaminant can be introduced (bioaugmentation) (Goel et al. 2008; Dash et al. 2021; Singh et al. 2021). The goal of bioremediation is to reduce pollution to safe or undetectable levels. As a result of rigorous agriculture and discharge of industrial waste, a number of xenobiotic compounds have been discharged into the environment. Crop production has been hampered by a lack of clean water and soil degradation due to the unchecked accumulation of toxic waste (Kamaludeen et al. 2003). This method involves encouraging the spread of specific microflora that are indigenous to the polluted areas and are capable of carrying out the desired tasks. Bioremediation techniques can be employed for both soil and water environments using *in-situ* and *ex-situ* methods. This section provides a concise overview of various bioremediation techniques tailored to address HMs and radionuclides as listed in Figures 16.2 and 16.3.

1) ***In-situ* bioremediation**
   These methods are preferred above others because they treat the area in question rather than affecting the toxins elsewhere (U.S. EPA 1990, 1996). The effective depth of *in-situ* treatment is a limiting factor in the practice. While depths of 60 cm and beyond have been successfully treated, the effective oxygen diffusion range for optimal rates of bioremediation in many soils is only a few centimeters to approximately 30 cm.

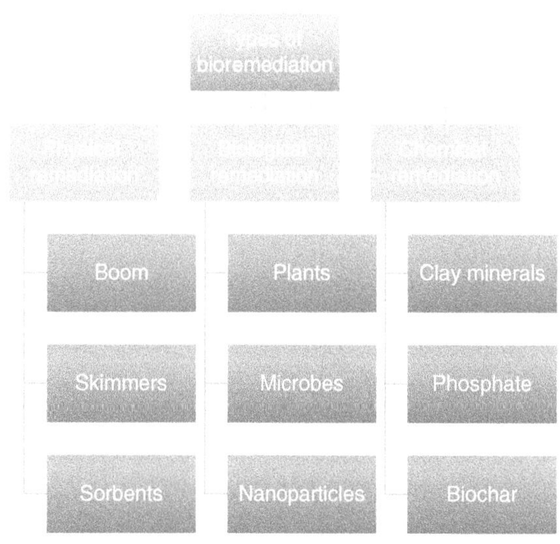

**Figure 16.1** Types of bioremediation.

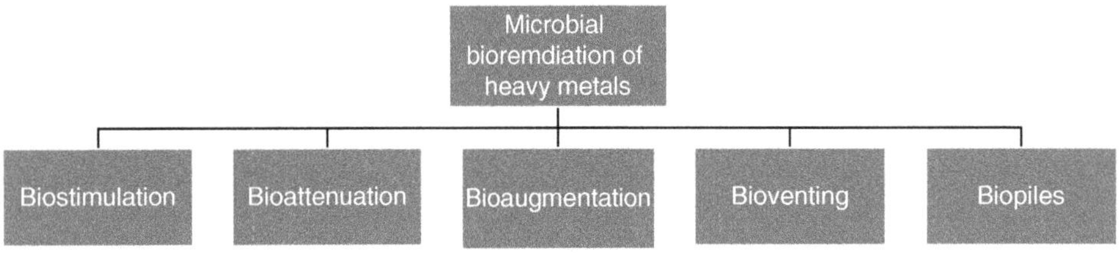


**Figure 16.2** Types of microbial bioremediation techniques for various contaminants.

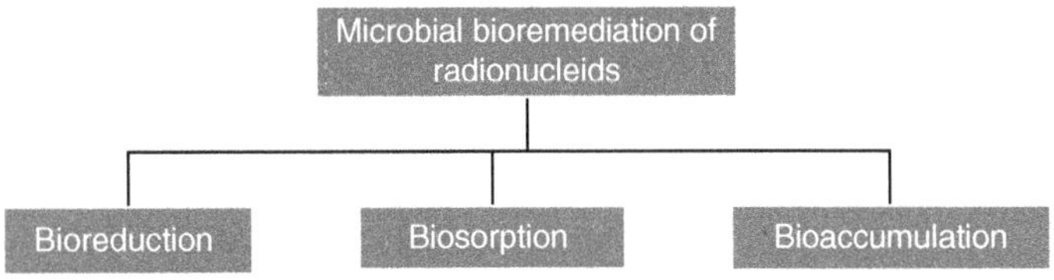


**Figure 16.3** Approaches for microbial bioremediation of radionuclides.

2) ***Ex-situ* bioremediation**
   Such procedures necessitate the remediation of soil contaminated with pollutants by excavation. There are numerous treatment technologies and methods that fall under the umbrella of bioremediation approaches. The keystones of bioremediation are biostimulation, bioattenuation, bioaugmentation, venting, and heaps.

### 16.4.1 Biostimulation

The introduction of certain improvements to the location such as soil or groundwater to encourage the growth of native microorganisms is the key to this unlocking mechanism. The emphasis is on reestablishing natural communities of microbes and other forms of life. The absorption rate and process can be accelerated further by providing additional natural requirements including pH, temperature, and oxygen (Adams et al. 2015; Kumar et al. 2011). A little concentration of toxins can also act as a stimulant, activating the genes responsible for bioremediation chemicals. The production of improvements and oxygen to aid native microorganisms typically continues in this important manner. These enhancements lay down foundations of life, enabling bacteria to develop properties such as imperativeness, cellular biomass, and impetuses to eradicate the poison (Madhavi and Mohini 2015).

### 16.4.2 Bioattenuation

Toxic material epicenters can be destroyed through bioattenuation, often known as gradual weakening. Natural cycles (including combustion and anaerobic decomposition, plant and animal uptake, and evaporation and condensation) and physical miracles (including changes in climate, dispersal, debilitation, spread, volatilization, and sorption/desorption) complete the task. Within the broader basic choke definition, terms such as inalienable remediation or biotransformation are merged (Mulligana and Yong 2004). There are four distinct ways in which nature might clean up after human-made pollution of the planet (Li et al. 2010):

1) Soil and groundwater bacteria and other tiny critters eat two different manmade mixtures. When they have finished digesting man-made innovations, they can transform them into harmless gases and liquids.
2) Chemicals can be fixed in place because they can adhere to or be absorbed by the soil. The created compounds are not cleaned up, but at least the groundwater and the surrounding area are protected from contamination.
3) Impure water can become mixed in with pure water as pollution travels through the ground. The level of impureness is diminished as a result of this.
4) Oil and solvents are two synthetic materials that can evaporate, suggesting that they undergo a phase transition from liquid to gaseous within the soil.

Such gases may be destroyed by the sun if they rise from the surface of the earth and enter the atmosphere. Bioremediation can be enhanced with biostimulation or bioaugmentation if the standard tightening is insufficient.

### 16.4.3 Bioaugmentation

It is a type of biodegradation system. Bioaugmentation is the process of engineering toxin-degrading microorganisms (proprietary/rare/intended) to increase the biodegradative capabilities of native microbes in the polluted zone. Improve debasement that ordinarily feeds on toxic sites and promote the rapid spread of signature microorganisms throughout human populations. Assembled microorganisms are then refined autonomously, undergo genetic engineering, and are released back into the remediation site. For example, all essential microorganisms, including those present in tetrachloroethylene and trichloroethylene, can be found in regions where chlorinated ethenes have contaminated the soil and groundwater. It is utilized to make sure the native microbes can eradicate the problem and convert the contaminants into harmless ethylene and chloride (Niu et al. 2009). Bioaugmentation is a strategy for memorizing designs utilizing microorganisms within a system as bioremediators to efficiently and comprehensively eliminate complex hazardous elements. Furthermore, newly discovered genetically edited microorganisms have proven the capacity to accelerate the breakdown of a wide spectrum of naturally hazardous chemicals. A less transient and more benign end result can be achieved by altering one's metabolic profile (Malik and Ahmed, 2012; Gomez et al. 2017).

Genetically produced life forms spread as distinct poisons far faster than the typical species and compete heavily with native species, hunters, and other remarkable abiotic causes, while regular species are unable to isolate certain mixtures without being innately modified through DNA control. Bioremediation of contaminated soil, water, and sludge using genetically engineered microbes has been recognized (Sayler and Ripp 2000; Bijay et al. 2012); these results reflect the most recent tainting capacity of a comprehensive assessment of manufactured and actual toxins.

### 16.4.4 Bioventing

Venting oxygen via soil, or "bioventing," is useful in energetically degradable mixes because it promotes the growth of native and invasive species of plants, animals, and parasites in the soil. During bioventing, minimal airflow is used to provide just enough oxygen to promote microbial growth. Well-drilling techniques often include injecting a mixture of air and oxygen directly into the contaminated soil below. Biodegradation of absorbed fuel residues and odd mixtures occurs gradually as exhaust diffuses across ordinarily robust soil. Several specialists have demonstrated that bioventing may be used to bioremediate oil-contaminated soil efficiently (Lee et al. 2006; Samuel and Ganiyu 2015).

### 16.4.5 Biopiles

"Biopiles" are a method for treating contaminated soil that has been exposed, namely, polluted hydrocarbon-contaminated soil. Oil toxins in exposed soils can be reduced in quantity through biodegradation using bio-piles (otherwise known as biocells, bioheaps, biomounds, and manure Piles) (Gomez 2014). By means of a series of ducts and siphons, air is recirculated through the biopile system, either by forcing into the mass of air at a predetermined pressure or by sucking air out of the stack at a negative pressure (Delille et al. 2008). Adsorbed oil pollution degradation was found to be high owing to enhanced microbial activity caused by microbial breath. (Emami et al. 2012).

## 16.5 Approaches for the Bioremediation of Radionuclide

Bacterial remediation refers to the process whereby bacteria use energy transfer to convert radioactive nuclides into stable isotopes. Microbes secrete reducing and oxidizing substances that alter the pH and oxidation state of radioactive nuclides, respectively. Natural microbe-catalyzed inorganic–organic material conversion can be harnessed to effectively solubilize or immobilize radioactive nuclides and hazardous metals in waste streams (Lloyd and Renshaw 2005; Ilyas and Lee 2014).

Oxidation–reduction reactions, altering formal charges and their solubility, facilitating complex formation, bioaccumulation, and biosorption, are the foundation of microbial bioremediation of radionuclides. You et al. (2020) and Iyas et al. (2012) have reported the ability to receive and donate electrons influences microbial activity significantly. In anaerobic environments, nitrate, sulfate, metals, and carbon dioxide are commonly employed as electron acceptors for microbes, but in oxygen-deprived environments, oxygen is typically used as the electron acceptor (Naseem et al. 2016). A greater degree of radioactive nuclide precipitation and solubilization occurs under anaerobic conditions. The formation of complexes connected to physical, biological, and chemical processes is critical to the efficient bioremediation of radioactive nuclides. The primary mechanisms that may attenuate toxicity and enhance radioactive nuclide transfer throughout the biogeochemical cycle are dissolution, oxidation, precipitation, reduction, sorption, and leaching. Such essential mechanisms are accomplished by microorganisms, and the bioprocesses actively involved in radioactive nuclide bioremediation are designed to immobilize these microbes in order to accelerate radionuclide recovery from waste streams (Elias et al. 2003; North et al. 2004). The degradation of concrete at nuclear waste storage facilities has been the subject of research by Turick and Berry (2016). The "BioRadBase" database, presented by Reena et al. (2012), contains information on all possible bacteria and fungi that could be employed in the cleanup of radioactive nuclide waste. High specificity, reusability, efficiency, cost effectiveness, little contamination, and proper elimination of contaminant are just few of the advantages of utilizing microorganisms for the cleanup of radioactive waste, as mentioned by Zhong et al. (2021).

Radionuclide transformation occurs via a number of bacterial remediation processes, including bioreduction, biosorption, bioaccumulations, and biomineralization (Akob et al. 2007). Directly or indirectly, bacteria can use enzymes to immobilize the radioactive nuclide. It entails bioreduction, biosorption, biomineralization, and bioaccumulation, to name a few of the fundamental processes.

### 16.5.1 Bioreduction

In the process of bioreduction, soluble radionuclide is converted into an insoluble form by using redox potential and a reduction by bacterial species. Due to its simplicity, safety, and efficiency, bioreduction has been widely adopted as a viable waste management option (Phillips et al. 2002). There are both direct and indirect approaches to bioreduction. Anaerobic bacteria are responsible for the dire ct enzymatic action that changes a high oxidation state to a reduced oxidation state. Radionuclides readily bind to bacterial surfaces during anaerobic respiration, when they serve as electron acceptors in the reduction of ethyl lactate. Direct bioreduction, on the other hand, employs sulfate-reducing bacteria to directly decrease transforming radionuclides into less soluble and less hazardous species. Bioreduction bacteria such as *Geobacter sulfurreducens*, *Shewanella putrefaciens*, *Desulfovibrio desulfuricans*, and *Desulfovibrio vulgaris* serve as illustrative examples.

### 16.5.2 Biosorption

Due to their shared relationship with radioactive nuclide and cell surface, biosorption, bioaccumulation, and biomineralization can be thought of as closely related processes (Koul and Adlakha 2021). These are all dynamic processes that rely on active energy transfer mechanisms. In biosorption, radionuclides with positive charges bind to dead or living biomass that has a negative charge (Prakash et al. 2013). The radioactive nuclide is immobilized through a biosorption process that is species-specific. Environmental factors such as temperature, pH, and oxygen content can all have an impact on biosorption, cell development phase, exopolymer secretion, and metabolite composition. Metal ions, electrostatic attraction, functional groups, and extracellular biopolymers all have a role in biosorption (Dobrowolski et al. 2017). *Citrobacter freudii* and Firmicutes are two examples of bacteria with impressive biosorption capabilities. One of the most active locations of biosorption, the carboxyl group in *C. freudii*'s cell wall, has a significant impact on sorption (Xie et al. 2008). They also discovered that functional groups other than the cell's biological activity influenced the process, meaning that dead cells had a greater sorption capability. Biosorption's primary drawback is that it quickly fills up on nuclide molecules, which might lead to competitive desorption when other cations try to occupy the cell's binding site (Jabbar and Wallner 2015). Since this is the case, biosorption is frequently utilized to remove radionuclides with low concentrations from effluent (Selenska-Pobell 2002).

### 16.5.3 Bioaccumulation

To produce a cell complex, positively charged radioactive nuclides adhere to negatively charged cellular components, forming small granules or precipitates (Tabak et al. 2005). This process is known as bioaccumulation. Bioaccumulation occurs when a radioactive nuclide binds to a ligand, such as phosphate, hydroxide, or sulfate, generating an insoluble material that may be transported out of the cell and disposed of without difficulty. Less environmental damage is done by these insoluble forms (Haferburg and Kothe 2007). One organism that is known to bioaccumulate strontium in the form of surface complexes is *Micrococcus luteus* (Phillips et al. 2002). Bioaccumulation can be either active or passive. Because of food constraints, passive bioaccumulation is preferable to the more time- and energy-consuming active bioaccumulation, which may or may not rely on metabolism. In comparison to active processes, this one takes far less time and energy to complete (Velásquez and Dussan 2009). Accumulation can occur either inside or outside of cells. When the presence or absence of a particular radioactive nuclide affects the permeability of the cell membrane, intracellular accumulation of radionuclides becomes more pronounced (Selenska-Pobell et al. 2002).

### 16.5.4 Bioremediation's Benefits and Drawbacks

It is a trademark cycle, and it works quite quickly as a means of treating waste from soiled materials like soil. Prepared microorganisms can multiply rapidly in the presence of pollution, turning it into a hazardous byproduct. When a toxin becomes contaminated, the number of biodegradable organisms and humans diminishes. Water, carbon dioxide, and cellular biomass are just a few of the often-harmless substances generated for treatment. It requires less effort and may frequently be completed nearby, without considerably complicating regular workouts. This also obviates the necessity to transport waste away from the site and the accompanying health and safety issues. Because other typical processes (developments) are not used for hazardous waste cleanup, it is implemented in a down-to-earth cycle (Montagnolli et al. 2015). Additionally, it aids in the entire eradication of the toxins, as many potentially harmful mixtures can be converted to innocuous substances, and this factor eliminates the potential for potential risks ahead linked to the treatment and removal of contaminated substances. There are no potentially harmful manufactured components in it. Microbial activity is sped up and made more dynamic thanks to the addition of enhancements, particularly fertilizers that are commonly applied on lawns and plant nurseries. Bioremediation converts dangerous manufactured products into innocuous gases and water, effectively crushing the ruined designed innovations (Shilpi 2012). Considering their inherent capabilities on Earth, they are straightforward, labor-saving, authentic, and humble.

The toxins are eliminated, not only shifted to another medium. Low-impact, taking into accounts the ability for future site uses (Kumar et al. 2011). Method that is both efficacious and sustainable for cleaning up polluted organic foundations. One drawback of bioremediation is that it can only be used on biodegradable mixtures. Some combinations can actually slow down or stop the degradation process. Biodegradation raises concerns since its byproducts may be more poisonous or tenacious than the original chemical. There is usually a lot of clarity in biological cycles. The presence of metabolically active microbial populations, low-cost ecological development settings, and adequate levels of nutrients and poisons are all crucial website variables necessary for development. It is difficult to extrapolate from study on a smaller scale (such as a seat or a pilot) to full-scale field operations. Through study, bioremediation innovations tailored to places with complex mixtures of contaminants that noticeably spread throughout the ecosystem are expected to be produced. Pollutants can exist in the form of solids, liquids, and gases. Bioremediation generally requires a greater amount of time than alternative treatment such as digging up and transporting away polluted soil or putting it on fire. Inadequate bioremediation implementation models remain a regulatory shortcoming. The term "clean" is not universally accepted, evaluating bioremediation projects is challenging, and bioremediation medications lack adequate objectives (Kensa 2011).

## 16.6 Novel Technologies in Bioremediation

### 16.6.1 Genetic Engineering

Incorporating genes into the bacterial genome to enhance metal regulation, resilience in the face of both environmental and biological stress factors, biodegradative enzyme production, metal-binding molecules and transporters, regulatory genes for metal uptake, and genes for reducing potential risks is a new and promising bioremediation method (Singh et al. 2011). The capability of these bacteria to tolerate and even thrive in environments contaminated with toxic metals is a major benefit. Foreign genes added to a species can make it more susceptible to the effects of several stressors. Furthermore, it can become a major issue when genetically created organisms and native species compete for food and other resources. Local species may fare better because they have evolved to thrive in polluted locations, and resistant strains might be selected consequently of prolonged exposure to HMs. Conversely, inclusion of certain genes gives the edge over the native strains, allowing genetically engineered bacterial species to prevail. In order to produce the recombinant bacteria, it is preferable to use preexisting microbes since they will share the genetic background of naturally occurring species but will be fortified by the introduction of resistance genes through genetic engineering.

Cloning and expressing the chromate efflux gene chrA from a chromate-resistant environmental isolate of *Bacillus pseudomycoides* improved *Bacillus subtilis* survival (Chromiková et al. 2019). *Ralstonia eutropha* CH34 has its bioremediation of cadmium enhanced in the laboratory when its metallothionein gene expression was altered (Valls et al. 2000). Effectively addressing arsenic contamination in drinking water and groundwater through bioremediation was traced back to changes in the metalloregulatory protein ArsR in an *Escherichia coli* strain (Kostal et al. 2004). Under laboratory conditions, the enhanced resistance to mercury was generated via a modified $Hg^{2+}$ transporter in *E. coli* JM109 (Zhao et al. 2005). Researchers found that *Pseudomonas fluorescens* OS8, *E. coli* MC1061, *Bacillus subtilis* BR161, and *Staphylococcus aureus* RN4220 all modified genes encoding proteins involved in cadmium, lead, zinc, and mercury biosensing in water-suspensions and soil extracts (Bondarenko et al. 2008).

### 16.6.2 Nanotechnology

Nanotechnology can be used for HM remediation. Bioremediation is aided by microorganisms, and many types of nanoparticles (iron, gold, silver, etc.) have been shown to do so (Dixit et al. 2015). Nanomaterial applications in bioremediation are hindered, however, by the usage of harmful substances in their physical and chemical manufacture. Some of the advantages of employing nanoparticles produced by bacterial enzymatic activities include increased enhanced catalytic reactivity, increased specific surface area, and enhanced enzyme–metal interaction (Pandey et al. 2015). When bacteria take in metal ions from their environment, they convert them into elemental metals through enzymatic activities, creating nanoparticles in the process. Nanoparticles can be biosynthesized either inside the cell, after metal ions have been taken up, or outside the cell, after the ions have been trapped on the cell surface and transformed by enzymes. *Shewanella oneidensis* synthesizes magnetite extracellularly (Perez-Gonzalez et al. 2010), and Rhodococcus sp. can produce Au nanoparticles intracellularly (Ahmad et al. 2003). *Bacillus cereus* can produce Ag nanoparticles intra-cellularly (Ganesh Babu and Gunasekaran 2009). Nanoparticles produced by microbial activity are common and can be encountered in a diverse array of environments. These nanoparticles have the potential to act as HM chelators. Using nanoparticles such as iron, poisonous hexavalent chromium can be transformed into nontoxic trivalent chromium (Ash et al. 2011; Abdeen et al. 2016). Higher adsorption capability and removal of $As^{3+}$ and $As^{5+}$ have been observed in bimetal oxide magnetic nanomaterials ($MnFe_2O_4$ and CoFe2$O_4$) than in $Fe_3O_4$ (Zhang et al. 2010). Perhaps because bimetal oxide nanoparticles have more surface hydroxyl groups. Adsorption onto -$Fe_2O_3$ nanoparticles has also been shown to rapidly remove $Pb^{2+}$ from wastewater (Cheng et al. 2012).

### 16.6.3 Metagenomics

Keeping an eye on the make-up of microbe populations is crucial during bioremediation. Only a tiny percentage of species found in soil (less than 1%) is culturable under laboratory conditions (Kaeberlein et al. 2002), making it a big concern in the past. As a result of recent scientific developments, metagenomics, a technique based on the examination of whole DNA extracted from soil samples, has evolved (Daniel 2005). The term "soil metagenome" refers to DNA recovered from soil that is thought to reflect the DNA of all native microorganisms present in the sample (Handelsman et al. 1998; Rondon et al. 1999). Several steps make up the standard protocol for metagenomic studies, including (i) sample enrichment, (ii) genomic DNA extraction, (iii) DNA library creation, (iv) target gene screening, and (v) target gene product expression (Kumar et al. 2020).

## 16.7 Future Perspectives and Conclusions

Despite the fact that utilizing enzymes produced by microorganisms for waste bioremediation has been the topic of countless research efforts, it is critical to optimize the technique to provide a safer and more sustainable environment. More study is needed to identify new bacteria that can bioremediate various contaminants, notably those produced by industry, quickly and efficiently. Maybe the innovative bacteria and their enzymes can bioremediate toxins more effectively than the conventional methods. More research is needed to create novel rapid detection technologies to disclose the growth of environmental pollutants or to aid in the confirmation of total biodegradation. Bioremediation bacteria can also be subjected to genetic modification to enhance enzyme production, improving their biodegradability. A more effective strategy for bioremediation would be to use a consortium of microorganisms rather than just one, as this would result in the presence of organisms that feed on a wider variety of substrates. Most of the microorganisms are used to breakdown organic substrates while inorganic contaminants that persist are ignored. Thus, more effort needs to be put into finding bacteria capable of breaking down inorganic (synthetic) contaminants. Nuclear waste from laboratories, hospitals, fuel processing plants, and power stations has persisted as a major contributor to environmental degradation around the world. Therefore, bioremediation of nuclear wastes should seriously examine the employment of bacteria and microbial enzymes. The bacteria we rely on to clean up our environments can sometimes be the cause of pollution. An instance of this is when algal blooms are caused by microbial biostimulation.

Thus, in order to guarantee a sustainable ecosystem, measures to prevent this should be developed. In addition, different chemicals degrade at different temperatures in the wild, and different microorganisms have varying temperature requirements for survival. More fieldwork is required to ascertain the natural optimum temperature for the degradation of specific compounds. It is also critical to achieve a balance between the average temperature of the environment and the lowest temperature required for specific bacteria to survive. This could help to minimize failed bioremediation operations that involve the introduction of new microorganisms from outside sources. It is critical to perform more research to understand the consequences of the microbes recruited for bioremediation, it is crucial to consider that certain organisms introduced to an environment may potentially lead to pollution due to mutations or other factors. As a result, there must be regulatory authorities that monitor the risks posed by germs in specific environments.

## References

Abdeen, M., Sabry, S., Ghozlan, H. et al. (2016). Microbial-physical synthesis of Fe and $Fe_3O_4$ magnetic nanoparticles using *Aspergillus niger* YESM1 and supercritical condition of ethanol. *Journal of Nanomaterials* 2016: Article ID 9174891.

Adams, G.O., Fufeyin, P.T., Okoro, S.E., and Ehinomen, I. (2015). Bioremediation, biostimulation and bioaugmentation: a review. *International Journal of Environmental Bioremediation & Biodegradation* 3: 28–39.

Ahmad, A., Senapati, S., Khan, M.I. et al. (2003). Intracellular synthesis of gold nanoparticles by a novel alkalotolerant actinomycete, *Rhodococcus* species. *Nanotechnology* 14: 824–828.

Akob, D.M., Mills, H.J., and Kostka, J.E. (2007). Metabolically active microbial communities in uranium-contaminated subsurface sediments. *FEMS Microbial Ecology* 59: 95–107.

Alexander, M. (1994). *Biodegradation and Bioremediation*. San Diego, CA: Academic.

AL-Huqail, A.A., Kumar, P., Eid, E.M. et al. (2022). Risk assessment of heavy metals contamination in soil and two rice (*Oryza sativa* L.) varieties irrigated with paper mill effluent. *Agriculture* 12 (11): 18–64.

Alvarez, L.M., Ruberto, L.A.M., Gurevich, J., and Mac Cormack, W.P. (2020). Environmental factors affecting reproducibility of bioremediation field assays in Antarctica. *Cold Regions Science and Technology* 169: 102915.

Aragao, W.A.B., Teixeira, F.B., Fagundes, N.C.F. et al. (2018). Hippocampal dysfunction provoked by mercury chloride exposure: evaluation of cognitive impairment, oxidative stress, tissue injury and nature of cell death. *Oxidative Medicine and Cellular Longevity* 1: 1–11.

Ash, A., Revati, K., and Pandey, B.D. (2011). Microbial synthesis of iron-based nanomaterials – a review. *Bulletin of Materials Science* 34 (2): 191–198.

Ayangbenro, A.S. and Babalola, O.O. (2018). Metal(loid) bioremediation: strategies employed by microbial polymers. *Sustainability* 10 (9): 3028.

Ayangbenro, A.S. and Babalola, O.O. (2020). Genomic analysis of *Bacillus cereus* NWUAB01 and its heavy metal removal from polluted soil. *Scientific Reports* 10 (1): 1–12.

Ayilara, M.S., Olanrewaju, O.S., Babalola, O.O., and Odeyemi, O. (2020). Waste management through composting: challenges and potentials. *Sustainability* 12 (11): 44–56.

Bala, S., Garg, D., Thirumalesh, B.V. et al. (2022). Recent strategies for bioremediation of emerging pollutants: a review for a green and sustainable environment. *Toxics* 10 (8): 484.

Balali-Mood, M., Naseri, K., Tahergorabi, Z. et al. (2021). Toxic mechanisms of five heavy metals: mercury, lead, chromium, cadmium, and arsenic. *Frontiers in Pharmacology* 12: 227.

Bhadouria, R., Das, S., Kumar, A. et al. (2020). Mycoremediation of agrochemicals. In: *Agrochemicals Detection, Treatment and Remediation* (ed. M. Narasimha Vara Prasad), 593–620. Oxford, United Kingdom: Butterworth-Heinemann https://doi.org/10.1016/B978-0-08-103017-2.00022-2.

Bhatt, P., Rene, E.R., Kumar, A.J. et al. (2020). Binding interaction of allethrin with esterase: bioremediation potential and mechanism. *Bioresource Technology* 315: 13845.

Bhatt, P., Bhatt, K., Sharma, A. et al. (2021a). Biotechnological basis of microbial consortia for the removal of pesticides from the environment. *Critical Reviews in Biotechnology* 41 (3): 317–338.

Bhatt, P., Joshi, T., Bhatt, K. et al. (2021b). Binding interaction of glyphosate oxidoreductase and C-P lyase: molecular docking and molecular dynamics simulation studies. *Journal of Hazardous Materials* 5 (409): 124–927.

Bijay, T., Ajay, K.C., and Anish, G. (2012). An audit on bioremediation of oil hydrocarbon contaminants in soil. Kathmandu University. *Diary of Science, Building And Innovation* 8: 164–170.

Bondarenko, O., Rõlova, T., Kahru, A., and Ivask, A. (2008). Bioavailability of Cd, Zn and Hg in soil to nine recombinant luminescent metal sensor bacteria. *Sensors* 8: 6899–6923.

Callaghan, A.V. (2013). Metabolomic investigations of anaerobic hydrocarbon-impacted environments. *Current Opinion in Biotechnology* 24: 506–515.

Cheng, Z., Tan, A.L.K., Tao, Y. et al. (2012). Synthesis and characterization of iron oxide nanoparticles and applications in the removal of heavy metals from industrial wastewater. *International Journal of Photoenergy* 2012: Article ID 608298.

Chromiková, Z., Čučak, D., Kučerová, K. et al. (2019). Microbial tolerance to heavy metal stress. In: *Abstract Book Bacell* (ed. I. Mandic-Mulec, T. Danevcic, and P. Stefanic). Ljubljana: University of Ljubljana Biotechnical Faculty.

Daniel, R. (2005). The metagenomics of soil. *Nature Reviews. Microbiology* 3: 470–478.

Dash, B., Sahu, N., Singh, A.K. et al. (2021). Arsenic efflux in *Enterobacter cloacae* RSN3 isolated from arsenic-rich soil. *Folia Microbiologica* 66: 189–196.

de Souza, M.B., de Souza Santos, L.R., Borges, R.E. et al. (2020). Current status of ecotoxicological studies of bats in Brazil. *Bulletin of Environmental Contamination and Toxicology* 104: 393–399.

Debbarma, P., Raghuwanshi, S., Singh, J. et al. (2017). Comparative in situ biodegradation studies of polyhydroxybutyrate film composites. *Biotech* 7 (178): 1–9.

Delille, D., Duval, A., and Pelletier, E. (2008). Highly efficient pilot biopiles for on location preparation treatment of diesel oil-tainted sub-Antarctic soil. *Cold Regions Science and Technology* 54: 7–18.

Dhankhar, R. and Hooda, A. (2011). Fungal biosorption – an alternative to meet the challenges of heavy metal pollution in aqueous solutions. *Environmental Technology* 32 (5): 467–491.

Dixit, R., Wasiullah Malaviya, D., Pandiyan, K. et al. (2015). Bioremediation of heavy metals from soil and aquatic environment: an overview of principles and criteria of fundamental processes. *Sustainability* 7: 2189–2212.

Dobrowolski, R., Szcześ, A., and Czemierska, M. (2017). Studies of cadmium (II), lead (II), nickel (II), cobalt (II) and chromium (VI) sorption on extracellular polymeric substances produced by *Rhodococcus opacus* and *Rhodococcus rhodochrous*. *Bioresource Technology* 225: 113–120.

Elias, D.A., Krumholz, L.R., and Wong, D. (2003). Characterization of microbial activities and U reduction in a shallow aquifer contaminated by uranium mill tailings. *Microbial Ecology* 46: 83–91.

Emami, S., Pourbabaee, A.A., and Alikhani, H.A. (2012). Bioremediation principles and techniques on petroleum hydrocarbon contaminated soil. *Special Journal of Engineering and Applied Sciences* 2: 320–323.

Fashola, M.O., Ngole-Jeme, V.M., and Babalola, O.O. (2016). Heavy metal pollution from gold mines: environmental effects and bacterial strategies for resistance. *International Journal of Environmental Research and Public Health* 13 (11): 1047.

Fashola, M.O., Ngole-Jeme, V.M., and Babalola, O.O. (2020). Physicochemical properties, heavy metals, and metal-tolerant bacteria profiles of abandoned gold mine tailings in Krugersdorp, South Africa. *Canadian Journal of Soil Science* 100 (3): 217–233.

Ganesh Babu, M.M. and Gunasekaran, P. (2009). Production and structural characterization of crystalline silver nanoparticles from *Bacillus cereus* isolate. *Colloids and Surfaces. B, Biointerfaces* 74: 191–195.

Giri, K., Rai, J.P.N., Pandey, S. et al. (2017a). Performance evaluation of isoproturon-degrading indigenous bacterial isolates in soil microcosm. *Chemistry and Ecology* 33 (9): 817–825.

Giri, K., Suyal, D.C., Mishra, G. et al. (2017b). Biodegradation of isoproturon by *Bacillus pumilus* K1 isolated from foothill agroecosystem of north west Himalaya. *Proceedings of the National Academy of Sciences, India Section B: Biological Sciences* 87 (3): 39–848.

Giri, D.D., Singh, S.K., Giri, A. et al. (2021). Bioremediation potential of methylotrophic bacteria. In: *Microbe Mediated Remediation of Environmental Contaminants* (ed. A. Kumar, V.K. Singh, P. Singh, and V.K. Mishra). Cambridge, MA: Woodhead Publishing.

Goel, R., Zaidi, M.G.H., Soni, R., and Kusumlata, S.Y.S. (2008). Implication of *Arthrobacter* and *Enterobacter* species for polycarbonate degradation. *International Biodeterioration and Biodegradation* 61 (2): 167–172.

Gomez, S.M. (2014). Optimization of field scale biopiles for bioremediation of oil hydrocarbon defiled soil at low temperature conditions by reaction surface approach (RSM). *International Biodeterioration & Biodegradation* 89: 103–109.

Gomez, M.G., Anderson, C.M., Graddy, C.M. et al. (2017). Large-scale comparison of bioaugmentation and biostimulation approaches for biocementation of sands. *Journal of Geotechnical and Geoenvironmental Engineering* 143 (5): 04016124.

Gond, D.P., Jha, S.S., Kumar, A., and Singh, S.K. (2021). Plant growth promoting bacteria and its role in green remediation. In: *Sustainable Environmental Clean-Up* (ed. V.K. Mishra and A. Kumar), 149–163. Cambridge, MA: Elsevier https://doi.org/10.1016/B978-0-12-823828-8.00007-4.

Gupta, A., Singh, S.K., Singh, V.K. et al. (2020). Endophytic microbe approaches in bioremediation of organic pollutants. In: *Microbial Endophytes: Prospects of Sustainable Agriculture* (ed. A. Kumar and V.K. Singh), 157–174. Cambridge, MA: Elsevier https://doi.org/10.1016/B978-0-12-818734-0.00007-3.

Haferburg, G. and Kothe, E. (2007). Microbes and metals: interactions in the environment. *Journal of Basic Microbiology* 47: 453–467.

Handelsman, J., Rondon, M.R., Brady, S.F. et al. (1998). Molecular biological access to the chemistry of unknown soil microbes: a new frontier for natural products. *Chemistry & Biology* 5 (10): 245–249.

Huang, Y., Pan, H., Wang, Q. et al. (2019). Enrichment of the soil microbial community in the bioremediation of a petroleum-contaminated soil amended with rice straw or sawdust. *Chemosphere* 224: 265–271.

Ilyas, S. and Lee, J. (2014). Bioleaching of metals from electronic scrap in a stirred tank reactor. *Hydrometallurgy* 149: 50–62.

Imam, A., Suman, S.K., Ghosh, D., and Kanaujia, P.K. (2019). Analytical approaches used in monitoring the bioremediation of hydrocarbons in petroleum contaminated soil and sludge. *TrAC Trends in Analytical Chemistry* 118: 50–64.

Imron, M.F., Kurniawan, S.B., Ismail, N.I., and Abdullah, S.R.S. (2020). Future challenges in diesel biodegradation by bacteria isolates: a review. *Journal of Cleaner Production* 251: 119716.

Isakovski, M.K., Maletić, S., Tamindžija, D. et al. (2020). Impact of hydrochar and biochar amendments on sorption and biodegradation of organophosphorus pesticides during transport through Danube alluvial sediment. *Journal of Environmental Management* 274: 111156.

Iyas, S., Ru'an, C.H.I., and Lee, J.C. (2012). One step bioleaching of sulphide ore with low concentration of arsenic by *Aspergillus niger* and Taguchi orthogonal array optimization. *Chinese Journal of Chemical Engineering* 20: 923–929.

Jabbar, T. and Wallner, G. (2015). Biotransformation of radionuclides: trends and challenges. *Radionuclides in the Environment* 2015: 169–184.

Jacob, J.M., Karthik, C., Saratale, R.G. et al. (2018). Biological approaches to tackle heavy metal pollution: a survey of literature. *Journal of Environmental Management* 217: 56–70.

Jafari, A., Ghaderpoori, M., Kamarehi, B., and Abdipour, H. (2019). Soil pollution evaluation and health risk assessment of heavy metals around Douroud cement factory. *The Iranian Journal of Earth Sciences* 78 (8): 1–9.

Kaeberlein, T., Lewis, K., and Epstein, S.S. (2002). Isolating "uncultivable" microorganisms in pure culture in a simulated natural environment. *Science* 296 (80): 1127–1130.

Kamaludeen, S.P.B., Arunkumar, K.R., Avudainayagam, S., and Ramasamy, K. (2003). Bioremediation of chromium contaminated environments. *Indian Journal of Experimental Biology* 41: 972–985.

Kebede, G., Tafese, T., Abda, E.M. et al. (2021). Factors influencing the bacterial bioremediation of hydrocarbon contaminants in the soil: mechanisms and impacts. *Journal of Chemistry* 2021: 1–17.

Kensa, V.M. (2011). Bioremediation – an overview. I. Control. *Pollution* 27 (2): 161–168.

Kharangate-Lad, A. and D'Souza, N.C. (2021). Current approaches in bioremediation of toxic contaminants by application of microbial cells; biosurfactants and bioemulsifiers of microbial origin. *Rhizobiont in Bioremediation of Hazardous Waste* 217–263.

Kostal, J., Yang, R., Wu, C.H. et al. (2004). Enhanced arsenic accumulation in engineered bacterial cells expressing ArsR. *Applied and Environmental Microbiology* 70 (8): 4582–4587.

Kour, D., Kaur, T., Devi, R. et al. (2021). Beneficial microbiomes for bioremediation of diverse contaminated environments for environmental sustainability: present status and future challenges. *Environmental Science and Pollution Research* 28: 24917–24939.

Koul, B. and Adlakha, K. (2021). Bioremediation of radionuclides by plant-microbe system: current progress and challenges. *Microbe Mediated Remediation of Environmental Contaminants* 303–320.

Kumar, A., Bisht, B.S., Joshi, V.D., and Dhewa, T. (2011). Review on bioremediation of polluted environment: a management tool. *Global Diary of Ecological Sciences* 1 (5): 1079–1093.

Kumar, A.M., Ravindran, B., Sarsaiya, S. et al. (2020). Metagenomics for taxonomy profiling: tools and approaches. *Bioengineered* 11 (1): 356–374.

Kumari, V. and Tripathi, A. (2020). Remediation of heavy metals in pharmaceutical effluent with the help of *Bacillus cereus*-based green-synthesized silver nanoparticles supported on alumina. *Applied Nanoscience* 10 (6): 1709–1719.

Lee, T.H., Byun, I.G., Kim, Y.O. et al. (2006). Monitoring biodegradation of diesel fuel in bioventing measures utilizing in situ breath rate. *Water Science and Technology* 53: 263–272.

Li, C.H., Wong, Y.S., and Tam, N.F. (2010). Anaerobic biodegradation of polycyclic sweet-smelling hydrocarbons with correction of iron (III) in mangrove dregs slurry. *Bioresource Technology* 101: 8083–8092.

Liu, X., Shi, H., Bai, Z. et al. (2020). Heavy metal concentrations of soils near the large opencast coal mine pits in China. *Chemosphere* 244: 125–360.

Lloyd, J.R. and Renshaw, J.C. (2005). Microbial transformations of radionuclides: fundamental mechanisms and biogeochemical implications. *Metal Ions in Biological Systems* 44: 253–260.

Madhavi, G.N. and Mohini, D.D. (2015). Review paper on parameters influencing bioremediation. *Worldwide Diary of Life Science and Pharma Research* 2: 77–80.

Mahmood, A., Bilal, B., Naeem, Z., Iram, S. 2021. Physical, chemical, and biological remediation techniques for textile effluents in context with developed and developing countries. In: *Rhizobiont in Bioremediation of Hazardous Waste* (ed. V. Kumar, R. Prasad, and M. Kumar), 409–44. Springer.

Mahmoud, G.A.E. (2021). Microbial scavenging of heavy metals using bioremediation strategies. In: *Rhizobiont in Bioremediation of Hazardous Waste* (ed. K. Vivek, P. Ram, and K. Manoj), 265–289. Singapore: Springer.

Malik, Z.A. and Ahmed, S. (2012). Degradation of oil hydrocarbons by oil field disconnected bacterial consortium. *African Journal of Biotechnology* 11: 650–658.

Medfu Tarekegn, M., Zewdu Salilih, F., and Ishetu, A.I. (2020). Microbes used as a tool for bioremediation of heavy metal from the environment. *Cogent Food & Agriculture* 6 (1): 1783174.

Methneni, N., Morales-Gonzalez, J.A., Jaziri, A. et al. (2021). Persistent organic and inorganic pollutants in the effluents from the textile dyeing industries: ecotoxicology appraisal via a battery of biotests. *Environmental Research* 196: 110956.

Mishra, M., Singh, S.K., and Kumar, A. (2021). Environmental factors affecting the bioremediation potential of microbes. In: *Woodhead Publishing Series in Food Science, Technology and Nutrition, Microbe Mediated Remediation of Environmental Contaminants* (ed. A. Kumar, V.K. Singh, P. Singh, and V.K. Mishra), 47–58. Cambridge, MA: Elsevier, Woodhead Publishing https://doi.org/10.1016/B978-0-12-821199-1.00005-5.

Montagnolli, R.N., Matos Lopes, P.R., and Bidoia, E.D. (2015). Assessing *Bacillus subtilis* biosurfactant impacts on the biodegradation of oil based commodities. *Environmental Monitoring and Assessment* 187: 1–17.

Mulligana, C.N. and Yong, R.N. (2004). Natural lessening of defiled soils. *Environment International* 30: 587–601.

Naseem, Z., Bhatti, H.N., and Sadaf, S. (2016). Sorption of uranium (VI) by *Trapa bispinosa* from aqueous solution: effect of pretreatments and modeling studies. *Desalination and Water Treatment* 57: 11121–11132.

Ndeddy Aka, R.J. and Babalola, O.O. (2017). Identification and characterization of Cr-, Cd-, and Ni-tolerant bacteria isolated from mine tailings. *Bioremediation Journal* 21 (1): 1–19.

Niu, G.L., Zhang, J.J., Zhao, S. et al. (2009). Bioaugmentation of a 4-chloronitrobenzene debased soil with *Pseudomonas putida* ZWL73. *Environmental pollution* 57: 763–771.

North, N.N., Dollhopf, S.L., and Petrie, L. (2004). Change in bacterial community structure during in situ biostimulation of subsurface sediment co-contaminated with uranium and nitrate. *Applied and Environmental Microbiology* 70: 4911–4920.

Ogunlaja, A., Ogunlaja, O.O., Okewole, D.M., and Morenikeji, O.A. (2019). Risk assessment and source identification of heavy metal contamination by multivariate and hazard index analyses of a pipeline vandalised area in Lagos state, Nigeria. *Science of the Total Environment* 651: 2943–2952.

Pandey, S., Kumari, M., Singh, S.P. et al. (2015). Bioremediation via nanoparticles: an innovative microbial approach. In: *Handbook of Research on Uncovering New Methods for Ecosystem Management Through Bioremediation*, 491–515. IGI Global.

Pandey, K.D., Singh, S.K., Shukla, L. et al. (2023). Mechanistic approaches and factors regulating microalgae mediated heavy metal remediation from the aquatic ecosystem. In: *Advances in Chemical Pollution, Environmental Management and Protection*, vol. 9 (ed. L.F.R. Ferreira, A. Kumar, and M. Bilal), 285–299. Cambridge, United States: Elsevier, Academic Press https://doi.org/10.1016/bs.apmp.2022.10.011.

Patel, A.B., Jain, K.R., Manvar, T. et al. (2022). Enriched bacterial community efficiently degrade polycyclic aromatic hydrocarbons in soil ecosystem: insights from a mesocosms study. *Biochemical Engineering Journal* 185: 108516.

Perez-Gonzalez, T., Jimenez-Lopez, C., Neal, A.L. et al. (2010). Magnetite biomineralization induced by *Shewanella oneidensis*. *Geochimica et Cosmochimica Acta* 74: 967–979.

Phillips, R.W., Wiegel, J., and Berry, C.J. (2002). *Kineococcusradiotolerans sp. nov.*, a radiation-resistant, gram-positive bacterium. *International Journal of Systematic and Evolutionary Microbiology* 52: 933–938.

Prabagar, S., Dharmadasa, R.M., Lintha, A. et al. (2021). Accumulation of heavy metals in grape fruit, leaves, soil and water: a study of influential factors and evaluating ecological risks in Jaffna, Sri Lanka. *Environmental and Sustainability Indicators* 12: 100–147.

Prakash, D., Gabani, P., and Chandel, A.K. (2013). Bioremediation: a genuine technology to remediate radionuclides from the environment. *Microbial Biotechnology* 6: 349–360.

Reena, R., Majhi, M.C., and Arya, A.K. (2012). BioRadBase: a database for bioremediation of radioactive waste. *African Journal of Biotechnology* 11: 8718–8721.

Rehman, A.U., Nazir, S., Irshad, R. et al. (2021). Toxicity of heavy metals in plants and animals and their uptake by magnetic iron oxide nanoparticles. *Journal of Molecular Liquids* 321: 114–455.

Ren, X., Zeng, G., Tang, L. et al. (2018). The potential impact on the biodegradation of organic pollutants from composting technology for soil remediation. *Waste Management* 72: 138–149.

Rondon, M.R., Goodman, R.M., and Handelsman, J. (1999). The Earth's bounty: assessing and accessing soil microbial diversity. *Trends in Biotechnology* 17: 403–409.

Samuel, A. and Ganiyu, K.L. (2015). Biodegradation of diesel oil in soil and its improvement by use of bioventing and revision with distillery squander effluents as biostimulation-bioaugmentation operators. *Diary of Ecological Engineering* 16: 82–91.

Sayler, G.S. and Ripp, S. (2000). Field utilizations of hereditarily built microorganisms for bioremediation measures. *Current Opinion in Biotechnology* 11: 286–289.

Selenska-Pobell, S. (2002). Diversity and activity of bacteria in uranium waste piles. *Radioactivity Environment* 2: 225–254.

Sharma, P., Dutta, D., Udayan, A. et al. (2022). Role of microbes in bioaccumulation of heavy metals in municipal solid waste: impacts on plant and human being. *Environmental Pollution* 305: 119248.

Shilpi, S. (2012). Bioremediation: features, strategies and applications. *Asian Journal of Biological and Life sciences* 2: 202–213.

Singh, J.S., Abhilash, P.C., Singh, H.B. et al. (2011). Genetically engineered bacteria: an emerging tool for environmental remediation and future research perspectives. *Gene* 1 (480): 1–9.

Singh, S.K., Singh, P.P., Gupta, A. et al. (2019). Tolerance of heavy metal toxicity using PGPR strains of *Pseudomonas* species. In: *PGPR Amelioration in Sustainable Agriculture* (ed. Amit Kishore Singh, Ajay Kumar, and Pawan Kumar Singh), 239–252. Cambridge, MA: Woodhead Publishing https://doi.org/10.1016/B978-0-12-815879-1.00012-4.

Singh, M., Singh, D., Rai, P. et al. (2021). Fungi in remediation of hazardous wastes: current status and future. In: *Recent Trends in Mycological Research, Fungal Biology, Environmental Industrial Perspective*, vol. 2 (ed. A.N. Yadav), 195–224.

Singh, S.K., Singh, M.K., Singh, V.K. et al. (2021a). Microbial enzymes and their exploitation in remediation of environmental contaminants. In: *Microbe Mediated Remediation of Environmental Contaminants* (ed. A. Kumar, V.K. Singh, P. Singh, and V.K. Mishra), 59–71. Cambridge, MA: Woodhead Publishing https://doi.org/10.1016/B978-0-12-821199-1.00006-7.

Siric, I., Eid, E.M., El-Morsy, M.H. et al. (2022a). Health risk assessment of hazardous heavy metals in two varieties of mango fruit (*Mangifera indica* l var. dasheri and langra). *Horticulturae* 8 (9): 832.

Strong, P.J. and Burgess, J.E. (2008). Treatment methods for wine-related ad distillery wastewaters: a review. *Bioremediation Journal* 12: 70–87.

Sun, L., Guo, D., Liu, K. et al. (2019). Levels, sources, and spatial distribution of heavy metals in soils from a typical coal industrial city of Tangshan, China. *Catena* 175: 101–109.

Tabak, H.H., Lens, P., and van Hullebusch, E.D.L. (2005). Developments in bioremediation of soils and sediments polluted with metals and radionuclides. *Reviews in Environmental Science and Biotechnology* 4: 115–156.

Tang, C.Y., Criddle, C.S., and Leckie, J.O. (2007). Effect of flux (trans membrane pressure) and membranes properties on fouling and rejection of reverse osmosis and nano-filtration membranes treating perfl uorooctane sulfonate containing waste water. *Environmental Science & Technology* 41: 2008–2014.

Turick, C.E. and Berry, C.J. (2016). Review of concrete biodeterioration in relation to nuclear waste. *Journal of Environmental Radioactivity* 151: 12–21.

U.S. EPA. Handbook on In Situ Treatment of Hazardous Waste Contaminated Soils. 1990. EPA/540/2-90/002.

U.S. EPA (1996). *Bioremediation of Hazardous Waste Sites: Practical Approaches to Implementation (EPA/625/K-96/001).* Washington, DC: U.S. Environmental Protection Agency.

Valls, M., Atrian, S., De Lorenzo, V., and Fernández, L.A. (2000). Engineering a mouse metallothionein on the cell surface of *Ralstonia eutropha* CH34 for immobilization of heavy metals in soil. *Nature Biotechnology* 18: 661–665.

Velásquez, L. and Dussan, J. (2009). Biosorption and bioaccumulation of heavy metals on dead and living biomass of *Bacillus sphaericus*. *Journal of Hazardous Materials* 167: 713–716.

Vidali, M. (2001). Bioremediation. An overview. *Pure and Applied Chemistry* 73: 1163–1172.

Xie, S., Yang, J., and Chen, C. (2008). Study on biosorption kinetics and thermodynamics of uranium by *Citrobacter freudii*. *Journal of Environmental Radioactivity* 99: 126–133.

You, J., Solongo, S.K., and Gomez-Flores, A. (2020). Intensified bioleaching of chalcopyrite concentrate using adapted mesophilic culture in continuous stirred tank reactors. *Bioresource Technology* 307: 123181.

Zeng, F., Qiu, B., Wu, X. et al. (2012). Glutathione mediated alleviation of chromium toxicity in rice plants. *Biological Trace Element Research* 148: 255–263.

Zhang, S., Niu, H., Cai, Y. et al. (2010). Arsenite and arsenate adsorption on coprecipitated bimetal oxide magnetic nanomaterials: $MnFe_2O_4$ and $CoFe_2O_4$. *Chemical Engineering Journal* 158: 599–607.

Zhang, C., Gan, C., Ding, L. et al. (2020). Maternal inorganic mercury exposure and renal effects in the Wanshan mercury mining area, southwest China. *Ecotoxicology and Environmental Safety* 189: 109987.

Zhao, X.W., Zhou, M.H., Li, Q.B. et al. (2005). Simultaneous mercury bioaccumulation and cell propagation by genetically engineered *Escherichia coli*. *Process Biochemistry* 40: 1611–1616.

Zhong, L., Lei, J., Deng, J. et al. (2021). Existing and potential decontamination methods for radioactively contaminated metals-A review. *Progress in Nuclear Energy.* 139: 103854.

Zhou, Y., Liu, X., and Wang, J. (2019). Characterization of microplastics and the association of heavy metals with microplastics in suburban soil of central China. *Science of the Total Environment* 694: 133798.

Zubair, M., Shakir, M., Ali, Q. et al. (2016). Rhizobacteria and phytoremediation of heavy metals. *Environmental Technology Reviews* 5 (1): 112–119.

Zwolak, A., Sarzyńska, M., Szpyrka, E., and Stawarczyk, K. (2019). Sources of soil pollution by heavy metals and their accumulation in vegetables: a review. *Water, Air, and Soil Pollution* 230 (7): 1–9.

# 17

## Chapter Role of Microbial Biofilms in Bioremediation: Current Perspectives

*Sahaya Nadar*[1,2] *and Tabassum Khan*[1]

[1] *Department of Pharmaceutical Chemistry, SVKMs Dr. Bhanuben Nanavati College of Pharmacy, Vile-Parle, Mumbai, Maharashtra, India*
[2] *Department of Pharmaceutical Chemistry, St. John Institute of Pharmacy and Research, Palghar, Maharashtra, India*

## 17.1 Introduction

Biofilms are a fascinating and complex facet of microbiology, playing a pivotal role in numerous natural and engineered environments. They represent intricate microbial communities, such as archaea, bacteria, fungi, and algae that attach to surfaces and are encapsulated within a self-formed extracellular matrix. This extracellular matrix, made of various substances like polysaccharides and proteins, bestows biofilms with their distinct characteristics, which include increased resilience to environmental challenges, such as antimicrobial agents and fluctuations in nutrient availability (Flemming et al. 2023).

Biofilms are not limited to a single domain of life; they can be found in diverse habitats, ranging from natural ecosystems like oceans and rivers to artificial settings such as water distribution systems and medical devices (Nadar et al. 2022). The ubiquity of biofilms is not only a testament to their adaptability but also to their relevance in various fields, including medicine, industry, and environmental science (Yin et al. 2019). Understanding the formation, structure, and function of biofilms is a subject of growing importance. Biofilm research has expanded our knowledge of microbial life, revealing intricate social behaviors, resource-sharing strategies, and complex signaling mechanisms that underpin biofilm development and maintenance (Edel et al. 2019). Biofilms have a substantial impact in diverse domains, including medicine, industry, and environmental science. Within the medical context, biofilms can develop on medical equipment, giving rise to infections that are difficult to manage. In industrial applications, biofilms can cause fouling and corrosion, impacting the efficiency and safety of processes. Additionally, in environmental contexts, biofilms participate in bioremediation and nutrient cycling (Kato et al. 2023).

Biofilms play a crucial role in the field of bioremediation, offering a promising avenue for the cleanup of contaminated environments. Bioremediation, the process of using biological organisms to mitigate and eliminate pollutants from various ecosystems, has gained significant attention due to its eco-friendly and cost-effective nature (Bhadouria et al. 2020; Giri et al. 2021; Gond et al. 2021). Within the realm of bioremediation, biofilms represent a critical biological strategy for remediation processes (Hlihor and Cozma 2023).

A biofilm is a complex community of microorganisms, such as bacteria, is often referred to as the "slime" or "matrix," and provides structural unification to the biofilm, as well as a protective environment for the resident archaea, algae, and fungi, that cling to surfaces and are encapsulated within a self-produced matrix of extracellular polymeric substances (EPSs) (Costa et al. 2018; Wingender et al. 1999). These matrix microorganisms within biofilms are embedded in a three-dimensional network and can interact in intricate ways, promoting diverse metabolic activities and adaptability in the face of environmental challenges.

The use of biofilms in bioremediation is gaining recognition due to their unique capabilities. Biofilms are resilient and can thrive in diverse environmental conditions, making them well-suited for the remediation of contaminated sites. Their ability to degrade or immobilize a wide range of pollutants, including organic and inorganic contaminants, has piqued the interest of researchers and environmental engineers alike. Moreover, biofilms have the potential to effectively remove toxic substances from water, soil, and air, making them a versatile tool in the ongoing efforts to combat environmental pollution (Kour et al. 2021; Yan and Bassler 2019).

*Microbes Based Approaches for the Management of Hazardous Contaminants*, First Edition. Edited by Ajay Kumar, Livleen Shukla, Joginder Singh, and Luiz Fernando R. Ferreira.

In this context, it is essential to explore the intricacies of biofilms in bioremediation comprehensively. This exploration includes understanding the formation, composition, and structure of biofilms, as well as their interactions with contaminants and their surrounding environments. It also involves investigating the genetic and physiological aspects of the microorganisms within biofilms and how they can be harnessed to enhance bioremediation processes.

Biofilms serve as natural habitats where bacterial cells engage in important interactions, including the exchange of genetic material, signaling molecules, and metabolites. Leveraging the power of biofilms in bioremediation allows for efficient and environmentally friendly methods to mitigate the impact of xenobiotic substances on ecosystems (Bhatt et al. 2022; Nicolella et al. 2005; Gupta et al. 2018, 2020). This approach not only facilitates the degradation of pollutants but also promotes the sharing of valuable resources and information among microbial communities, making it a holistic and effective solution for environmental cleanup.

## 17.2 Formation of Biofilm

Biofilm formation is a crucial process in microbial proliferation, distinct from the growth of planktonic organisms in the environment. Biofilms develop through various methods and typically require a moist surface, which can be artificial, inanimate, or living. Beneficial biofilm communities have been found in diverse habitats, ranging from wastewater treatment facilities, rivers, and streams to the mammalian digestive tracts. Biofilms can be seen in various types of interfaces, including air–solid, liquid–liquid, and solid–liquid. For instance, biofilms growing at liquid–liquid interfaces have been utilized in degrading hydrocarbons such as fuels, oils, and industrial coolants. The architectural integrity of the biofilm throughout its lifecycle is primarily managed by physicochemical bonding forces that hold together the components of the EPS (Arnaouteli et al. 2021; O'Toole et al. 2000; Rabin et al. 2015). During the initial stages of biofilm development, free-floating bacteria attach to substrates through formation of van der Waals and electrostatic bonds. Subsequently, hydrogen bonds or ionic interactions are employed to enhance bacterial adhesion. The biofilm's structure goes through four phases, commencing with the adhesion of planktonic bacteria to a moist, non-sterile substrate and concluding with detachment. Between these stages, the three-dimensional structure of biofilm and the formation of microcolonies play a role. Microcolonies consist of individual or grouped bacterial cells that exhibit various morphologies, including even or uneven, filamentous and flat, fluffy or mushroom shaped. They are enclosed by interstitial spaces occupied with water (Flowers et al. 2023; Mampel et al. 2006; Schembri et al. 2002). The amount of EPSs produced within a small colony is contingent on the bacterial population within it. It has been observed that microcolonies typically comprise 70–90% EPS contents, with the left portion consisting of cells. Each of these stages is elaborated upon in the following sections.

### 17.2.1 Reversible Adhesion

In conducive environments, a solitary planktonic cell undergoes migration and establishes a reversible attachment to a surface, marking the initial stage of the biofilm formation process. This attachment is characterized by weak interactions such as electrostatic, van der Waals, or hydrophobic forces, and it is of a reversible nature. Cellular projections such as pili, fimbriae, or flagella contribute to the strength and adhesion of the attachment to the surface. Once attached, the cells become progressively enveloped within EPSs (Greenman et al. 2020; Solano et al. 2014) (Figure 17.1).

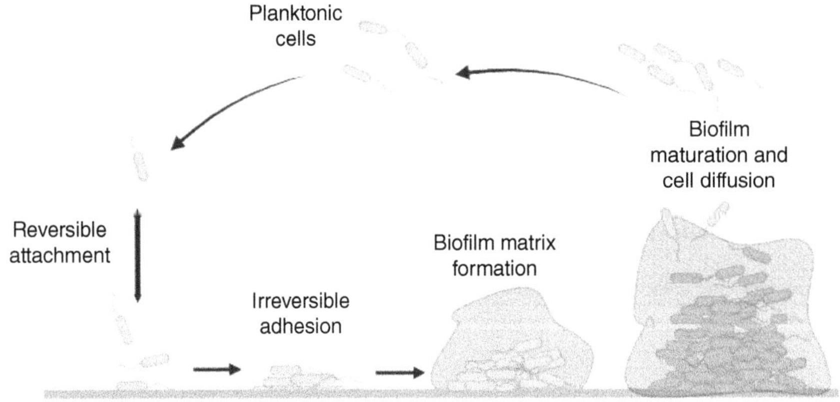


**Figure 17.1** Development and formation of biofilm.

### 17.2.2 Irreversible Adhesion

During this phase, the planktonic cells undergo a distinct transition, adopting a more stratified arrangement and establishing a structured microcolony with aqueous channels, resulting in an irreversible attachment. Colonization, a defining characteristic of biofilms, holds significant importance in their dormancy and ability to cause disease. As the cells firmly attach to a suitable surface, numerous microorganisms accumulate and release EPSs to serve as a sealant, effectively anchoring the microorganisms. These coordinated processes culminate in the formation of microcolonies (Dunne 2002; Tolker-Nielsen 2015; van Loosdrecht et al. 1990).

### 17.2.3 Formation of Biofilm Matrix

The EPS secreted by adherent cells develops a matrix within which the cells form their community and reach their maximum cell density. The EPS that envelops cells within a biofilm comprises a mixture of components, including extracellular DNA, auto-sensing substances, lipids, proteins, persister cells, and polysaccharides.

Polysaccharides found in this matrix play a crucial role in providing support for the cells inside the biofilm, facilitating adhesion, offering protection, and contributing to stability of the structure. The aggregative polysaccharides, acting as an adhesive, support the colonization process and provide defense against physical stresses caused by the movement of fluids that might otherwise deprive the cells of essential nutrients (Karygianni et al. 2020).

Different organisms inherit various types of polysaccharides, each playing a significant role in sustaining the integrity of the biofilm. Nucleic acids, such as e-DNA, extracellular RNA, and ribosomal DNA, engage with various constituents of the EPS, ensuring the availability of nutrients, contributing to structural organization, and protecting against gene transfer into the biofilm. Proteins within the matrix impart integrity to the biofilm. Additionally, dormant cells called persister cells display exceptional resistance to antimicrobial agents (Muthami et al. 2022; Tai et al. 2022).

### 17.2.4 Biofilm Maturation and Diffusion

Under conducive environment and in the availability of sufficient nutrients, the cells proliferate and undergo specialization, resulting in the formation of mature biofilms characterized by a distinct spatial structure. These well-developed biofilms take on the appearance of a cohesive community governed by chemical signaling modulators released by the bacterial residents in the biofilm. Individual planktonic cells are released with the maturation of the biofilm, having the capacity to migrate to new surfaces, thereby contributing to the spread of bacterial colonization (Gilbert et al. 2002).

## 17.3 Microbes Forming Biofilm

### 17.3.1 Bacteria

Bacterial biofilms are microbial colonies vital for endurance in challenging environments. Biofilms represent self-regulating communities of bacteria that adhere to surfaces, organizing into a well-structured community with a polymeric matrix self-formed, that promotes their growth within the environment.

The competency of bacterial biofilms is influenced by a range of factors, including capsular polysaccharides, surface proteins, autolysin, adhesins, carbon dioxide levels, anaerobic conditions, osmotic conditions, availability of glucose, temperature, pH, ionic concentrations, nutritional resources, and the existence of surface-active agents (Sharma et al. 2019; van Wolferen et al. 2018).

Mature bacterial biofilms exhibit geographical, temporal, and dynamic heterogeneity, adopting different structures based on the encircling environment, such as pH, osmolarity, temperature, and presence of nutrients, as well as the composition of the microbial community.

The composition of polysaccharides plays a pivotal role in shaping bacterial biofilms, involving various biomolecules such as DNA, proteins, lipids, and organic compounds alongside polysaccharides. Research by Andersson and colleagues delved into characteristics of bacterial strains in terms of biofilm formation and adhesion in both pure and mixed populations of bacterial strains (Andersson et al. 2008; Chen et al. 2019). Turki et al. explored the structural diversity of bacterial population in a pilot testing facility using the rotating biological contactor approach. This study identified dominant bacterial species, including *Enterobacter agglomerans*, *Pantoea agglomerans,* and *Cronobacter sakazakii,* known for their capability for bioremediation, while *Salmonella* was frequently detected, suggesting limited impact from the rotating biodisk system on *Salmonella* presence (Chattopadhyay et al. 2022; Turki et al. 2017).

### 17.3.2 Algae

Algal biofilms are ecosystems primarily dominated by various types of algae. These biofilms colonize surfaces exposed to light, provided there is moisture and a source of nutrients available. Similar to bacterial biofilms, algal biofilms can establish themselves as colonies on surfaces and may detach in clusters or as a single entity. They are also capable of responding to environmental changes.

Despite being widespread in various environmental settings, research on algal biofilms has predominantly focused on understanding their structural characteristics. This emphasis on structural analysis is often motivated by concerns related to safety and economic considerations. Researchers aim to develop strategies to control or prevent the expansion of algal biofilms, particularly when they can lead to issues such as fouling or unwanted growth in industrial and aquatic systems (Häubner et al. 2006; Hoffmann 1998; Leadbeater and Callow 1992).

Li et al. conducted a study in which microalgal biofilms were utilized to recover zinc from effluents and leachates emanating from a mining dump. They utilized two types of microalgae, *Halochlorella rubescens* and the *trebouxiophyceae* to create microalgal biofilms (Li et al. 2015).

The selection of *Streptomyces bacillaris* strains was based on their recognized efficiency as adsorbers of heavy metals, while *H. rubescens* had previously demonstrated effectiveness in bilayer photobioreactors for wastewater treatment. The results of the study indicated that the *S. bacillaris* strain displays maximum resilience to zinc (Shi et al. 2014).

Interestingly, the *H. rubescens* and *S. bacillaris* 1896 B strains were obtained in quasi-natural environments. Notably, these species were originally found in the drainage stream of a metal waste site in Decazeville, France. It is known that microalgae exposed to toxic substances in polluted environments can develop a certain degree of tolerance to those substances. Additionally, previous studies have indicated that algal strains obtained from contaminated environments settings often have a heightened capacity to remove heavy metals, making them valuable candidates for bioremediation efforts (Doshi et al. 2008; Skowroński et al. 1985; Tran and Lee 2023).

### 17.3.3 Fungi

Fungi, much like bacteria, share the fundamental prerequisites for the formation of biofilm, which involve the capability to attach to surfaces or to each other and to be encapsulated in EPSs. Recent discoveries have shown that filamentous fungi possess hydrophobic proteins of an amphipathic nature, which facilitate biofilm growth in fungi by enabling the fungus to attach to surfaces. Hydrophobins like these serve as morphogenetic signals allowing spores to cling to hydrophobic surfaces, whether they are natural or artificial. *Aspergillus fumigatus* secretes hydrophobins with an eight-cysteine residue pattern, creating strong attachment forces during spore adhesion to surfaces (Sheppard and Howell 2016).

Various fungal species, including *Penicillium rubrum, Neocosmopora, Candida albicans, Acremonium,* and *Fusarium species*, are known to form biofilms. Yet, in various microbial biofilm, communities found in clean water delivery systems, fungal diversity is typically lower when compared to bacterial diversity (Roudbary et al. 2022).

Researchers like Howell and Shepperd have conducted comparative studies, comparing the biofilms of *C. albicans* and *C. fumigatus* with bacterial biofilms. The goal of such studies is to acquire a more comprehensive understanding of the biological functions, biosynthesis, and structure of fungal biofilms. Despite differences in the homology of glycan and the use of different biosynthetic enzymes between fungi and bacteria, both groups are capable of developing strategies for antibiotic resistance that involve evading the immune system and biofilm formation (Douterelo et al. 2018; Pérez Mendoza et al. 2011; Gulati and Nobile 2016).

In the context of treating skin diseases, various treatment techniques are available. However, enzyme-based therapeutic techniques that aim to inhibit and break down biofilm elements are becoming potential alternatives in the future for more effective management of these conditions. These enzyme-based approaches hold promise in addressing biofilm-related skin diseases.

## 17.4 Biofilms in Bioremediation

### 17.4.1 Biofilms in Polycyclic Aromatic Hydrocarbons Remediation

Polycyclic aromatic hydrocarbons (PAHs), which belong to a group of hazardous aromatic compounds with fused rings, are organic pollutants that can be found widely within the ecosystem. They accumulate as a result of both human activities and natural processes. Notably, these substances possess substantial molecular weight and can endure within the ecosystem for extended time frame. These pollutants are primarily generated through the partial burning of fossil fuels, petroleum products, and various industrial activities. Additionally, natural disasters like volcanic eruptions and forest fires also have an impact on the accumulation of these pollutants in the ecosystem. PAHs, recognized for their mutagenic and carcinogenic characteristics, present substantial health hazards for mankind. In recognition of these risks, they are designated as priority pollutants (Qin et al. 2019; Wei et al. 2017). These compounds can penetrate the human system through various pathways, including breathing polluted air, consumption of contaminated food and water, skin contact with polluted soil, and occupational exposure (Wattiau 2002).

Additionally, PAHs can infiltrate water supplies from multiple origins, such as industrial sources and domestic, urban drainage, and discharges from vehicles. The need for the elimination of these particular pollutants from the atmosphere has garnered considerable attention due to the broad spectrum of damage they cause, affecting not only human health but also the ecosystem, including marine and terrestrial fauna, and agricultural farmland. However, removing PAHs from soil presents a unique challenge due to their insolubility in water and slow degradation (Aralu et al. 2023; Choi et al. 2010).

Biofilm-based bioremediation exhibits significant potential for the detoxification of persistent organic contaminants like PAHs. To fully explore biofilms capabilities in cleanup of the environment, a comprehension of their physiological and genetic makeup is essential. PAHs, a class of aromatic hydrocarbons, are of particular concern due to their toxic effects on living organisms including carcinogenesis and mutagenesis and amid the several available methods for decontamination, emerging bioremediation technologies hold promise. Biological methods are favored over chemical procedures because they are environmentally friendly and cost-effective (Gevao et al. 2007; Li et al. 2023; Mishra et al. 2021, Pandey et al. 2023).

While planktonic cells grown in monitored laboratory conditions have been widely investigated for bioremediation, in nature, they are mostly found in biofilms. These biofilms develop on solid surfaces where nutrient gradients exist. Bacterial biofilms are valuable entities for PAH bioremediation, thanks to their resilient structures that can withstand various environmental challenges. Therefore, optimizing bioremediation processes to suit real-world conditions necessitates a deep understanding of biofilm structural integrity and the interactions of microbiome with contaminants and various environmental factors (Kumari et al. 2022; Singh et al. 2019b). This chapter provides insights into the existing understanding and the potential use of biofilms for PAH bioremediation.

Chauhan et al. worked on the deterioration of various aromatic compounds by a range of bacteria, including *Nocardioides* species, *Comamonas testosteroni*, *Mycobacterium*, *Sphingomonas paucimobilis*, *Pseudomonas stutzeri*, and *Alcaligenes faecalis*. These bacteria demonstrated the ability to degrade naphthalene, anthracene, benzo[*b*]fluoranthene, pyrene, benzo[*a*]pyrene, 1-nitropyrene, and phenanthrene (Chauhan et al. 2008; Lahiri et al. 2022).

Similarly, Annweiler et al. conducted a study in which they isolated *Bacillus thermoleovorans* from contaminated compost. They found that this bacterium exhibited the capability to degrade naphthalene compounds (Annweiler et al. 2000). In another study, Chaudhary and coworkers identified *Mycobacterium* species, *Haemophilus* species, *Pseudomonas* species, and *Rhodococcus* species isolated from soil, all of which showed the capacity to degrade phenanthrene, naphthalene, anthracene, pyrene, and benzo[*a*]pyrene (Chaudhary et al. 2015). Mangwani and team reported the isolation of specific microbes from various environmental sources, all of which demonstrated the ability to degrade PAHs compounds such as phenanthrene and pyrene (Mangwani et al. 2017). The isolated microbes include *Pseudomonas aeruginosa* from Paradeep port and *A. faecalis* and *Stenotrophomonas acidaminiphila* from Chilika lagoon, and *Pseudomonas mendocina* from Rushikulya estuary (Vandana and Das 2021). These findings highlight the potentiality of these organisms in contributing to the remediation of PAH-contaminated environments.

### 17.4.2 Biofilms in Chlorinated Compounds Remediation

Chlorinated compounds are exceptionally stubborn chemicals that frequently occur in effluents from various chemical industries. They have the capacity to rapidly migrate through soil, making them a challenging issue. These compounds are categorized as a few of the most prevalent soil and aquifer contaminants and pose a cancer-causing risk even at very low levels. In an effort to eliminate 2,4-dichlorophenol (DCP) from synthetic eluents, Kargi and team implemented a biofilm reactor equipped with a rotating perforated tube. This innovative system contained mixed microbial species derived from activated sludge, which was boosted with DCP-degrading *Pseudomonas putida*. Impressively, this approach led to the degradation of nearly 100% of the DCP present in the wastewater (Dilaver and Kargi 2009; Eker and Kargi 2008).

Biofilms formed by *Burkholderia species* NK8 and *P. aeruginosa* PA01 play a crucial role in the bioremediation of chlorinated benzoates. This bioremediation process is significant for reducing the impact of chlorinated benzoates on the environment (Chattopadhyay et al. 2022; Yoshida et al. 2009).

### 17.4.3 Biofilms in Petroleum Remediation

Petroleum is widely regarded as a critical resource that is essential for the functioning of every nation. It serves as a vital cornerstone for contemporary society, particularly as a main energy source, and is found in various forms, such as petrol, kerosine, crude oil, paraffin oil, and asphalt. When petroleum hydrocarbons (PHs) are released into a pristine environment, they pose significant risks not only to human health but also to the balance of the ecosystem (Dasgupta et al. 2013; Xu et al. 2018).

The EPS content within bacterial biofilms plays a crucial role in the degradation of toxic inorganic substances like PAHs. This is because EPS has the capability to absorb, trap, and break down these environmental contaminants. Moreover, it aids bacteria in thriving within oil-contaminated areas, helping them withstand challenging environmental conditions (Vishwakarma 2020).

Bacterial biofilms are diverse communities that bring together aerobic and anaerobic microorganisms, heterotrophic and nitrifying species, sulfate-reducing and sulfate-oxidizing organisms (Dang and Lovell 2015; Davey and O'toole 2000; Le Borgne and Baquerizo 2019). This is made possible by the structural dimensionality of the biofilm matrix, where oxygen levels decrease toward the center. This arrangement enhances the efficient and swift breakdown of pollutants, including PHs. Individual microbes reside on rock surfaces to acquire nutrients and shield themselves from harsh environmental circumstances.

Previous findings have shown that bioaugmentation and biostimulation methods are more effective in remediating PH oil (Myazin et al. 2021; Sayed et al. 2021; Udume et al. 2023; Wu et al. 2019). Additionally, there is a strong positive relationship between the EPS content of bacterial biofilms and the degradation of PAHs. The composition and specific three-dimensional framework of the biofilm matrix enhance its interaction with various hydrophobic compounds, including PAHs. This is attributed to the presence of charged, hydrophobic components in bacterial EPS, which demonstrates their binding ability with specific PAH compounds.

Biofilms can function as a valuable tool for the detoxification of hazardous chemicals. In nature, microbial biotransformation facilitated by the production of glycoconjugates can lead to the creation of less harmful byproducts without the need for genetic modification of microorganisms. During the stationary growth phase of microorganisms, they naturally produce glycoconjugates with surfactant properties. These amphiphilic compounds are created on the cell surfaces of microorganisms and have both hydrophilic and lipophilic counterparts effectively reducing interfacial and surface tension. Glycoconjugates come in various forms, including lipopolysaccharides, glycopeptides, glycosides, glycolipids, glycoproteins, and peptidoglycans. These glycoconjugates facilitate the aggregation and further bioremediation of the organic pollutants. The bioremediation of various contaminants relies on a wide array of metabolizing enzymes, showcasing a remarkable diversity. This process takes place through enzymatic reactions, which generate a range of intermediate metabolites by following specific metabolic pathways (Ayilara and Babalola 2023; Bala et al. 2022; Bhatt et al. 2021; Filote et al. 2021) (Figure 17.2).

Polyhydrocarbon-degrading bacteria have either naturally existed or adapted to thrive in response to the escalating levels of petroleum contamination, which can alter the population of hydrocarbon-remedying microbial species in soil due to oil spills. Dasgupta and colleagues discovered that *Pseudomonas* species found in oil-spilled water in the Bay of Bengal possess the capability to break down hydrocarbons with complex structures (Bakaeva et al. 2020; Dasgupta et al. 2013; Pacwa-Płociniczak et al. 2014; Patowary et al. 2017). This degradation is facilitated by their biofilms, a conclusion reinforced by the significant biofilm formation observed at the oil–water juncture.

In marine environments, oil spills can be remediated through the use of hydrocarbon-degrading bacteria, such as *Rhodococcus*, *Arthrobacter*, *Bacillus*, *Pseudomonas*, *Alcanivorax*, and *Cycloclasticus* species belonging to the gammaproteobacteria class. To address the remediation of PHs through degradation, research has demonstrated that a microbial consortium consisting of *Bacillus subtilis* and *Acinetobacter radioresistens*, combined with a surfactant-forming strain, proves

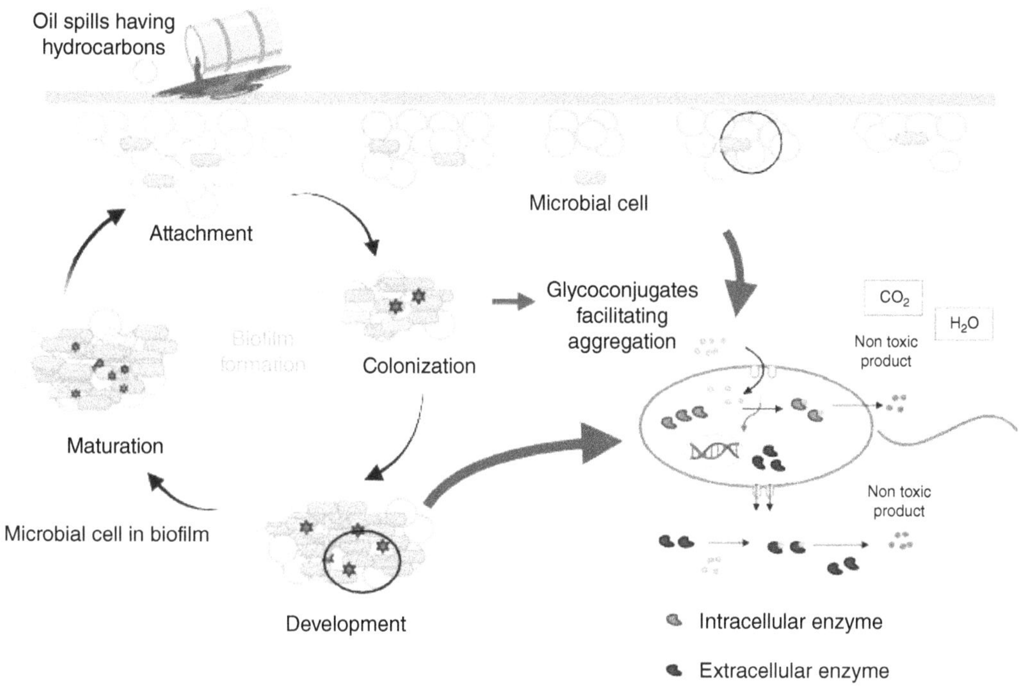


**Figure 17.2** Degradation of oil spills by microbial glycoconjugates.

more effective than a single bacterial strain (Bacosa et al. 2022; Perdigão et al. 2021; Xu et al. 2018). This highlights the effectiveness of a synergistic association of bacteria in PH degradation.

*P. putida OUS82* and *P. aeruginosa*, as well as *P. stutzeri*, are recognized for their capacity to degrade PHs (Koshlaf et al. 2017; Liu et al. 2021; Seo et al. 2009; Van Hamme et al. 2003). However, their usage primarily focuses on decontaminating already polluted soil and water. Concerns regarding their direct contact with humans and potential inclusion in the food chain are not typically considered, which places a greater emphasis on their ability to break down PH than on their pathogenic potential. Moreover, there is limited available literature on this topic.

*Bacillus pumilus KS2*, *Bacillus cereus*, *Bacillus simplex*, and *B. pumilus* are also documented for their capability to degrade PAHs. Frequently, these bacteria are seen to coexist in a consortium along with *Pseudomonas* (Bibi et al. 2018; Nzila et al. 2023; Patowary et al. 2015).

### 17.4.4 Biofilms in Heavy Metals Remediation

Heavy metals, known for their persistence and toxicity across all life forms, pose a significant environmental challenge. The continued growth in industrialization, urbanization, and unsustainable agricultural practices contributes to the ongoing introduction of heavy metals in the ecosystem, resulting in pollution of water and soil. Heavy metals are notorious for endangering human well-being, often resulting in a range of adverse effects. Contact with these metals can lead to skin irritation, respiratory issues, fatigue, abdominal cramps, headaches, diarrhea, anemia, and even permanent damage to vital organs such as the brain and the kidney (Ayangbenro and Babalola 2017; Jaishankar et al. 2014; Ojuederie and Babalola 2017; Zang et al., 2023). Acknowledging the harmful impact of these heavy metals on the ecosystem, various approaches have been formulated for their cleanup. However, it is crucial to recognize that several of these methods come with inherent constraints and limitations of their own. Microbial bioremediation of metals offers an efficient, cost-effective, and environmentally friendly approach to detoxify metal pollutants (Abbas et al. 2018; Chibuike and Obiora 2014; Edelstein and Ben-Hur 2018; Jalmi et al. 2018). Microorganisms possess the ability to utilize metal contaminants as an energy source and convert them into less toxic forms.

Biosorption is a dynamic phenomenon involving the nonspecific binding of metal ions to the polysaccharides and proteins found on the surface of cells. This process is not limited to living cells but also extends to the binding sites offered by deceased microbial biomass, allowing these sites to effectively capture heavy metals, even from extremely diluted solutions. The presence of a peptidoglycan layer in bacteria contributes to their capacity for metal adsorption (Ali Redha 2020;

Anastopoulos and Kyzas 2015; Liu et al. 2019; Naz et al. 2016). Sahmoune's research revealed that *Streptomyces rimosus* demonstrates a strong ability to bind with metals, including lead and iron (Kirova et al. 2021; Sahmoune 2018). In a separate investigation by Rahman and colleagues, it was unequivocally demonstrated that a bacterium resistant to lead, *Staphylococcus hominis* strain AMB-2, is highly effective for biosorbing lead and cadmium (Albert et al. 2018; Rahman et al. 2019). Moreover, coral-associated solubilizing bacteria, such as *Cronobacter muytjensii KSCAS2*, exhibited the capacity to biosorb multiple heavy metals. In the bacterial cell membrane, the mechanisms that lead to the bioaccumulation of heavy metals encompass ion channels, complex permeation, lipid permeation carrier-mediated transport, and endocytosis (Saranya et al. 2018).

In a recent report, various gram-negative bacterial species were identified due to their proven efficacy in the detoxification of specific heavy metals. *Enterobacter* sp. demonstrated the capacity for copper detoxification, while *Stenotrophomonas* sp. and *Providencia* sp. were found to be proficient in bioremediating cadmium. Additionally, *Chryseobacterium* sp. and *Comamonas* sp. were documented for their ability to remove cobalt, and *Ochrobactrum* sp. was recognized for its role in chromium bioremediation. Notably, these bacterial strains exhibited resistance to concentrations of 275 mg/l of copper, 320 mg/l of cadmium, 140 mg/l of cobalt, and 29 mg/l of chromium within a wastewater context (Al-Dhabi et al. 2019; Bao et al. 2023; Fathollahi et al. 2021; Syed and Chinthala 2015).

Ahemad et al. involved bioaccumulation studies of various metals such as mercury, lead, silver, cadmium, and nickel (Ahemad 2019; Pande et al. 2022). Goel investigated cadmium using transmission electron microscopy and identified intracellular and periplasmic accumulation of the metal in *P. putida 62 BN* (Rani and Goel 2009). Srivastava reported that both *Monodictys pelagica* and *Aspergillus niger* can bioaccumulate cadmium, chromium, and lead (Sher and Rehman 2019; Srivastava et al. 2011). In a more recent study, Animesh et al. demonstrated that approximately 20% of nickel(II) accumulates intracellularly in cells of *B. cereus M116* (Naskar et al. 2020). In some documented cases, microorganisms have been used to alter the state of metals. *Acinetobacter* sp. and *Micrococcus* sp. were able to convert toxic As(III) into a less harmful and less soluble form, effectively reducing its toxicity (Basu et al. 2023; Medfu Tarekegn et al. 2020). Similarly, Das and colleagues presented findings demonstrating that *Bacillus* sp. SFC 500-1E, resistant to Cr(VI), can utilize NADH-dependent reductase to transform toxic Cr(VI) into Cr(III), which is less toxic. Moreover, the more soluble and mobile Cr(VI) was converted into a considerably less soluble Cr(IV) (Dhal et al. 2010; Thatoi et al. 2014; Zhou et al. 2022). Another report highlighted the remarkable capabilities of urease-producing bacteria, including *Variovorax boronicumulans*, *Sporosarcina pasteurii*, and *Stenotrophomonas rhizophila*. These bacteria were isolated from calcareous soils in Iranian mines and were found to be instrumental in biomineralizing a range of heavy metals (Jalilvand et al. 2020). An overview of a few bacterial and fungal species that aids in bioremediation of several heavy metals is given in Table 17.1.

### 17.4.5 Other Applications

Biofilms offer versatile applications in the realm of bioremediation. They are particularly valuable in addressing challenges associated with nonaqueous phase liquids (NAPLs) and other environmental issues. NAPLs in contaminated soil and sediments can create localized variations in the distribution of contaminants, making traditional remediation challenging (Gupta and Gandhi 2023; Mukherji and Chavan 2012). Law suggested that biofilms, through chemotaxis, can enhance cell density at interfaces where chemical attractants desorb or dissipate (Law and Aitken 2003). This, in turn, can lead to increased rates of degradation of the attractant, aiding in the remediation of NAPLs. Biofilms exhibit a crucial role in addressing the prevailing environmental problem of acid-mine drainage resulting from mining activities. They contribute to the removal of acidic compounds and the oxidation of sulfide particles within tailings, waste rock heaps, and open-pit walls. In environments where oxygen transport is limited, bacterial biofilms significantly improve the availability of oxygen, thereby enhancing the rates of remediation.

These applications highlight the adaptability and effectiveness of biofilms in addressing a range of environmental challenges and contaminants, making them valuable tools in bioremediation efforts.

## 17.5 Emerging Opportunities

The field of genetic engineering has made significant strides, enabling the modification of microbes with specific traits and capabilities. These engineered microbes are designed to possess desired characteristics, including the ability to withstand metal stress, overexpress proteins chelating these metal, and collect the metals efficiently.

**Table 17.1** Bioremediation of heavy metal by resistant microorganisms.

| Bacteria/fungi | Heavy metals | References |
|---|---|---|
| *Acinetobacter junii* | Pb(II) | Kushwaha et al. (2017) and Singh et al. (2019a) |
| *Arthrobacter viscosus* | Cr(IV) | Hlihor et al. (2017), Lameiras et al. (2008), and Silva et al. (2012) |
| *Geobacillus thermodenitrificans* | Cu(II), Pb(II), Zn(II) | Chatterjee et al. (2010), Jeong et al. (2019), and Sen (2023) |
| *Oceanobacillus profundus* | Pb (II), Zn(II) | Mwandira et al. (2020) |
| *Bacillus thuringiensis* | Pb (II), Cd(II) | Goo et al. (2014); Li et al. (2019), and Zhou et al. (2018) |
| *Enterobacter cloacae* KJ-47, *E. cloacae* KJ-46, *Sporosarcina soli* B-22, *Viridibacillus arenosi* B-21 | Pd(II), Cu(II), Cd(II) | Kang et al. (2016) and Saha et al. (2021) |
| *Turbinaria vulgaris* | Cr(VI) | Boddu et al. (2022) and Khan et al. (2022) |
| *Chlorella vulgaris ZBS1* | Cr(VI) | Tattibayeva et al. (2022) |
| *Ochrobactrum* sp. *POC9* | Cd(II) | Chen et al. (2016) and Poszytek et al. (2018) |
| *Pseudomonas aeruginosa* | As(III), Pb(II), Hg(II), Cu(II), Cd(II) | Al-Ansari et al. (2021), Chellaiah (2018), Taran et al. (2019), Tariq et al. (2019), Teitzel and Parsek (2003), and Zhang et al. (2016) |
| *Rhodobacter capsulatus* | Co(II), Zn(II) | Dhar et al. (2023), Giotta et al. (2006), Hussain et al. (2022), and Tarfeen et al. (2022) |
| *Sporosarcina pasteurii* | Pb(II), Cd(II), Zn(II) | Ghorbanzadeh et al. (2022), Huang et al. (2022), and Zhang et al. (2023b) |
| *Bacillus licheniformis* | Cu(II), Cr(VI), Fe(II) | Alotaibi et al. (2021), Kapahi and Sachdeva (2019), Pande et al. (2022), and Wróbel et al. (2023) |
| *Stenotropho monasmaltophilia* | Cr(VI), Cu(II) | Ghosh and Saha (2013), Raman et al. (2018), and Said et al. (2023) |
| *Variovorax boronicumulans* | Pb(II), Cd(II), Zn(II) | Elnabi et al. (2023) and del Vargas-García et al., (2012) |
| *Aspergillus niger* | Pb(II), Cd(II) | Acosta-Rodríguez et al. (2018), Kapoor et al. (1999), Qiu et al. (2021), and Zhang et al. (2023a) |
| *Aspergillus fumigatus* | Cd(II), Cr(VI) | Dey et al. (2022) and Sharma et al. (2020) |
| *Aspergillus flavus* | Cr(VI), Pb(II), Zn(II), Cd(II), Mg(IV) | Akar and Tunali (2006) and Anupong et al. (2022) |
| *Acremonium* sp. | Al(III), Mn(III), Cu(II), Pb(II), Zn(III), Fe(II) | Kaur and Balomajumder (2020) and Li et al. (2020) |
| *Cellulosimicrobium* species | Cu(II), Pb(II), Cd(II), Ni(II), Zn(II) | Bhati et al. (2019) and Kranthi Raj et al. (2018) |
| *Termitomyces clypeatus* | Cr(VI) | Das and Guha (2009), Ramrakhiani et al. (2011), and Zapana-Huarache et al. (2019) |
| *Penicillium chrysogenum* | Pb(II), Cr(VI) | Deng and Ting (2005), García-Hernández et al. (2017), and Leitão (2009) |
| *Trichoderma* sp. | Cd(II) | Mohsenzadeh and Shahrokhi (2014) |
| *Trichoderma viride* | Cr(IV), Pb(II) | Joshi et al. (2011) and Luo et al. (2021, 2022) |

For instance, Woodruff et al. reported their successful engineering of microorganisms to produce trehalose, a natural sugar, which was then used to demonstrate its capacity to reduce 1 mM of Cr(VI) to Cr(III) (O'Neill et al. 2017; Woodruff and Frederick 2013). This reduction process is significant because it transforms a highly toxic form of chromium into a less harmful form.

In another example, bioengineered *Chlamydomonas reinhardtii* displayed a substantial increase in resistance to cadmium toxicity. Furthermore, these engineered microorganisms exhibited enhanced capabilities for the accumulation of Cd, contributing to their potential role in phytoremediation efforts aimed at removing this toxic metal from the environment (Samadani et al. 2018, p. 121; Thiriet-Rupert et al. 2021).

These advancements in genetic engineering of microbes are crucial for developing more eco-friendly and efficient methods for the restoration of metal-polluted sites and other environmental challenges.

In recent years, biosurfactants have garnered increasing attention and interest, primarily due to their low toxicity, biodegradability, and remarkable diversity. These unique properties make biosurfactants a valuable and sustainable resource in various applications, ranging from environmental remediation to industrial processes and beyond (Karlapudi et al. 2018; Sharma et al. 2022; Singh et al. 2020).

## 17.6 Challenges in Bioremediation Using Biofilms

When assessing a biofilm, it is important to consider several significant variables. Two key factors include the biofilm's thickness and cell density. Additionally, the pore space within the matrix, which impacts the biofilm structure and capabilities, is a crucial parameter to take into account. Research has shown that thin biofilms tend to have fewer pores and a higher cell density compared to thick biofilms. These pores within the biofilm matrix can serve as substrates for solid–liquid partitioning. Due to this, the biofilm's ability to degrade contaminants is closely linked to its thickness. Consequently, the thickness of a biofilm is proposed as a viable and efficient approach for bioremediation. By optimizing biofilm thickness, it is possible to enhance its efficiency in degrading specific pollutants, making it a valuable approach for environmental cleanup efforts.

In the field of biofilm research, several areas require further exploration:

**Recovery of aged biofilms**: There is a requirement for the advancement of cost-effective and environmentally friendly methods to recover aged biofilms. This is important for the maintenance and reactivation of biofilm-based systems, especially in applications such as wastewater treatment (Chen et al. 2020; Jing et al. 2019).

**Advanced monitoring and molecular insights**: Exploring the characteristics of biofilms from a further microscopic perspective is crucial. *In situ* monitoring and microbiological techniques, such as metagenomics and macroproteomics, can give deeper comprehensions into the composition and dynamics of biofilm microcolonies. Understanding the genetic and proteomic profiles of biofilms can lead to more targeted and effective strategies for biofilm management (Funari and Shen 2022; Joo and Otto 2012).

**Shortening biofilm formation process**: Developing reliable and reproducible methods for expediting biofilm formation is of interest. In certain industrial and bioremediation processes, quicker biofilm establishment may be desirable to achieve efficient pollutant removal or other objectives. Finding ways to accelerate the biofilm formation process without compromising its stability and functionality is an area for potential research and innovation.

These research directions aim to advance our understanding of biofilms and their applications, making them more efficient and environmentally friendly solutions for various challenges, from pollution control to industrial processes.

## 17.7 Conclusions

The build-up of harmful xenobiotic substances in the ecosystem is on the rise, underscoring the critical need to bring down the concentration of these substances through various methods. Microbe-based removal of environmental contaminants is an eco-friendly and sustainable approach. Among the various strategies available, biofilm-based bioremediation stands out as a promising approach for reducing pollutant levels in the environment.

There is a growing and well-founded public disquiet regarding the elimination of noxious pollutants that have been instigated into the environment due to various human activities. Bioremediation, a technology-centered around biological systems, is emerging as a novel and highly credible approach in the domain of pollution management. In comparison to various physiochemical methods, bioremediation is considered a viable and cost-effective means of waste disposal. Continued efforts to identify new bioresources are essential for the successful implementation of bioremediation and the preservation of our natural environment. Investigating the impact of microbes, either individually or in combinations, on a wide range of pollutants has become an urgent priority. Furthermore, utilization of genetically modified microorganisms capable of breaking down a wider range of contaminants represents a significant step forward in this field. The enzymes that play a pivotal role in the bioremediation process can be harnessed by overexpression, purification, and practical utilization. Deeper insights into the mechanisms underlying microbial-mediated bioremediation are essential for advancing this technology. Raising awareness and providing education to the public about the vital function of microbial populations in environmental cleanup is equally crucial. Conducting field trials to demonstrate the efficiency and effectiveness of bioremediation technology is a key aspect of its validation. Moreover, metagenomics presents a valuable approach for studying the microbial community within polluted area, enabling the identification of genes that can enhance the remediation capabilities of microbial strains. In conclusion, there is immense capacity for the advancement and refinement of bioremediation processes using microbial resources. This holds promise for addressing environmental pollution challenges while safeguarding the health of our ecosystems.

## References

Abbas, G., Murtaza, B., Bibi, I. et al. (2018). Arsenic uptake, toxicity, detoxification, and speciation in plants: physiological, biochemical, and molecular aspects. *International Journal of Environmental Research and Public Health* 15 (1): 59. https://doi.org/10.3390/ijerph15010059.

Acosta-Rodríguez, I., Cárdenas-González, J.F., Rodríguez Pérez, A.S. et al. (2018). Bioremoval of different heavy metals by the resistant fungal strain *Aspergillus niger*. *Bioinorganic Chemistry and Applications* 2018: e3457196. https://doi.org/10.1155/2018/3457196.

Ahemad, M. (2019). Remediation of metalliferous soils through the heavy metal resistant plant growth promoting bacteria: paradigms and prospects. *Arabian Journal of Chemistry* 12 (7): 1365–1377. https://doi.org/10.1016/j.arabjc.2014.11.020.

Akar, T. and Tunali, S. (2006). Biosorption characteristics of *Aspergillus flavus* biomass for removal of Pb(II) and Cu(II) ions from an aqueous solution. *Bioresource Technology* 97 (15): 1780–1787. https://doi.org/10.1016/j.biortech.2005.09.009.

Al-Ansari, M.M., Benabdelkamel, H., AlMalki, R.H. et al. (2021). Effective removal of heavy metals from industrial effluent wastewater by a multi metal and drug resistant *Pseudomonas aeruginosa* strain RA-14 using integrated sequencing batch reactor. *Environmental Research* 199: 111240. https://doi.org/10.1016/j.envres.2021.111240.

Albert, Q., Leleyter, L., Lemoine, M. et al. (2018). Comparison of tolerance and biosorption of three trace metals (Cd, Cu, Pb) by the soil fungus *Absidia cylindrospora*. *Chemosphere* 196: 386–392. https://doi.org/10.1016/j.chemosphere.2017.12.156.

Al-Dhabi, N.A., Esmail, G.A., Mohammed Ghilan, A.-K., and Valan Arasu, M. (2019). Optimizing the management of cadmium bioremediation capacity of metal-resistant *Pseudomonas* sp. strain Al-Dhabi-126 isolated from the industrial city of Saudi Arabian environment. *International Journal of Environmental Research and Public Health* 16 (23): 4788. https://doi.org/10.3390/ijerph16234788.

Ali Redha, A. (2020). Removal of heavy metals from aqueous media by biosorption. *Arab Journal of Basic and Applied Sciences* 27 (1): 183–193. https://doi.org/10.1080/25765299.2020.1756177.

Alotaibi, B.S., Khan, M., and Shamim, S. (2021). Unraveling the underlying heavy metal detoxification mechanisms of *Bacillus* species. *Microorganisms* 9 (8): Article 8 https://doi.org/10.3390/microorganisms9081628.

Anastopoulos, I. and Kyzas, G.Z. (2015). Progress in batch biosorption of heavy metals onto algae. *Journal of Molecular Liquids* 209: 77–86. https://doi.org/10.1016/j.molliq.2015.05.023.

Andersson, S., Kuttuva Rajarao, G., Land, C.J., and Dalhammar, G. (2008). Biofilm formation and interactions of bacterial strains found in wastewater treatment systems. *FEMS Microbiology Letters* 283 (1): 83–90. https://doi.org/10.1111/j.1574-6968.2008.01149.x.

Annweiler, E., Richnow, H.H., Antranikian, G. et al. (2000). Naphthalene degradation and incorporation of naphthalene-derived carbon into biomass by the thermophile *Bacillus thermoleovorans*. *Applied and Environmental Microbiology* 66 (2): 518–523.

Anupong, W., Jutamas, K., On-Uma, R. et al. (2022). Bioremediation competence of *Aspergillus flavus* DDN on pond water contaminated by mining activities. *Chemosphere* 304: 135250. https://doi.org/10.1016/j.chemosphere.2022.135250.

Aralu, C.C., Okoye, P.-A.C., Abugu, H.O. et al. (2023). Characterization, sources, and risk assessment of PAHs in borehole water from the vicinity of an unlined dumpsite in Awka, Nigeria. *Scientific Reports* 13: 9688. https://doi.org/10.1038/s41598-023-36691-3.

Arnaouteli, S., Bamford, N.C., Stanley-Wall, N.R., and Kovács, Á.T. (2021). *Bacillus subtilis* biofilm formation and social interactions. *Nature Reviews. Microbiology* 19 (9): 600–614. https://doi.org/10.1038/s41579-021-00540-9.

Ayangbenro, A.S. and Babalola, O.O. (2017). A new strategy for heavy metal polluted environments: a review of microbial biosorbents. *International Journal of Environmental Research and Public Health* 14 (1): 94. https://doi.org/10.3390/ijerph14010094.

Ayilara, M.S. and Babalola, O.O. (2023). Bioremediation of environmental wastes: the role of microorganisms. *Frontiers in Agronomy* 5: https://www.frontiersin.org/articles/10.3389/fagro.2023.1183691.

Bacosa, H.P., Ancla, S.M.B., Arcadio, C.G.L.A. et al. (2022). From surface water to the deep sea: a review on factors affecting the biodegradation of spilled oil in marine environment. *Journal of Marine Science and Engineering* 10 (3): Article 3. https://doi.org/10.3390/jmse10030426.

Bakaeva, M., Kuzina, E., Vysotskaya, L. et al. (2020). Capacity of *Pseudomonas* strains to degrade hydrocarbons, produce auxins and maintain plant growth under normal conditions and in the presence of petroleum contaminants. *Plants* 9 (3): 379. https://doi.org/10.3390/plants9030379.

Bala, S., Garg, D., Thirumalesh, B.V. et al. (2022). Recent strategies for bioremediation of emerging pollutants: a review for a green and sustainable environment. *Toxics* 10 (8): 484. https://doi.org/10.3390/toxics10080484.

Bao, L., Cui, Y., Wu, H. et al. (2023). Breeding, biosorption characteristics, and mechanism of a lead-resistant strain. *Toxics* 11 (5): 412. https://doi.org/10.3390/toxics11050412.

Basu, S., Dutta, A., Mukherjee, S.K., and Hossain, S.T. (2023). Isolation and characterization of an As(III) oxidizing bacterium, *Acinetobacter* sp. TMKU4 from paddy field for possible arsenic decontamination. *Journal of Hazardous Materials Advances* 10: 100289. https://doi.org/10.1016/j.hazadv.2023.100289.

Bhadouria, R., Das, S., Kumar, A. et al. (2020). Mycoremediation of agrochemicals. In: *Agrochemicals Detection, Treatment and Remediation* (ed. M. Narasimha Vara Prasad), 593–620. Oxford, United Kingdom: Butterworth-Heinemann https://doi.org/10.1016/B978-0-08-103017-2.00022-2.

Bhati, T., Gupta, R., Yadav, N. et al. (2019). Assessment of bioremediation potential of *Cellulosimicrobium* sp for treatment of multiple heavy metals. *Microbiology and Biotechnology Letters* 47 (2): 269–277. https://doi.org/10.4014/mbl.1808.08006.

Bhatt, P., Verma, A., Gangola, S. et al. (2021). Microbial glycoconjugates in organic pollutant bioremediation: recent advances and applications. *Microbial Cell Factories* 20 (1): 72. https://doi.org/10.1186/s12934-021-01556-9.

Bhatt, P., Bhatt, K., Huang, Y. et al. (2022). Biofilm formation in xenobiotic-degrading microorganisms. *Critical Reviews in Biotechnology* 1–21. https://doi.org/10.1080/07388551.2022.2106417.

Bibi, N., Hamayun, M., Khan, S.A. et al. (2018). Anthracene biodegradation capacity of newly isolated rhizospheric bacteria *Bacillus cereus* S13. *PLoS ONE* 13 (8): e0201620. https://doi.org/10.1371/journal.pone.0201620.

Boddu, S., Alugunulla, V.N., Dulla, J.B. et al. (2022). Estimation of biosorption characteristics of chromium (VI) from aqueous and real tannery effluents by treated *T. vulgaris*: experimental assessment and statistical modelling. *International Journal of Environmental Analytical Chemistry* 102 (16): 4842–4861. https://doi.org/10.1080/03067319.2020.1789617.

Chatterjee, S.K., Bhattacharjee, I., and Chandra, G. (2010). Biosorption of heavy metals from industrial waste water by *Geobacillus thermodenitrificans*. *Journal of Hazardous Materials* 175 (1–3): 117–125. https://doi.org/10.1016/j.jhazmat.2009.09.136.

Chattopadhyay, I., Rajesh, J.B., Usman, T.M.M., and Varjani, S. (2022). Exploring the role of microbial biofilm for industrial effluents treatment. *Bioengineered* 13 (3): 6420–6440. https://doi.org/10.1080/21655979.2022.2044250.

Chaudhary, P., Sahay, H., Sharma, R. et al. (2015). Identification and analysis of polyaromatic hydrocarbons (PAHs) – biodegrading bacterial strains from refinery soil of India. *Environmental Monitoring and Assessment* 187 (6): 391. https://doi.org/10.1007/s10661-015-4617-0.

Chauhan, A., Fazlurrahman, Oakeshott, J.G., and Jain, R.K. (2008). Bacterial metabolism of polycyclic aromatic hydrocarbons: strategies for bioremediation. *Indian Journal of Microbiology* 48 (1): 95–113. https://doi.org/10.1007/s12088-008-0010-9.

Chellaiah, E.R. (2018). Cadmium (heavy metals) bioremediation by *Pseudomonas aeruginosa*: a minireview. *Applied Water Science* 8 (6): 154. https://doi.org/10.1007/s13201-018-0796-5.

Chen, C., Lei, W., Lu, M. et al. (2016). Characterization of Cu(II) and Cd(II) resistance mechanisms in *Sphingobium* sp. PHE-SPH and *Ochrobactrum* sp. PHE-OCH and their potential application in the bioremediation of heavy metal-phenanthrene co-contaminated sites. *Environmental Science and Pollution Research* 23 (7): 6861–6872. https://doi.org/10.1007/s11356-015-5926-0.

Chen, Q., Xie, S., Lou, X. et al. (2019). Biofilm formation and prevalence of adhesion genes among *Staphylococcus aureus* isolates from different food sources. *MicrobiologyOpen* 9 (1): e00946. https://doi.org/10.1002/mbo3.946.

Chen, X., Thomsen, T.R., Winkler, H., and Xu, Y. (2020). Influence of biofilm growth age, media, antibiotic concentration and exposure time on *Staphylococcus aureus* and *Pseudomonas aeruginosa* biofilm removal in vitro. *BMC Microbiology* 20 (1): 264. https://doi.org/10.1186/s12866-020-01947-9.

Chibuike, G.U. and Obiora, S.C. (2014). Heavy metal polluted soils: effect on plants and bioremediation methods. *Applied and Environmental Soil Science* 2014: e752708. https://doi.org/10.1155/2014/752708.

Choi, H., Harrison, R., Komulainen, H., and Saborit, J.M.D. (2010). Polycyclic aromatic hydrocarbons. In: *WHO Guidelines for Indoor Air Quality: Selected Pollutants*. World Health Organization https://www.ncbi.nlm.nih.gov/books/NBK138709/.

Costa, O.Y.A., Raaijmakers, J.M., and Kuramae, E.E. (2018). Microbial extracellular polymeric substances: ecological function and impact on soil aggregation. *Frontiers in Microbiology* 9: https://www.frontiersin.org/articles/10.3389/fmicb.2018.01636.

Dang, H. and Lovell, C.R. (2015). Microbial surface colonization and biofilm development in marine environments. *Microbiology and Molecular Biology Reviews: MMBR* 80 (1): 91–138. https://doi.org/10.1128/MMBR.00037-15.

Das, S.K. and Guha, A.K. (2009). Biosorption of hexavalent chromium by *Termitomyces clypeatus* biomass: kinetics and transmission electron microscopic study. *Journal of Hazardous Materials* 167 (1): 685–691. https://doi.org/10.1016/j.jhazmat.2009.01.037.

Dasgupta, D., Ghosh, R., and Sengupta, T.K. (2013). Biofilm-mediated enhanced crude oil degradation by newly isolated *Pseudomonas* species. *ISRN Biotechnology* 2013: 250749. https://doi.org/10.5402/2013/250749.

Davey, M.E. and O'toole, G.A. (2000). Microbial biofilms: from ecology to molecular genetics. *Microbiology and Molecular Biology Reviews* 64 (4): 847–867.

del Vargas-García, M.C., López, M.J., Suárez-Estrella, F., and Moreno, J. (2012). Compost as a source of microbial isolates for the bioremediation of heavy metals: in vitro selection. *The Science of the Total Environment* 431: 62–67. https://doi.org/10.1016/j.scitotenv.2012.05.026.

Deng, S. and Ting, Y.P. (2005). Polyethylenimine-modified fungal biomass as a high-capacity biosorbent for Cr(VI) anions: sorption capacity and uptake mechanisms. *Environmental Science & Technology* 39 (21): 8490–8496. https://doi.org/10.1021/es050697u.

Dey, P., Malik, A., Singh, D.K. et al. (2022). Insight into the molecular mechanisms underpinning the mycoremediation of multiple metals by proteomic technique. *Frontiers in Microbiology* 13: 872576. https://doi.org/10.3389/fmicb.2022.872576.

Dhal, B., Thatoi, H., Das, N., and Pandey, B.D. (2010). Reduction of hexavalent chromium by *Bacillus* sp. isolated from chromite mine soils and characterization of reduced product. *Journal of Chemical Technology & Biotechnology* 85 (11): 1471–1479. https://doi.org/10.1002/jctb.2451.

Dhar, K., Venkateswarlu, K., and Megharaj, M. (2023). Anoxygenic phototrophic purple non-sulfur bacteria: tool for bioremediation of hazardous environmental pollutants. *World Journal of Microbiology and Biotechnology* 39 (10): 283. https://doi.org/10.1007/s11274-023-03729-7.

Dilaver, M. and Kargi, F. (2009). 2,4-Dichlorophenol (DCP) containing wastewater treatment using a hybrid-loop bioreactor. *Bioresource Technology* 100 (3): 1459–1462. https://doi.org/10.1016/j.biortech.2008.07.054.

Doshi, H., Seth, C., Ray, A., and Kothari, I.L. (2008). Bioaccumulation of heavy metals by green algae. *Current Microbiology* 56 (3): 246–255. https://doi.org/10.1007/s00284-007-9070-z.

Douterelo, I., Fish, K.E., and Boxall, J.B. (2018). Succession of bacterial and fungal communities within biofilms of a chlorinated drinking water distribution system. *Water Research* 141: 74–85. https://doi.org/10.1016/j.watres.2018.04.058.

Dunne, W.M. (2002). Bacterial adhesion: seen any good biofilms lately? *Clinical Microbiology Reviews* 15 (2): 155–166. https://doi.org/10.1128/CMR.15.2.155-166.2002.

Edel, M., Horn, H., and Gescher, J. (2019). Biofilm systems as tools in biotechnological production. *Applied Microbiology and Biotechnology* 103 (13): 5095–5103. https://doi.org/10.1007/s00253-019-09869-x.

Edelstein, M. and Ben-Hur, M. (2018). Heavy metals and metalloids: sources, risks and strategies to reduce their accumulation in horticultural crops. *Scientia Horticulturae* 234: 431–444. https://doi.org/10.1016/j.scienta.2017.12.039.

Eker, S. and Kargi, F. (2008). Biological treatment of 2,4-dichlorophenol containing synthetic wastewater using a rotating brush biofilm reactor. *Bioresource Technology* 99 (7): 2319–2325. https://doi.org/10.1016/j.biortech.2007.05.016.

Elnabi, M.K.A., Elkaliny, N.E., Elyazied, M.M. et al. (2023). Toxicity of heavy metals and recent advances in their removal: a review. *Toxics* 11 (7): https://doi.org/10.3390/toxics11070580.

Fathollahi, A., Khasteganan, N., Coupe, S.J., and Newman, A.P. (2021). A meta-analysis of metal biosorption by suspended bacteria from three phyla. *Chemosphere* 268: 129290. https://doi.org/10.1016/j.chemosphere.2020.129290.

Filote, C., Roșca, M., Hlihor, R.M. et al. (2021). Sustainable application of biosorption and bioaccumulation of persistent pollutants in wastewater treatment: current practice. *Processes* 9 (10): Article 10. https://doi.org/10.3390/pr9101696.

Flemming, H.-C., van Hullebusch, E.D., Neu, T.R. et al. (2023). The biofilm matrix: multitasking in a shared space. *Nature Reviews. Microbiology* 21 (2): 70–86. https://doi.org/10.1038/s41579-022-00791-0.

Flowers, S., Holder, K.H., and Gardner, S.M. (2023). Here is the biology, now what is the mechanism? Investigating biology undergraduates' mechanistic reasoning within the context of biofilm development. *Journal of Microbiology & Biology Education* 24 (2): e00221–e00222. https://doi.org/10.1128/jmbe.00221-22.

Funari, R. and Shen, A.Q. (2022). Detection and characterization of bacterial biofilms and biofilm-based sensors. *ACS Sensors* 7 (2): 347–357. https://doi.org/10.1021/acssensors.1c02722.

García-Hernández, M.A., Villarreal-Chiu, J.F., and Garza-González, M.T. (2017). Metallophilic fungi research: an alternative for its use in the bioremediation of hexavalent chromium. *International journal of Environmental Science and Technology* 14 (9): 2023–2038. https://doi.org/10.1007/s13762-017-1348-5.

Gevao, B., Al-Bahloul, M., Zafar, J. et al. (2007). Polycyclic aromatic hydrocarbons in indoor air and dust in Kuwait: implications for sources and nondietary human exposure. *Archives of Environmental Contamination and Toxicology* 53 (4): 503–512. https://doi.org/10.1007/s00244-006-0261-6.

Ghorbanzadeh, N., Ghanbari, Z., Farhangi, M.B., and Khalili Rad, M. (2022). Zinc bioremediation in soil by two isolated L-asparaginase and urease producing bacteria strains. *Applied Geochemistry* 140: 105271. https://doi.org/10.1016/j.apgeochem.2022.105271.

Ghosh, A. and Saha, P.D. (2013). Optimization of copper bioremediation by *Stenotrophomonas maltophilia* PD2. *Journal of Environmental Chemical Engineering* 1 (3): 159–163. https://doi.org/10.1016/j.jece.2013.04.012.

Gilbert, P., Maira-Litran, T., McBain, A.J. et al. (2002). The physiology and collective recalcitrance of microbial biofilm communities. *Advances in Microbial Physiology* 46: 202–256.

Giotta, L., Agostiano, A., Italiano, F. et al. (2006). Heavy metal ion influence on the photosynthetic growth of *Rhodobacter sphaeroides*. *Chemosphere* 62 (9): 1490–1499. https://doi.org/10.1016/j.chemosphere.2005.06.014.

Giri, D.D., Singh, S.K., Giri, A. et al. (2021). Bioremediation potential of methylotrophic bacteria. In: *Microbe Mediated Remediation of Environmental Contaminants* (ed. A. Kumar, V.K. Singh, P. Singh, and V.K. Mishra). Cambridge, MA: Woodhead Publishing.

Gond, D.P., Jha, S.S., Kumar, A., and Singh, S.K. (2021). Plant growth promoting bacteria and its role in green remediation. In: *Sustainable Environmental Clean-Up* (ed. V.K. Mishra and A. Kumar), 149–163. Cambridge, MA: Elsevier https://doi.org/10.1016/B978-0-12-823828-8.00007-4.

Goo, B.G., Hwang, Y.J., and Park, J.K. (2014). *Bacillus thuringiensis*: a specific gamma-cyclodextrin producer strain. *Carbohydrate Research* 386: 12–17. https://doi.org/10.1016/j.carres.2013.12.005.

Greenman, J., Hewett, K., and Saad, S. (2020). Discovery, development and exploitation of steady-state biofilms. *Journal of Breath Research* 14 (4): 044001. https://doi.org/10.1088/1752-7163/abb765.

Gulati, M. and Nobile, C.J. (2016). *Candida albicans* biofilms: development, regulation, and molecular mechanisms. *Microbes and Infection* 18 (5): 310–321. https://doi.org/10.1016/j.micinf.2016.01.002.

Gupta, P.K. and Gandhi, M. (2023). Bioremediation of organic pollutants in soil–water system: a review. *Biotech* 12 (2): 36. https://doi.org/10.3390/biotech12020036.

Gupta, A., Singh, M., Pandey, K.D., and Kumar, A. (2018). Biodegradation of xenobiotics using cyanobacteria. *Journal of Scientific Research* 62: 35–40.

Gupta, A., Singh, S.K., Singh, V.K. et al. (2020). Endophytic microbe approaches in bioremediation of organic pollutants. In: *Microbial Endophytes: Prospects of Sustainable Agriculture* (ed. A. Kumar and V.K. Singh), 157–174. Cambridge: Elsevier https://doi.org/10.1016/B978-0-12-818734-0.00007-3.

Häubner, N., Schumann, R., and Karsten, U. (2006). Aeroterrestrial microalgae growing in biofilms on facades – response to temperature and water stress. *Microbial Ecology* 51 (3): 285–293. https://doi.org/10.1007/s00248-006-9016-1.

Hlihor, R.M. and Cozma, P. (2023). Microbial bioremediation of environmental pollution. *Processes* 11 (5): Article 5. https://doi.org/10.3390/pr11051543.

Hlihor, R.M., Figueiredo, H., Tavares, T., and Gavrilescu, M. (2017). Biosorption potential of dead and living *Arthrobacter viscosus* biomass in the removal of Cr(VI): batch and column studies. *Process Safety and Environmental Protection* 108: 44–56. https://doi.org/10.1016/j.psep.2016.06.016.

Hoffmann, J.P. (1998). Wastewater treatment with suspended and nonsuspended algae. *Journal of Phycology* 34 (5): 757–763. https://doi.org/10.1046/j.1529-8817.1998.340757.x.

Huang, X., Zhang, R., Cui, M., and Lai, H. (2022). Experimental investigation on bioremediation of heavy metal contaminated solution by *Sporosarcina pasteurii* under some complex conditions. *Water* 14 (4): Article 4. https://doi.org/10.3390/w14040595.

Hussain, S., Khan, M., Sheikh, T.M.M. et al. (2022). Zinc essentiality, toxicity, and its bacterial bioremediation: a comprehensive insight. *Frontiers in Microbiology* 13: 900740. https://doi.org/10.3389/fmicb.2022.900740.

Jaishankar, M., Tseten, T., Anbalagan, N. et al. (2014). Toxicity, mechanism and health effects of some heavy metals. *Interdisciplinary Toxicology* 7 (2): 60–72. https://doi.org/10.2478/intox-2014-0009.

Jalilvand, N., Akhgar, A., Alikhani, H.A. et al. (2020). Removal of heavy metals zinc, lead, and cadmium by biomineralization of urease-producing bacteria isolated from Iranian mine calcareous soils. *Journal of Soil Science and Plant Nutrition* 20 (1): 206–219. https://doi.org/10.1007/s42729-019-00121-z.

Jalmi, S.K., Bhagat, P.K., Verma, D. et al. (2018). Traversing the links between heavy metal stress and plant signaling. *Frontiers in Plant Science* 9: https://www.frontiersin.org/articles/10.3389/fpls.2018.00012.

Jeong, S.-W., Kim, H.K., Yang, J.E., and Choi, Y.J. (2019). Removal of Pb(II) by pellicle-like biofilm-producing *Methylobacterium hispanicum* EM2 strain from aqueous media. *Water 11* (10): Article 10. https://doi.org/10.3390/w11102081.

Jing, X., Huang, X., Haapasalo, M. et al. (2019). Modeling oral multispecies biofilm recovery after antibacterial treatment. *Scientific Reports* 9: 804. https://doi.org/10.1038/s41598-018-37170-w.

Joo, H.-S. and Otto, M. (2012). Molecular basis of in vivo biofilm formation by bacterial pathogens. *Chemistry & Biology* 19 (12): 1503–1513. https://doi.org/10.1016/j.chembiol.2012.10.022.

Joshi, P.K., Swarup, A., Maheshwari, S. et al. (2011). Bioremediation of heavy metals in liquid media through fungi isolated from contaminated sources. *Indian Journal of Microbiology* 51 (4): 482–487. https://doi.org/10.1007/s12088-011-0110-9.

Kang, C.-H., Kwon, Y.-J., and So, J.-S. (2016). Bioremediation of heavy metals by using bacterial mixtures. *Ecological Engineering* 89: 64–69. https://doi.org/10.1016/j.ecoleng.2016.01.023.

Kapahi, M. and Sachdeva, S. (2019). Bioremediation options for heavy metal pollution. *Journal of Health & Pollution* 9 (24): 191203. https://doi.org/10.5696/2156-9614-9.24.191203.

Kapoor, A., Viraraghavan, T., and Cullimore, D.R. (1999). Removal of heavy metals using the fungus *Aspergillus niger*. *Bioresource Technology* 70 (1): 95–104. https://doi.org/10.1016/S0960-8524(98)00192-8.

Karlapudi, A.P., Venkateswarulu, T.C., Tammineedi, J. et al. (2018). Role of biosurfactants in bioremediation of oil pollution – a review. *Petroleum* 4 (3): 241–249. https://doi.org/10.1016/j.petlm.2018.03.007.

Karygianni, L., Ren, Z., Koo, H., and Thurnheer, T. (2020). Biofilm matrixome: extracellular components in structured microbial communities. *Trends in Microbiology* 28 (8): 668–681. https://doi.org/10.1016/j.tim.2020.03.016.

Kato, H., Okino, N., Kijitori, H. et al. (2023). Analysis of biofilm and bacterial communities in the towel environment with daily use. *Scientific Reports* 13 (1): 7611. https://doi.org/10.1038/s41598-023-34501-4.

Kaur, P. and Balomajumder, C. (2020). Bioremediation process optimization and effective reclamation of mixed carbamate-contaminated soil by newly isolated *Acremonium* sp. *Chemosphere* 249: 125982. https://doi.org/10.1016/j.chemosphere.2020.125982.

Khan, A.A., Mukherjee, S., Mondal, M. et al. (2022). Assessment of algal biomass towards removal of Cr(VI) from tannery effluent: a sustainable approach. *Environmental Science and Pollution Research International* 29 (41): 61856–61869. https://doi.org/10.1007/s11356-021-16102-8.

Kirova, G., Velkova, Z., Stoytcheva, M., and Gochev, V. (2021). Tetracycline removal from model aqueous solutions by pretreated waste *Streptomyces fradiae* biomass. *Biotechnology & Biotechnological Equipment* 35 (1): 953–963. https://doi.org/10.1080/13102818.2021.1938677.

Koshlaf, E., Ball, A.S., Koshlaf, E., and Ball, A.S. (2017). Soil bioremediation approaches for petroleum hydrocarbon polluted environments. *AIMS Microbiology* 3 (1): 25–49. https://doi.org/10.3934/microbiol.2017.1.25.

Kour, D., Kaur, T., Devi, R. et al. (2021). Beneficial microbiomes for bioremediation of diverse contaminated environments for environmental sustainability: present status and future challenges. *Environmental Science and Pollution Research International* 28 (20): 24917–24939. https://doi.org/10.1007/s11356-021-13252-7.

Kranthi Raj, K., Sardar, U.R., Bhargavi, E. et al. (2018). Advances in exopolysaccharides based bioremediation of heavy metals in soil and water: a critical review. *Carbohydrate Polymers* 199: 353–364. https://doi.org/10.1016/j.carbpol.2018.07.037.

Kumari, M., Qureshi, K.A., Jaremko, M. et al. (2022). Deciphering the role of endophytic microbiome in postharvest diseases management of fruits: opportunity areas in commercial up-scale production. *Frontiers in Plant Science* 13: 4390. https://doi.org/10.3389/fpls.2022.1026575.

Kushwaha, A., Rani, R., Kumar, S. et al. (2017). A new insight to adsorption and accumulation of high lead concentration by exopolymer and whole cells of lead-resistant bacterium *Acinetobacter junii* L. Pb1 isolated from coal mine dump. *Environmental Science and Pollution Research International* 24 (11): 10652–10661. https://doi.org/10.1007/s11356 017-8752-8.

Lahiri, D., Nag, M., Dey, A. et al. (2022). Biofilm mediated degradation of petroleum products. *Geomicrobiology Journal* 39 (3–5): 389–398. https://doi.org/10.1080/01490451.2021.1968979.

Lameiras, S., Quintelas, C., and Tavares, T. (2008). Biosorption of Cr(VI) using a bacterial biofilm supported on granular activated carbon and on zeolite. *Bioresource Technology* 99 (4): 801–806. https://doi.org/10.1016/j.biortech.2007.01.040.

Law, A.M.J. and Aitken, M.D. (2003). Bacterial chemotaxis to naphthalene desorbing from a nonaqueous liquid. *Applied and Environmental Microbiology* 69 (10): 5968–5973. https://doi.org/10.1128/AEM.69.10.5968-5973.2003.

Le Borgne, S. and Baquerizo, G. (2019). Microbial ecology of biofiltration units used for the desulfurization of biogas. *ChemEngineering 3* (3): Article 3. https://doi.org/10.3390/chemengineering3030072.

Leadbeater, B.S.C. and Callow, M.E. (1992). Formation, composition and physiology of algal biofilms. In: *Biofilms – Science and Technology* (ed. L.F. Melo, T.R. Bott, M. Fletcher, and B. Capdeville), 149–162. Netherlands: Springer https://doi.org/10.1007/978-94-011-1824-8_15.

Leitão, A.L. (2009). Potential of *Penicillium* species in the bioremediation field. *International Journal of Environmental Research and Public Health* 6 (4): 1393–1417. https://doi.org/10.3390/ijerph6041393.

Li, T., Lin, G., Podola, B., and Melkonian, M. (2015). Continuous removal of zinc from wastewater and mine dump leachate by a microalgal biofilm PSBR. *Journal of Hazardous Materials* 297: 112–118. https://doi.org/10.1016/j.jhazmat.2015.04.080.

Li, Z., Wang, P., Yue, X. et al. (2019). Effects of *Bacillus thuringiensis* HC-2 combined with biochar on the growth and Cd and Pb accumulation of radish in a heavy metal-contaminated farmland under field conditions. *International Journal of Environmental Research and Public Health* 16 (19): 3676. https://doi.org/10.3390/ijerph16193676.

Li, Q., Liu, J., and Gadd, G.M. (2020). Fungal bioremediation of soil co-contaminated with petroleum hydrocarbons and toxic metals. *Applied Microbiology and Biotechnology* 104 (21): 8999–9008. https://doi.org/10.1007/s00253-020-10854-y.

Li, Z., Feng, Q., Dang, J. et al. (2023). Types and source apportionment of polycyclic aromatic hydrocarbons (PAHs) in soil-groundwater of a closed coking plant in Shanxi Province, China. *Water* 15 (11): Article 11. https://doi.org/10.3390/w15112002.

Liu, L., Liu, J., Liu, X. et al. (2019). Kinetic and equilibrium of U(VI) biosorption onto the resistant bacterium *Bacillus amyloliquefaciens*. *Journal of Environmental Radioactivity* 203: 117–124. https://doi.org/10.1016/j.jenvrad.2019.03.008.

Liu, Y., Hu, H., Zanaroli, G. et al. (2021). A *Pseudomonas* sp. strain uniquely degrades PAHs and heterocyclic derivatives via lateral dioxygenation pathways. *Journal of Hazardous Materials* 403: 123956. https://doi.org/10.1016/j.jhazmat.2020.123956.

Luo, D., Geng, R., Wang, W. et al. (2021). *Trichoderma viride* involvement in the sorption of Pb(II) on muscovite, biotite and phlogopite: batch and spectroscopic studies. *Journal of Hazardous Materials* 401: 123249. https://doi.org/10.1016/j.jhazmat.2020.123249.

Luo, D., Qiang, S., Geng, R. et al. (2022). Mechanistic study for mutual interactions of $Pb^{2+}$ and *Trichoderma viride*. *Ecotoxicology and Environmental Safety* 233: 113310. https://doi.org/10.1016/j.ecoenv.2022.113310.

Mampel, J., Spirig, T., Weber, S.S. et al. (2006). Planktonic replication is essential for biofilm formation by *Legionella pneumophila* in a complex medium under static and dynamic flow conditions. *Applied and Environmental Microbiology* 72 (4): 2885–2895. https://doi.org/10.1128/AEM.72.4.2885-2895.2006.

Mangwani, N., Kumari, S., and Das, S. (2017). Marine bacterial biofilms in bioremediation of polycyclic aromatic hydrocarbons (PAHs) under terrestrial condition in a soil microcosm. *Pedosphere* 27 (3): 548–558. https://doi.org/10.1016/S1002-0160(17)60350-3.

Medfu Tarekegn, M., Zewdu Salilih, F., and Ishetu, A.I. (2020). Microbes used as a tool for bioremediation of heavy metal from the environment. *Cogent Food & Agriculture* 6 (1): 1783174. https://doi.org/10.1080/23311932.2020.1783174.

Mishra, M., Singh, S.K., and Kumar, A. (2021). Environmental factors affecting the bioremediation potential of microbes. In: *Woodhead Publishing Series in Food Science, Technology and Nutrition, Microbe Mediated Remediation of Environmental Contaminants* (ed. A. Kumar, V.K. Singh, P. Singh, and V.K. Mishra), 47–58. Cambridge, MA: Elsevier, Woodhead Publishing https://doi.org/10.1016/B978-0-12-821199-1.00005-5.

Mohsenzadeh, F. and Shahrokhi, F. (2014). Biological removing of cadmium from contaminated media by fungal biomass of *Trichoderma* species. *Journal of Environmental Health Science and Engineering* 12: 102. https://doi.org/10.1186/2052-336X-12-102.

Mukherji, S. and Chavan, A. (2012). Treatment of aqueous effluents containing non-aqueous phase liquids in rotating biological contactor with algal bacterial biofilm. *Chemical Engineering Journal* 200–202: 459–470. https://doi.org/10.1016/j.cej.2012.06.076.

Muthami, J.M., Fernández-García, L., Tomás, M., and Wood, T.K. (2022). What is the fate of the biofilm matrix? *Environmental Microbiology* 24 (10): 4495–4499. https://doi.org/10.1111/1462-2920.16097.

Mwandira, W., Nakashima, K., Kawasaki, S. et al. (2020). Biosorption of Pb(II) and Zn(II) from aqueous solution by *Oceanobacillus profundus* isolated from an abandoned mine. *Scientific Reports* 10: 21189. https://doi.org/10.1038/s41598-020-78187-4.

Myazin, V.A., Korneykova, M.V., Chaporgina, A.A. et al. (2021). The effectiveness of biostimulation, bioaugmentation and sorption-biological treatment of soil contaminated with petroleum products in the Russian Subarctic. *Microorganisms* 9 (8): Article 8. https://doi.org/10.3390/microorganisms9081722.

Nadar, S., Khan, T., Patching, S.G., and Omri, A. (2022). Development of antibiofilm therapeutics strategies to overcome antimicrobial drug resistance. *Microorganisms* 10 (2): Article 2. https://doi.org/10.3390/microorganisms10020303.

Naskar, A., Majumder, R., and Goswami, M. (2020). Bioaccumulation of Ni(II) on growing cells of *Bacillus* sp.: response surface modeling and mechanistic insight. *Environmental Technology & Innovation* 20: 101057. https://doi.org/10.1016/j.eti.2020.101057.

Naz, T., Khan, M.D., Ahmed, I. et al. (2016). Biosorption of heavy metals by *Pseudomonas* species isolated from sugar industry. *Toxicology and Industrial Health* 32 (9): 1619–1627. https://doi.org/10.1177/0748233715569900.

Nicolella, C., Zolezzi, M., Rabino, M. et al. (2005). Development of particle-based biofilms for degradation of xenobiotic organic compounds. *Water Research* 39 (12): 2495–2504. https://doi.org/10.1016/j.watres.2005.04.016.

Nzila, A., Musa, M.M., Afuecheta, E. et al. (2023). Benzo[*a*]pyrene biodegradation by multiple and individual mesophilic bacteria under axenic conditions and in soil samples. *International Journal of Environmental Research and Public Health* 20 (3): 1855. https://doi.org/10.3390/ijerph20031855.

Ojuederie, O.B. and Babalola, O.O. (2017). Microbial and plant-assisted bioremediation of heavy metal polluted environments: a review. *International Journal of Environmental Research and Public Health* 14 (12): 1504. https://doi.org/10.3390/ijerph14121504.

O'Neill, M.K., Piligian, B.F., Olson, C.D. et al. (2017). Tailoring trehalose for biomedical and biotechnological applications. *Pure and Applied Chemistry. Chimie Pure et Appliquee* 89 (9): 1223–1249. https://doi.org/10.1515/pac-2016-1025.

O'Toole, G., Kaplan, H.B., and Kolter, R. (2000). Biofilm formation as microbial development. *Annual Review of Microbiology* 54: 49–79. https://doi.org/10.1146/annurev.micro.54.1.49.

Pacwa-Płociniczak, M., Płaza, G.A., Poliwoda, A., and Piotrowska-Seget, Z. (2014). Characterization of hydrocarbon-degrading and biosurfactant-producing *Pseudomonas* sp. P-1 strain as a potential tool for bioremediation of petroleum-contaminated soil. *Environmental Science and Pollution Research International* 21 (15): 9385–9395. https://doi.org/10.1007/s11356-014-2872-1.

Pande, V., Pandey, S.C., Sati, D. et al. (2022). Microbial interventions in bioremediation of heavy metal contaminants in agroecosystem. *Frontiers in Microbiology* 13: 824084. https://doi.org/10.3389/fmicb.2022.824084.

Pandey, K.D., Singh, S.K., Shukla, L. et al. (2023). Mechanistic approaches and factors regulating microalgae mediated heavy metal remediation from the aquatic ecosystem. In: *Advances in Chemical Pollution, Environmental Management and Protection*, vol. 9 (ed. L.F.R. Ferreira, A. Kumar, and M. Bilal), 285–299. Cambridge: Elsevier, Academic Press https://doi.org/10.1016/bs.apmp.2022.10.011.

Patowary, K., Saikia, R.R., Kalita, M.C., and Deka, S. (2015). Degradation of polyaromatic hydrocarbons employing biosurfactant-producing *Bacillus pumilus* KS2. *Annals of Microbiology 65* (1): Article 1. https://doi.org/10.1007/s13213-014-0854-7.

Patowary, K., Patowary, R., Kalita, M.C., and Deka, S. (2017). Characterization of biosurfactant produced during degradation of hydrocarbons using crude oil as sole source of carbon. *Frontiers in Microbiology* 8: https://www.frontiersin.org/articles/10.3389/fmicb.2017.00279.

Perdigão, R., Almeida, C.M.R., Magalhães, C. et al. (2021). Bioremediation of petroleum hydrocarbons in seawater: prospects of using lyophilized native hydrocarbon-degrading bacteria. *Microorganisms* 9 (11): 2285. https://doi.org/10.3390/microorganisms9112285.

Pérez-Mendoza, D., Coulthurst, S.J., Sanjuán, J., and Salmond, G.P.C. (2011). *N*-Acetylglucosamine-dependent biofilm formation in *Pectobacterium atrosepticum* is cryptic and activated by elevated c-di-GMP levels. *Microbiology (Reading, England)* 157 (Pt 12): 3340–3348. https://doi.org/10.1099/mic.0.050450-0.

Poszytek, K., Karczewska-Golec, J., Ciok, A. et al. (2018). Genome-guided characterization of *Ochrobactrum* sp. POC9 enhancing sewage sludge utilization – biotechnological potential and biosafety considerations. *International Journal of Environmental Research and Public Health* 15 (7): https://doi.org/10.3390/ijerph15071501.

Qin, Z., Zhao, Z., Xia, L. et al. (2019). The dissipation and risk alleviation mechanism of PAHs and nitrogen in constructed wetlands: the role of submerged macrophytes and their biofilms-leaves. *Environment International* 131: 104940. https://doi.org/10.1016/j.envint.2019.104940.

Qiu, J., Song, X., Li, S. et al. (2021). Experimental and modeling studies of competitive Pb (II) and Cd (II) bioaccumulation by *Aspergillus niger*. *Applied Microbiology and Biotechnology* 105 (16–17): 6477–6488. https://doi.org/10.1007/s00253-021-11497-3.

Rabin, N., Zheng, Y., Opoku-Temeng, C. et al. (2015). Biofilm formation mechanisms and targets for developing antibiofilm agents. *Future Medicinal Chemistry* 7 (4): 493–512. https://doi.org/10.4155/fmc.15.6.

Rahman, Z., Thomas, L., and Singh, V.P. (2019). Biosorption of heavy metals by a lead (Pb) resistant bacterium, *Staphylococcus hominis* strain AMB-2. *Journal of Basic Microbiology* 59 (5): 477–486. https://doi.org/10.1002/jobm.201900024.

Raman, N.M., Asokan, S., Shobana Sundari, N., and Ramasamy, S. (2018). Bioremediation of chromium(VI) by *Stenotrophomonas maltophilia* isolated from tannery effluent. *International journal of Environmental Science and Technology* 15 (1): 207–216. https://doi.org/10.1007/s13762-017-1378-z.

Ramrakhiani, L., Majumder, R., and Khowala, S. (2011). Removal of hexavalent chromium by heat inactivated fungal biomass of *Termitomyces clypeatus*: surface characterization and mechanism of biosorption. *Chemical Engineering Journal* 171 (3): 1060–1068. https://doi.org/10.1016/j.cej.2011.05.002.

Rani, A. and Goel, R. (2009). Strategies for crop improvement in contaminated soils using metal-tolerant bioinoculants. In: *Microbial Strategies for Crop Improvement* (ed. M.S. Khan, A. Zaidi, and J. Musarrat), 85–104. Springer https://doi.org/10.1007/978-3-642-01979-1_5.

Roudbary, M., Vahedi-Shahandashti, R., Santos, A.L.S.D. et al. (2022). Biofilm formation in clinically relevant filamentous fungi: a therapeutic challenge. *Critical Reviews in Microbiology* 48 (2): 197–221. https://doi.org/10.1080/1040841X.2021.1950121.

Saha, L., Tiwari, J., Bauddh, K., and Ma, Y. (2021). Recent developments in microbe–plant-based bioremediation for tackling heavy metal-polluted soils. *Frontiers in Microbiology* 12: 731723. https://doi.org/10.3389/fmicb.2021.731723.

Sahmoune, M.N. (2018). Performance of *Streptomyces rimosus* biomass in biosorption of heavy metals from aqueous solutions. *Microchemical Journal* 141: 87–95. https://doi.org/10.1016/j.microc.2018.05.009.

Said, M.S., Tirthani, E., and Lesho, E. (2023). *Stenotrophomonas maltophilia*. In: *StatPearls*. StatPearls Publishing http://www.ncbi.nlm.nih.gov/books/NBK572123/.

Samadani, M., Perreault, F., Oukarroum, A., and Dewez, D. (2018). Effect of cadmium accumulation on green algae *Chlamydomonas reinhardtii* and acid-tolerant *Chlamydomonas* CPCC 121. *Chemosphere* 191: 174–182. https://doi.org/10.1016/j.chemosphere.2017.10.017.

Saranya, K., Sundaramanickam, A., Shekhar, S. et al. (2018). Biosorption of multi-heavy metals by coral associated phosphate solubilising bacteria *Cronobacter muytjensii* KSCAS2. *Journal of Environmental Management* 222: 396–401. https://doi.org/10.1016/j.jenvman.2018.05.083.

Sayed, K., Baloo, L., and Sharma, N.K. (2021). Bioremediation of total petroleum hydrocarbons (TPH) by bioaugmentation and biostimulation in water with floating oil spill containment booms as bioreactor basin. *International Journal of Environmental Research and Public Health* 18 (5): 2226. https://doi.org/10.3390/ijerph18052226.

Schembri, M.A., Givskov, M., and Klemm, P. (2002). An attractive surface: gram-negative bacterial biofilms. *Science's STKE: Signal Transduction Knowledge Environment* 2002 (132): re6. https://doi.org/10.1126/stke.2002.132.re6.

Sen, T.K. (2023). Agricultural solid wastes based adsorbent materials in the remediation of heavy metal ions from water and wastewater by adsorption: a review. *Molecules* 28 (14): Article 14. https://doi.org/10.3390/molecules28145575.

Seo, J.-S., Keum, Y.-S., and Li, Q.X. (2009). Bacterial degradation of aromatic compounds. *International Journal of Environmental Research and Public Health* 6 (1): 278–309. https://doi.org/10.3390/ijerph6010278.

Sharma, D., Misba, L., and Khan, A.U. (2019). Antibiotics versus biofilm: an emerging battleground in microbial communities. *Antimicrobial Resistance and Infection Control* 8: 76. https://doi.org/10.1186/s13756-019-0533-3.

Sharma, R., Talukdar, D., Bhardwaj, S. et al. (2020). Bioremediation potential of novel fungal species isolated from wastewater for the removal of lead from liquid medium. *Environmental Technology & Innovation* 18: 100757. https://doi.org/10.1016/j.eti.2020.100757.

Sharma, N., Lavania, M., and Lal, B. (2022). Biosurfactant: a next-generation tool for sustainable remediation of organic pollutants. *Frontiers in Microbiology* 12: https://www.frontiersin.org/articles/10.3389/fmicb.2021.821531.

Sheppard, D.C. and Howell, P.L. (2016). Biofilm exopolysaccharides of pathogenic fungi: lessons from bacteria. *The Journal of Biological Chemistry* 291 (24): 12529–12537. https://doi.org/10.1074/jbc.R116.720995.

Sher, S. and Rehman, A. (2019). Use of heavy metals resistant bacteria-a strategy for arsenic bioremediation. *Applied Microbiology and Biotechnology* 103 (15): 6007–6021. https://doi.org/10.1007/s00253-019-09933-6.

Shi, J., Podola, B., and Melkonian, M. (2014). Application of a prototype-scale twin-Layer photobioreactor for effective N and P removal from different process stages of municipal wastewater by immobilized microalgae. *Bioresource Technology* 154: 260–266. https://doi.org/10.1016/j.biortech.2013.11.100.

Silva, B., Figueiredo, H., Quintelas, C. et al. (2012). Improved biosorption for Cr(VI) reduction and removal by *Arthrobacter viscosus* using zeolite. *International Biodeterioration & Biodegradation* 74: 116–123. https://doi.org/10.1016/j.ibiod.2012.05.026.

Singh, P.K., Kushwaha, A., Hans, N. et al. (2019a). Evaluation of the cytotoxicity and interaction of lead with lead resistant bacterium *Acinetobacter junii* Pb1. *Brazilian Journal of Microbiology* 50 (1): 223–230. https://doi.org/10.1007/s42770-019-00041-1.

Singh, S.K., Singh, P.P., Gupta, A. et al. (2019b). Tolerance of heavy metal toxicity using PGPR strains of *Pseudomonas* species. In: *PGPR Amelioration in Sustainable Agriculture*, 239–252. Cambridge, Massachusetts, USA: Woodhead Publishing https://doi.org/10.1016/B978-0-12-815879-1.00012-4.

Singh, S., Kumar, V., Singh, S. et al. (2020). 16 – Biosurfactant-based bioremediation. In: *Bioremediation of Pollutants* (ed. V.C. Pandey and V. Singh), 333–358. Elsevier https://doi.org/10.1016/B978-0-12-819025-8.00016-8.

Singh, S.K., Singh, M.K., Singh, V.K. et al. (2021). Microbial enzymes and their exploitation in remediation of environmental contaminants. In: *Microbe Mediated Remediation of Environmental Contaminants* (ed. A. Kumar, V.K. Singh, P. Singh, and V.K. Mishra), 59–71. Cambridge, Massachusetts USA: Woodhead Publishing https://doi.org/10.1016/B978-0-12-821199-1.00006-7.

Skowroński, T., Jakubowski, M., and Pawlik, B. (1985). Cadmium toxicity to green alga *Stichococcus bacillaris*. *Acta Microbiologica Polonica* 34 (3–4): 309–312.

Solano, C., Echeverz, M., and Lasa, I. (2014). Biofilm dispersion and quorum sensing. *Current Opinion in Microbiology* 18: 96–104. https://doi.org/10.1016/j.mib.2014.02.008.

Srivastava, P.K., Vaish, A., Dwivedi, S. et al. (2011). Biological removal of arsenic pollution by soil fungi. *The Science of the Total Environment* 409 (12): 2430–2442. https://doi.org/10.1016/j.scitotenv.2011.03.002.

Syed, S. and Chinthala, P. (2015). Heavy metal detoxification by different *Bacillus* species isolated from solar salterns. *Scientifica* 2015: 319760. https://doi.org/10.1155/2015/319760.

Tai, J.-S.B., Mukherjee, S., Nero, T. et al. (2022). Social evolution of shared biofilm matrix components. *Proceedings of the National Academy of Sciences of the United States of America* 119 (27): e2123469119. https://doi.org/10.1073/pnas.2123469119.

Taran, M., Fateh, R., Rezaei, S., and Gholi, M.K. (2019). Isolation of arsenic accumulating bacteria from garbage leachates for possible application in bioremediation. *Iranian Journal of Microbiology* 11 (1): 60–66.

Tarfeen, N., Nisa, K.U., Hamid, B. et al. (2022). Microbial remediation: a promising tool for reclamation of contaminated sites with special emphasis on heavy metal and pesticide pollution: a review. *Processes* 10 (7): Article 7. https://doi.org/10.3390/pr10071358.

Tariq, A., Ullah, U., Asif, M., and Sadiq, I. (2019). Biosorption of arsenic through bacteria isolated from Pakistan. *International Microbiology: The Official Journal of the Spanish Society for Microbiology* 22 (1): 59–68. https://doi.org/10.1007/s10123-018-0028-8.

Tattibayeva, Z., Tazhibayeva, S., Kujawski, W. et al. (2022). Peculiarities of adsorption of Cr (VI) ions on the surface of *Chlorella vulgaris* ZBS1 algae cells. *Heliyon* 8 (9): e10468. https://doi.org/10.1016/j.heliyon.2022.e10468.

Teitzel, G.M. and Parsek, M.R. (2003). Heavy metal resistance of biofilm and planktonic *Pseudomonas aeruginosa*. *Applied and Environmental Microbiology* 69 (4): 2313–2320. https://doi.org/10.1128/AEM.69.4.2313-2320.2003.

Thatoi, H., Das, S., Mishra, J. et al. (2014). Bacterial chromate reductase, a potential enzyme for bioremediation of hexavalent chromium: a review. *Journal of Environmental Management* 146: 383–399. https://doi.org/10.1016/j.jenvman.2014.07.014.

Thiriet-Rupert, S., Gain, G., Jadoul, A. et al. (2021). Long-term acclimation to cadmium exposure reveals extensive phenotypic plasticity in *Chlamydomonas*. *Plant Physiology* 187 (3): 1653–1678. https://doi.org/10.1093/plphys/kiab375.

Tolker-Nielsen, T. (2015). Biofilm development. *Microbiology Spectrum* 3 (2): MB-0001-2014. https://doi.org/10.1128/microbiolspec.MB-0001-2014.

Tran, T.M. and Lee, J.-U. (2023). Biosorption of Cd by an indigenous Cd-resistant bacterium isolated from soil contaminated with Cd. *Geosciences Journal* https://doi.org/10.1007/s12303-023-0031-8.

Turki, Y., Mehri, I., Lajnef, R. et al. (2017). Biofilms in bioremediation and wastewater treatment: characterization of bacterial community structure and diversity during seasons in municipal wastewater treatment process. *Environmental Science and Pollution Research International* 24 (4): 3519–3530. https://doi.org/10.1007/s11356-016-8090-2.

Udume, O.A., Abu, G.O., Stanley, H.O. et al. (2023). Biostimulation of petroleum-contaminated soil using organic and inorganic amendments. *Plants (Basel, Switzerland)* 12 (3): 431. https://doi.org/10.3390/plants12030431.

Van Hamme, J.D., Singh, A., and Ward, O.P. (2003). Recent advances in petroleum microbiology. *Microbiology and Molecular Biology Reviews* 67 (4): 503–549. https://doi.org/10.1128/MMBR.67.4.503-549.2003.

van Loosdrecht, M.C., Norde, W., and Zehnder, A.J. (1990). Physical chemical description of bacterial adhesion. *Journal of Biomaterials Applications* 5 (2): 91–106. https://doi.org/10.1177/088532829000500202.

van Wolferen, M., Orell, A., and Albers, S.-V. (2018). Archaeal biofilm formation. *Nature Reviews. Microbiology* 16 (11): 699–713. https://doi.org/10.1038/s41579-018-0058-4.

Vandana and Das, S. (2021). Structural and mechanical characterization of biofilm-associated bacterial polymer in the emulsification of petroleum hydrocarbon. *3 Biotech* 11 (5): 239. https://doi.org/10.1007/s13205-021-02795-8.

Vishwakarma, V. (2020). Impact of environmental biofilms: industrial components and its remediation. *Journal of Basic Microbiology* 60 (3): 198–206. https://doi.org/10.1002/jobm.201900569.

Wattiau, P. (2002). Microbial aspects in bioremediation of soils polluted by polyaromatic hydrocarbons. In: *Biotechnology for the Environment: Strategy and Fundamentals*, vol. Vol. 3A (ed. S.N. Agathos and W. Reineke), 69–89. Netherlands: Springer https://doi.org/10.1007/978-94-010-0357-5_5.

Wei, R., Ni, J., Li, X. et al. (2017). Dissipation and phytoremediation of polycyclic aromatic hydrocarbons in freshly spiked and long-term field-contaminated soils. *Environmental Science and Pollution Research International* 24 (9): 7994–8003. https://doi.org/10.1007/s11356-017-8459-x.

Wingender, J., Neu, T.R., and Flemming, H.-C. (1999). What are bacterial extracellular polymeric substances? In: *Microbial Extracellular Polymeric Substances* (ed. J. Wingender, T.R. Neu, and H.-C. Flemming), 1–19. Berlin, Heidelberg: Springer https://doi.org/10.1007/978-3-642-60147-7_1.

Woodruff, P. J., & Frederick, T. M. (2013). Bacteria with Increased Trehalose Production and Method for Using the Same in Bioremediation (United States Patent US8481026B1), https://patents.google.com/patent/US8481026B1/en

Wróbel, M., Śliwakowski, W., Kowalczyk, P. et al. (2023). Bioremediation of heavy metals by the genus *Bacillus*. *International Journal of Environmental Research and Public Health* 20 (6): 4964. https://doi.org/10.3390/ijerph20064964.

Wu, M., Wu, J., Zhang, X., and Ye, X. (2019). Effect of bioaugmentation and biostimulation on hydrocarbon degradation and microbial community composition in petroleum-contaminated loessal soil. *Chemosphere* 237: 124456. https://doi.org/10.1016/j.chemosphere.2019.124456.

Xu, X., Liu, W., Tian, S. et al. (2018). Petroleum hydrocarbon-degrading bacteria for the remediation of oil pollution under aerobic conditions: a perspective analysis. *Frontiers in Microbiology* 9: https://www.frontiersin.org/articles/10.3389/fmicb.2018.02885.

Yan, J. and Bassler, B.L. (2019). Surviving as a community: antibiotic tolerance and persistence in bacterial biofilms. *Cell Host & Microbe* 26 (1): 15–21. https://doi.org/10.1016/j.chom.2019.06.002.

Yin, W., Wang, Y., Liu, L., and He, J. (2019). Biofilms: the microbial "protective clothing" in extreme environments. *International Journal of Molecular Sciences* 20 (14): 3423. https://doi.org/10.3390/ijms20143423.

Yoshida, S., Ogawa, N., Fujii, T., and Tsushima, S. (2009). Enhanced biofilm formation and 3-chlorobenzoate degrading activity by the bacterial consortium of Burkholderia sp. NK8 and *Pseudomonas aeruginosa* PAO1. *Journal of Applied Microbiology* 106 (3): 790–800. https://doi.org/10.1111/j.1365-2672.2008.04027.x.

Zapana-Huarache, S.V., Romero-Sánchez, C.K., Gonza, A.P.D. et al. (2019). Chromium (VI) bioremediation potential of filamentous fungi isolated from Peruvian tannery industry effluents. *Brazilian Journal of Microbiology* 51 (1): 271–278. https://doi.org/10.1007/s42770-019-00209-9.

Zhang, Z., Yin, N., Cai, X. et al. (2016). Arsenic redox transformation by *Pseudomonas* sp. HN-2 isolated from arsenic-contaminated soil in Hunan, China. *Journal of Environmental Sciences* 47: 165–173. https://doi.org/10.1016/j.jes.2015.11.036.

Zhang, L., Yang, X., Li, S. et al. (2023a). A contrast of Pb(II), Cd(II), and Cu(II) toxicities to *Aspergillus niger* through biochemical, morphological, and genetic investigations. *Journal of Hazardous Materials* 446: 130691. https://doi.org/10.1016/j.jhazmat.2022.130691.

Zhang, W., Zhang, H., Xu, R. et al. (2023b). Heavy metal bioremediation using microbially induced carbonate precipitation: Key factors and enhancement strategies. *Frontiers in Microbiology* 14: https://www.frontiersin.org/articles/10.3389/fmicb.2023.1116970.

Zhou, X., Li, H., Liu, D. et al. (2018). Effects of toxin from *Bacillus thuringiensis* (Bt) on sorption of Pb (II) in red and black soils: equilibrium and kinetics aspects. *Journal of Hazardous Materials* 360: 172–181. https://doi.org/10.1016/j.jhazmat.2018.07.114.

Zhou, Z., Zhu, L., Dong, Y. et al. (2022). Identification of a novel chromate and selenite reductase FesR in *Alishewanella* sp. WH16-1. *Frontiers in Microbiology* 13: https://www.frontiersin.org/articles/10.3389/fmicb.2022.834293.

# 18

# Green Nanoparticles for Textile Wastewater Treatment: The Current Insights

*Irfan Haidri[1], Aneeza Ishfaq[1], Muhammad Shahid[2], Tanvir Shahzad[1], Sabir Hussain[1], and Faisal Mahmood[1]*

[1] *Department of Environmental Sciences, Government College University, Faisalabad, Pakistan*
[2] *Department of Bioinformatics & Biotechnology, Government College University, Faisalabad, Pakistan*

## 18.1 Introduction

Wastewater from textile industries consists of high pollutant levels, including dyes, heavy metals, and organic chemicals. This untreated or inadequately treated textile wastewater creates substantial environmental and health risks by contaminating water bodies. It endangers aquatic life and affects human health through the consumption of contaminated water and agricultural products (Paul et al. 2012). Consequently, there is an urgent need for effective wastewater treatment methods to mitigate the adverse impacts of textile industry effluents. Conventional wastewater treatment technologies, such as chemical coagulation, sedimentation, and activated sludge processes, have limitations in their ability to treat textile wastewater due to the complex and recalcitrant nature of the pollutants involved. These methods often demand significant quantities of chemicals, generate substantial sludge, and consume large amounts of energy. Exploring alternative and sustainable wastewater treatment approaches, like green synthesized nanoparticles, is emerging as a promising technology to mitigate this issue (Hassanien et al. 2019).

Green nanoparticles, synthesized from natural resources using plant extracts, microbes, or algae, provide a sustainable and eco-friendly alternative as compared to traditional methods of chemical coagulants and flocculants for wastewater treatment. Nanoparticles harness the power of nature to efficiently remove contaminants from water, representing a significant step toward a more environmentally conscious and effective wastewater treatment process. Nanoparticles exhibit unique physicochemical properties compared to chemically synthesized nanoparticles, including a high surface area, surface charge, and reactivity. They render them highly effective in adsorbing, flocculating, or catalytically degrading various pollutants in textile wastewater (Abdel-Shafy and Mansour 2018). Green nanoparticles are derived from renewable resources and are biodegradable, significantly reducing their environmental negative impacts compared to nanoparticles synthesized through chemicals. By substituting traditional coagulants with green nanoparticles, the textile industry can decrease its reliance on harmful chemicals and minimize the generation of toxic sludge. This contributes to a more sustainable and environmentally responsible manufacturing process. This shift toward green synthesized nanoparticles in wastewater treatment addresses the pollution concerns associated with textile production and aligns with broader efforts to promote greener technologies across various industries (Bahrodin et al. 2021).

The synthesis of green nanoparticles is often cost-effective, particularly when utilizing locally available natural resources. This economic advantage further bolsters the viability of green nanoparticle-based wastewater treatment methods. Moreover, their effectiveness in removing pollutants can lower operational costs in wastewater treatment plants. These economic advantages make green nanoparticles appealing to large-scale textile industries and small-scale dyeing units in developing countries. Some green nanoparticles, such as those derived from algae or microbes, have the potential to accumulate and recover valuable metals and dyes from textile wastewater. This not only aids in pollutant removal but also presents an opportunity for resource recovery and recycling within the textile production cycle, aligning with circular economy principles (Castro et al. 2014).

*Microbes Based Approaches for the Management of Hazardous Contaminants*, First Edition. Edited by Ajay Kumar, Livleen Shukla, Joginder Singh, and Luiz Fernando R. Ferreira.

The chapters highlight the use of green nanoparticles to treat textile wastewater. It also discussed the current state of knowledge and research in using green nanoparticles to treat textile industry wastewater, focusing on their efficacy in pollutant removal, sustainability, and potential for resource recovery. Additionally, the chapter discussed various biological synthesis methods characterization employed for different nanoparticles. Evolving trends, challenges, and prospects within this critical environmental science and technology field.

## 18.2 Sources and Composition of Textile Wastewater

Textile wastewater is generated across various stages of the textile production process, resulting in a complex mixture of contaminants. Wastewater is produced from processes such as dyeing, printing, finishing, and washing, each contributing to the diverse composition of the wastewater. Synthetic dyes, heavy metals, organic chemicals, surfactants, and suspended solids are the main pollutants present in textile wastewater (Kishor et al. 2021; Singh et al. 2023a,b; Yadav et al. 2023). Textile wastewater consists of synthetic and natural dyes, imparting vibrant colors to textiles but posing significant environmental challenges due to their persistent nature and potential toxicity. Heavy metals, frequently employed in dye fixation and as catalysts in textile processes, present another substantial concern. In addition, the presence of organic chemicals, such as solvents, detergents, and fabric treatment agents, further adds to the complexity of the wastewater. This intricate composition underscores the pressing need to develop and implement effective treatment strategies to alleviate the negative environmental impacts associated with using textile wastewater for agricultural purpose (Khan and Malik 2014).

## 18.3 Environmental Effects of Textile Wastewater

The environmental effects of textile wastewater are far-reaching and complex, impacting both natural surroundings and local communities. Pollution of natural water sources is one of the significant issues, such as rivers, lakes, and groundwater reservoirs. Textile wastewater, comprising various pollutants such as dyes, heavy metals, and organic compounds, infiltrates into aquatic systems and disturbs aquatic ecosystems. Synthetic dyes, in particular, can alter water color and reduce light penetration, disrupting photosynthesis and harming aquatic plants. Excessive heavy metals threaten aquatic life, accumulating in organisms and leading to toxicity and biodiversity loss. Additionally, organic chemicals can deplete oxygen levels, creating oxygen-depleted "dead zones" harmful to fish and other aquatic species (Testa et al. 2010). On terrestrial ecosystems, textile wastewater is used for irrigation purpose, which infiltrates into the soil and contaminate agricultural lands. This contamination may lead to uptake of harmful chemicals by agricultural crops, ultimately affecting the food chain (Haidri et al. 2023). This not only poses a hazard to the health of the crops themselves but also has wider consequences throughout the food chain. Animals that consume these contaminated crops may experience adverse effects, and ultimately, this contamination can impact human populations that rely on these crops and animals for nutrition (Figure 18.1). Thus, the pollution of water bodies due to textile wastewater has far-reaching consequences that not only affect the aquatic as well as terrestrial ecosystems (Table 18.1) (Pereao et al. 2020).

## 18.4 Nanotechnology in Environmental Pollution Remediation

Nanotechnology can revolutionize multiple industries, providing innovative solutions to environmental pollution. Employing nanoscale materials in environmental remediation represents a promising approach for mitigating pollution, restoring contaminated areas, and protecting ecological health. Nanotechnology enhances pollutant elimination efficiency through nanomaterials like nanoparticles and nanocomposites. Nanoparticles possess specific features such as high surface area-to-volume ratios and high reactivity, which increase contaminant adsorption, degradation, or immobilization. For example, nanoscale zero-valent iron (nZVI) effectively remediates groundwater contaminated with heavy metals and chlorinated solvents (Carroll et al. 2013) Nano sorbents, including graphene-based materials and carbon nanotubes, offer environmentally conscious and efficient alternatives for addressing oil spills and water purification. Furthermore, nanocatalysts like titanium dioxide ($TiO_2$) and zinc oxide (ZnO) harness solar energy to

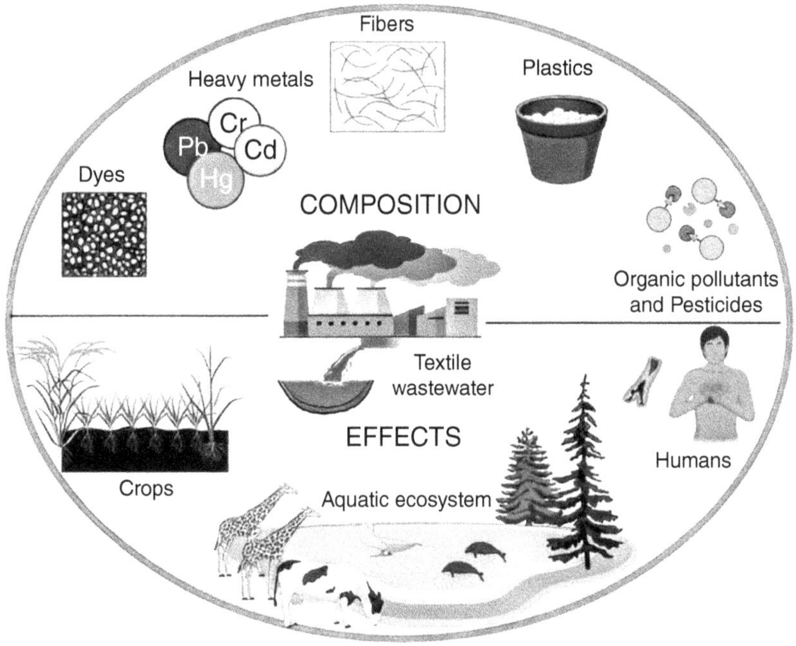


**Figure 18.1** Composition and effects of textile wastewater.

**Table 18.1** Physicochemical properties of textile wastewater.

| Country | pH | BOD (mg/l) | COD (mg/l) | TDS (mg/l) | TSS (mg/l) | Color | References |
|---|---|---|---|---|---|---|---|
| Johor, Malaysia | 6.5–8.5 | 7.1–85 | 55–294 | N/A | 69–205 | 17–140 | Bakar et al. (2020) |
| Tanzania | 6.5–8.5 | 366.7 | 2133.8 | N/A | N/A | 1685.9Pt-Co color | Al-Gheethi et al. (2022) |
| India | 8.5 ± 1 | 224.6 ± 40.7 | 433.7 ± 35 | 2570 ± 490 | 244 ± 76 | 456 ± 22.6 | Nawaz and Ahsan (2014) |
| India | 6–10 | 100–4000 | 150–10 000 | 1800–6000 | 100–5000 | – | Upadhye and Joshi (2012) |
| India | 6–10 | 80–6000 | 150–12 000 | 2900–3100 | 15–8000 | 50–2500 | Ghaly et al. (2014) |
| India | – | 141 | 1090 | – | 1004 | – | Watharkar et al. (2015) |
| Malaysia | 6.95–11.8 | 80–6000 | 150–30 000 | 2900–3100 | 15–8000 | 50–2500 | Gomathi et al. (2017) |

catalyze pollutant degradation, mitigate air pollution, and decontaminate water sources Overall, using nanoscale materials in environmental remediation holds immense potential for addressing pollution challenges and promoting the overall well-being of ecosystems (Kanan et al. 2020).

## 18.5 Types of Biologically Synthesized Nanoparticles Used in the Treatment of Textile Wastewater

Green synthesis of nanoparticles involves using natural, eco-friendly sources like plants, microbes, fungi, and polymers to reduce metal ions and create nanoparticles. These biologically synthesized nanoparticles have garnered significant attention for their potential in treating textile wastewater due to their environmentally friendly attributes and cost effectiveness (Abdel-Shafy and Mansour 2018). The unique properties of these nanoparticles can effectively treat the contamination presence in textile wastewater, ultimately contributing to the preservation of ecosystems. Different biological synthesized nanoparticles possess different functions like silver nanoparticles (Ag-NPs), renowned for their solid antimicrobial properties, which combat microbial contamination. Iron nanoparticles (Fe-NPs) have the ability to remove heavy metals and organic dyes, while titanium dioxide nanoparticles ($TiO_2$NPs) are highly regarded for their photocatalytic capabilities under UV light. Zinc oxide nanoparticles (ZnO-NPs) prove effective in disinfection and the removal of organic contaminants, and copper nanoparticles (Cu-NPs) exhibit antimicrobial properties (Das et al. 2018). Among nanoparticles for treating textile wastewater, Ag-NPs and ZnO-NPs have demonstrated their ability to enhance various properties of finished fabrics, including antibacterial characteristics, self-cleaning properties, and UV-blocking capabilities. ZnO-NPs have found application in the textile industry to improve UV blocking and antibacterial properties, with inorganic NPs like $TiO_2$. ZnO-NPs being preferred over organic NPs for UV blocking due to their chemical stability and non-toxicity when exposed to UV radiation and high temperatures. NPs offer significant advantages over bulk materials in UV-blocking efficacy due to their large surface area-to-volume ratio (Huh and Kwon 2011).

Gold nanoparticles (Au-NPs) are more prominent in the removal of heavy metals and dyes, similar to nano zero-valent iron, which reduces heavy metal ions in textile wastewater (O'Carroll et al. 2013). Magnetic nanoparticles play a significant role in efficient separation, while nanostructured metal oxides such as manganese and cerium oxide serve as catalysts. These nanoparticles present a promising and sustainable approach to textile wastewater treatment, aligning with the principles of green chemistry and environmental safety concerns. They not only mitigate water pollution but also hold the potential to recover valuable resources from textile wastewater, thereby enhancing both environmental and economic benefits (Gupta et al. 2021).

## 18.6 Green Synthesis Methods

Green synthesis methods play a crucial role in textile wastewater treatment using nanoparticles. One approach is plant-mediated synthesis, where plant extracts are involved in facilitating the formation of nanoparticles. Additionally, microbial synthesis employs microorganisms to reduce metal ions effectively, a critical step in nanoparticle production (Haidri et al. 2023). Enzymatic synthesis offers precise control over nanoparticle characteristics through specific enzymes. Biopolymer-assisted synthesis utilizes natural biopolymers to stabilize nanoparticles in an environmentally friendly manner. The adoption of eco-friendly solvents like water aids in minimizing the overall environmental footprint of the synthesis process. Furthermore, solar or microwave irradiation methods harness sustainable energy sources to power nanoparticle synthesis, aligning with eco-friendly practices. Waste biomass conversion provides a means of transforming waste materials from various sources into valuable nanoparticles. Natural extracts from fruits and spices serve as effective reducing agents, contributing to sustainable and green nanoparticle synthesis (Prasad et al. 2016).

### 18.6.1 Plant Synthesis of Nanoparticles

Plant synthesis of nanoparticles, also known as plant-mediated or green synthesis, is a sustainable and eco-friendly method for producing nanoparticles. It involves using various parts of plants, such as leaves, stems, roots, or extracts, to facilitate the reduction and stabilization of metal ions, leading to the formation of nanoparticles. This process typically occurs in an aqueous environment under mild conditions, without the need for high temperatures, pressure, or toxic chemicals (Shafqat et al. 2023). The utilization of plant extracts for the biologically synthesized nanoparticles. Particularly in the context of textile wastewater treatment, this method offers distinct advantages compared to other biological synthesis methods. For instance, it stands out from methods involving microorganisms, which entail the complex task of maintaining microbial cultures. Notably, one of the key advantages of plant-mediated nanoparticle synthesis lies in its kinetics, which is significantly more favorable. This stands in contrast to alternative biosynthetic approaches, including chemical nanoparticle preparation. This enhanced kinetic efficiency underscores the potential of plant-based methods for nanoparticle production. Furthermore, various components of plants, ranging from fruits and leaves to stems and roots, have gained widespread utilization in green nanoparticle synthesis. This is attributed to their ability to yield a diverse range of valuable phytochemicals, rendering them invaluable resources in the pursuit of environmentally friendly wastewater treatment solutions (de Jesus et al. 2021).

Green tea leaves, with their abundance of catechins, notably epigallocatechin gallate (EGCG), have shown remarkable potential in facilitating the synthesis of nanoparticles with specific properties. Neem leaves are renowned for their rich content of bioactive compounds like azadirachtin. Nimbin and quercetin offer a sustainable resource for nanoparticle synthesis with antimicrobial and other beneficial properties. Turmeric rhizomes, which contain the potent bioactive compound curcumin as their primary component, have played a crucial role in the development of nanoparticles with diverse applications (Shafqat et al. 2023). The exploration of various plant species underscores the adaptability and versatility of

Plant-Mediated Synthesis. This enables researchers to customize nanoparticle synthesis according to specific requirements, be it for the treatment of textile wastewater or for other innovative applications (Table 18.2) (Kharwar et al. 2020).

### 18.6.2 Microbial Synthesis of Nanoparticles

Microbial synthesis of nanoparticles, also known as biogenic synthesis or green synthesis, is a process in which microorganisms, such as bacteria, fungi, algae, and yeast, are used to produce nanoparticles (Figure 18.2). This eco-friendly approach utilizes the biological machinery of microorganisms to reduce metal ions and produce nanoparticles tailored for the treatment of textile wastewater (Haidri et al. 2023). Among land-dwelling bacterial species, *Escherichia coli, Bacillus spp.*, and *Pseudomonas aeruginosa* emerge as notable candidates. This is attributed to their extensively reactive enzymatic systems and their high effectiveness in reducing metal ions. These bacteria are adept at producing nanoparticles with controlled characteristics, making them suitable for various applications in wastewater treatment. In the fungal kingdom, species such as *Aspergillus, Fusarium,* and *Trichoderma* have garnered significant attention for their nanoparticle synthesis capabilities. These fungi exhibit unique enzymatic pathways, including reductases and extracellular enzymes, that are pivotal in reducing and stabilizing metal ions. As a result, they contribute to emerging nanoparticles with specific properties optimized for textile wastewater remediation (Banerjee and Rai 2017).

**Table 18.2** Plant synthesis of nanoparticles.

| Name of plant | Nanoparticles | Size (nm) | Shape of NPs | Applications of nanoparticles | References |
|---|---|---|---|---|---|
| *Asparagus racemosus* | Iron oxide | 32–40 | Spherical shape | Photodegradation of textile wastewater | Kumara et al. (2018) |
| *Phyllanthus niruri* | Zinc oxide | 25.62 | Hexagonal | Photodegradation of methylene blue in textile wastewater | Anbuvannan et al. (2015) |
| *Plectranthus amboinicus* | Zinc oxide | 50–180 | Rod shape | Photocatalytic activity for textile wastewater | Fu and Fu (2015) |
| *Artocarpus gomezianus* | Zinc oxide | 11.52 | Spherical, hexagonal shape | Photocatalytic degradation of dyes | Suresh et al. (2015) |
| *Allium sativum* | Zinc oxide | 14–70 | Hexagonal shape | Photodegradation of methylene blue dyes | Stan et al. (2015) |
| *Artocarpus heterophyllus* | Zinc oxide | 15–25 | Hexagonal shape | Photocatalytic degradation of dyes | Vidya et al. (2016) |

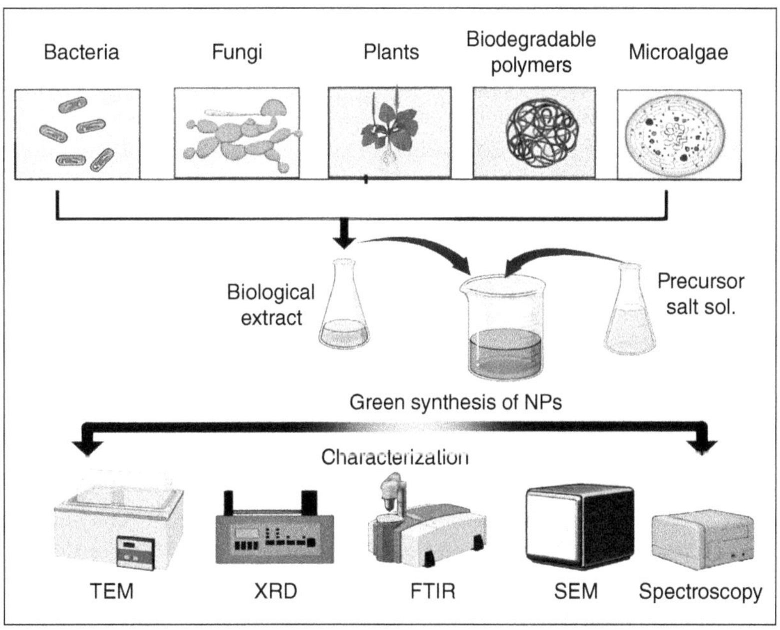


**Figure 18.2** Green synthesis and characterization nanoparticles using different biological sources.

**Table 18.3** Microbial synthesis of nanoparticles.

| Microbial source | NPs | Metal Ion Source | Size | Shape | Studied applications | References |
|---|---|---|---|---|---|---|
| Algae | Zinc oxide | Zinc acetate dihydrate | 21 | Hexagonal wurtzite shape | Photocatalytic activity | Priyadharshini et al. (2014) |
| Algae | Zinc oxide | Zinc acetate | 10–50 | Sponge-like asymmetrical | Photocatalytic activity | Ishwarya et al. (2018) |
| Bacteria | | | | | | |
| Bacteria | Zinc oxide | Zinc sulfate | 100–120 | Hexagonal shape | Photocatalytic degradation of malachite green | Kundu et al. (2014) |
| Bacteria | Titanium oxide | Metatitanic acid | 14–17 | Spherical shape | Textile wastewater treatment | Fulekar et al. (2018) |
| Fungi | Cesium oxide | Cerous chloride heptahydrate | 5–20 | Cubic and spherical shape | Antibacterial treatment of wastewater | Gopinath et al. (2015) |

Moreover, microalgae, such as Chlorella and Spirulina, are emerging as promising candidates for nanoparticle synthesis. These microorganisms not only participate in reducing metal ions but also offer the advantage of sustainable biomass production. Harnessing the photosynthetic capabilities of microalgae, researchers are exploring their potential in the green synthesis of nanoparticles, further enhancing the eco-friendly aspects of this approach (Table 18.3) (Tang et al. 2020).

### 18.6.3 Synthesis of Nanoparticles by Using Biodegradable Polymers

Biopolymer synthesis of nanoparticles is a sustainable approach that utilizes natural biopolymers, such as chitosan, cellulose, starch, and alginate, as integral components in the synthesis of nanoparticles for textile wastewater treatment. These biodegradable polymers not only act as environmentally friendly reducing and stabilizing agents but also enhance the overall eco-friendliness of the nanoparticle production process. Chitosan, for example, is derived from chitin, a biopolymer found abundantly in the shells of crustaceans like shrimp and crabs. Its cationic nature plays a pivotal role in nanoparticle synthesis by facilitating electrostatic interactions with metal ions, thereby aiding in the formation of stable nanoparticles (Philibert et al. 2016).

Cellulose, a widely explored biodegradable polymer, offers distinct advantages owing to its abundant availability from plant-based sources and remarkable biocompatibility. Similarly, starch, derived from diverse plant sources, is notable for its eco-friendly attributes and capacity to act as a reducing agent in the synthesis of nanoparticles. Additionally, alginate, sourced from brown algae, is esteemed for its biocompatibility and effective role as a stabilizing agent in nanoparticle production. These biopolymers play a pivotal role in the green synthesis of nanoparticles, significantly reducing the environmental impact of the process while enhancing the resulting nanoparticles' biocompatibility. This eco-friendly approach aligns perfectly with sustainable practices and is significant in addressing the challenges associated with textile wastewater treatment (Samuel et al. 2022). Utilizing these biodegradable polymers not only aids in mitigating pollution but also underscores our commitment to environmental stewardship in the quest for a greener, more sustainable future (Zinge and Kandasubramanian 2020).

### 18.6.4 Characterization of Biologically Synthesized Nanoparticles

Spectroscopy techniques are essential for characterizing biologically synthesized nanoparticles for textile wastewater treatment. These methods offer valuable insights into the nanoparticles' composition, structure, and optical properties. The position and intensity of these peaks provide insights into the size and shape of the nanoparticles. Smaller, spherical nanoparticles typically display absorption peaks around 400 nm, which are distinguished by their narrow spectral profiles. X-ray photoelectron spectroscopy (XPS) plays a pivotal role in determining the elemental composition and chemical states of the nanoparticles. This aids in identifying the oxidation states of metal nanoparticles and their surface chemistry, providing crucial insights into reactivity and stability (Haidri et al. 2023). Fourier Transform Infrared Spectroscopy (FTIR) analyzes the chemical composition and functional groups on the surface of nanoparticles. It is useful for surface modification studies (Negrescu et al. 2022).

**Table 18.4** Characterization techniques of biologically synthesized nanoparticles for wastewater treatment.

| Types of catalysts used | Characterization | Conditions for treatment of textile wastewater | Degradation of dyes in textile wastewater | References |
|---|---|---|---|---|
| ZnO-MFs | TEM, SEM and thermogravimetric | 300 min at 150 mg dosage, 10 ppm concentration dye, pH 7 | 99.5% | Isa et al. (2020) |
| $TiO_2$ | Chromatography and FTIR | UV/at pH 3 for 60 min | 93.1%/90.2% | Inderyas et al. (2020) |
| ZnO-NPs | FTIR, XRD, UV–VIS, and SEM | 0.24 g/l Catalyst, concentration 20 ppm dye, pH 9. for one h | 90.0% of the dye | Devi et al. (2020) |
| ZnO-NPs | SEM, FTIR, XRD, TEM and EDX | 1.0 g/l catalyst for 159 min under solar irradiation | 91.0% of the dye | Khan et al. (2019) |

Several techniques are employed for precise particle size analysis. Dynamic light scattering (DLS) measures the hydrodynamic size of nanoparticles within liquid suspensions, providing valuable insights into particle size distribution, polydispersity, and zeta potential. DLS proves particularly useful for assessing nanoparticle stability and aggregation behavior in aqueous solutions. Nanoparticle tracking analysis (NTA) takes a real-time approach to determine nanoparticle size distribution and concentration in liquid samples, tracking individual nanoparticles based on their Brownian motion and delivering high-resolution size data. Static light scattering (SLS) measures the absolute size of nanoparticles by scrutinizing the scattering of laser light, making it well-suited for precise size measurements, especially in the case of monodisperse nanoparticles. These combined techniques collectively empower researchers to thoroughly characterize the size-related properties of biologically synthesized nanoparticles. This ensures their suitability for efficient and environmentally sustainable treatment of textile wastewater (Table 18.4) (Cho and Hackley 2010).

## 18.7 Treatment of Textile Wastewater by Different Process

Biologically synthesized nanoparticles are essential in treating textile wastewater through various mechanisms. These mechanisms encompass adsorption, where nanoparticles like Ag-NPs, ZnO-NPs, and Fe-NPs exhibit high adsorption capacity, effectively attracting and binding pollutants like heavy metals and organic dyes. Additionally, nanoparticles possess antimicrobial solid action, combating microbial contamination and preventing foul odors, as seen with silver nanoparticles. Photocatalysis is another mechanism, with nanoparticles like $TiO_2$-NPs and ZnO-NPs catalyzing the degradation of organic pollutants when exposed to UV light, reducing the organic contamination in wastewater (AlAbduljabbar et al. 2021). Furthermore, nanoparticles contribute to disinfection, as ZnO-NPs and Cu-NPs, with their antibacterial properties ensuring water safety emitting from hospital and other sources. Their role extends to heavy metal removal, where nano zero-valent iron effectively reduces heavy metal ions like chromium and arsenic, facilitating precipitation and immobilization. Magnetic nanoparticles facilitate separation and filtration processes, enabling efficient recovery and reuse of nanoparticles while ensuring clean treated water discharge (Mohammadzadeh et al. 2022).

### 18.7.1 Treatment of Textile Wastewater by Adsorption Process

Adsorption is a widely used method for the treatment of textile wastewater. It involves the adherence of dissolved or suspended pollutants onto the surface of a solid material, known as an adsorbent. This process is highly effective in removing various contaminants from wastewater streams. Activated carbon nanoparticles, known for their high surface area and porosity, effectively remove organic dyes, pharmaceuticals, and volatile organic compounds from wastewater. For instance, due to their substantial adsorption capacity, activated carbon nanoparticles can eliminate stains like methylene blue from textile wastewater (Jadhav and Jadhav 2021). Zero-valent iron nanoparticles (nZVI) exhibit strong reducing properties, making them ideal for removing heavy metals such as arsenic, chromium, and chlorinated solvents from contaminated groundwater and soil. In textile wastewater, nZVI can convert hexavalent chromium into less toxic trivalent chromium. Titanium dioxide nanoparticles capitalize on their photocatalytic activity under UV light to degrade organic dyes, pollutants, and disinfect water (Ainali et al. 2021). Silica nanoparticles, known for their extensive surface area and functional capacity, exhibit a selective adsorption capability for heavy metals such as lead (Pb) when modified with thiol groups. Graphene oxide nanoparticles, with exceptional surface area and strong adsorption affinity, efficiently remove dyes such as Congo red from textile wastewater. Magnetic nanoparticles, like iron oxide nanoparticles, are easily separable using a magnetic field, allowing for recovery and potential reuse after removing various contaminants, including heavy metals and organic compounds. As an example, iron oxide nanoparticles enveloped in chitosan can extract heavy metals like copper from textile wastewater and can be easily separated for subsequent regeneration. Silver nanoparticles, possessing antibacterial and antiviral properties, disinfect textile wastewater by inhibiting microbial growth. The selection of the appropriate nanoparticle depends on the specific contaminants present, and careful optimization of process conditions, such as pH and temperature, is essential for maximizing efficiency while adhering to safety measures during handling and disposal (Bishayee et al. 2022).

### 18.7.2 Photocatalytic Degradation of Textile Wastewater

Photodegradation of textile wastewater involves the utilization of nanoparticles. These nanoparticles are typically semiconducting materials like $TiO_2$ or ZnO, activated by ultraviolet or visible light to catalyze the degradation of organic pollutants and dyes within the wastewater. This process, known as photocatalytic degradation, offers a promising approach for treating textile wastewater. Initially, nanoparticles are dispersed in the textile wastewater. When exposed to UV or visible light, they generate electron–hole pairs, creating highly reactive species on their surfaces (Haidri et al. 2023). These electron–hole pairs then interact with the organic contaminants, including dyes and other pollutants in the wastewater, leading to the degradation of these substances into smaller, less harmful molecules. The photocatalytic process sometimes goes beyond fragmenting contaminants. It mineralizes them into harmless end products, carbon dioxide ($CO_2$) and water ($H_2O$). This method effectively addresses the challenge of removing vividly colored dyes and associated odors in textile wastewater by breaking down the responsible chromophoric groups (Pandey et al. 2020).

The process of photocatalysis accelerates the activity of biologically synthesized nanoparticles, due to the presence of multiple functional groups on their surface. This augmentation enhances their reactivity and adsorption capacity. When these nanoparticles are dispersed in the textile wastewater and exposed to UV or visible light, they not only generate electron–hole pairs on their surfaces but also offer several advantages that can improve the overall efficiency of the process. These functional groups can act as active sites for the adsorption of organic contaminants, providing more binding sites for pollutants present in the wastewater (Figure 18.3). This increased adsorption capacity facilitates a higher concentration of contaminants near the nanoparticles, promoting more effective interactions during photocatalytic degradation. The functional groups can also act as catalysts or co-catalysts, further accelerating the degradation reactions (Bahnemann et al. 2018). They can enhance the generation of reactive oxygen species (ROS) and facilitate the conversion of contaminants into smaller, less harmful molecules. This catalytic activity can lead to more efficient pollutant degradation and mineralization (Abdal Dayem et al. 2017).

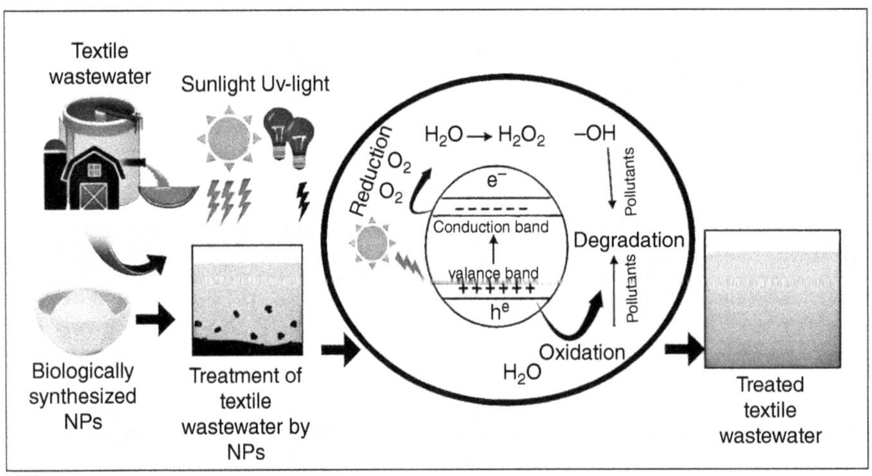


**Figure 18.3** Photocatalytic treatment of textile wastewater by green synthesized nanoparticles.

### 18.7.3 Filtration and Membrane Processes

Filtration and membrane processes depend on specialized membranes or filters that selectively remove contaminants. Filtration involves passing wastewater through a physical barrier or filter medium that traps particles and contaminants. Biologically synthesized nanoparticles with a higher density of functional groups offer enhanced efficiency in improving nanofiltration, nano-enhanced filtration, and various membrane processes for wastewater treatment. They increase surface functionalization, incorporating hydroxyl (–OH) or amino (–$NH_2$) groups that expand active sites for contaminant interactions, ultimately boosting adsorption and separation efficiency (Lari et al. 2021). Additionally, their adsorption capacity allows them to effectively capture more significant quantities of contaminants from wastewater, resulting in increased removal rates. The functional groups on these nanoparticles enhance their selectivity, making them more proficient in targeting specific pollutants, such as heavy metals or organic dyes. Certain functional groups also exhibit antifouling properties, which are valuable in membrane processes. They prevent the buildup of contaminants on membrane surfaces, consequently reducing fouling and ensuring sustained, efficient filtration over extended periods. Biologically synthesized nanoparticles often demonstrate synergistic effects when integrated with membrane processes. Their functional groups complement the mechanisms of membrane separation, ultimately enhancing overall performance (Yang et al. 2021).

## 18.8 Degradation of Dyes by Green Synthesized Nanoparticles

Different biologically synthesized nanoparticles were used to treat different pollutants present in textile wastewater. Textile wastewater mainly compromises dyes, so the researcher used biologically synthesized nanoparticles to treat the dyes present in textile wastewater. Zinc nanoparticles have been synthesized using various sources for their application in the degradation of different dyes. *Artocarpus heterophyllus*-synthesized zinc nanoparticles effectively degraded Rose Bengal dye, with particle sizes ranging from 15 to 25 nm and achieving nearly 85% dye degradation (Vidya et al. 2016). *Momordica cochinchinensis* was utilized as a source for Au nanoparticle synthesis, resulting in nanoparticles ranging in size from 10 to 80 nm. These nanoparticles exhibited a swift decolorization of dyes, including Congo red, Methylene blue, Rhodamine B, Eosin Y, and Methyl Orange, within a short time of 4–8 minutes (Paul et al. 2016). *Aspergillus* sp. NJP02-derived zinc nanoparticles, sized between 40 and 80 nm, demonstrated remarkable photocatalytic degradation of Methylene Blue dye (Bhargava et al. 2016). The synthesis of gold nanoparticles using various sources has shown remarkable efficiency in degrading dyes in wastewater treatment. Similarly, palladium (Pd) nanoparticles obtained from *Catharanthus roseus* proved highly efficient in degrading Phenol Red (Kalaiselvi et al. 2015). Zirconium oxide nanoparticles synthesized using *Lagerstroemia speciosa* demonstrated the ability to degrade Methyl Orange up to 94.58% in textile wastewater (Saraswathi et al. 2017). Titanium dioxide ($TiO_2$) nanoparticles, synthesized from *Bacillus amyloliquefaciens*, effectively removed Reactive Red 31 from textile wastewater, achieving a 90.98% degradation rate with platinum doping (*Morinda tinctoria* produced larger Au nanoparticles, ranging from 79 to 96 nm, effectively deleting Methylene blue (Vanaja et al. 2014; Khan and Fulekar 2016). *Coccinia grandis*-synthesized silver nanoparticles (20–30 nm) demonstrated good photocatalytic activity, degrading Coomassie Brilliant Blue 250-G in 90 minutes (Arunachalam et al. 2012). These findings underscore the potential of biologically synthesized nanoparticles as efficient agents for dye degradation in textile wastewater treatment, offering various sizes and degradation rates to address different wastewater challenges (Table 18.5).

## 18.9 Removal Efficiency of Green Synthesized Nanoparticles for the Treatment of Textile Wastewater

The removal efficiency of green synthesized nanoparticles for textile wastewater treatment can vary widely due to multiple influencing factors. The choice of nanoparticle type, its size, and surface properties can significantly influence pollutant interactions. Factors such as treatment time, pH, and temperature conditions further modulate removal efficiency, emphasizing the need for comprehensive optimization. These factors depend on the type and concentration of nanoparticles utilized, the physicochemical properties of wastewater itself, and the specific treatment conditions used. Green synthesized nanoparticles, produced through environmentally friendly methods, often involving natural extracts or plant-based materials, have increased their sustainability. These nanoparticles are used to treat different contaminants in textile wastewater, including dyes, heavy metals, and organic compounds. The choice of nanoparticle type, size, and surface properties can influence pollutant interactions, while factors like treatment time and conditions, including pH and temperature, play a

**Table 18.5** Treatment of textile wastewater by using different nanoparticles.

| Source for nanoparticles synthesis | Nanoparticles | Size range | Dye degradation in textile wastewater | Degradation time | References |
|---|---|---|---|---|---|
| *Leuconostoc lactis* | Silver nanoparticles | 30–200 nm | Methyl orange, congo red | 240 min | Saravanan et al. (2017) |
| Cladosporium oxysporum | Gold nanoparticles | 72.31 ± 21.81 nm | Rhodamine B | 7 min | Bhargava et al. (2016) |
| *Aspergillum* sp. | Gold nanoparticles | 10–100 nm (H2O), 12–48 nm (PBS) | Various dyes | 7 min (varied) | Qu et al. (2017) |
| *Azadirachta indica* | Zinc nanoparticles | 9.6–25.4 nm | Methylene blue | Efficient degradation | Bhuyan et al. (2015) |
| *Azadirachta Indica* | Copper nanoparticles | Varies | Methyl orange | 60 min (92%)/30 min (98%) | Nagar and Devra (2017) |
| *Tinospora cordifolia* | Copper oxide nanoparticles | Varies | Methylene blue | 80% removal rate | Nethravathi et al. (2015) |
| *Plectranthus amboinicus* | Zinc nanoparticles | 51–180 nm | Methyl red | Up to 92.44% in 180 min | Fu and Fu (2015) |
| *Cassia fistula* | Zinc nanoparticles | 5–15 nm | Methylene blue | pH-dependent efficiency | Suresh, Nethravathi et al. (2015) |
| *Sterculia acuminata* | Gold nanoparticles | 9.37–37.12 nm | Methyl orange, methylene-blue, direct blue 24 | 12 min (methyl orange, methylene blue) | Bogireddy et al. (2015) |
| *Piper longum* | Gold nanoparticles | 56 nm | Methylene blue, methyl -red, crystal violet, acridine orange | Varied (up to 64% after 28 h) | Nakkala et al. (2016) |
| *Gmelina arborea* | Silver nanoparticles | 8–32 nm | Complete degradation of methylene blue | 10 min | Saha et al. (2017) |

role in removal efficiency (Lade and Shanware 2020). The relationship between dye concentration, light, and catalysts assumes paramount significance in photocatalytic processes. The specific target pollutant also plays a pivotal role in determining the efficiency of photocatalysis due to their different reactivity with nanoparticles., Overall the efficiency of wastewater treatment by green synthesized nanoparticles highly depends on different parameters including the type of catalyst, pollutant concentration, light intensity, and wavelength.(Bahnemann et al. 2018).

## 18.10 Toxicity and Safety Considerations for the Treatment of Textile Wastewater Using Green Synthesized Nanoparticles

The application of nanotechnology, particularly nanostructures, has found relevance across various sectors, including textile wastewater treatment, where their enhanced surface area offers unique advantages (Haidri et al. 2023). However, this heightened surface area has raised concerns about potential toxicity, a matter not always as discernible as it is with bulk materials. In this situation, direct ingestion and inhalation of nanomaterials rarely happen. However, it is crucial to conduct thorough research to understand how nanomaterials might affect the metabolism when they come into direct contact with stored food items. The current research lacks specific regulations governing the production and utilization of nanosized particles, largely attributable to the limited understanding of the associated risks and their implications for human health (Bardos et al. 2011).

Currently, the regulatory framework concerning the production and application of nanosized particles remains undefined, leaving a void in safeguarding against potential harm. The concerns about nanomaterials in food packaging persist, various food products containing nano-scale additives, such as ZnO in cereals, Fe in soft drinks, and phytochemicals in oils, have already entered the market, raising further questions about their safety for human consumption. The regulatory bodies, including the Food and Drug Administration (FDA), should take proactive measures to ensure the safe utilization of nanoparticles, not only in food products but also in emerging environmental treatment applications like textile wastewater treatment. This necessitates the implementation of stringent safety evaluations and risk assessments to comprehensively understand and mitigate potential hazards. Such measures are essential and effective for the textile wastewater treatment and preservation of the health of humans and ecosystem (Yadav et al. 2017).

## 18.11 Cost-effectiveness

The application of green-synthesized and biogenic nanocatalysts and nanomaterials offers a promising solution for the cost-effective and efficient treatment of textile wastewater. It treats a wide range of pollutants including inorganic, organic, pharmaceutical, and heavy metal contaminants in textile wastewater. Green-synthesized nanoparticles offer a more cost-effective solution because of their sustainable synthesis processes, often using readily available, low-cost, green raw materials and energy-efficient methods. The reduced need for costly infrastructure and maintenance also contributes to their cost-effectiveness. The efficiency of green synthesis nanoparticles in treating contaminants from textile wastewater is the most feasible. Their high surface area and unique surface properties enable them to effectively adsorb and degrade various pollutants. This results in a more efficient treatment process, reducing the time and resources required to achieve desired water quality standards. The enhanced performance of green-synthesized nanoparticles also means that smaller quantities of these materials are needed, further lowering operational costs. The environmentally friendly nature of green synthesized nanoparticles aligns with sustainability goals. Traditional chemical treatments often involve using hazardous substances, which can generate harmful byproducts and contribute to pollution. In contrast, green-synthesized nanoparticles typically produce minimal secondary toxic pollution, making them more eco-friendly for textile wastewater treatment (Singh et al. 2021).

## 18.12 Challenges and Limitations

### 18.12.1 Long-Term Environmental Impact

The treatment of textile wastewater using biologically synthesized nanoparticles is an excellent choice for mitigating the textile industry's environmental impact. Treating wastewater requires careful consideration of both short-term and long-term ecological impacts. The potential long-term effects, including the toxicity of nanoparticles and their accumulation in soil and water ecosystems, are becoming threats to terrestrial and aquatic life. Continuous monitoring is necessary while treating textile wastewater by nanoparticles because their presence in water bodies causes bioaccumulation. This bioaccumulation of nanoparticles ultimately disturbs the food chain, becoming a threat to living organisms. The release of nanoparticles in the ecosystem during wastewater treatment leads to unseen ecological consequences due to their persistence and reactivity. When nanoparticles enter the soil, they may alter microbial communities, affecting nutrient cycling and ecosystem stability and prompting potential regulatory adjustments. Comprehensive environmental assessments, ongoing monitoring, and development are essential to reduce the adverse effects. Treating textile wastewater with biologically synthesized nanoparticles poses a complex challenge due to the numerous transformation pathways, especially within wastewater treatment plants. These pathways complicate the assessment of their environmental risk. These transformations include oxidative dissolution, photoreduction, sulfidation, and chlorination. Understanding these transformation processes is critical for a comprehensive assessment of the environmental implications of ENP exposure in aquatic systems, particularly within the context of wastewater treatment plants treating textile wastewater (Zhang et al. 2022).

### 18.12.2 Scalability and Commercial Viability

In developing countries like Pakistan, managing industrial wastewater poses unique challenges, particularly in treating textile wastewater. A critical concern in this context is the safe disposal of loaded adsorbents, especially when employing adsorption techniques. Currently, the predominant method for dealing with textile sludge in the industrial sector is landfilling, a traditional approach that has prompted recent explorations into more sustainable and economically viable alternatives. In the context of developing nations, the feasibility and scalability of treating textile wastewater with green synthesized nanoparticles require necessary considerations. Implementing biologically synthesized nanoparticles for textile wastewater treatment depends on various factors, including efficient production methods, treatment capacity, infrastructure development, adherence to regulatory standards, cost-effectiveness, market demand, environmental sustainability, and community acceptance. In essence, the adoption of biologically synthesized nanoparticles presents a promising solution for addressing the challenge of safely disposing of loaded adsorbents in textile wastewater treatment. Still, its practicality in countries like Pakistan depends on a thorough evaluation encompassing technical, economic, regulatory, and market-related aspects. Rigorous research, pilot studies, and comprehensive cost-benefit analyses are imperative steps to determine its potential for success in these settings (Baskar et al. 2022).

## 18.13 Future Trends and Research Directions

The future trends in using biologically synthesized nanoparticles for treating textile wastewater involve adopting green or sustainable technologies on an industrial scale to provide efficient and cost-effective solutions, especially for developing countries. This includes exploring hybrid treatment systems integrating advanced techniques like photo and electro-Fenton, UV photolysis, and ozone treatment with biological treatment technologies. Exploring emerging nanoparticle technologies holds great offers as it allows the creation of new nanoparticles that can capture more selective pollutants. Another trend involves combining existing wastewater treatment methods with physical techniques like ultrasound, adsorption, and gamma radiation to improve the treatment process. Developing innovative nanoparticle-based systems for treating textile wastewater, scaling up to pilot projects, establishing regulatory frameworks, and promoting knowledge sharing and education within the industry are all integral to harnessing the full potential of biologically synthesized nanoparticles for sustainable textile wastewater treatment. Furthermore, there will be a continued emphasis on investigating the degradation processes associated with substances like formic acid and acetic acid, which is crucial for developing relevant models and improving treatment strategies. There is a need to conduct laboratory experiments and tests, and results are applied in real-world pilot projects, set up rules and guidelines, share knowledge, and educate people in the industry.

## 18.14 Conclusion

Small and medium-sized textile companies face wastewater treatment before being released in an open environment. The urgent challenge faced by textile companies with the discharge of untreated wastewater into municipal treatment plants is due to the financial constraints associated with the high cost and complex process of conventional treatment technologies. Using green-synthesized nanoparticles for treating textile wastewater offers a promising and sustainable solution to the textile industries due to low cost and sustainability. By merging nanotechnology principles with eco-friendly synthesis methods, this innovative approach efficiently mitigates the detrimental effects of textile wastewater on the environment and public health. Green nanoparticles contribute to environmental sustainability by reducing the industry's ecological footprint, reducing pollutant emissions, and conserving valuable resources. Moreover, their impressive efficiency and cost-effectiveness compared to conventional treatment methods translate into substantial long-term savings for textile manufacturers. This adoption also paves the way for reduced chemical usage, aligning with the global emphasis on sustainability, while continued research promises further innovations and regulatory compliance assurance. Embracing green-synthesized nanoparticles enhances a company's reputation and leads to cost savings, compliance with evolving standards, and a meaningful contribution to resource conservation. In essence, this technology represents a pivotal step toward sustainable and responsible textile production, where environmental consciousness aligns with economic viability, promising a brighter and more environmentally conscious future for the industry.

## References

Abdal Dayem, A., Hossain, M.K., Lee, S.B. et al. (2017). The role of reactive oxygen species (ROS) in the biological activities of metallic nanoparticles. *International Journal of Molecular Sciences* 18 (1): 120.

Abdel-Shafy, H.I. and Mansour, M.S.M. (2018). *Green Synthesis of Metallic Nanoparticles from Natural Resources and Food Waste and Their Environmental Application*, 321–385. Wiley https://doi.org/10.1002/9781119418900.ch11.

Ainali, N.M., Kalaronis, D., Evgenidou, E. et al. (2021). Insights into biodegradable polymer-supported titanium dioxide photocatalysts for environmental remediation. *Macromolecules* 1 (3): 201–233. https://doi.org/10.3390/macromol1030015.

Al-Gheethi, A.A., Azhar, Q.M., Kumar, P.S. et al. (2022). A review of sustainable approaches for removing Rhodamine B dye using agricultural waste adsorbents. *Chemosphere* 287: 132080.

AlAbduljabbar, F.A., Haider, S., Ali, F.A.A. et al. (2021). Efficient photocatalytic degradation of organic pollutant in wastewater by electrospun functionally modified polyacrylonitrile nanofibers membrane anchoring {TiO}2 nanostructured. *Membranes* 11 (10): 785. https://doi.org/10.3390/membranes11100785.

Anbuvannan, M., Ramesh, M., Viruthagiri, G. et al. (2015). Synthesis, characterization and photocatalytic activity of ZnO nanoparticles prepared by biological method. *Spectrochimica Acta Part A: Molecular and Biomolecular Spectroscopy* 143: 304–308.

Arunachalam, R., Dhanasingh, S., Kalimuthu, B. et al. (2012). Photosynthesis of silver nanoparticles using *Coccinia grandis* leaf extract and its application in photocatalytic degradation. *Colloids and Surfaces B: Biointerfaces* 94: 226–230. https://doi.org/10.1016/j.colsurfb.2012.01.040.

Bahnemann, D., Cunningham, J., Fox, M.A. et al. (2018). Photocatalytic treatment of water. In: *Aquatic and Surface Photochemistry* (ed. G.R. Helz, R.G. Zepp, and D.G. Crosby), 261–316. CRC Press.

Bahrodin, M.B., Zaidi, N.S., Hussein, N. et al. (2021). Recent advances on coagulation-based wastewater treatment: transition from chemical to natural coagulant. *Current Pollution Reports* 7 (3): 379–391. https://doi.org/10.1007/s40726-021-00191-7.

Bakar, N.A., Othman, N., Yunus, Z.M. et al. (2020). Physico-chemical water quality parameters analysis on textile. *IOP Conference Series: Earth and Environmental Science* 498 (1): 12077.

Banerjee, K. and Rai, V.R. (2017). A review on mycosynthesis, mechanism, and characterization of silver and gold nanoparticles. *BioNanoScience* 8 (1): 17–31. https://doi.org/10.1007/s12668-017-0437-8.

Bardos, P., Bone, B., Elliott, D. et al. (2011). A risk/benefit approach to the application of iron nanoparticles for the remediation of contaminated sites in the environment. *Defra Research Project Final Report.* .

Baskar, A.V., Bolan, N., Hoang, S.A. et al. (2022). Recovery, regeneration and sustainable management of spent adsorbents from wastewater treatment streams: a review. *Science of the Total Environment* 822: 153555. https://doi.org/10.1016/j.scitotenv.2022.153555.

Bhargava, A., Jain, N., Khan, M.A. et al. (2016). Utilizing metal tolerance potential of soil fungus for efficient synthesis of gold nanoparticles with superior catalytic activity for degradation of rhodamine B. *Journal of Environmental Management* 183: 22–32.

Bhuyan, T., Mishra, K., Khanuja, M. et al. (2015). Biosynthesis of zinc oxide nanoparticles from *Azadirachta indica* for antibacterial and photocatalytic applications. *Materials Science in Semiconductor Processing* 32: 55–61.

Bishayee, B., Chatterjee, R.P., Ruj, B. et al. (2022). Strategic management of nitrate pollution from contaminated water using viable adsorbents: an economic assessment-based review with possible policy suggestions. *Journal of Environmental Management* 303: 114081. https://doi.org/10.1016/j.jenvman.2021.114081.

Bogireddy, N.K.R., Anand, K.K.H., and Mandal, B.K. (2015). Gold nanoparticles – synthesis by Sterculia acuminata extract and its catalytic efficiency in alleviating different organic dyes. *Journal of Molecular Liquids* 211: 868–875. https://doi.org/10.1016/j.molliq.2015.07.027.

Carroll, D., Sleep, B., Krol, M. et al. (2013). Nanoscale zero valent iron and bimetallic particles for contaminated site remediation. *Advances in Water Resources* 51: 104–122. https://doi.org/10.1016/j.advwatres.2012.02.005.

Castro, L., Blázquez, M.L., Muñoz, J. et al. (2014). Mechanism and applications of metal nanoparticles prepared by bio-mediated process. *Reviews in Advanced Sciences and Engineering* 3 (3): 199–216. https://doi.org/10.1166/rase.2014.1064.

Cho, T.J. and Hackley, V.A. (2010). Fractionation and characterization of gold nanoparticles in aqueous solution: asymmetric-flow field flow fractionation with MALS, DLS, and UV–Vis detection. *Analytical and Bioanalytical Chemistry* 398: 2003–2018.

Das, S., Chakraborty, J., Chatterjee, S., and Kumar, H. (2018). Prospects of biosynthesized nanomaterials for the remediation of organic and inorganic environmental contaminants. *Environmental Science: Nano* 5 (12): 2784–2808. https://doi.org/10.1039/c8en00799c.

de Jesus, R.A., de Assis, G.C., de Oliveira, R.J. et al. (2021). Environmental remediation potentialities of metal and metal oxide nanoparticles: mechanistic biosynthesis, influencing factors, and application standpoint. *Environmental Technology & Innovation* 24: 101851. https://doi.org/10.1016/j.eti.2021.101851.

Devi, S.A., Singh, K.J., and Devi, K.N. (2020). A comparative study on the photocatalytic activity of eucalyptus leaf assisted green synthesized ZnO and chemically synthesized ZnO towards the degradation of malachite green dye. *Integrated Ferroelectrics* 205 (1): 38–51.

Fu, L. and Fu, Z. (2015). Plectranthus amboinicus leaf extract–assisted biosynthesis of ZnO nanoparticles and their photocatalytic activity. *Ceramics International* 41 (2): 2492–2496.

Fulekar, J., Dutta, D.P., Pathak, B., and Fulekar, M.H. (2018). Novel microbial and root mediated green synthesis of $TiO_2$ nanoparticles and its application in wastewater remediation. *Journal of Chemical Technology & Biotechnology* 93 (3): 736–743.

Ghaly, A.E., Ananthashankar, R., Alhattab, M., and Ramakrishnan, V.V. (2014). Production, characterization and treatment of textile effluents: a critical review. *Journal of Chemical Engineering & Process Technology* 5 (1): 1–19.

Gomathi, E., Rathika, G., and Santhini, E. (2017). Physico-chemical parameters of textile dyeing effluent and its impacts with casestudy. *International Journal of Research in Chemistry and Environment (IJRCE)* 7 (1): 17–24.

Gopinath, K., Karthika, V., Sundaravadivelan, C. et al. (2015). Mycogenesis of cerium oxide nanoparticles using *Aspergillus niger* culture filtrate and their applications for antibacterial and larvicidal activities. *Journal of Nanostructure in Chemistry* 5: 295–303.

Gupta, K., Joshi, P., Gusain, R., and Khatri, O.P. (2021). Recent advances in adsorptive removal of heavy metal and metalloid ions by metal oxide-based nanomaterials. *Coordination Chemistry Reviews* 445: 214100. https://doi.org/10.1016/j.ccr.2021.214100.

Haidri, I., Shahid, M., Hussain, S. et al. (2023). Efficacy of biogenic zinc oxide nanoparticles in treating wastewater for sustainable wheat cultivation. *Plants* 12 (17): 3058.

Hassanien, R., Abed-Elmageed, A.A.I., and Husein, D.Z. (2019). Eco-friendly approach to synthesize selenium nanoparticles: photocatalytic degradation of sunset yellow azo dye and anticancer activity. *ChemistrySelect* 4 (31): 9018–9026. https://doi.org/10.1002/slct.201901267.

Huh, A.J. and Kwon, Y.J. (2011). "Nanoantibiotics": a new paradigm for treating infectious diseases using nanomaterials in the antibiotics-resistant era. *Journal of Controlled Release* 156 (2): 128–145. https://doi.org/10.1016/j.jconrel.2011.07.002.

Inderyas, A., Bhatti, I.A., Ashar, A. et al. (2020). Synthesis of immobilized ZnO over polyurethane and photocatalytic activity evaluation for the degradation of azo dye under UV and solar light irardiation. *Materials Research Express* 7 (2): 25033.

Isa, E.D.M., Jusoh, N.W.C., Hazan, R., and Shameli, K. (2020). Photocatalytic degradation of methyl orange using pullulan-mediated porous zinc oxide microflowers. *Environmental Science and Pollution Research* 28 (5): 5774–5785. https://doi.org/10.1007/s11356-020-10939-1.

Ishwarya, R., Vaseeharan, B., Kalyani, S. et al. (2018). Facile green synthesis of zinc oxide nanoparticles using Ulva lactuca seaweed extract and evaluation of their photocatalytic, antibiofilm and insecticidal activity. *Journal of Photochemistry and Photobiology B: Biology* 178: 249–258.

Jadhav, A.C. and Jadhav, N.C. (2021). Treatment of textile wastewater using adsorption and adsorbents. In: *Sustainable Technologies for Textile Wastewater Treatments* (ed. S. Senthil1kannan Muthu), 235–273. Elsevier https://doi.org/10.1016/b978-0-323-85829-8.00008-0.

Kalaiselvi, A., Roopan, S.M., Madhumitha, G. et al. (2015). Synthesis and characterization of palladium nanoparticles using *Catharanthus roseus* leaf extract and its application in the photo-catalytic degradation. *Spectrochimica Acta Part A: Molecular and Biomolecular Spectroscopy* 135: 116–119. https://doi.org/10.1016/j.saa.2014.07.010.

Kanan, S., Moyet, M.A., Arthur, R.B., and Patterson, H.H. (2020). Recent advances on $TiO_2$-based photocatalysts toward the degradation of pesticides and major organic pollutants from water bodies. *Catalysis Reviews* 62 (1): 1–65.

Khan, R. and Fulekar, M.H. (2016). Biosynthesis of titanium dioxide nanoparticles using *Bacillus amyloliquefaciens* culture and enhancement of its photocatalytic activity for the degradation of a sulfonated textile dye Reactive Red 31. *Journal of Colloid and Interface Science* 475: 184–191. https://doi.org/10.1016/j.jcis.2016.05.001.

Khan, S. and Malik, A. (2014). Environmental and health effects of textile industry wastewater. In: *Environmental Deterioration and Human Health: Natural and Anthropogenic Determinants* (ed. A. Malik, E. Grohmann, and R. Akhtar), 55–71. Dordrecht: Springer Netherlands.

Khan, Z.U.H., Sadiq, H.M., Shah, N.S. et al. (2019). Greener synthesis of zinc oxide nanoparticles using *Trianthema portulacastrum* extract and evaluation of its photocatalytic and biological applications. *Journal of Photochemistry and Photobiology B: Biology* 192: 147–157.

Kharwar, R.N., Sharma, V.K., Mishra, A. et al. (2020). Harnessing the phytotherapeutic treasure troves of the ancient medicinal plant azadirachta indica (Neem) and associated endophytic microorganisms. *Planta Medica* 86 (13/14): 906–940. https://doi.org/10.1055/a-1107-9370.

Kishor, R., Purchase, D., Saratale, G. D., et al. (2021). Ecotoxicological and health concerns of persistent coloring pollutants of textile industry wastewater and treatment approaches for environmental safety. *Journal of Environmental Chemical Engineering*, 9(2), 105012. /https://doi.org/10.1016/j.jece.2020.105012.

Kumara, P., Pammi, S.V.N., and Shameem, U. (2018). A Green approach for the synthesis of iron oxide nanoparticles by using roots of *A. racemosus* and its degradation of dye methyl orange. *International Journal of Pharmaceutics and Drug Analysis* 6: 22–28.

Kundu, D., Hazra, C., Chatterjee, A. et al. (2014). Extracellular biosynthesis of zinc oxide nanoparticles using *Rhodococcus pyridinivorans* NT2: multifunctional textile finishing, biosafety evaluation and in vitro drug delivery in colon carcinoma. *Journal of Photochemistry and Photobiology B: Biology* 140: 194–204.

Lade, B.D. and Shanware, A.S. (2020). Phytonanofabrication: methodology and factors affecting biosynthesis of nanoparticles. In: *Smart Nanosystems for Biomedicine, Optoelectronics and Catalysis* (ed. T. Shabatina and V. Bochenkov). IntechOpen https://doi.org/10.5772/intechopen.90918.

Lari, S., Parsa, S.A.M., Akbari, S. et al. (2021). Fabrication and evaluation of nanofiltration membrane coated with amino-functionalized graphene oxide for highly efficient heavy metal removal. *International Journal of Environmental Science and Technology* 19 (6): 4615–4626. https://doi.org/10.1007/s13762-021-03464-2.

Mohammadzadeh, V., Barani, M., Amiri, M.S. et al. (2022). Applications of plant-based nanoparticles in nanomedicine: a review. *Sustainable Chemistry and Pharmacy* 25: 100606. https://doi.org/10.1016/j.scp.2022.100606.

Nagar, N. and Devra, V. (2017). Activation of peroxodisulfate and peroxomonosulfate by green synthesized copper nanoparticles for Methyl Orange degradation: a kinetic study. *Journal of Environmental Chemical Engineering* 5 (6): 5793–5800.

Nakkala, J.R., Mata, R., and Sadras, S.R. (2016). The antioxidant and catalytic activities of green synthesized gold nanoparticles from *Piper longum* fruit extract. *Process Safety and Environmental Protection* 100: 288–294.

Nawaz, M.S. and Ahsan, M. (2014). Comparison of physico-chemical, advanced oxidation and biological techniques for the textile wastewater treatment. *Alexandria Engineering Journal* 53 (3): 717–722.

Negrescu, A.M., Killian, M.S., Raghu, S.N.V. et al. (2022). Metal oxide nanoparticles: review of synthesis, characterization and biological effects. *Journal of Functional Biomaterials* 13 (4): 274. https://doi.org/10.3390/jfb13040274.

Nethravathi, P.C., Kumar, M.A.P., Suresh, D. et al. (2015). Tinospora cordifolia mediated facile green synthesis of cupric oxide nanoparticles and their photocatalytic, antioxidant and antibacterial properties. *Materials Science in Semiconductor Processing* 33: 81–88.

O'Carroll, D., Sleep, B., Krol, M. et al. (2013). Nanoscale zero valent iron and bimetallic particles for contaminated site remediation. *Advances in Water Resources* 51: 104–122. https://doi.org/10.1016/j.advwatres.2012.02.005.

Pandey, S., Mandari, K.K., Kim, J. et al. (2020). Recent advancement in visible-light-responsive photocatalysts in heterogeneous photocatalytic water treatment technology. *Photocatalysts in Advanced Oxidation Processes for Wastewater Treatment* 167–196.

Paul, B., Bhuyan, B., Purkayastha, D.D. et al. (2016). One-pot green synthesis of gold nanoparticles and studies of their anticoagulative and photocatalytic activities. *Materials Letters* 185: 143–147. https://doi.org/10.1016/j.matlet.2016.08.121.

Paul, S.A., Chavan, S.K., and Khambe, S.D. (2012). Studies on characterization of textile industrial waste water in Solapur city. *International Journal of Chemical Sciences* 10 (2): 635–642.

Pereao, O., Opeolu, B., and Fatoki, O. (2020). Microplastics in aquatic environment: characterization, ecotoxicological effect, implications for ecosystems and developments in South Africa. *Environmental Science and Pollution Research* 27 (18): 22271–22291. https://doi.org/10.1007/s11356-020-08688-2.

Philibert, T., Lee, B.H., and Fabien, N. (2016). Current status and new perspectives on chitin and chitosan as functional biopolymers. *Applied Biochemistry and Biotechnology* 181 (4): 1314–1337. https://doi.org/10.1007/s12010-016-2286-2.

Prasad, R., Pandey, R., and Barman, I. (2016). Engineering tailored nanoparticles with microbes: quo vadis? *Wiley Interdisciplinary Reviews: Nanomedicine and Nanobiotechnology* 8 (2): 316–330.

Priyadharshini, R.I., Prasannaraj, G., Geetha, N., and Venkatachalam, P. (2014). Microwave-mediated extracellular synthesis of metallic silver and zinc oxide nanoparticles using macro-algae (Gracilaria edulis) extracts and its anticancer activity against human PC3 cell lines. *Applied Biochemistry and Biotechnology* 174: 2777–2790.

Qu, Y., Pei, X., Shen, W. et al. (2017). Biosynthesis of gold nanoparticles by *Aspergillum* sp. WL-Au for degradation of aromatic pollutants. *Physica E: Low-dimensional Systems and Nanostructures* 88: 133–141.

Saha, J., Begum, A., Mukherjee, A., and Kumar, S. (2017). A novel green synthesis of silver nanoparticles and their catalytic action in reduction of Methylene Blue dye. *Sustainable Environment Research* 27 (5): 245–250.

Samuel, M.S., Ravikumar, M., John, A.J. et al. (2022). A review on green synthesis of nanoparticles and their diverse biomedical and environmental applications. *Catalysts* 12 (5): 459. https://doi.org/10.3390/catal12050459.

Saraswathi, V.S., Kamarudheen, N., BhaskaraRao, K.V., and Santhakumar, K. (2017). Phytoremediation of dyes using *Lagerstroemia speciosa* mediated silver nanoparticles and its biofilm activity against clinical strains *Pseudomonas aeruginosa*. *Journal of Photochemistry and Photobiology B: Biology* 168: 107–116. https://doi.org/10.1016/j.jphotobiol.2017.02.004.

Saravanan, C., Rajesh, R., Kaviarasan, T. et al. (2017). Synthesis of silver nanoparticles using bacterial exopolysaccharide and its application for degradation of azo-dyes. *Biotechnology Reports* 15: 33–40.

Shafqat, U., Hussain, S., Shahzad, T. et al. (2023). Elucidating the phytotoxicity thresholds of various biosynthesized nanoparticles on physical and biochemical attributes of cotton. *Chemical and Biological Technologies in Agriculture* 10 (1): 1–15.

Singh, N.B., Susan, M.A.B.H., and Guin, M. (2021). Applications of green synthesized nanomaterials in water remediation. *Current Pharmaceutical Biotechnology* 22 (6): 733–761. https://doi.org/10.2174/1389201021666201027160029.

Singh, R.P., Yadav, P., Gupta, R.K. et al. (2023a). Pathogenic microbes in wastewater: Identification and characterization. In: *Advances in Chemical Pollution, Environmental Management and Protection*, vol. 9 (ed. L.F.R. Ferreira, A. Kumar, and M. Bilal), 247–262. Cambridge: Elsevier, Academic Press https://doi.org/10.1016/bs.apmp.2022.10.010.

Singh, S.K., Shukla, L., Singh, R.P. et al. (2023b). Microalgae mediated wastewater treatment and its production for biofuels and bioproducts. In: *Advances in Chemical Pollution, Environmental Management and Protection*, vol. 9 (ed. L.F.R. Ferreira, A. Kumar, and M. Bilal), 153–165. Cambridge: Elsevier, Academic Press https://doi.org/10.1016/bs.apmp.2022.10.007.

Stan, M., Popa, A., Toloman, D. et al. (2015). Enhanced photocatalytic degradation properties of zinc oxide nanoparticles synthesized by using plant extracts. *Materials Science in Semiconductor Processing* 39: 23–29.

Suresh, D., Nethravathi, P.C., Rajanaika, H. et al. (2015). Green synthesis of multifunctional zinc oxide (ZnO) nanoparticles using Cassia fistula plant extract and their photodegradative, antioxidant and antibacterial activities. *Materials Science in Semiconductor Processing* 31: 446–454.

Suresh, D., Shobharani, R.M., Nethravathi, P.C. et al. (2015). *Artocarpus gomezianus* aided green synthesis of ZnO nanoparticles: luminescence, photocatalytic and antioxidant properties. *Spectrochimica Acta Part A: Molecular and Biomolecular Spectroscopy* 141: 128–134.

Tang, D.Y.Y., Khoo, K.S., Chew, K.W. et al. (2020). Potential utilization of bioproducts from microalgae for the quality enhancement of natural products. *Bioresource Technology* 304: 122997. https://doi.org/10.1016/j.biortech.2020.122997.

Testa, J., Gurbisz, C., Murray, L. et al. (2010). Investigating aquatic dead zones. *The Science Teacher* 77 (2): 29.

Upadhye, V.B. and Joshi, S.S. (2012). Advances in wastewater treatment. a review. *International Journal of Chemical Sciences and Applications* 3 (2): 264–268.

Vanaja, M., Paulkumar, K., Baburaja, M. et al. (2014). Degradation of methylene blue using biologically synthesized silver nanoparticles. *Bioinorganic Chemistry and Applications* 2014: 1–8. https://doi.org/10.1155/2014/742346.

Vidya, C., Prabha, M.N.C., and Raj, M.A.L.A. (2016). Green mediated synthesis of zinc oxide nanoparticles for the photocatalytic degradation of Rose Bengal dye. *Environmental Nanotechnology, Monitoring & Management* 6: 134–138.

Watharkar, A.D., Khandare, R.V., Waghmare, P.R. et al. (2015). Treatment of textile effluent in a developed phytoreactor with immobilized bacterial augmentation and subsequent toxicity studies on *Etheostoma olmstedi* fish. *Journal of Hazardous Materials* 283: 698–704.

Yadav, K.K., Singh, J.K., Gupta, N., and Kumar, V. (2017). A review of nanobioremediatsssssion technologies for environmental cleanup: a novel biological approach. *Journal of Materials and Environmental Science* 8 (2): 740–757.

Yadav, P., Singh, R.P., Gupta, R.K. et al. (2023). Contamination of soil and food chain through wastewater application. In: *Advances in Chemical Pollution, Environmental Management and Protection*, vol. 9 (ed. L.F.R. Ferreira, A. Kumar, and M. Bilal), 109–132. Cambridge: Elsevier, Academic Press https://doi.org/10.1016/bs.apmp.2022.11.001.

Yang, H., Liu, H.-B., Tang, Z.-S. et al. (2021). Synthesis, performance, and application of molecularly imprinted membranes: a review. *Journal of Environmental Chemical Engineering* 9 (6): 106352. https://doi.org/10.1016/j.jece.2021.106352.

Zhang, F., Wang, Z., Peijnenburg, W.J.G.M., and Vijver, M.G. (2022). Review and prospects on the ecotoxicity of mixtures of nanoparticles and hybrid nanomaterials. *Environmental Science & Technology* 56 (22): 15238–15250. https://doi.org/10.1021/acs.est.2c03333.

Zinge, C. and Kandasubramanian, B. (2020). Nanocellulose based biodegradable polymers. *European Polymer Journal* 133: 109758. https://doi.org/10.1016/j.eurpolymj.2020.109758.

# 19

# Microbial Inoculants: Application in the Management of Metal Stress

*Poulomi Ghosh*[1] *and Saprativ P. Das*[2]

[1] *Department of Biotechnology, Institute of Genetic Engineering (affiliated to MAKAUT), Kolkata, West Bengal, India*
[2] *Department of Chemical Engineering, Indian Institute of Technology Bombay, Mumbai, Maharashtra, India*

## 19.1 Introduction

Metal stress, arising from the accumulation of heavy metals in the environment, poses a serious threat to ecosystems and human well-being. The rapid industrialization and improper disposal of metal-containing wastes have led to widespread contamination of soils, water bodies, and air, causing adverse effects on biodiversity, agricultural productivity, and human health. Conventional methods of remediating metal-contaminated sites often involve expensive and time-consuming processes, making them less viable for large-scale applications (Siqueira and Santos 2023). In recent years, biotechnology deploying microbial inoculants has materialized as a favorable resolution for the management of metal stress. Microbial inoculants, poised with various microbes, *viz*., bacteria, fungi, and algae, exhibit unique competencies to interact with toxic heavy metals, convert them into less harmful forms, or facilitate their removal, thereby modifying their fate in the environment. Harnessing the potential of these natural micro-organisms offers an eco-friendly and sustainable strategy to combat metal pollution (El-Nahal et al. 2022).

Microbial inoculants represent a promising frontier in the fight against metal stress (Sammauria et al. 2020). Understanding the gravity of the metal stress issue is essential for recognizing the urgency and relevance of employing novel approaches to bioremediation. This piece of work explores the usage of microbial inoculants in the management of metal stress, focusing on their mechanisms of action and potential benefits in remediating metal-contaminated environments. An overview of metal stress, discussing the sources of metal contamination, its impacts on ecosystems, and the potential risks it poses to human health, is presented. Subsequently, microbial inoculants, along with their numerous arrayed mechanisms to tolerate and detoxify heavy metals, shed light on the scientific basis of their effectiveness are being explored. The practical applications of microbial inoculants in the management of metal stress, describing their role in bioremediation, aiding in the transformation of metal contaminants, and making them less hazardous to the environment have been explored. Despite the potential benefits of microbial inoculants, several challenges hinder their widespread implementation (Siqueira and Santos 2023) Identification of these obstacles with potential strategies to overcome them has been elucidated. By delving into the mechanisms and practical applications, this literature survey aims to contribute to the growing body of knowledge in the field of environmental biotechnology and pave the way for more effective metal stress management strategies.

## 19.2 Microbial Inoculants

Microbial inoculants, also known as microbial bio-fertilizers or microbial bio-pesticides, refer to formulations that contain live beneficial microorganisms, such as bacteria, fungi, algae, or archaea. These microbes are applied to plants, soil, or other environmental substrates to enhance nutrient uptake, plant growth, and pest resistance. In the context of metal stress management, they play a crucial role in remediating metal-contaminated sites and mitigating the harmful effects of heavy metals (Singh et al. 2016) (Table 19.1).

*Microbes Based Approaches for the Management of Hazardous Contaminants*, First Edition. Edited by Ajay Kumar, Livleen Shukla, Joginder Singh, and Luiz Fernando R. Ferreira.

**Table 19.1** Array of microbial inoculants targeted toward different metal stress remediation.

| Type of heavy metal | Bacteria | Fungi | Algae | References |
|---|---|---|---|---|
| Copper | *Bacillus* sp., *Corynebacterium glutamicum*, *Desulfovibrio desulfuricans*, *Geobacter metallireducens*, *Klebsiella planticola* | *Aspergillus flavus*, *Aspergillus fumigatus*, *Paecilomyces* sp., *Cladosporium* sp., *Mucor* sp., *Trichoderma viride* | *Chlamydomonas reinhardtii*, *Chlorella vulgaris*, *Jania rubens*, *Pterocladia capillacea*, *Cladophora fascicularis* | Romera et al. (2007); El-Kassas and El-Taher (2009); Cardenas-Gonzalez et al. (2017) |
| Chromium | *Arthrobacter* sp., *Corynebacterium glutamicum*, *Pseudomonas aeruginosa*, *Bacillus licheniformis* | *Aspergillus fumigatus*, *Paecilomyces* sp., *Cladosporium* sp., *Trichoderma viride* | *Asparagopsis* sp., *Codium* sp., *Spirogyra* sp., *Chondrus* sp., *Fucus* sp., *Ascophyllum* sp., *Chlorella vulgaris*, *Scenedesmus armatus* | Sari and Tuzen (2008); El-Kassas and El-Taher (2009); Hasan and Srivastava (2009); Cardenas-Gonzalez et al. (2017) |
| Lead | *Arthrobacter* sp., *Geobacter sulfurreducens*, *Pseudomonas aeruginosa*, *Bacillus licheniformis* | *Agaricus bisporus*, *Aspergillus flavus*, *Paecilomyces* sp., *Cladosporium* sp. | *Chlorella kessleri*, *Corallina mediterranea*, *Galaxaura oblongata*, *Caulerpa lentillifera*, *Caulerpa fastigiata* | Cardenas-Gonzalez et al. (2017); El-Kassas and El-Taher (2009); Sari and Tuzen (2008) |
| Mercury | *Arthrobacter* sp., *Corynebacterium glutamicum*, *Geobacter sulfurreducens*, *Pseudomonas aeruginosa*, *Bacillus licheniformis* | *Agaricus bisporus*, *Pleurotus ostreatus*, *Trichoderma viride*, *Aspergillus* sp., *Rhizopus* sp., *Schizophyllum commune* | *Asparagopsis* sp., *Codium* sp., *Chondrus* sp., *Fucus* sp., *Ascophyllum* sp., *Ceramium virgatum* | El-Kassas and El-Taher (2009); Romera et al. (2007); Sari and Tuzen (2008) |
| Cadmium | *Arthrobacter* sp., *Bacillus* sp., *Corynebacterium glutamicum*, *Geobacter sulfurreducens* | *Agaricus bisporus*, *Pleurotus ostreatus*, *Trichoderma viride*, *Pleurotus ferulae*, *Pleurotus florida* | *Chlorella kessleri*, *Jania rubens*, *Pterocladia capillacea*, *Cladophora fascicularis* | Deng et al. (2007); El-Kassas and El-Taher (2009) |

### 19.2.1 Bacterial Inoculants

Bacterial inoculants are among the most common types of microbial bio-fertilizers used in metal stress management (Singh et al. 2016). They include various strains of beneficial bacteria, such as nitrogen-fixing bacteria (e.g., *Rhizobium*, *Azotobacter* sp.) that assist in nitrogen fixation, phosphorus-solubilizing bacteria (e.g., *Pseudomonas*, *Bacillus* sp.) that augment phosphorus availability, and plant growth-promoting rhizobacteria (PGPR) that yield phytohormones and facilitate nutrient uptake. Some bacterial inoculants also possess metal-tolerance capabilities and can promote metal detoxification in contaminated environments.

### 19.2.2 Fungal Inoculants

Fungi-based inoculants consist of beneficial fungi, including mycorrhizal fungi and *Trichoderma* sp. Mycorrhizal fungi form symbiotic relationships with plant roots, extending their reach into the soil and enhancing the plant's ability to access nutrients, including metals. These fungi can also aid in the immobilization and sequestration of heavy metals, reducing their bioavailability and toxicity. *Trichoderma* sp., on the other hand, acts as a biocontrol agent against plant pathogens, thereby contributing to overall plant health and stress tolerance (Sammauria et al. 2020).

### 19.2.3 Algal Inoculants

Algae-based inoculants are less commonly used but have demonstrated potential in metal stress management. Certain algal species possess metal-binding capabilities and can bioaccumulate heavy metals from the environment. These algal inoculants can be employed in conjunction with other bioremediation techniques to improve metal removal efficiency from contaminated sites.

### 19.2.4 Archaeal Inoculants

Archaeal inoculants, specifically halophilic archaea, have shown aptitude in metal stress management in saline and hyper-saline environments. These microorganisms can thrive in high-salt conditions and contribute to metal immobilization and reduction in hyper-saline soils, preventing metal leaching into nearby water bodies.

Each type of microbial inoculant offers unique capability and benefit, making them suitable for specific environmental conditions and metal stress scenarios. When applied properly, these microbial bio-fertilizers can significantly contribute to sustainable approaches of metal-polluted sites' management. Their ability to promote plant growth, enhance nutrient uptake, and mitigate the toxic effects of heavy metals presents a promising alternative to conventional remediation methods. However, successful application requires careful consideration of environmental factors, proper strain selection, and optimization of application techniques to achieve the desired outcomes in metal stress management (Singh et al. 2016).

## 19.3 Factors Influencing Microbial Inoculants' Efficacy

The efficacy of microbial inoculants can be influenced by several factors, which determine their ability to successfully carry out their intended functions. These factors can impact the survival, activity, and performance of the introduced microbes, affecting the overall effectiveness of the microbial inoculation (Singh et al. 2019). Certain key factors influencing the efficacy of microbial inoculants include the following:

- **Environmental conditions:** The physical and chemical characteristics of the target environment, such as moisture content, oxygen levels, temperature, pH, and nutrient availability, play a critical role in determining the success of microbial inoculation. Microbes need specific environmental supplies for growth and activity, and insurance of favorable conditions is crucial for their survival and functionality.

The efficacy of microbial inoculants, which consist of beneficial microorganisms introduced into ecosystems to confer various advantages, such as improved plant growth and soil health, is intricately shaped by a multitude of environmental factors. Temperature, a critical determinant, significantly affects microbial growth and metabolic activities. Microorganisms have specific temperature ranges at which they function optimally. Extremes in temperature, either too high or too low, can suppress their growth or even lead to their demise (Khan et al. 2023). Therefore, the success of microbial inoculants hinges on upholding temperatures within the appropriate range for the specific microbial inoculant.

Moisture levels also play a pivotal role in the efficacy of microbial inoculants. Adequate moisture is essential for microbial survival, nutrient uptake, and enzymatic processes. In environments with unsatisfactory moisture, microbial activity can be severely limited, leading to decreased efficiency of the inoculant (Xiang et al. 2022).

The pH of the environment, signifying its acidity or alkalinity, is another critical factor influencing microbial inoculant ability. Microorganisms exhibit varying pH preferences; inoculants containing specific microbial strains may only function optimally within a certain pH range (Khan et al. 2023). Therefore, understanding the pH needs of the inoculated microorganisms and ensuring compatibility with the pH of the target environment is essential for achieving the desired outcomes.

The composition of the soil and its native microbial community significantly impact the success of microbial inoculants. Different soil types possess distinct nutrient levels, organic matter content, and physical properties. These factors can influence how effectively introduced microbes establish themselves and interact with the native soil microbiota. Competition with native microorganisms for resources and space can hinder the colonization and function of the introduced microbes (Xiang et al. 2022). Furthermore, nutrient availability in the environment is crucial for microbial growth and activity. Microorganisms require a balanced supply of essential nutrients, including macronutrients *viz.*, nitrogen, phosphorus, and potassium, as well as micronutrients. In environments with limited nutrient availability, the growth and metabolic processes of inoculated microbes can be compromised, leading to reduced efficacy (Khan et al. 2023).

- **Microbial strain selection:** Choice of microbial strains is a critical factor in determining the efficacy of microbial inoculants. Different strains possess varying abilities to solubilize phosphates, fix nitrogen, degrade pollutants, and promote plant growth. Selecting the most suitable strains for the specific remediation or agricultural purpose is essential to achieve the desired outcomes (Singh et al. 2019).

Selecting strains that have a natural affinity for the plant species in question can greatly enhance the inoculant's effectiveness. The efficacy of microbial inoculants, designed to enhance soil quality and plant growth, is profoundly influenced by the careful selection of microbial strains. Microbial strain selection is a pivotal factor in determining the success of these inoculants (Xiang et al. 2022). Different strains possess unique traits and capabilities that can either complement or hinder the intended objectives of the inoculation strategy. Some microorganisms have evolved to form symbiotic relationships with certain plant species, stimulating nutrient uptake and disease resistance.

Furthermore, the metabolic capabilities of microbial strains are crucial. Some strains are adept at fixing atmospheric nitrogen and converting it into forms that plants can utilize (Xiang et al. 2022). Others excel at breaking down complex organic matter into simpler compounds, enriching the soil with nutrients. The selection of strains with appropriate metabolic traits aligns with the specific nutrient needs of the plants, optimizing their growth potential. Microbial strain compatibility also extends to the environmental conditions of the target site. By choosing strains adapted to the local environment, the likelihood of successful establishment and persistence of the inoculated microorganisms is significantly improved. In addition, microbial diversity within the inoculant can contribute to its overall efficacy. A mixture of compatible strains can provide a broader spectrum of benefits, such as enhanced disease suppression, improved nutrient cycling, and increased soil structure. This diversity can lead to more resilient and productive ecosystems.

- **Compatibility with indigenous microbiota:** The presence of indigenous microorganisms in the target environment can influence the success of microbial inoculation. Competition for resources and ecological niches among the introduced and indigenous microorganisms can impact the establishment and dominance of the inoculated strains (Khan et al. 2023).

The efficacy of microbial inoculants, designed to augment soil health and plant growth by introducing beneficial microbes, is significantly influenced by their compatibility with the indigenous microbiota of the target environment. The interactions between introduced microbial strains and the existing native microorganisms play a pivotal role in determining the success of the inoculation strategy (Khan et al. 2023). The indigenous microbiota of a specific ecosystem has evolved over time to acclimatize to local conditions, including soil type, climate, and available nutrients. When introducing new microbial strains, their compatibility with the existing community is crucial. Incompatible strains may struggle to establish themselves or could be outcompeted by native microorganisms, reducing the overall effectiveness of the inoculant. In cases where introduced strains are closely related to native species or have similar ecological functions, they may integrate more seamlessly into the existing microbiota. This can lead to a smoother transition and more effective cooperation between the introduced and native microorganisms.

- **Delivery system and application method:** The method of introducing microbial inoculants to the environment can influence their survival and activity along with distribution (Singh et al. 2019; Xiang et al. 2022). Factors such as the type of carrier material, application technique (e.g., seed coating, soil drenching), and timing of application can affect the inoculant's efficacy.

The effectiveness of microbial inoculants, designed to improve soil quality and foster plant growth through beneficial microorganisms, is significantly impacted by the chosen delivery system and application method. The delivery system refers to the physical form in which the microbial inoculants are provided, such as liquid suspensions, granules, powders, or carrier-based formulations. Each delivery system has its own advantages and limitations. Liquid formulations are typically easy to distribute and can provide direct contact with soil and plant surfaces. Granules and powders, on the other hand, offer better protection and a longer shelf life for the microorganisms.

Equally important is the application method, which involves how the microbial inoculants are introduced into the soil or onto plant surfaces. Common methods include seed coating, soil drenching, foliar spraying, and root dipping. Seed coating ensures direct contact between the microorganisms and germinating seeds, giving the inoculants a head start in establishing a symbiotic relationship. Soil drenching and root dipping methods target the root zone, allowing the microorganisms to interact with the root system and influence nutrient uptake and disease resistance. Foliar spraying applies microorganisms directly to the plant's aerial parts, potentially aiding in disease prevention and enhancing nutrient absorption through the leaves (Khan et al. 2023). Balancing the desire to introduce new beneficial microorganisms with the need to maintain the stability of the existing ecosystem is a critical consideration. An excessively dominant introduction of non-native strains might upset the ecological balance, potentially leading to unintended consequences such as shifts in nutrient cycling, disease dynamics, or overall ecosystem functions.

- **Inoculant viability and shelflife:** The viability and shelflife of microbial inoculants are critical considerations for their successful application (Xiang et al. 2022). Ensuring that the inoculant remains viable and active during storage and transportation is crucial to achieve utmost efficacy upon application.

The feasibility of the microorganisms is affected by various factors, together with the formulation of the inoculant, the deployed specific strains, and the storage conditions. Formulations that provide protection and nutrients for the microorganisms can improve their survival rates. Additionally, the choice of microbial strains matters; some strains are naturally more resilient and capable of withstanding adverse conditions, while others might be more sensitive.

Shelf life, which refers to the period during which an inoculant remains viable and effective, is closely related to viability. Extending the shelf life of an inoculant can provide flexibility in terms of application timing and reduce wastage (Xiang et al. 2022). Encapsulation of specific microbes in protective matrices or spore forms can help preserve their viability for longer periods. Significantly observing the viability and shelflife of microbial inoculants can vary between individual microorganisms and among different products. Appropriate storage practices, including refrigeration and protection from extreme conditions, are essential to conserving the viability of the inoculant until its application.

- **Target interactions:** The ability of microbial inoculants to interact with and degrade specific contaminants or form beneficial symbiotic relationships with plants affect their overall remediation or agricultural potential. Understanding these communications is crucial for optimizing the inoculation strategy along with the specific targets'interactions they are intended to benefit (Khan et al. 2023).

These target interactions distinguish between a range of complex relationships among the introduced microorganisms, the host plants, and the surrounding environment. One key aspect of target interactions is the compatibility between the introduced microorganisms and the host plants. Microorganisms that establish symbiotic relationships with specific plants, such as mycorrhizal fungi aiding in nutrient uptake or nitrogen-fixing bacteria enhancing plant nitrogen availability, demonstrate compatibility and can significantly contribute to the host plant's health and growth. Additionally, the physiological and ecological characteristics of the target plants play a role in the efficiency of microbial inoculants. Some plants have structures or mechanisms that facilitate the colonization and establishment of introduced microbes, while others might have traits that hinder such interactions (Singh et al. 2019).

Environmental conditions also shape target interactions. The physical and chemical properties of the soil, as well as climatic factors, influence the formation and enactment of introduced microbes. Microorganisms that are well-suited to the prevailing conditions can establish themselves more effectively and contribute positively to the ecosystem. Interactions with the native microbiota are another crucial consideration (Xiang et al. 2022). Beneficial microorganisms from the inoculant must compete with or cooperate with existing soil microbes. Some introduced microorganisms might enhance the existing microbial community's functions, while others may face challenges in establishing themselves due to competition or antagonistic interactions, which in turn benefits plant health (Singh et al. 2019).

- **Competitors:** The introduced microbial inoculants may face competition from other microbes present in the environment. Understanding the microbial ecology of the site and managing potential competitors can enhance the success of the inoculation.

Competitors refer to native microbes potentially present in the soil or on plant surfaces that might interact with the introduced microorganisms, either positively or negatively. Native microorganisms can either outcompete with or complement the introduced microorganisms, shaping their success and effectiveness. Competition for resources such as nutrients, space, and energy can limit the establishment and growth of introduced microorganisms (Xiang et al. 2022).

However, not all interactions with the competitors are detrimental. In some cases, introduced microorganisms can form synergistic relationships with native microbes. These synergies might involve cooperation in nutrient cycling, disease suppression, or other ecological processes.

The extent of competition or cooperation depends on various factors. The specific traits of the introduced and native microorganisms, such as their metabolic capabilities, resource utilization strategies, and growth rates, influence the nature of their interactions along with their dynamics. Strategies to enhance the efficiency of microbial inoculants often involve considering these competitor interactions. Additionally, providing the right conditions, such as optimizing nutrient availability or minimizing stress factors, can tip the balance in favor of the introduced microorganisms (Khan et al. 2023).

- **Regulatory and ethical considerations:** Compliance with regulatory guidelines and ethical considerations surrounding the use of genetically modified organisms (GMOs) or genetically engineered strains may influence the selection and deployment of microbial inoculants (Xiang et al. 2022).

The potential for genetically modified organisms (GMOs) in microbial inoculants adds another layer of complexity to the regulatory and ethical landscape. Regulatory considerations involve compliance with governmental regulations, guidelines, and protocols that oversee the release and application of microbial products. Regulatory agencies evaluate the safety, efficacy, and potential environmental impacts of microbial inoculants before granting approval for commercial use. Stringent analysis and documentation are required to ensure that these products do not pose risks to human health, ecosystems, or nontarget organisms. Adhering to regulatory standards is essential to gain market access and ensure that microbial inoculants meet established safety criteria (Xiang et al. 2022).

Ethical considerations encompass a range of factors, including potential risks, unintended consequences, and an equitable distribution of benefits. There may be ethical concerns related to the preface of GMOs, the eventuality of ecological dislocation, and the impact on biodiversity. Ethical discussions also include questions about transparency in labeling, informed consent from stakeholders, and the broader societal implications of using microbial inoculants. Debates focus on issues similar as the eventuality of unintended consequences, the rights of unborn generations, and the balance between invention and palladium. Transparent communication between scientists, policymakers, assiduity stakeholders, and the public is vital to address enterprises, build trust, and form informed opinions.

To ensure the efficacy of microbial inoculants, thorough research, observation, and alteration of the inoculation approach based on site-specific conditions are indispensable. Consideration and addressing of these influencing factors can optimize microbial inoculants to achieve their intended goals in bioremediation and other environmental applications.

## 19.4 Sources of Heavy Metals

Heavy metals can originate from both natural and anthropogenic sources.

### 19.4.1 Natural Sources

The Earth's geology is one of the main sources of heavy metals due to its composition. Heavy metals are naturally occurring elements of the earth's crust. They exist in varied concentrations in soils, rocks, minerals, and water bodies, with natural geological and environmental processes contributing to their release into the environment. Over time, these rocks undergo weathering and erosion, releasing heavy metal ions into the soil and water. Geological activities like volcanic eruptions and earthquakes can expose previously trapped heavy metals, contributing to their dispersion in the environment (Vardhan et al. 2019).

Weathering and erosion processes, including physical, chemical, and biological weathering, break down rocks and minerals containing heavy metals. These release metal ions into soils and sediments. Additionally, water percolating through rocks can dissolve and transport heavy metals, leading to their presence in groundwater and surface water bodies (Arora and Khosla 2022).

Furthermore, volcanic activity is a significant natural source of heavy metals. Volcanic emissions release heavy metal-laden gases and particles into the atmosphere. These materials eventually reach the earth's surface and settle, where they may be incorporated into soils and water bodies, which contribute to the natural distribution of heavy metals. Additionally, certain geological formations, such as mineral deposits or ore bodies, contain inherently high levels of heavy metals. Certain geological formations, such as mineral deposits and ore bodies, contain concentrated amounts of heavy metals (Dhaliwal et al. 2020).

Mining and erosion of these formations can release heavy metals into the environment. However, while these processes have been going on for millions of years; human activities, particularly industrialization and mining, have accelerated the release of heavy metals above natural levels leading to environmental pollution.

### 19.4.2 Anthropogenic Sources

Anthropogenic sources refer to human activities that release heavy metals into the environment at levels significantly higher than those caused by natural processes (Alengebawy et al. 2021). These sources play an important role in the increased presence of heavy metals in soil, water, and air. Industrial processes are among the foremost anthropogenic sources of heavy metals. Activities such as mining, smelting, manufacturing, and metal production release large amounts of heavy metals into the environment (Ali et al. 2019). These metals can be released as particles, gases, or effluents. Mining metals such as lead, mercury, and cadmium from ores can spread these metals into surrounding ecosystems, leading to soil and water contamination (Vardhan et al. 2019).

Agricultural practices also contribute to heavy metal contamination. The application of fertilizers, pesticides, and sewage sludge that may contain elevated levels of heavy metals can result in soil accumulation (Okereafor et al. 2020). These metals can then enter the food chain through crops and affect human health. Waste disposal and improper management are significant sources of heavy metal pollution. Improper disposal of hazardous materials can release heavy metals like lead, mercury, and cadmium into the environment. Landfills, if not properly lined or managed, can leach heavy metals into groundwater, contaminating nearby water sources (Okereafor et al. 2020). Traffic and vehicular emissions are also sources of heavy metals in urban environments. When fossil fuels burn, heavy metals, such as lead, are released into the atmosphere as particulate matter. These particles can settle onto soils and water bodies, affecting both terrestrial and aquatic ecosystems (Arora and Khosla 2022). In addition, construction activities, deforestation, and the use of fossil fuels for energy also contribute to heavy metal emissions. These activities release metal-containing particles and gases into the atmosphere, which can then be transported over long distances before deposition.

### 19.4.3 Electronic Waste (*e-Waste*) Generating Heavy Metals

Electronic waste, commonly known as "*e-waste,*" is a significant contributor to heavy metal pollution. *E-waste* comprises discarded electronic and electrical equipment, such as computers, smartphones, televisions, refrigerators, and other consumer electronics. When improperly managed, e-waste can release toxic heavy metals into the environment.

Some of the major heavy metals and their potential impacts are presented in Table 19.2.

Heavy metals are elements with high atomic weights and densities, including metals like lead (Pb), mercury (Hg), cadmium (Cd), and arsenic (As).

**Table 19.2** Acute effects of different heavy metals leading to soil contamination and health issues.

| Type of heavy metal | Sources of exposure | Effects | Permissible limits (mg/l) | References |
|---|---|---|---|---|
| Cadmium | Stabilizers, Electroplating, Paints, Fertilizers | Acute: Pneumonitis,<br>Chronic: Proteinuria, Osteomalcia, Lung cancer | 0.06 | Alengebawy et al. (2021); Nazir et al. (2015) |
| Arsenic | Ore mining, Pesticides, Fungicides, Smelting, Wood reserves | Acute: Neuropathy, Encephalopathy, Nausea, Diarrhea,<br>Chronic: Diabetes, Hypopigmentation/ Hyperkeratosis, Cancer in skin, lungs and bladder. | 0.01 | Rahman and Singh (2019) |
| Chromium | Steel industries, Tanneries, Fly Ash | Acute: Renal failure, Gastro-intestinal Hemorrhage, Hemolysis<br>Chronic: Pulmonary Fibrosis, Lung cancer | 0.05 | Chen et al. (2009); Nazir et al. (2015) |
| Mercury | Medical waste, Coal combustion, Au-Ag mining | Acute: Elemental-Fever, Vomiting, Diarrhea, Acute Lung Injury<br>Chronic: Neurasthenia, Nephritic syndrome, Hypersentivity | 0.01 | Romera et al. (2007); Sari and Tuzen (2008) |
| Lead | Herbicides, Battery wastes, Insecticides,<br>Leaded Fuel | Acute: Vomiting, Diarrhea, Abdominal pain<br>Chronic: Osteoporosis, Anemia, Copper deficiency, Neurologic degeneration | 0.1 | Lakherwal (2014) |
| Copper | Ore refinery, Pesticides, Fertilizers | Acute: Gastro-intestinal irritation/ Hemorrhage,<br>Blue Vomiting, Multi-organ dysfunction<br>Chronic: Wilson disease, Insomnia, Rheumatoid arthritis, Postpartum psychosis, Autism | 0.1 | Cardenas-Gonzalez et al. (2017) |
| Nickel | Automobiles, Batteries, Surgical instruments, Kitchen appliances | Acute: Myocarditis, Encephalopathy, Dermatitis<br>Chronic: Pulmonary fibrosis, Nasopharyngeal tumors, Oligospermia | 0.2 | Nazir et al. (2015) |

Circuit boards often contain lead (Pb) in solder joints. Lead solder was traditionally used to connect components on these boards due to its conductivity and ease of use. When e-waste is improperly disposed of or processed, these circuit boards can break down, releasing lead particles into the environment (Houessionon et al. 2021). This can lead to soil and water contamination, with potential health risks to both humans and ecosystems.

Another heavy metal commonly found in e-waste is mercury (Hg). Compact fluorescent lamps (CFLs) and certain types of batteries, often found in electronic devices, contain mercury. When these items are discarded and not properly managed, mercury can be released into the environment, posing risks to aquatic systems and entering the food chain through bioaccumulation (Rahman and Singh 2019).

Cadmium (Cd) is yet another heavy metal of concern in *e-waste*. Cd is used in rechargeable batteries and semiconductors, which are prevalent in various electronic devices. When e-waste is incinerated or disposed of in landfills, cadmium can leach into soils and water bodies, negatively affecting the environment and potentially causing health problems in humans (Saha et al. 2021).

Moreover, certain electronic components are coated with chromium (Cr), which can be released into the environment if not managed properly. Chromium can have detrimental effects on both human health and aquatic ecosystems.

The improper recycling and disposal of e-waste, often practiced in informal recycling operations in some regions, can exacerbate the release of heavy metals. Burning *e-waste* to recover valuable materials can release heavy metal-containing particulates into the air, contributing to air pollution. Leaching of heavy metals from landfills where *e-waste* is improperly disposed of can contaminate groundwater, affecting local water sources (Arora and Khosla 2022).

Many countries have implemented regulations and guidelines for the proper handling and recycling of *e-waste* to address the growing concerns related to heavy metal pollution from electronic waste. It is important to note that heavy metal contamination can occur as a result of the interaction of these sources (Alengebawy et al. 2021). Once heavy metals enter the environment, they can persist for long periods and pose significant risks to soil, plants, wildlife, human health, and the entire ecosystem. Proper waste management, pollution control measures, and the adoption of cleaner technologies are essential to minimize heavy metal pollution and protect the environment (Alengebawy et al. 2021).

## 19.5 Effects of Heavy Metals

Heavy metals, even at low concentrations, can have noteworthy hostile effects on plants, microorganisms, and human health (Figure 19.1). Their toxic nature stems from their ability to accumulate in living organisms and disrupt essential physiological processes.

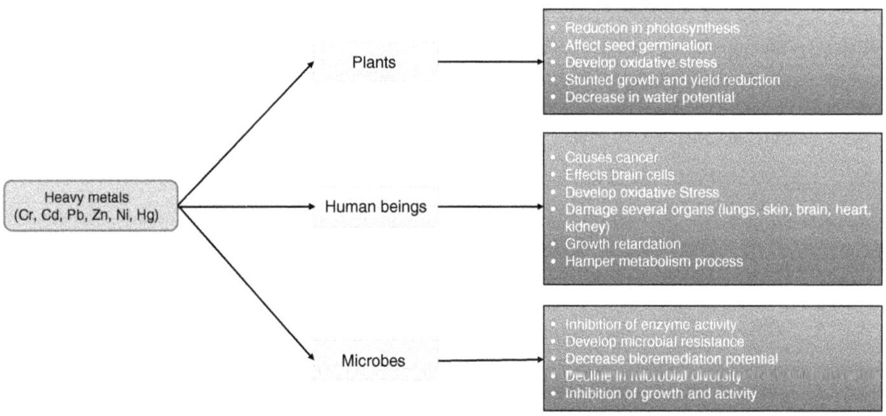


**Figure 19.1** Effect of heavy metals on ecosystem. Heavy metals disturb plants, causing growth inhibition, nutrient imbalance, oxidative stress, and root damage, leading to accumulation in tissues and affecting the overall food chain. In human beings, gathering of heavy metals results in neurological, respiratory, carcinogenic, renal, hepatic, reproductive, and developmental issues. Drastic consequences of heavy metals on microbes include the hindering of their growth and activity along with their communities' modification, resulting in a lessened prospect for bioremediation coupled with the release of harmful metabolites.

### 19.5.1 Plants

Heavy metals can have profound effects on plants, impacting their growth, development, and overall health. When absorbed by plant roots, heavy metals can disrupt essential physiological processes, *viz.*, heavy metals can interfere with nutrient uptake, as they can be mistaken for essential minerals and transported into the plant's vascular system. This can lead to nutrient imbalances and deficiencies. Additionally, heavy metals can inhibit enzyme activities involved in prime biochemical pathways, disrupting metabolic processes like photosynthesis and respiration. Accumulation of heavy metals in plant tissues can also result in oxidative stress, generating reactive oxygen species (ROS) that damage cell structures and impair plant growth (Alengebawy et al. 2021).

Furthermore, heavy metals can alter the soil's microbial community, affecting nutrient cycling and the availability of beneficial microbes. The visible effects of heavy metal toxicity in plants include stunted growth, chlorosis (yellowing of leaves), necrosis (cell death), and reduced reproductive output. The impacts of heavy metals can extend to higher trophic levels, as heavy metal-contaminated plants can affect herbivores and organisms in the ecological food chain (Alengebawy et al. 2021; Diaconu et al. 2021).

Heavy metals can undergo bioaccumulation within plant tissues. As plants take up heavy metals from the soil, these metals can accumulate in plant cells and tissues over time. This accumulation not only affects the plants themselves but also poses risks to organisms higher up the food chain that consume these contaminated plants (El-Meihy et al. 2019). This phenomenon, known as biomagnification, can lead to the transfer of heavy metals to herbivores, predators, and ultimately to humans, potentially causing health issues.

The presence of heavy metals in the environment can lead to a range of detrimental effects on plants. Heavy metals can interfere with cell division, elongation, and differentiation, leading to stunted root and shoot growth (Diaconu et al. 2021).

Moreover, heavy metals can disrupt nutrient uptake and assimilation processes, causing nutrient imbalances. Some heavy metals can compete with essential minerals for uptake, causing deficiencies in vital nutrients like calcium, magnesium, and iron, which are necessary for various physiological functions. This disruption in nutrient balance further compromises plant health (El-Meihy et al. 2019).

Another adverse impact of heavy metals is the induction of oxidative stress. Heavy metals can trigger the production of ROS within plant cells. The excessive accumulation of ROS can damage cell membranes, proteins, and DNA, resulting in celldeath and tissue damage. This oxidative stress hampers the plant's ability to perform crucial metabolic processes and impairs overall growth and development (Alengebawy et al. 2021).

### 19.5.2 Microorganisms

Heavy metals are known to have profound effects on microbes, exerting both direct and indirect influences on their growth, behavior, and ecosystem functions. These metals, including lead, mercury, cadmium, chromium, and arsenic, are released into the environment primarily through industrial processes, mining activities, and improper management (Raklami et al. 2022). Their presence in ecosystems can lead to various detrimental impacts on microbes. The growth of microorganisms is significantly affected by heavy metal exposure. High concentrations of heavy metals can disrupt key cellular processes such as DNA replication, transcription, translation, and enzyme activity. Metal ions can replace essential ions in enzymatic reactions, leading to decreased enzyme efficiency or complete inactivation (Raklami et al. 2021).

Consequently, these disruptions can impair cellular metabolism, hinder energy production, and ultimately inhibit microbial growth. One critical consequence of heavy metal contamination is the diminished bioremediation potential of microbial communities. Bioremediation relies on the metabolic capabilities of microbes to degrade or transform pollutants, thereby aiding in the clean-up of contaminated environments (Thakare et al. 2021). However, heavy metals can directly inhibit the enzymatic activities involved in bioremediation processes. Microorganisms adapted for bioremediation may exhibit reduced efficiency or even complete inability to detoxify and degrade pollutants in the presence of heavy metals. This not only slows down the remediation process but also limits the effectiveness of using microbial communities as a means of environmental restoration (Thakare et al. 2021).

Furthermore, heavy metal exposure can lead to alterations in microbial communities themselves. Some microbes have developed mechanisms to tolerate or resist heavy metals through various strategies, such as metal efflux pumps, metal-binding proteins through bioaccumulation. As a result, heavy metal-tolerant species might outcompete sensitive species in contaminated environments, leading to a shift in microbial community composition. Such changes can disrupt the delicate balance of ecosystem functions and nutrient cycling carried out by microorganisms, potentially cascading through the entire food web (Sun et al. 2020; Thakare et al. 2021).

In addition to altering microbial communities, heavy metals can also trigger the discharge of toxic metabolites. As microorganisms struggle to cope with heavy metal stress, they may divert resources away from normal metabolic pathways (Sun et al. 2020). This redirection can lead to the accumulation and release of harmful by-products, which may include ROS and other toxic compounds (Alengebawy et al. 2021). These metabolites can contribute to oxidative stress, DNA damage, and cell death in nearby organisms, including higher trophic levels, thus further disrupting ecosystem health.

### 19.5.3 Human Health

The effects of heavy metals on microorganisms extend beyond environmental impacts and can also have significant implications for human health. Heavy metals, such as lead, mercury, cadmium, and arsenic, can accumulate in the food chain and reach humans through the consumption of contaminated water, crops, and animals. Once ingested, these metals can have detrimental effects on human health due to their ability to disrupt cellular processes and interact with biomolecules (Alengebawy et al. 2021).

When heavy metals interact with microorganisms in the environment, they can be transformed into various chemical forms, some of which are more easily absorbed by humans. For instance, methylmercury is a highly toxic form of mercury that can bioaccumulate in aquatic food chains and pose a serious risk to human health when consumed through contaminated fish and seafood (Bala et al. 2022). Similarly, cadmium can accumulate in crops and plants, making its way into the human diet and potentially causing adverse health effects.

In the context of human health, heavy metals can disrupt essential biological processes. These metals can interfere with enzyme activities, alter protein structures, and affect cellular signaling pathways. Lead, is known to disrupt neurodevelopment in children, leading to cognitive and behavioral impairments. It can replace calcium ions in biological processes, affecting neurotransmission and neuronal function. Chronic exposure to lead can result in learning disabilities, decreased IQ, and behavioral problems (Mitra et al. 2022).

Mercury is another heavy metal with well-documented human health effects. Methylmercury exposure, primarily through the consumption of contaminated fish, can lead to neurological and developmental disorders in fetuses and infants. It can damage the central nervous system, impair cognitive function, and cause motor control issues. Additionally, long-term exposure to cadmium has been associated with kidney damage, bone density reduction, and an increased risk of certain cancers, such as lung and prostate cancer (Mishra et al. 2019).

Arsenic, often found in drinking water in some regions, is a carcinogenic heavy metal that can lead to skin, lung, and bladder cancer (Rahman and Singh 2019). It interferes with DNA repair mechanisms and disrupts cellular pathways, contributing to the development of cancerous cells. Chronic exposure to arsenic can also lead to various non-cancer health issues, including cardiovascular diseases, diabetes, and skin lesions. Heavy metal pollution poses significant ecological and public health concerns (Bala et al. 2022). Diminishing heavy metal contamination through appropriate waste management is vital to safeguard the environment.

## 19.6 Microbial Mechanisms of Metal Tolerance and Remediation

Bioremediation involves harnessing the natural abilities of microbes (bacteria, fungi, and algae) to shrink, degrade, transform, or immobilize the heavy metal content in contaminated sites, making it an ecological approach to environmental clean-up. However, it is vital to judiciously achieve bioremediation processes that avert unintentional consequences, such as the discharge of mobilized heavy metals into diverse environments. The mechanisms employed by microbial inoculants to tolerate and detoxify heavy metals are critical to their effectiveness (Singh et al. 2016). These metal-metabolizing microbes, often referred to as metal-resistant or metal-tolerant microorganisms, have evolved specialized enzymes and cellular processes to deal with heavy metal pollutants. This section explores various strategies employed by microorganisms to withstand metal stress, including biosorption (binding and accumulation), bioaccumulation (active uptake and storage), and enzymatic transformations. This approach can be applied *insitu* (on-site) or *exsitu* (off-site), depending on the specific contamination scenario.

### 19.6.1 Biosorption

Biosorption is a passive process where microbes bind and accumulate heavy metal ions on their cell surfaces or within their cell walls. This mechanism is mainly facilitated by functional groups on the microbial cell, such as carboxyl, amino, and phosphate groups, which have a high affinity for metal ions (Yaashikaa et al. 2022). The metal ions are effectively sequestered and immobilized, reducing their bioavailability and toxicity in the surrounding environment. Bacterial cell walls and extracellular polymeric substances (EPS) produced by certain microbes are particularly instrumental in biosorption.

### 19.6.2 Bioaccumulation

Bioaccumulation involves the active uptake and accumulation of heavy metals by microbes within their cellular structures. In this process, metal ions are transported through the microbial cell membrane by specific transporter proteins. The accumulated metals can then be stored intracellularly, usually in vacuoles or specialized compartments, minimizing their harmful effects on cellular functions. Certain metal-accumulating bacteria and fungi, such as *Pseudomonas* and *Saccharomyces* sp., have been extensively studied for their bioaccumulation potential (Kumar et al. 2016). *Microbacterium esteraromaticum* harbors the gene "*mepA*," which encodes for a metal-binding protein. This protein facilitates the sequestration and accumulation of heavy metals such as lead and cadmium within the microbial cells through phytostabilization, thereby reducing the bioavailability and potential toxicity of these metals to other organisms in the ecosystem (Mushtaq 2021).

### 19.6.3 Biomineralization

Biomineralization is a unique mechanism employed by some microorganisms to convert soluble metal ions into insoluble mineral forms. Certain microorganisms can induce the precipitation of heavy metals inside (intracellular precipitation) or outside their cells (extracellular precipitation). They do this by altering the pH or redox conditions in their immediate environment, causing the metal ions to precipitate in less toxic or insoluble forms. Some microbes may accumulate heavy metals within their cells, sequestering the metals in specific organelles or vacuoles. This compartmentalization helps shield the rest of the cell from the toxic effects of heavy metals. The microorganisms can produce metal-specific precipitates, such as metal sulfides, carbonates, or oxalates, which are less toxic and less mobile than their soluble counterparts (Malik et al. 2019). This process not only detoxifies the environment but also contributes to the formation of stable metal precipitates, potentially aiding in long-term metal immobilization.

### 19.6.4 Enzymatic Transformations

Certain microorganisms possess metal-transforming enzymes that facilitate the conversion of toxic metal ions into less harmful forms. For instance, some bacteria and fungi produce specific metal detoxification enzymes and metal-reducing enzymes, such as metallothioneins (MTs) and superoxide dismutase, which can mitigate the toxic effects of ROS generated by metal stress. MTs are a group of minor, cysteine-rich proteins originating in a wide range of organisms, including bacteria, fungi, plants, and animals, including humans. These proteins have a high affinity for binding heavy metal ions and play a crucial role in metal homeostasis and detoxification, which can bind and sequester heavy metals, preventing their harmful effects on cellular components.

Additionally, metal-transforming enzymes can participate in redox reactions, leading to the reduction or oxidation of metal ions to less toxic or less mobile states. Some microorganisms can transform the oxidation state of heavy metals through reduction or oxidation reactions. For example, certain bacteria can transform toxic forms of chromium (Cr(VI)) into less toxic and less mobile forms like chromium (Cr(III)) (Singh et al. 2016).

### 19.6.5 Chelation and Complexation

Microorganisms can synthesize and secrete organic molecules, such as siderophores, exopolysaccharides, and organic acids, that have a high affinity for metal ions. These molecules act as chelators or metal-binding ligands, forming stable complexes with the metal ions, thereby reducing their toxicity and mobility. The metal-complexed compounds can also serve as a source of essential nutrients for the microorganisms, contributing to their growth and proliferation in metal-contaminated environments (Yaashikaa et al. 2022). Some microbes have the ability to methylate certain heavy metals, like mercury (Hg), through methylation. Methylated forms of heavy metals are often less toxic but can become more mobile and may enter the food chain. Additionally, certain microbes can facilitate the volatilization of certain heavy metals into the atmosphere, reducing their impact on the soil or water.

Thus, by employing each mechanism or mixed mechanisms, microbial inoculants can alter the heavy metals in metal-polluted surroundings, either by sequestering them in their cellular structures, converting them into less toxic forms, or facilitating their immobilization (Malik et al. 2019). However, successful applications in real-world scenarios require a thorough understanding of the specific microbial strains and their metal tolerance mechanisms, as well as careful consideration of environmental factors that may influence their effectiveness.

## 19.7 Other Remediation Approaches

As heavy metal contamination continues to be a significant environmental concern, various approaches for remediation have been developed and implemented to mitigate their adverse effects (Zhu et al. 2019). These remediation strategies aim to either remove or immobilize heavy metals from the environment, reducing their bioavailability and potential for harm (Kapahi and Sachdeva 2019) (Figure 19.2). Certain current approaches include:

### 19.7.1 Phytoremediation

Phytoremediation is a plant-based approach that utilizes specific plant species to uptake, accumulate, and sequester heavy metals from contaminated soil and water. Certain plants, known as hyperaccumulators, have the natural ability to absorb and tolerate high concentrations of heavy metals. Phytoremediation can be applied through techniques, *viz*., phytoextraction (where metal-enriched plants are harvested and disposed of), phytostabilization (reducing metal mobility in soil through root uptake and immobilization), and phytovolatilization (plants volatilize metals, which can then be collected). The gene "*zntA*," found in metal-resistant bacteria like *Pseudomonas fluorescens*, encodes for a zinc transporter protein that helps in the internal sequestration of zinc ions, promoting phytostabilization and preventing zinc toxicity in the soil environment. This method is eco-friendly and can be used in combination with other remediation strategies (Chugh et al. 2022; Saravanan et al. 2023).

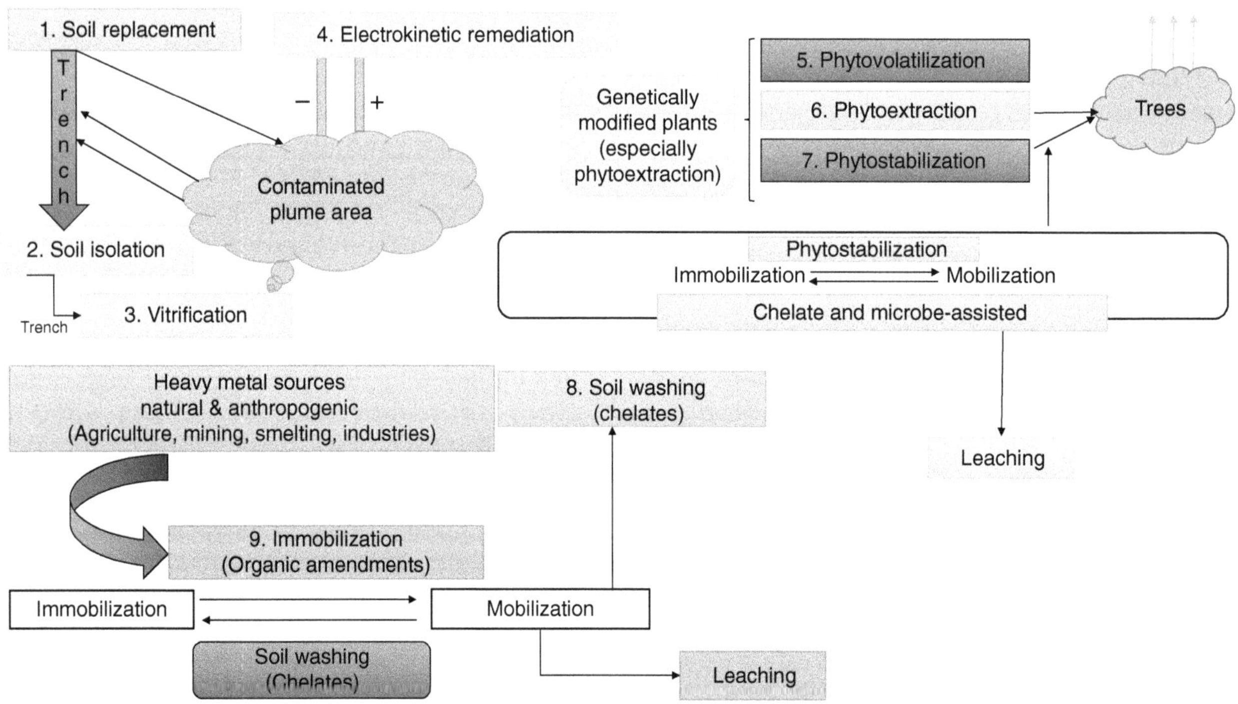


**Figure 19.2** Diverse approaches to heavy metals' remediation in contaminated sites. Remediation approaches intent in immobilizing heavy metals from the atmosphere, dropping bioavailability and destruction probability. Vitrification involves a thermal process of destructing heavy metal waste present in contaminated soils. Electrokinetic remediation exploits the electrical fields in the soil for transportation and consequent elimination of metal ions. Phytoextraction, phytostabilization, and phytovolatilization are the various application techniques of phytoremediation involving specific plant species in heavy metal removal. Soil washing is a reversible physical-chemical process encompassing the application of specific solvents to the contaminated soil to eradicate heavy metals.

### 19.7.2 Rhizoremediation

In the context of rhizoremediation, microbes interact with plant roots to enhance metal uptake and tolerance. For example, the gene "*acdS*" in *Pseudomonas putida* codes for an enzyme called 1-aminocyclopropane-1-carboxylate (ACC) deaminase, which facilitates plant growth and stress tolerance by dropping the levels of stress-induced ethylene in the plant, aiding in the plant's ability to cope with metal stress.

### 19.7.3 Chemical Precipitation

Chemical precipitation involves the addition of chemicals that can react with heavy metal ions in contaminated water, forming insoluble precipitates (Chugh et al. 2022). The formed precipitates can then be easily separated from the water. Lime (calcium hydroxide) and other metal hydroxides are commonly used for this purpose. Chemical precipitation is effective in removing heavy metals from water bodies but requires careful monitoring to prevent secondary pollution.

### 19.7.4 Ion Exchange

Ion exchange involves the use of ion exchange resins or materials that can selectively bind heavy metal ions and release other nontoxic ions in exchange. This method is commonly used for treating industrial wastewater containing heavy metals.

### 19.7.5 Electrokinetic Remediation

Electrokinetic remediation utilizes electrical fields to transport heavy metal ions through the soil toward an electrode for subsequent removal. This method is particularly useful in clayed soils, where traditional methods may have limited efficiency.

### 19.7.6 Soil Washing

Soil washing is a physical-chemical process that involves the washing of contaminated soil with specific solvents or surfactants to remove heavy metals. The extracted metals can be recovered and treated further, while the cleaned soil can be returned to the site.

### 19.7.7 Nanotechnology

Nanotechnology-based approaches involve the use of nanoparticles that can selectively adsorb heavy metals. Nanoparticles, such as zero-valent iron (nZVI) and carbon nanotubes, have shown promising outcomes in the removal of heavy metals from water and soil.

### 19.7.8 Constructed Wetlands

Constructed wetlands utilize natural processes in wetland ecosystems to remove heavy metals from water. Wetland plants and microorganisms can aid in the uptake and transformation of heavy metals, while sediments act as sinks for metal removal.

It is important to note that the choice of remediation approach depends on the specific characteristics of the contaminated site, the type of heavy metals present, and the desired remediation goals. Often, a combination of multiple techniques (a hybrid approach) may be more effective in achieving comprehensive heavy metal remediation. Additionally, ongoing research and development are focused toward the improvement of the existing techniques and explore innovative solutions for the management of heavy metal pollution (Chugh et al. 2022).

## 19.8 Metal Remediation in Co-contaminated Soils

Metal remediation in co-contaminated soils is an ongoing area of research that presents both challenges and opportunities. Recent key findings have shed light on the complex interactions between different contaminants in co-contaminated soils. Studies have exposed synergistic effects where the presence of certain organic pollutants can influence the behavior of heavy metals in soils and *vice-versa*. Certain co-contaminants can enhance or inhibit metal mobility, bioavailability, and transformation, leading to intricate interactions that disturb the efficacy of remediation strategies. Additionally, research has explored the practice of biochar amendments as a promising soil treatment option for co-contaminated sites. Biochar, a carbon-rich material created from the pyrolysis of organic matter, has presented the capability to effectually adsorb both heavy metals and organic pollutants, thereby dropping their movement and bioavailability in soils (Sharma et al. 2021).

However, several challenges exist in metal remediation in co-contaminated soils. The complexity of co-contaminated sites is a major hurdle. The presence of multiple pollutants with variable behaviors and remediation techniques necessitates tailored and site-specific approaches. Moreover, contaminant interactions can be unpredictable, and in some cases, co-contaminants may hinder or inhibit the usefulness of traditional remediation approaches. Careful assessment and understanding of the specific interactions are crucial to design prosperous remediation strategies. Despite these challenges, the potential viability of metal remediation in co-contaminated soils is promising with the right approach. Integrated remediation strategies that combine physical, chemical, and biological methods have demonstrated the potential in efficiently addressing co-contaminated sites. For instance, a combination of phytoremediation for metal uptake with bioremediation for organic pollutant degradation can offer a synergistic approach (Sharma et al. 2021). Advances in nanotechnology have also opened up likelihoods for using nanomaterials to concurrently aim heavy metals and organic pollutants. Moreover, understanding the dynamics of the microbial community in co-contaminated soils is acute, as microbes play a key part in the transformation and contaminants' breakdown.

Various microbes are cast off for metal remediation in co-contaminated soils. Metal-resistant bacteria are frequently engaged for their ability to endure and gather heavy metals, binding and sequestering them to lessen their bioavailability. These microbial inoculants encompass a diverse array of microbial species with specialized genes and activities. *Cupriavidus metallidurans* has been extensively studied for its metal-tolerant properties (O'Callaghan et al. 2022). This bacterium possesses a unique genomic feature, the chromosomal cluster "*cnr*," which encodes for the "*Cnr*" system. The "*Cnr*" system includes an efflux pump capable of expelling toxic metal ions from the bacterial cell, reducing intracellular metal concentrations, and preventing metal-induced cytotoxicity. Metal-transforming microbes also play a pivotal role in co-contaminated soil management. Notably, *Stenotrophomonas maltophilia* has been found to possess the gene "*nfsA*," encoding for a nitro-reductase enzyme. This enzyme is crucial in the reduction of toxic heavy metal ions like chromium (VI) to their less toxic counterparts, such as chromium (III), thereby mitigating the harmful effects of hexavalent chromium in the environment. The collective efforts of these microbial inoculants pave the way for innovative and sustainable solutions to tackle the challenges posed by heavy metal contamination and its detrimental impacts on ecosystems and human health (Singh et al. 2019).

Mycorrhizal fungi that form favorable associations with plant roots can benefit in the uptake and translocation of heavy metals, paying to phytoremediation efforts. Certain algal species and cyanobacteria can also store heavy metals and organic pollutants, making them suitable candidates for bioremediation applications. However, the selection of microorganisms should be based on factors such as environmental conditions, nutrient availability, and the presence of inhibitory substances in the soil (Sharma et al. 2023). As research in this field progresses, the development of pioneering and competent strategies for metal remediation in co-contaminated soils holds assurance for addressing intricate environmental bottlenecks.

## 19.9 Concomitant Strategies for Metal Stress Management

### 19.9.1 Biosurfactant-Based Metal Remediation

Biosurfactants are naturally produced surface-active compounds that are synthesized by microbes, *viz*., bacteria, fungi, and yeast. These compounds have amphiphilic properties, having both hydrophilic (water-loving) and hydrophobic (water-repelling) regions, which facilitate them to interact with both water and organic compounds. Biosurfactants have gained significant attention in recent years owing to their potential uses in various industries, along with playing a crucial role in environment-friendly metal remediation (Yadav et al. 2023). Several advantages include:

- **Metal solubilization:** Biosurfactants can solubilize and chelate heavy metals from contaminated soils and water. The hydrophilic portion of the biosurfactant molecules can interact with metal ions, making them more soluble in water and facilitating their removal. They stimulate the growth and activity of metal-metabolizing microorganisms through metal solubility and accessibility, in turn accelerating the biodegradation or bioaccumulation of metals.

- **Enriched metal mobilization:** Biosurfactants can heighten the movement of heavy metals in the soil, making them more accessible for remediation techniques. They can help to reduce the adsorption of metals to soil particles, increasing metal bioavailability for uptake by remediation-targeted organisms or other processes.
- **Soil washing and extraction:** Biosurfactants can be used in soil washing and extraction processes to eradicate heavy metals from contaminated soils. In this method, biosurfactant solutions are applied to the soil, and the metals are subsequently washed out with water or extracted using other techniques.
- **Phytoextraction enhancement:** Biosurfactants can expand phytoextraction, a phytoremediation technique where plants are used to take up heavy metals from soils. Biosurfactants can boost metal solubilization and translocation within the soil, making metals more accessible for plant uptake (Yadav et al. 2023).
- **Diminished environmental impact:** Compared to synthetic surfactants, biosurfactants are often biodegradable, nontoxic, and ecological friendly. Deployment in metal remediation can reduce the environmental impact related to the application of harsh chemical agents.

While biosurfactants offer great potential for metal remediation, some challenges persist. The production of biosurfactants on a large scale can be cost-intensive, and their efficiency may vary depending on the specific site conditions and contaminants involved. Additionally, the choice of appropriate biosurfactant-generating microorganisms and the optimization of their growth conditions are critical for successful metal remediation (Yadav et al. 2023). In recent years, research in this area has been focused on exploring different types of biosurfactants, understanding their action mechanisms, and evaluating their effectiveness in real-world metal-contaminated sites. As the field of biosurfactant research continues to progress, it holds promise for contributing to more sustainable and effective metal remediation approaches and providing greener alternatives to traditional remediation methods.

### 19.9.2 Biofilms-Based Metal Remediation

Biofilms play a significant role in metal remediation and have been extensively studied for their potential in environmental clean-up. Biofilms are communities of microorganisms that adhere to surfaces and are encased in a matrix of EPS (Manju et al. 2023). These biofilms can form on various substrates, including soil particles, sediments, rocks, and metal surfaces. Due to their inherent stability and resistance to environmental fluctuations, biofilms are suitable candidates for bioremediation strategies. They can be employed for biostimulation, bioaugmentation, and phytoremediation to enhance metal remediation efficiency. Biofilms offer several gains in metal remediation through various processes:

- **Metal binding and sequestration:** Biofilms have a high capacity to bind and sequester heavy metals. The EPS matrix contains functional groups (e.g., carboxyl, amino, and phosphate) that can bind metal ions, effectually dropping their mobility and bioavailability in the environment. This immobilization of metals reduces their potential for leaching into water sources and prevents their uptake by plants or animals (Manju et al. 2023).
- **Metal accumulation:** Certain microbial biofilms possess the ability to accumulate and tolerate greater concentrations of heavy metals. These metal-tolerant or metal-resistant microorganisms can actively take up metals from the surrounding environment and store them intracellularly or in the EPS matrix (Sharma 2022).
- **Bioreduction and biomineralization:** Some metal-metabolizing microorganisms within biofilms can facilitate the reduction of certain metal ions to less toxic and less mobile forms. For example, some bacteria can reduce toxic forms of chromium (Cr(VI)) to less toxic Cr(III) compounds, which are less soluble and less mobile in the environment (Sharma 2022).
- **Enhanced microbial activity:** The close proximity and interactions within biofilm communities create microenvironments with specific physico-chemical settings. These microenvironments can enrich microbial activity, stimulating breakdown of organic pollutants and, thereby, metal contaminants' transformation.
- **Safeguard from environmental stressors:** The EPS matrixes of biofilms provide protection to the microbes against environmental stressors, such as fluctuations in temperature, pH and importantly, metal toxicity. This protection can enhance the resilience of metal-metabolizing microorganisms and help them withstand their activity in challenging environments.

Despite their benefits, several bottlenecks are posed by biofilms in metal remediation. The complex and heterogeneous nature of biofilms makes their study and manipulation perplexing. Additionally, the presence of biofilms in certain industrial settings, such as wastewater treatment plants, can lead to metal biofouling, where metals accumulate and negatively affect the system's performance (Manju et al. 2023). Researchers continue to explore biofilm-based remediation strategies to optimize their use in metal-contaminated environments. Understanding the dynamics of biofilm formation, microbial interactions, and the role of the EPS matrix can lead to the development of innovative and sustainable biotechnological solutions for metal remediation.

### 19.9.3 Chemotaxis-Based Metal Remediation

Chemotaxis is the process by which cells, including many microorganisms, respond to chemical gradients in their environment by moving toward or away from the chemical signal. In the context of metals, chemotaxis plays a crucial role in the behavior of metal-metabolizing microbes. Unlike negative chemotaxis for survival strategy, some microorganisms exhibit positive chemotaxis toward certain metals, indicating that they are attracted to higher concentrations of these metals. This attributes them to efficiently locate and accumulate metals in their surroundings, which is especially important in metal-contaminated environments. The process permits these microorganisms to concentrate metals within their cells or in biofilms, where they can promote the interaction with the metals through various detoxification or bioaccumulation mechanisms (Yadav et al. 2023). Understanding chemotaxis in metal-metabolizing microorganisms can provide valuable insights into their behavior and response to metal contamination, posing potential applications in bioremediation and environmental monitoring efforts (Yadav et al. 2023).

### 19.9.4 Melanin-Based Metal Remediation

Melanin is a naturally occurring pigment found in various organisms, including plants, fungi, and some bacteria. Catechol-melanin is predominantly found in plants. However, apart from its role in pigmentation, melanin has also been recognized for its importance in metal detoxification and protection against metal toxicity (Tran-Ly et al. 2020). These include:

- **Metal binding:** Melanin has strong metal-binding properties due to its chemical structure, which encompasses numerous functional groups, such as phenolic and amine groups. These functional groups can form stable complexes with metal ions, effectively sequestering the metals and reducing their bioavailability and toxicity.
- **Drop in reactive oxygen species (ROS) formation:** Melanin acts as an antioxidant by reducing the formation of ROS in cells. ROS are highly reactive molecules that can damage cellular components, including DNA, proteins, and lipids. In the presence of heavy metals, ROS production can increase due to metal-induced oxidative stress. Melanin's antioxidant properties help mitigate this damage, protecting cells from metal-induced oxidative stress (Tran-Ly et al. 2020).
- **Metal transformation:** Melanin-producing organisms, such as certain fungi and bacteria, can enzymatically transform metal ions into less toxic or less bioavailable forms. This process, known as biotransformation, is facilitated by melanin's redox-active nature and can contribute to the detoxification of metals in the environment.

Overall, melanin's importance in metal detoxification lies in its ability to bind, sequester, and reduce the toxicity of heavy metals (Tran-Ly et al. 2020). This natural pigment provides plants with an inherent defense against metal-induced stress and damage, helping them to thrive in metal-contaminated environments.

## 19.10 Challenges, Impending Visions, and Conclusions

While microbial inoculants offer significant potential in bioremediation, there are numerous challenges to conquer. Various regulatory aspects for the usage of microbial inoculants and potential ecological risks coupled with the need for effective monitoring and evaluation need to be addressed (O'Callaghan et al. 2022)

- **Regulatory considerations:** Regulatory hurdles are encountered, particularly when dealing with genetically modified organisms (GMOs). The approval process for using genetically engineered strains can be lengthy and complex, requiring comprehensive risk assessments to ensure their safety and environmental impact.
- **Ecological risks:** Introducing foreign microbes into an ecosystem conveys the threat of unintended ecological consequences. The introduced microbial inoculants may compete with indigenous microorganisms or disrupt existing ecological balances, leading to unforeseen impacts on biodiversity along with ecosystem functions.
- **Survival and persistence:** Ensuring the survival and persistence of introduced microbial inoculants in the target environment can be arduous. Microbes may face harsh environmental conditions, competition from indigenous species, or a lack of compatible nutrient sources, potentially reducing their effectiveness over time.

- **Delivery and distribution:** Homogenous and effective distribution of microbial inoculants across large and diverse contaminated sites can be technically demanding. Developing efficient delivery systems to ensure the appropriate spatial coverage is critical for optimal bioremediation outcomes.
- **Monitoring and evaluation:** Precise and timely observation of microbial activity and biodegradation progress is critical for assessing the efficacy of microbial inoculants. Establishment of appropriate biomarkers and monitoring techniques can be complex, needing on-going research and innovation (El-Nahal et al. 2022).

Embracing on-going research and adopting cutting-edge technologies are the prerequisites for the prospective future of microbial inoculants' application in bioremediation. Through continuous research and innovative applications, we can unlock the full potential of microbial inoculants as invaluable allies in the fight against metal stress and its associated environmental challenges (Deb et al. 2020).

- **Molecular biology advancements:** On-going advancements in molecular techniques, such as genomics and metagenomics, will provide deeper insights into microbial communities' composition and function, will aid in the selection of more effective microbial strains, and help predict their behavior in different environmental conditions.
- **Microbiome-based inoculants:** Moving beyond single-strain inoculants, microbiome-based inoculants harnessing the power of diverse microbial communities may offer greater stability, resilience, and functionality in bioremediation. Understanding microbial interactions and synergies can lead to more efficient and robust inoculant formulations. (Saravanan et al. 2023).
- **Synthetic biology and genetic engineering:** Development of synthetic biology tools and genetic engineering will allow for the design and engineering of microbial strains with enhanced capabilities for bioremediation. Customized strains with specific metabolic pathways can aim a broader range of contaminants and optimize biodegradation.
- **Smart delivery systems:** Innovations in delivery systems, such as nanotechnology-based carriers and controlled-release mechanisms, can improve the survivability and targeted delivery of microbial inoculants, enhancing their performance and longevity.
- **Bioinformatics and data analytics:** Utilization of bioinformatics and data analytics tools to process large-scale microbial datasets can aid in predicting the efficacy of different inoculant formulations and optimizing bioremediation strategies for specific environmental conditions.
- **International collaboration and standardization:** International collaboration along with protocols' standardization for the assessment of microbial inoculants' safety and efficacy can streamline the regulatory approval process and facilitate knowledge sharing among researchers and practitioners.

As a whole, conducting comprehensive environmental risk assessments prior to the deployment of microbial inoculants will ensure a proactive approach to detect and, thereby, mitigate probable ecological risks related to their practice. With careful consideration and continuous innovation, microbial inoculants can continue to play a crucial role in sustainable environmental clean-up and eco-system restoration efforts (Mushtaq 2021).

In the pursuit of sustainable environmental management, microbial inoculants offer an eco-friendly and cost-effective alternative to traditional remediation methods. Microbial inoculants offer a range of valuable applications in the management of metal stress, providing encouraging resolutions to address the challenges posed by heavy metal pollution. The unique capabilities of various microbial species, such as metal resistance, transformation, accumulation, and interaction with plants, make them powerful associates in remediating metal-contaminated environments. Embracing these valuable microbes can lead us toward a cleaner and healthier environment, safeguarding natural ecosystems and human well-being from the detrimental magnitudes of heavy metal contamination.

## References

Alengebawy, A., Abdelkhalek, S.T., Qureshi, S.R., and Wang, M.Q. (2021). Heavy metals and pesticides toxicity in agricultural soil and plants: ecological risks and human health implications. *Toxics* 9 (3): 42.

Ali, H., Khan, E., and Ilahi, I. (2019). Environmental chemistry and ecotoxicology of hazardous heavy metals: environmental persistence, toxicity and bioaccumulation. *Journal of Chemistry* 2019: 6730305.

Arora, V., Khosla, B., 2022. Conventional and contemporary techniques for removal of heavy metals from soil. In: Mendes, K.F., de Sousa, R.N., and Mielke, K.C. (Eds.), Biodegradation Technology of Organic and Inorganic Pollutants. IntechOpen. doi: https://doi.org/10.5772/intechopen.98569.

Bala, S., Garg, D., Thirumalesh, B.V. et al. (2022). Recent strategies for bioremediation of emerging pollutants: a review for a green and sustainable environment. *Toxics* 10 (8): 484.

Cardenas-Gonzalez, J.F., Acosta-Rodriguez, I., Teran-Figueroa, Y., and Rodriguez-Perez, A.S. (2017). Bioremoval of arsenic (V) from aqueous solutions by chemically modified fungal biomass. *3 Biotech* 7 (3): 226.

Chen, T.L., Wise, S.S., Kraus, S. et al. (2009). Particulate hexavalent chromium is cytotoxic and genotoxic to the North Atlantic right whale (*Eubalaena glacialis*) lung and skin fibroblasts. *Environmental and Molecular Mutagenesis* 50: 387–393.

Chugh, M., Kumar, L., Bhardwaj, D., and Bharadvaja, N. (2022). Bioaccumulation and detoxification of heavy metals: an insight into the mechanism. In: *Development in Wastewater Treatment Research and Processes* (ed. M.P. Shah, S. Rodriguez-Couto, and R.T. Kapoor), 243–264. Elsevier.

Deb, V.K., Rabbani, A., Upadhyay, S. et al. (2020). Microbe-assisted phytoremediation in reinstating heavy metal-contaminated sites: concepts, mechanisms, challenges, and future perspectives. In: *Microbial Technology for Health and Environment. Microorganisms for Sustainability*, vol. 22 (ed. P. Arora), 161–189. Singapore: Springer.

Deng, L., Su, Y., Su, H. et al. (2007). Sorption and desorption of lead (II) from wastewater by green algae *Cladophora fascicularis*. *Journal of Hazardous Materials* 143 (1-2): 220–225.

Dhaliwal, S.S., Singh, J., Taneja, P.K., and Mandal, A. (2020). Remediation techniques for removal of heavy metals from the soil contaminated through different sources: a review. *Environmental Science and Pollution Research* 27: 1319–1333.

Diaconu, M., Pavel, L.V., Hlihor, R.M. et al. (2021). Characterization of heavy metal toxicity in some plants and microorganisms – a preliminary approach for environmental bioremediation. *New Biotechnology* 56: 130–139.

El-Kassas, H.Y. and El-Taher, E.M. (2009). Optimization of batch process parameters by response surface methodology for mycoremediation of chrome-VI by a chromium resistant strain of marine *Trichoderma viride*. *Journal of Agriculture and Environmental Sciences* 5 (5): 676–681.

El-Meihy, R.M., Abou-Aly, H.E., Youssef, A.M. et al. (2019). Efficiency of heavy metals-tolerant plant growth promoting bacteria for alleviating heavy metals toxicity on sorghum. *Environmental and Experimental Botany* 162: 295–301.

El-Nahal, A.S., El-Saadony, M.T., Saad, A.M. et al. (2022). The use of microbial inoculants for biological control, plant growth promotion, and sustainable agriculture: a review. *European Journal of Plant Pathology* 162 (4): 759–792.

Hasan, S.H. and Srivastava, P. (2009). Batch and continuous biosorption of Cu($2^{+}$) by immobilized biomass of *Arthrobacter* sp. *Journal of Environmental Management* 90 (11): 3313–3321.

Houessionon, M.K., Ouendo, E.M.D., Bouland, C. et al. (2021). Environmental heavy metal contamination from electronic waste (e-waste) recycling activities worldwide: a systematic review from 2005 to 2017. *International Journal of Environmental Research and Public Health* 18 (7): 3517.

Kapahi, M. and Sachdeva, S. (2019). Bioremediation options for heavy metal pollution. *Journal of Health and Pollution* 9 (24): 191203.

Khan, N., Humm, E.A., Jayakarunakaran, A., and Hirsch, A.M. (2023). Reviewing and renewing the use of beneficial root and soil bacteria for plant growth and sustainability in nutrient-poor, arid soils. *Frontiers in Plant Science* 14: 1147535.

Kumar, R., Mishra, R.K., Mishra, V. et al. (2016). Detoxification and tolerance of heavy metals in plants. In: *Plant Metal Interaction* (ed. P. Ahmad), 335–359. Elsevier.

Lakherwal, D. (2014). Adsorption of heavy metals: a review. *The International Journal of Environmental Research and Development* 4: 41–48.

Malik, L.A., Bashir, A., Qureashi, A., and Pandith, A.H. (2019). Detection and removal of heavy metal ions: a review. *Environmental Chemistry Letters* 17: 1495–1521.

Manju, Verma, R.K., Bohra, Y. et al. (2023). Role of microbial biofilms in bioremediation. In: *Developments in Applied Microbiology and Biotechnology, Microbial Inoculants* (ed. V.K. Sharma, A. Kumar, M.R.Z. Passarini, et al.), 253–276. Academic Press.

Mishra, S., Bharagava, R.N., More, N. et al. (2019). Heavy metal contamination: an alarming threat to environment and human health. In: *Environmental Biotechnology: For Sustainable Future* (ed. R. Sobti, N. Arora, and R. Kothari), 103–125. Singapore: Springer.

Mitra, S., Chakraborty, A.J., Tareq, A.M. et al. (2022). Impact of heavy metals on the environment and human health: novel therapeutic insights to counter the toxicity. *Journal of King Saud University, Science* 34 (3): 101865.

Mushtaq, Z. (2021). PGPR: present role, mechanism of action and future prospects along bottlenecks in commercialization. *Journal of Environmental Quality* 41: 9–15.

Nazir, R., Khan, M., Masab, M. et al. (2015). Accumulation of heavy metals (Ni, Cu, Cd, Cr, Pb, Zn, Fe) in the soil, water and plants and analysis of physico-chemical parameters of soil and water collected from Tanda dam kohat. *Journal of Pharmaceutical Sciences and Research* 7: 89–97.

O'Callaghan, M., Ballard, R.A., and Wright, D. (2022). Soil microbial inoculants for sustainable agriculture: limitations and opportunities. *Soil Use and Management* 38 (3): 1340–1369.

Okereafor, U., Makhatha, M., Mekuto, L. et al. (2020). Toxic metal implications on agricultural soils, plants, animals, aquatic life and human health. *International Journal of Environmental Research and Public Health* 17 (7): 2204.

Rahman, Z. and Singh, V.P. (2019). The relative impact of toxic heavy metals (THMs) (arsenic (As), cadmium (Cd), chromium (Cr)(VI), mercury (Hg), and lead (Pb)) on the total environment: an overview. *Environmental Monitoring and Assessment* 191: 1–21.

Raklami, A., Tahiri, A.I., Bechtaoui, N. et al. (2021). Restoring the plant productivity of heavy metal-contaminated soil using phosphate sludge, marble waste and beneficial microorganisms. *Journal of Environmental Sciences* 99: 210–221.

Raklami, A., Meddich, A., Oufdou, K., and Baslam, M. (2022). Plants - Microorganisms-based bioremediation for heavy metal clean-up: recent developments, phytoremediation techniques, regulation mechanisms, and molecular responses. *International Journal of Molecular Sciences* 23 (9): 5031.

Romera, E., Gonzalez, F., Ballester, A. et al. (2007). Comparative study of biosorption of heavy metals using different types of algae. *Bioresource Technology* 98 (17): 3344–3353.

Saha, L., Kumar, V., Tiwari, J. et al. (2021). Electronic waste and their leachates impact on human health and environment: global ecological threat and management. *Environmental Technology and Innovation* 24: 102049.

Sammauria, R., Kumawat, S., Kumawat, P. et al. (2020). Microbial inoculants: potential tool for sustainability of agricultural production systems. *Archives of Microbiology* 202 (4): 677–693.

Saravanan, A., Swaminaathan, P., Senthil Kumar, P. et al. (2023). A comprehensive review on immobilized microbes-biochar and their environmental remediation: mechanism, challenges and future perspectives. *Environmental Research* 116723.

Sari, A. and Tuzen, M. (2008). Biosorption of cadmium (II) from aqueous solution by red algae (*Ceramium virgatum*): equilibrium, kinetic and thermodynamic studies. *Journal of Hazardous Materials* 157 (2-3): 448–454.

Sharma, P., Pandey, A.K., Udayan, A., and Kumar, S. (2021). Role of microbial community and metal-binding proteins in phytoremediation of heavy metals from industrial wastewater. *Bioresource Technology* 326: 124750.

Sharma, P. (2022). Role and significance of biofilm-forming microbes in phytoremediation – a review. *Environmental Technology and Innovation* 25: 102182.

Sharma, P., Bano, A., Singh, S.P., and Tong, Y.W. (2023). Metal-tolerant microbial inoculants for improved phytoextraction. In: *Developments in Applied Microbiology and Biotechnology, Microbial Inoculants* (ed. V.K. Sharma, A. Kumar, M.R.Z. Passarini, et al.), 75–98. Academic Press.

Singh, J.S., Koushal, S., Kumar, A. et al. (2016). Book review: microbial inoculants in sustainable agricultural productivity – Vol. II: functional application. *Frontiers in Microbiology* 7: 2105.

Singh, D.P., Prabha, R., and Gupta, V.K. (2019). Microbial inoculants for sustainable crop management. In: *Microbial Interventions in Agriculture and Environment* (ed. D. Singh, V. Gupta, and R. Prabha), 1–35. Singapore: Springer.

Siqueira, A.H. and Santos, R.C.B. (2023). Microbial inoculants for the management of herbicide toxicity in plants. In: *Developments in Applied Microbiology and Biotechnology, Microbial Inoculants* (ed. V.K. Sharma, A. Kumar, M.R.Z. Passarini, et al.), 149–159. Academic Press.

Sun, Y., Lan, J., Du, Y. et al. (2020). Efficient removal of heavy metals by synergistic actions of microorganisms and waste molasses. *Bioresource Technology* 302: 122797.

Thakare, M., Sarma, H., Datar, S. et al. (2021). Understanding the holistic approach to plant-microbe remediation technologies for removing heavy metals and radionuclides from soil. *Current Research in Biotechnology* 3: 84–98.

Tran-Ly, A.N., Ribera, J., Schwarze, F.W.M.R. et al. (2020). Fungal melanin-based electrospun membranes for heavy metal detoxification of water. *Sustainable Materials and Technologies* 23: e00146.

Vardhan, K.H., Kumar, P.S., and Panda, R.C. (2019). A review on heavy metal pollution, toxicity and remedial measures: current trends and future perspectives. *Journal of Molecular Liquids* 290: 111197.

Xiang, L., Harindintwali, J.D., Wang, F. et al. (2022). Integrating biochar, bacteria, and plants for sustainable remediation of soils contaminated with organic pollutants. *Environmental Science & Technology* 56 (23): 16546–16566.

Yaashikaa, P.R., Kumar, P.S., Jeevanantham, S., and Saravanan, R. (2022). A review on bioremediation approach for heavy metal detoxification and accumulation in plants. *Environmental Pollution* 301: 119035.

Yadav, R., Singh, G., Santal, A.R., and Singh, N.P. (2023). Omics approaches in effective selection and generation of potential plants for phytoremediation of heavy metal from contaminated resources. *Journal of Environmental Management* 336: 117730.

Zhu, Y., Fan, W., Zhou, T., and Li, X. (2019). Removal of chelated heavy metals from aqueous solution: a review of current methods and mechanisms. *Science of The Total Environment* 678: 253–266.

# 20

## Harnessing *In Silico* Techniques for Bioremediation Solutions

*Nischal Pradhan and Ajay Kumar*

*Amity Institute of Biotechnology, Amity University, Noida, India*

### 20.1 Introduction

In the past century, the world has observed a daunting increase in overall global pollution arising due to rapid industrialization and urbanization. The sudden disregard for environmental safety has ultimately led to humanity being gravely affected, resulting in the depletion of our natural resources. becoming contaminated and the occurrence of many phenomena such as biomagnification, where toxins accumulate and intensify, which has become a critical concern, reaching alarming levels over the past few decades (Philp et al. 2014). This contamination, whether originating from natural sources or man-made, encompasses a wide range of organic and inorganic substances, all of which pose serious threats to our biological systems. Due to the unchecked addition of new and large influx of various organic and inorganic compounds, which include pesticides, insecticides, herbicides, fungicides, and various other compounds and their nonbiodegradable and persistent nature have consequently resulted in the environment being flooded with chemical contaminants (Varshney 2019).

"Remediate," which refers to the process of solving or reversing a problem; similarly, "bio-remediate" refers to the process of solving environmental problems by harnessing microbes to clean up contamination discreetly by eliminating harmful contaminants from the environment.

As seen in Figure 20.1, we present a visual representation of the fundamental principle underlying bioremediation. Figure 20.1 shows the basic principle regarding bioremediation. The bacterial degradation of sewage organics provided a foundation for applying microbes to other persistent pollutants, like petroleum products (Sonawdekar 2012). This process depends on the type of contaminates, encompasses various strategies, including partial or complete breakdown, alteration, or detoxification of pollutants through microbial action (Varshney 2019).

The process of bioremediation owes its inception to George M. Robinson, who made significant contributions in the 1960s by pioneering the concept of crafting custom mixtures of dried bacteria cultures for commercial use. These custom "bug-brew" recipes gained recognition and prominence through various well-publicized demonstrations (Sonawdekar 2012). According to Hoff (1993), the earliest instances of research on bioremediation can be traced back to the 1970s. During this period many articles documenting the process of microbial degradation of oil, both in laboratory settings and field trials, were published. This documentation included several review papers that discussed the mechanisms of biodegradation, laying the foundation for further advancements in the field of bioremediation. However, a milestone in the history of bioremediation occurred in 1972 when the first commercial in situ bioremediation system was installed to clean up a Sun Oil pipeline spill in Ambler, Pennsylvania. Another major breakthrough in the field of bioremediation occurred in 1979, when Anand Mohan Chakrabarty, an American scientist of Indian origin, obtained a strain of *Pseudomonas putida* containing the *XYL* and *NAH* plasmids, as well as a hybrid plasmid created by recombining parts of *CAM* and *OCT*, with the particular strain exhibiting remarkable efficiency in metabolizing hydrocarbons found in crude oil (Sonawdekar 2012). This identification of bacteria that are capable of breaking down hydrocarbons present in crude oil and refined petroleum products opened the door to using bioremediation for these ubiquitous pollutants (Das and Chandran 2011), marking a significant breakthrough in bioremediation. Although research on bioremediation continued during the 1980s with the discovery of bacterial degradation of sewage organics. The topic remained relatively unknown outside of academic and applied research

*Microbes Based Approaches for the Management of Hazardous Contaminants*, First Edition. Edited by Ajay Kumar, Livleen Shukla, Joginder Singh, and Luiz Fernando R. Ferreira.

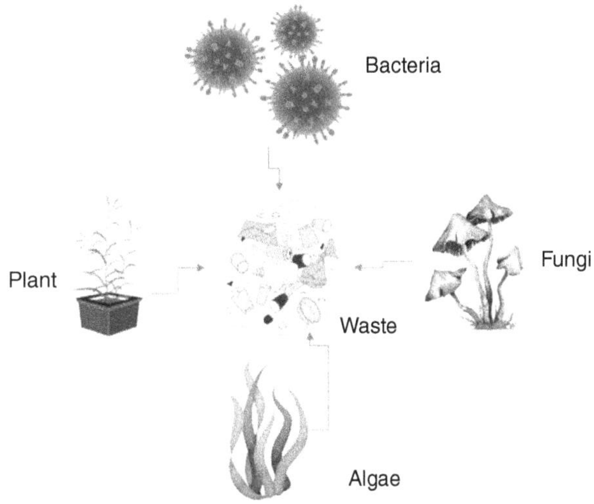


**Figure 20.1** Principle illustration of bioremediation.

communities, with only a small fraction of individuals being familiar with bioremediation before the infamous Exxon Valdez spill (Hoff 1993).

The Exxon Valdez spill, which occurred in March 1989, had a significant impact on the field of bioremediation. The immediate aftermath of the spill saw bioremediation being recognized as one of the essential methods for combating environmental contamination. This led to the creation of the Environmental Protection Agency (EPA), which established bioremediation protocols based on various case studies on the subject (Hoff 1993). Consequently, by the 1990s, scientists began to place greater reliance on natural microorganisms and techniques to enhance their performance in bioremediation efforts (Sonawdekar 2012).

Current bioremediation techniques encompass mineralization, co-metabolism, and the use of microorganisms to fully transform contaminants into harmless inorganic compounds (Varshney 2019). Traditional bioremediation research relied on laboratory experiments to isolate microbes capable of metabolizing contaminants and test their efficiency. However, in recent times, *in silico* methods have emerged, incorporating comparative genomics, genome-scale metabolic modeling, and machine learning. These methods have become indispensable for identifying novel biodegradation pathways, analyzing intricate microbial communities, and uncovering new microorganisms with bioremediation potential in a high-throughput manner. With the continued advancement in computational power and technology, bioremediation has made substantial progress.

## 20.2 Emergence of *In Silico* Approaches

In recent times, there has been a notable rise in the use of computational techniques for bio remedial processes; these methods are termed as "*in silico* approaches" within the field of bioremediation (Figure 20.2).

This methodology has proven invaluable tool in improving our understanding and utilizing the application of microorganisms and plants to solve various environmental problems by providing a cost-effective technique. The development of these computational techniques is driven by advances in genomics, systems biology, and computational modeling, which allow accurate simulation of complex systems.

In the early 2000s, the field of bioremediation started incorporating *in silico* techniques for bioremediation. During the late 1990s and early 2000s, researchers had already turned to computer-based pathway modeling, employing metabolic pathway analysis tools. This innovative approach aimed to gain deeper insights into the catabolic pathways and substrate preferences of pollutant-degrading bacteria and fungi (Patil et al. 2005), For instance, *Pseudomonas putida* was engineered to exhibit enhanced degradation capabilities for substances like toluene and other aromatic compounds (Loh and Cao 2008). During this time, there was also an increased acceptance of genome-scale metabolic models that covered all metabolic reactions in a specific microorganism.

The subsequent decade, the 2010s, brought forth the emergence of high-throughput "omics" technologies such as genomics, transcriptomics, proteomics, and metabolomics. These technologies significantly expanded the wealth of

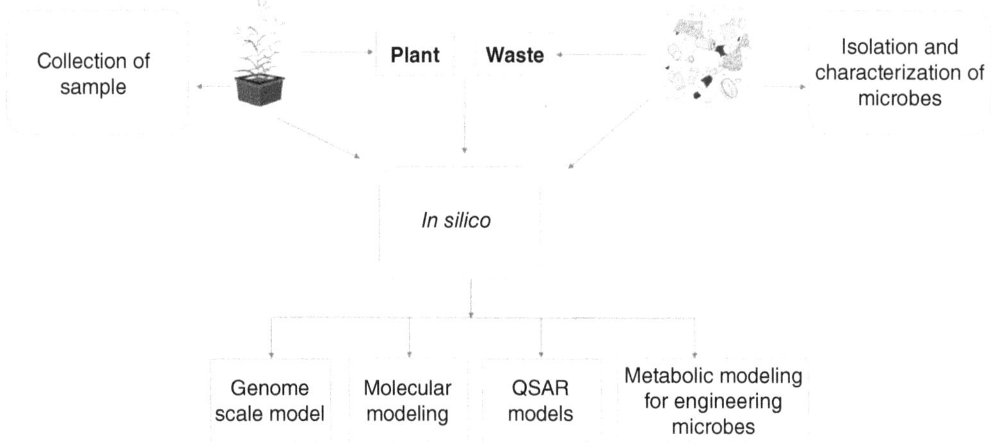


**Figure 20.2** *In silico* approaches to bioremediation.

biological data available for modeling bioremediation microbes (Yergeau et al. 2012). Instead of relying solely on known metabolic reactions, researchers could now integrate comprehensive multi-omics datasets into genome-scale models to gain a more comprehensive understanding of organism function. Researchers leveraged the powers of systems biology modeling to study microbial communities and their interactions during bioremediation (Song et al. 2015). Community metabolic network models were built by integrating individual organism models, providing insights into syntrophic and competitive interactions (Song et al. 2015; Zengler and Palsson 2012).

*In silico* methods bring a multitude of advantages when compared to traditional experimental approaches. They facilitate high-throughput screening of microbes and metabolic pathways, allowing for the examination of microbial interactions and providing valuable insights into the dynamics of microbial ecology.

## 20.3 Genome-Scale Models

Genome-scale metabolic models (GEMs) are computational representations of an organism's entire metabolism, capturing the genotype–phenotype relationships linking genes to metabolic reactions and growth capabilities. GEMs have emerged as powerful tools for *in silico* modeling of microbes involved in bioremediation (Figure 20.3).

Recently, Nogales et al. (2019). constructed a high-quality GEM called iJN1462 for the biotechnologically important bacterium *Pseudomonas putida KT2440. P. putida* is renowned for its metabolic versatility and robustness in degraded environments, with recognized applications in the bioremediation of aromatic hydrocarbons, heavy metals, and other pollutants.

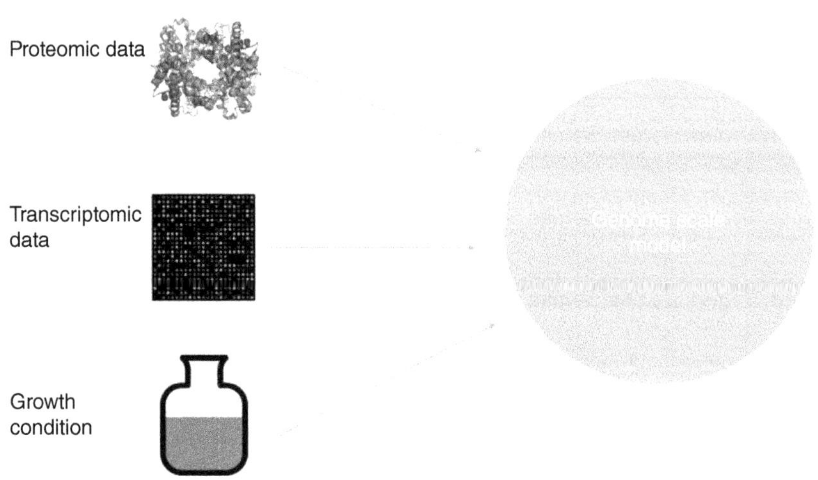


**Figure 20.3** Basic genome scale model illustration.

**Table 20.1** Basic summary comparison study table.

| Study | Nogales et al. (2019) | Lewis, Nagarajan, and Palsson (2012) |
|---|---|---|
| Focus | Genome-scale metabolic modeling | Understanding the genotype–phenotype relationship |
| Significance | Contribution to the field of metabolic modeling | Contribution to understanding the genotype–phenotype relationship |
| Metabolic coverage | Highlights the broad metabolic capabilities of *Pseudomonas putida* through high-quality genome-scale metabolic modeling | Describes the evolution and diversity of constraint-based reconstruction and analysis (COBRA) methods for studying the genotype–phenotype relationship |
| Bio remedial application | Suggests that *Pseudomonas putida*'s broad metabolic capabilities make it a promising candidate for bioremediation applications, although specific applications are not discussed. | Indicates that COBRA methods can contribute to improving bioremediation strategies by providing insights into the metabolic pathways and phenotypic states of microorganisms involved in bioremediation processes. |

The iJN1462 model significantly expands the metabolic coverage of *P. putida* compared to previous GEMs, incorporating decades of legacy knowledge from the literature. It contains 1501 reactions not found in other GEMs, including pathways for degradation of aromatic compounds like lignin monomers that are relevant to bioremediation. Extensive experimental validation was performed to assess the accuracy of iJN1462 for predicting *P. putida* metabolism. Growth phenotypes on over 100 novel carbon and nitrogen sources showed 79–84% accuracy. Flux measurements and gene essentiality data further demonstrated the high predictive ability of iJN1462 (Nogales et al. 2019). A key benefit of high-quality GEMs is their utility as scaffolds for reconstructing other strains. The validated iJN1462 model was leveraged to efficiently reconstruct over 80 draft GEMs of different *P. putida* strains by identifying shared and unique gene content. Despite being drafts, these models predicted diverse and broad bioremediation capabilities across strains, including the potential for growth on lignin monomers. The multi-strain modeling illuminated both shared and specialized metabolic abilities, providing targets to assess environmental niches and guide bioprocessing applications of *P. putida* strains (Nogales et al. 2019). Similar to a complementary study, (Lewis et al. 2012) offered the COBRA framework, for understanding the mechanistic basis of cellular functions and predicting phenotypic states in biological systems.

The evidence Table 20.1 demonstrates the comparison between the bioremediation potential of microbial species at the genome scale, as well as the potential of genotype–phenotype relationship in contributing to better bioremediation strategies. Since bioremediation research will increasingly improve, it will employ omics datasets and GEMs, providing a critical computational platform for contextualizing this data and guiding engineering of designer bioremediators. High-quality GEMs like iJN1462 for key microbes enable *in silico* modeling that complements and expands benchtop bioremediation research.

## 20.4 Molecular Modeling

Molecular modeling techniques are methods used to analyze biomolecular structure and interactions, which are used in bioremediation research to analyze as well as to elucidate contaminant degradation pathways and engineer microbial biocatalysts. These computational methods, which are part of the broader field of *in silico* analysis, provide valuable insights into the behavior and capabilities of microorganisms such as *Pseudomonas putida* KT2440, The comprehensive and validated reconstruction of the metabolic network of *P. putida* KT2440, known as iJN1462, greatly expands the computational predictions of its metabolic states and allows for the identification of new nutrients supporting growth (Nogales et al. 2019).

Additionally, the multi-strain modeling of *P. putida*, based on gene ortholog analysis, enables the identification of strains of interest without the need for wet-lab experiments. These advancements in molecular modeling contribute to a better understanding of the metabolic capabilities of environmental bacteria and have significant implications for bioremediation research and biotechnology applications (Dangi et al. 2018). A recent study by Zhang et al. (2022) demonstrates the value of molecular modeling in characterizing pollutant-degrading enzymes. The authors isolated the *Sphingobium strain SHPJ-2* from a polyaromatic hydrocarbon (PAH)-degrading consortium. Metabolite profiling showed SHPJ-2 could degrade various high-molecular-weight PAHs. Although Docking studies with different PAH substrates revealed two RHDs,

*PhnA1B1* and *PhnA2B2*, had favorable binding to pyrene and chrysene (Zhang et al. 2022), heterologously expressed *PhnA1B1* and *PhnA2B2* confirmed their predicted HMW-PAH metabolism capabilities (Zhang et al. 2022). Screening *PhnA1B1* and *PhnA2B2* mutant libraries *in silico* can also guide engineering efforts to alter their substrate scope (Zhang et al. 2022). This study exemplifies molecular modeling's utility in determining biodegradation pathways and optimizing catalytic efficiency. As computational power increases, molecular modeling will likely become integral to elucidating contaminant metabolism mechanisms and designing bioremediating enzymes.

## 20.5 QSAR Models

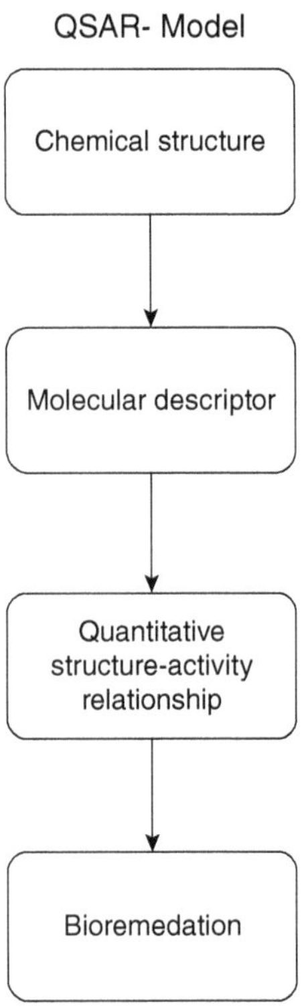


**Figure 20.4** General protocol used in QSAR model for bioremediation.

QSAR/QSPR models, also known as "Quantitative Structure–Activity/Property Relationship," are statistical models used in estimating the physicochemical characteristics of organic pollutants. These model's application is utilized in computational chemistry to predict properties such as biodegradability and volatility (Figure 20.4), which are essential for assessing environmental fate and remediation parameters (Fatemi and Chahi 2012). Figure 20.4 shows the basic work flow that is followed for predicting these properties.

Recent studies done by Fatemi and Chahi (2012) indicate that QSAR/QSPR models can play a major role in designing remediation strategies for organic pollutants, such as emerging contaminants, where the experimental data available are limited. The model was successfully able to identify specific molecular descriptors, such as lopping centric index, mean topological charge index of order, and many other attributes that may contribute to the calculation precision and were found to be important in determining the half-lives of PCB congeners (Fatemi and Chahi 2012). Longer estimated half-lives suggest that a contaminant will be more resistant to bioremediation. The study by Fatemi and Chahi (2012) also estimated the depuration half-lives of polychlorinated biphenyl (PCB) congeners in juvenile rainbow trout using Quantitative structure-activity relationships also known as QSAR methodology. The study used multiple linear regression, multilayer perceptron neural networks, and support vector regression techniques to develop robust and predictive QSAR models for prediction. Similarly, in a modeling study done by Bordás, Bélai, and Koomíves (2011), developed statistically significant and predictive linear regression models for the BCF values of PCDDs/PCDFs and PCBs in zucchini based on retrospective, The models incorporated a set of 1660 DRAGON, 150 VolSurf, and 11 quantum chemical descriptors, including properties like hydrophobicity, molecular volume, and number of chlorines. Table 20.2 shows an overview of the different descriptors.

Although the applications of these models in the domain of designing remediation techniques remain relatively unexplored, predictive models can be trained and put to use in the

**Table 20.2** Different types of descriptors predicted.

| Types of descriptors | Number of descriptors | Description | Source |
|---|---|---|---|
| DRAGON | 1660 | Molecular descriptors providing information on hydrophobicity, molecular size, shape, and electronic properties | Todeschini and Consonni (2009) |
| VOLSurf | 150 | Molecular descriptors capturing information about molecular surface properties like hydrophobicity, polarity, and shape | Puzyn et al. (2009) |
| Quantum | 11 | Descriptors related to electronic structure | Karelson et al. (1996) |

initial phases of developing remediation processes, which will be able to provide valuable insights that extend from bench-scale assessments to pilot-scale and full-scale implementation. However, since these models for biodegradation are created based on experimental biodegradation data and structures of the corresponding chemicals. The advantages of QSAR models lie in providing statistically significant predictions and implying the physicochemical properties of the respective chemicals.

Even though QSAR modeling is still in its early stages of bioremediation technology design, following this with experimental optimization, an efficient strategy or tool can be developed in bioremediation. Through these applications, researchers are able to gain a deeper comprehension of the effectiveness of various remediation technologies in addressing specific pollutants, thereby enabling more efficient and targeted remediation efforts.

## 20.6 Metabolic Modeling for Engineering Microbes

As we know, microbial communities are ubiquitous in nature and play critical roles in the bio-geochemical cycles and bioremediation of pollutants. However, it is challenging to elucidate the complex interactions within microbial communities using experimental techniques alone. In this regard, genome-scale metabolic modeling has emerged as a useful computational approach to analyze complex communities and predict the fate of pollutants (Ang et al. 2018). Metabolic modeling starts with literature and data mining after the identification of the required data, followed by target identification and modeling, which is validated using an experimental process (Figure 20.5).

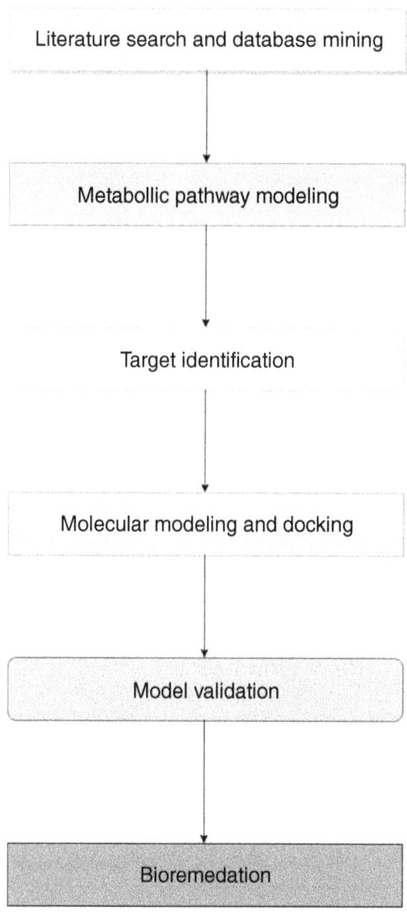


**Figure 20.5.** General overview of the protocols used during metabolic modeling approaches to bioremediation.

Metabolic modeling approaches are broadly classified based on the level of segregation between member species. In compartmentalized models, genome-scale models of individual species are first built and then integrated into a combined community model. This allows tracing metabolite exchanges between modeled species. However, it requires high-quality genome annotations and models for member species. In contrast, lumped models do not differentiate between species. Instead, community metabolic potential is represented as a single combined reaction network based on metagenomic data. Although lumped models cannot capture species-specific interactions, they allow studying communities with poor characterization (Ang et al. 2018).

Once formed, the community metabolic model is solved by techniques such as flux balance analysis (FBA) to predict steady-state flux distributions that optimize a biologically relevant objective like community growth rate. The computed fluxes can provide insights into inter-species metabolite exchanges, structural robustness, and biosynthetic capabilities. Dynamic community behaviors can be analyzed using extensions like dynamic Flux Balance Analysis (FBA), which integrates metabolite kinetics with stoichiometric models (Ang et al. 2018).

Ang et al. (2018) used compartmentalized modeling to elucidate syntrophic interactions in methanogenic communities, sulfate-reducing bacteria, and microbial oxidation of sulfide minerals. Lumped models have characterized microbial mats and soil communities. Dynamic FBA modeling revealed metabolic competition between *Geobacter* and *Rhodoferax* species during in situ uranium bioremediation. The model predicted conditions favoring uranium-reducing *Geobacter* growth over *Rhodoferax*. Further, multi-objective community optimization frameworks like OptCom can examine cooperation, competition, and emergent capabilities in communities (Ang et al. 2018).

Combined with high-throughput cloning, such advanced modeling promises more robust bio design. As demonstrated by metabolic modeling remains a powerful *in silico* tool for engineering microbes' degradative abilities and expanding applications in bioremediation.

## 20.7 Development of *In Silico* Platforms for Bioremediation Research

As we witnessed, *in silico* analysis is an essential part in predicting degradative pathways and reactions within microorganisms before their practical application in environmental cleanup.

One of the main challenges in developing an *in silico* platform for bioremediation research is the accurate representation of complex biological processes. As it involves complex interactions between various microorganisms, pollutants, and environmental factors, the process of capturing these interactions in computational models requires a deep understanding of the underlying biological mechanisms and their integration into mathematical algorithms. In addition, scalability *in silico* platforms is an important concern (Varjani et al. 2017).

Developing models that can effectively bridge these scales and provide meaningful predictions at larger scales is a challenge. In addition, the computational resources required to run complex models at larger scales can be a limitation, especially for researchers with limited access to high-performance computing facilities. Data availability and quality pose challenges to developing *in silico* platforms. Since these studies are based on extensive experimental data such as microbial kinetics, pollutant degradation rates, and environmental parameters, therefore acquiring a comprehensive and reliable database can be a time-consuming and expensive process. In addition, data variability, such as the lack of standard data formats and the heterogeneity of data sources, makes data integration and model validation difficult (Varjani et al. 2017).

The table below shows some important structured databases and tools that are used in computational bioremediation.

| S. No. | Name | Utility | URL | References |
|---|---|---|---|---|
| 1 | NCBI | NCBI's extensive databases, including GenBank (®) and PubMed, provide vital biological data for bioremediation research. | https://www.ncbi.nlm.nih.gov/ | NCBI Resource Coordinators (2015) |
| 3 | BioSurfDB | BioSurfDB is a database that contains various data on advancing bioremediation efforts and harnessing the full potential of biosurfactants. | https://www.biosurfdb.org/#/ | Oliveira et al. (2015) |
| 4 | MetaCyc | This database offers comprehensive information on metabolic pathways, including related metabolites, enzymes, reactions, and genes spanning all domains of life. | https://metacyc.org/ | Caspi et al. (2019) |
| 5 | FastQC | This tool is employed to assess the quality of sequencing data produced by different next-generation sequencing platforms. | https://www.bioinformatics.babraham.ac.uk/projects/fastqc/ | Andrews S (2010) |
| 6 | Trimmomatic | It is a freely available program for trimming next-generation sequencing data. | http://www.usadellab.org/cms/index.php?page=trimmomatic | Bolger et al. (2014) |
| 7 | BLAST2GO | This software is used for functional annotation, visualization, and analysis of high-throughput sequencing data. | https://www.biobam.com/ | Conesa et al. (2005) |
| 8 | AutoDock | It is well known and widely used software for molecular docking and protein-ligand interaction study. | https://autodock.scripps.edu/ | Morris et al. (2009) |

## 20.8 Challenges and Limitations

While promising, some key challenges and limitations remain in effectively harnessing *in silico* approaches for bioremediation. Many tools are designed for generic applications and not optimized for bioremediation-specific queries.

*In silico* bioremediation studies have many challenges and limitations that must be addressed for effective implementation. The selection and screening of suitable bacteria and enzymes to reduce bacterial contamination is a major challenge. This process is time-consuming and labor-intensive, as it involves identifying bacteria with the desired enzyme capacity and testing their effectiveness in neutralizing the bacteria (Singh, Chowdhary and Raj 2020).

Another limitation is the application of *in silico* bioremediation technologies at scale at contaminated sites. Although *in silico* methods provide good information about degradation pathways and possible enzyme effects, their application in real situations requires *in vitro* and *in vivo* validation and testing (Singh, Chowdhary, and Raj 2020).

## 20.9 Conclusion

Although *In silico* approaches have emerged as powerful computational tools to advance the field of bioremediation. The development of various bioinformatics techniques, such as comparative genomics, metabolic modeling, molecular docking, and QSAR models, has enabled comprehensive system-level analysis of biodegrading microbes and microbial communities. These *in silico* methods allow high-throughput screening and discovery that complement traditional laboratory experiments. Genome-scale metabolic models in particular provide scaffolding to characterize biodegradation potentials across microbial strains and species. Specialized databases and analytical software tailored for bioremediation play a key role in contextualizing the growing volumes of omics data. Structured curation efforts are needed to leverage knowledge from the literature and enrich datasets with domain-specific metadata. Integrated platforms that seamlessly link databases and analytical tools will further empower researchers by facilitating workflows from data retrieval to downstream applications.

While challenges remain in optimizing currently available tools for bioremediation queries, the application potential of *in silico* methods is immense. Future directions include developing ontologies to standardize exchange formats and advancing community-level multi-omic modeling. Enhanced accessibility of web-deployed resources and intuitive interfaces will also promote widespread use. With continued collaboration between computational and domain scientists, *in silico* approaches hold promise to revolutionize bioremediation research through their unmatched abilities for high-throughput discovery, hypothesis generation, and optimization of bioremediation strategies.

## References


Andrews S (2010) FastQC: A Quality Control Tool for High Throughput Sequence Data. http://www.bioinformatics.babraham.ac.uk/projects/fastqc

Ang, K.S., Lakshmanan, M., Lee, N.-R., and Lee, D.-Y. (2018). Metabolic modeling of microbial community interactions for health, environmental and biotechnological applications. *Current Genomics* 19 (8): 712–722. https://doi.org/10.2174/1389202919666180911144055.

Bolger, A., Lohse, M., and Usadel, B. (2014). Trimmomatic: a flexible trimmer for Illumina sequence data. *Bioinformatics* 30 (15): 2114–2120. https://doi.org/10.1093/bioinformatics/btu170.

Bordás, B., Bélai, I., and Koomíves, T. (2011). Theoretical molecular descriptors relevant to the uptake of persistent organic pollutants from soil by zucchini. A QSAR Study. *Journal of Agricultural and Food Chemistry* 59 (7): 2863–2869. https://doi.org/10.1021/jf1038772.

Caspi, R., Billington, R., Keseler, I.M. et al. (2019). The MetaCyc database of metabolic pathways and enzymes – a 2019 update. *Nucleic Acids Research* 48 (D1): D445–D453. https://doi.org/10.1093/nar/gkz862.

Conesa, A., Götz, S., García-Gómez, J.M. et al. (2005). Blast2GO: a universal tool for annotation, visualization and analysis in functional genomics research. *Bioinformatics* 21 (18): 3674–3676. https://doi.org/10.1093/bioinformatics/bti610.

Dangi, A.K., Sharma, B., Hill, R.T., and Shukla, P. (2018). Bioremediation through microbes: systems biology and metabolic engineering approach. *Critical Reviews in Biotechnology* 39 (1): 79–98. https://doi.org/10.1080/07388551.2018.1500997.

Das, N. and Chandran, P. (2011). Microbial degradation of petroleum hydrocarbon contaminants: an overview. *Biotechnology Research International* 2011: 1–13. https://doi.org/10.4061/2011/941810.

Fatemi, M. and Chahi, Z.G. (2012). QSPR-based estimation of the half-lives for polychlorinated biphenyl congeners. *SAR and QSAR in Environmental Research* 23 (1–2): 155–168. https://doi.org/10.1080/1062936x.2011.645876.

Hoff, R.Z. (1993). Bioremediation: an overview of its development and use for oil spill cleanup. *Marine Pollution Bulletin* 26 (9): 476–481. https://doi.org/10.1016/0025-326x(93)90463-t.

Karelson, M., Lobanov, V.S., and Katritzky, A.R. (1996). Quantum-chemical descriptors in QSAR/QSPR studies. *Chemical Reviews* 96 (3): 1027–1044. https://doi.org/10.1021/cr950202r.

Lewis, N.E., Nagarajan, H., and Palsson, B.Ø. (2012). Constraining the metabolic genotype–phenotype relationship using a phylogeny of in silico methods. *Nature Reviews Microbiology* 10 (4): 291–305. https://doi.org/10.1038/nrmicro2737.

Loh, K. and Cao, B. (2008). Paradigm in biodegradation using *Pseudomonas putida* – a review of proteomics studies. *Enzyme and Microbial Technology* 43 (1): 1–12. https://doi.org/10.1016/j.enzmictec.2008.03.004.

Morris, G.M., Huey, R., Lindstrom, W.M. et al. (2009). AutoDock4 and AutoDockTools4: automated docking with selective receptor flexibility. *Journal of Computational Chemistry* 30 (16): 2785–2791. https://doi.org/10.1002/jcc.21256.

NCBI Resource Coordinators (2015). Database resources of the national center for biotechnology information. *Nucleic Acids Research* 44 (D1): D7–D19. https://doi.org/10.1093/nar/gkv1290.

Nogales, J., Mueller, J., Gudmundsson, S. et al. (2019). High-quality genome-scale metabolic modelling of *Pseudomonas putida* highlights its broad metabolic capabilities. *Environmental Microbiology* 22 (1): 255–269. https://doi.org/10.1111/1462-2920.14843.

Oliveira, J.S., Araújo, W.J., Sales, A.I.L. et al. (2015). BioSurfDB: knowledge and algorithms to support biosurfactants and biodegradation studies. *Database* 2015: https://doi.org/10.1093/database/bav033.

Patil, K.R., Rocha, I., Förster, J., and Nielsen, J. (2005). Evolutionary programming as a platform for in silico metabolic engineering. *BMC Bioinformatics* 6 (1): https://doi.org/10.1186/1471-2105-6-308.

Philp, J., Bamforth, S.M., Singleton, I., and Atlas, R.M. (2014). *Environmental Pollution and Restoration: A Role for Bioremediation*, 1–48. ASM Press eBooks https://doi.org/10.1128/9781555817596.ch1.

Puzyn, T., Leszczynski, J., and Cronin, M.T. (2009). *Recent Advances in QSAR Studies: Methods and Applications*. Springer.

Singh, A.K., Chowdhary, P., and Raj, A. (2020). In silico bioremediation strategies for removal of environmental pollutants released from paper mills using bacterial ligninolytic enzymes. In: *Elsevier eBooks*, 249–285. 10.1016/b978-0-12-819001-2.00013-9.

Sonawdekar, S. (2012). Bioremediation: A boon to hydrocarbon degradation. *International Journal of Environmental Sciences* 2: 2408–2424.

Song, H., Renslow, R.S., Fredrickson, J.K., and Lindemann, S.R. (2015). Integrating ecological and engineering concepts of resilience in microbial communities. *Frontiers in Microbiology* 6: https://doi.org/10.3389/fmicb.2015.01298.

Todeschini, R. and Consonni, V. (2009). *Molecular Descriptors for Chemoinformatics, 2 volume set: Volume I: Alphabetical Listing / Volume II: Appendices, References*. Wiley-VCH https://doi.org/10.1002/9783527628766.

Varjani, S.J., Agarwal, A.K., Gnansounou, E., and Gurunathan, B. (2017). *Bioremediation: Applications for Environmental Protection and Management*. Springer.

Varshney, K.A. (2019). Bioremediation of pesticide waste at contaminated sites. *Journal of Emerging Technologies and Innovative Research* [Preprint]. https://www.jetir.org/view?paper=JETIRCH06013.

Yergeau, É., Sanschagrin, S., Beaumier, D., and Greer, C.W. (2012). Metagenomic analysis of the bioremediation of diesel-contaminated Canadian high arctic soils. *PLoS One* 7 (1): e30058. https://doi.org/10.1371/journal.pone.0030058.

Zengler, K. and Palsson, B.Ø. (2012). A road map for the development of community systems (CoSy) biology. *Nature Reviews Microbiology* 10 (5): 366–372. https://doi.org/10.1038/nrmicro2763.

Zhang, L., Liu, H., Dai, J. et al. (2022). Unveiling degradation mechanism of PAHs by a *Sphingobium* strain from a microbial consortium. *mLife* 1 (3): 287–302. https://doi.org/10.1002/mlf2.12032.

# 21

# Microbial Inoculants and Their Potential Application in Bioremediation

*Ankita Agrawal, Jitesh Kumar Maharana, and Amiya Kumar Patel*

*Department of Biotechnology and Bioinformatics, Sambalpur University, Burla, Odisha, India*

## 21.1 Introduction

The healthiness of the environment influences the living style of all creatures on earth. Tremendous increase in environmental pollution owing to rapid urbanization and increased industrial and agricultural activities accumulate huge number of hazardous contaminants. These detrimental contaminants, such as pesticides, polycyclic aromatic hydrocarbons (PAHs), crude oils, chlorophenols, and heavy metals, pose significant threats to natural ecosystem and human health (Aparicio et al. 2022; Chen et al. 2015; Verma and Kuila 2019; Kour et al. 2022). The existence of pollutants is a major concern owing to their high toxicity, nonbiodegradable, carcinogenic, and mutagenic properties, leading to health hazards such as cancer, neurological disorders, allergies, and infertility (Das and Osborne 2018; Kumar et al. 2021). As a result, nature is continually developing and imparting fresh hurdles and threats (Sonune 2021). Therefore, ecosystem restoration by various remediation tactics, which must be effective, reliable, eco-friendly, and affordable, is essential for sustainable development (Aparicio et al. 2022). Conventional remediation strategies such as physical methods (excavation, landfilling) and chemical treatments (use of minerals, aluminum salts, and sulfide) are often expensive and energy-intensive and result in the production of secondary pollutants and incomplete degradation (Cheng et al. 2021), which implies biological approaches as the ideal choice (Das and Osborne 2018). There has been growing interest in bioremediation as a versatile, sustainable, and expenditure-friendly approach for environmental cleanup with low side effects over the traditional remediation methods that often rely on physical or chemical processes (Chaudhary et al. 2023; Sonune 2021). Bioremediation is the integration of various strategies to retrieve natural ecosystems that have been poisoned by anthropogenic activities through biodegradation (Ganguly et al. 2023; Yadav 2021). Moreover, bioremediation is a green trick that harnesses the natural abilities of microbes to mitigate toxic contaminants and improve soil fertility (Chaudhary and Shukla 2019). Management of environmental pollutants through bioremediation strategies promotes substantial development and ecological stability (Raj et al. 2022).

Microbial inoculants in bioremediation are based on the principle of employing viable microbes to restore contaminated areas. Diverse microbial inoculants with higher catabolic and adaptivity potential have emerged as a promising tool to enhance the rate and efficacy of bioremediation in polluted sites for eco-restoration (Adetunji and Anani 2021; Tyagi et al. 2011; Xu and Zhou 2017). These inoculants can be naturally occurring microbial strains (bacteria, fungi, or other microbial species) isolated from contaminated sites or genetically engineered microbes (GEMs) with enhanced pollutant degradation capabilities (Kour et al. 2022). Microbial inoculants such as *Pseudomonas* sp., *Bacillus* sp., *Flavobacterium* sp., *Aspergillus* sp., *Corynebacterium* sp., *Arthrobacter* sp., *Nocardia* sp., *Rhizopus* sp., and *Rhodococcus* sp. are considered as potent bioremediators (Nivetha et al. 2022; Verma and Kuila 2019). Besides, the biodegradation process can be accelerated by augmenting the degrading microbes, leading to faster and more efficient clean-up. Such inoculants are carefully selected based on their ability to degrade target contaminants and survival in extreme environmental conditions. One of the key benefits of using microbial inoculants is their potential to accelerate the degradation of contaminants that are recalcitrant to degradation by native microorganisms alone. Some carcinogenic pollutants, like PAHs and chlorinated compounds, are challenging to degrade due to their complex chemical structures, causing ecological dysfunction. However, certain

*Microbes Based Approaches for the Management of Hazardous Contaminants*, First Edition. Edited by Ajay Kumar, Livleen Shukla, Joginder Singh, and Luiz Fernando R. Ferreira.

bacterial strains and white-rot fungi have evolved specialized metabolic pathways to detoxify these carcinogenic compounds (Sayara et al. 2011). Further, microbial inoculants can promote the growth and activities of indigenous microbes, leading to improved overall efficiency of bioremediation. Also, these inoculants release various enzymes and bioactive compounds that facilitate the breakdown of complex pollutants, making them more accessible to native microbial communities. Additionally, microbial inoculants inhibit the growth of harmful microorganisms, preventing their proliferation and thereby promoting relative dominance of beneficial species. Such competitive exclusion creates a favorable environment for the degradation of contaminants. However, understanding and knowledge of the potential risks and benefits associated with the use of microbial inoculants in bioremediation is a prerequisite to evaluate their long-term impacts on environment and living creatures.

The chapter deals with in-depth exploration of microbial inoculants and their potential applications in bioremediation, outlining their advantages and limitations. Besides, it delves into the mechanism by which microbial inoculants contribute to the degradation of contaminants and the enhancement of native microbial communities. In addition, the chapter highlights the different microbial-mediated remediation strategies as well as their selection, optimization, and monitoring. Furthermore, the case studies and success stories have been presented with the objective to illustrate the practical applications of microbial inoculants in bioremediation. Nevertheless, future perspectives, research approaches, and recommendations to harness the full potential of microbial inoculants in the field of bioremediation are discussed.

## 21.2 Overview of Bioremediation

### 21.2.1 Definition and Principle

Bioremediation is an inventive and integrated microbiological approach utilizing diverse microbial communities or their enzymes to degrade, detoxify, immobilize, and reduce contaminants that are hazardous to environment in a comprehensive manner (Patel et al. 2022). Bioremediation represents a holistic approach based on the principle of microbial biodegradation (Abatenh et al. 2017; Abo-Alkasem et al. 2023), utilizing metabolic capabilities of microbes to break down complex pollutants into simpler or less toxic forms suitable for ecological restoration (Jain and Bajpai 2012).

### 21.2.2 Types of Bioremediation

The process of bioremediation can be categorized into (i) *in situ*; (ii) *ex situ*; (iii) other strategies; and (iv) emerging strategies (Figure 21.1). Bioremediation can be either *in situ*, where the contaminants are treated at the original site, or *ex situ*, where the contaminated materials are excavated from the site of origin and treated elsewhere (Nwankwegu et al. 2022). The choice between *in situ* and *ex situ* bioremediation largely depends upon the location of contamination site, the type and toxicity of pollutants, and economic feasibility (Patel et al. 2022). Both *in situ* and *ex situ* bioremediation depend upon the metabolism of microbial inoculants and functional diversity in terms of eco-restoration efficiency, with systematic and operational difficulties to distinguish efficacy of individual treatment strategy (Nwankwegu et al. 2022). The *in situ* bioremediation approach is sustainable, does not require excavation of pollutants, and is more reliable and cost-efficient compared to *ex situ* approach (Kour et al. 2022). In addition, *in situ* bioremediation is effective for the treatment of large amounts of contaminants utilizing native harmless microbial species and is commonly employed for the degradation of chlorinated hydrocarbons, anilines, and nitrobenzene in soil and groundwater (Sonune 2021).

i) ***In situ* bioremediation**
   *In situ* bioremediation can be classified into intrinsic (Bioattenuation), and engineered bioremediation based on the origin of microbial species (Kim et al. 2014; Roy et al. 2015). Engineered *in situ* bioremediation approaches include bioventing, biosparing, biostimulation, and bioaugmentation (Sayqal and Ahmed 2021; Vidali 2001).
   a) **Bioattenuation** (natural attenuation) relies on the degradation ability of native microbes to degrade contaminants from environment, which is a natural *in situ* degradation process without any external intervention where the polluted sites are passively remediated. The lack of external factors makes the approach cheapest *in situ* bioremediation. The limitation of this approach is longer time duration (Azubuike et al. 2016; Patel et al. 2022).
   b) **Bioventing** is an *in situ* detoxifying process involving controlled stimulation of air flow to provide limited supply of oxygen to sustain the activities of native microbial species for effective bioremediation while limit the process of volatilization (Vidali 2001). Besides, it is considered as effective approach for contaminated simple hydrocarbons

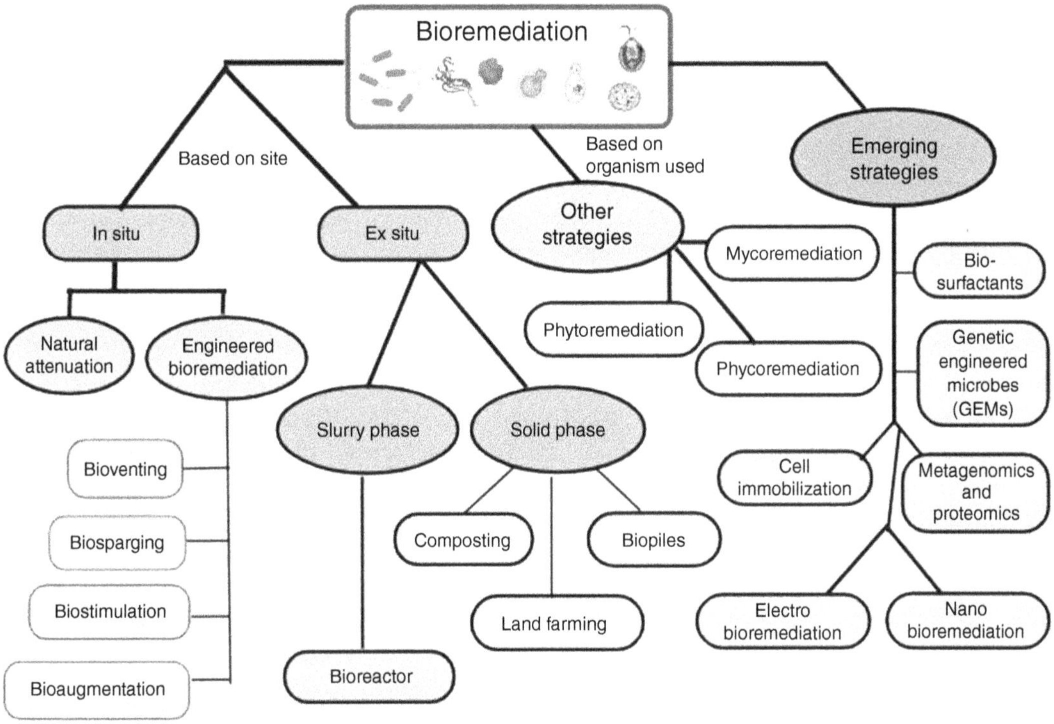


**Figure 21.1** Different strategies of bioremediation: (i) *in situ* bioremediation; (ii) *ex situ* bioremediation; (iii) other strategies of bioremediation; and (iv) emerging strategies of bioremediation.

and oil products in soil (Brown and Ulrich 2014; Chipasa and Mędrzycka 2006). However, the inability to deliver oxygen to polluted soil and insufficient aeration are major limitations of bioventing (Azubuike et al. 2016).

c) **Biosparging** is an easy and low-cost approach that involves injection of pressurized air beneath the water table to increase the oxygen content in groundwater, thereby enhancing the pace of bioremediation (Sayqal and Ahmed 2021). The effectiveness of the biosparging approach mostly depends on soil permeability and pollutant biodegradability (Sales da Silva et al. 2020). The biosparging approach is commonly used to remediate aquifers contaminated with oil derivatives (Kao et al. 2008).

d) **Biostimulation** is considered as one of the efficient *in situ* approach for commencing the bioremediation process through the addition of enriched nutrients, supplements, enzymes, metabolites, electron donors/acceptors (phosphorus, nitrogen, oxygen, carbon, and molasses), growth-promoting factors, and/or optimization of environmental conditions by aeration in order to enhance the microbial-mediated soil processes exhibited by resident microbes (Ayilara and Babalola 2023; Patel et al. 2022; Yadav 2021). Biostimulation accelerates the inherent degradation potential or metabolism of microbes for rapid cleanup of pollutants (Kanissery and Sims 2011). It is advantageous because bioremediation is carried out by the extant native microbes that are well suited to the inherent environmental settings and are geographically spread across the subsurface. It also constitutes one of the most effective strategies for detoxification of hydrocarbon, petroleum, and chlorinated pollutants (Patel et al. 2022). The key problem lies in the difficulties of delivering additives in a way that enables them to be easily accessible to subsurface bacteria (Adams et al. 2015). Moreover, it may lead to soil toxicity, eutrophication, and algal blooms (Nwankwegu et al. 2022).

e) **Bioaugmentation** is an effective and economical detoxification process that involves the inoculation of exogenous microbial strains with specific degradation capabilities in the contaminated site to augment biodegradative capacity of the existing microbial population (Gao et al. 2022; Ma et al. 2018). The efficacy of bioaugmentation largely depends upon the adaptability and degrading quality of the inoculated microbes (Adams et al. 2015; Patel et al. 2022). It holds the potential to improve the treatment process by supplementing it with various exogenous microbes such as *Bacillus thuringiensis*, *Aspergillus* sp., *Pseudomonas*, *Penicillium* sp., and *Arthrobacter* either individually or in combination with specialized metabolic activities for the precise degradation of specific pollutants at a speedy rate

(Ma et al. 2018). Several research revealed bioaugmentation as a primary bioremediation development trend in the future, as the selected microbes have a strong affinity for degrading impact, shorter remediation cycles, and broad range of adaptability (Gao et al. 2022; Ma et al. 2018; Patel et al. 2022).

ii) ***Ex situ* bioremediation**

*Ex situ* bioremediation is the excavation and transportation of pollutants from contaminated site to another site with special facilities for treatment and can be further categorized into (a) solid-phase and (b) slurry-phase systems based on the state of pollutant excavated. The solid phase system involves land farming, composting, and biopiles techniques used for the detoxification of agricultural, industrial, domestic, and municipal solid wastes, whereas slurry-based techniques involve bioreactors for degradation (Vidali 2001).

a) **Land farming** is a simplified, low-cost and low-footprint soil bioremediation process that entails removing the contaminated soil and distributing it over a thin surface to stimulate the native microbes for bioremediation (Sales da Silva et al. 2020). This technique is generally employed to remediate petroleum-contaminated soil (Sayqal and Ahmed 2021).
b) **Compositing** involves the integration of contaminated soil with non-hazardous organic amendments such as manure, straw, and agricultural wastes to accelerate microbial growth for successful degradation (Kour et al. 2022).
c) **Biopiles** is an economically feasible bioremediation process involved in detoxification of transported piles of solid pollutants (soil, silt, or solid wastes) above the dumping site using a forced aeration system to regulate microbial activities (Abatenh et al. 2017; Patel et al. 2022; Sayqal and Ahmed 2021) by converting the pollutants to carbon dioxide and water, which are relatively harmless to nature (Kour et al. 2022).
d) **Bioreactor** is the slurry-based *ex situ* bioremediation process that involves the excavation of contaminated soil followed by mixing with water supplemented with native microbial inoculants using a bioreactor with optimized environmental conditions for detoxification (Kour et al. 2022). Bioreactors are one of the finest ways to clean contaminated soil because the reactor environment includes optimized variables that can be easily regulated, allowing for enhanced microbial biodegradation activity (Sales da Silva et al. 2020).

iii) **Other remediation strategies**

The remediation strategies based on the organism can be divided into three different types: phytoremediation, mycoremediation, and phycoremediation. These approaches involve combination of microbial inoculants with other organisms to achieve synergistic effect, leading to more efficient and comprehensive pollutant removal. These strategies are more advantageous in terms of remediation efficiency by targeting specific pollutants of interest through a multifaceted approach using a combination of microbial inoculants.

a) **Phytoremediation** is an emerging technology that utilizes plants and their associated microorganisms to degrade contaminants in soil and water. Moreover, the introduction of microbial inoculant in phytoremediation enhances pollutant degradation and plant growth. Microbes facilitate the breakdown of complex pollutants, improve nutrient availability and promote the plant–microbe interactions, leading to enhanced remediation efficiency (Patel et al. 2022; Sayqal and Ahmed 2021).
b) **Mycoremediation** utilizes microbial inoculants containing fungi or its derivatives (such as white rot) to degrade toxic pollutants including heavy metals, pharmaceutical wastes, and polynuclear aromatic hydrocarbons in soil and water (Akhtar and Mannan 2020).
c) **Phycoremediation** involves the utilization of microorganisms such as microalgae, macroalgae, and cyanobacteria to detoxify pollutants and derive toxic products from waste water using eco-friendly approach (Priyadharshini et al. 2021).

iv) **Emerging remediation techniques**

Array of advanced techniques have arisen that overcome the limitations of conventional bioremediation strategies that hastened the process of remediation to many folds. Integration of microbial inoculants with emerging techniques (biosurfactants, cell immobilization, GEMs, nanobioremediation, electrobioremediation, and metagenomics) offers a comprehensive and multidimensional approach to bioremediation.

a) **Biosurfactants** are surface-active amphipathic molecules produced by aerobic microbes that are potentially used for soil and water bioremediation. Reduction in the hydrophobic nature of pollutants makes them easily accessible to microbes for rapid degradation, which is the principle of biosurfactants. Usually, the microbial surfactants are nontoxic, biodegradable with environmental compatibility, low production cost, and unlimited applicability for long-term remediation process (Mulligan 2021). They are primarily utilized for biodegradation of hydrocarbons and heavy metals in polluted areas, as well as to improve oil recovery and oil spills (Silva et al. 2014; Sonune 2021).
b) **Cell immobilization** is one of the most potent bioremediation approaches that uses microbial inoculants multiple times, has a relatively lower cost, higher degradation efficiency, resistance to environmental

heterogeneity, is tolerant to high pollutant concentrations, and has a large biomass (Bayat et al. 2015). Immobilization involves adsorption, covalent binding, entrapment, and encapsulation of microbial inoculants (*Pseudomonas* sp., *Xanthomonas* sp., *Arthrobacter* sp., and *Acinetobacter* sp.) that have been employed for bioremediation of crude oil and waste water effluents (Bayat et al. 2015; Srivastava and Gupta 2021).

c) **Genetically engineered microbes (GEMs)** with superior catabolic activities are widely used as potential tool for the restoration of soil, groundwater, and activated sludge (Janssen and Stucki 2020). Bioremediation using the GEMs accelerates the recovery rate of polluted sites and increases substrate degradation, which provides a sustainable ecosystem that is environmentally feasible. Besides, GEMs have been widely employed for bioremediation of heavy metals such as Ni, Hg, Cd, Cu, Fe, and As (Bruschi and Goulhen 2006). However, studies on landscape-scale fields are prerequisites to gather information in order to establish the overall efficacy and consequences of GEMs introduction into natural ecosystems.
d) **Nanobioremediation** is a promising bioremediation approach that uses nanoparticles for detoxification of pollutants, either *in situ* or *ex situ* approaches. Owing to particle size, they can penetrate through small spaces and move faster compared to large-size particles, which offers a wider spectrum for managing pollutants in ground water, hydrocarbons, heavy metals, organic or inorganic compounds by reducing intermediates at a minimal cost, facilitating chemical reduction, and mitigating negative environmental impacts (Bharagava et al. 2020; Bhatt et al. 2022).
e) **Electrobioremediation** is considered as an integrated remediation technique that integrates the principle of bioremediation as well as electrokinetic approaches for detoxification of hydrophobic organic pollutants (Jain et al. 2020).
f) **Metagenomics** approach is one of the beneficial techniques that is useful to detect potent microbes involved in detoxification of specific contaminants and to pinpoint the catabolic gene expression patterns solely accountable for the catabolism of specific pollutant. In the same breath, proteomics approach is also equally capable of identifying the beneficial proteins involved in detoxification process. Hence, both culture-independent approaches are significantly used for bioremediation (Sonune 2021). However, it is very difficult to detoxify all the pollutants at contamination site due to higher genetic diversity and ever-changing dynamics. Novel applications of advanced techniques and their integration might overcome the existing obstacles by minimizing the risk associated with bioremediation.

### 21.2.3 Advantages and Challenges

Bioremediation is a promising alternative to conventional remediation approaches, as evident by several scientific investigations (Abatenh et al. 2017; Vidali 2001; Yadav 2021). Bioremediation utilizes potent microbes with inherent biological activities that completely degrade the contaminants into nontoxic end products (water, carbon dioxide, cell biomass) without any leftover (Agrawal et al. 2018). Further, it is a natural, eco-friendly approach that reduces the use of harsh and toxic chemicals considered as environmental pollutants. In comparison to other remediation approaches, this strategy is more economical and feasible. Besides, it can be applied to varieties of contaminants in diverse environmental settings with less technical assistance, making it easier to execute with higher public acceptance. Furthermore, bioremediation is nonintrusive and could potentially allow for continued site usage.

However, bioremediation confronts various challenges, depending on the nature and degree of pollutants, site-specific environmental conditions, and regulatory requirements. For instance, it only works on a limited range of pollutants, that are biodegradable. Some pollutants, like persistent organic pollutants are resistant to microbial degradation. Additionally, the success of bioremediation depends on various environmental factors like pH, temperature, oxygen demand, and nutrient availability. It is also a time-consuming process, especially for large-scale contaminated sites. Further, secondary soil variables that indirectly influence the quantum of bioremediation need to be explored. Moreover, it might be difficult to fulfill the regulatory requirements and obtain necessary clearance from the statutory bodies for the implementation of bioremediation strategies. In light of mitigating such bioremediation issues, microbial inoculants offer a potential solution to improve the bioremediation process.

## 21.3 Microbial Inoculants: Concept and Types

Microbial inoculants (nature's recycling agents) are defined as living microbes that are deliberately introduced into contaminated site to conduct absorption, precipitation, oxidation, and reduction of pollutants to enhance bioremediation (Bumbra et al. 2022). Microbial inoculants have metabolic pathways that utilize diverse pollutants as energy sources for their growth and proliferation and assure the retrieval of ecological balance (Ayangbenro and Babalola 2017;

Sayqal and Ahmed 2021). Microbial inoculants include algae, bacteria, fungi, and archaea, which are unlimited resources used for the biodegradation of various environmental pollutants. The concept behind microbial inoculants is to introduce potent microbial inoculants with novel deterioration capacity to boost the degradation capabilities of the existing microbial communities for complete mineralization and thus improve the overall efficiency of bioremediation (Table 21.1).

### 21.3.1 Microbes Used as Bioremediation Drivers

Different microbial species, including bacteria, fungi, and algae, have the potential to utilize broad variety of substrates as carbon source in variety of environments (Jabbar et al. 2022; Nivetha et al. 2022). This inherent ability of the microbes makes them ideally suited for bioremediation process. Moreover, their capacity to adopt and thrive under extreme circumstances boosts their effectiveness in many folds. Utilization of different bacterial, fungal, and algal strains for bioremediation of diverse contaminants has been tabulated (Table 21.1).

Microbial inoculants possess several key characteristics that necessitate their utility as potential tools in bioremediation. Diverse metabolic pathways, high compatibility with broad range of environmental settings, effective colonization in contaminated sites, adaptability, interactions with native microbial communities, and the capacity to develop bioactive compounds (biosurfactants, siderophores) make them the principal agents to enhance the bioremediation process (Abatenh et al. 2017; Sayqal and Ahmed 2021; Singh 2014; Sonune 2021). Further, the microbial inoculants are capable of biofilm production that protect microenvironment, improves their resilience, attachment to contaminated surfaces, degradation efficiency, and thereby overall efficiency of bioremediation (Sonune 2021).

### 21.3.2 Types of Microbial Inoculants

The microbial inoculants utilized in bioremediation can be classified based on their attributes and composition. The categories of microbial inoculants encompass a range of options, such as pure culture, mixed culture, genetically engineered microorganism (GEMs), consortia of indigenous microbes, and encapsulated or immobilized inoculants (Mukred et al. 2008). Pure cultures include specific microbial strains sourced from the contaminated sites or chosen for their degradation capabilities and cultured in controlled laboratory settings to ensure the desired metabolic pathways and eliminate competition (Nwankwegu et al. 2022). While providing targeted degradation, pure cultures might struggle with survival in the field. In contrast, mixed cultures comprise multiple microbial strains forming consortia with complementary capabilities that enable broader contaminant degradation and greater adaptability to varying environmental conditions, which exhibit relatively enhanced stability and synergistic metabolic activities compared to pure cultures compensating for individual limitations (Daniel et al. 2019; Kumar and Philip 2006). GEMs developed by genetic engineering express specific enzymes and have upgraded metabolic pathways for rapid degradation of diverse contaminants (Kumar et al. 2018). Consortia of indigenous microbes include microbes from the contamination site to improve degradation abilities. Besides, encapsulation or immobilization techniques protect and relatively improve the performance of microbial inoculants through immobilization matrix such as resins, beads, as well as biodegradable polymers. Such diverse inoculant types offer adaptable solutions for specific bioremediation scenarios, with selection depending on the nature and chemical complexity of the contaminants, site circumstances, and desired outcome.

### 21.3.3 Selection Criteria of Microbial Inoculants

The selection of microbial inoculants for bioremediation involves precise evaluation of several criteria to ensure their efficiency in degrading or transforming contaminants. The key selection criteria include the inoculants specificity to the target contaminants, their compatibility with environmental conditions of polluted site, including several factors (pH, temperature, salinity, oxygen availability, nutrient availability, and presence of chemical constituents), and their interactions with native microbial communities (Abatenh et al. 2017; Medfu Tarekegn et al. 2020). The resilience and adaptability of microbial inoculants are important, as they should be able to endure challenges induced by a contaminated environment, such as high contaminant concentrations, nutrient limitations, or fluctuations in environmental variables (Zhang and Zhang 2022). Safety consideration is crucial, and therefore the selected inoculants should be nonpathogenic, nontoxic, and harmless to human health and the environment (Chaudhary et al. 2023). Besides, availability and culturable criteria should also be considered, and the chosen inoculants should be readily obtainable in sufficient quantities for large-scale

**Table 21.1** Microbial inoculants used for bioremediation of contaminated environment.

| Pollutant | Microorganism | Category | References |
|---|---|---|---|
| Anthracene | *Cladosporium herbarum* | Fungi | Krivobok et al. (1998) |
| | *Cryphonectria parasitica* | Fungi | Krivobok et al. (1998) |
| | *Mycobacterium vanbaalenii* | Bacteria | Biache et al. (2017) |
| Pyrene | *Alcaligenes faecalis* | Bacteria | Mangwani et al. (2017) |
| | *Pseudomonas mendocina* | Bacteria | Mangwani et al. (2017) |
| | *Rhodococcus* sp. | Bacteria | Kunihiro et al. (2005) |
| Steroids | *Escherichia coli* | Bacteria | Ye et al. (2019) |
| Benzo(a)pyrene | *Pleurotus ostreatus* | Fungi | Li et al. (2012) |
| | *Rhizobium tropici* | Bacteria | Yessica et al. (2013) |
| As | *Microcystis aeruginosa* | Algae | Wang et al. (2018b) |
| | *Chlamydomonas reinhardtii* | Algae | Nivetha et al. (2022) |
| | *Saccharomyces cerevisiae* | Fungi | Verma and Kuila (2019) |
| | *Ascomycota consortium* | Fungi | Akpasi et al. (2023) |
| | *Microbacterium* sp. | Bacteria | Bhakat et al. (2019) |
| | *Shigella* sp. | Bacteria | Bhakat et al. (2019) |
| Zn | *Pseudomonas aeruginosa* | Bacteria | Mohamed and Abo-amer (2012) |
| | *Penicillium chrysogenum* | Fungi | Bwapwa et al. (2017) |
| | *Stigeoclonium* sp. | Algae | del Carmen Vargas-García et al. (2012) |
| | *Fucus vesiculosus* | Algae | Brinza et al. (2020) |
| Cd | *Aerococcus* sp. | Bacteria | Sravya and Sangeetha (2022) |
| | *Cladosporium* sp. | Fungi | del Carmen Vargas-García et al. (2012) |
| | *Graphium putredinis* | Fungi | |
| | *Fusarium solani* | Fungi | Joshi et al. (2011) |
| | *Aspergillus flavus* | Fungi | Barapatre and Jha (2016) |
| | *Penicillium chrysogenum* | Fungi | Akpasi et al. (2023) |
| | *Scenedesmus incrassatulus* | Algae | Dubey et al. (2023) |
| Hg | *Aspergillus* sp. *A31* | Fungi | Pietro-Souza et al. (2020) |
| | *Curvularia geniculate* | Fungi | Pietro-Souza et al. (2020) |
| | *Cunninghamella elegans* | Fungi | Duc et al. (2021) |
| | *Cyclotella cryptica* | Algae | Shah and Jain (2020) |
| Pb | *Azotobacter chroococcum* | Bacteria | Ndeddy Aka and Babalola (2016) |
| | *Enterobacter cloacae* | Bacteria | Kang et al. (2015) |
| | *Variovorax boronicumulans* | Bacteria | Nivetha et al. (2022) |
| | *Penicillium simplicissimum* | Fungi | Iskandar et al. (2011) |
| | *TalaromycesIslandicus* | Fungi | Akpasi et al. (2023) |
| Ni | *Bacillus subtilis* | Bacteria | Abou-Shanab et al. (2019) |
| | *Kluyvera ascorbata* | Bacteria | Ndeddy Aka and Babalola (2016) |
| Fe | *Chlorococcum humicola* | Algae | Chugh et al. (2022) |
| | *Pseudomonas fluorescence* | Bacteria | Širić et al. (2022) |
| | *Ascomycota consortium* | Fungi | Akpasi et al. (2023) |
| Ce | *Fusarium solani* | Fungi | del Carmen Vargas-García et al. (2012) |
| | *Penicillium crysogenum* | Fungi | |
| Cr | *Bacillus firmus* | Bacteria | Bachate et al. (2013) |
| | *Aspergillus awamori* | Fungi | Joshi et al. (2011) |
| | *Azotobacter chroococcum* | Bacteria | Abo-amer et al. (2014) |
| U | *Microbacterium oxydans* | Bacteria | Sánchez-Castro et al. (2017) |
| | *Pseudomonas aeruginosa* | Bacteria | Choudhary and Sar (2011) |

(*Continued*)

**Table 21.1** (Continued)

| Pollutant | Microorganism | Category | References |
|---|---|---|---|
| Phenol | *Corynebacterium propinquum* | Bacteria | Gaur et al. (2018) |
| | *Bacillus subtilis* | Bacteria | Gaur et al. (2018) |
| | *Chlorella* sp. | Algae | Wu et al. (2022) |
| | | Algae | Wu et al. (2022) |
| Naphthalene | *Alcaligenes faecalis* | Bacteria | John et al. (2012) |
| | *Bacillus thermoleovorans* | Bacteria | Annweiler et al. (2000) |
| Phenanthrene | *Phomopsis liquidambari* | Fungi | Fu et al. (2018) |
| Crude oil | *Pseudomonas aeruginosa* | Bacteria | Mukjang et al. (2022) |
| | *Alcaligenes* sp. | Bacteria | Jabbar et al. (2022) |
| | *Achremonium* sp. | Fungi | Jabbar et al. (2022) |
| | *Candida glabrata* | Fungi | Burghal et al. (2016) |
| Aliphatic hydrocarbons | *Aspergillus* sp. | Fungi | Li et al. (2020) |
| | *Penicillium* sp. | Fungi | Li et al. (2020) |
| Pesticides and Herbicides | *Pseudomonas* sp. | Bacteria | Malakar et al. (2020) |
| | *Brevundimonas diminuta* | Bacteria | Thakur et al. (2019) |
| | *Penicillum verruculosum* | Fungi | Akhtar and Mannan (2020) |
| | *Botryospaeria laricina* | Fungi | Akhtar and Mannan (2020) |
| Palm oil | *Bacillus pumilus* | Bacteria | Saranya et al. (2019) |
| Congo red dye | *Oudemansiella canarii* | Fungi | Iark et al. (2019) |
| Azo dye | *Lysinibacillus sphaericus* | Bacteria | Lu et al. (2020) |
| Dibenzofurans | *Coprinellus radians* | Fungi | Aranda et al. (2010) |
| Benzopyrene | *Candida viswanathii* | Fungi | Hesham et al. (2012) |
| Diesel oil | *Pseudomonas cepecla* | Bacteria | Kehinde and Isaac (2016) |
| | *Bacillus cereus A* | Bacteria | Maliji et al. (2013) |
| Polychlororinated biphenyls (PCBs) | *Pseudomonas fluorescens* | Bacteria | Toussaint et al. (2012) |
| | *Rhodococcus ruber* | Bacteria | Wang et al. (2018a) |
| Acenaphthene | *Aeromonas hydrophila* | Bacteria | Alegbeleye et al. (2017) |
| Fluorene | *Raoultellaornithinolytica* | Bacteria | Alegbeleye et al. (2017) |
| Polyethene | *Trichoderma harzianum* | Fungi | Akpasi et al. (2023) |
| Plastics | *Streptococcus* sp. | Bacteria | Sharma (2018) |
| | *Pseudomonas* sp. | Bacteria | |
| | *Bacillus* sp. | Bacteria | |

application. Further, prior knowledge of microbial inoculants in similar bioremediation, as applications evidenced by field trials or case studies, provides valuable insights to aid in the selection process. The meticulous examination of these criteria leads the researchers to identify potent microbial inoculants for specific bioremediation, ensuring their high efficiency in the degradation or transformation of target pollutants in the environment.

## 21.4 Mode of Action of Microbial Inoculants in Bioremediation

The interaction of microbial flora with diverse pollutants is essential for effective bioremediation strategies. However, the interaction varies with the nature and complexity of the pollutant and microbial strains used for detoxification (Tayang and Songachan 2021). As a result, the mode of action of microbial inoculants varies and solely depends on their nature of interactions with environmental pollutants, which can be grouped into several key aspects.

### 21.4.1 Biosorption

Biosorption is an interactive remediation technique that removes toxic pollutants from aqueous solutions rapidly by binding nonspecifically with the cell surface of bacteria, fungi and algae (biosorbent). The biosorbent used should be cheap, reusable, have faster kinetics, and be recovered efficiently (Vijayaraghavan and Yun 2008). Microbial strains such as *Bacillus anthracis*, *Rhodococcus erythropolis*, *Streptomyces* sp., *Bacillus subtilis*, *Rhizopus arrhizus*, *Saccharomyces cerevisiae,* and microalgae are potent biosorbents applied for biological cleanup of heavy metals and other pollutants (Pande et al. 2022; Wang and Chen 2009).

### 21.4.2 Bioleaching

Bioleaching is an economical and eco-friendly process that uses acidophiles to transform solid chemicals within the contaminated soil into soluble substances through extraction and thereby recover pollutants (mostly iron and sulfur metals). This process specifically involves iron and sulfur oxidizing bacteria, including heterotrophic bacterial strains (*Aspergillus* and *Penicillium* sp.), acidophiles (*Acidithiobacillus ferroxidans* and *Acidithiobacillus thioxidans*), and thermophiles (*Sulfobacillus thermosulfidooxidans*), which generate an acidic environment to solubilize the heavy metals (Medfu Tarekegn et al. 2020; Roy and Roy 2015).

### 21.4.3 Bioprecipitation

Bioprecipitation involves the conversion of harmful pollutants into precipitates or crystals to reduce toxicity using suitable bioreactor so that stable metal ions are recovered (Ayilara and Babalola 2023; Sharma et al. 2022). Microbial strains such as *B. subtilis* and *Oceanobacillus indicireducens* are mostly used to precipitate heavy metals from environmental pollutants (Maity et al. 2019; Willis and Donati 2017).

### 21.4.4 Bioaccumulation

Bioaccumulation is one of the potent strategies used for bioremediation with import-storage system that stores toxic compounds in the intracellular portion of microbes through endocytosis, ion channels, protein channels, and carrier-mediated transports (Geva et al. 2016; Pande et al. 2022). The applications of bioaccumulation for detoxification include the accumulation of cadmium by *Pseudomonas putida* (Rani and Goel 2009), thorium and lead accumulation by *Aspergillus niger* (Naskar et al. 2020; Sharma et al. 2022), chromium accumulation by *Monodictys pelagic* (Sher and Rehman 2019), as well as *R. arrhizus* (Sharma et al. 2022).

### 21.4.5 Biotransformation

The process of biotransformation uses microbial inoculants with enzymatic machinery necessary for the breakdown of contaminants supplemented with heavy metals like selenium, mercury, and chromium. Mechanisms of enzyme-mediated actions involve oxidative, reductive, condensation, ring opening, and hydrolysis that act on specific pollutants for bioremediation (Pande et al. 2022). Microbes such as *Acinetobacter* sp., *Micrococcus* sp., and *Bacillus* sp. produce hydrocarbon-degrading enzymes like alkane monooxygenases or aromatic ring-cleaving dioxygenases, whereas fungal strains such as *Pycnoporus* sp. secrete ligninolytic enzymes like peroxidases and laccases for heavy metal degradation (Gangola et al. 2018). These enzymes enable existing microbial inoculants to initiate and accelerate the degradation process of contaminants, thereby converting them into less harmful or readily biodegradable forms. However, the enzyme-mediated transformation exhibited by the microbial inoculants is an expensive and time-consuming process (Narayanan et al. 2023).

## 21.5 Applications of Microbial Inoculants

Microbial inoculants have immense potential for various bioremediation applications due to their unique features and mechanisms of action in dealing with different contaminants and environmental toxins. Microbial inoculants are instrumental in detoxification and cleanup of a wide variety of contaminated environments, including petroleum hydrocarbons in soil, water, and oil spills (Nwankwegu et al. 2022), chlorinated compounds (Dutta et al. 2022), toxic heavy metal pollutants (Sayqal and Ahmed 2021; Tayang and Songachan 2021), pesticides and herbicides (Odukkathil and Vasudevan 2013), organic pollutants/solvents (PAHs, polyethene, crude oil), and emerging contaminants such as hospital wastes, pharmaceuticals, personal care products, and endocrine disrupting compounds, as well as complex matrices (Ayilara and Babalola 2023; Wang et al. 2023) (Table 21.1).

Additionally, microbial bioremediation is utilized for the degradation of plastic and polyethylene pollutants (Sharma 2018). Moreover, microbial inoculants find application in remediating agricultural pollutants to enhance soil quality, crop yield, plant productivity, and overall agricultural output (Kour et al. 2022; Raj et al. 2022). Consequently, microbes are regarded as potential agents for detoxifying and degrading a diverse array of pollutants that endanger the environment and ecosystem functionality, thus contributing to improved environmental health and human well-being.

## 21.6 Process Optimization for Enhanced Bioremediation

### 21.6.1 Optimization of Microbial Inoculants and Environmental Parameters

The optimization of microbial inoculants and environmental parameters is crucial in ensuring their efficacy in bioremediation processes. The achievement of restoration initiatives hinges on the careful consideration of multiple factors. Firstly, microbial inoculants possessing unique enzymatic capabilities and enzyme-mediated metabolic pathways should be selected, targeting the specific environmental pollutant of interest for efficient breakdown. Secondly, optimization of environmental conditions for a specific contaminated site is essential for implementing appropriate bioremediation strategies (Sayqal and Ahmed 2021). The key environmental determinants for optimization include the optimal range of pH (5.5–8.8), moisture (25–28%), temperature (15–45°C), oxygen availability (usually aerobic), and nutrient levels (carbon, nitrogen, and phosphate) essential for maximizing the survival and activity of microbial inoculants supplemented at target contaminated site (Vidali 2001; Medfu Tarekegn et al. 2020). The rate of response and microbial growth are influenced by temperature and pH; aerobic or anaerobic bioremediation is determined by the availability of oxygen; microbial growth and metabolism are influenced by the availability of nutrients; ensuring that there is enough oxygen for aerobic degradation and controlling the moisture level to generate the ideal circumstances to increase the degradation efficiency (Sayqal and Ahmed 2021). Thirdly, the adoption of microbial inoculants with relatively high genetic and functional diversity enhances their overall degradation potential and resilience (Das and Dash 2014). Moreover, it is important to take into account the interplay between introduced microbial inoculants and the preexisting indigenous microbial communities in the polluted area. This consideration seeks to ensure compatibility and advantageous interactions, ultimately amplifying the overall degradation effectiveness (Abatenh et al. 2017). Nonetheless, strategies for formulation and delivery, including carrier materials or protective matrices, are being devised to enhance the resilience of microbial inoculants during storage and application. These approaches aim to guarantee their successful dispersal and establishment within the polluted site. Regular monitoring of the performance and activities exhibited by microbial inoculants during bioremediation is essential to assess their viability, survival, growth patterns, and degradation efficiency over time, which facilitates precise adjustment based on the nature of application or environmental heterogeneity to optimize their overall performance (Ayilara and Babalola 2023). Besides, scalability considerations, including their relative abundance and distribution and economic feasibility, are vital to ensure successful large-scale application of the chosen microbial inoculants (Yadav et al. 2022). With careful consideration and optimization of these factors, the application of microbial inoculants can be maximized, resulting in successful bioremediation efforts.

### 21.6.2 Genetic Modification in Microbial Inoculants

Genetic modification and engineering methods are being explored as viable means to amplify the effectiveness of microbial inoculants engaged in bioremediation processes (Rafeeq et al. 2023). These approaches involve the manipulation of microbial genome to introduce specific traits or augment their existing characteristics (Arunraja et al. 2023). Diverse tactics have been utilized, encompassing gene transfer, a process where particular genes responsible for the degradation of specific contaminants are inserted into microbial inoculants (Ikuma and Gunsch 2012); metabolic pathway engineering, which entails the introduction or enhanced expression of genes encoding enzymes related to the decomposition of targeted pollutants; genetic regulation involves the incorporation of the inducible promoters, repressor elements or other regulatory elements into the genome of the microbial inoculants that activate or repress target genes in response to specific environmental signals or conditions, thereby regulating the enzyme production and enzyme-mediated metabolic activities and synthetic biology approaches used for designing and constructing synthetic materials incorporated in microbial systems (Janssen and Stucki 2020). Besides, synthetic biology techniques facilitate the *de novo* design and assembly of genetic components to achieve the desired functions and optimize the performance of bioremediation. Several researchers have proposed different genetic alterations in microbial strains, such as improved biofilm formation that facilitates microbial colonization and persistence in contaminated environments (Peacock et al. 2004; Singh et al. 2006); improvement of stress tolerance in microbial inoculants to promote stress tolerance capabilities; and quorum sensing manipulation to increase the synchronization of biodegradation of environmental pollutants (French et al. 2020). These methods involved genetic alterations in microbial inoculants, which are crucial in light of biosafety, regulatory, and moral considerations. In order to guarantee the security and responsible application of genetically modified microbial inoculants in bioremediation, proper risk assessment and monitoring are equally important.

## 21.7 Challenges and Future Prospects of Microbial Inoculants

Microbial inoculants show immense potential in the context of bioremediation but face certain challenges and limitations that require considerable attention. Firstly, environmental complexities arise due to diverse pollutants and fluctuations in environmental conditions. It is difficult to design microbial inoculants that successfully target and destroy various pollutants in such complicated settings (Alori et al. 2022). Besides, microbial inoculants must endure and persist in the contaminated environment to carry out effective bioremediation. However, competition with indigenous microbes, adverse conditions, and limited nutrient availability hinders their survival and persistence (Rafeeq et al. 2023). Further, effective delivery and distribution of microbial inoculants to target sites pose practical challenges especially in heterogeneous environments (Vassilev et al. 2020). Additionally, the involvement of microbial inoculants in sterile and harsh environments with higher salinity, extreme temperature, and low oxygen levels necessitates the development of specialized inoculants (Kochhar et al. 2022; Trabelsi and Mhamdi 2013). Furthermore, scaling up microbial inoculants for large-scale bioremediation is problematic due to comparatively higher cost, restricted scalability, and requirements of specialized facilities (Helmy et al. 2015). Also, the periodic monitoring and assessment of the efficiency of microbial inoculants is crucial to determine the progress of bioremediation, which entails many sophisticated procedures (Chikere 2013). In addition, the microbial inoculants are also subjected to various regulatory constraints.

Therefore, to fully capitalize on their promises for bioremediation, new avenues of improvement must be explored. Designing multifunctional inoculants capable of degrading a wide range of contaminants with adaptation to dynamic environmental changes (Rebello et al. 2021). Enhancement in survival and persistence of microbial strains, improving scalability, developing reliable monitoring and assessment techniques, and microbial inoculants can significantly contribute to effective and sustainable environmental cleanup (Das and Dash 2014; Qiu et al. 2019). Besides, developing innovative delivery methods like encapsulation techniques, aerosolization, and biofilm carriers improves the uniformity and efficiency of microbial inoculants toward bioremediation (Catania et al. 2020). Further, integration with other remediation strategies has synergistic effects for microbial inoculants with significant tolerance and resilience to environmental extremes, capable of mitigating heavy metal toxicity and converting complex pollutants into biodegradable form. Also, researchers, regulatory agencies, and stakeholders must work together to produce clear standards and rules that support the responsible and safe use of microbial inoculants. By pursuing innovative solutions, microbes can act as powerful tools for bioremediation that offer great promise for the ecological restoration of contaminated environments (Alori et al. 2022).

## 21.8 Ecological Consequences

Although microbial inoculants play a significant role in bioremediation, as it is critical to examine the potential ecological ramifications of their use. Some of the important environmental issues linked with the microbial remediation strategies are listed below.

### 21.8.1 Disruption of Indigenous Microbial Communities

Application of microbial inoculants into contaminated environment could potentially disrupt the existing indigenous microbial communities, leading to a decline in microbial biodiversity and associated ecological processes (Trabelsi and Mhamdi 2013; Manfredini et al. 2021). Therefore, to maintain ecosystem stability and diversity, careful selection of microbial inoculants that are compatible with resident microbial communities is crucial.

### 21.8.2 Shift in Microbial Community Structure

Introducing the microbial inoculants into polluted environments has the potential to disturb the established native microbial communities, resulting in the decline of microbial diversity and its related ecological functions. Such variation in their relative distribution and abundance has a cascade effect that alters biogeochemical cycling, nutrient turnover, organic matter decomposition, biotransformation, and other ecosystem processes. The long-term consequences of such alternations are still not well documented and need further study to assess the ecological impacts of bioremediation (Trabelsi and Mhamdi 2013).

### 21.8.3 Gene Transfer and Modification

The potential risks of genetically modified organisms (GMOs) used as bioremediates include horizontal transfer of recombinant genes, genetic contamination, ecological damage, and negative impacts on human health (Perera and Hemamali 2022). Additionally, it could have a lot of undiscovered and indirect environmental effects. Thereby, the development of ecologically safe genetically modified microbial inoculants for bioremediation with minimal adverse effects is highly recommended.

### 21.8.4 Ecological Niche Alteration

Introduction of microbial inoculants alters the ecological niches within the contaminated environment. Inoculants may compete with indigenous microbes for resources and occupy niches that were occupied by other microbial populations, which disrupts natural ecological balance and interactions within ecosystem. Careful consideration toward compatibility of inoculants with the existing ecological niches is necessary to minimize ecological disturbances.

### 21.8.5 Potency for Unintended Effects

Use of microbial inoculants for bioremediation results in unforeseen outcomes that are not immediately evident. Microbial-mediated unintended ecological ramifications include modifications in plant–microbe relationships, pressure on nontarget microbes, or disruptions in ecosystem dynamics (Perera and Hemamali 2022). Comprehensive risk assessment is the need of the hour to detect and address ecological problems linked with their application.

## 21.9 Assessment and Implementation of Microbial Inoculants

The assessment of microbial inoculants is not only useful for the evaluation of their performance and behavior in specific environment but also for the implementation of appropriate remediation strategies that deal with the importance of monitoring and implementation, highlighting the use of various techniques (Andreoni and Gianfreda 2007; Manfredini et al. 2021). Effective monitoring techniques allow the real-time tracking of the survival and activities of microbial inoculants, their interactions with contaminants, as well as existing native microbes (Chandran et al. 2020), which provide appropriate insights into the efficacy and long-term impact of microbial inoculants in bioremediation.

Metagenomics study is useful for identification, quantification, and characterization of specific microbial strains or genetic markers associated with the inoculants (Awasthi et al. 2022). Polymerase chain reaction, fluorescent *in situ* hybridization (FISH), next-generation sequencing, terminal restriction fragment length polymorphism (T-RFLP), and stable isotope probing are some of the techniques used for monitoring bioinoculants (Chandran et al. 2020; Manfredini et al. 2021; Schmidt and Eickhorst 2014). Besides, metaproteomic and metabolomics studies involve the analysis of proteins or metabolites present in microbial communities, which provide insights into the functional activities and metabolic profiles of microbial inoculants. By comparing proteome or metabolomic profiles of environmental samples before and after the application of microbial inoculants, the changes in their relative abundance and microbial activities can be assessed. Utilizing omics approaches (metagenomics, metaproteomic, and metabolomics), the researchers can identify, quantify, and track the inoculant strains, their abundance, and functional activities in complex environmental samples (Rodríguez et al. 2020) that provide insights into the behavior, persistence, and effectiveness of microbial inoculants and analyze the functional genes responsible for catabolism of pollutants, aiding in the optimization and success of bioremediation (Sharma et al. 2022).

Biomarkers are specific molecules used as indicators for determining the presence, activities, and abundance of microbial strains useful for monitoring microbial inoculants (Jansson et al. 2000). Several biomarkers used for monitoring microbial inoculants involved in bioremediation include fatty acid methyl esters, phospholipid fatty acid analysis, stable isotope probing, and genetic markers (Pandey et al. 2009). Biomarker analysis is one of the potent tools used to assess the presence, activity, and persistence of microbial inoculants in specific environment. Such biomarker-based approach is useful to track the survival and efficacy of microbial inoculants, their interactions with existing microbial communities, and their overall performance in bioremediation applications (Jansson et al. 2000).

Moreover, establishing a well-planned and efficient monitoring protocol for microbial inoculants is essential for evaluating their effectiveness, longevity, and potential consequences for the intended environment (Manfredini et al. 2021). The well-designed monitoring program ensures the effectiveness and safety of microbial inoculants throughout the bioremediation process, which delves into the essential components and stages encompassing the development and execution of a robust monitoring program for microbial inoculants. The primary emphasis is on setting clear monitoring goals, encompassing tasks like quantifying native microorganisms, observing alterations in microbial community composition, gauging pollutant degradation rates, and assessing relevant environmental factors. Secondly, method selection is pivotal, necessitating techniques that exhibit sensitivity, specificity, and reliability to provide accurate data on the presence, activities, and relative ratios of inoculants (Schmidt and Eickhorst 2014). An effective sampling strategy can be devised for encapsulating both spatial and temporal dimensions, tailored to the specific contaminated site. Further, determining optimal frequency and timing of sampling and synchronizing with key bioremediation stages is vital to assess the persistence of the inoculants over time. Additionally, aseptic methodologies to thwart contamination and preserve samples in appropriate ways to maintain viability during transport are crucial (Sharma et al. 2022). The subsequent phase entails data analysis and interpretation, entailing meticulous statistical procedures and pattern recognition (Foran et al. 2015). Integrating an adaptive management strategy guided by monitoring results improves the effectiveness of bioremediation (Foran et al. 2015). Continual evaluation and refinement of the monitoring program, adapting objectives, methodologies, and sampling strategies in response to emerging data, optimize monitoring endeavors, and elevate the overall performance of microbial inoculants (Sharma et al. 2022). Regular monitoring, coupled with adaptive management strategies, bolsters bioremediation success, ensuring both the safe utilization and efficiency of microbial inoculants.

## 21.10 Case Studies and Success of Restoration Efforts

Some of the case studies and success stories of microbial inoculants involved in bioremediation, demonstrating their potency and effectiveness in remediating environmental contamination challenges, are discussed.

### 21.10.1 Oil Spill Remediation

Oil spills pose significant environmental catastrophes that have severe impacts on ecosystem and human health. As evident from case studies, microbial inoculants are considered as one of the most effective bioremediation agents for oil-contaminated environment. Leveraging the inherent capabilities of oil-degrading microbes (oleophilic), oil biodegradation can be enhanced, mitigating the ecological impacts of oil spills. Nevertheless, ongoing research and development remain imperative to optimize their role in bolstering oil spill bioremediation strategies. The viability of this approach is illustrated by the following instances: Following the Exxon Valdez oil spill in Alaska in 1989, various oil-degrading bacterial strains, including *P. putida*, *Rhodococcus* sp., and Mycobacterium sp., have demonstrated their efficacy as effective bioremediators. These strains utilize metabolic pathways to enhance the breakdown of hydrocarbons and mitigate the consequences of oil spills (Arora et al. 2022; Tyagi et al. 2011). Likewise, studies on the Deepwater Horizon oil spill in 2010 reported that hydrogen-degrading bacterial inoculants (*Alcanivorax borkumensis*, *Marinobacter hydrocarbonoclasticus*, and *P. putida*) accelerate the breakdown of petroleum through bioaugmentation, hastening ecological recovery (Geng et al. 2022). Furthermore, investigations have highlighted the enrichment of *Alteromonas*, *Pseudoalteromonas, and Cycloclasticus* strains in response to oil spills. These strains have demonstrated the capability to degrade various aliphatic and aromatic hydrocarbons, rendering them well-suited for oil bioremediation purposes (Gutierrez et al. 2018). Another study by Bôto et al. (2021) highlighted that indigenous microbialstrains like *Pseudomonas* sp., *Alcanivorax*, and *Thalassospira* hold significant promise for the degradation of aromatic compounds, thereby offering substantial potential for mitigating the impact of oil spills and contributing to the successful restoration of the Iberian coastal ecosystem. Furthermore, a study reported that oleophilic microbes (*Pseudomonas aeruginosa*) break down the spilled oil through the process of composting (Mukjang et al. 2022). According to Pineda-Flores et al. (2004), consortium of bacterial strains encompassing *Corynebacterium*, *Bacillus*, *Brevibacillus,* and *Staphylococcus* species utilizes asphaltene as sole carbon source and can be used as potent detoxification agents in oil spills. Furthermore, studies have reported that genetically modified microbes (*Pseudomonas fluorescens*) with enhanced naphthalene degradation ability have been implemented for oil bioremediation (Adams et al. 2015). Furthermore, the utilization of microbial consortia has been employed to enhance the reclamation process of oil sand tailings (Chegounian et al. 2020). In each of these cases, microbial inoculants showcased their ability to work in tandem with natural processes, effectively mitigating the impacts of oil spills and facilitating the process of environmental recovery.

### 21.10.2 Heavy Metal Contaminated Sites

The application of various microbial inoculants holds significant potential for the bioremediation of sites contaminated with hazardous heavy metals such as cadmium, mercury, arsenic, lead, and uranium (Kour et al. 2022; Tanmoy and Saha 2019; Zeraatkar et al. 2016). Several case studies demonstrate the efficacy of specific microbes with metal-resistant and metal-transforming abilities, in resolving heavy metal contamination and improving the restoration process (Naz et al. 2022). For example, in a cadmium-contaminated site, several researchers reported the use of microbial species like *Aspergillus*, *Pseudomonas*, *Stenotrophomonas*, *Paecilomyces*, *Cladosporium*, *Trichoderma,* and *Bacillus* for effective detoxification (Ayilara and Babalola 2023; Kour et al. 2022; Unuofin et al. 2021). These metal-resistant bacteria sequester and immobilize cadmium, significantly reducing its levels and restoring soil fertility (Aibeche et al. 2022; Kumar et al. 2021; Li et al. 2023). In mercury-contaminated sediment, microbial inoculants (*Pseudomonas monteilii*) have demonstrated effectiveness in mercury remediation that form consortia for enhanced volatilization and methylation processes (Kumari et al. 2020; Yadav et al. 2022). Subsequently, microalgae (*Microcystis aeruginosa*) not only have high arsenic-tolerance capacity but also have high tolerance range for pH changes and are considered as a potent microalga for arsenic bioremediation (Wang et al. 2018b).

A recent study conducted by Song et al. (2022) demonstrated the detoxification of the carcinogenic heavy metal lead using *Enterobacter cloacae*. This microbial strain exhibited the capability to immobilize and transform lead, resulting in a decrease in the bioavailability and mobility of the metal. Similarly, in India, a study focused on chromium-contaminated soil employed a consortium of chromium-tolerant bacteria (*Strenotrophomonas maltophilia*) that effectively transforms mutagenic, teratogenic, and carcinogenic chromium into less toxic forms (Bakshi and Panigrahi 2022). Further, Jalilvand et al. (2020) reported the application of urease-producing bacterial isolates (*Stenotrophomonas rhizophila*, *Variovorax boronicumulans,* and *Sporosarcina pasteurii*) for biomineralization of various heavy metals (lead, cadmium, and zinc) with higher stability and efficacy. These instances collectively underscore the potential use of microbial inoculants to address various heavy metal contaminations and offer promising remediation solutions.

In addition to heavy metals, microbial inoculants are also used for the bioremediation of radionuclides like uranium, neptunium, technetium, etc. Various microbial species, such as *Serratia*, *Halomonas*, *Pseudomonas*, *Pseudanabaena*, *Desulfovibrio*, and *Denitromonas*, can directly interact with the radionuclides through the processes of biosorption, biomineralization, biooxidation, bioreduction, and bioaccumulation, thereby offering a more sustainable and efficient approach to bioremediation (Williamson et al. 2023).

### 21.10.3 Pesticide and Herbicide Degradation

The utilization of microbial inoculants for pesticide and herbicide degradation is gaining tremendous interest for sustainable agricultural practices (Bhatt et al. 2023). Prominent case studies and research findings underscore the potential use of specialized microbial inoculants in the degradation of pesticides and herbicides, contributing to soil detoxification and mitigating their adverse effects. Effective microbial strains such as *Alcaligenes*, *Flavobacterium*, *Actinomyces*, *Nocardia*, *Streptomyces*, *Cladosporium*, *Aspergillus*, *Penicillium*, *Trichoderma*, *Pseudomonas*, *Mucor*, and *Bacillus* have been recognized for their capability in degrading various herbicides and pesticides such as parathion, endosulfan, alachlor, methyl parathion, DDT, and atrazine. This microbial action contributes significantly to diminishing the risk of contamination in essential food and water sources (Kour et al. 2022). Atrazine is a predominantly used herbicide that persists in the soil and poses risks to environment. Research has identified that bacteria such as *Arthrobacter* sp., *Pseudomonas* sp., and *B. subtilis*, which possess atrazine-degrading capabilities, can effectively facilitate the complete degradation of atrazine. This microbial action plays a pivotal role in substantially reducing the persistence of atrazine in soil (Abd Rani et al. 2022). Similarly, DDT is a persistent pesticide that not only imposes adverse effects on ecosystem and nontarget creatures but also has a carcinogenic nature. Ito et al. (2021) reported that gram-positive bacterial strains *Streptomyces achromogenes* can potentially degrade the DDT into harmless compounds and prevent their accumulation in the soil. Subsequently, microbial strains like *Alcaligenes* sp., *Rhodococcus* sp., *Chryseobacterium* sp., *Sphingobacterium* sp., and *Stenotrophomonas* sp. have also been recognized as DDT-degrading bacteria that can effectively promote DDT degradation, leading to decline in its concentration and reducing its environmental impact (Ito et al. 2021). Organophosphate pesticides (like malathion, ethion, chlorpyrifos, and methyl parathion, dimethoate) are extensively used in agriculture and contaminates water bodies and pose toxic effects. Researchers revealed that fungal isolates (*Aspergillus fumigatus*, *Tricholoma giganteum*, and *Aspergillus terreus*) have inherent potential for the breakdown of organophosphate pesticides in the environment (Mali et al. 2023). Additionally, several other microbial inoculants including bacterial, fungal, and algal strains (*Pseudomonas*, *Flavobacterium*, *Chlorella, Penicillium*, *Tricholoma*, *Arthrobacter*, *Nostoc*, *Anabaena*, and *Spirulina*) also aid in the degradation and removal of organophosphate pesticides, which significantly declines their concentration and thereby minimizes their adverse impacts on aquatic ecosystem (Mali et al. 2023). Subsequently, mixed cultures of *Bacillus* and *Staphylococcus* species have been reported for endosulfan detoxification (Kumar and Philip 2006).

Similarly, the bioremediation of persistent organic pollutants (glyphosate) involves the use of potent glyphosate-resistant microbes like *Bacillus cereus*, *Agrobacterium tumefaciens*, *E. cloacae*, *P. putida*, *P. fluorescens*, *Chryseobacterium* sp., *Micrococcus* sp., *Corynebacterium* sp., *Achromobacte* sp., *Rhodococcus* sp., and *Streptomyces* sp. that efficiently degrade and metabolize the glyphosate in the environment and reduce its negative effects (Masotti et al. 2023). Another study reveals the combined utilization of a fungal strain (*Arbuscular mycorrhizal*) and a bacterial strain (*Hansschlegelia zhihuaiae*), which has a synergistic impact on the detoxification of an herbicide (bensulfuron-methyl) for soil cleanup (Bhatt et al. 2023).

### 21.10.4 Industrial Waste Treatment

The industrial sector stands as a significant contributor of toxic pollutants, giving rise to severe environmental threats and health-related concerns. Moreover, the harmful industrial effluents not only cause soil and water pollution but also severely affects human health. Microbial degradation using specific detoxifying microbes holds the assurance for efficient remediation of industrial waste for environmental cleanup. For instance, hydrocarbon degraders such as *Proteobacteria*, *Actinobacteria*, *Firmicutes*, *Bacteroidetes,* and *Chlamydiae* isolated from petroleum hydrocarbon-contaminated soil were proven to be efficient for bioremediation of petroleum refinery effluents (Chirwa and Bezza 2015). Similarly, augmented microbial consortium of *Galactomyces geotrichum*, *Penicillium simplicissimum*, and *Bacillus* sp. were able to detoxify the textile effluents contaminated with heavy metals, recalcitrant dyes, and xenobiotics (Patil et al. 2020). Further, Khan and Malik (2018) employed indigenous dye-degrading *Arthrobacter* sp. for the degradation of textile effluents. Furthermore, microbial strains like *Ochrobactrum intermedium*, *Staphylococcus aureus*, *P. aeruginosa*, *Bacillus fusiformis*, *Acinetobacter junii,* and *Micrococcus luteus* have been reported as potent tools for the detoxification of textile dyes (Khalid et al. 2016; Kour et al. 2022).

Likewise, within the realm of food processing, a combination of bacteria known as a consortium, encompassing amylolytic and cellulolytic strains like *B. thuringiensis*, *Bacillus licheniformis*, *Brevibacillus borstelensis*, and *B. licheniformis*, has been harnessed to break down food waste (Awasthi et al. 2018). Also, fungal strains such as *S. cerevisiae* have been employed for the removal and detoxification of heavy metal pollutants in the food industry. Further, fungal strain *Aspergillus sclerotiorum*, with its high lipolytic activity, has been used as potent detoxifying agent for effluents produced by diary industry (de Moura Dickel et al. 2023). Moreover, fungal varieties like *Aspergillus niger*, *A. oryzae*, and *Trichoderma viride* are recognized as highly effective candidates for the bioremediation of winery wastewater (Pandey et al. 2009).

Similarly, the paper and pulp industry generate substantial quantities of hazardous effluents, serving as a primary contributor to aquatic pollution. Research findings have documented the utilization of various bacterial isolates, including *P. putida*, *Bacillus megaterium*, *P. aeruginosa*, *Serratia marcescens*, *Pannonibacter* sp., and *Ochrobactrum* sp., not only for the decolorization but also for the reduction of both chemical and biochemical oxygen demand in the effluents

(Garg et al. 2012; Sonkar et al. 2021). Similarly, Sonkar et al. (2020) reported the efficacy of bacterial consortium consisting of *Pseudomonas* sp., *Pseudomonas stutzeri*, and *Bacillus* sp. for the detoxification of paper mill sludge. Moreover, another investigation has reported the utilization of three white rot fungi (*Trametes versicolor*, *Phanerochaete chrysosporium,* and *Corius versicolor*) and a saprophytic fungus, *Gliocladium virens* as novel strains for color detoxification in paper and pulp mill effluents (Malaviya and Rathore 2007). Other than that, improper handling or accidental spills in chemical industry release chlorinated hydrocarbons into the nature causing water and soil pollution that endanger human health. Various bacterial species, including *Dehalogenimonas*, *Desulfoluna*, *Desulfomonile*, and *Dehalococcoides*, recognized as proficient dechlorinators, have been harnessed for the breakdown of chlorinated compounds (Dutta et al. 2022).

## 21.11 Conclusion

Increased environmental pollution owing to changing industrial activities poses significant threats to mankind. Bioremediation, a green approach, emerged as the potential solution to resolve the environmental issues. Microbial bioremediation for detoxification of contaminated environments is an eco-friendly and economical approach. These specialized microbial inoculants offer unique advantages for degradation, transformation, elimination, and detoxification of diverse pollutants, which lead to effective remediation solution for sustainable ecological restoration. Indigenous as well as genetically modified microbes have demonstrated successful bioremediation applications in oil spills, heavy metal pollutants, pesticide/herbicide degradation, and industrial effluents. Mechanisms including enzymatic degradation, bioaccumulation, biofilm formation, and metabolic transformations by microbial inoculants enable efficient breakdown of complex pollutants and mitigate their harmful effects on environment. However, challenges and limitations still exist. Monitoring programs incorporating molecular tools, biomarker analysis, and real-time systems are essential for the periodic assessment of performance and efficacy of microbial inoculants in bioremediation. Further, detailed accounts of ecological consequences and regulatory aspects related to the use of microbes are essential to ensure their safe and responsible deployment. Therefore, in-depth understanding of the bioremediation mechanism and awareness about ecological implications of microbial-mediated remediation approaches are significantly important.

## 21.12 Future Perspectives

To unlock the full potential of microbial inoculants, intensive research is required to explore and integrate emerging techniques like genetic engineering, synthetic biology, metagenomics, nanotechnology, and microbiome engineering into bioremediation approaches. Integrated approaches will provide multifaceted and comprehensive strategy for long-term sustainable development. In addition, these will improve capabilities of microbial inoculants, enabling tailored and targeted applications at specific contaminated site. With further, progress in metagenomic methods, new and novel detoxifying genes will be identified, accelerating the bioremediation process. Additionally, the use of genetically modified microbes with diverse metabolic activities and overexpression of enzyme-producing genes will be a better approach to increase the rate of bioremediation. Thus, continued research to unravel new and potent microbial strains and their field trials to demonstrate their efficacy for bioremediation of toxic pollutants is important. This entails that by harnessing the remarkable potential of microbial inoculants with advanced technology, a more practical, efficient, economical, and ecologically sound solution for bioremediation and revitalization of the contaminated environment can be achieved. Moreover, public acceptance and awareness of the benefits and safety of microbial inoculants are crucial for their widespread adoption. Global collaboration between industries, government agencies, and communities is vital for promoting the utilization of microbes in bioremediation for sustainable development.

## Acknowledgment

The author, Miss. Ankita Agrawal, would like to acknowledge the Department of Science and Technology (DST), Govt. of India, New Delhi, for providing fellowship through DST/INSPIRE FELLOWSHIP/2020/IF200208, and to the Department of Biotechnology and Bioinformatics, Sambalpur University, for providing necessary scientific knowledge.

## References

Abatenh, E., Gizaw, B., Tsegaye, Z., and Wassie, M. (2017). The role of microorganisms in bioremediation – a review. *Open Journal of Environmental Biology* 2 (1): 038–046.

Abd Rani, N.F., Ahmad Kamil, K., Aris, F. et al. (2022). Atrazine-degrading bacteria for bioremediation strategy: a review. *Biocatalysis and Biotransformation* 40 (4): 233–247.

Abo-Alkasem, M.I., Hassan, N.M.H., and Abo Elsoud, M.M. (2023). Microbial bioremediation as a tool for the removal of heavy metals. *Bulletin of the National Research Centre* 47 (1): 1–15.

Abo-Amer, A.E., Abu-Gharbia, M.A., Soltan, E.S.M., and Abd El-Raheem, W.M. (2014). Isolation and molecular characterization of heavy metal-resistant *Azotobacter chroococcum* from agricultural soil and their potential application in bioremediation. *Geomicrobiology Journal* 31 (7): 551–561.

Abou-Shanab, R.A., El-Sheekh, M.M., and Sadowsky, M.J. (2019). *Role of Rhizobacteria in Phytoremediation of Metal-Impacted Sites. Emerging and Eco-Friendly Approaches for Waste Management*, 299–328. Springer.

Adams, G.O., Fufeyin, P.T., Okoro, S.E., and Ehinomen, I. (2015). Bioremediation, biostimulation and bioaugmentation: a review. *International Journal of Bioremediation Biodegradation* 3 (1): 28–39.

Adetunji, C.O. and Anani, O.A. (2021). Bioaugmentation: a powerful biotechnological technique for sustainable ecorestoration of soil and groundwater contaminants. *Microbial Rejuvenation of Polluted Environment* 1: 373–398.

Agrawal, N., Verma, P., and Shahi, S.K. (2018). Degradation of polycyclic aromatic hydrocarbons (phenanthrene and pyrene) by the ligninolytic fungi *Ganoderma lucidum* isolated from the hardwood stump. *Bioresources and Bioprocessing* 5 (1): 1–9.

Aibeche, C., Selami, N., Zitouni-Haouar, F.E.H. et al. (2022). Bioremediation potential and lead removal capacity of heavy metal-tolerant yeasts isolated from Dayet Oum Ghellaz Lake water (northwest of Algeria). *International Microbiology* 25: 1–13.

Akhtar, N. and Mannan, M.A.U. (2020). Mycoremediation: expunging environmental pollutants. *Biotechnology Reports* 26: e00452.

Akpasi, S.O., Anekwe, I.M.S., Tetteh, E.K. et al. (2023). Mycoremediation as a potentially promising technology: current status and prospects – a review. *Applied Sciences* 13 (8): 4978.

Alegbeleye, O.O., Opeolu, B.O., and Jackson, V. (2017). Bioremediation of polycyclic aromatic hydrocarbon (PAH) compounds: (acenaphthene and fluorene) in water using indigenous bacterial species isolated from the Diep and Plankenburg rivers, Western Cape, South Africa. *Brazilian Journal of Microbiology* 48 (2): 314–325.

Alori, E.T., Gabasawa, A.I., Elenwo, C.E., and Agbeyegbe, O.O. (2022). Bioremediation techniques as affected by limiting factors in soil environment. *Frontiers in Soil Science* 2: 937186.

Andreoni, V. and Gianfreda, L. (2007). Bioremediation and monitoring of aromatic-polluted habitats. *Applied Microbiology and Biotechnology* 76: 287–308.

Annweiler, E., Richnow, H.H., Antranikian, G. et al. (2000). Naphthalene degradation and incorporation of naphthalene-derived carbon into biomass by the thermophile *Bacillus thermoleovorans*. *Applied and Environmental Microbiology* 66 (2): 518–523.

Aparicio, J.D., Raimondo, E.E., Saez, J.M. et al. (2022). The current approach to soil remediation: a review of physicochemical and biological technologies, and the potential of their strategic combination. *Journal of Environmental Chemical Engineering* 10 (2): 107141.

Aranda, E., Ullrich, R., and Hofrichter, M. (2010). Conversion of polycyclic aromatic hydrocarbons, methyl naphthalenes and dibenzofuran by two fungal peroxygenases. *Biodegradation* 21 (2): 267–281.

Arora, S., Saxena, S., Sutaria, D., and Sethi, J. (2022). Bioremediation: an ecofriendly approach for the treatment of oil spills. In: *Advances in Oil-Water Separation* (ed. P. Das, S. Manna, and J.K. Pandey), 353–373. Elsevier.

Arunraja, D., Romauld, S.I., Devi, P.B. et al. (2023). Genetically engineered microbes for bioremediation and phytoremediation of contaminated environment. In: *Metagenomics to Bioremediation* (ed. V. Kumar, M. Bilal, S.K. Shahi, and V.K. Garg), 709–721. Academic Press.

Awasthi, M.K., Wong, J.W., Kumar, S. et al. (2018). Biodegradation of food waste using microbial cultures producing thermostable α-amylase and cellulase under different pH and temperature. *Bioresource Technology* 248: 160–170.

Awasthi, M.K., Ravindran, B., Sarsaiya, S. et al. (2022). Metagenomics for taxonomy profiling: tools and approaches. *Bioengineered* 11: 356.

Ayangbenro, A.S. and Babalola, O.O. (2017). A new strategy for heavy metal polluted environments: a review of microbial biosorbents. *International Journal of Environmental Research and Public Health* 14 (1): 94.

Ayilara, M.S. and Babalola, O.O. (2023). Bioremediation of environmental wastes: the role of microorganisms. *Frontiers in Agronomy* 5: 1183691.

Azubuike, C.C., Chikere, C.B., and Okpokwasili, G.C. (2016). Bioremediation techniques–classification based on site of application: principles, advantages, limitations and prospects. *World Journal of Microbiology and Biotechnology* 32: 1–18.

Bachate, S.P., Nandre, V.S., Ghatpande, N.S., and Kodam, K.M. (2013). Simultaneous reduction of Cr(VI) and oxidation of As(III) by *Bacillus firmus* TE7 isolated from tannery effluent. *Chemosphere* 90 (8): 2273–2278.

Bakshi, A. and Panigrahi, A.K. (2022). Chromium contamination in soil and its bioremediation: an overview. In: *Advances in Bioremediation and Phytoremediation for Sustainable Soil Management: Principles, Monitoring and Remediation*, 229–248. Springer.

Barapatre, A. and Jha, H. (2016). Decolourization and biological treatment of pulp and paper mill effluent by lignin-degrading fungus *Aspergillus flavus* strain F10. *International Journal of Current Microbiology and Applied Sciences* 5 (5): 19–32.

Bayat, Z., Hassanshahian, M., and Cappello, S. (2015). Immobilization of microbes for bioremediation of crude oil polluted environments: a mini review. *Open Microbiology Journal* 9: 48.

Bhakat, K., Chakraborty, A., and Islam, E. (2019). Characterization of arsenic oxidation and uranium bioremediation potential of arsenic resistant bacteria isolated from uranium ore. *Environmental Science and Pollution Research* 26: 12907–12919.

Bharagava, R.N., Saxena, G., and Mulla, S.I. (2020). Introduction to industrial wastes containing organic and inorganic pollutants and bioremediation approaches for environmental management. In: *Bioremediation of Industrial Waste for Environmental Safety. I: Industrial Waste and Its Management*, 1–18. Springer.

Bhatt, P., Pandey, S.C., Joshi, S. et al. (2022). Nanobioremediation: a sustainable approach for the removal of toxic pollutants from the environment. *Journal of Hazardous Materials* 427: 128033.

Bhatt, P., Chen, S., and Murphy, C. (2023). Recent advancements in microbe-pesticide interaction: a smart-soil bioremediation approach. *Frontiers in Microbiology* 14: 1206179.

Biache, C., Ouali, S., Cébron, A. et al. (2017). Bioremediation of PAH-contained soils: consequences on formation and degradation of polar-polycyclic aromatic compounds and microbial community abundance. *Journal of Hazardous Materials* 329: 1–10.

Bôto, M.L., Magalhães, C., Perdigão, R. et al. (2021). Harnessing the potential of native microbial communities for bioremediation of oil spills in the Iberian Peninsula NW coast. *Frontiers in Microbiology* 12: 633659.

Brinza, L., Geraki, K., Cojocaru, C. et al. (2020). Baltic *Fucus vesiculosus* as potential bio-sorbent for Zn removal: mechanism insight. *Chemosphere* 238: 124652.

Brown, L.D. and Ulrich, A.C. (2014). Bioremediation of oil spills on land. In: *Handbook of Oil Spill Science and Technology*, 395–406. Wiley.

Bruschi, M. and Goulhen, F. (2006). New bioremediation technologies to remove heavy metals and radionuclides using Fe (III)-, sulfate-and sulfur-reducing bacteria. In: *Environmental Bioremediation Technologies* (ed. S.N. Singh and R.D. Tripathi), 35–55. Berlin, Heidelberg: Springer Berlin Heidelberg.

Bumbra, P., Tanwar, N., Arora, V. et al. (2022). Microbial inoculants: an invasive approach for the bioremediation of soil contaminated with heavy metals. In: *Microbial and Biotechnological Interventions in Bioremediation and Phytoremediation* (ed. J.A. Malik), 259–276. Cham: Springer International Publishing.

Burghal, A.A., Abu-Mejdad, N.M.J.A., and Al-Tamimi, W.H. (2016). Mycodegradation of crude oil by fungal species isolated from petroleum contaminated soil. *International Journal of Innovative Research in Science, Engineering and Technology* 5 (2): 1517–1524.

Bwapwa, J.K., Jaiyeola, A.T., and Chetty, R. (2017). Bioremediation of acid mine drainage using algae strains: a review. *South African Journal of Chemical Engineering* 24: 62–70.

Catania, V., Lopresti, F., Cappello, S. et al. (2020). Innovative, ecofriendly biosorbent-biodegrading biofilms for bioremediation of oil-contaminated water. *New Biotechnology* 58: 25–31.

Chandran, H., Meena, M., and Sharma, K. (2020). Microbial biodiversity and bioremediation assessment through omics approaches. *Frontiers in Environmental Chemistry* 1: 570326.

Chaudhary, T. and Shukla, P. (2019). Bioinoculants for bioremediation applications and disease resistance: innovative perspectives. *Indian Journal of Microbiology* 59: 129–136.

Chaudhary, P., Ahamad, L., Chaudhary, A. et al. (2023). Nanoparticle-mediated bioremediation as a powerful weapon in the removal of environmental pollutants. *Journal of Environmental Chemical Engineering* 11: 109591.

Chegounian, P., Zerriffi, H., and Yadav, V.G. (2020). Engineering microbes for remediation of oil sands tailings. *Trends in Biotechnology* 38 (11): 1192–1196.

Chen, M., Xu, P., Zeng, G. et al. (2015). Bioremediation of soils contaminated with polycyclic aromatic hydrocarbons, petroleum, pesticides, chlorophenols and heavy metals by composting: applications, microbes and future research needs. *Biotechnology Advances* 33 (6): 745–755.

Cheng, Y., Sun, H., Yang, E. et al. (2021). Distribution and bioaccessibility of polycyclic aromatic hydrocarbons in industrially contaminated site soils as affected by thermal treatment. *Journal of Hazardous Materials* 411: 125129.

Chikere, C.B. (2013). Application of molecular microbiology techniques in bioremediation of hydrocarbons and other pollutants. *British Biotechnology Journal* 3 (1): 90.

Chipasa, K.B. and Mędrzycka, K. (2006). Behavior of lipids in biological wastewater treatment processes. *Journal of Industrial Microbiology & Biotechnology* 33 (8): 635–645.

Chirwa, E.M.N. and Bezza, F.A. (2015). Petroleum hydrocarbon spills in the environment and abundance of microbial community capable of biosurfactant production. *Journal of Petroleum & Environmental Biotechnology* 6 (237): 2.

Choudhary, S. and Sar, P. (2011). Uranium biomineralization by a metal resistant *Pseudomonas aeruginosa* strain isolated from contaminated mine waste. *Journal of Hazardous Materials* 186 (1): 336–343.

Chugh, M., Kumar, L., Shah, M.P., and Bharadvaja, N. (2022). Algal Bioremediation of heavy metals: an insight into removal mechanisms, recovery of by-products, challenges, and future opportunities. *Energy Nexus* 100129.

Daniel, D., Jegathambal, P., and Bevers, B. (2019). In situ bioremediation of textile dye effluent-contaminated soils using mixed microbial culture. *International Journal of Civil Engineering* 17: 1527–1536.

Das, S. and Dash, H.R. (2014). Microbial bioremediation: a potential tool for restoration of contaminated areas. In: *Microbial Biodegradation and Bioremediation* (ed. S. Das), 1–21. Elsevier.

Das, A. and Osborne, J.W. (2018). Bioremediation of heavy metals. In: *Nanotechnology, Food Security and Water Treatment*, 277–311. Springer.

de Moura Dickel, J.D., Carvalho, J.K., Silveira, M.A.D. et al. (2023). *Aspergillus sclerotiorum* lipolytic activity and its application in bioremediation of high-fat dairy wastewater environments. *Environmental Science and Pollution Research* 30 (13): 35517–35527.

del Carmen Vargas-García, M., López, M.J., Suárez-Estrella, F., and Moreno, J. (2012). Compost as a source of microbial isolates for the bioremediation of heavy metals: in vitro selection. *Science of the Total Environment* 431: 62–67.

Dubey, S., Chen, C.W., Haldar, D. et al. (2023). Advancement in algal bioremediation for organic, inorganic, and emerging pollutants. *Environmental Pollution* 317: 120840.

Duc, H.D., Hung, N.V., and Oanh, N.T. (2021). Anaerobic degradation of endosulfans by a mixed culture of *Pseudomonas* sp. and *Staphylococcus* sp. *Applied Biochemistry and Microbiology* 57 (3): 327–334.

Dutta, N., Usman, M., Ashraf, M.A. et al. (2022). A critical review of recent advances in the bio-remediation of chlorinated substances by microbial dechlorinators. *Chemical Engineering Journal Advances* 12: 100359.

Foran, C.M., Baker, K.M., Grosso, N.R., and Linkov, I. (2015). An enhanced adaptive management approach for remediation of legacy mercury in the South River. *PLoS One* 10 (2): e0117140.

French, K.E., Zhou, Z., and Terry, N. (2020). Horizontal 'gene drives' harness indigenous bacteria for bioremediation. *Scientific Reports* 10 (1): 15091.

Fu, W., Xu, M., Sun, K. et al. (2018). Biodegradation of phenanthrene by endophytic fungus *Phomopsis liquidambari in-vitro* and *in-vivo*. *Chemosphere* 203: 160–169.

Gangola, S., Bhatt, P., Chaudhary, P. et al. (2018). Bioremediation of industrial waste using microbial metabolic diversity. In: *Microbial Biotechnology in Environmental Monitoring and Cleanup*, 1–27. IGI Global.

Ganguly, D., Prasanna, K.L.V., Neelapu, S., and Goswami, G. (2023). Role of microbes in bioremediation. In: *Industrial Microbiology and Biotechnology: Emerging Concepts in Microbial Technology* (ed. P. Verma), 549–584. Singapore: Springer Nature Singapore.

Gao, D., Zhao, H., Wang, L. et al. (2022). Current and emerging trends in bioaugmentation of organic contaminated soils: a review. *Journal of Environmental Management* 320: 115799.

Garg, S.K., Tripathi, M., Kumar, S. et al. (2012). Microbial dechlorination of chloroorganics and simultaneous decolorization of pulp–paper mill effluent by *Pseudomonas putida* MTCC 10510 augmentation. *Environmental Monitoring and Assessment* 184: 5533–5544.

Gaur, N., Narasimhulu, K., and PydiSetty, Y. (2018). Recent advances in the bio-remediation of persistent organic pollutants and its effect on environment. *Journal of Cleaner Production* 198: 1602–1631.

Geng, X., An, C., Lee, K., and Boufadel, M.C. (2022). Modeling oil biodegradation and bioremediation within beaches. *Current Opinion in Chemical Engineering* 35: 100751.

Geva, P., Kahta, R., Nakonechny, F. et al. (2016). Increased copper bioremediation ability of new transgenic and adapted *Saccharomyces cerevisiae* strains. *Environmental Science and Pollution Research* 23: 19613–19625.

Gutierrez, T., Morris, G., Ellis, D. et al. (2018). Hydrocarbon-degradation and MOS-formation capabilities of the dominant bacteria enriched in sea surface oil slicks during the deepwater horizon oil spill. *Marine Pollution Bulletin* 135: 205–215.

Helmy, Q., Laksmono, R., and Kardena, E. (2015). Bioremediation of aged petroleum oil contaminated soil: from laboratory scale to full scale application. *Procedia Chemistry* 14: 326–333.

Hesham, A.E.L., Khan, S., Tao, Y. et al. (2012). Biodegradation of high molecular weight PAHs using isolated yeast mixtures: application of meta-genomic methods for community structure analyses. *Environmental Science and Pollution Research* 19: 3568–3578.

Iark, D., dos Reis Buzzo, A.J., Garcia, J.A.A. et al. (2019). Enzymatic degradation and detoxification of azo dye Congo red by a new laccase from *Oudemansiella canarii*. *Bioresource Technology* 289: 121655.

Ikuma, K. and Gunsch, C.K. (2012). Genetic bioaugmentation as an effective method for in situ bioremediation: functionality of catabolic plasmids following conjugal transfers. *Bioengineered* 3 (4): 236–241.

Iskandar, N.L., Zainudin, N.A.I.M., and Tan, S.G. (2011). Tolerance and biosorption of copper (Cu) and lead (Pb) by filamentous fungi isolated from a freshwater ecosystem. *Journal of Environmental Sciences* 23 (5): 824–830.

Ito, K., Mahmood, A., Kataoka, R., and Takagi, K. (2021). Dichlorodiphenyltrichloroethane (DDT) degradation by Streptomyces sp. isolated from DDT contaminated soil. *Bioremediation Journal* 25 (2): 148–158.

Jabbar, N.M., Alardhi, S.M., Mohammed, A.K. et al. (2022). Challenges in the implementation of bioremediation processes in petroleum-contaminated soils: a review. *Environmental Nanotechnology, Monitoring & Management* 18: 100694.

Jain, P.K. and Bajpai, V. (2012). Biotechnology of bioremediation – a review. *International Journal of Environmental Sciences* 3 (1): 535–549.

Jain, K., Johnson, J., Devpura, N. et al. (2020). Emerging bioremediation technologies for the treatment of wastewater containing synthetic organic compounds. In: *Emerging Technologies in Environmental Bioremediation* (ed. M.P. Shah, S. Rodriguez-Couto, and S.S. Sengor), 131–150. Elsevier.

Jalilvand, N., Akhgar, A., Alikhani, H.A. et al. (2020). Removal of heavy metals zinc, lead, and cadmium by biomineralization of urease-producing bacteria isolated from Iranian mine calcareous soils. *Journal of Soil Science and Plant Nutrition* 20: 206–219.

Janssen, D.B. and Stucki, G. (2020). Perspectives of genetically engineered microbes for groundwater bioremediation. *Environmental Science. Processes & Impacts* 22 (3): 487–499.

Jansson, J.K., Björklöf, K., Elvang, A.M., and Jørgensen, K.S. (2000). Biomarkers for monitoring efficacy of bioremediation by microbial inoculants. *Environmental Pollution* 107 (2): 217–223.

John, R.C., Essien, J.P., Akpan, S.B., and Okpokwasili, G.C. (2012). Polycyclic aromatic hydrocarbon-degrading bacteria from aviation fuel spill site at Ibeno, Nigeria. *Bulletin of Environment Contamination and Toxicology* 88: 1014–1019.

Joshi, P.K., Swarup, A., Maheshwari, S. et al. (2011). Bioremediation of heavy metals in liquid media through fungi isolated from contaminated sources. *Indian Journal of Microbiology* 51: 482–487.

Kang, C.H., Oh, S.J., Shin, Y. et al. (2015). Bioremediation of lead by ureolytic bacteria isolated from soil at abandoned metal mines in South Korea. *Ecological Engineering* 74: 402–407.

Kanissery, R.G. and Sims, G.K. (2011). Biostimulation for the enhanced degradation of herbicides in soil. *Applied and Environmental Soil Science* 2011: 843450.

Kao, C.M., Chen, C.Y., Chen, S.C. et al. (2008). Application of in situ biosparging to remediate a petroleum-hydrocarbon spill site: field and microbial evaluation. *Chemosphere* 70 (8): 1492–1499.

Kehinde, F.O. and Isaac, S.A. (2016). Effectiveness of augmented consortia of *Bacillus coagulans*, *Citrobacter koseri* and *Serratia ficaria* in the degradation of diesel polluted soil supplemented with pig dung. *African Journal of Microbiology Research* 10 (39): 1637–1644.

Khalid, T., Fatima, A., Shafiq, A. et al. (2016). Microbial decolorization of textile effluent. *RADS Journal of Biological Research & Applied Sciences* 7 (1): 28–34.

Khan, S. and Malik, A. (2018). Toxicity evaluation of textile effluents and role of native soil bacterium in biodegradation of a textile dye. *Environmental Science and Pollution Research* 25: 4446–4458.

Kim, S., Krajmalnik-Brown, R., Kim, J.O., and Chung, J. (2014). Remediation of petroleum hydrocarbon-contaminated sites by DNA diagnosis-based bioslurping technology. *Science of the Total Environment* 497: 250–259.

Kochhar, N., Shrivastava, S., Ghosh, A. et al. (2022). Perspectives on the microorganism of extreme environments and their applications. *Current Research in Microbial Sciences* 3: 100134.

Kour, D., Kaur, T., Devi, R. et al. (2022). Beneficial microbiomes for bioremediation of diverse contaminated environments for environmental sustainability: present status and future challenges. *Environmental Science and Pollution Research* 28: 24917–24939.

Krivobok, S., Miriouchkine, E., Seigle-Murandi, F., and Benoit-Guyod, J.L. (1998). Biodegradation of anthracene by soil fungi. *Chemosphere* 37 (3): 523–530.

Kumar, M. and Philip, L. (2006). Enrichment and isolation of a mixed bacterial culture for complete mineralization of endosulfan. *Journal of Environmental Science and Health Part B* 41 (1): 81–96.

Kumar, M., Singh, D., Singh, R., and Singh, R.P. (2018). Genetic engineering strategies for enhancing phytoremediation of contaminated environments. *Biotechnology Letters* 40 (3): 341–352.

Kumar, A., Subrahmanyam, G., Mondal, R. et al. (2021). Bio-remediation approaches for alleviation of cadmium contamination in natural resources. *Chemosphere* 268: 128855.

Kumari, S., Jamwal, R., Mishra, N., and Singh, D.K. (2020). Recent developments in environmental mercury bioremediation and its toxicity: a review. *Environmental Nanotechnology, Monitoring & Management* 13: 100283.

Kunihiro, N., Haruki, M., Takano, K. et al. (2005). Isolation and characterization of *Rhodococcus* sp. strains TMP2 and T12 that degrade 2,6,10,14-tetramethylpentadecane (pristane) at moderately low temperatures. *Journal of Biotechnology* 115 (2): 129–136.

Li, X., Wu, Y., Lin, X. et al. (2012). Dissipation of polycyclic aromatic hydrocarbons (PAHs) in soil microcosms amended with mushroom cultivation substrate. *Soil Biology and Biochemistry* 47: 191–197.

Li, Q., Liu, J., and Gadd, G.M. (2020). Fungal bioremediation of soil co-contaminated with petroleum hydrocarbons and toxic metals. *Applied Microbiology and Biotechnology* 104: 8999–9008.

Li, G., Yan, L., Chen, X. et al. (2023). Phytoremediation of cadmium from soil, air and water. *Chemosphere* 320: 138058.

Lu, J., Zhang, C., Leong, H.Y. et al. (2020). Overproduction of lipoxygenase from *Pseudomonas aeruginosa* in *Escherichia coli* by auto-induction expression and its application in triphenylmethane dyes degradation. *Journal of Bioscience and Bioengineering* 129 (3): 327–332.

Ma, H., Wang, A., Zhang, M. et al. (2018). Compared the physiological response of two petroleum-tolerant contrasting plants to petroleum stress. *International Journal of Phytoremediation* 20 (10): 1043–1048.

Maity, J.P., Chen, G.S., Huang, Y.H. et al. (2019). Ecofriendly heavy metal stabilization: microbial induced mineral precipitation (MIMP) and biomineralization for heavy metals within the contaminated soil by indigenous bacteria. *Geomicrobiology Journal* 36 (7): 612–623.

Malakar, N., Mitra, S., Toppo, P., and Mathur, P. (2020). Understanding the functional attributes of different microbial enzymes in bioremediation. *NBU Journal of Plant Sciences* 12: 58–69.

Malaviya, P. and Rathore, V.S. (2007). Bioremediation of pulp and paper mill effluent by a novel fungal consortium isolated from polluted soil. *Bioresource Technology* 98 (18): 3647–3651.

Mali, H., Shah, C., Raghunandan, B.H. et al. (2023). Organophosphate pesticides an emerging environmental contaminant: pollution, toxicity, bioremediation progress, and remaining challenges. *Journal of Environmental Sciences* 127: 234–250.

Maliji, D., Olama, Z., and Holail, H. (2013). Environmental studies on the microbial degradation of oil hydrocarbons and its application in Lebanese oil polluted coastal and marine ecosystem. *International Journal of Current Microbiology and Applied Sciences* 2 (6): 1–18.

Manfredini, A., Malusà, E., Costa, C. et al. (2021). Current methods, common practices, and perspectives in tracking and monitoring bioinoculants in soil. *Frontiers in Microbiology* 12: 698491.

Mangwani, N., Kumari, S., and Surajit, D.A.S. (2017). Marine bacterial biofilms in bioremediation of polycyclic aromatic hydrocarbons (PAHs) under terrestrial condition in a soil microcosm. *Pedosphere* 27 (3): 548–558.

Masotti, F., Garavaglia, B.S., Gottig, N., and Ottado, J. (2023). Bioremediation of the herbicide glyphosate in polluted soils by plant-associated microbes. *Current Opinion in Microbiology* 73: 102290.

Medfu Tarekegn, M., Zewdu Salilih, F., and Ishetu, A.I. (2020). Microbes used as a tool for bioremediation of heavy metal from the environment. *Cogent Food & Agriculture* 6 (1): 1783174.

Mohamed, R.M. and Abo-Amer, A.E. (2012). Isolation and characterization of heavy-metal resistant microbes from roadside soil and phylloplane. *Journal of Basic Microbiology* 52 (1): 53–65.

Mukjang, N., Chitov, T., Mhuantong, W. et al. (2022). Bacterial communities associated with crude oil bioremediation through composting approaches with indigenous bacterial isolate. *Life* 12 (11): 1712.

Mukred, A.M., Hamid, A.A., Hamzah, A., and Yusoff, W.M.W. (2008). Development of three bacteria consortium for the bioremediation of crude petroleum-oil in contaminated water. *Online Journal of Biological Sciences* 8 (4): 73–79.

Mulligan, C.N. (2021). Sustainable remediation of contaminated soil using biosurfactants. *Frontiers in Bioengineering and Biotechnology* 9: 635196.

Narayanan, M., Ali, S.S., and El-Sheekh, M. (2023). A comprehensive review on the potential of microbial enzymes in multipollutant bioremediation: mechanisms, challenges, and future prospects. *Journal of Environmental Management* 334: 117532.

Naskar, A., Majumder, R., and Goswami, M. (2020). Bioaccumulation of Ni (II) on growing cells of *Bacillus* sp.: response surface modeling and mechanistic insight. *Environmental Technology and Innovation* 20: 101057.

Naz, M., Raza, M.A., Zaib, S. et al. (2022). Major groups of microorganisms employed in bioremediation. In: *Microbial Bioremediation: Sustainable Management of Environmental Contamination* (ed. R.A. Bhat, M. Butnariu, G.H. Dar, and K.R. Hakeem), 141–159. Cham: Springer International Publishing.

Ndeddy Aka, R.J. and Babalola, O.O. (2016). Effect of bacterial inoculation of strains of *Pseudomonas aeruginosa*, *Alcaligenes feacalis* and *Bacillus subtilis* on germination, growth and heavy metal (Cd, Cr, and Ni) uptake of *Brassica juncea*. *International Journal of Phytoremediation* 18 (2): 200–209.

Nivetha, N., Srivarshine, B., Sowmya, B. et al. (2022). A comprehensive review on bio-stimulation and bio-enhancement towards remediation of heavy metals degeneration. *Chemosphere* 312: 137099.

Nwankwegu, A.S., Zhang, L., Xie, D. et al. (2022). Bioaugmentation as a green technology for hydrocarbon pollution remediation. Problems and prospects. *Journal of Environmental Management* 304: 114313.

Odukkathil, G. and Vasudevan, N. (2013). Toxicity and bioremediation of pesticides in agricultural soil. *Reviews in Environmental Science and Bio/Technology* 12: 421–444.

Pande, V., Pandey, S.C., Sati, D. et al. (2022). Microbial interventions in bioremediation of heavy metal contaminants in agroecosystem. *Frontiers in Microbiology* 13: 824084.

Pandey, J., Chauhan, A., and Jain, R.K. (2009). Integrative approaches for assessing the ecological sustainability of in situ bioremediation. *FEMS Microbiology Reviews* 33 (2): 324–375.

Patel, A.K., Singhania, R.R., Albarico, F.P.J.B. et al. (2022). Organic wastes bioremediation and its changing prospects. *Science of the Total Environment* 824: 153889.

Patil, S.M., Suryavanshi, M.V., Chandanshive, V.V. et al. (2020). Regeneration of textile wastewater deteriorated microbial diversity of soil microcosm through bioaugmentation. *Chemical Engineering Journal* 380: 122533.

Peacock, A.D., Chang, Y.J., Istok, J.D. et al. (2004). Utilization of microbial biofilms as monitors of bioremediation. *Microbial Ecology* 47: 284–292.

Perera, I.C. and Hemamali, E.H. (2022). Genetically modified organisms for bioremediation: current research and advancements. *Bioremediation of Environmental Pollutants: Emerging Trends and Strategies* 1: 163–186.

Pietro-Souza, W., de Campos Pereira, F., Mello, I.S. et al. (2020). Mercury resistance and bioremediation mediated by endophytic fungi. *Chemosphere* 240: 124874.

Pineda-Flores, G., Boll-Argüello, G., Lira-Galeana, C., and Mesta-Howard, A.M. (2004). A microbial consortium isolated from a crude oil sample that uses asphaltenes as a carbon and energy source. *Biodegradation* 15: 145–151.

Priyadharshini, S.D., Babu, P.S., Manikandan, S. et al. (2021). Phycoremediation of wastewater for pollutant removal: a green approach to environmental protection and long-term remediation. *Environmental Pollution* 290: 117989.

Qiu, Z., Egidi, E., Liu, H. et al. (2019). New frontiers in agriculture productivity: optimised microbial inoculants and *in situ* microbiome engineering. *Biotechnology Advances* 37 (6): 107371.

Rafeeq, H., Afsheen, N., Rafique, S. et al. (2023). Genetically engineered microorganisms for environmental remediation. *Chemosphere* 310: 136751.

Raj, A., Jhariya, M.K., Khan, N., and Banerjee, A. (2022). Bioremediation of problematic soil for sustainability. In: *Microbial Biotechnology: Role in Ecological Sustainability and Research*, 201–223. Wiley.

Rani, A. and Goel, A.G.R. (2009). In situ boremediation effect of *Pseudomonas putida* KNP9 strain on cadmium and lead toxicity. *Biotechnology Emerging Trends* 379.

Rebello, S., Nathan, V.K., Sindhu, R. et al. (2021). Bioengineered microbes for soil health restoration: present status and future. *Bioengineered* 12 (2): 12839–12853.

Rodríguez, A., Castrejón-Godínez, M.L., Salazar-Bustamante, E. et al. (2020). Omics approaches to pesticide biodegradation. *Current Microbiology* 77: 545–563.

Roy, S. and Roy, M. (2015). Bioleaching of heavy metals by sulfur oxidizing bacteria: a review. *International Research Journal of Environmental Science* 4: 75–79.

Roy, M., Giri, A.K., Dutta, S., and Mukherjee, P. (2015). Integrated phytobial remediation for sustainable management of arsenic in soil and water. *Environment International* 75: 180–198.

Sales da Silva, I.G., Gomes de Almeida, F.C., da Rocha, P. et al. (2020). Soil bioremediation: overview of technologies and trends. *Energies* 13 (18): 4664.

Sánchez-Castro, I., Amador-García, A., Moreno-Romero, C. et al. (2017). Screening of bacterial strains isolated from uranium mill tailings porewaters for bioremediation purposes. *Journal of Environmental Radioactivity* 166: 130–141.

Saranya, P., Selvi, P.K., and Sekaran, G. (2019). Integrated thermophilic enzyme-immobilized reactor and high-rate biological reactors for treatment of palm oil-containing wastewater without sludge production. *Bioprocess and Biosystems Engineering* 42: 1053–1064.

Sayara, T., Borràs, E., Caminal, G. et al. (2011). Bioremediation of PAHs-contaminated soil through composting: influence of bioaugmentation and biostimulation on contaminant biodegradation. *International Biodeterioration and Biodegradation* 65 (6): 859–865.

Sayqal, A. and Ahmed, O.B. (2021). Advances in heavy metal bioremediation: an overview. *Applied Bionics and Biomechanics* 2021: 9871378.

Schmidt, H. and Eickhorst, T. (2014). Detection and quantification of native microbial populations on soil-grown rice roots by catalyzed reporter deposition-fluorescence in situ hybridization. *FEMS Microbiology Ecology* 87 (2): 390–402.

Shah, H. and Jain, S. (2020). Bioremediation: an approach for environmental pollutants detoxification. *Waste to Energy: Prospects and Applications*, 2020 121–142.

Sharma, S.R. (2018). Bioremediation of polythenes and plastics: a microbial approach. In: *Approaches in Bioremediation: The New Era of Environmental Microbiology and Nanobiotechnology*, 97–114. Springer.

Sharma, P., Dutta, D., Udayan, A. et al. (2022). Role of microbes in bioaccumulation of heavy metals in municipal solid waste: impacts on plant and human being. *Environmental Pollution* 305: 119248.

Sharma, P., Singh, S.P., Iqbal, H.M., and Tong, Y.W. (2022). Omics approaches in bioremediation of environmental contaminants: an integrated approach for environmental safety and sustainability. *Environmental Research* 211: 113102.

Sher, S. and Rehman, A. (2019). Use of heavy metals resistant bacteria – a strategy for arsenic bioremediation. *Applied Microbiology and Biotechnology* 103: 6007–6021.

Silva, R.D.C.F., Almeida, D.G., Rufino, R.D. et al. (2014). Applications of biosurfactants in the petroleum industry and the remediation of oil spills. *International Journal of Molecular Sciences* 15 (7): 12523–12542.

Singh, R. (2014). Microorganism as a tool of bioremediation technology for cleaning environment: a review. *Proceedings of the International Academy of Ecology and Environmental Sciences* 4 (1): 1.

Singh, R., Paul, D., and Jain, R.K. (2006). Biofilms: implications in bioremediation. *Trends in Microbiology* 14 (9): 389–397.

Širić, I., Eid, E.M., Taher, M.A. et al. (2022). Combined use of spent mushroom substrate biochar and PGPR improves growth, yield, and biochemical response of cauliflower (*Brassica oleracea* var. botrytis): a preliminary study on greenhouse cultivation. *Horticulturae* 8 (9): 830.

Song, P., Xu, D., Yue, J. et al. (2022). Recent advances in soil remediation technology for heavy metal contaminated sites: a critical review. *Science of the Total Environment* 838: 156417.

Sonkar, M., Kumar, V., and Dutt, D. (2020). Use of paper mill sludge and sewage sludge powder as nitrogen and phosphorus sources with bacterial consortium for the treatment of paper industry wastewater. *Biocatalysis and Agricultural Biotechnology* 30: 101843.

Sonkar, M., Kumar, V., Kumar, P. et al. (2021). Bioaugmentation with existing potent microorganisms to accelerate the treatment efficacy of paper industry wastewater pollutants. *Journal of Environmental Chemical Engineering* 9 (5): 105913.

Sonune, N. (2021). Microbes: a potential tool for bioremediation. *Rhizobiont in bioremediation of hazardous waste* 391–407.

Sravya, K. and Sangeetha, S. (2022). Feasibility study on bioremediation techniques to contaminated soils. *Materials Today Proceedings* 51: 2556–2560.

Srivastava, S. and Gupta, B. (2021). Application of immobilization techniques in heavy metal and metalloid remediation. In: *Immobilization Strategies: Biomedical, Bioengineering and Environmental Applications*, 581–595. Springer.

Tanmoy, P.A.U.L. and Saha, N.C. (2019). Environmental arsenic and selenium contamination and approaches towards its bioremediation through the exploration of microbial adaptations: a review. *Pedosphere* 29 (5): 554–568.

Tayang, A. and Songachan, L.S. (2021). Microbial bioremediation of heavy metals. *Current Science* 120 (6): 00113891.

Thakur, M., Medintz, I.L., and Walper, S.A. (2019). Enzymatic bioremediation of organophosphate compounds – progress and remaining challenges. *Frontiers in Bioengineering and Biotechnology* 7: 289.

Toussaint, J.P., Pham, T.T.M., Barriault, D., and Sylvestre, M. (2012). Plant exudates promote PCB degradation by a rhodococcal rhizobacteria. *Applied Microbiology and Biotechnology* 95: 1589–1603.

Trabelsi, D. and Mhamdi, R. (2013). Microbial inoculants and their impact on soil microbial communities: a review. *BioMed Research International* 2013: 1–11.

Tyagi, M., da Fonseca, M.M.R., and de Carvalho, C.C. (2011). Bioaugmentation and biostimulation strategies to improve the effectiveness of bioremediation processes. *Biodegradation* 22: 231–241.

Unuofin, J.O., Falade, A.O., and Aladekoyi, O.J. (2021). Applications of microbial laccases in bioremediation of environmental pollutants: potential issues, challenges, and prospects. *Bioremediation for Environmental Sustainability* 519–540.

Vassilev, N., Vassileva, M., Martos, V. et al. (2020). Formulation of microbial inoculants by encapsulation in natural polysaccharides: focus on beneficial properties of carrier additives and derivatives. *Frontiers in Plant Science* 11: 270.

Verma, S. and Kuila, A. (2019). Bioremediation of heavy metals by microbial process. *Environmental Technology and Innovation* 14: 100369.

Vidali, M. (2001). Bioremediation. an overview. *Pure and Applied Chemistry* 73 (7): 1163–1172.

Vijayaraghavan, K. and Yun, Y.S. (2008). Bacterial biosorbents and biosorption. *Biotechnology Advances* 26 (3): 266–291.

Wang, J. and Chen, C. (2009). Biosorbents for heavy metals removal and their future. *Biotechnology Advances* 27 (2): 195–226.

Wang, H., Hu, J., Xu, K. et al. (2018a). Biodegradation and chemotaxis of polychlorinated biphenyls, biphenyls, and their metabolites by *Rhodococcus* spp. *Biodegradation* 29: 1–10.

Wang, Z., Luo, Z., Yan, C. et al. (2018b). Biokinetics of arsenate accumulation and release in *Microcystis aeruginosa* regulated by common environmental factors: practical implications for enhanced bioremediation. *Journal of Cleaner Production* 199: 112–120.

Wang, Z., Wang, H., Nie, Q. et al. (2023). Pb(II) bioremediation using fresh algal-bacterial aerobic granular sludge and its underlying mechanisms highlighting the role of extracellular polymeric substances. *Journal of Hazardous Materials* 444: 130452.

Williamson, A.J., Binet, M., and Sergeant, C. (2023). Radionuclide biogeochemistry: from bioremediation toward the treatment of aqueous radioactive effluents. *Critical Reviews in Biotechnology* 31: 1–19.

Willis, G. and Donati, E.R. (2017). Heavy metal bioprecipitation. In: *Heavy Metals in the Environment: Microorganisms and Bioremediation*, 114. CRC Press.

Wu, P., Zhang, Z., Luo, Y. et al. (2022). Bioremediation of phenolic pollutants by algae – current status and challenges. *Bioresource Technology* 350: 126930.

Xu, Y. and Zhou, N.Y. (2017). Microbial remediation of aromatics-contaminated soil. *Frontiers of Environmental Science & Engineering* 11: 1–9.

Yadav, S.K. (2021). Microbial indicators of bioremediation: potential and success. *Microbial Rejuvenation of Polluted Environment* 1: 85–100.

Yadav, A.N., Suyal, D.C., Kour, D. et al. (2022). Bioremediation and waste management for environmental sustainability. *Journal of Applied Biology & Biotechnology* 10 (2): 1–5.

Ye, X., Peng, T., Feng, J. et al. (2019). A novel dehydrogenase 17β-HSDx from *Rhodococcus* sp. P14 with potential application in bioremediation of steroids contaminated environment. *Journal of Hazardous Materials* 362: 170–177.

Yessica, G.P., Alejandro, A., Ronald, F.C. et al. (2013). Tolerance, growth and degradation of phenanthrene and benzo [*a*] pyrene by *Rhizobium tropici* CIAT 899 in liquid culture medium. *Applied Soil Ecology* 63: 105–111.

Zeraatkar, A.K., Ahmadzadeh, H., Talebi, A.F. et al. (2016). Potential use of algae for heavy metal bioremediation: a critical review. *Journal of Environmental Management* 181: 817–831.

Zhang, T. and Zhang, H. (2022). Microbial consortia are needed to degrade soil pollutants. *Microorganisms* 10 (2): 261.

# 22

## Microbial Inoculant Approaches for Disease Management

*S.T.M. Aravindharajan[1], Sivaprakasam Navarasu[2], Velmurugan Shanmugam[3], S.S. Deepti Varsha[1], D. Vijaysri[1], Sandeep Kumar Singh[1], and Livleen Shukla[1]*

[1] Division of Microbiology, ICAR-Indian Agricultural Research Institute, New Delhi, India
[2] Department of Plant Pathology, Tamil Nadu Agricultural University, Coimbatore, India
[3] Division of Plant Pathology, ICAR-Indian Agricultural Research Institute, New Delhi, India

## 22.1 Introduction

Agricultural production is the source of 90% of calories and 80% of protein to humankind. In the recent past, agricultural crop production has significantly reduced due to various parameters. Among such, plant disease stands out as a serious threat in agriculture and is caused by different micro-organisms such as fungi, bacteria, protozoa, viruses, mycoplasma, spiroplasma, and so on. It has the potential to destroy the standing crops completely, which depends on the tools of the disease triangle, that is, a virulent pathogen, a susceptible host, and favorable environmental conditions. The interaction and alignment of these factors can contribute to the development and spread of crop diseases and, ultimately, significant yield loss. In the absence of any one component, disease does not appear. In history, the first visualized impact of plant diseases created a cumbersome situation for human life. As a result, 1 million people lost their lives due to starvation, and 1.5 million people migrated to other countries due to the infection of *Phytophthora infestans* in potatoes in 1845 in Ireland. This disease epidemic provoked the entire world to find out the reason. The significant contribution by Anton de Bary revealed the role of microorganisms in causing plant disease. Later, in 1943, another noteworthy Bengal famine reported the deaths of three million people, which was caused by *Helminthosporium oryzae* in rice. Whereas, the world's population is anticipated to reach 10 billion by 2050, posing a significant agricultural challenge in terms of food production and distribution. Also, an investigation by Jaiswal et al. (2022) alarmed that biotic stress causes an annual crop loss of around 20–40%, in terms of money american dollar values of $40 billion in economic loss. In this situation, using agrochemicals for disease management is not advisable because it has a lot of negative environmental consequences. Under these circumstances, we must have necessitated to find alternatives to tackle the plant diseases that should be safe, eco-friendly, and reliable, and also restore the depleted resources. To overcome this crisis, using microbial inoculants, that is, biocontrol agents, is a potential option, like "hitting two birds with a single stone," resulting in disease management and plant growth promotions.

"Biocontrol" means the "reduction of the amount of inoculum or disease-producing activity of a pathogen accomplished by or through one or more organisms other than man" (Cook and Baker 1983). Large-scale application of biofertilizers has been possible in recent decades through plant protection training in India, Brazil, and China, where research and application are both classical and supplemental. In this part, we discuss the different mechanisms used by the biocontrol agents to tackle plant pathogens from infection.

*Microbes Based Approaches for the Management of Hazardous Contaminants*, First Edition. Edited by Ajay Kumar, Livleen Shukla, Joginder Singh, and Luiz Fernando R. Ferreira.

## 22.2 Approaches of Various Microbial Inoculants for Controlling the Economically Important Disease

### 22.2.1 Fungi

Fungal plant pathogens pose a significant threat to global agriculture, causing substantial yield losses and economic damage (Jayawardena et al. 2021). One-third of all food crops are thought to be lost each year due to fungal pathogens (Fisher et al. 2012).

Many of the fungal genera, such as *Aspergillus*, *Penicillium*, *Alternaria*, *Fusarium*, and *Claviceps*, produce toxins that also cause health hazards to plant and animal health (Adeyeye and Samuel 2016). The beneficial fungal agents have numerous strategies that were identified to combat plant pathogens. There was a division into direct mechanisms, such as competition, parasitism, and antibiosis, and indirect mechanisms, such as the induction of systemic resistance (Figure 22.1) (Pal and Gardener 2006).

### 22.2.2 Competition

Fungi utilize a diverse array of carbohydrates, comprising disaccharides, monosaccharides, and nitrogenous sources; thus, rivalry for resources and habitat is one of the primary modes of operation in the majority of microbial antagonists (Edward-Rajanayagam et al. 2023). As it can limit the nutrients and space available to the pathogen, making it more difficult for it to grow and reproduce, this method of controlling pathogenic infections may be effective (Omidvari et al. 2023). Root exudates serve as crucial chemical messengers, mediating communication and signaling between microorganisms and plants also they serve as an important nutrient source for beneficial and harmful microbes (Turrà and Di Pietro 2015). Many fungal biocontrol agents are applied before and after planting, which is a potential competitor for nutrients to manage harmful plant pathogens. *Trichoderma* species have been observed to display antagonistic behavior toward a wide range of microorganisms (López-Bucio et al. 2015). These fungi are highly recognized for their fast growth and competitive nature, which allows them to effectively invade substrates and displace slower pathogens.

Additionally, recent research has investigated the significance of endophytic properties in promoting plant health (Oszust et al. 2020; Gupta et al. 2020b; Verma et al. 2021). *Pythium oligandrum* is a soil-dwelling oomycete that directly combats pathogens; it can compete with target organisms for space and nutrients (Bělonožníková et al. 2020; Campbell 1989). The basidiomycetes yeast *Papiliotrema terrestris* acts as a potential biocontrol agent against postharvest diseases of fruits and vegetables. They exhibit faster colonization on the fruit surface, and wounds occur during the handling and harvesting process, as well as depletion of available nutrients for the survival of postharvest pathogens and withstand oxidative and cellular stress (Castoria et al. 2021).

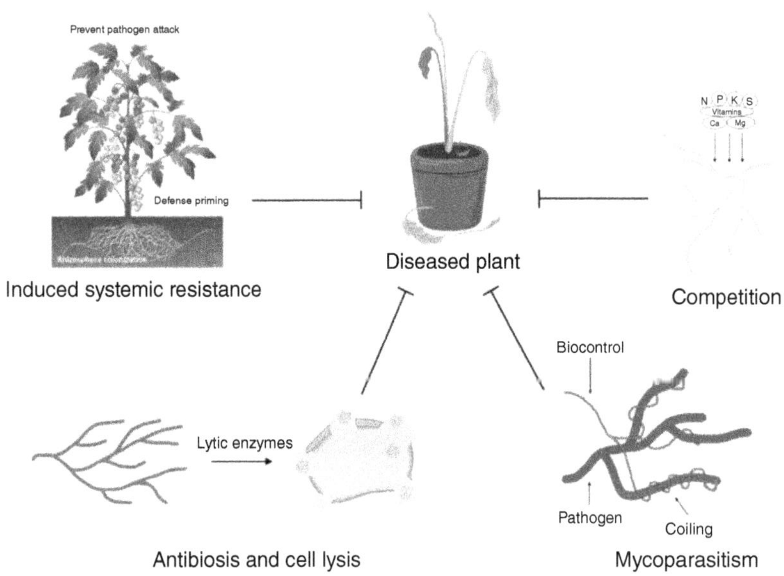


**Figure 22.1** Various aspects of fungal biocontrol mechanisms.

Edward-Rajanayagam et al. (2023) reported that yeast mixtures such as *Meyerozyma guilliermondii*, *Pseudozyma* spp., and *Saccharomyces cerevisiae* have potential for managing fungal rots in postharvest lemon fruit, viz., green/blue mold (*Penicillium* spp.), fusarium rot (*F. oxysporum, F. solani, and F. proliferatum*), and anthracnose (*Colletotrichum* spp.). *Leucosporidium scottii* and *Cryptococcus Laurentii*, two low-temperature yeasts isolated from Antarctic and Tibet soil samples, were very successful in their biocontrol activities on tomato and apple inoculations with *Botrytis cinerea* (Vero et al. 2013; Hu et al. 2017).

### 22.2.3 Parasitism

Mycoparasites are fungi that can attack and parasitize other fungi. By parasitizing and consuming plant pathogens, mycoparasites can help suppress the spread and impact of these harmful plant pathogenic fungi. *Trichoderma* spp., a widely used and explored mycoparasite stimulates the immune system of plants, promoting systemic resistance and plant's growth, vitality, and health. *Clonostachys rosea* parasitizes the plant pathogenic fungi by coiling around their hyphae and penetrating them. It has a broad range of host fungus, including species from the genera *Botrytis, Fusarium,* and *Rhizoctonia* (Meng et al. 2022; Gimeno et al. 2021; Salamone et al. 2018). *Stachybotrys elegans* is a mycoparasite that attacks and destroys plant pathogenic fungi *Rhizoctonia solani* (Aydin 2022). *Ampelomyces* is one of the first mycoparasites employed to manage powdery mildew fungus because of their ability to effectively eradicate mycelial growth and decrease the mycohosts overwintering inoculums (Prahl et al. 2023). In addition, *Coniothyrium minitans* is an example of a parasite of sclerotia-forming plant pathogens. Numerous mycoparasites have been used still to manage fungal phytopathogens; some important are listed below in Table 22.1.

### 22.2.4 Antibiosis

Antibiosis refers to the phenomenon where one organism inhibits the growth or survival of another through the release of toxic compounds or antimicrobial substances (Hutchings et al. 2019). Numerous biocontrol agents have been identified for their capacity to suppress phytopathogens by releasing antibiotics. *Trichoderma* spp. is widely distributed in soil and

**Table 22.1** List of Mycoparasites against fungal plant pathogens.

| Parasite | Pathogen | References |
|---|---|---|
| *Trichoderma* spp. | Wide range of pathogens | Palmieri et al. (2022) |
| *Ampelomyces quisqualis* | Powdery mildew pathogens | Huth et al. (2021) |
| *Acremonium alternatum* | *Plasmodiophora brassicae* | Auer and Ludwig-Müller (2015) |
| *Calonectria hemileiae* | *Hemileia vastatrix* | Salcedo-Sarmiento et al. (2021) |
| *Sphaerodes mycoparasitica* | *Fusarium graminearum* | Kim et al. (2020), Kim and Vladimir (2022) |
| *Paecilomyces fumosoroseus* | *Sphaerotheca fuliginea* | Li et al. (2020) |
| *Coniothyrium minitans* | *Sclerotinia sclerotiorum* | Zhao et al. (2020) |
| *Lecanicillium lecanii* | *Hemileia vastatrix, Podosphaera fuliginea, Cercospora arachidicola,* and *Phaeoisariopsis personata* | Tounwendsida et al. (2022)<br>Hajian-Forooshani et al. (2023) |
| *Clonostachys rosea* | *Sclerotinia sclerotiorum* | Venkatesan et al. (2023) |
| *Stachybotrys elegans* | *Rhizoctonia solani* | Chamoun et al. (2015) |
| *Arthrobotrys eudermata* | *Rhizoctonia solani* | Kumar and Gouda (2018) |
| *Talaromyces pinophilus* | *Botrytis cinerea*<br>*Botryotinia fuckeliana* | Abdel-Rahim and Kamal (2018) |
| *Beauveria bassiana* strain 11–98 | *Rhizoctonia solani*<br>*Pythium myriotylum* | Jaber (2015) |
| *Alternaria alternata* | *Puccinia striiformis* | Zheng et al. (2017) |
| *Chaetomium globosum* and *C. cupreum* | *Fusarium, Phytophthora,* and *Pythium* | Sharma (2015) |

synthesizes a number of volatile and nonvolatile chemicals. They create substances that can stop fungus from growing into mycelium, such as hydrogen cyanide, aldehydes, ethylene (ET), alcohols, and ketones, among other volatile and nonvolatile substances like peptides. Numerous antifungal drugs, including viridin, heptelidic acid, gliovirin, gliotoxin, viridiol, and valinotrocin, have been linked to the biocontrol fungus *Gliocladium virens* (Tariq et al. 2020). For certain BCAs, the release of killer toxins and VOC generation operate in synergy. Killer toxin production by yeasts was originally documented in the 1960s, and each strain's capacity to produce more than one killer toxin was found. *Wickerhamomyces anomalus* BS91, *Debaryomyces hansenii* KI2a, and *D. hansenii* MI1a were able to generate killer toxins, which were checked under *in vitro* and *in vivo* against *Monilinia fructigena* and *M. fructicola*.

### 22.2.5 Bacteria

Bacteria in agriculture are known for their role in causing plant diseases, biological nitrogen fixation, and biocontrol. Yet there are more than 200 species of bacteria that are pathogenic in this group (Buttimer et al. 2017). The bacteria with a beneficial role in agriculture are called as plant growth promoting bacteria (PGPB) (Figure 22.2) (Ramakrishna et al. 2019; Gond et al. 2021; Singh et al. 2020). The use of PGPB is preferred to achieve sustainability by reducing the ill effects of hazardous chemicals used for plant disease management and plant growth (Raaijmakers et al. 2002; Gupta et al. 2020a). Different mechanisms of bacterial biocontrol agents are the same as those discussed in the previous section. Other than the above-mentioned mechanisms, lytic enzymes (chitinases, cellulases, glucanases, lipases, and proteases), volatile organic compounds, bacteriocins, and siderophores of bacterial origin efficiently reduced the pathogen inoculum and thus plant disease. Recent advancements in the biocontrol approach include the use of plant-associated microbes such as endophytes, phyllosphere, and rhizosphere microbes to manage plant disease; as a whole, we can say the plant microbiome (Singh et al. 2022a; Anand et al. 2023). Understanding the structure and functions of plant microbiomes opens up new green strategies for plant disease management (Ab Rahman et al. 2018; Kumar et al. 2020a, 2020b). Some of the successful bacterial biocontrol agents are *Pseudomonas fluorescence, Bacillus* spp., *Streptomyces* spp., *Agrobacterium radiobacter, Pantoea agglomerans*, and *Burkholderia cepacian,* as listed in Table 22.2 (Pal and Gardener 2006).

### 22.2.6 Nematode

Nematodes are small, tiny, thread-like organisms with a free-living and parasitic lifestyle. Nematology has been studied from the perspective of plant damage, but nematodes are successfully used in managing a few insects' pests. The use of nematodes in managing plant disease is little explored. The nematodes that feed on fungi are called fungivorous/mycophagous nematodes. *Aphelenchus, Aphelenchoides, Ditylenchus,* and *Tylenchus* are examples of fungi-feeding nematodes that act as biological controls in managing plant disease. For the first time, *Aphelenchus avenae* was used to control the Pythium root rot of corn (Rhoades and Linford 1959). *Apehelchus avenae* is also able to control other pathogenic fungi like *Rhizoctonia solani*

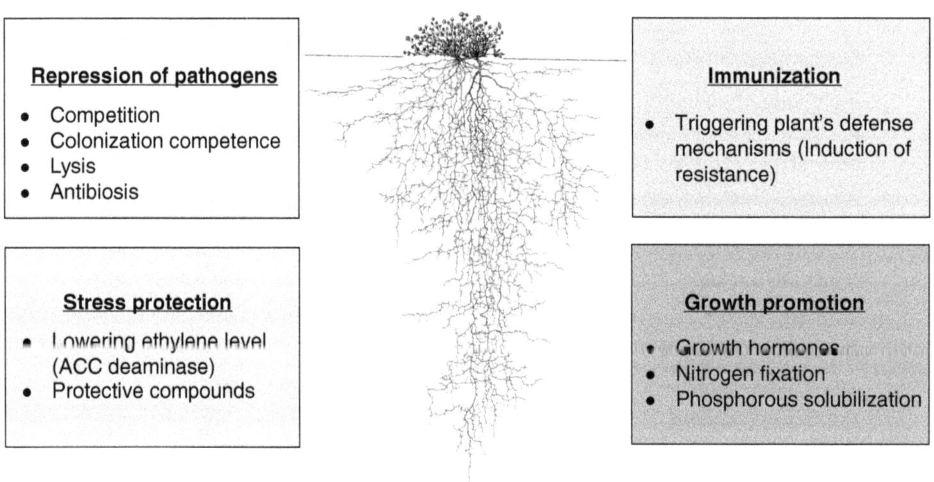


**Figure 22.2** Different roles of plant growth promoting bacteria (PGPB).

**Table 22.2** List of bacterial biocontrol agents and their mode of actions.

| Bacterial isolates | Mode of action | Pathogen | Disease and host | References |
|---|---|---|---|---|
| *Pseudomonas putida* A1 | Hydrolysis | *Xanthomonas oryzae* pv. *oryzae*<br>*X. oryzae* pv. *oryzicola*<br>*X. citri* subsp. *citri*<br>*Ralstonia solanecearum* | Rice BLB, Rice leaf streak disease, Citrus canker<br>Tomato wilt | Sun et al. (2017) |
| *Paenibacillus elgii* JCK-5075 | Lytic enzymes, antibacterial | *Botrytis cinerea, Puccinia triticina, Pectobacterium carotovorum* subsp. *carotovorum*, *Ralstonia solanacearum, Xanthomonas euvesicatoria* | Kimchi Cabbage soft rot, Wheat leaf rust, Tomato bacterial wilt, Tomato gray mold | Kim et al. (2020) and Le et al. (2020) |
| *Bacillus amyloliquefaciens* CHB 310 | Antibiosis, induced systemic resistance | *Xanthomonas perforans* | Tomato bacterial spot | Chien and Huang (2020) |
| *Xenorhabdus budapestensis* C72 | Lytic enzymes, induced systemic resistance | *Bipolaris maydis* | Southern corn leaf blight | Li et al. (2021) |
| *Pseudomonas mosselii* (CR 14), *Bacillus tequilensis* (CR 42), *Streptomyces hawaiiensis* (SE 4), *Pseudomonas koreensis* (KB 19), *Bacillus pumilus* (SKR 6) | Direct antagonism | *Bipolaris oryzae*<br>*Magnaporthe oryzae*<br>*Rhizoctonia solani*<br>*X. oryzae* | Brown spot, Rice blast, Sheath blight, Bacterial leaf blight | Jasrotia et al. (2021) and Brindhaa (2019) |
| *Bacillus pumilis, B. amyloliquefaciens*<br>*Bacillus subtilis* | Biofilm formation, antifungal lipopetides, lytic enzymes, VOCs | *Penicillium digitatum, Penicillium italicum* | Green and blue mold of citrus fruits | Hammami et al. (2022) |

and *Fusarium solani* in alfa alfa, *Rhizoctonia solani, Fusarium solani*, and *Fusarium oxysporum* in beans or peas in lab conditions and field conditions (Barnes 1981; Hong and Estey, 1985). *A. avenae* and *Aphelenchoides* spp. reduced the cauliflower damping off by suppressing *R. solani* (Lagerlöf et al. 2011).

The combined use of *Aphelenchus* and *Trichoderma* resulted in good control of *Pythium* damping off of radish seedlings, which demonstrates the synergism (Jun and Kim, 2004). The fungivorous nematode brings indirect control over plant pathogenic fungi, for example, *Aphelenchoides hylurgi* spreads the hypovirulent strains of the fungi *Cryphonectria parasitica* and reduces the infection by virulent strain (Griffin et al. 2009). *A. avenae* controls the *Ditylenchus destructor* by competing for the food resource. *D. destructor,* during the early stage of its lifecycle, fed on the fungi associated with the garlic plant due to food deprivation by *A. avenae,* the population was reduced, and further infection was prevented (Haraguchi and Yoshiga 2020). The bacterivorous nematode, which does not consume the bacterial pathogen, can invite the antagonistic bacteria that kill the pathogens (Li et al. 2023). The role of soil-dwelling nematodes in plant health is overlooked. As the nematode and plant are coevolved together, further explorations have to be done to harness the potential of nematodes in managing soil-borne plant pathogens.

### 22.2.7 Virus

Nearly 10% of the yield loss in agriculture is caused by plant disease (Strange and Scott, 2005). Though viruses are of particular importance in causing plant disease in recent days, the viruses that infect bacteria (bacteriophages) are widely used in controlling plant diseases caused by bacteria. Bacteriophages (phages) are at least 10 times greater than the total number of prokaryotic cells in the ecosystem and are the most prevalent biological species in the natural world, likely a total of $10^{31}$ (Whitman et al. 1998; Wommack and Colwell 2000), and are specific forms of bacteria-infecting viruses that interfere with their hosts' metabolic processes to grow and multiply (Żaczek et al. 2015). They also have a high degree of host specification and lytic activity toward their primary hosts (de Jonge et al. 2019). The AgriPhage™ products are EPA-registered biopesticides developed by the USA-based company Omnilytics that prevent and control harmful bacteria on tomatoes, peppers, apples, citrus, peaches, cherries, walnuts, almonds, and hazelnuts (OmniLytics 2006).

### 22.2.8 Transgenic Micro-Organisms

Despite employing random genetic manipulation strategies such as mutagenesis, genome shuffling, or protoplast fusion to enhance biocontrol traits of microorganisms, genetic engineering of native strains with the targeted introduction of desired genes for novel and efficient secondary metabolites, antibiotics, and hydrolytic enzymes is a highly efficient approach in the field of phytopathogen biocontrol (Kashyap et al. 2023; Kumari et al. 2022). The widespread and commercial application of native biocontrol strains is becoming less prospective due to their inconsistency in performance, narrow spectrum of activity, and the question of cell survivability under diverse environmental conditions (Weller and Thomashow 2015; Pathak et al. 2022). Several engineering strategies have been reported successfully in *Bacillus* (Xiao et al. 2023), *Trichoderma* (Wang et al. 2022), and Actinomycetes species (Figure 22.3) (Baltz 2016). It mainly focusing on the enhanced synthesizing of different hydrolytic enzymes such as cellulases and the overproduction of novel secondary metabolites like antibiotics, lipopeptides, non-ribosomal peptides, and polyketides encoded by biosynthetic gene clusters.

Traditional recombination techniques based on restriction endonucleases and ligases followed by genetic transformation such as electroporation, Agrobacterium-mediated transformation, polyethylene glycol-mediated transformation, and biolistic approaches are being used to construct engineered strains in *Trichoderma* and *Bacillus* (Li et al. 2017). Other recombination techniques based on transposons and lambda red recombinase system are gaining popularity. The Tn5 transposon system is commonly used in the development of transgenic strains of *Pseudomonas* (Domrose et al. 2015). The lambda red recombination system is employed for homologous recombination of foreign DNA in bacterial strains that express lambda phage proteins Exo, Beta, and Gam. The desired gene is introduced into the host bacterium through electroporation. Gam protein inhibits the endogenous nucleases that otherwise digest the introduced DNA. Exonuclease (Exo protein) degrades one of the two strands creating a ssDNA, and Beta protein anneals the ssDNA to the complementary strand and incorporates it into the host genome (Murphy 2016).

Employing artificially engineered nucleases is a recent and efficient approach in genetic engineering that enables to attain gene replacements, genome-wide gene disruptions, and precise gene editing for the development of novel engineered strains (Zheng et al. 2017). Among the gene editing strategies such as clustered regularly interspaced short palindromic repeats (CRISPR) RNA-guided Cas9 (CRISPR-Cas9) and transcription activator-like effector nucleases (TALENs), CRISPR-Cas9 is the most widely used and convenient tool for genetic manipulation. CRISPR-Cas9 involves the introduction of two components: Cas genes that encode nucleases and guide RNAs, which guide the Cas protein-RNA complex to bind to the complementary DNA site (Singh et al. 2022b).

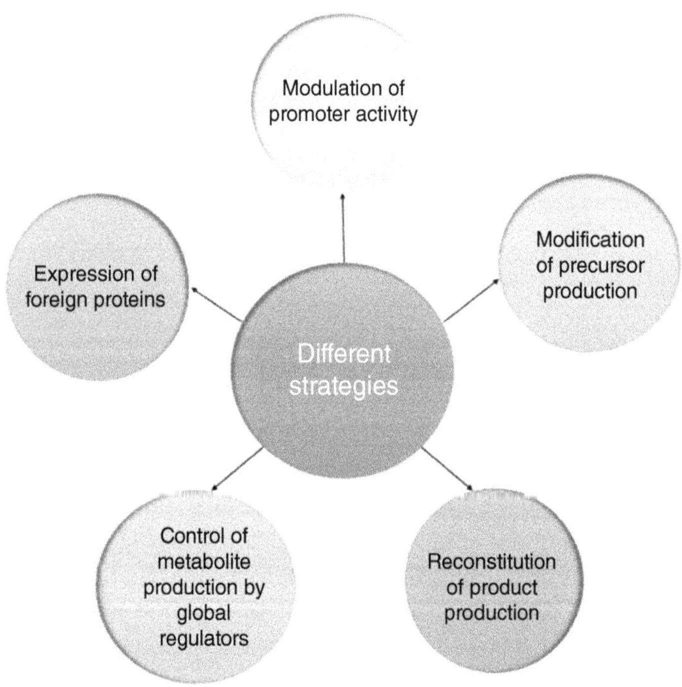


**Figure 22.3** Different techniques of genetic engineering for biocontrol agents.

After that, it introduces double-stranded breaks, which are subsequently repaired by either nonhomologous end joining, resulting in insertions, deletions, and rare substitutions near cleavage site, or by homology-directed repair leading to precise sequence alterations with the endogenous repair pathways (Liu and Fan 2014). The CRISPR-Cas system is introduced into the host by classical transformation methods. It can introduce gene edits at multiple sites of the genome to enhance their expression of antimicrobial-related genes through the deletion of repressors, introduction of heterologous genes, and overexpression of activators (Liu et al. 2015). Another improved strategy referred to as scarless Cas9-assisted recombineering, uses lamba red system for recombination and Cas9 enzyme to cleave dsDNA of antibiotic resistance marker genes in the host after recombination (Reisch and Prather 2015).

Despite the nonacceptance of transgenic biocontrol agents in today's agriculture, they are sure to be adopted as prospective means of plant disease control shortly (del Pilar Martφnez-Diz et al. 2021). iGEM (International Genetically Engineered Machine) Wageningen is an independent, nonprofit organization that supports projects on developing transgenic microorganisms for plant disease biocontrol, such as BananaGuard. It contains engineered *Pseudomonas putida* for control of Panama disease in cavendish plantations caused by a novel virulent strain of *F. oxysporum* sp. *cubense (Foc),* tropical race 4 (TR4), renamed as *F. odoratissimum* (Maymon et al. 2020). The recombinant *P. putida* expresses fungal growth inhibitors such as 2,4-diacetylphloroglucinol (2,4-DAPG), dimethyldisulfide (DMDS), dimethyltrisulfide (DMTS), pyoverdine siderophores, and chitinase (Gross and Loper, 2009) only upon detection of Fusarium in soil by sensing fusaric acid as well as self-destruction in the absence of the pathogen. Moreover, some of the transgenic biocontrol agents are listed in Table 22.3.

## 22.3 Central Role of Micro Organisms to Induced the Innate Immunity

Plants encounter different plant pathogens using induced resistance through several defensive molecules such as salicylic acid (SA), jasmonic acid (JA) and ET produced during different signaling pathways (Van Wees et al. 2008; Sidharthan 2017). These defensive molecules may act either independently or jointly using host defensive mechanisms such as systemic acquired resistance (SAR) or induced systemic resistance (ISR) (Subramanian et al. 2011; Maithani et al. 2021). In SAR, preinfection of phytopathogens triggers plants to enhance disease resistance by synthesizing SA, which acts as a protective agent against different pathogens (Ryals et al. 1996). In ISR, nonpathogenic and beneficial microbes incite the plants to produce ET and JA (Pieterse et al. 1998). Both mechanisms paved the way for broad-spectrum resistance against different pathogens (Subramanian et al. 2011). As compare to SAR, ISR enabled plant defense via antioxidant enzymes, which deeply rely on signal transduction pathways and will be discussed on the following topics.

### 22.3.1 ISR Induction

Elicitors are capable of eliciting a series of signaling events to activate defense response against biotic and abiotic stress (Figure 22.4) (Kesawat et al. 2023). Microorganisms can produce microbe-associated molecular pattern (MAMP) or pathogen-associated molecular pattern (PAMP) molecules such as chitin, flagellin, lipopolysaccharides, and other molecules recognized by pattern recognition receptors (PRR). There are numerous other unknown receptors found in plant cells that recognize the hormones, siderophores, volatile molecules, and chemical molecules produced by microbes (Zehra et al. 2021).

MAMP act as an elicitor that connects with the plant cell's receptors and are known as PRR. These signals were transferred and activated by the PTI (PAMP-triggered immunity) and MYB72 transcription factors, which are important for the transfer of signals through systematic movement throughout the plant system (Pieterse et al. 2014). The transcriptional factor (TF) and MYC2 genes are activated by the NPR 1 (Non-expressor of Pathogenesis Related genes 1) gene, a master regulator of induced resistance, and they aid defense reactions by producing JA and ET hormones (Zhai and Li 2019).

These hormones were expressed during the interaction with the beneficial microorganisms. Further, JA, ET and nitric oxide hormones interact with transcription factors that induce defense activities against harmful microbes, such as the production of reactive oxygen species (ROS) and activation of defense-related genes, hormones, and chemical compounds (Meena et al. 2022). During the ISR plants were primed against the plant pathogen to confer the resistance (Martínez et al. 2016). Alkaline pathogenesis-related (PR) gene and other proteins such as defensins, proteinase inhibitors, and thionins are related to ISR (Thomma et al. 2002).

**Table 22.3** Details of transgenic biocontrol agents.

| Parent strain | Donor strain | Gene | Strategy | Trait | Pathogen control | References |
|---|---|---|---|---|---|---|
| *T. atroviride* P1 | *S. cerevisiae* BY4741 | *ERG20* | Cotransformation with pAN52-1 NotI plasmid and pRMLex_30 plasmid | FPP synthase (an enzyme involved in mevalonate pathway), DPM synthase activity responsible for enhanced biocontrol activity by O- and N-glycosylation of secretory hydrolases | *P. ultimum* | Graczyk et al. (2020) |
| *P. chlororaphis* HT66 | *Nocardiopsis* sp. 13–12–13 | *NaphzNO1* | No SCAR deletion strategy (scarless Cas9-assisted recombineering) | Synthesis of novel phenazine-N oxides | Not proved | Guo et al. (2020) |
| | *P. aeruginosa* PAO1 | *PhzS* | | | | |
| *B. velezensis* FZB42 | *B. velezensis* 916 | *loc* gene (*loc DABC*) cluster | Lambda Red recombination system | Production of locillomycins | Not proved | Luo et al. (2019) |
| *P. fluorescens* HC1–07 | *P. synxantha* 2–79 | *phzABCDEFG* | Tn5 transposon | Phenazine biosynthetic operon | *Gaeumannomyces graminis* var. *tritici* | Yang et al. (2017) |
| *E. amylovora* phage Y2 | *Vibrio harveyi* phage L1 | Depolymerase encoding *dpoL1* | Transformation with recombinant pBlueScript plasmid by electroporation | Degradation of EPS of *Erwinia* | *E. amylovora* | Born et al. (2017) |
| | | *luxAB* | | Luciferase detection | | |
| *T. harzianum* | *T. arundinaceum* | *tri5* and *tri4* | Transformation with recombinant plasmid pb1Tatri4 | Production of trichothecenes | *F. graminearum* | Cardoza et al. (2015) |

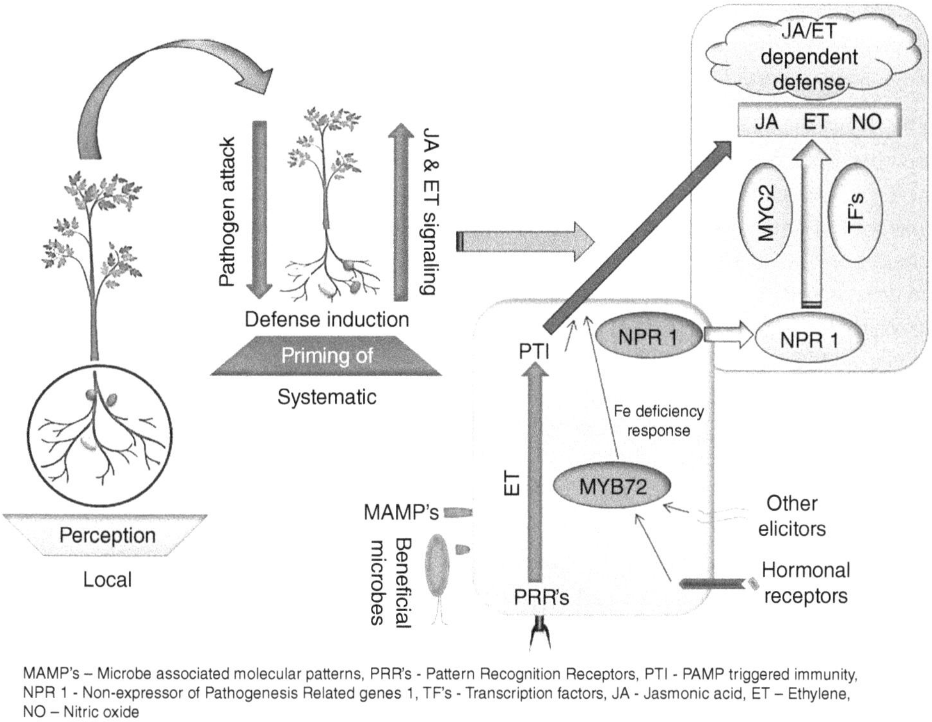


**Figure 22.4** Signal transduction pathway of ISR.

Whereas, the co-inoculation of *Trichoderma asperellum* and *T. harzianum* effectively inhibits the anthracnose in chili caused by *Colletotrichum truncatum*. During seed biopriming with pepper, this antagonist induces six defense genes such as *CaPDF1.2, SOD, APx, GPx, PR-2,* and *PR-5* in the plant system (Yadav et al. 2023). *Rhizoctonia solani* was effectively controlled by *Trichoderma pubescens,* which also promotes plant growth and systemic resistance. It also stimulates the production of antioxidants by influencing the three defense-related genes, including PAL, CHS, and HQT (Behiry et al. 2023).

### 22.3.2 Role of Antioxidant Against Plant Pathogens

In ISR, defense-related genes are directed to synthesize the JA and ET molecules that encode some precursors of antioxidant enzymes, namely, polyphenol oxidase (PPO), peroxidase (PO), tyrosine ammonia lysine (TAL), oxidative phenolics and phenyl alanine lyase (PAL) (Avdiushko et al. 1993). Such enzymes employ defensive reactions in plants and were categorized into two parts, that is, nonenzymatic and enzymatic antioxidants will be discussed in the next section (He and Hannon 2004). Host defense system follows different strategies to combat host diseases including (i) reinforcement of cell wall; (ii) enhanced production of antioxidants, PR protein, phytoalexins, proline, lignin deposition, and defensive enzyme; (iii) physiological tolerance; and (iv) adjustment or modulation of phenols to develop antioxidant and antimicrobial properties (Jain and Gould, 2015).

#### 22.3.2.1 Nonenzymatic Antioxidants: Phenolics

In general, phenolics are often synthesized by specialized cells in plants and stored in their vacuoles. Upon an injury, vacuoles burst, and rapid oxidation of the phenolic content ensues in suberization and lignification of cells. This results in pathogen cell death or injury at the point of penetration. If there is a failure of the defense mechanism, lignified phenolics promote the production of IAA and ET. This ensures the host has taken into account secondary metabolism, which triggers the peridermal defense (Beckman, 2000).

#### 22.3.2.2 Enzymatic Antioxidants During Stress

As a kind of defense reactions, plants have produced different enzymes such as phenylalanine ammonia lyase (PAL), PPO, PO, chitinase, and -1,3-glucanase (Prasannath and De Costa 2015).

**Superoxide dismutase (SOD):** SOD (E.C.1.15.1.1) is an important plant stress hormone that, in the face of diverse stress situations, acts as the plant's first line of defense against oxidative damage (Sagar et al. 2020). It is a metalloenzyme that converts the superoxide radical ($O_2^{\cdot -}$) into either oxygen ($O_2$) or hydrogen peroxide ($H_2O_2$) in aerobic organisms (Hatem et al. 2010). It is the strongest intracellular enzymatic antioxidant in the cell and is found in all aerobic species, primarily where the electron transport chain takes place, i.e., mitochondria, chloroplast, glyoxysomes, and cytosol (Elstner 1991). SOD can be grouped into three types based on metals present at their active site: primarily, (Fe) SOD, manganese (Mn) SOD, and copper/zinc iron (Cu/Zn) SOD. Among these, (Fe) SOD is seen in prokaryotes, (Mn) SOD is present in eukaryotic mitochondria and prokaryotes, and (Cu/Zn) SOD is found in chloroplasts, cytosols of eukaryotes, and some species of bacteria (Mruk et al. 2002). In terms of enzymatic perception, "one unit SOD represents the amount of the enzyme required to inhibit 50% of the autoxidation of pyrogallol." It is associated to generate the systemic resistance in tissues and is essential to preventing cells against superoxide toxicity.

**Catalase (CAT):** Catalase (E.C. 1.11.1.6) an oxidoreductase enzyme plays a central role in the subsequent breakdown of non-radical ROS such as hydrogen peroxide ($H_2O_2$) into oxygen ($O_2$) and water ($H_2O$) via a catalytic reaction (Beers and Sizer 1952; Ighodaro and Akinloye 2018). It has a tetrameric Fe-containing porphyrin enzyme with each chain being 500 amino acids long. It maintains cellular redox homeostasis. It is a first-line antioxidant defense enzyme that is photosensitive and leads to photoinactivation and subsequent degradation (Feierabend et al. 1992). Catalase utilizes $H_2O_2$ for peroxidative oxidation of various organic substrates. It is usually located in peroxisomes and the inner mitochondrial membrane of almost all aerobic organisms (Mruk et al. 2002). Depending on sequence homology and assembly, it has been categorized into three groups: (i) monofunctional heme-containing catalase is present in all aerobic organisms, (ii) bifunctional heme-containing catalase-peroxidase is less abundant and closely related to plant PO, and (iii) Mn-containing catalase groups lack heme group (Glorieux and Calderon, 2017). One unit of catalase means 1 M of $H_2O_2$ decomposed per minute (Sagar et al. 2020). It plays a significant role in reducing the concentration and removal of hydrogen peroxide, which is also involved in the signaling process, and against several oxidative stress-related infections as a therapeutic agent (Nandi et al. 2019). Under biotic and abiotic stress, ROS such as hydrogen peroxide and superoxide anions accelerate the oxidative damage and eventually cell death (Fimognari et al. 2020). Hence, it acts as an antioxidant and protects the cell against oxidative stress.

**Peroxidases (PO)**: PO (E.C.1.11.1.x) has been known for its defense-related functions like lignification, cross-linking of phenolics and glycoproteins, hypersensitive response, suberization, and phytoalexin production (Nicholson and Hammerschmidt 1992; Wojtaszek 1997). PO catalyzes the oxidization of phenolics into quinones with the release of hydrogen peroxide. The gene expression of PO in higher plants can be induced by bacteria (Prasannath et al. 2014), fungi (Sasaki et al. 2004), viruses (Diaz-Vivancos et al. 2006), and viroids (Vera et al. 1993). $H_2O_2$ released is an antimicrobial compound that generates large amount of ROS derivatives, which kill the pathogens by oxidative burst (Halfeld-Vieira et al. 2006). It also enhances the phenolic compounds polymerization into lignin substances, which helps to strengthen the cell walls leads impede the pathogen's ability to proliferate and evolve further (Agrios 2005).

**Polyphenol oxidase (PPO's)**: Polyphenol oxidases (E.C.1.14.18.1) (PPOs) convert the phenolics to quinone, which has antimicrobial activity (Figure 22.5) (Chunhua et al. 2001). In response to pathogens, plants instantly get higher PPO, which has been seen in various plant taxa, including monocots and dicots. (Chen et al. 2000; Deborah et al. 2001). In a report by Saravanan et al. (2022), it was noticed that, PPO was increased when pathogen was infected by Banana plants. Then, it was treated with *Psuedomonas fluorescens* to combat the same disease, which showed a startling rise in PPO activity (Saravanan et al. 2004). The outcome correlates an increase in the induction of defense enzymes like PPO with the protective action of rhizosphere bacteria.

**Phenylalanine ammonia lyase (PAL)**: PAL (E.C.4.1.3.5) is the first enzyme in the phenylpropanoid pathway and flavonoid pathway. It catalyzes the conversion of L-phenylalanine to *trans*-cinnamic acid and ammonia, which in turn leads to the biosynthesis of polyphenol compounds such as flavonoid, phenylpropanoid, and lignin in plants (Zhang and Liu 2015). It belongs to the ammonia-lyase family and has a catabolic role in breaking carbon–nitrogen bonds. *trans*-Cinnamic acid is the direct precursor of SA. SA is a signal molecule and elicits a SAR response as a mechanism to protect against pathogen invasion (Kavitha et al. 2012). Phenolic metabolism in plants plays important roles in structural support, chemical attractants or repellents, defense-related compounds, UV protection, providing aromatic amino acids, and color (Jun et al. 2018).

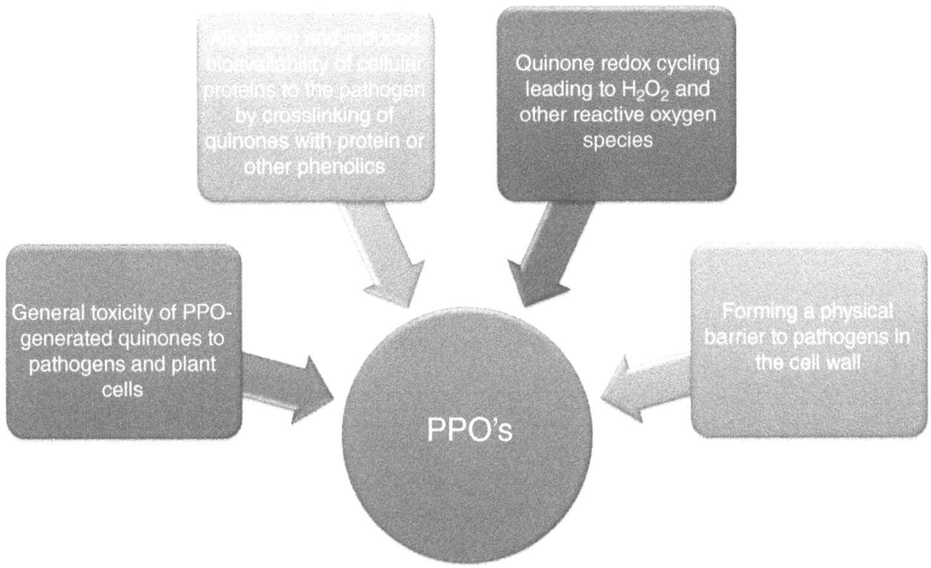


**Figure 22.5** Different roles of PPO in plant defensive mechanisms.

## 22.4 Synthetic Microbial Communities in Plant Disease Management

Synthetic microbial communities (Syncoms) also known as microbiome engineering or microbial consortia have emerged as a promising approach in plant disease management. These communities consist of carefully selected and assembled beneficial microorganisms that interact with plants and each other to enhance plant health and suppress the growth of pathogens (Wang et al. 2023). These offer greater flexibility in designing and tailoring specific beneficial traits and interactions, allowing for more targeted and effective plant disease management compared to traditional bioinoculants (Sarrocco 2023). Syncoms can be engineered to include multiple microbial species that work synergistically to provide enhanced disease suppression, nutrient cycling, and plant growth promotion, providing a more comprehensive and robust solution for plant health (Vannier et al. 2019).

This technique uses top-down and bottom-up approaches to tailoring better artificial microbial communities. In the top-down approach, microbial communities are assembled by starting with a complex community that manipulates specific microorganisms to achieve desired functions. It allows for the exploration of potential microbial communities against plant diseases. In the bottom-up approach, microbial communities are built from the beginning by adding individual microbes by species or genus level with known beneficial traits of the artificial communities (Figure 22.6).

This approach provides a greater impact on the compositions and interactions within the community, enabling the improvement of plant health (Ke et al. 2021). These can involve both culturable and nonculturable microbes, expanding the diversity and functionality of the community beyond the culturable microbes (de Souza et al. 2016).

Nonculturable microbes are identified through a metagenomic and molecular technique that contributes to the complexity and functionality of artificial complex microbes to enhance their potential for plant disease management and other beneficial interactions in the ecosystem (Oberhardt et al. 2015, Chen et al. 2007). The possible mechanism of Syncoms may be a cry for help hypothesis. An adaptive technique of the plant for enhancing plant survival under biotic stress is the recruitment of advantageous microorganisms and suppressive soil mechanism during protection from biotic and abiotic stress. The cry for help refers to the mechanism by which plants release chemical signals or volatile organic compounds when they are under stress or attacked by pathogens. The signals are often attracted and perceived by the beneficial microorganisms and trigger defense responses to protect themselves or activate beneficial interactions (Wang et al. 2022). Suppressive soil mechanism refers to the natural ability of certain soils to suppress diseases through diverse microbial communities and beneficial interactions, enhancing plant health and reducing pathogen impact (Du et al. 2021). In recent years, this approach has been promising in plant disease management by utilizing the collective power of beneficial microorganisms and plant health.

Investigation by Prigigallo et al. (2022) reported that Banana Fusarium wilt pathogen *Fusarium oxysporum* f. sp. *cubense* tropical race 4 was successfully controlled by Syncom. The cross-kingdom rhizomicrobiome of tomato induces the suppressive effect of the Fusarium wilt pathogen. Also, Sym of various microbial communities like bacteria and fungi showed

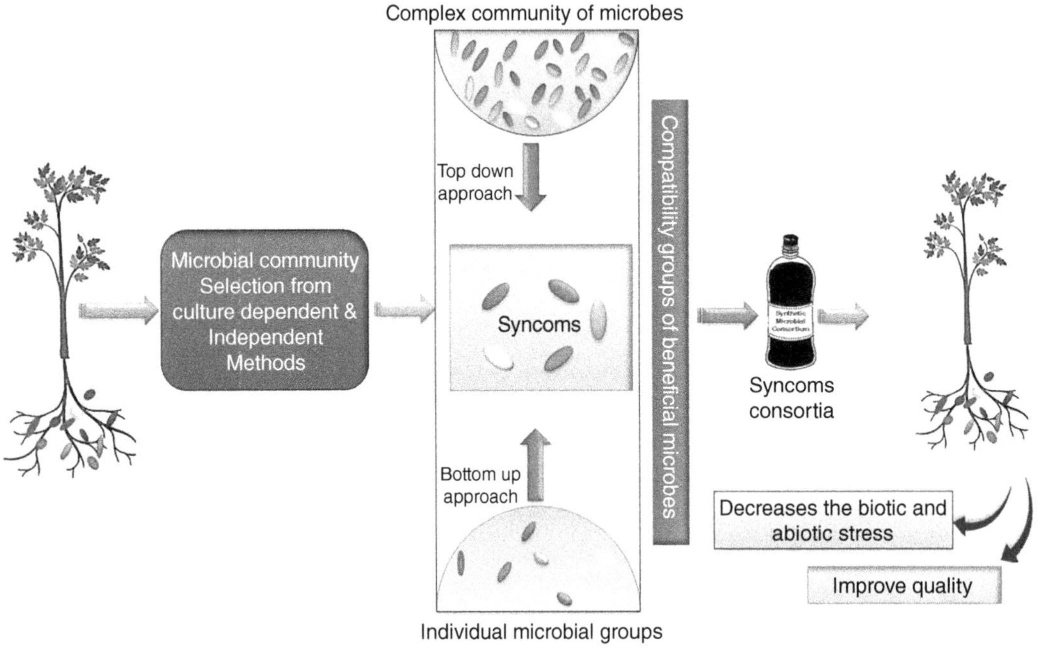


**Figure 22.6** Approaches of SynComs in plant stress management.

excellent disease reduction and enhanced innate immunity of plant system (Zhou et al. 2022). For example, Syncom made up of *Trichoderma virens* T2C1.4, *Pseudomonas chlororaphis* subsp. *piscium* PS5, and *Bacillus velezensis* BN8.2, reducing disease severity and outperforming commercial biological control agents. The Syncoms are efficient in significant disease control, both in the root zone and on the foliage of tomato plants (Minchev et al. 2021).

## 22.5 Recent Trends of Biocontrol Agent

Global agriculture is heading toward the employment of biocontrol strategies for plant disease control with the gradual reduction in the use of pesticide chemicals. The value of the global biocontrol market reached 4.0 billion dollars in 2020 and it is expected to reach 10.6 billion dollars by 2027. Several fungal biocontrol agents based on *Trichoderma* including *T. asperellum, T. harzianum, T. viride* and *T. virens* have been commercially developed for the treatment of soil-borne fungal pathogens, since the release of the first commercial *Trichoderma*-based biocontrol agent Bunab-*T. Trichoderma* isolate T22, a protoplast fused strain developed in the 1990s, it still holds a place in plant disease management (Harman 2006). Apart from *Trichoderma*, strains of *Clonostachys roseum*, *Talaromyces flavus, Ulocladium oudemansii*, and *Coniothyrium minitans* gain commercial significance as biocontrol formulations for the control of soil-borne fungal pathogens, including *Sclerotinia sclerotiarum*, *Phytophthora,* and *Pythium* (Rashad and Moussa 2020). Powdery mildew-targeting fungus *Ampelomyces quisqualis* (Carbó et al. 2020) and mycorrhiza-like endophytic fungus *Serendipita indica* (identified from the Indian desert, popularized as "Rootonic" (Singhal et al. 2017) that protects plants against biotic and abiotic stresses occupy prime importance as antimicrobials. The bacterial biocontrol agents based on *Bacillus subtilis, B. amyloliquefaciens,* and *B. pumilus* have become commercially popular with the development of improved strains and formulations with severe antagonism toward fungal pathogens (Borriss 2015; Fira et al. 2018). In addition to *Bacillus* sp., various *Pseudomonas* species have outperformed in inhibiting a lot of plant pathogens, mainly *P. fluorescens* for *Erwinia amylovora*, nonpathogenic *Pseudomonas syringae* for postharvest diseases, *P. chlororaphis* for diseases of turf grass and ornamental crops, and *Pseudomonas aureofaciens* against soil-borne fungi are commercially successful crop specific biocontrol products (Lahlali et al. 2022).

Several yeast-based biocontrol agents are employed to commercially inhibited the several postharvest pathogens of fruits and vegetables, including Candifruit™ (*Candida sake*), Shemer™ (*Metschnikowia fructicola*), Boni-Protect® (*Aureobasidium pullulans*), and Nexy (*C. oleophila*) (Spadaro and Droby 2016). Non-toxigenic *Aspergillus* strains have been employed for the control of aflatoxin contamination, such as Aflasafe and Aflaguard (Chang et al. 2021). Several phage-based bioproducts such as Agriphage against bacterial diseases such as citrus canker, tomato bacterial speck, fire blight, and peach blight (Nakayinga et al. 2021) and XylPhiPD™® (Wagemans et al. 2022), against *Xylella fastidiosa*-caused Pierce's diseases have been commercially developed. The recombinant *P. putida* expresses fungal growth inhibitors such as 2,4-DAPG, DMDS, DMTS, pyoverdine siderophores, and chitinase (Gross and Loper 2009) only upon detection of *Fusarium* in soil by sensing fusaric acid as well as self-destruction in the absence of the pathogen.

## 22.6 Conclusion

In a nutshell, pressure from the various biotic and abiotic stressors significantly reduced crop productivity. In these circumstances, microbial inoculants are greatly helped to exit the all-agricultural constraints and hope to achieve Sustainable Development Goal 2 by 2030. Currently, various mechanisms related to plant growth and biocontrol of different pathogens are being primarily explored. Keeping in view, a biocontrol agent cum plant growth promoter is a successful and active candidate for all aspects of a sustainable agriculture system. Moreover, further exploration of interactional studies between plant and biocontrol agents helped to grasp more knowledge of unknown functions during signaling pathways under stresses. In future, plant–microbe interactions and their traits will be clear to understand for the optimal performance of particular crops. Also, more research is required for efficient delivery system for long-term survival and actions of microbial inoculants, which should ensure the sustainable agriculture (Table 22.4).

**Table 22.4** Commercially available products of biocontrol agents.

| Commercial product name | Biocontrol strain | Pathogen/disease | Countries approved | References |
|---|---|---|---|---|
| 1) **Bacteria** | | | | |
| Serenade (Bayer Crop Science) | *Bacillus amyloliquefaciens* QST 713 | *Pythium*, clubroot, yellow rust | Global | Reiss and Jørgensen (2017) |
| Avogreen | *Bacillus subtilis* B246 | *Cercospora, Collelotrichum, Botrytis, Penicillium* | USA | Demoz and Korsten (2006) |
| Kodiak (Gustafson, USA) | *B. subtilis* GB03 | Cotton wilts caused by *Rhizoctonia* and *Fusarium*, Take all disease | USA, Canada | Wang et al. (2018) |
| Galltroll, (Becker Underwood) | *Agrobacterium radiobacter* (now *Rhizobium rhizogenes*) K84 | Crown gall (*A. tumefaciens*) | Australia, USA, Turkey | Kerr and Bullard (2020) |
| Nogall | *A. radiobacter* K1026 | Crown gall | Australia, Turkey, USA | Kerr and Bullard (2020) |
| Biosave 10L (Jet Harvest Solutions, USA) | *Pseudomonas syringae* ESC 10 | *Penicillium, Botrytis, Mucor* | USA, Canada, Moldova | Carmona-Hernandez (2019) |
| BlightBan (Frost Technology Corporation, USA) | *Pseudomonas fluorescens* A506 | Fire blight, bunch rot | USA, Canada | Stockwell et al. (2010) |
| Tivano (AEC Global) | *Lactobacillus casei* strain LPT-111 | Angular leaf spot caused by *Xanthomonas fragaria* | Canada | Dubois et al. (2017) |
| 2) **Fungi** | | | | |
| Radix soil (Isagro), Remedier (Gowan) | *Trichoderma asperullum* ICCO12 + *Trichoderma gamsii* ICC080 | Soilborne pathogens and grapevine trunk diseases | EU | Gerin et al. (2018) and Martínez-Diz et al. (2020) |
| Trichodex (BioWorks, USA) | *Trichoderma harzianum* T39 | *Botrytis cinerea*, Downy mildew | USA, South Africa, India | Roatti et al. (2013) and Harel et al. (2014) |

(*Continued*)

**Table 22.4** (Continued)

| Commercial product name | Biocontrol strain | Pathogen/disease | Countries approved | References |
|---|---|---|---|---|
| Root Shield (Bioworks), Trianum-P (Koppert) | *T. harzianum* T22 | Root diseases | USA, EU | Blaya et al. (2013) |
| Sporidex (Ecogen Inc.) | *Pseudozyma flocculosa* PF-A22UL | Powdery mildew on wheat, barley, grapevines, apple, and vegetables | USA, Canada | Laur et al. (2018) |
| Rootonic: SOM Phytopharma | *Serendipita indica* (*Piriformonspora indica*) | Soilborne pathogens | India | Varma et al. (2013) |
| AQ10 BioFungicide Ecogen (Grand Junction, USA) | *Ampelomyces quisqualis* M-10 | Powdery mildew | USA, UK | Caffi et al. (2022) |
| Contans WG (Prophyta Biologischer) | *Coniothyrium minitans* CON/M/91–08 | *Sclerotinia sclerotiorum, S. minor* | Germany, UK | Hammond et al. (2008) and Zeng et al. (2012) |
| GlioGard (Thermo Trilogy Corporation, USA) | *Gliocladium virens* GL-21 | Soilborne pathogens | Finland, USA | Jensen et al. (2021) |
| Tough Block (Idemitsu Kosan Co., Ltd.) | *Talaromyces flavus* SAY-Y-94-01 | Anthracnose caused by *Glomerella cingulata* and *Colletotrichum acutatum* | Japan | Ishikawa (2013) |
| 3) **Post harvest biocontrol agents** | | | | |
| Aspire (Ecogen, USA) | *Candida oleophila* I-182 | *Botrytis, Penicillium, Monilinia* | USA, EU, California | Sui et al. (2020) |
| Candifruit (IRTA, Spain) | *Candida sake* CPA 1 | *Penicillium, Botrytis, Rhizopus* | USA | Freimoser et al. (2019) |
| Shemer (Bayer, Netherlands) | *Metschnikowia fructicola NRRL Y-27328* | *Botrytis, Penicillium, Rhizopus, Aspergillus* | UK, EU, USA, California | Prodorutti et al. (2006) |
| BoniProtect (Bio-fern, Austria) | *Aureobasidium pullulans* DSM 14940 and DSM 14941 | *Botrytis, Penicillium, Monilinia* | USA, Canada, Australia, California | Freimoser et al. (2019) |
| 4) **Virus** | | | | |
| Biolyse-PB (APS biocontrol) | Consortium of bacteriophage | Postharvest soft rot of potato | UK | https://www.apsbiocontrol.com/ |
| XylPhi-PD (Wilbur-Ellis) | Bacteriophage cocktail | Pierce's disease on vine (*Xylella fastidiosa*) | California, USA | https://www.wilburellisagribusiness.com/product/xylphi-pd |

Information from Pesticide Properties Database, University of Hertfordshire. The Manual of Biocontrol Agents (Loughlin 2014).

## References


Ab Rahman, S.F.S., Singh, E., Pieterse, C.M., and Schenk, P.M. (2018). Emerging microbial biocontrol strategies for plant pathogens. *Plant Science* 267: 102–111.

Abdel-Rahim, I.R. and Kamal, A.M. (2018). *Talaromyces pinophilus* strain AUN-1 as a novel mycoparasite of *Botrytis cinerea*, the pathogen of onion scape and umbel blights. *Microbiological Research* 212: 1–9.

Adeyeye, M. and Samuel, A.O. (2016). Fungal mycotoxins in foods: a review. *Cogent Food & Agriculture* 2 (1): 1213127.

Agrios, G.N. (2005). *Plant Pathology*, 5e, 11. San Diego: Elsevier-Academic Press.

Anand, U., Pal, T., Yadav, N. et al. (2023). Current scenario and future prospects of endophytic microbes: promising candidates for abiotic and biotic stress management for agricultural and environmental sustainability. *Microbial Ecology* 86: 1455–1486. https://doi.org/10.1007/s00248-023-02190-1.

Auer, S. and Ludwig-Müller, J. (2015). Biological control of clubroot (*Plasmodiophora brassicae*) by the endophytic fungus *Acremonium alternatum*. *Journal of Endocytobiosis and Cell Research* 26: 43–49.

Avdiushko, S.A., Ye, X.S., and Kuc, J. (1993). Detection of several enzymatic activities in leaf prints of cucumber plants. *Physiological and Molecular Plant Pathology* 42 (6): 441–454.

Aydin, M.H. (2022). *Rhizoctonia solani* and its biological control. *Türkiye Tarımsal Araştırmalar Dergisi* 9 (1): 118–135.

Baltz, R.H. (2016). Genetic manipulation of secondary metabolite biosynthesis for improved production in *Streptomyces* and other actinomycetes. *Journal of Industrial Microbiology and Biotechnology* 43 (2): 343–370.

Barnes, G. (1981). *Aphelenchus avenae*, a potential biological control agent for root rot fungi. *Plant Disease* 65: 423.

Beckman, C.H. (2000). Phenolic-storing cells: keys to programmed cell death and periderm formation in wilt disease resistance and in general defence responses in plants? *Physiological and Molecular Plant Pathology* 57 (3): 101–110.

Beers, R.F. and Sizer, I.W. (1952). A spectrophotometric method for measuring the breakdown of hydrogen peroxide by catalase. *Journal of Biological Chemistry* 195 (1): 133–140.

Behiry, S., Soliman, S.A., Massoud, M.A. et al. (2023). *Trichoderma pubescens* elicit induced systemic resistance in tomato challenged by *Rhizoctonia solani*. *Journal of Fungi* 9 (2): 167.

Bělonožníková, K., Vaverová, K., Vaněk, T. et al. (2020). Novel insights into the effect of pythium strains on rapeseed metabolism. *Microorganisms* 8 (10): 1472.

Blaya, J., López-Mondéjar, R., Lloret, E. et al. (2013). Changes induced by *Trichoderma harzianum* in suppressive compost controlling *Fusarium* wilt. *Pesticide Biochemistry and Physiology* 107 (1): 112–119.

Born, Y., Fieseler, L., Thöny, V. et al. (2017). Engineering of bacteriophages Y2 *dpoL1-C* and Y *luxAB* for efficient control and rapid detection of the fire blight pathogen, *Erwinia amylovora*. *Applied and Environmental Microbiology* 83 (12): e00341–e00317.

Borriss, R. (2015). *Bacillus*, a plant beneficial bacterium. In: *Principles of Plant-Microbe Interactions. Microbes for Sustainable Agriculture* (ed. B. Lugtenberg), 379–391. Berlin: Springer.

Brindhaa, N. T. 2019. *Biocontrol potential of indigenous fluorescent pseudomonads against banded leaf and sheath blight of maize incited by Rhizoctonia solani f. sp. Sasakii*. Thesis, M. Sc. GB Pant University of Agriculture and Technology, Pantnagar.

Buttimer, C., McAuliffe, O., Ross, R.P. et al. (2017). Bacteriophages and bacterial plant diseases. *Frontiers in Microbiology* 8: 34.

Caffi, T., Legler, S.E., Furiosi, M. et al. (2022). *Application of Ampelomyces Quisqualis in the Integrated Management of Grapevine Powdery Mildew*. CABI Books. CABI International.

Campbell, R. (1989). Biocontrol of postharvest diseases – introduction. *Biological Control of Foliar and Post-Harvest Diseases* 16: 93–94.

Carbó, A., Torres, R., Usall, J. et al. (2020). Biocontrol potential of *Ampelomyces quisqualis* strain CPA-9 against powdery mildew: conidia production in liquid medium and efficacy on zucchini leaves. *Scientia Horticulturae* 267: 109337.

Cardoza, R.E., McCormick, S.P., Malmierca, M.G. et al. (2015). Effects of trichothecene production on the plant defense response and fungal physiology: overexpression of the *Trichoderma arundinaceum* tri4 gene in *T. harzianum*. *Applied and Environmental Microbiology* 81 (18): 6355–6366.

Carmona-Hernandez, S., Reyes-Pérez, J.J., Chiquito-Contreras, R.G. et al. (2019). Biocontrol of postharvest fruit fungal diseases by bacterial antagonists: a review. *Agronomy* 9 (3): 121.

Castoria, R., Miccoli, C., Barone, G. et al. (2021). Molecular tools for the yeast *Papiliotrema terrestris* LS28 and identification of Yap1 as a transcription factor involved in biocontrol activity. *Applied and Environmental Microbiology* 87 (7): e02910–e02920.

Chamoun, R., Aliferis, K.A., and Jabaji, S. (2015). Identification of signatory secondary metabolites during mycoparasitism of *Rhizoctonia solani* by *Stachybotrys elegans*. *Frontiers in Microbiology* 6: 353.

Chang, P.K., Chang, T.D., and Katoh, K. (2021). Deciphering the origin of *Aspergillus flavus* NRRL21882, the active biocontrol agent of Afla-Guard®. *Letters in Applied Microbiology* 72 (5): 509–516.

Chen, C., Belanger, R.R., Benhamou, N., and Paulitz, T.C. (2000). Defense enzymes induced in cucumber roots by treatment with plant growth-promoting rhizobacteria (PGPR) and *Pythium aphanidermatum*. *Physiological and Molecular Plant Pathology* 56 (1): 13–23.

Chen, A.H., Chai, Y.R., Li, J.N., and Chen, L. (2007). Molecular cloning of two genes encoding cinnamate 4-hydroxylase (C4H) from oilseed rape (*Brassica napus*). *Journal of Biochemistry and Molecular Biology Research* 40: 247–260.

Chien, Y.C. and Huang, C.H. (2020). Biocontrol of bacterial spot on tomato by foliar spray and growth medium application of *Bacillus amyloliquefaciens* and *Trichoderma asperellum*. *European Journal of Plant Pathology* 156 (4): 995–1003.

Chunhua, S., Ya, D., Bingle, X. et al. (2001). The purification and spectral properties of PPO I from Nicotianan tababcum. *Plant Molecular Biology* 19: 301–314.

Cook, R.J. and Baker, K.F. (1983). *The Nature and Practice of Biological Control of Plant Pathogens*, 539. Saint Paul: Minnesota's, American Phytopathological Society.

Deborah, S.D., Palaniswami, A., Vidhyasekaran, P. et al. (2001). Time-course study of the induction of defense enzymes, phenolics and lignin in rice in response to infection by pathogen and non-pathogen/Zeitlicher Verlauf der Induktion von Abwehrsystemen, phenolischen Verbindungen und Lignin als Reaktion auf die Infektion mit einem Pathogen oder Nichtpathogen. *Zeitschrift für Pflanzenkrankheiten und Pflanzenschutz/Journal of Plant Diseases and Protection* 204–216.

Demoz, B.T. and Korsten, L. (2006). *Bacillus subtilis* attachment, colonization, and survival on avocado flowers and its mode of action on stem-end rot pathogens. *Biological Control* 37 (1): 68–74.

de Souza, R.S.C., Okura, V.K., Armanhi, J.S.L. et al. (2016). Unlocking the bacterial and fungal communities assemblages of sugarcane microbiome. *Scientific Reports* 6 (1): 28774.

Díaz-Vivancos, P., Rubio, M., Mesonero, V. et al. (2006). The apoplastic antioxidant system in Prunus: response to long-term plum pox virus infection. *Journal of Experimental Botany* 57 (14): 3813–3824.

Domrose, A., Klein, A.S., Hage-Hülsmann, J. et al. (2015). Efficient recombinant production of prodigiosin in *Pseudomonas putida*. *Frontiers in Microbiology* 6: 972.

Du, J., Li, Y., Mukhtar, I. et al. (2021). Synergistically promoting plant health by harnessing synthetic microbial communities and prebiotics. *Iscience* 24 (8): 102918.

Dubois, C., Arsenault-Labrecque, G., and Pickford, J. (2017). Evaluation of a new biopesticide against angular leaf spot in a commercial operation system. *Acta Horticulturae* 1156: 757–764.

Edward-Rajanayagam, R.M.A., Narváez-Zapata, J.A., del Socorro Ramírez-González, M. et al. (2023). Yeast mixtures for postharvest biocontrol of diverse fungal rots on *Citrus limon* var Eureka. *Horticulturae* 9 (5): 573.

Elstner, E.F. (1991). Oxygen radicals—biochemical basis for their efficacy. *Klinische Wochenschrift* 69 (21–23): 949–956.

Feierabend, J., Schaan, C., and Hertwig, B. (1992). Photoinactivation of catalase occurs under both high-and low-temperature stress conditions and accompanies photoinhibition of photosystem II. *Plant Physiology* 100 (3): 1554–1561.

Fimognari, L., Dölker, R., Kaselyte, G. et al. (2020). Simple semi-high throughput determination of activity signatures of key antioxidant enzymes for physiological phenotyping. *Plant Methods* 16 (1): 1–19.

Fira, D., Dimkić, I., Berić, T. et al. (2018). Biological control of plant pathogens by *Bacillus* species. *Journal of Biotechnology* 285: 44–55.

Fisher, M.C., Henk, D.A., Briggs, C.J. et al. (2012). Emerging fungal threats to animal, plant and ecosystem health. *Nature* 484 (7393): 186–194.

Freimoser, F.M., Rueda-Mejia, M.P., Tilocca, B., and Migheli, Q. (2019). Biocontrol yeasts: mechanisms and applications. *World Journal of Microbiology and Biotechnology* 35 (10): 154.

Gerin, D., Pollastro, S., Raguseo, C. et al. (2018). A ready-to-use single- and duplex-TaqMan-qPCR assay to detect and quantify the biocontrol agents *Trichoderma asperellum* and *Trichoderma gamsii*. *Frontiers in Microbiology* 9: 2073.

Gimeno, A., Leimgruber, M., Kägi, A. et al. (2021). UV protection and shelf life of the biological control agent *Clonostachys rosea* against *Fusarium graminearum*. *Biological Control* 158: 104600.

Glorieux, C. and Calderon, P.B. (2017). Catalase, a remarkable enzyme: targeting the oldest antioxidant enzyme to find a new cancer treatment approach. *Biological Chemistry* 398 (10): 1095–1108.

Gond, D.P., Jha, S.S., Kumar, A., and Singh, S.K. (2021). Plant growth promoting bacteria and its role in green remediation. In: *Sustainable Environmental Clean-up* (ed. V.K. Mishra and A. Kumar), 149–163. Cambridge, MA: Elsevier https://doi.org/10.1016/B978-0-12-823828-8.00007-4.

Graczyk, S., Perlińska-Lenart, U., Górka-Nieć, W. et al. (2020). Increased activity of the sterol branch of the mevalonate pathway elevates glycosylation of secretory proteins and improves antifungal properties of *Trichoderma atroviride*. *Fungal Genetics and Biology* 137: 103334.

Griffin, G.J., Eisenback, J.D., Yancey, M.M., and Templeton, J. (2009). *Aphelenchoides hylurgi* as a carrier of white, hypovirulent *Cryphonectria parasitica* and its possible role in hypovirulence spread on blight-controlled American chestnut trees. *Journal of Nematology* 41 (4): 267.

Gross, H. and Loper, J.E. (2009). Genomics of secondary metabolite production by *Pseudomonas* spp. *Natural Product Reports* 26: 1408–1446.

Guo, S., Liu, R., Wang, W. et al. (2020). Designing an artificial pathway for the biosynthesis of a novel phenazine N-oxide in *Pseudomonas chlororaphis* HT66. *ACS Synthetic Biology* 9 (4): 883–892.

Gupta, A., Singh, S.K., Singh, M.K. et al. (2020a). Plant growth–promoting rhizobacteria and their functional role in salinity stress management. In: *Abatement of Environmental Pollutants* (ed. P. Singh, A. Kumar, and A. Borthakur), 151–160. Cambridge, MA: Elsevier https://doi.org/10.1016/B978-0-12-818095-2.00007-2.

Gupta, A., Singh, S.K., Singh, V.K. et al. (2020b). Endophytic microbe approaches in bioremediation of organic pollutants. In: *Microbial Endophytes: Prospects of Sustainable Agriculture* (ed. A. Kumar and V.K. Singh), 157–174. Cambridge, MA: Elsevier https://doi.org/10.1016/B978-0-12-818734-0.00007-3.

Hajian-Forooshani, Z., Perfecto, I., and Vandermeer, J. (2023). Novel community assembly and the control of a fungal pathogen in coffee agroecosystems. *Biological Control* 177: 105099.

Halfeld-Vieira, B.A., Vieira, J.R. Jr., Romeiro, R.S. et al. (2006). Introduction of systemic resistance in tomato by the autochthonous phylloplane resident *Bacillus cereus*. *Pesquisa Agropecuária Brasileira* 41 (8): 1247–1252.

Hammami, R., Oueslati, M., Smiri, M. et al. (2022). Epiphytic yeasts and bacteria as candidate biocontrol agents of green and blue molds of citrus fruits. *Journal of Fungi* 8 (8): 818.

Hammond, C.N., Cummings, T.F., and Johnson, D.A. (2008). Deposition of Ascospores of *Sclerotinia sclerotiorum* in and near potato fields and the potential to impact efficacy of a biocontrol agent in the Columbia Basin. *American Journal of Potato Research* 353–360.

Haraguchi, S. and Yoshiga, T. (2020). Potential of the fungal feeding nematode *Aphelenchus avenae* to control fungi and the plant parasitic nematode *Ditylenchus destructor* associated with garlic. *Biological Control* 143: 104203.

Harel, Y.M., Mehari, Z.H., Rav-David, D., and Elad, Y. (2014). Systemic resistance to gray mold induced in tomato by benzothiadiazole and *Trichoderma harzianum* T39. *Phytopathology* 104 (2): 150–157.

Harman, G.E. (2006). Overview of mechanisms and uses of *Trichoderma* spp. *Phytopathology* 96 (2): 190–194.

Hatem, M.E., Saleh, A.B., Abdulaziz, A.A., and Samah, G. (2010). Microbial superoxide dismutase enzyme as therapeutic agent and future gene therapy. In: *Current Research, Technology and Education Topics in Applied Microbiology and Microbial Biotechnology*. Singapore: World Scientific.

He, L. and Hannon, G.J. (2004). MicroRNAs: small RNAs with a big role in gene regulation. *Nature Reviews Genetics* 5 (7): 522–531.

Hong, C.P. and Estey, R.H. (1985). Control of the damping-off disease of pea by *Aphelenchus avenae*. *Indian Journal of Nematology* 15 (1): 1–4.

Hu, H., Wisniewski, M.E., Abdelfattah, A., and Zheng, X. (2017). Biocontrol activity of a cold-adapted yeast from Tibet against gray mold in cherry tomato and its action mechanism. *Extremophiles* 21: 789–803.

Hutchings, K., Matthew, I., Andrew, W. et al. (2019). Antibiotics: past, present and future. *Current Opinion in Microbiology* 51: 72–80.

Huth, L., Ash, G.J., Idnurm, A. et al. (2021). The "bipartite" structure of the first genome of *Ampelomyces quisqualis,* a common hyperparasite and biocontrol agent of powdery mildews, may point to its evolutionary origin from plant pathogenic fungi. *Genome Biology and Evolution* 13 (8): 182.

Ighodaro, O.M. and Akinloye, O.A. (2018). First line defence antioxidants superoxide dismutase (SOD), catalase (CAT) and glutathione peroxidase (GPX): their fundamental role in the entire antioxidant defence grid. *Alexandria Journal of Medicine* 54 (4): 287–293.

Ishikawa, S. (2013). Integrated disease management of strawberry anthracnose and development of a new biopesticide. *Journal of General Plant Pathology* 79: 441–443.

Jaber, L.R. (2015). Grapevine leaf tissue colonization by the fungal entomopathogen *Beauveria bassiana sl* and its effect against downy mildew. *BioControl* 60: 103–112.

Jain, G. and Gould, K.S. (2015). Are betalain pigments the functional homologues of anthocyanins in plants? *Environmental and Experimental Botany* 119: 48–53.

Jaiswal, D.K., Gawande, S.J., Soumia, P.S. et al. (2022). Biocontrol strategies: an eco-smart tool for integrated pest and diseases management. *BMC Microbiology* 22 (1): 1–5.

Jasrotia, S., Salgotra, R.K., and Sharma, M. (2021). Efficacy of bioinoculants to control of bacterial and fungal diseases of rice (*Oryza sativa* L.) in northwestern Himalaya. *Brazilian Journal of Microbiology* 52: 687–704.

Jayawardena, R.S., Hyde, K.D., de Farias, A.R.G. et al. (2021). What is a species in fungal plant pathogens? *Fungal Diversity* 109 (1): 239–266.

Jensen, D., Dubey, M., Jensen, B., and Karlsson, M. (2021). *Clonostachys rosea* to control plant diseases. In: *Microbial Bioprotectants for Plant Disease Management* (ed. J. Köhl and W. Ravensberg), 429–471. Cambridge, UK: Burleigh Dodds Science Publishing.

Jun, O.K. and Kim, Y.H. (2004). *Aphelenchus avenae* and antagonistic fungi as biological control agents of Pythium spp. *The Plant Pathology Journal* 20 (4): 271–276.

Jun, S.Y., Sattler, S.A., Cortez, G.S. et al. (2018). Biochemical and structural analysis of substrate specificity of a phenylalanine ammonia-lyase. *Plant Physiology* 176 (2): 1452–1468.

de Jonge, P.A., Nobrega, F.L., Brouns, S.J.J., and Dutilh, B.E. (2019). Molecular and evolutionary determinants of bacteriophage host range. *Trends in Microbiology* 27 (1): 51–63.

Kashyap, N., Singh, S.K., Yadav, N. et al. (2023). Biocontrol screening of endophytes: applications and limitations. *Plants* 12 (13): 2480. https://doi.org/10.3390/plants12132480.

Kavitha, K., Nakkeeran, S., and Chandrasekar, G. (2012). Rhizobacterial-mediated induction of defense enzymes to enhance the resistance of turmeric (*Curcuma longa* L) to *Pythium aphanidermatum* causing rhizome rot. *Archives of Phytopathology and Plant Protection* 45 (2): 199–219.

Ke, J., Wang, B., and Yoshikuni, Y. (2021). Microbiome engineering: synthetic biology of plant-associated microbiomes in sustainable agriculture. *Trends in Biotechnology* 39 (3): 244–261.

Kerr, A. and Bullard, G. (2020). Biocontrol of crown gall by *Rhizobium rhizogenes*: challenges in biopesticide commercialisation. *Agronomy* 10 (8): 1126.

Kesawat, M.S., Satheesh, N., Kherawat, B.S. et al. (2023). Regulation of reactive oxygen species during salt stress in plants and their crosstalk with other signaling molecules—current perspectives and future directions. *Plants* 12 (4): 864. https://doi.org/10.3390/plants12040864.

Kim, S.H. and Vladimir, V. (2022). ATP-binding cassette (ABC) transporters in *Fusarium* specific mycoparasite *Sphaerodes mycoparasitica* during biotrophic mycoparasitism. *Applied Sciences* 12 (15): 7641.

Kim, J., Le, K.D., Yu, N.H. et al. (2020). Structure and antifungal activity of pelgipeptins from *Paenibacillus elgii* against phytopathogenic fungi. *Pesticide Biochemistry and Physiology* 163: 154–163.

Kumar, D. and Gouda, S.A. (2018). Evaluation of mycoparasitic efficacy of nematode-trapping fungi against *Rhizoctonia solani* inciting sheath blight disease in rice (*Oryza sativa* L.). *Biological Control* 122: 31–40.

Kumar, A., Droby, S., Singh, V.K. et al. (2020a). Entry, colonization, and distribution of endophytic microorganisms in plants. In: *Microbial Endophytes, Functional Biology and Applications* (ed. A. Kumar and E.K. Radhakrishnan), 1–33. Cambridge, MA: Woodhead Publishing, Elsevier https://doi.org/10.1016/B978-0-12-819654-0.00001-6.

Kumar, A., Droby, S., White, J.F. et al. (2020b). Endophytes and seed priming: agricultural applications and future prospects. In: *Microbial Endophytes, Functional Biology and Applications* (ed. A. Kumar and E.K. Radhakrishnan), 107–124. Cambridge, MA: Woodhead Publishing, Elsevier https://doi.org/10.1016/B978-0-12-819654-0.00005-3.

Kumari, M., Qureshi, K.A., Jaremko, M. et al. (2022). Deciphering the role of endophytic microbiome in postharvest diseases management of fruits: opportunity areas in commercial up-scale production. *Frontiers in Plant Science* 13: 4390. https://doi.org/10.3389/fpls.2022.1026575.

Lagerlöf, J., Insunza, V., Lundegårdh, B., and Rämert, B. (2011). Interaction between a fungal plant disease, fungivorous nematodes and compost suppressiveness. *Acta Agriculturae Scandinavica Section B Soil and Plant Science* 61 (4): 372–377.

Lahlali, R., Ezrari, S., Radouane, N. et al. (2022). Biological control of plant pathogens: a global perspective. *Microorganisms* 10 (3): 596.

Laur, J., Ramakrishnan, G.B., LabbÚ, C. et al. (2018). Effectors involved in fungal–fungal interaction lead to a rare phenomenon of hyperbiotrophy in the tritrophic system biocontrol agent–powdery mildew–plant. *New Phytologist* 217 (2): 713–725.

Le, K.D., Kim, J., Yu, N.H. et al. (2020). Biological control of tomato bacterial wilt, kimchi cabbage soft rot, and red pepper bacterial leaf spot using *Paenibacillus elgii* JCK-5075. *Frontiers in Plant Science* 11: 775.

Li, D., Tang, Y., Lin, J., and Cai, W. (2017). Methods for genetic transformation of filamentous fungi. *Microbial Cell Factories* 16: 168.

Li, X.Q., Xu, K., Liu, X.M., and Zhang, P. (2020). A systematic review on secondary metabolites of *Paecilomyces* species: chemical diversity and biological activity. *Planta Medica* 86 (12): 805–821.

Li, B., Kong, L., Qiu, D. et al. (2021). Biocontrol potential and mode of action of entomopathogenic bacteria *Xenorhabdus budapestensis* C72 against *Bipolaris maydis*. *Biological Control* 158: 104605.

Li, G., Liu, T., Whalen, J.K., and Wei, Z. (2023). Nematodes: an overlooked tiny engineer of plant health. *Trends in Plant Science* 29 (1): 52–63.

Liu, L. and Fan, X.D. (2014). CRISPR-Cas system: a powerful tool for genome engineering. *Plant Molecular Biology* 85: 209–218.

Liu, R., Chen, L., Jiang, Y. et al. (2015). Efficient genome editing in filamentous fungus *Trichoderma reesei* using the CRISPR/Cas9 system. *Cell Discovery* 1: 15007.

López-Bucio, J., Pelagio-Flores, R., and Herrera-Estrella, A. (2015). Trichoderma as biostimulant: exploiting the multilevel properties of a plant beneficial fungus. *Scientia Horticulturae* 196: 109–123.

Loughlin, D. (2014). The manual of biocontrol agents. *International Pest Control* 56 (4): 224.

Luo, C., Chen, Y., Liu, X. et al. (2019). Engineered biosynthesis of cyclic lipopeptide locillomycins in surrogate host *Bacillus velezensis* FZB42 and derivative strains enhance antibacterial activity. *Applied Microbiology and Biotechnology* 103: 4467–4481.

Maithani, D., Singh, H., and Sharma, A. (2021). Stress alleviation in plants using SAR and ISR: current views on stress signaling network. In: *Microbes and Signaling Biomolecules Against Plant Stress. Rhizosphere Biology* (ed. A. Sharma). Singapore: Springer.

Martínez, J.L., Petranovic, D., and Nielsen, J. (2016). Heme metabolism in stress regulation and protein production: From Cinderella to a key player. *Bioengineered* 7 (2): 112–115.

Martínez-Diz, M.D.P., Díaz-Losada, E., Díaz-Fernández, A. et al. (2020). Protection of grapevine pruning wounds against *Phaeomoniella chlamydospora* and *Diplodia seriata* by commercial biological and chemical methods. *Crop Protection* 143: 105465.

Maymon, M., Sela, N., Shpatz, U. et al. (2020). The origin and current situation of *Fusarium oxysporum* f. sp. *cubense* tropical race 4 in Israel and the Middle East. *Scientific Reports* 10 (1): 1590.

Meena, M., Yadav, G., Sonigra, P. et al. (2022). Role of elicitors to initiate the induction of systemic resistance in plants to biotic stress. *Plant Stress* 5: 100103.

Meng, F., Lv, R., Cheng, M. et al. (2022). Insights into the molecular basis of biocontrol of *Botrytis cinerea* by *Clonostachys rosea* in tomato. *Scientia Horticulturae* 291: 110547.

Minchev, Z., Kostenko, O., Soler, R., and Pozo, M.J. (2021). Microbial consortia for effective biocontrol of root and foliar diseases in tomato. *Frontiers in Plant Science* 12: 756368.

Mruk, D.D., Silvestrini, B., Mo, M.Y., and Cheng, C.Y. (2002). Antioxidant superoxide dismutase – a review: its function, regulation in the testis, and role in male fertility. *Contraception* 65 (4): 305–311.

Murphy, K.C. (2016). λ Recombination and recombineering. *EcoSal Plus* 7 (1): 10–1128.

Nakayinga, R., Makumi, A., Tumuhaise, V., and Tinzaara, W. (2021). *Xanthomonas* bacteriophages: a review of their biology and biocontrol applications in agriculture. *BMC Microbiology* 21 (1): 1–20.

Nandi, A., Yan, L.J., Jana, C.K., and Das, N. (2019). Role of catalase in oxidative stress-and age-associated degenerative diseases. *Oxidative Medicine and Cellular Longevity* 2019: 9613090.

Nicholson, R.L. and Hammerschmidt, R. (1992). Phenolic compounds and their role in disease resistance. *Annual Review of Phytopathology* 30 (1): 369–389.

Oberhardt, M.A., Zarecki, R., Gronow, S. et al. (2015). Harnessing the landscape of microbial culture media to predict new organism–media pairings. *Nature Communications* 6 (1): 8493.

Omidvari, M., Abbaszadeh-Dahaji, P., Hatami, M., and Kariman, K. (2023). Biocontrol: a novel eco-friendly mitigation strategy to manage plant diseases. *Plant Stress Mitigators* 5: 27–56.

Omnilytics, (2006). What makes bacteriophage safe? http://www.omnilytics.com/documents/What%20Makes%20 Bacteriophage%20Safe.pdf. (accessed 4 August 2007.

Oszust, K., Cybulska, J., and Frąc, M. (2020). How do *Trichoderma* genus fungi win a nutritional competition battle against soft fruit pathogens? A report on niche overlap nutritional potentiates. *International Journal of Molecular Sciences* 21 (12): 4235.

Pal, K.K. and Gardener, B.M. (2006). Biological control of plant pathogens. *The Plant Health Instructor* 2: 1117–1142.

Palmieri, D., Ianiri, G., Del Grosso, C. et al. (2022). Advances and perspectives in the use of biocontrol agents against fungal plant diseases. *Horticulturae* 8 (7): 577.

Pathak, P., Rai, V.K., Can, H. et al. (2022). Plant-endophyte interaction during biotic stress management. *Plants* 11 (17): 2203. https://doi.org/10.3390/plants11172203.

Pieterse, C.M., Van Wees, S.C., Van Pelt, J.A. et al. (1998). A novel signaling pathway controlling induced systemic resistance in Arabidopsis. *The Plant Cell* 10 (9): 1571–1580.

Pieterse, C.M., Zamioudis, C., Berendsen, R.L. et al. (2014). Induced systemic resistance by beneficial microbes. *Annual Review of Phytopathology* 52: 347–375.

del Pilar Martínez-Diz, M., Díaz-Losada, E., Andrés-Sodupe, M. et al. (2021). Field evaluation of biocontrol agents against black-foot and Petri diseases of grapevine. *Pest Management Science* 77 (2): 697–708.

Prahl, R.E., Khan, S., and Deo, R.C. (2023). *Ampelomyces* mycoparasites of powdery mildews – a review. *Canadian Journal of Plant Pathology* 1–14.

Prasannath, K. and De Costa, D.M. (2015). Induction of peroxidase activity in tomato leaf tissues treated with two crop management systems across a temperature gradient. *Proceedings of the International Conference on Dry Zone Agriculture 2015*. Faculty of Agriculture, University of Jaffna, Sri Lanka. 15th & 16th.

Prasannath, K., De Costa, D. M. and Hemachandra, K. S. (2014). Quantification of peroxidase activity in Chilli tissues grown under two crop protection systems across a temperature gradient. *Proceedings of the HETC Symposium*.

Prigigallo, M.I., Gómez-Lama Cabanás, C., Mercado-Blanco, J., and Bubici, G. (2022). Designing a synthetic microbial community devoted to biological control: the case study of *Fusarium* wilt of banana. *Frontiers in Microbiology* 13: 967885.

Prodorutti, D., Ferrari, A., and Pertot, I. (2006). Efficacy of Shemer (*Metschnikowia fructicola*) against postharvest rot of small fruits. In: *Giornate Fitopatologiche, Riccione (RN)*, 27-29 marzo 2006. Atti, volume secondo 2006 (ed. A. Brunelli, A. Canova, and M. Collina), 423–428.

Raaijmakers, J.M., Vlami, M., and De Souza, J.T. (2002). Antibiotic production by bacterial biocontrol agents. *Antonie Van Leeuwenhoek* 81: 537–547.

Ramakrishna, W., Yadav, R., and Li, K. (2019). Plant growth promoting bacteria in agriculture: two sides of a coin. *Applied Soil Ecology* 138: 10–18.

Rashad, Y.M. and Moussa, T.A.A. (2020). Biocontrol agents for fungal plant diseases management. In: *Cottage Industry of Biocontrol Agents and their Applications* (ed. N. El-Wakeil, M. Saleh, and M. Abu-hashim), 337–363. Cham: Springer.

Reisch, C. and Prather, K. (2015). The no-SCAR (Scarless Cas9 Assisted Recombineering) system for genome editing in *Escherichia coli*. *Science Reports* 5: 15096.

Reiss, A. and Jørgensen, L.N. (2017). Biological control of yellow rust of wheat (*Puccinia striiformis*) with Serenade® ASO (*Bacillus subtilis* strain QST713). *Crop Protection* 93: 1–8.

Rhoades, H.L. and Linford, M.B. (1959). Control of pythium root rot by the nematode *Aphelenchus avenae*. *Plant Disease Report* 43: 323–328.

Roatti, B., Perazzolli, M., Gessler, C., and Pertot, I. (2013). Abiotic stresses affect *Trichoderma harzianum* T39-induced resistance to downy mildew in grapevine. *Phytopathology* 103 (12): 1227–1234.

Ryals, J.A., Neuenschwander, U.H., Willits, M.G. et al. (1996). Systemic acquired resistance. *The Plant Cell* 8 (10): 1809.

Sagar, A., Sayyed, R.Z., Ramteke, P.W. et al. (2020). ACC deaminase and antioxidant enzymes producing halophilic *Enterobacter* sp. PR14 promotes the growth of rice and millets under salinity stress. *Physiology and Molecular Biology of Plants* 26 (9): 1847–1854.

Salamone, A.L., Gundersen, B., and Inglis, D.A. (2018). *Clonostachys rosea*, a potential biological control agent for *Rhizoctonia solani* AG-3 causing black scurf on potato. *Biocontrol Science and Technology* 28 (9): 895–900.

Salcedo-Sarmiento, S., Aucique-Pérez, C.E., Silveira, P.R. et al. (2021). Elucidating the interactions between the rust *Hemileia vastatrix* and a *Calonectria* mycoparasite and the coffee plant. *Iscience* 24 (4): 345–456.

Saravanan, T., Bhaskaran, R., and Muthusamy, M. (2004). *Pseudomonas fluorescens* induced enzymological changes in banana roots (Cv. Rasthali) against *Fusarium* wilt disease. *Plant Pathology Journal* 234 (5): 234–256.

Saravanan, R., Nakkeeran, S., Saranya, N. et al. (2022). Biohardening of Banana cv. Karpooravalli (ABB; Pisang Awak) With *Bacillus velezensis* YEBBR6 promotes plant growth and reprograms the innate immune response against *Fusarium oxysporum* f. sp. *cubense*. *Frontiers in Sustainable Food Systems* 6: 845512.

Sarrocco, S. (2023). Biological disease control by beneficial (micro) organisms: selected breakthroughs in the past 50 years. *Phytopathology* 113 (4): 732–740.

Sasaki, K., Iwai, T., Hiraga, S. et al. (2004). Ten rice peroxidases redundantly respond to multiple stresses including infection with rice blast fungus. *Plant and Cell Physiology* 45 (10): 1442–1452.

Sharma, P. (2015). Bio fungicides: their role in plant disease management. In: *New Paradigms in Diseases Management: Conventional and Molecular Approaches for Rapeseed-Mustard Production*. Indian Council of Agricultural Research, Winter School training manual, 113 p.

Sidharthan, K. (2017). *Selection of a compatible biocontrol consortium effective against Fusarium wilt of tomato*. Thesis, M. Sc. New Delhi: Division of Plant Pathology, ICAR-Indian Agricultural Research Institute.

Singh, D., Singh, S.K., Singh, V.K. et al. (2020). Plant growth–promoting bacteria and their role in environmental management. In: *Abatement of Environmental Pollutants* (ed. P. Singh, A. Kumar, and A. Borthakur), 161–175. Cambridge, MA: Elsevier https://doi.org/10.1016/B978-0-12-818095-2.00008-4.

Singh, R., Pandey, K.D., Singh, M. et al. (2022a). Isolation and characterization of endophytes bacterial strains of *Momordica charantia* L. and their possible approach in stress management. *Microorganisms* 10 (2): 290. https://doi.org/10.3390/microorganisms10020290.

Singh, S.K., Kumar, A., Yadav, M.K. et al. (2022b). Cyanobacterial availability for CRISPR-based genome editing: current and future challenges. In: *Cyanobacterial Lifestyle and its Applications in Biotechnology*, 231–252. Academic Press.

Singhal, U., Prasad, R., and Varma, A. (2017). *Piriformospora indica* (*Serendipita indica*): the novel symbiont. In: *Mycorrhiza-Function, Diversity, State of the Art* (ed. A. Varma, R. Prasad, and N. Tuteja), 349–364. Cham: Springer.

Spadaro, D. and Droby, S. (2016). Development of biocontrol products for postharvest diseases of fruit: the importance of elucidating the mechanisms of action of yeast antagonists. *Trends in Food Science & Technology* 47: 39–49.

Stockwell, V.O., Johnson, K.B., Sugar, D., and Loper, J.E. (2010). Control of fire blight by *Pseudomonas fluorescens* A506 and *Pantoea vagans* C9-1 applied as single strains and mixed inocula. *Phytopathology* 100 (12): 1330–1339.

Strange, R.N. and Scott, P.R. (2005). Plant disease: a threat to global food security. *Annual Review of Phytopathology* 43: 83–116.

Subramanian, S., Sangha, J.S., Gray, B.A. et al. (2011). Extracts of the marine brown macroalga, *Ascophyllum nodosum*, induce jasmonic acid dependent systemic resistance in *Arabidopsis thaliana* against *Pseudomonas syringae* pv. *tomato* DC3000 and *Sclerotinia sclerotiorum*. *European Journal of Plant Pathology* 131: 237–248.

Sui, Y., Wisniewski, M., Droby, S. et al. (2020). Genome sequence, assembly, and characterization of the antagonistic yeast *Candida oleophila* used as a biocontrol agent against post-harvest diseases. *Frontiers in Microbiology* 11: 295.

Sun, D., Zhuo, T., Hu, X. et al. (2017). Identification of a *Pseudomonas putida* as biocontrol agent for tomato bacterial wilt disease. *Biological Control* 114: 45–50.

Tariq, M., Khan, A., Asif, M. et al. (2020). Biological control: a sustainable and practical approach for plant disease management. *Acta Agriculturae Scandinavica Section B Soil and Plant Science* 70 (6): 507–524.

Thomma, B.P., Cammue, B.P., and Thevissen, K. (2002). Plant defensins. *Planta* 216: 193–202.

Tounwendsida, A.N., Adama, Z.O.N.G.O., Bawomon, F.N., and Philippe, S.A.N.K.A.R.A. (2022). Assessing the effects of *Lecanicillium lecanii* in the biological control of early and late leaf spot of peanut in vitro (Burkina Faso, West Africa). *African Journal of Agricultural Research* 18 (1): 1–7.

Turrà, D. and Di Pietro, A. (2015). Chemotropic sensing in fungus–plant interactions. *Current Opinion in Plant Biology* 26: 35–140.

Van Wees, S.C., Van der Ent, S., and Pieterse, C.M. (2008). Plant immune responses triggered by beneficial microbes. *Current Opinion in Plant Biology* 11 (4): 443–448.

Vannier, N., Agler, M., and Hacquard, S. (2019). Microbiota-mediated disease resistance in plants. *PLoS Pathogens* 15 (6): e1007740.

Varma, A., Bajaj, R., Agarwal, A. et al. (2013). *Memoirs of 'Rootonic'-the Magic Fungus*. Noida: Amity University Press, India.

Venkatesan, R.M., Muthusamy, K., Iruthayasamy, J. et al. (2023). First report of *Clonostachys rosea* as a Mycoparasite on *Sclerotinia sclerotiorum* causing head rot of cabbage in India. *Plants* 12 (1): 199.

Vera, P., Tornero, P., and Conejero, V. (1993). Cloning and expression analysis of a viroid-induced peroxidase from tomato plants. *Molecular Plant - Microbe Interactions* 6 (6): 790–794.

Vero, S., Garmendia, G., González, M.B. et al. (2013). Evaluation of yeasts obtained from Antarctic soil samples as biocontrol agents for the management of postharvest diseases of apple (Malus× domestica). *FEMS Yeast Research* 13 (2): 189–199.

Verma, H., Kumar, D., Kumar, V. et al. (2021). The potential application of endophytes in Management of Stress from drought and salinity in crop plants. *Microorganisms* 9 (8): 1729. https://doi.org/10.3390/microorganisms9081729.

Wagemans, J., Holtappels, D., Vainio, E. et al. (2022). Going viral: virus-based biological control agents for plant protection. *Annual Review of Phytopathology* 60: 21–42.

Wang, X.Q., Zhao, D.L., Shen, L.L. et al. (2018). Application and mechanisms of *Bacillus subtilis* in biological control of plant disease. *Role of Rhizospheric Microbes in Soil:: Stress Management and Agricultural Sustainability* 1: 225–250.

Wang, Y., Chen, H., Ma, L. et al. (2022). Use of CRISPR-Cas tools to engineer *Trichoderma* species. *Microbial Biotechnology* 15 (10): 2521–2532.

Wang, Z., Hu, X., Solanki, M.K., and Pang, F. (2023). A synthetic microbial community of plant core microbiome can be a potential biocontrol tool. *Journal of Agricultural and Food Chemistry* 71 (13): 5030–5041.

Weller, D.M. and Thomashow, L.S. (2015). Phytosanitation and the development of transgenic biocontrol agents. In: *OECD Biosafety and the Environmental Uses of Micro-Organisms: Conference Proceedings*, 35–43. Paris: OECD publishing 220 p.

Whitman, W.B., Coleman, D.C., and Wiebe, W.J. (1998). Prokaryotes: the unseen majority. *Proceedings of the National Academy of Sciences of the United States of America* 95 (12): 6578–6583.

Wojtaszek, P. (1997). Mechanisms for the generation of reactive oxygen species in plant defence response. *Acta Physiologiae Plantarum* 19 (4): 581–589.

Wommack, K.E. and Colwell, R.R. (2000). Virioplankton: viruses in aquatic ecosystems. *Microbiology and Molecular Biology Reviews* 64 (1): 69–114.

Xiao, Z., Zhao, Q., Li, W. et al. (2023). Strain improvement of *Trichoderma harzianum* for enhanced biocontrol capacity: strategies and prospects. *Frontiers in Microbiology* 14: 1146210.

Yadav, M., Divyanshu, K., Dubey, M.K. et al. (2023). Plant growth promotion and differential expression of defense genes in chilli pepper against *Colletotrichum truncatum* induced by *Trichoderma asperellum* and *T. harzianum*. *BMC Microbiology* 23 (1): 54.

Yang, M., Mavrodi, D.V., Mavrodi, O.V. et al. (2017). Construction of a recombinant strain of *Pseudomonas fluorescens* producing both phenazine-l-carboxylic acid and cyclic lipopeptide for the biocontrol of take-all disease of wheat. *European Journal of Plant Pathology* 149: 683–694.

Żaczek, M., Weber-Dąbrowska, B., and Górski, A. (2015). Phages in the global fruit and vegetable industry. *Journal of Applied Microbiology* 118 (3): 537–556.

Zehra, A., Raytekar, N.A., Meena, M., and Swapnil, P. (2021). Efficiency of microbial bio-agents as elicitors in plant defense mechanism under biotic stress: a review. *Current Research in Microbial Sciences* 2: 100054.

Zeng, W., Wang, D., Kirk, W., and Hao, J. (2012). Use of *Coniothyrium minitans* and other microorganisms for reducing *Sclerotinia sclerotiorum*. *Biological Control* 60: 225–232.

Zhang, X. and Liu, C.J. (2015). Multifaceted regulations of gateway enzyme phenylalanine ammonia-lyase in the biosynthesis of phenylpropanoids. *Molecular Plant* 8 (1): 17–27.

Zhai, Q. and Li, C. (2019). The plant Mediator complex and its role in jasmonate signaling. *Journal of Experimental Botany* 70 (13): 3415–3424.

Zhao, H., Zhou, T., Xie, J. et al. (2020). Mycoparasitism illuminated by genome and transcriptome sequencing of *Coniothyrium minitans*, an important biocontrol fungus of the plant pathogen *Sclerotinia sclerotiorum*. *Microbial Genomics* 6 (3): 456–459.

Zheng, Y.M., Lin, F.L., Gao, H. et al. (2017). Development of a versatile and conventional technique for gene disruption in filamentous fungi based on CRISPR-Cas9 technology. *Scientific Reports* 7 (1): 9250.

Zhou, Q., Ma, K., Hu, H. et al. (2022). Extracellular vesicles: their functions in plant–pathogen interactions. *Molecular Plant Pathology* 23 (6): 760–771.

# 23

# Impact of Microbial Inoculants on the Secondary Metabolites Production of Medicinal Plants

*Haralakal Keerthi Kumari[1], D. Vijaysri[2], T. Chethan[3], Swati[1], and V. Mamtha[1]*

[1] *Department of Agricultural Microbiology, University of Agricultural Sciences, Bangalore, India*
[2] *Division of Microbiology, ICAR-Indian Agricultural Research Institute, New Delhi, India*
[3] *Department of Agricultural Entomology, University of Agricultural Sciences, GKVK, Bengaluru, India*

## 23.1 Introduction

Living microorganisms, known as microbial inoculants, which have a variety of effects and mechanisms on plants, are used as biostimulants or biocontrol agents (Grobelak et al. 2018). This connection, which is as old as plant colonization of the planet, is a complicated, dynamic, and ongoing process. Different chemical signaling routes are used by plants and microorganisms to communicate with one another. Positive or negative interactions between plants and microbes can be distinguished based on the results of their interactions. Figure 23.1 shows a schematic illustration of the relationship between plants and microbes. Dynamic changes in both partners' genomes, metabolisms, and signaling networks can be seen in both interactions. Changes in plants can also be caused by several external sources. In fact, some microorganisms, including fungi, bacteria, algae, yeast, and viruses, can alter how a plant functions, grows, or reproduces in either a positive or negative way (Aria 2021).During this interaction, microbes can secrete some chemical substances, which can promote the primary synthesis of secondary phytochemicals in medicinal plants. Microbial inoculants area class of biotic triggers that comprise microorganisms, or polysaccharides, and are biologically derived. Microbial elicitors, which are classified as MAMPs, connect the cell wall of the host to cause effector-triggered immunity (ETI), via exchanging signal molecules with bacteria and plants. Contingent upon PTI and ETI, this interaction triggers a number of signaling pathways that can produce a number of transcriptional factors (secondary messengers), hence triggering particular physiological reactions in plants through regulating gene expression (Chamkhi et al. 2021; Kesawat et al. 2023). The bacteria generate and secrete elicitor molecules that trigger the host plant to respond to biological or abiotic threats by creating secondary compounds in the plant–microbe relationship.

## 23.2 Biosynthesis of Plant Secondary Phytochemicals and Their Classification

Secondary phytochemicals as an idea were first described by the 1910 Nobel Prize winner for physiology or medicine, Albrecht Kossel. Secondary metabolites are low-molecular-weight substances having less than 3000 kDa of molecular weight, metabolites with several functions that are frequently used by plants for communication and defense in their environment. Additionally, these are linked to the color, taste, and scent of the plants by taking part in several reactions occurring in them. The production of secondary compounds by the primary compounds (Carbohydrates, lipid, orproteins) that are synthesized by plants during photosynthesis. The primary metabolites are called precursors; they are typically created in the citric acid cycle and glycolytic cycle. Depending on their chemical makeup, purpose, and distribution in plants, both main and secondary metabolites can be distinguished from each other. The shikimic acid cycle, which yields tannins, phenols, and aromatic alkaloids, is one of the major pathways in the biosynthesis of these metabolites. Other important pathways include the malonic acid pathway, which yields terpenes, alkaloids, and steroids; the acetate-malonate pathway,

*Microbes Based Approaches for the Management of Hazardous Contaminants*, First Edition. Edited by Ajay Kumar, Livleen Shukla, Joginder Singh, and Luiz Fernando R. Ferreira.

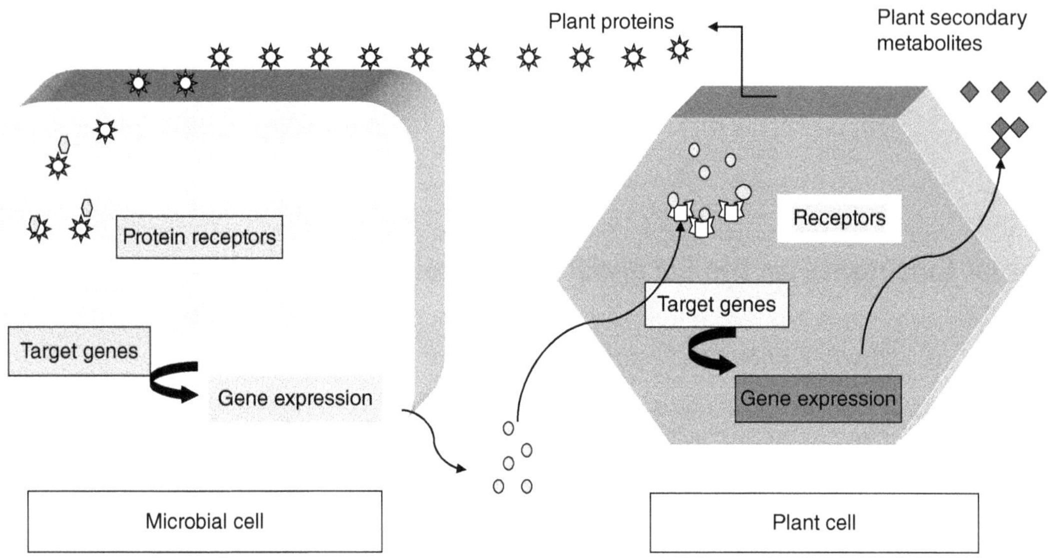


**Figure 23.1** Microbe–plant interaction to produce secondary metabolites.

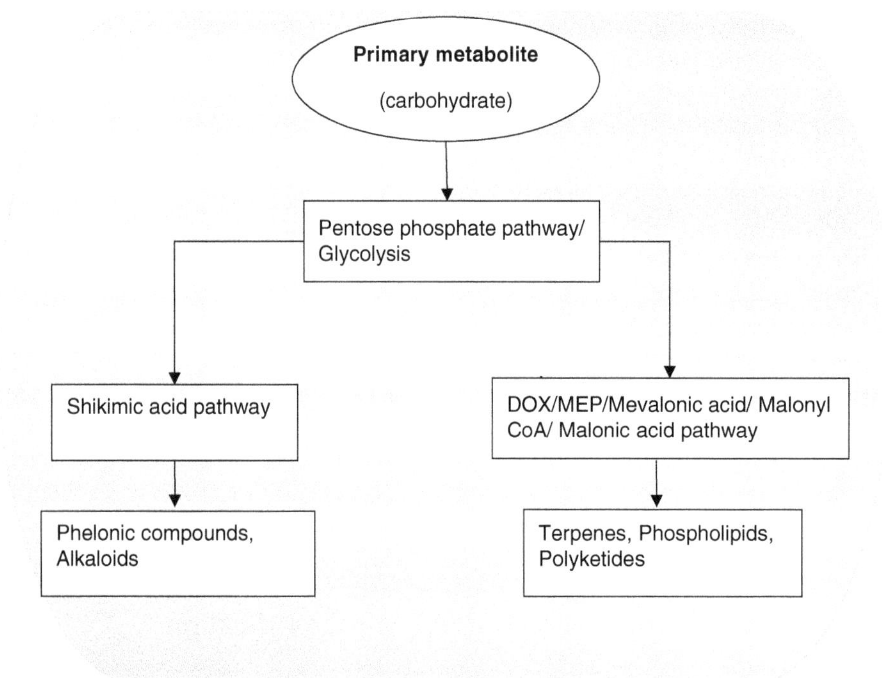


**Figure 23.2** Illustration of the routes used in the synthesis of secondary metabolites in plants.

which yields alkaloids, lipids, and phenols; and the hexose monophosphate shunt, which yields glycosides (Khare et al. 2020). Figure 23.2 depicts the metabolic process that results in the creation of secondary metabolites.

The forerunner of the shikimate route, shikimic acid, is produced by mixing phosphoenolpyruvate obtained through the Embden Meyerhof Parnas (EMP) pathway with erythrose 4-phosphate from the hexose monophosphate pathway. The shikimate pathway results in the formation of phenylalanine, tyrosine, and tryptophan, which are essential amino acids for protein synthesis and frequently used as precursors for secondary metabolites in plants such as phenolics and nitrogen-containing compounds. Phosphoenolpyruvate gives rise to pyruvic acid production, which is a substrate for terpenoid/terpenes production through deoxy D-xylose phosphate (DOX)/mavelonic-independent pathway (MEP), whereas acetyl CoA also aids in the production of terpenoids through mavelonic acid pathway. Acetyl CoA, in combination with malonyl CoA, produces lipids and phospholipids. In addition, sugars are converted into myo-inositol to produce complex polysaccharides, viz., cell walls (Parankusamet al. 2020).

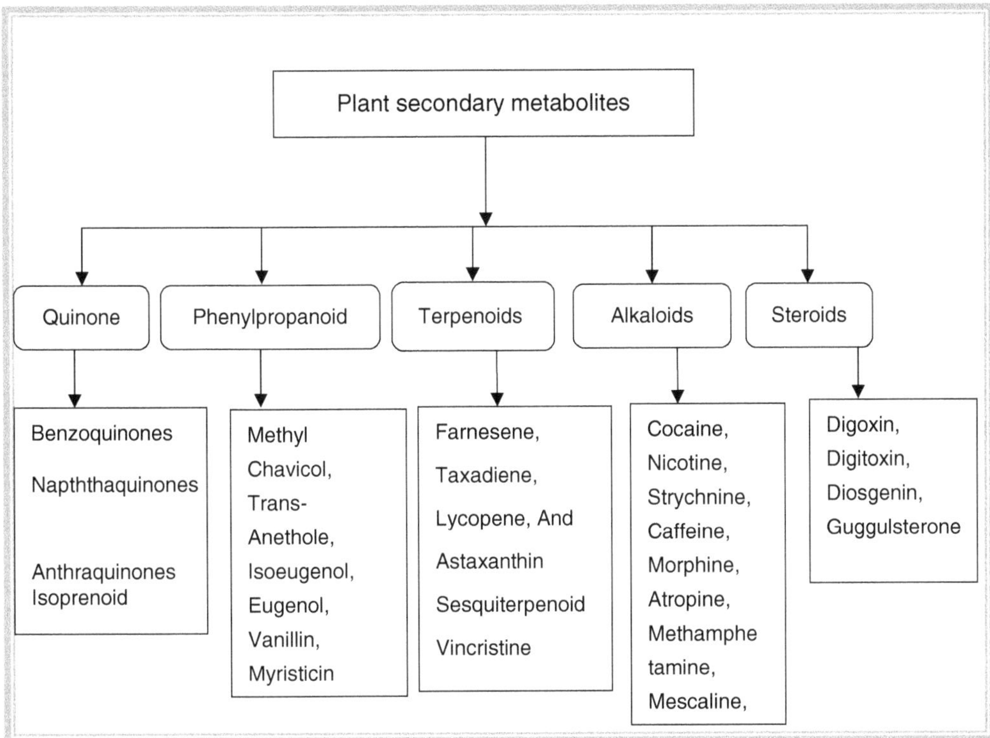


**Figure 23.3** Flow diagram for the secondary phytochemical classification of plants.

Additional classification of plant secondary metabolites into five major classes, such as terpenes, phenylpropenoids, steroids, quinones, and alkaloids, is categorized according to their chemical make-up (Jamwal et al. 2018). Classes of plant secondary metabolites with examples are listed in Figure 23.3. These diverse classes of secondary metabolites are crucial to the pharmaceutical and cosmetic spheres since they are employed in agrochemicals, flavorings, essential oils, biopesticides, and food additives. Many biotic compounds, such as caffeine, ephedrine, papervine, quinine, and vincristine, are highly sought-after on the global market but typically only occur in a single genus or species of plant. The amount of secondary metabolites that plants naturally produce – less than 1% of the dry weight of the plant – depends on the physiological and the plant's level of development as well as the ambient factors.

## 23.3 General Mechanism of Microbial Inoculants-Induced Production of Secondary Compounds

The microbial inoculants create and produce elicitor molecules that cause the macrosymbiont (plant) to produce secondary metabolites in response to biotic or abiotic stress (Anand et al. 2023; Kesawat et al. 2023; Meena et al. 2023). The plant cell receptors are bound by the microbial elicitors, which activate signaling pathways. As a result of this activation, some genes are expressed as a result of the activation of transcription factors and the creation of secondary messengers (Bhaskar et al. 2021). Contact between microbial elicitors and plants is made possible via ligand–receptor relationships between elicitors (ligands) and the numerous external or internal receptors within cells, such as receptors connected to G-proteins and $Ca^{2+}$ circuit sensors. When membrane receptors linked to G-protein become active, inositol triphosphate (IP3), phospholipase C (PLC), and cyclase adenylate (CA) activities are triggered. When $Ca^{2+}$ channels are opened by CA and IP3, intracellular calcium levels rise, and they attach to particular receptors that release $Ca^{2+}$ from the endoplasmic reticulum. PLC and elevated intracellular $Ca^{2+}$ ion concentrations have the ability to trigger various phosphate kinase actions. The oxidative burst process, which leads to the creation of hydrogen peroxide ($H_2O_2$) and hydroxy radicals (OH), is likewise carried out by PLC. These free radical molecules cause the creation of reactive oxygen compounds (ROS), nitric oxide(NO), and ethylene, which activate transcriptional factors vital for the host plant's ability to produce metabolites that are secondary in nature (Chamkhi et al. 2021).

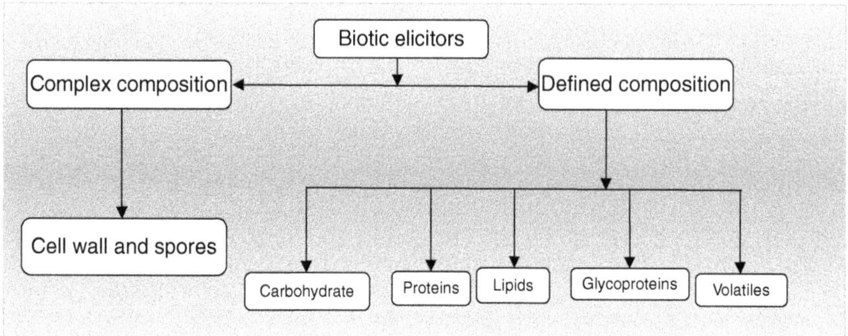


**Figure 23.4** Classification of microbial-derived elicitors.

Microbial inoculants, viz., fungi, bacteria, algae, yeast, and viruses, produce elicitors that fall under biotic elicitors class (Figure 23.4). Cell walls, spores, sugars, lipids, proteins, glycoproteins, volatiles, and additional elements are among them (Radman et al. 2003).

## 23.4 Determinants of Secondary Phytochemical Synthesis

- **Environmental factors:** Medicinal herbs communicate with their surroundings as they grow and develop, coming into contact with a variety of biotic and abiotic elements like microbes, insects, animals, humans, water, light, temperature, soil, and chemicals, respectively. Adverse biotic and abiotic variables, such as deteriorating soil or a buildup of harmful substances, surges of temperatures and light, drought, or flooding, the presence of pathogens, etc., create secondary stressors, and these cause alterations in the process of biosynthesis of metabolites that are secondary in nature (Verma and Shukla 2015; Meena et al. 2023; Singh et al. 2022).
- **Genetic elements:** The secondary metabolite chemicals are created in plants using a variety of biosynthetic pathways, each of which is unique to a particular class of chemical compounds. Any modification to the metabolic route, single gene, or group of genes has an influence on the synthesis and buildup of these metabolites. The pathway can be linear or branched in operation. It is well known that altering the expression of the relevant genes, either naturally or artificially, causes changes in the enzymatic processes involved in the creation of secondary metabolites. Recombinant pathway gene expression and its relationship to the increase of secondary metabolites in plants have made significant strides in the literature (Jadaun et al. 2023).

## 23.5 Ideal Characteristics of Microbial Inoculants

- Should be more competent than a pathogen
- Should be capable of targeting a specific metabolite and triggering the respective gene
- Amenable to genetic modification
- Possess a stable genetic construction
- Easy to grow and maintain
- Should not cause the disease in host
- Capable of establishing symbiotic relation with hostplant
- It should not cause any negative effect on plant growth and development

Let us study about microbial inoculants, their mechanism of action, and examples affecting the generation of secondary phytochemicals.

## 23.6 Fungi

The most significant and popular organisms involved in the commercial generation of plant bioactive compounds are fungi, both free-living and endophytic. Carbohydrate, chitosan, protein, culture filtrate and mycelial homogenate are examples of fungal-derived elicitors that can be utilized in synthesis of secondary metabolites. The most important and commonly employed biotic elicitors for the manufacture of industrial compounds are fungi, both free-living and endophytic. The creation of many chemical families of secondary metabolites, such as terpenoids, coumarin derivatives, alkaloids, and flavonoids, has increased as a result of this approach. According to a report, andrographolide production by *Aspergillusniger* and *Penicilliumexpansum* elicitors in suspension cultures of cells of *Andrographis paniculata* has increased (Vakil and Mendhulkar 2013).

The manufacture of phytoalexins, lignin, and ethylene is all induced by an oligosaccharide that comes from the cell wall of a fungus. This oligosaccharide is also known to cause a number of particular biochemical changes that are linked to resistance. Fungi mediate a mechanism that promotes plant phytoalexin and results in plant hypersensitivity or a self-defense reaction during the physiological processes of plant disease resistance (Sangeetha and Sarada 2015). The elicitor signal transduction pathway's initial committed step is signal perception, which is followed by the detection of appropriate receptors in plant plasma membranes. Fungal inoculants have the ability to produce a variety of elicitors, including chitin, xyloglucans, chitosan, glucan, and oligogalacturonide, which have stimulation action across numerous plant species and greatly boost phytoalexins in plants. Different plants also share some common receptors for detecting these signals. The majority of the protein-based or lipid stimulants and saccharides interact with their sensors in a way that is highly selective, bidirectional, and displays saturable binding, demonstrating that they are interacting with the receptor and ligand (Zhao et al. 2005). According to Algar et al. (2012), metabolites, accumulated materials, fermented liquids, and breakdown products of fungi are all examples of fungal elicitors. Oligosaccharides, proteins, and polyunsaturated fatty acids can be applied to categorize the chemical makeup of a fungal elicitor.

## 23.7 Mechanism of Fungal Elicitors

Fungal inoculants mediate a signal transduction process that is very complex and involve series of reaction, which activates host DNA transcription and end up in the initiation of plant defense mechanisms by triggering the cycles responsible for the production of plant secondary compounds. Nitrogen oxide (NO), abscisic acid (ABA), $Ca_2$, cyclic AMP, ethylene, G protein, IP3, $O_2$, ROS, $H_2O_2$, and JA are important signaling chemicals. Signal transduction is a complex process in which several molecules of signal communicate with one another over a variety of circumstances. Normally, fungal stimulants result in the opening of ion channels along with the coupling of G protein through IP3 as intracellular transcriptional elements. Oxidative burst is a common and ubiquitous method of producing reactive oxygen species (ROS) ($O_2$, $H_2O_2$, and NO). Medicinal plants have diverse signal pathways that are mediated by the reciprocal transformation of $O_2$ and $H_2O_2$, and NO can send the signal through either a dependent or an independent oxidative burst signaling cycle. Salicylic acid (SA) can accumulate more quickly when nitrogen oxide is present, which can slow down the generation of jasmonic acid (JA) in plant cells (Zhai et al. 2016). Even though there is a reciprocal restriction between the SA and JA signaling pathways, there is also a unique coordination that is complementary to it. Inspite of our knowledge about all these signals, the complete pathway involved is still unclear.

In signaling, the primary response by plants is ions flow, in and out of the cell. Signal from fungal inoculant activates the receptor-coupled effectors, such as G-proteins or protein-based kinases, which further activate the $K^+/H^+$, $Cl^-/Ca^{2+}$ ion fluxes, among which the $Ca^{2+}$ flux is major since it plays key function in a number of communication systems (Almagro et al. 2012;Du et al. 2015). G-proteins and their GTPase activity will be altered by some particular toxins, like cholera, which lead to the transformation of signals to jasmonate pathway to produce the desired metabolite. Ion ($Ca^{2+}$) binding sensors can also result in specific pathways to produce secondary metabolites and $Ca^{2+}$sensors that fall into two categories. One is calcium-dependent protein kinases (CDPK), which have protein kinases and calcium ions coupled in their structural domain. ROS have the ability to directly activate defense genes and the production of phytoalexin, a transcriptional element that will result in the formation of tiny molecules like JA signal. By acting as lipoxygenase's substrate, this, in turn, stimulates and regulates the firing of subsequent defense genes (Hao et al. 2014). Hydrogen peroxide ($H_2O_2$), one of the ROS, is crucial as a signaling molecule in reaction to a variety of stressors and is involved in the buildup of secondary metabolites.

Even though fungi are competent microbial inoculum in synthesis of secondary compounds in medicinal herbs, not all fungi will be beneficial and suitable in producing secondary phytochemicals. Some plant-harmful fungi elicit hypersensitive cell reactions and control cell death, while helpful fungi might trigger a general defense response or have null effect at all. Some of the examples of fungal inoculants in metabolite production are given in Table 23.1.

**Table 23.1** Fungal inoculants in generation of secondary phytochemicals.

| Medicinal plant | Fungal species | Secondary metabolite | References |
|---|---|---|---|
| *Rauwolfia serpentina* | *Aspergillus niger* (protein) | Ajmaline and ajmalicine | Srivastava et al. (2016) |
| *Solanum khasianum* | *Saccharomyces cerevisiae* (carbohydrate) | Solasodine and α-solanine | Srivastava et al. (2016) |
| *Echinacea purpurea L. moench* | *Aspergillus niger*<br>*Fusarium oxysporum* | Total phenolics and total flavonoids | Elshahawy et al. (2022) |
| *Salvia miltiorrhiza* | *Mucor fragilis* | Salvianolic acid B and oleic acid | Wenjuan et al. (2021) |
| *Tripterigeum wilfordii* | *Fusarium subglutinans* | Subglutinols | Lee et al. (1995) |
| *Dysoxylum binectariferum* | *Fusarium proliferatum* | Rohitukine | Mohana et al. (2012) |
| *Camptotheca acuminate* | *Fusarium solani* | Camptothecin | Puri et al. (2005) |
| *Dioscorea* spp. | *Glomus clarum, G. etunicatum, G. fasciculatum, Gigaspora* sp., *G. mosseae*, and *Acaulospora* sp. | Polyphenols, anthocyanin, flavonoids | Lu et al. (2015) |

### 23.7.1 Bacteria

A variety of bacteria were identified by Tamilarasi et al. (2008) from the rhizosphere of 50 medicinal herbs, with Bacillus emerging as the leading species, in addition to Enterobacter, Micrococcus, Pseudomonas, Corynebacterium, and Serratia. The exchange of plant metabolites may be the primary cause of microbes' preferences for particular therapeutic plants. The main way that metabolic processes for functional chemical synthesis in medicinal herbs/plants are positively affected is by the rise in the concentration of secondary phytochemicals (Ghorbanpour et al. 2013). The quantity and quality of the bioactive compounds and their effective applications in the fields of agriculture, pharmaceuticals, and medicine are also influenced by indigenous bacteria of these medicinal plants. Additionally, it may affect these therapeutic plants' bioactivity and metabolic rate. The composition of medicinal plants was the primary focus of research in the past, but, more recently, the architecture and behavior of various medicinal plant microbiomes have come into focus.

Interestingly, endophytes, a type of bacteria associated with plants, as well as the plants themselves, were able to produce chemicals having phytotherapeutic properties. It is assumed that the complex community of the endomicrobiome, in particular, has a direct or indirect role in the generation of bioactive phytochemicals in numerous medicinal plants (Koberl et al. 2013). Some examples of bacterial inoculants involved in improving medicinal plants are given in Table 23.2.

### 23.7.2 Algae

Microalgae are microscopic (prokaryotic or eukaryotic) photosynthetic organisms that can thrive in a variety of aquatic environments, such as ponds, rivers, lakes, seas, wastewater, and even moist soil (Pandey et al. 2023). In coastal areas, algae has long been utilized as soil fertilizer, and there have been reports of benefits from spraying seaweed extracts on crop

**Table 23.2** Bacterial inoculants involved in secondary metabolite production.

| Medicinal plant | Bacteria | Secondary metabolites | References |
|---|---|---|---|
| *Hyoscyamus niger* | *Pseudomonas putida* | Tropane alkaloids | Ghorbanpouret al. (2013) |
| *H. niger* | *Pseudomonas fluorescens* | Tropane alkaloids | Ghorbanpouret al. (2013) |
| *Ocimum basilicum* | *Bacillus subtilis* GB03 | Eugenol and A-terpineol | Banchio et al. (2009) |
| *Atractylodes lancea* | *P. fluorescens* ALEB7B | Volatileoil | Zhou et al. (2016) |

**Table 23.3** Algae in generation of secondary phytochemical.

| Algae | Plant | Mechanisms | References |
|---|---|---|---|
| *Ascophyllum nodosum* | *Piper nigrum* | Salicylic acid and jasmonic acid | Ali et al. (2019) |
| *Caulerpa cylindracea* | *Nicotiana benthamiana* | Unidentified | Alburquerque et al. (2019) |
| *Ulva Lactuca* (Ulvan) | *Olea europaea* | Increase in PAL activity and polyphenol content | Salah et al. (2018) |
| *Kappaphycus alvarezii* | *Picrorhizakurroa Royleex Benth* | Picroside-Iproduction | Sharma et al. (2014) |
| *Botryococcus braunii* | *Capsicum frutescens* L. | Vanillylamineand Capsaicin | Sharma et al. (2014) |

plants. As a result, when different plants were treated with seaweed extracts, better fruit shelf life, higher yields, increased resistance to pests and diseases, and improved seed germination were all seen, along with the generation of secondary phytochemicals (Mercier et al. 2008). Algal extracts have been postulated to act as stimulants of plant defense responses because of their actions as plant protectants; some examples are given in Table 23.3.

Their use as useful products is nonetheless constrained by a lack of understanding of their mode of action. There are well-established mechanisms by which plants react to infections and their elicitors in order to prevent penetration and development into their tissues. Elicitors function as signal molecules, and when they are detected at the cell surface and then transmitted, they cause the activation of plant defense genes. The induction of defense responses on a local and systemic level is mediated by elicitor effects through signaling pathways, and SA, JA, and ethylene, either separately or together, play important roles in this process (Reymond and Farmer 1998).

### 23.7.3 Yeast

Yeast extract is a particularly intriguing naturally occurring plant inducer due to its capacity to function as a biostimulant of plant growth or in the manufacture of plant pigments and several other bioactive molecules, as well as its role in the production of numerous growth regulators. Chitin, glucan, glycopeptides, and ergosterol are only a few of the many substances found in yeast extract that are involved in plant defense responses. Yeast extract is used as a microbial inoculant derivative to induce production of plant secondary metabolites by following similar mechanisms, like bacteria, fungi, etc. (Baenas et al. 2014). List of medicinal plants and their associated secondary metabolites influenced by yeast extract is given in Table 23.4.

### 23.7.4 Virus

Although plant viruses are typically associated with causing illness in crops, multiple studies have shown that their interactions with macrosymbiont (plant) can actually prolong the lives of untamed plant hosts. Since viruses depend on host

**Table 23.4** Plants with their secondary metabolites influenced by yeast.

| Yeast | Plant | Secondary metabolite | References |
|---|---|---|---|
| Yeast extract | *Solanum hainanense* | Solasodine | Loc et al. (2014) |
| *Saccharomyces cerevisiae* (yeast extract) | *Lactuca sativa* L. | Total phenols | Zlotek and Swieca (2015) |
| Yeast extract | *Ammi visnaga* L. | Khellol glucoside, khellin, and visnagin | Akbary and Golkar (2023) |
| Yeast extract | *Azadirachta indica* | Azadirachtin | Farjaminezhad and Garoosi (2021) |
| Yeast extract | *Astragalus membranaceus* | Astragalosides | Park et al. (2021) |
| Yeast extract | *Puerariacandollei* var. *mirifica*. | Daidzein and genistein | Rani et al. (2020) |

**Table 23.5** Viruses improving secondary phytochemicals creation in medicinal plants.

| Virus | Medicinal herbs | Plant metabolite | References |
|---|---|---|---|
| *Cucumber mosaic virus* | *Passiflora edulis* | Flavonoids and Polyphenols | Lan et al. (2020) |
| *Saffron latent virus* | *Crocus sativus* | Crocetin, picrocrocin, and kaempferols | Parizad et al. (2019) |
| *Telosma mosaic virus* | *P. edulis* | Phenols | Chen et al. (2019) |
| Grapevine red blotch-associated virus | *Vitis vinifera* | Biosynthesis of anthocyanin and flavonoid | Barbara et al. (2017) |
| *Grapevine leaf roll-associated virus* 3 | *V. vinifera* | Flavonols | Montero et al. (2016) |

resources for their own growth and spread, the general consensus is that virus infections are damaging to the host. This model, however, only offers a partial understanding of the interactions between viruses and hosts. Some viruses also aid their hosts' survival and procreation. Large number of viruses can influence plant secondary metabolite production; some of them are listed in Table 23.5.

## 23.8 Advantages of Microbial Inoculants over Chemical Inoculants for Metabolite Production

- Environmentally safe to use
- Cost-effective compared to chemical inoculants
- Easy to maintain and reusable
- Nontoxic to plant ecosystem
- No chemical residue will be added to the soil and in food chain

## 23.9 Applications of Plant Secondary Metabolites

- Used in various dye preparations
- Used in pharmaceutical industry
- To prepare perfumes and cosmetics
- Exploited in making agrochemicals like insecticides
- They can also be used as food color and flavoring agent

## 23.10 Conclusion

Through a comprehensive analysis of the available literature, it is evident that microbial inoculants can significantly enhance the biosynthesis of secondary metabolites in medicinal plants. This effect is attributed to the intricate interactions between plants and beneficial microorganisms, which can stimulate various metabolic pathways and trigger the synthesis of valuable compounds. The application of microbial inoculants not only increases the yield of secondary metabolites but also enhances their quality, making medicinal plants more potent and effective. However, it is essential to acknowledge that the effectiveness of microbial inoculants can vary depending on various factors, such as plant species, microbial strains, environmental conditions, and cultivation practices. As our understanding of these complex plant–microbe interactions continues to grow, we can look forward to harnessing the full potential of microbial inoculants to enhance the quality and quantity of secondary metabolites in medicinal plants, thus contributing to the advancement of healthcare and sustainable agriculture. Therefore, future research should focus on optimizing the selection and application of specific microbial inoculants tailored to different medicinal plant species and growing conditions.

## References

Akbary, R. and Golkar, P. (2023). Elicitation of medicinally-valuable secondary metabolites, enzymatic, and antioxidant activity using chitin and yeast extract in callus cultures of *Ammi visnaga* L. *Plant Cell, Tissue and Organ Culture* 154: 689–702.

Alburquerque, N., Faize, L., Faize, M. et al. (2019). Towards the valorization of the invasive seaweeds *Caulerpacylindracea* and *Asparagopsistaxiformis* in the Mediterranean Sea: applications for *in vitro*plant regeneration and crop protection. *Journal of Applied Phycology* 31 (2): 1403–1413.

Algar, E., Gutierrez-Manero, F.J., Bonilla, A. et al. (2012). *Pseudomonasfluorescens* N21.4 metabolites enhance secondary metabolism isoflavones in soybean (*Glycinemax*) calli cultures. *Journal of Agricultural and Food Chemistry* 60: 11080–11087.

Ali, O., Ramsubhag, A., and Jayaraman, J. (2019). Biostimulatory activities of *Ascophyllumnodosum* extract in tomato and sweet pepper crops in a tropical environment. *Public Library of Science* 14 (5): e0216710.

Almagro, L., Bru, R., Pugin, A. et al. (2012). Early signaling network in tobacco cells elicited with methyl jasmonate and cyclodextrins. *Plant Physiology and Biochemistry* 51: 1–9.

Anand, U., Pal, T., Yadav, N. et al. (2023). Current scenario and future prospects of endophytic microbes: promising candidates for abiotic and biotic stress management for agricultural and environmental sustainability. *Microbial Ecology* 86: 1455–1486. https://doi.org/10.1007/s00248-023-02190-1.

Aria, D. (2021). Plant microbe interaction. *Biology* 10 (1): 15.

Baenas, N., García-Viguera, C., and Moreno, D.A. (2014). Elicitation: a tool for enriching the bioactive composition of foods. *Molecules* 19: 13541–13563.

Banchio, E., Xie, X., Zhang, H. et al. (2009). Soil bacteria elevate essential oil accumulation and emissions in sweet basil. *Journal of Agricultural and Food Chemistry* 57 (2): 653–657.

Barbara, B., Helene, H., Rosa, F.B. et al. (2017). Red blotch disease alters grape berry development and metabolism by interfering with the transcriptional and hormonal regulation of ripening. *Journal of Experimental Botany* 5 (68): 1225–1238.

Bhaskar, R., Louretta, S., Eruva, X. et al. (2021). Biotic elicitors: a boon for the in vitro production of plant secondary metabolites. *Plant Cell Tissue and Organ Culture* 149 (1–2): 1–18.

Chamkhi, I., Benali, T., Aanniz, T. et al. (2021). Plant-microbial interaction: the mechanism and the application of microbial elicitor induced secondary metabolites biosynthesis in medicinal plants. *Plant Physiology and Biochemistry* 167: 269–295.

Chen, S., Yu, N., Yang, S. et al. (2019). Identification of Telosma mosaic virus infection in *Passifloraedulis* and its impact on phytochemical contents. *Virology Journal* 15 (1): 168–169.

Du, W., Liang, J., Han, Y. et al. (2015). Nitric oxide mediates hypocrellin accumulation induced by fungal elicitor in submerged cultures of *Shiraiabambusicola*. *Biotechnology Letters* 37: 153–159.

Elshahawy, A.M.O., Mohamed, E.F.Z., Hamza, A.M. et al. (2022). Improving the production of total phenolics and flavonoids and the antioxidant capacity of *Echinaceapurpurea*callus through biotic elicitation. *Egyptian Journal of Chemistry* 12 (65): 137–149.

Farjaminezhad, R. and Garoosi, G. (2021). Improvement and prediction of secondary metabolites production under yeast extract elicitation of *Azadirachtaindica* cell suspension culture using response surface methodology. *AMB Express* 11: 43–45.

Ghorbanpour, M., Mehrnaz, H., and Khavazi, K. (2013). Role of plant growth promoting rhizobacteria on antioxidant enzyme activities and tropane alkaloid production of *Hyoscyamus niger* under water deficit stress. *Turkish Journal of Biology* 37: 350–360.

Grobelak, A., Paulina, K., Hutchison, D. et al. (2018). Plant growth-promoting rhizobacteria as an alternative to mineral fertilizers in assisted bioremediation-sustainable land and waste management. *Journal of Environmental Management* 227: 1–9.

Hao, W., Guo, H., Zhang, J. et al. (2014). Hydrogen peroxide is involved in salicylic acid-elicited rosmarinic acid production in *Salviamiltiorrhiza* cell cultures. *Science World Journal* 3: 843764–843765.

Jadaun, J.S., Yadav, R., Yadav, N. et al. (2023). Influence of genetics on the secondary metabolites of plants. In: *Natural Secondary Metabolites* (ed. M. Carocho, S.A. Heleno, and L. Barros). Cham: Springer https://doi.org/10.1007/978-3-031-18587-8_12.

Jamwal, K., Bhattacharya, S., and Puri, S. (2018). Plant growth regulator mediated consequences of secondary metabolites in medicinal plants. *Journal of Applied Research on Medicinal and Aromatic Plants* 9: 26–38.

Kesawat, M.S., Satheesh, N., Kherawat, B.S. et al. (2023). Regulation of reactive oxygen species during salt stress in plants and their crosstalk with other signaling molecules – current perspectives and future directions. *Plants* 12 (4): 864. https://doi.org/10.3390/plants12040864.

Khare, S., Singh, N.B., Singh, A. et al. (2020). Plant secondary metabolites synthesis and their regulations under biotic and abiotic constraints. *Journal of Plant Biology* 63: 203–216.

Koberl, M., Schmidt, R., Ramadan, E.M. et al. (2013). The microbiome of medicinal plants: diversity and importance for plant growth, quality and health. *Frontiers in Microbiology* 400 (4): 1–9.

Lan, H., Lai, B., Zhao, P. et al. (2020). Cucumber mosaic virus infection modulated the phytochemical contents of *Passifloraedulis*. *Microbial Pathogenesis* 138: 103828.

Lee, J.C., Lobkovsky, E., Pliam, N.B. et al. (1995). Subglutinols A and B: immunosuppressive compounds from the endophytic fungus *Fusarium subglutinans*. *Journal of Organic Chemistry* 60: 7076–7077.

Loc, N.H., Anh, N.H.T., Khuyen, L.T.M. et al. (2014). Effects of yeast extract and methyl jasmonate on the enhancement of solasodine biosynthesis in cell cultures of *Solanumhainanense*hance. *Journal of BioScience and Biotechnology* 3: 1–6.

Lu, F.C., Lee, C.Y., and Wang, C.L. (2015). The influence of arbuscular mycorrhizal fungi inoculation on yam (*Dioscorea* spp.) tuber weights and secondary metabolite content. *PeerJ* 3: e1266.

Meena, M., Singh, S.K., Swapnil, P., and Kumari, P. (2023). The role of PGPRs in medicinal plants under abiotic stress. In: *Medicinal Plants: Their Response to Abiotic Stress* (ed. A. Husen and M. Iqbal), 267–285. Singapore: Springer Nature Singapore https://doi.org/10.1007/978-981-19-5611-9_10.

Mercier, L., Lafitte, C., Borderies, G. et al. (2008). The algal polysaccharide carrageenans can act as an elicitor of plant defence. *New Phytologist* 149: 43–51.

Mohana, K.P., Zuehlke, S., Priti, V. et al. (2012). *Fusarium proliferatum*, an endophytic fungus from *Dysoxylum binectariferum* Hook.f, produces rohitukine, a chromane alkaloid possessing anticancer activity. *Antonie van Leeuwenhoek* 101: 323–239.

Montero, R., Perez-Bueno, M., Baron, M. et al. (2016). Alterations in primary and secondary metabolism in *Vitisvinifera* 'Malvasía de Banyalbufar' upon infection with grapevine leafroll-associated virus 3. *Physiologia Plantarum* 157 (4): 442–452.

Pandey, K.D., Singh, S.K., Shukla, L. et al. (2023). Mechanistic approaches and factors regulating microalgae mediated heavy metal remediation from the aquatic ecosystem. In: *Advances in Chemical Pollution, Environmental Management and Protection, Volume 9* (ed. L.F.R. Ferreira, A. Kumar, and M. Bilal), 285–299. Cambridge: Elsevier, Academic Press https://doi.org/10.1016/bs.apmp.2022.10.011.

Parankusam, S., Hemalatha, S., Gundavarapu, S. et al. (2020). *Nitric Oxide as a Signal in Inducing Secondary Metabolites During Plant Stress*, 593–562. Cham: Springer.

Parizad, S., Dizadji, A., Habibi, K. et al. (2019). The effects of geographical origin and virus infection on the saffron (*Crocussativus* L.) quality. *Food Chemistry* 295: 387–394.

Park, Y.J., Kim, J.K., and Park, S.U. (2021). Yeast extract improved biosynthesis of astragalosides in hairy root cultures of *Astragalusmembranaceus*. *Preparative Biochemistry and Biotechnology* 51 (5): 467–474.

Puri, S.C., Verma, V., Amna, T. et al. (2005). An endophytic fungus from *Nothapodytes foetida* that produces camptothecin. *Journal of Natural Products* 68: 1717–1719.

Radman, R., Teresa, S., Christopher, B. et al. (2003). Elicitation of plants and microbial cell systems. *Biotechnology and Applied Biochemistry* 37: 91–102.

Rani, D., Meelaph, T., De-Eknamkul, W. et al. (2020). Yeast extract elicited isoflavonoid accumulation and biosynthetic gene expression in *Puerariacandollei* var. mirifica cell cultures. *Plant Cell, Tissue and Organ Culture* 141: 661–667.

Reymond, P. and Farmer, E.E. (1998). Jasmonate and salicylate as global signals for defense gene expression. *Current Opinion in Plant Biology* 1: 404–411.

Salah, B.I., Aghrouss, S., Douira, A. et al. (2018). Seaweed polysaccharides as bio-elicitors of natural defenses in olive trees against verticillium wilt of olive. *Journal of Plant Interactions* 13 (1): 248–255.

Sangeetha, S. and Sarada, D.V.L. (2015). Phenyl derivative of pyranocoumarin precludes *Fusariumoxysporum*f.sp.*Lycopersici* infection in *Lycopersiconesculentum* via induction of enzymes of the phenylpropanoid pathway. *Applied Biochemistry and Biotechnology* 175: 1168–1180.

Sharma, H.S., Fleming, C., Selby, C. et al. (2014). Plant biostimulants: a review on the processing of macroalgae and use of extracts for crop management to reduce abiotic and biotic stresses. *Journal of Applied Phycology* 26 (1): 465–490.

Singh, S.K., Singh, V.K., Kumar, D. et al. (2022). Endophytes of medicinal plants: diversity and bioactivity. In: *Bacterial Endophytes for Sustainable Agriculture and Environmental Management* (ed. A.K. Singh, V. Tripathi, A.K. Shukla, and P. Kumar), 117–128. Singapore: Springer Singapore https://doi.org/10.1007/978-981-16-4497-9_6.

Srivastava, M., Swati, S., and Pratibha, M. (2016). Elicitation based enhancement of secondary metabolites in *Rauwolfia serpentina* and *Solanum khasianum*hairy root cultures. *Pharmacognosy Magazine* 12: 15–20.

Tamilarasi, S., Nanthakumar, K., Karthikeyan, K. et al. (2008). Diversity of root associated microorganisms of selected medicinal plants and influence of rhizo microorganisms on the antimicrobial property of *Coriandrumsativum*. *Journal of Environmental Biology* 29 (1): 127–134.

Vakil, M.M.A. and Mendhulkar, V.D. (2013). Enhanced synthesis of andrographolide by *Aspergillusniger* and *Penicilliumexpansum* elicitors in cell suspension culture of *Andrographispaniculata* (Burm. f.) Nees. *Botanical Studies* 54 (1): 49.

Verma, N. and Shukla, S. (2015). Impact of various factors responsible for fluctuation in plant secondary metabolites. *Journal of Applied Research on Medicinal and Aromatic Plants* 2 (4): 105–113.

Wenjuan, X., Xiaoyan, J., Miao, Y. et al. (2021). Primary and secondary metabolites produced in *Salviamiltiorrhiza* hairy roots by an endophytic fungal elicitor from *Mucorfragilis*. *Plant Physiology and Biochemistry* 160: 404–412.

Zhai, X., Jia, M., Chen, L. et al. (2016). The regulatory mechanism of fungal elicitor-induced secondary metabolite biosynthesis in medical plants. *Critical Reviews in Microbiology* 43: 238–261.

Zhao, J., Davis, L.C., and Verpoorte, R. (2005). Elicitor signal transduction leading to production of plant secondary metabolites. *Biotechnology Advances* 23: 283–333.

Zhou, Z., Xia, L., Zheng, J.Y., and Dai, C.C. (2016). Volatiles released by endophytic *Pseudomonasfluorescens* promoting the growth and volatile oil accumulation in *Atractylodeslancea*. *Plant Physiology and Biochemistry* 101: 132–140.

Zlotek, U. and Swieca, M. (2015). Elicitation effect of *Saccharomycescerevisiae* yeast extract on main health-promoting compounds and antioxidant and anti-inflammatory potential of butter lettuce (*Lactucasativa* L.). *Journal of the Science of Food and Agriculture* 96 (7): 2565–2572.

# 24

## Bioremediation of High Molecular Weight Polycyclic Aromatic Hydrocarbons

*Fahad S. Alotaibi, Abdullah Alrajhi, and Saif Alharbi*

*King Abdulaziz City for Science and Technology, Riyadh, Saudi Arabia*

### 24.1 Introduction

Polycyclic aromatic hydrocarbons (PAHs) are perilous organic compounds known for their mutagenic, carcinogenic, and genotoxic impacts. As outlined by the United States Environmental Protection Agency (USEPA), a total of 16 PAHs are classified as primary pollutants with toxicity to living organisms (Mrozik et al. 2003). Scientific investigations into the biological breakdown of PAHs have revealed that bacteria, fungi, and algae harbor catabolic capabilities that can be harnessed for PAH-contaminated soil and water restoration. The efficacy of bioremediation has been evidenced in addressing soil pollution caused by PAHs with low molecular weights (LMWs) (Juhasz and Naidu 2000). The global utilization and dispersion of PAHs result in their introduction into the environment and subsequent bioaccumulation in the food cycle (Kotoky et al. 2018). Owing to their enduring attributes and pronounced hydrophobic characteristics, PAH molecules attach to the particles, rendering them inaccessible for biodegradation (Weber et al. 2017). The 1989 Exxon Valdez oil spill sparked a global surge of interest in the study of PAH degradation. Three decades on, innovative tactics and emerging tools are introducing fresh perspectives to address this challenge (Pandey et al. 2021). While various approaches have been proposed to eliminate PAHs, i.e., burning, oxidation, and fixation, what is frequently overlooked is the significance of enhancing their bioavailability. Common techniques occasionally fail to entirely eradicate PAHs, leading to the creation of fractional by-products that can be more carcinogenic than the original products (Gan et al. 2009). Bacteria possess diverse mechanisms for breaking down PAHs and other xenobiotic compounds (Kotoky and Pandey 2020). Among the well-examined PAH-degrading bacteria are *Pseudomonas aeruginosa*, *P. fluorescens*, various *Mycobacterium* species, *Haemophilus* species, *Rhodococcus* species, and *Paenibacillus* species (He et al. 2019; Singh et al. 2019; Kumar et al. 2017). A crucial microbial community involved in this process is biosurfactant-producing microbes, which play a vital role in rendering contaminants accessible for biodegradation (Singha and Pandey 2021).

The duration of PAHs' existence in their surroundings relies on their physical and chemical attributes. While photo-oxidation and chemical oxidation contribute to the breakdown of PAHs (Abdel-Shafy and Mansour 2016), it's likely that their primary mode of reduction occurs through biological transformation (Byrne et al. 2018). The resistance of PAHs to microbial breakdown becomes more pronounced as their molecular weight and octanol: water partition coefficient (log Kow) increase (Huang et al. 2018). Recent developments in genomics, proteomics, and transcriptomics have significantly enhanced our understanding of PAH degradation and shed light on the diverse metabolic pathways at play. Techniques include microarray analysis, isotope probing, single-cell genomics, and qPCR expression analysis. Given the intricate nature of PAH degradation pathways, genome sequencing plays a crucial role in unraveling potential catabolic processes and facilitating the informed design of consortia (Kotoky and Pandey 2021). The utilization of high-throughput Illumina sequencing has seen widespread application in identifying non-cultivable microorganisms engaged in PAH breakdown (Wang et al. 2018).

Additionally, two invaluable techniques, metaproteomic and metabolomics, have proven effective in pinpointing the proteins and metabolites generated throughout the biodegradation process. Soil management recommended an approach

*Microbes Based Approaches for the Management of Hazardous Contaminants*, First Edition. Edited by Ajay Kumar, Livleen Shukla, Joginder Singh, and Luiz Fernando R. Ferreira.

for large-scale PAH biodegradation, where the native microbial community is activated. Furthermore, alternative techniques like bioaugmentation, composting, biopiling, phytoremediation, vermi-remediation, and the application of plant growth-promoting bacteria (PGPB) in polluted sites have proven remarkably effective, contributing significantly to PAH removal (Zaki et al. 2013). Rhizoremediation is a bioremediation strategy that harnesses plant-microbe interactions to eliminate higher-molecular-weight PAHs (Singha and Pandey 2021). To comprehensively address the status of our lives and environment, it is imperative to maintain up-to-date literature. This holds true even when considering the pressing issue of pollution facilitated by PAHs, which warrants particular attention. This review offers an inclusive overview of the latest insights into how microorganisms degrade and transform PAHs. Additionally, it delves due to the helps of bacteria, fungi, and algae in degrading PAHs, explores the factors influencing this process, highlights recent advancements in PAH bioremediation techniques, and discusses their practical applications.

## 24.2 Polycyclic Aromatic Hydrocarbons (PAHs): Sources, Pollution, and Exposure Routes

PAHs represent a class of organic pollutants characterized by the presence of two or more bonded aromatic rings composed of hydrogen (H) and carbon (C) molecules. These compounds typically exist in the form of colorless, white, or pale-yellow solids (Suman et al. 2016). The spatial configuration of these aromatic rings can vary, ranging from linear to angular or clustered arrangements (Abdel-Shafy and Mansour 2016). PAHs are categorized based on the number of rings within their molecular structure. They are categorized into two groups: LMW-PAHs, which consist of two or three aromatic rings, and HMW-PAHs, comprising four or more aromatic rings. Their emission characteristics differ according to their molecular weight, with LMW-PAHs being released in gaseous form and HMW-PAHs existing as particulate matter (Lee 2010). Additionally, PAHs can be further categorized based on the arrangement of their aromatic rings. Alternant PAHs, for instance, consist solely of the fusion of 6C benzene rings.

In contrast, non-alternant PAHs, such as fluorene, feature the fusion of 6C benzene rings along with an extra ring containing fewer than 6C atoms (Gupte et al. 2016). The presence of dense $\pi$ electrons on these aromatic rings contributes to the biochemical persistence of PAHs, rendering them (Haritash and Kaushik 2009). In 1983, the USEPA designated 16 PAHs as top-name pollutants due to their high concentrations, increased exposure, resistance to degradation, and toxicity (Mojiri et al. 2019). PAHs recorded specific characteristics: high melting and boiling points, low solubility, and vapor pressure; however, they are diverse due to their molecular composition. PAHs with HMW tend to have reduced water solubility and increased lipophilicity, making them less degradable (Okere and Semple 2012).

PAH pollution sources can be primarily categorized into two groups (Mojiri et al. 2019). Sources in nature, such as volcanic outbursts and wildfires, played a less significant role (Abdel-Shafy and Mansour 2016). However, the main contributors are anthropogenic sources divided into four types: agricultural sources, domestic industry, and mobile sources (Dat and Chang 2017). Similarly, incomplete combustion includes primary PAH emissions from waste incineration, industrial processes, steel production, aluminum production, coal-tar pitch production, asphalt, fungicide and insecticide, and refinery exhaust (Mallah et al. 2022). Other industrial sources include coal gasification, electric arc furnaces, oxygen furnaces, diesel engines, and gasoline-powered engines used in large machinery (Moka et al. 2014). Mobile emission sources encompass exhaust emissions from various vehicles, including aircraft, ships, and trains, as well as heavyweight and lightweight off-road vehicles (Lawrence et al. 2015). Domestic emissions and other forms of residential heating would also contribute to atmospheric pollutants (Gupte et al. 2016). Agricultural emissions arise from open biomass burning and the burning of agricultural waste when combustion occurs under less-than-optimal conditions (Nayal et al. 2016). PAH concentrations fluctuate throughout the year, with the highest recorded at low temperatures relative to hot temperatures (Miura et al. 2019).

PAH sources can be classified based on their production origins into three distinct categories: pyrogenic, petrogenic, and biogenic (Miura et al. 2019). Pyrogenic PAHs are generated because of unintentional incomplete combustion of organic substances at extremely high temperatures (350–1200°C) in the absence or presence of minimal $O_2$ (Zhou et al. 2010). Some deliberate pyrolytic processes, such as degrading complex petroleum compounds into low hydrocarbons (LHCs) and coal distillation to produce coal tar and coke, also yielded pyrogenic PAHs. Typically, pyrogenic concentrations of PAHs are elevated in urban areas (Mojiri et al. 2019). Petrogenic PAHs are mainly composed of petroleum and its by-product (Patel et al. 2020), pyrogenic sources are formed from HMW-PAHs, and petrogenic sources comprise LMW-PAHs (Marris et al. 2020).

On the other hand, biogenic PAHs are produced by biological organisms, i.e., algae, microorganisms, phytoplankton, and algae, which are significant during the continuing biological conversion of organic matter (Miura et al. 2019).

## 24.3 Biodegradation Pathways

One of the key sections of the review explores the diverse biodegradation pathways employed by microorganisms to break down HMW PAHs. The enzymatic mechanisms involved in the degradation process are elucidated, highlighting the roles of dioxygenases, monooxygenases, and other enzymes. The section emphasizes the importance of microbial communities and their synergistic activities in efficiently degrading these complex compounds (Amobonye et al. 2021). Biological approaches have garnered significant attention in recent times for remediating PAHs, thanks to their eco-friendly nature. This shift in focus is a response to various shortcomings associated with physical and chemical methods, such as high costs, procedural intricacies, regulatory challenges, and the inability to achieve complete degradation (Patel et al. 2020). Bioremediation techniques fall into two broad categories: *ex-situ* and *in-situ* methods. *Ex-situ* methods include physically relocating polluted contaminants, such as excavating soil or pumping water, to another specified area, usually within the same site, for treatment, relative to *in-situ* methods, which treat the polluted material directly at its original location. The public favors the latter approach due to its cost-effectiveness, minimal disruption to the site, and the potential for permanent waste elimination (Hussain et al. 2022).

Kuppusamy et al. (2017) indicated a notable focus on employing biological approaches for PAH remediation, closely followed by a combined process involving chemical oxidation and physical techniques in terms of interest. Among the various biological approaches for PAH remediation, facilitating PAH degradation, bioaugmentation, and biostimulation are the utmost methods, accounting for approximately 33% of usage. Bioreactors and phytoremediation/rhizoremediation are also commonly applied, representing around 22% of the methods used. Composting is employed in about 13% of cases, while biopiles, enzyme-mediated bioremediation, vermi-remediation, and other methods have a lower utilization rate of 4% or less (Kuppusamy et al. 2017). As discussed above, integrated methods are a more significant choice to remediate the soil. It contributes 42% to biological, 27% to chemical and biological, and 21% to physical and chemical. However, 5% contributed using the thermal chemical method (Kuppusamy et al. 2017). Microbial remediation involves biostimulation, and bioaugmentation encompasses the separate or combined use of specific organisms in the form of algae, bacteria, archaea, and fungi. Nonetheless, extensive research has been conducted on bacteria- and fungi-assisted degradation (Nwankwegu et al. 2022).

Bacteria exhibit remarkable metabolic versatility when breaking down PAH pollutants in the soil (Alegbeleye et al. 2017). In the process of aerobic PAH degradation by bacteria, $O_2$ represents as acceptor and co-substrate for hydroxylation, and $O_2$ reconciles aromatic ring breakdown (Aydin et al. 2017). Conversely, when bacteria engage in anaerobic PAH degradation, they adopt a distinct strategy for dismantling and opening the aromatic ring. This strategy depends on the type of reductive reaction and alternative final electron acceptors (Dhar et al. 2020). Bacterial PAH degradation depends purely on aerobic oxygenase-facilitated metabolism followed by monooxygenase and dioxygenase enzymes. The preliminary process involves hydroxylation of aromatic rings; these enzymes form *cis*-dihydrodiols in aerobic degradation. Dehydrogenase enzyme breaks down the diols into different intermediate diol compounds. These intermediate diols breakdown divided the components into intra/extra diol rings following an ortho-cleavage or meta-cleavage pathway. It involves catechol, gentistic acid, and protocatechuic acid transformed into tricarboxylic acid (TCA) cycle intermediates (Choi et al. 2021). The dioxygenase enzyme is multifunctional, comprising reductase, ferredoxin, and terminal oxygenase subunits (Wu et al. 2020). Bacteria also employ the cytochrome P450-assisted pathway for PAH degradation, forming trans-dihydrodiols. Additionally, under nitrate conditions with sulfate reduction, bacteria have the capacity for anaerobic degradation of PAHs (Zhang et al. 2019).

The traditional and preferred approach is aerobic PAH degradation. In contemporary times, there is a growing focus on anaerobic PAH degradation, especially in environments with anoxic conditions like the phreatic zone, deep aquatic sediment, and water-flooded soil (Dhar et al. 2020). Extensive literature reports the involvement of aerobic and anaerobic bacterial species in the degradation of both LMW and HMW-PAHs. Enhanced or complete PAH degradation is achievable through co-cultures and bacterial conglomerates, leveraging the cumulative catabolic performances of their constituents and the potential presence of diverse degradation pathways. Consequently, recent research has significantly emphasized PAH degradation assisted by mixed bacterial cultures and consortia (Feng et al. 2021). Additionally, using immobilized and genetically modified bacteria (GMB) represents a viable method to enhance the process (Sharma et al. 2022). In the context of soil and sediment, one of the significant challenges in PAH degradation is the dispersion of inoculum. This process is relatively direct for surface soil but poses challenges for subsurface soil due to restricted microbial movement, given that bacterial cells strongly adhere to soil organic matter.

### 24.3.1 Algae, Co-Culture

Algae, the primary producers in coastal and estuarine ecosystems, play a crucial role in the bioremediation of PAHs within aquatic environments. Algae effectively eliminate PAHs through cellular biodegradation and/or bioaccumulation (Subashchandrabose et al. 2019). PAH biodegradation involves both monooxygenase and dioxygenase enzymatic processes, resulting in the production of hydroxylated and dehydroxylated intermediates. The specific outcomes depend on the type of algae (Tomar and Jajoo 2021); microalgae, such as Cyanobacteria, are single-celled green algae found in freshwater environments (Barone et al. 2020). They have garnered significant attention because they are found to be extensively accessible to cultivate and have remarkable efficiency in degrading HMW-PAHs (García and de Llasera 2021). While numerous studies on PAH eradication through alga have been conducted under controlled conditions or on a microcosm scale, there remains a gap in understanding large-scale or field-level application remediation. Addressing this gap is essential and necessitates a concentrated scientific focus to devise effective strategies.

In natural ecosystems, bacterial-fungal interactions play a crucial role. Co-cultivation methods, such as bacterial-fungal co-cultures, fungal–algal collaborations, and bacterial-algal partnerships, have been established as highly effective bioremediation approaches in laboratory experiments and large-scale applications (Harry-Asobara and Kamei 2019). Algae play a critical role in aerobic degradation processes by supplying oxygen, thereby enhancing the degradation process. Notably, bacterial-algal synergy offers more advantages than bacterial consortia and bacterial-fungal co-cultures, as algae contribute additional polymeric and lightweight compounds. These compounds, in turn, promote the growth of bacteria and/or fungi, thereby significantly boosting the degradation of PAHs (Pozdnyakova et al. 2022). Combine application enhanced the mycelium's dry weight and increased the number of bacterial colonies relative to single. The co-culture product converts the PAHs into their respective quinones. The corresponding degradation also enhanced the production of extracellular fungal laccase and peroxidase. In addition, indole-3-acetic acid (IAA) production augmented PAH degradation and enzyme production (Pozdnyakova et al. 2022).

### 24.3.2 Fungi

In recent years, there has been extensive research on mycoremediation of PAHs, involving various fungal species. Unlike bacteria, not all fungi use PAHs as their primary carbon source (Bhadouria et al. 2020). Instead, they co-metabolize PAHs, yielding a range of oxidized by-products, including $CO_2$. This degradation process is facilitated by monooxygenase enzymes (Udom et al. 2023). Two main categories of fungi have been documented in the literature for the bioremediation of LMW- and HMW-PAHs: ligninolytic fungi (white-rot fungi) and non-ligninolytic fungi. Ligninolytic fungi produce enzymes like lignin peroxidase, manganese peroxidase, and laccases, which not only break down lignin found in wood but also oxidize PAHs and convert into diphenol derivatives that eventually become quinones (Aydin et al. 2017). The ligninolytic enzymes break down aromatic compounds, producing polar, water-soluble products that can be utilized by fungal metabolism and the adjacent soil microflora (Gupta and Pathak 2020).

Conversely, non-ligninolytic fungi produce enzymes resembling cytochrome P450 monooxygenases. These enzymes oxidize PAHs, resulting in the formation of arene oxides and water. Subsequently, arene oxides undergo nonenzymatic rearrangement to form phenols, which can conjugate with xylose, gluconeric acid, and glucose (Ghosal et al. 2016). Some fungal species have the additional capability of producing biosurfactants to enhance the degradation of less soluble HMW PAHs (Ojha et al. 2019). While myco-remediation shows promise, there is a limited understanding of the specific mechanisms and pathways involved in PAH breakdown through this process (Pozdnyakova 2012). Applying fungi directly in the field presents various challenges, including insufficient biomass growth, difficulties in handling large quantities of biomass, a lack of standardized application methods, and the need for substantial enzyme production. These challenges can be addressed through the application of oxidative fungal enzymes for PAH bioremediation (Deshmukh et al. 2016).

### 24.3.3 Bioremediation Strategies

Microbial bioremediation involves the utilization of enzymes isolated from microorganisms used for PAH removal (Giri et al. 2021; Gond et al. 2021). This process exhibits peak efficiency at higher rates and catalyzes the reaction across varying degrees of pH and temperature. Different kinds of enzymes, i.e., phenol oxidase, lignin peroxidases, manganese peroxidase, laccases, and phenol oxidases, are used for oxidation (Thacharodi et al. 2023). Fungal oxidative enzymes are particularly adept in this context due to their comparatively lower substrate specificity (Gupta and Pathak 2020).

Zhang et al. (2020) conducted the isolation of a novel manganese peroxidase gene from *Cerrena unicolor* BBP6, subsequently cloning it into *Pichia pastoris*. This genetic modification resulted in an enzyme with diverse dye-decolorizing capabilities, achieving 80% and 91% degradation of fluorene and phenanthrene within a 24-hour timeframe. Notably, the highest expression of the recombinant enzyme reached 154.5 units per liter (unit $l^{-1}$). Nevertheless, it is important to acknowledge that the primary limitation associated with this approach pertains to the costs incurred during the production, extraction, and purification of these enzymes (Kuppusamy et al. 2017).

Land farming represents a cost-efficient and secure remediation method for contaminated soil, wherein the indigenous microbiome at the contaminated site is activated to enhance PAH degradation through the optimization of aeration, moisture, and nutrient content, thereby fostering improved microbial interactions with both pollutants and nutrients (Sarkar et al. 2020). The reduction rate is notably greater for LMW-PAHs (2–3 rings) to (4–6 rings) in HMW, and typically, this technique is employed on a shallow layer of soil (Mundo et al. 2022). The current method demands little ongoing maintenance, involves minimal cleanup efforts, and requires only minor monitoring. Nonetheless, it comes with certain constraints, such as a decline in degradation speed following the initial rapid phase, largely attributed to variations in pollutant concentration. Furthermore, its effectiveness is confined to the top 10–35 cm of accessible soil and is substantially impacted by uncontrollable and unforeseen factors like heavy rainfall (Certini et al. 2021). Similarly, another method called facilitating the PAH degradation method boosts the degradation capabilities of the native microbiome by enhancing aeration, moisture, and nutrient levels. When applied to contaminated land, it is a noteworthy illustration of the land farming approach (Aparicio et al. 2022). A remarkable 98% degradation of total petroleum hydrocarbons and an 85% reduction in PAH were reported by promoting increased facilitation of PAH degradation within a large-scale bioremediation project to address crude oil pollution. This enhanced facilitation of PAH degradation was achieved by incorporating nutrients at a ratio of 2:1:1 (N:P:K), tilling the soil, periodically irrigating with water, and intermittently turning the soil to ensure uniform aeration (Guarino et al. 2017). Biostimulation and bioaugmentation are frequently utilized in bioremediation approaches. Biostimulation involves adding nutrients like nitrogen (N), phosphorus (P), potassium (K), or other alternate treatments such as fertilizers, municipal sewage sludge, compost, organic waste, and humic substances, aiming to boost the performance of local microbial populations (Bianco et al. 2020). On the other hand, bioaugmentation entails introducing native or externally sourced PAH-degrading bacteria into polluted areas to improve the effectiveness of the remediation process (Haleyur et al. 2019).

### 24.3.4 Factors Influencing Bioremediation

This section discusses the various factors that impact the efficiency of bioremediation processes, encompassing environmental conditions like temperature, pH, oxygen availability, and nutrient levels (Gupta et al. 2018, 2020). It also delves into the complex interplay between microbial communities, illustrating how optimizing these parameters can augment biodegradation rates. The efficiency of PAH bioremediation hinges on a multitude of factors. These encompass the properties of PAHs, characteristics of the contaminated material, environmental conditions, and microbial ecology. Equally abiotic and biotic factors, such as salinity, temperature, pH, humidity, nutrient availability, oxygen levels, bioavailability of pollutants, pollutant concentrations, presence of native microflora, their degradation capabilities, microbial substrate preferences, prior exposure to pollutants, production or addition of biosurfactants, and the availability of other carbon sources, may vary from site to site, significantly influencing the bioremediation process (Kebede et al. 2021; Srivastava et al. 2014). The key considerations in soil PAH bioremediation pertain to soil/sediment characteristics through microbial activity (Mehta 2019; Mishra et al. 2021; Pandey et al. 2023). The effectiveness of microbial remediation is influenced by the dispersion of pollutants, microbial distribution, and the metabolic capacity of microbes (Kebede et al. 2021; Singh et al. 2020, 2021). The composition of PAHs showed a critical point in determining the choice of degradation methods. Thermodynamically, the angular configuration represents the most stable form of PAHs; however, enzymatic degradation is more likely to occur in their bay regions. This accounts for the higher biodegradation rates in angular structures compared to linear and clustered ones. Conversely, linear and clustered PAHs undergo rapid degradation through processes such as photooxidation and chemical oxidation (Rathankumar et al. 2020). The degradation of pyrene occurs at a faster rate compared to phenanthrene through photochemical or chemical oxidation. This behavior was elucidated in the study by (Peluffo et al. 2018), where they observed a higher degradation of pyrene (96% of 2800 mg $kg^{-1}$) in comparison to phenanthrene (91% of 1200 mg $kg^{-1}$) within a seven-day period using a thermal (65°C for six hour)-activated persulfate oxidation method in a batch reactor containing aged, polluted soil at room temperature. The significant influence of soil type on PAH bioremediation was determined through bioslurry experiments conducted at 28°C in the dark and under continuous shaking (120 rpm). The source of soil type is from three different sites: a gas plant site, a coking plant, and an active wood treatment plant (Patel et al. 2020).

### 24.3.5 Emerging Technologies

In recent times, breakthrough technologies like genomics, metagenomics, and synthetic biology have brought about a transformative shift in the realm of bioremediation. In this section, we illuminate how these tools offer a glimpse into the genetic capacity of microorganisms, empowering scientists to create and optimize strains with improved abilities to degrade PAHs. Studying the microbial community that degrades PAHs, the genes involved in degradation, i.e., regulatory genes, intermediates, enzyme systems, and final products of degradation pathways, as well as the indigenous microbial diversity at contaminated sites, is crucial for gaining a comprehensive understanding of PAH bioremediation (Wang et al. 2020). Effective design for PAH bioremediation can be developed by strategizing based on knowledge of native microbes at contaminated sites, their metabolic adaptability, and their ability to thrive in challenging environmental conditions (Sakshi and Haritash 2020). Regrettably, obtaining this specific knowledge isn't readily accessible, but it can be acquired through the "genomic era," utilizing culture-independent methods like metagenomics (Sakshi and Haritash 2020), meta-transcriptomics (Aydin et al. 2017), meta-proteomics (Malik et al. 2021), and metabolomics (Li et al. 2019). The limitation of traditional culture-dependent methods lies in the fact that only a tiny fraction, ranging from 0.1% to 1.0%, of microbes in the environment can be cultured. The vast majority, constituting 99.0% to 99.9%, remains unculturable due to insufficient knowledge about their specific culturing requirements (Zafra-Rojas et al. 2016).

Similarly, metagenomic omics encompasses two approaches: sequence-based and function-based. The sequence-based approach is significant for exploring new degrading genes involved in degrading pathway evolution among different microbial diversity. On the contrary, the function-based approach is beneficial for identifying novel degradative genes (Malla et al. 2018). Microbial communities' functional capacities can be assessed across various tiers, including meta-transcriptomics, meta-proteomics, and metabolomics. Meta-transcriptomics entails unraveling mRNA profiles to understand the up- or down-regulation of degradative microbes and associated genes. Meta-proteomics involves monitoring pollutants induced by proteins from degrading microbes (Lu et al. 2019). Metabolomics analyzes metabolites produced by degradative microbes to help them cope with pollutants (Miglani et al. 2022). Meta-transcriptomic approaches are involved in predicting the functions of abundant members even at low-level cell populations (Marcatili et al. 2016). Similarly, amplicon metagenomics helps to find microbial taxa for precision bioremediation in polluted soil. The identified microbial taxa exhibit degradation abilities in the form of bio-stimulants, abundant growth, cultivation ease, and ability to thrive and metabolize high pollutant concentration, using bioaugmentation and promote sustainable approach for remediation (Redfern et al. 2019). *Geobacter* spp. as targets for biostimulation, while identifying *Mycobacterium* spp. and *Sphingomonads* spp. as suitable targets for bioaugmentation in creosote-polluted soil, employing a metagenomic approach. In addition, the correlation matrix index analysis results suggested that taxa showing a positive correlation can be harnessed in PAH bioremediation due to their resistance to PAHs, while conversely, taxa displaying a negative correlation are sensitive to PAHs (Redfern et al. 2019). Another study suggested alterations in bacterial communities when comparing original seawater to a mixed PAH-degrading consortium. The results utilized high-throughput 16S rRNA gene sequencing. The findings indicated a heightened abundance of proteobacteria and a reduced bacteroidetes within the seawater consortium. Additionally, the Shannon and Simpson diversity indices were observed to be lower. Additionally, the research delved into the microbial diversity within a consortium capable of diesel degradation. It also investigated essential genes linked to degrading enzymes, employing 16S rRNA amplicon and shotgun metagenome sequencing (García-Sánchez et al. 2018).

Interesting results were also obtained from their research utilizing sediment artificially polluted with pyrene. This study encompassed investigations into bacterial synergism, metabolic profiles of bacteria under pyrene-induced stress, metagenome-derived genomes (MAGs) assembly, and the interplay of aromatic hydrocarbon degrading genes with diverse metabolic pathways involved in the process. They assembled a collective total of 56 (MAGs) attributed to Gamma- and proteobacteria. A notable portion of these genomes were neither cultivable nor classified. Within this set, 20 MAGs were identified to possess PAH degradation genes (Zhang et al. 2019). Particular genes responsible for degrading PAHs, such as PAH ring hydroxylating dioxygenases (RHDs), can serve as effective biomarkers for identifying PAH-degrading microorganisms. The gene-targeted metagenomics approach focusing on PAH-RHDs has been demonstrated to be a suitable and promising method for studying the diversity of PAH-degrading bacteria in both oil-contaminated soil and mangrove sediment (Liang et al. 2019). Storey et al. (2018) suggested a 10,000-fold increase in PAH-RHD phenanthrene genes in polluted soil relative to control. Omics strategies provide a significant base for diagnosing the effectiveness and progress of bioremediation. In addition, it addresses vital inquiries related to the destiny of native degrading microorganisms, differential expression genes related to degrading pathways, and their involvement in uncultivable microorganisms (Zafra and Valdivieso Quintero 2016).

Computational studies, including techniques like homology modeling and molecular docking of the P450 monooxygenase (CYP protein), can prove highly effective in elucidating the docking relationship between protein structure and substrates. This provides comprehensive insights into substrate binding and the subsequent metabolic degradation of the substrates (Suresh Kumar et al. 2018). PAHs engage with the active sites of CYP through various intermolecular interactions like hydrogen bonding, hydrophobic bonding, $\pi$–$\pi$ interactions, and van der Waals forces. Consequently, a structural elucidation study affirms that the microalgae *Parachlorella kessleri* demonstrates a higher efficiency in remediating HMW than other microorganisms. The possible utilization of microalgae for degrading aromatic compounds is explored through molecular docking, employing the modeled structure of the enzyme CYP450 monooxygenase as the target (Suresh Kumar et al. 2018).

## 24.4 Challenges and Future Directions

The review extensively explores persistent challenges encountered in the bioremediation of HMW and PAHs. These challenges encompass critical factors such as substrate specificity, toxicity of metabolites, and an incomplete understanding of specific biodegradation pathways. The contamination by harmful, bio-toxic HMW-PAHs from diverse sources, as mentioned earlier, has significantly impacted both natural settings and human well-being. In response to these challenges, the review underscores substantial advancements in the development of diverse technologies and strategies aimed at remediating PAH pollution in the environment. Furthermore, it emphasizes the importance of interdisciplinary collaboration among microbiologists, environmental scientists, engineers, and geneticists to devise comprehensive and sustainable solutions for addressing these intricate pollutants. In summary, the review offers a comprehensive perspective on the current status of bioremediation concerning HMW-PAH. It serves as a valuable asset for researchers, practitioners, and policymakers keen on comprehending the potential of bioremediation approaches in mitigating the environmental and health ramifications of these persistent pollutants. Nonetheless, several unresolved challenges persist, making the remediation of PAHs at sites a formidable task. Both laboratory and pilot/reactor-scale studies have furnished abundant evidence, offering intricate mechanistic insights into the degradation of LMW and HMW-PAHs. Notably, LMW-PAHs were relatively easy to degrade, contrasting the HMW-PAHs, which are more carcinogenic due to their compact ring structures, rendering their degradation highly challenging.

Furthermore, extensive research spanning several years has indicated that replicating lab-scale results in field-scale applications is insufficient. Bioremediation approaches present a viable solution for PAH remediation. Recent advancements in traditional landform compositing have augmented phytoremediation, rejuvenating the PAH treatment process. The effective leveraging of synergistic interactions between microorganisms and plants has proven beneficial for *in situ* bioremediation of PAHs. Emerging integrated approaches, such as eco-engineered bioremediation methodologies, have enhanced microbial degradation, and the immobilization of microbial consortiums in macro-capsules has furthered the degradation process in polluted areas. Understanding and identifying the intricacies of PAH degradation in soil and marine environments pose significant challenges requiring focused attention in future research endeavors. In the soil ecosystem, PAHs often become trapped within black-clayish-carbon particles and coal tar, substantially reducing their bioavailability. This poses an important obstacle to the successful bioremediation of PAHs. The efficacy of PAH remediation approaches is influenced by several factors, including optimal dissolved oxygen concentration, nutrient availability, co-existing pollutants such as heavy metals and phenolics, as well as various other abiotic factors. Considering the natural environment functions as an interactive and interdependent system, where microbial communities play a pivotal role in niche and nutrient cycling, understanding the synergistic mechanisms of hydrocarbon degradation becomes vital for successfully applying and scaling up *in situ* bio-stimulation and bioaugmentation strategies. Specifically, exploring recent progress in creating value-added by-products during PAH remediation represents a practical approach warranting further investigation in future studies.

## List of Abbreviations

| | |
|---|---|
| ATP | adenosine triphosphate |
| BBP6 | *Cerrena unicolor* BBP6 |
| $CO_2$ | carbon dioxide |
| CYP | cytochrome P450 monooxygenase |
| GMB | genetically modified bacteria |
| HMW | high molecular weight |
| IAA | indole-3-acetic acid |
| K | potassium |
| LHC | low hydrocarbons |
| LMW | low molecular weight |
| MAGs | metagenome-derived genomes |
| N | nitrogen |
| $O_2$ | oxygen |
| P | phosphorus |
| PAHs | polycyclic aromatic hydrocarbons |
| RHDs | ring hydroxylating dioxygenases |
| TCA | tricarboxylic acid |
| USEPA | United States Environmental Protection Agency |

## References


Abdel-Shafy, H.I. and Mansour, M.S. (2016). A review on polycyclic aromatic hydrocarbons: source, environmental impact, effect on human health and remediation. *Egyptian Journal of Petroleum* 25 (1): 107–123.

Alegbeleye, O.O., Opeolu, B.O., and Jackson, V.A. (2017). Polycyclic aromatic hydrocarbons: a critical review of environmental occurrence and bioremediation. *Environmental Management* 60: 758–783.

Amobonye, A., Bhagwat, P., Singh, S., and Pillai, S. (2021). Plastic biodegradation: frontline microbes and their enzymes. *Science of the Total Environment* 759: 143536.

Aparicio, J.D., Raimondo, E.E., Saez, J.M. et al. (2022). The current approach to soil remediation: a review of physicochemical and biological technologies, and the potential of their strategic combination. *Journal of Environmental Chemical Engineering* 10 (2): 107141.

Aydin, S., Karaçay, H.A., Shahi, A. et al. (2017). Aerobic and anaerobic fungal metabolism and omics insights for increasing polycyclic aromatic hydrocarbons biodegradation. *Fungal Biology Reviews* 31 (2): 61–72.

Barone, G.D., Ferizović, D., Biundo, A., and Lindblad, P. (2020). Hints at the applicability of microalgae and cyanobacteria for the biodegradation of plastics. *Sustainability* 12 (24): 10449.

Bhadouria, R., Das, S., Kumar, A. et al. (2020). Mycoremediation of agrochemicals. In: *Agrochemicals Detection, Treatment and Remediation* (ed. M. Narasimha Vara Prasad), 593–620. Oxford: Butterworth-Heinemann https://doi.org/10.1016/B978-0-08-103017-2.00022-2.

Bianco, F., Race, M., Papirio, S., and Esposito, G. (2020). Removal of polycyclic aromatic hydrocarbons during anaerobic biostimulation of marine sediments. *Science of the Total Environment* 709: 136141.

Byrne, G., Dimitrov, D., Monostori, L. et al. (2018). Biologicalisation: biological transformation in manufacturing. *CIRP Journal of Manufacturing Science and Technology* 21: 1–32.

Certini, G., Moya, D., Lucas-Borja, M.E., and Mastrolonardo, G. (2021). The impact of fire on soil-dwelling biota: a review. *Forest Ecology and Management* 488: 118989.

Choi, I., Son, H., and Baek, J.-H. (2021). Tricarboxylic acid (TCA) cycle intermediates: regulators of immune responses. *Life* 11 (1): 69.

Dat, N.-D. and Chang, M.B. (2017). Review on characteristics of PAHs in atmosphere, anthropogenic sources and control technologies. *Science of the Total Environment* 609: 682–693.

Deshmukh, R., Khardenavis, A.A., and Purohit, H.J. (2016). Diverse metabolic capacities of fungi for bioremediation. *Indian Journal of Microbiology* 56: 247–264.

Dhar, K., Subashchandrabose, S.R., Venkateswarlu, K. et al. (2020). Anaerobic microbial degradation of polycyclic aromatic hydrocarbons: a comprehensive review. *Reviews of Environmental Contamination and Toxicology* 251: 25–108.

Feng, L., Jiang, X., Huang, Y. et al. (2021). Petroleum hydrocarbon-contaminated soil bioremediation assisted by isolated bacterial consortium and sophorolipid. *Environmental Pollution* 273: 116476.

Gan, S., Lau, E., and Ng, H.K. (2009). Remediation of soils contaminated with polycyclic aromatic hydrocarbons (PAHs). *Journal of Hazardous Materials* 172 (2–3): 532–549.

García, M.M. and de Llasera, M.P.G. (2021). A review on the enzymes and metabolites identified by mass spectrometry from bacteria and microalgae involved in the degradation of high molecular weight PAHs. *Science of the Total Environment* 797: 149035.

García-Sánchez, M., Košnář, Z., Mercl, F. et al. (2018). A comparative study to evaluate natural attenuation, mycoaugmentation, phytoremediation, and microbial-assisted phytoremediation strategies for the bioremediation of an aged PAH-polluted soil. *Ecotoxicology and Environmental Safety* 147: 165–174.

Ghosal, D., Ghosh, S., Dutta, T.K., and Ahn, Y. (2016). Current state of knowledge in microbial degradation of polycyclic aromatic hydrocarbons (PAHs): a review. *Frontiers in Microbiology* 7: 1369.

Giri, D.D., Singh, S.K., Giri, A. et al. (2021). Bioremediation potential of methylotrophic bacteria. In: *Microbe Mediated Remediation of Environmental Contaminants* (ed. A. Kumar, V.K. Singh, P. Singh, and V.K. Mishra), 199–207. Cambridge, MA: Woodhead Publishing https://doi.org/10.1016/B978-0-12-821199-1.00017-1.

Gond, D.P., Jha, S.S., Kumar, A., and Singh, S.K. (2021). Plant growth promoting bacteria and its role in green remediation. In: *Sustainable Environmental Clean-Up* (ed. V.K. Mishra and A. Kumar), 149–163. Cambridge, MA: Elsevier https://doi.org/10.1016/B978-0-12-823828-8.00007-4.

Guarino, C., Spada, V., and Sciarrillo, R. (2017). Assessment of three approaches of bioremediation (natural attenuation, landfarming and bioagumentation–assisted landfarming) for a petroleum hydrocarbons contaminated soil. *Chemosphere* 170: 10–16.

Gupta, S. and Pathak, B. (2020). Mycoremediation of polycyclic aromatic hydrocarbons. In: *Abatement of Environmental Pollutants* (ed. A. Borthakur, P. Singh, and A. Kumar), 127–149. Elsevier.

Gupta, A., Singh, M., Pandey, K.D., and Kumar, A. (2018). Biodegradation of xenobiotics using cyanobacteria. *Journal of Scientific Research* 62: 35–40.

Gupta, A., Singh, S.K., Singh, V.K. et al. (2020). Endophytic microbe approaches in bioremediation of organic pollutants. In: *Microbial Endophytes: Prospects of Sustainable Agriculture* (ed. A. Kumar and V.K. Singh), 157–174. Cambridge, MA: Elsevier https://doi.org/10.1016/B978-0-12-818734-0.00007-3.

Gupte, A., Tripathi, A., Patel, H. et al. (2016). Bioremediation of polycyclic aromatic hydrocarbon (PAHs): a perspective. *The Open Biotechnology Journal* 10 (1): 363–378.

Haleyur, N., Shahsavari, E., Jain, S.S. et al. (2019). Influence of bioaugmentation and biostimulation on PAH degradation in aged contaminated soils: response and dynamics of the bacterial community. *Journal of Environmental Management* 238: 49–58.

Haritash, A. and Kaushik, C. (2009). Biodegradation aspects of polycyclic aromatic hydrocarbons (PAHs): a review. *Journal of Hazardous Materials* 169 (1–3): 1–15.

Harry-Asobara, J.L. and Kamei, I. (2019). Growth management of white-rot fungus *Phlebia brevispora* improved degradation of high-molecular-weight polycyclic aromatic hydrocarbons. *3 Biotech* 9 (11): 403.

He, Z., Zhang, R., Chen, S. et al. (2019). FLT1 hypermethylation is involved in polycyclic aromatic hydrocarbons-induced cell transformation. *Environmental Pollution* 252: 607–615.

Huang, Y., Xiao, L., Li, F. et al. (2018). Microbial degradation of pesticide residues and an emphasis on the degradation of cypermethrin and 3-phenoxy benzoic acid: a review. *Molecules* 23 (9): 2313.

Hussain, A., Rehman, F., Rafeeq, H. et al. (2022). In-situ, ex-situ, and nano-remediation strategies to treat polluted soil, water, and air – a review. *Chemosphere* 289: 133252.

Juhasz, A.L. and Naidu, R. (2000). Bioremediation of high molecular weight polycyclic aromatic hydrocarbons: a review of the microbial degradation of benzo[*a*]pyrene. *International Biodeterioration & Biodegradation* 45 (1–2): 57–88.

Kebede, G., Tafese, T., Abda, E.M. et al. (2021). Factors influencing the bacterial bioremediation of hydrocarbon contaminants in the soil: mechanisms and impacts. *Journal of Chemistry* 2021: 1–17.

Kotoky, R. and Pandey, P. (2020). Rhizosphere assisted biodegradation of benzo(a)pyrene by cadmium resistant plant-probiotic *Serratia marcescens* S2I7, and its genomic traits. *Scientific Reports* 10 (1): 5279.

Kotoky, R. and Pandey, P. (2021). The genomic attributes of Cd-resistant, hydrocarbonoclastic *Bacillus subtilis* SR1 for rhizodegradation of benzo(a)pyrene under co-contaminated conditions. *Genomics* 113 (1): 613–623.

Kotoky, R., Rajkumari, J., and Pandey, P. (2018). The rhizosphere microbiome: Significance in rhizoremediation of polyaromatic hydrocarbon contaminated soil. *Journal of Environmental Management* 217: 858–870.

Kumar, A., Verma, H., Singh, V.K. et al. (2017). Role of *Pseudomonas* sp. in sustainable agriculture and disease management. In: *Agriculturally Important Microbes for Sustainable Agriculture: Vol. 2: Applications in Crop Production and Protection*, 195–215. Springer.

Kuppusamy, S., Thavamani, P., Venkateswarlu, K. et al. (2017). Remediation approaches for polycyclic aromatic hydrocarbons (PAHs) contaminated soils: technological constraints, emerging trends and future directions. *Chemosphere* 168: 944–968.

Lawrence, M.J., Stemberger, H.L., Zolderdo, A.J. et al. (2015). The effects of modern war and military activities on biodiversity and the environment. *Environmental Reviews* 23 (4): 443–460.

Lee, B.-K. (2010). Sources, distribution and toxicity of polyaromatic hydrocarbons (PAHs) in particulate matter. In: *Air Pollution* (ed. V. Villanyi). IntechOpen.

Li, X., Qu, C., Bian, Y. et al. (2019). New insights into the responses of soil microorganisms to polycyclic aromatic hydrocarbon stress by combining enzyme activity and sequencing analysis with metabolomics. *Environmental Pollution* 255: 113312.

Liang, C., Huang, Y., Wang, Y. et al. (2019). Distribution of bacterial polycyclic aromatic hydrocarbon (PAH) ring-hydroxylating dioxygenases genes in oilfield soils and mangrove sediments explored by gene-targeted metagenomics. *Applied Microbiology and Biotechnology* 103: 2427–2440.

Lu, C., Hong, Y., Liu, J. et al. (2019). A PAH-degrading bacterial community enriched with contaminated agricultural soil and its utility for microbial bioremediation. *Environmental Pollution* 251: 773–782.

Malik, G., Arora, R., Chaturvedi, R., and Paul, M.S. (2021). Implementation of genetic engineering and novel omics approaches to enhance bioremediation: a focused review. *Bulletin of Environmental Contamination and Toxicology* 108: 1–8.

Malla, M.A., Dubey, A., Yadav, S. et al. (2018). Understanding and designing the strategies for the microbe-mediated remediation of environmental contaminants using omics approaches. *Frontiers in Microbiology* 9: 1132.

Mallah, M.A., Changxing, L., Mallah, M.A. et al. (2022). Polycyclic aromatic hydrocarbon and its effects on human health: an overview. *Chemosphere* 296: 133948.

Marcatili, P., Nielsen, M.W., Sicheritz-Pontén, T. et al. (2016). A novel approach to probe host-pathogen interactions of bovine digital dermatitis, a model of a complex polymicrobial infection. *BMC Genomics* 17: 1–13.

Marris, C., Kompella, S.N., Miller, M. et al. (2020). Polyaromatic hydrocarbons in pollution: a heart-breaking matter. *The Journal of Physiology* 598 (2): 227–247.

Mehta, N. (2019). *Radium Cycling in Groundwater: Implications for Bioremediation and Tracer Studies*. Massachusetts Institute of Technology.

Miglani, R., Parveen, N., Kumar, A. et al. (2022). Degradation of xenobiotic pollutants: an environmentally sustainable approach. *Metabolites* 12 (9): 818.

Mishra, M., Singh, S.K., and Kumar, A. (2021). Environmental factors affecting the bioremediation potential of microbes. In: *Woodhead Publishing Series in Food Science, Technology and Nutrition, Microbe Mediated Remediation of Environmental Contaminants* (ed. A. Kumar, V.K. Singh, P. Singh, and V.K. Mishra), 47–58. Cambridge, MA: Elsevier, Woodhead Publishing https://doi.org/10.1016/B978-0-12-821199-1.00005-5.

Miura, K., Shimada, K., Sugiyama, T. et al. (2019). Seasonal and annual changes in PAH concentrations in a remote site in the Pacific Ocean. *Scientific Reports* 9 (1): 12591.

Mojiri, A., Zhou, J.L., Ohashi, A. et al. (2019). Comprehensive review of polycyclic aromatic hydrocarbons in water sources, their effects and treatments. *Science of the Total Environment* 696: 133971.

Moka, S., Pande, M., Rani, M. et al. (2014). Alternative fuels: an overview of current trends and scope for future. *Renewable and Sustainable Energy Reviews* 32: 697–712.

Mrozik, A., Piotrowska-Seget, Z., and Labuzek, S. (2003). Bacterial degradation and bioremediation of polycyclic aromatic hydrocarbons. *Polish Journal of Environmental Studies* 12 (1): 15–25.

Mundo, R., Matsunaka, T., Iwai, H. et al. (2022). Environmental processes and fate of PAHs at a shallow and enclosed bay: West Nanao Bay, Noto Peninsula, Japan. *Marine Pollution Bulletin* 184: 114105.

Nayal, F.S., Mammadov, A., and Ciliz, N. (2016). Environmental assessment of energy generation from agricultural and farm waste through anaerobic digestion. *Journal of Environmental Management* 184: 389–399.

Nwankwegu, A.S., Zhang, L., Xie, D. et al. (2022). Bioaugmentation as a green technology for hydrocarbon pollution remediation. Problems and prospects. *Journal of Environmental Management* 304: 114313.

Ojha, N., Mandal, S.K., and Das, N. (2019). Enhanced degradation of indeno (1, 2, 3-cd) pyrene using *Candida tropicalis* NN4 in presence of iron nanoparticles and produced biosurfactant: a statistical approach. *3 Biotech* 9: 1–13.

Okere, U. and Semple, K. (2012). Biodegradation of PAHs in 'pristine'soils from different climatic regions. *Journal of Bioremediation Biodegradation* 1 (2): 1–11.

Pandey, P., Kapley, A., and Brar, S.K. (2021). Biodegradation of high molecular weight polyaromatic hydrocarbons in different environments. *Frontiers in Microbiology* 2079.

Pandey, K.D., Singh, S.K., Shukla, L. et al. (2023). Mechanistic approaches and factors regulating microalgae mediated heavy metal remediation from the aquatic ecosystem. In: *Advances in Chemical Pollution, Environmental Management and Protection*, vol. 9 (ed. L.F.R. Ferreira, A. Kumar, and M. Bilal), 285–299. Cambridge, MA: Elsevier, Academic Press https://doi.org/10.1016/bs.apmp.2022.10.011.

Patel, A.B., Shaikh, S., Jain, K.R. et al. (2020). Polycyclic aromatic hydrocarbons: sources, toxicity, and remediation approaches. *Frontiers in Microbiology* 11: 562813.

Peluffo, M., Rosso, J.A., Morelli, I.S., and Mora, V.C. (2018). Strategies for oxidation of PAHs in aged contaminated soil by batch reactors. *Ecotoxicology and Environmental Safety* 151: 76–82.

Pozdnyakova, N.N. (2012). Involvement of the ligninolytic system of white-rot and litter-decomposing fungi in the degradation of polycyclic aromatic hydrocarbons. *Biotechnology Research International* 2012: 243217.

Pozdnyakova, N., Muratova, A., and Turkovskaya, O. (2022). Degradation of polycyclic aromatic hydrocarbons by co-culture of *Pleurotus ostreatus* Florida and *Azospirillum brasilense*. *Applied Microbiology* 2 (4): 735–748.

Rathankumar, A.K., Saikia, K., Ramachandran, K. et al. (2020). Effect of soil organic matter (SOM) on the degradation of polycyclic aromatic hydrocarbons using *Pleurotus dryinus* IBB 903-A microcosm study. *Journal of Environmental Management* 260: 110153.

Redfern, L.K., Gardner, C.M., Hodzic, E. et al. (2019). A new framework for approaching precision bioremediation of PAH contaminated soils. *Journal of Hazardous Materials* 378: 120859.

Sakshi and Haritash, A. (2020). A comprehensive review of metabolic and genomic aspects of PAH-degradation. *Archives of Microbiology* 202 (8): 2033–2058.

Sarkar, J., Roy, A., Sar, P., and Kazy, S.K. (2020). Accelerated bioremediation of petroleum refinery sludge through biostimulation and bioaugmentation of native microbiome. In: *Emerging Technologies in Environmental Bioremediation* (ed. M. Shah, S. Rodriguez-Couto, and S. Sengor), 23–65. Elsevier.

Sharma, P., Bano, A., Singh, S.P. et al. (2022). Engineered microbes as effective tools for the remediation of polyaromatic aromatic hydrocarbons and heavy metals. *Chemosphere* 306: 135538.

Singh, S.K., Singh, P.P., Gupta, A. et al. (2019). Tolerance of heavy metal toxicity using PGPR strains of *Pseudomonas* species. In: *PGPR Amelioration in Sustainable Agriculture* (ed. A.K. Singh, A. Kumar, and A.K. Singh), 239–252. Woodhead Publishing.

Singh, D., Singh, S.K., Singh, V.K. et al. (2020). Plant growth–promoting bacteria and their role in environmental management. In: *Abat. Environ. Pollut* (ed. P. Singh, A. Kumar, and A. Borthakur), 161–175. Cambridge, MA: Elsevier https://doi.org/10.1016/B978-0-12-818095-2.00008-4.

Singh, S.K., Singh, M.K., Singh, V.K. et al. (2021). Microbial enzymes and their exploitation in remediation of environmental contaminants. In: *Microbe Mediated Remediation of Environmental Contaminants* (ed. A. Kumar, V.K. Singh, P. Singh, and V.K. Mishra), 59–71. Cambridge, MA: Woodhead Publishing https://doi.org/10.1016/B978-0-12-821199-1.00006-7.

Singha, L.P. and Pandey, P. (2021). Rhizosphere assisted bioengineering approaches for the mitigation of petroleum hydrocarbons contamination in soil. *Critical Reviews in Biotechnology* 41 (5): 749–766.

Srivastava, J., Naraian, R., Kalra, S., and Chandra, H. (2014). Advances in microbial bioremediation and the factors influencing the process. *International journal of Environmental Science and Technology* 11: 1787–1800.

Storey, S., Ashaari, M.M., Clipson, N. et al. (2018). Opportunistic bacteria dominate the soil microbiome response to phenanthrene in a microcosm-based study. *Frontiers in Microbiology* 9: 2815.

Subashchandrabose, S.R., Venkateswarlu, K., Venkidusamy, K. et al. (2019). Bioremediation of soil long-term contaminated with PAHs by algal–bacterial synergy of Chlorella sp. MM3 and *Rhodococcus wratislaviensis* strain 9 in slurry phase. *Science of the Total Environment* 659: 724–731.

Suman, A., Yadav, A.N., and Verma, P. (2016). Endophytic microbes in crops: diversity and beneficial impact for sustainable agriculture. In: *Microbial Inoculants in Sustainable Agricultural Productivity: Vol. 1: Research Perspectives*, 117–143. Springer.

SureshKumar, P., Thomas, J., and Poornima, V. (2018). Structural insights on bioremediation of polycyclic aromatic hydrocarbons using microalgae: a modelling-based computational study. *Environmental Monitoring and Assessment* 190: 1–16.

Thacharodi, A., Hassan, S., Singh, T. et al. (2023). Bioremediation of polycyclic aromatic hydrocarbons: an updated microbiological review. *Chemosphere* 328: 138498.

Tomar, R.S. and Jajoo, A. (2021). Enzymatic pathway involved in the degradation of fluoranthene by microalgae *Chlorella vulgaris*. *Ecotoxicology* 30 (2): 268–276.

Udom, G.J., Frazzoli, C., Ekhator, O.C. et al. (2023). Pervasiveness, bioaccumulation and subduing environmental health challenges posed by polycyclic aromatic hydrocarbons (PAHs): a systematic review to settle a one health strategy in Niger Delta, Nigeria. *Environmental Research* 226: 115620.

Wang, W., Wang, L., and Shao, Z. (2018). Polycyclic aromatic hydrocarbon (PAH) degradation pathways of the obligate marine PAH degrader *Cycloclasticus* sp. strain P1. *Applied and Environmental Microbiology* 84 (21): e01261–e01218.

Wang, Q., Hou, J., Yuan, J. et al. (2020). Evaluation of fatty acid derivatives in the remediation of aged PAH-contaminated soil and microbial community and degradation gene response. *Chemosphere* 248: 125983.

Weber, W.J., Young, T.M., and Hillers, A. (2017). Microscale heterogeneities in soil properties and their effects on contaminant sorption and bioavailability. In: *Fundamentals and Applications of Bioremediation* (ed. S.K. Sikdar), 221–258. Routledge.

Wu, Y., Xu, Y., and Zhou, N. (2020). A newly defined dioxygenase system from *Mycobacterium vanbaalenii* PYR-1 endowed with an enhanced activity of dihydroxylation of high-molecular-weight polyaromatic hydrocarbons. *Frontiers of Environmental Science & Engineering* 14: 1–11.

Zafra, G. and Valdivieso Quintero, W. (2016). Use of molecular tools to monitor microbial communities during the bioremediation of polycyclic aromatic hydrocarbon-contaminated soils. *Journal of Industrial Pollution Control* 32: 534–543.

Zafra-Rojas, Q.Y., Cruz-Cansino, N.S., Quintero-Lira, A. et al. (2016). Application of ultrasound in a closed system: optimum condition for antioxidants extraction of blackberry (*Rubus fructicosus*) residues. *Molecules* 21 (7): 950.

Zaki, M.S., Fawzi, O.M., and Abd El-Zaher, M.F. (2013). Bioremediation of contaminants. *Life Science Journal* 10 (1): 3329–3332.

Zhang, K., Hu, Z., Zeng, F. et al. (2019). Biodegradation of petroleum hydrocarbons and changes in microbial community structure in sediment under nitrate-, ferric-, sulfate-reducing and methanogenic conditions. *Journal of Environmental Management* 249: 109425.

Zhang, S., Hu, Z., and Wang, H. (2019). Metagenomic analysis exhibited the co-metabolism of polycyclic aromatic hydrocarbons by bacterial community from estuarine sediment. *Environment International* 129: 308–319.

Zhang, H., Zhang, X., and Geng, A. (2020). Expression of a novel manganese peroxidase from Cerrena unicolor BBP6 in *Pichia pastoris* and its application in dye decolorization and PAH degradation. *Biochemical Engineering Journal* 153: 107402.

Zhou, Y., Stuart-Williams, H., Farquhar, G.D., and Hocart, C.H. (2010). The use of natural abundance stable isotopic ratios to indicate the presence of oxygen-containing chemical linkages between cellulose and lignin in plant cell walls. *Phytochemistry* 71 (8–9): 982–993.

# 25

## Microbial Indicators for Monitoring Pollution and Bioremediation

*Vijay Kumar[1], Ashok Chhetri[1], Joy Kumar Dey[2], and Ashim Debnath[3]*

[1] *Department of Plant Protection & Department of Horticulture, Multi-Technology Testing Centre & Vocational Training Centre, Central Agricultural University (Imphal Manipur), Agartala, Tripura, India*
[2] *Krishi Vigyan Kendra-Sepahijala, CAU (I), Latiacherra, Tripura, India*
[3] *Department of Genetics and Plant Breeding, Faculty of Agricultural Sciences, Rajiv Gandhi University, Doimukh, Arunachal Pradesh, India*

## 25.1 Introduction

A significant setback of our civilization's industrial and technical advancement is pollution. The environment has been neglected for many years in the pursuit of more wealth. The results are still concerning, and the chosen techniques are still ineffective, because of the lack of strict guidelines and rules for management and restriction of the production of polluting agents into the different ecosystems. Decision-makers and scientists are looking for more sustainable solutions as they become more aware of the sensitivity of the current pollution situation. Traditional methods for cleaning up environmental toxins typically use physical, chemical, and biological techniques (Khalid et al. 2017) to reduce contamination to a level that is not harmful to the ecosystem. Experts are adamant that an effective cleanup strategy should identify the contaminant and encourage the use of sufficient cleaning equipment. Microorganisms have evolved methods to break down the majority of the chemical and biological components that have emerged throughout the course of development in order to survive. There are different bioremediation methods which are employed to convert/decompose the toxic pollutants into harmless or beneficial products (Gond et al. 2021; Mishra et al. 2021; Giri et al. 2021). According to Kumar et al. (2019), Kulkarni et al. (2012), and Baez-Rogelio et al. (2017), the agriculture, textile, medical, all forms of oil or chemical industries, and others are using bioremediation to address pollution issues.

By utilizing plants, algae, fungi, and bacteria, several solutions were implemented to address various pollution scenarios (Gupta et al. 2020; Bhadouria et al. 2020; Bhattacharya et al. 2015; Yin et al. 2019). However, according to Quintella et al. (2019) and de Lima et al. (2018) some contaminants may be detected by specific biological markers, each of which is used to pinpoint a specific pollutant or group of pollutants. The use of microbial flora, higher plants, and animals to detect environmental pollutants is discussed in the current chapter. Here are the most important biological signs. The mentioned plant indicators include higher plants, while enzymes, bacteria, fungi, algae, lichens, and planktons are microbial origin indicators and animal poisons, toads, insects, earthworms, macroinvertebrates, and frogs are animal origin indicators.

According to Holt and Miller (2011) material which have the potential of degrading human health, reducing beneficial microbial flora, increase harmful microbial flora, and interfere with the normal ecosystem functioning fall under the pollution category. Few biologically active compounds may not have harmful effects on living things at some concentrations, but many of them may serve as pollutants (Butterworth et al. 2001). Polluting agents can be categorized by composition their physical condition, their qualities (such as, biodegradability, reactivity, and solubility), their sectors (such as air, aquatic, or terrestrial), or their sources (such as residential, industrial, or industrial combustion). Nearly all of these classifications are arbitrary, but the ones that take into account a pollutant's biological, chemical, and physical characteristics are the most common. On the other hand, classifying pollutants according to their functions is connected to the state of our understanding of biology and physiology. Generally speaking, the method of classification that is used clearly depends on the purpose for which the information is being gathered.

*Microbes Based Approaches for the Management of Hazardous Contaminants*, First Edition. Edited by Ajay Kumar, Livleen Shukla, Joginder Singh, and Luiz Fernando R. Ferreira.

### 25.1.1 Microbial Indicators for Air Pollution

Recently, networks for tracking air pollution have been set up in many nations. These networks typically use physicochemical techniques for analysis. Biological indicator systems are frequently employed to assess the quality of the air. According to the Arndt et al. (1985) some highly standardized techniques involve using lichens, tobacco, or grass cultures that collect heavy metals, which got reacted with particles of air pollution. A variety of green plant species and lichens are frequently employed in conjunction to identify the impacts of different air contaminants. These bio-indicators, which are living creatures, respond to a combination of airborne toxins, and as a result, they may provide useful information more pertinent to human health than those offered by just physicochemical measurements of polluting agents of the air. Most commonly used microbial agents are bacterial origin organisms for the testing of efficacy of different antimicrobial drugs. Ames's test developed by the Ames et al. (1975), shows that bacteria are also highly useful indicators of DNA-damaging chemicals.

### 25.1.2 Microbial Indicators for Terrestrial and Aquatic Pollution

Ecosystems on land and in water are complex heterogeneous biocolloidal environments where biological, chemical, and physical events are likely to interact, to create a dynamic equilibrium. The ecological sensitivity of a specific contaminant under diverse conditions directly controls its fate. Since the ecosystem serves as the primary repository for most pollutants, the ecological degradation undoubtedly primarily caused by the pollution. Pollution is undoubtedly one of the most important characteristics of ecological degradation (Jain et al. 2010). Pollution is typically a result of unhygienic habits, agricultural practises, improper waste management, and fallout from the atmosphere. To reduce the chances of harming the ecosystem, the dumping of waste into the ecosystem, such as terrestrial and aquatic, should be avoided unless it has been chemically and biologically analyzed.

According to Hoballah et al. (2014, 2015) as well as Saber et al. (2015), the majority of wastes contain significant amounts of disease-causing agents, toxic/poisonous substances, and polluting agents. Guideline should be formulated for reducing and a managing the pollution of terrestrial and aquatic ecosystem based on toxicity of the polluting agents. Ecological stability and resilience have been used to define ecological health in response to stress or disturbance. However, the strategies and the amount of time needed for the remediation could act as indicators of the health of an ecosystem (Khatri and Tyagi 2015). On a modern ecosystem paradigm, a thorough understanding of all the available microbial, animal, and higher plants indicators should be supported.

### 25.1.3 Microbial Indicators

Because of their rapid development, microorganisms are frequently utilized in assessing contamination in marine and coastal ecosystems (Gerhardt 2002). These organisms always respond to pollution, despite very low pollutant concentration and show clear evidence of ecosystem alterations. Microbial indicators are abundant in their occurrence in environment and easy to detect and access (Khatri and Tyagi 2015). For the selection of microbiological indicators, there are six distinct, clearly defined criteria. Microbial biotas are accurately and consistently leaving a variety of environments. Microbial toxins, especially in aquatic habitats, are simple to measure and can be used as a sign of contamination.

Despite the fact that Butterworth et al. (2001) indicated that microbiota is particularly easy to track in comparison to other conventional research, their regulation may merely result from the positive effects of toxins on the communities in which they reside. The microbial consortium is highly capable of adjusting the activity levels, biomass, and group makeup to control contaminants in the ecosystem. These factors are crucial in determining how well-developed a certain ecosystem is. Numerous applications for bioluminescent bacteria can be found in the identification of ecosystem poisons as microbial indicators.

### 25.1.4 Bio-Indicators of Bacterial Origin

The bacterial populations or species whose existence in an ecosystem, above a specific threshold, signals exposure to contaminants. The importance of bacteria in analyzing pollution issues in various ecosystems cannot be denied (Kalkan and Altuğ 2015). However, the relationship between the polluting agents and microbial flora raises concerns about the efficacious use and diagnostic accuracy (Sumampouw and Risjani 2014). The estimation of total bacterial counts is a crucial bacterial biological indicator, notwithstanding the substantial experimental error.

However, it should be remembered that an exact total bacterial count is almost never acquired because it is difficult to identify a single culture medium in which all bacteria in a given environment might successfully establish their colonies. The individual bacterial counts are measured by using the specific growing medium that is most appropriate in each situation. *Salmonella typhimurium* and *Clostridium* sp. counts among anaerobic mesophilic bacteria, as well as their abundance in a particular ecosystem, may be used as biological indicators for the presence of aquatic logging. Coliforms, which are a diverse group of facultative aerobic and anaerobic bacteria that can digest lactose with gas production in less than 48 hours at 37°C, are frequently utilized as biological indicators.

The coliforms known as fecal coliforms are those that can thrive on an average temperature of 44–45°C and help in the fermentation of lactose. They primarily consist of *E. coli* types I and II in high concentrations and are therefore helpful as biological indicators for contamination with a recent fecal origin (Saber et al. 2015). Fecal coliforms, which also comprise of bacterial species normally present on plants and in soil, are more effective as biological markers than total coliforms in recreational aquatic habitats, such as *Klebsiella pneumoniae*, although classified as a fecal coliform, is not of fecal origin.

In most ecosystems, bacteria are always present in large numbers and are simpler to sample than other biotas. The bacterial indicators are only important when cost is taken into consideration, and the current bacterial indicators do not meet these essential qualifications. Some bacteria generate novel stress proteins in response to exposure to specific pollutants, such cadmium or benzene, which may be utilized in advanced alerting system to tackle specific ecosystem polluting agents. Since a bacterial pollution indicator must not be a pathogen, the coliform bacteria must not last for an extended amount of time in the ecosystem after being ejected from the gut and their occurrence must be closely associated to polluted fecal organisms.

*Streptococcus bovis* and specific *Clostridia* spp. are examples of non-coliform bacteria that may also be employed as a measure of fecal contamination in terrestrial and aquatic habitats. Microbial functions such as respiration, carbon and nitrogen mineralization, fixation of nitrogen, and enzymes may be used to classify the microbial biomass. According to Aslam et al. (2012), some microbial behaviours, such biomass-specific breathing, appear to be more sensitive than their single action or population rate. Additionally, there may be some "global regulation" indicators, such as the biomass to organic matter ratio. Such techniques could be developed to examine whether poisons alter the natural ecosystem without using expensive and time-consuming field ways.

### 25.1.5 Fungal Indicators

The ecosystem, such as aquatic and terrestrial, the faunal communities are widely spread and play an essential role in their operations. Fungi such as *Candida albicans, Aspergillus niger, Aspergillus fumigatus, Trichoderma* species, *Stachybotrys* species, *Aspergillus versicolor*, *Exophiala* species, *Phialophora* species, *Fusarium* species, *Penicillium* species, *Ulocladium* species, and few yeasts are widely utilized fungal bio-indicators for detecting polluting agents. On the basis of their resistance to specific ecological fluctuation, many fungi indicator species are used to efficiently monitor ecosystem occurrences. Hasselbach et al. (2005) fungal indicators for potential toxic elements (PTEs) were a moss known as *Hylocomium splendens* which is found in a rare habitat of Northwest Alaska ecosystem. The results after the examination of moss tissue shows the PTE level high in moss growing near to roadside.

### 25.1.6 Algal Indicators

The concept of algal bio-indicators, given by the Hosmani (2013), found that agal species such as *Chlorella* sp. and *Euglena* sp. were utilized as an indicator for aquatic ecosystem pollutant. Most often, only one or two types of algae would be the most prevalent in polluted aquatic habitats (Pandey et al. 2023). According to Markert and Breure (2003) and Zancan et al. (2006) the polluted aquatic ecosystem has abundant diatom as a pollutant. There is insufficient amount of knowledge about the role of algae in terrestrial, aquas ecosystem as a pollutant indicator.

The difficulties of studying algae's presence in terrestrial ecosystems is one reason for the dearth of data on them, but it is anticipated that many algae could persist in them, especially in their resting stages (Singh et al. 2013). Under polluted conditions, both the pH value and the amount of nutrients in terrestrial ecosystems are subject to significant change. Decrease in nutrient level is indicative of low pH in the habitat. Blue-green algae, according to Markert and Breure (2003), could be utilized as biological indicators to detect pH changes in many environments, but their growth is slow when pH is under 5. According to Jain et al. (2010), a decline in the health of the marine ecosystem is always shown by a rise Population of *Phacus tortus*, *Euglena clastica,* and *Trachelon anas*. Algal blooms in water bodies are frequently utilized to show significant improvement in nitrate and phosphate levels.

The prevalence of PTEs is one of the main issues that is constantly associated with ecosystem pollution (Zaghloul 2014). Various methods have been used frequently to use algae as reliable biological markers of PTEs, but none of these methods had been refined to the point where they could be applied consistently. Algae have been used as a biological indicator in several PTE pollution. The challenging information may ultimately turn out to be simple, but much more knowledge is still required on the efficacy of beneficial algae which include *Nitella* spp., *Lemanea*, *Enteromorpha,* and *Cladophora* spp. against the PTEs. Microalgae have recently been demonstrated to be susceptible to contaminants in many habitats. The most well-known microalga, *Euglena gracilis*, is a flagellated, motile, freshwater, photosynthetic microalga that is vulnerable to pollution, acidity, and stressors, especially those brought in by PTEs and persistent organic pollutants. This microalga exhibits distinctive reactions by restricting mobility and altering orientation characteristics. It is simple to handle and effective in ecotoxicological evaluations to employ *E. gracilis*. Its gravitational orientation is particularly susceptible to pollution.

### 25.1.7 Microbial Indicators for Bioremediation

Traditional pathogen detection methods have relied on microbial markers, however today three types are recognized (Ashbolt 2001). As their name implies, microbial indicators cannot generate a quantitative measurement or remove the contaminants; they can only reveal the existence of pollutants (Singh et al. 2020). Recent developments in systems and synthetic biology have paved the way for fresh, creative methods to address this pressing issue by bringing in more sensitive, focused, and environmentally friendly solutions (de Lorenzo 2008; Nunes-Alves 2015). The two modules of these microbiological tools are detection and remediation. The efficient and precise detection of pollutants is the foundation of the first module.

The ability to identify molecules at extremely low concentrations has been demonstrated in a growing number of scientific studies (Trabelsi et al. 2018; Lin et al. 2018; Zhang et al. 2017; Fernandez-López et al. 2015). The first module's generic name is a "microbial indicator," also known as a "biosensor" in layman's terms. All the components necessary to remove the pollutants or lower their concentrations below the permitted limit are included in the second module. The majority of these components are bio-bricks that encode intricate biodegradation processes. To achieve the aim, a series of biological events are typically required. This chapter will describe the creation of such biosensors and their uses in remediation, detecting, and monitoring. Some cases will be used as examples to help readers gain a better understanding of these important tools.

## 25.2 Biosensors for Microbial Remediation

### 25.2.1 Biosensors Based on Nucleic Acid

Nucleic acid sequences, i.e., DNA or RNA with structural conformations that permit binding affinity to a target ligand are used to construct nucleic acid-based biosensors. The structure of the nucleic acid sequence is changed by the ligands binding to it. An output signal could be produced using this structural modification. This family of biosensors includes aptamers and riboswitches. Single-stranded DNA or RNA molecules that exhibit affinity for a target ligand are known as aptamers (Ellington and Szostak 1992; Tuerk and Gold 1990; Aghdam et al. 2016). The aptamer undergoes a conformational shift upon binding to the target ligand. According to Song et al. (2008), a signal could be produced using this modification of the secondary or tertiary structures (Song et al. 2008). Single-stranded RNA molecules called riboswitches have two connected domains. Following ligand interaction, the second domain produces a signal (Findeiß et al. 2017). The first domain binds to the target ligand.

### 25.2.2 Transcription Factor-Based Biosensors

The culmination of several years' worth of combined study on protein expression regulation and transcription led to the development of biosensors based on the behaviour. Interaction of transcription factors and target ligand factors to produce a specific transcription response. New biosensors can be developed from the naturally occurring transcription factors because they are all reasonably simple to engineer (Fernandez-López et al. 2015). The biosensors are a genetic circuit that uses a promoter-operator pair, a transcriptional regulator, and a signal output for the intended test. Heterologous components could be put together because the system is modular. The three key steps in the validation process are regulating the specificity of the binding ligand and optimizing the dose–response and biosensor behaviour prediction.

Scientists have developed de novo transcription factors that can bind additional ligands using protein engineering techniques in order to increase the limited number of transcription factor promoter pairs that are available (Snoek et al. 2020; Machado et al. 2019). Sensipath, a different computational method, was also applied to broaden the scope of molecule detection. By including one or two steps of catalytic reactions, a nondetectable chemical could be changed into a detectable one (Delépine et al. 2016). It could be difficult to employ transcription factors from various species. In fact, when the same system is implanted into an entirely different organism, a response that occurs in one species may be wholly different.

### 25.2.3 Biosensors Based on Transcription-Independent Protein

This group of biosensors is quite diverse and provides a wide range of output reactions and detecting techniques. Typically, a response domain and a receptor domain for a target ligand are produced as a fusion protein. A conformational change caused by the ligand's binding to the receptor domain is conveyed to the response domain and results in the production of a signal (Nagai et al. 2001; Guntas et al. 2004). Successful design requires a thorough understanding of the structural details of both the ligand-binding domain and the response domain. However, the design could be quite difficult if the necessary information is unavailable.

## References


Aghdam, E.M., Hejazi, M.S., and Barzegar, A. (2016). Riboswitches: from living biosensors to novel targets of antibiotics. *Gene* 592 (2): 244–259. https://doi.org/10.1016/j.gene.2016.07.035.

Ames, B.N., McCann, J., and Yamasaki, E. (1975). Methods for detecting carcinogens and mutagens with the *Salmonella/* mammalian-microsome mutagenicity test. *Mutatation Research (Netherlands)* 31: 347–364.

Arndt, U., Erhardt, W., Keitel, A. et al. (1985). Standardisierte Exposition von pflanzlichen Reaktionsindkatoren. *Staub - Reinhaltung der Luft* 45 (10): 481–483.

Ashbolt, N.J. (2001). Indicators of microbial water quality. In: *Water Quality: Guidelines, Standards and Health* (ed. L. Fewtrell and J. Bartram). IWA Publishing.

Aslam, M., Verma, D.K., Dhakerya, R. et al. (2012). Bioindicator: a comparative study on uptake and accumulation of heavy metals. *Research Journal of Environmental and Earth Sciences* 4 (12): 1060–1070.

Baez-Rogelio, A., Morales-García, Y.E., Quintero-Hernández, V., and Muñoz-Rojas, J. (2017). Next generation of microbial inoculants for agriculture and bioremediation. *Microbial Biotechnology* 10 (1): 19–21. https://doi.org/10.1111/1751-7915.12448.

Bhadouria, R., Das, S., Kumar, A. et al. (2020). Mycoremediation of agrochemicals. In: *Agrochemicals Detection, Treatment and Remediation* (ed. M. Narasimha Vara Prasad), 593–620. Oxford, United Kingdom: Butterworth-Heinemann https://doi.org/10.1016/B978-0-08-103017-2.00022-2.

Bhattacharya, P., Chakraborty, N., and Pal, R. (2015). Bioremediation of toxic metals using algae. In: *Algal Biorefinery: An Integrated Approach* (ed. D. Das), 439–462. Springer International Publication https://doi.org/10.1007/978-3-319-22813-6_19.

Butterworth, F., Gunatilaka, A., and Gonsebatt, M. (2001). Biomonitors and biomarkers of environmental change. In: *Presentation/Presentación and Abstracts. Revista Internacional de Contaminación Ambiental, 17(1)*, vol. II (ed. F. Butterworth, A. Gunatilaka, and M. Gonsebatt), 53–63. New York: Kluwer Academic/Plenum Publishing.

de Lima, D.P., dos Santos, E.D.A., Marques, M.R. et al. (2018). Fungal bioremediation of pollutant aromatic amines. *Current Opinion in Green and Sustainable Chemistry* 11: 34–44. https://doi.org/10.1016/j.cogsc.2018.03.012.

de Lorenzo, V. (2008). Systems biology approaches to bioremediation. *Current Opinion in Biotechnology* 19 (6): 579–589.

Delépine, B., Libis, V., Carbonell, P., and Faulon, J.L. (2016). SensiPath: computer-aided design of sensing-enabling metabolic pathways. *Nucleic Acids Research* 44 (W1): W226–W231. https://doi.org/10.1093/nar/gkw305.

Ellington, A.D. and Szostak, J.W. (1992). Selection in vitro of single-stranded DNA molecules that fold into specific ligand-binding structures. *Nature* 355 (6363): 850–852. https://doi.org/10.1038/355850a0.

Fernandez-López, R., Ruiz, R., de la Cruz, F., and Moncalián, G. (2015). Transcription factor-based biosensors enlightened by the analyte. *Frontiers in Microbiology* 6: 648. https://doi.org/10.3389/fmicb.2015.00648.

Findeiß, S., Etzel, M., Will, S. et al. (2017). Design of artificial riboswitches as biosensors. *Sensors* 17 (9): 1990. https://doi.org/10.3390/s17091990.

Gerhardt, A. (2002). Bioindicator species and their use in biomonitoring. *Environmental Monitoring* 1: 77–123.

Giri, D.D., Singh, S.K., Giri, A. et al. (2021). Bioremediation potential of methylotrophic bacteria. In: *Microbe Mediated Remediation of Environmental Contaminants* (ed. A. Kumar, V.K. Singh, P. Singh, and V.K. Mishra), 199–207. Cambridge, MA: Woodhead Publishing https://doi.org/10.1016/B978-0-12-821199-1.00017-1.

Gond, D.P., Jha, S.S., Kumar, A., and Singh, S.K. (2021). Plant growth promoting bacteria and its role in green remediation. In: *Sustainable Environmental Clean-up* (ed. V.K. Mishra and A. Kumar), 149–163. Cambridge, MA: Elsevier https://doi.org/10.1016/B978-0-12-823828-8.00007-4.

Guntas, G., Mitchell, S.F., and Ostermeier, M. (2004). A molecular switch created by in vitro recombination of nonhomologous genes. *Chemistry & Biology* 11 (11): 1483–1487. https://doi.org/10.1016/j.chembiol.2004.08.020.

Gupta, A., Singh, S.K., Singh, V.K. et al. (2020). Endophytic microbe approaches in bioremediation of organic pollutants. In: *Microbial Endophytes: Prospects of Sustainable Agriculture* (ed. A. Kumar and V.K. Singh), 157–174. Cambridge, MA: Elsevier https://doi.org/10.1016/B978-0-12-818734-0.00007-3.

Hasselbach, L., Ver Hoef, J.M., Ford, J. et al. (2005). Spatial patterns of cadmium and lead deposition on and adjacent to National Park Service lands in the vicinity of Red Dog Mine, Alaska. *Science of the Total Environment* 348 (1–3): 211–230.

Hoballah, E., Saber, M., Matter, I., and Zaghloul, A. (2014). Bioremediation of polychlorinated biphenyl (PCBs) in a sewaged soil by certain remediative amendments followed by phytoremediation. *Research Journal of Pharmaceutical, Biological and Chemical Sciences* 5 (2): 91–103.

Hoballah, E., Saber, M., Matter, I., and Zaghloul, A. (2015). Bioremediation of polyaromatic hydrocarbons (PAHs) in a sewaged soil by certain remediative amendments followed by phytoremediation. In: *8th Ann Inter Symp on Agric Res*, 13–16.

Holt, E.A. and Miller, S.W. (2011). Bioindicators: using organisms to measure. *Nature* 3: 8–13.

Hosmani, S.P. (2013). Fresh aquatic algae as indicators of aquatic quality. *Universal Journal of Environmental Research and Technology* 3: 473–482.

Jain, A., Singh, B.N., Sl, S. et al. (2010). Exploring biodiversity as biological indicators for aquatic pollution. In: *National Conf. on Biodiversity, Development and Poverty Alleviation* (ed. A. Jain and B.N. Singh). Lucknow, Uttar Pradesh (India): Uttar Pradesh State Biodiversity Board.

Kalkan, S. and Altuğ, G. (2015). Bio-indicator bacteria & environmental variables of the coastal zones: the example of the Güllük Bay, Aegean Sea, Turkey. *Marine Pollution Bulletin* 95 (1): 380–384.

Khalid, S., Shahid, M., Niazi, N.K. et al. (2017). A comparison of technologies for remediation of heavy metal contaminated soils. *Journal of Geochemical Exploration* 182: 247–268. https://doi.org/10.1016/J.GEXPLO.2016.11.021.

Khatri, N. and Tyagi, S. (2015). Influences of natural and anthropogenic factors on surface and groundwater quality in rural and urban areas. *Frontiers in Life Science* 8 (1): 23–39.

Kulkarni, S.V., Palande, A.S., and Deshpande, M.V. (2012). Bioremediation of petroleum hydrocarbons in soils. In: *Microorganisms in Environmental Management* (ed. T. Satyanarayana and B.N. Johri), 589–606. https://doi.org/10.1007/978-94-007-2229-3_26.

Kumar, M., Jaiswal, S., Sodhi, K.K. et al. (2019). Antibiotics bioremediation: perspectives on its ecotoxicity and resistance. *Environment International* 124: 448–461. https://doi.org/10.1016/j.envint.2018.12.065.

Lin, C., Jair, Y.C., Chou, Y.C. et al. (2018). Transcription factor-based biosensor for detection of phenylalanine and tyrosine in urine for diagnosis of phenylketonuria. *Analytica Chimica Acta* 1041: 108–113. https://doi.org/10.1016/j.aca.2018.08.053.

Machado, F.M.L., Currin, A., and Dixon, N. (2019). Directed evolution of the PcaV allosteric transcription factor to generate a biosensor for aromatic aldehydes. *Journal of Biological Engineering* 13 (1): 1–15. https://doi.org/10.1101/689232.

Markert, B.A. and Breure, A.M. (2003). Bryophytes. In: *Bioindicators and Biomonitors* (ed. H.G. Zechmeister, K. Grodzińska, and G. Szarek-Łukaszewska), 329. Elsevier.

Mishra, M., Singh, S.K., and Kumar, A. (2021). Environmental factors affecting the bioremediation potential of microbes. In: *Woodhead Publishing Series in Food Science, Technology and Nutrition, Microbe Mediated Remediation of Environmental Contaminants* (ed. A. Kumar, V.K. Singh, P. Singh, and V.K. Mishra), 47–58. Cambridge, MA: Elsevier, Woodhead Publishing https://doi.org/10.1016/B978-0-12-821199-1.00005-5.

Nagai, T., Sawano, A., Park, E.S., and Miyawaki, A. (2001). Circularly permuted green fluorescent proteins engineered to sense $Ca^{2+}$. *Proceedings of the National Academy of Sciences of the United States of America* 98 (6): 3197–3202. https://doi.org/10.1073/pnas.051636098.

Nunes-Alves, C. (2015). GMOs in lockdown. *Nature Reviews Microbiology* 13 (3): 125–125. https://doi.org/10.1038/nrmicro3443.

Pandey, K.D., Singh, S.K., Shukla, L. et al. (2023). Mechanistic approaches and factors regulating microalgae mediated heavy metal remediation from the aquatic ecosystem. In: *Advances in Chemical Pollution, Environmental Management and Protection*, vol. 9 (ed. L.F.R. Ferreira, A. Kumar, and M. Bilal), 285–299. Cambridge, MA: Elsevier, Academic Press https://doi.org/10.1016/bs.apmp.2022.10.011.

Quintella, C.M., Mata, A.M., and Lima, L.C. (2019). Overview of bioremediation with technology assessment and emphasis on fungal bioremediation of oil contaminated soils. *Journal of Environmental Management* 241: 156–166. https://doi.org/10.1016/j.jenvman.2019.04.019.

Saber, M., Abouziena, H., Hoballah, E. et al. (2015). Phytoremediation of potential toxic elements in contaminated sewaged soils by sunflower (Helianthus annuus) and corn (Zea mays L.) plants. In: *12th International Phytotechnology Conference, Manhattan, KS*, 27–30.

Singh, U.B., Ahluwalia, A.S., Sharma, C. et al. (2013). Planktonic indicators: a promising tool for monitoring water quality (early-warning signals). *Ecology, Environment and Conservation* 19 (3): 793–800.

Singh, D., Singh, S.K., Singh, V.K. et al. (2020). Plant growth–promoting bacteria and their role in environmental management. In: *Abatement of Environmental Pollutants* (ed. P. Singh, A. Kumar, and A. Borthakur), 161–175. Cambridge, MA: Elsevier https://doi.org/10.1016/B978-0-12-818095-2.00008-4.

Snoek, T., Chaberski, E.K., Ambri, F. et al. (2020). Evolution-guided engineering of small-molecule biosensors. *Nucleic Acids Research* 48 (1): e3–e3. https://doi.org/10.1101/601823.

Song, S., Wang, L., Li, J. et al. (2008). Aptamer-based biosensors. *TrAC Trends in Analytical Chemistry* 27 (2): 108–117. https://doi.org/10.1016/j.trac.2007.12.004.

Sumampouw, O.J. and Risjani, Y. (2014). Bacteria as indicators of environmental pollution. *Environment* 51 (52): 10–5923.

Trabelsi, H., Koch, M., and Faulon, J.L. (2018). Building a minimal and generalizable model of transcription factor–based biosensors: showcasing flavonoids. *Biotechnology and Bioengineering* 115 (9): 2292–2304. https://doi.org/10.1002/bit.26726.

Tuerk, C. and Gold, L. (1990). Systematic evolution of ligands by exponential enrichment: RNA ligands to bacteriophage T4 DNA polymerase. *Science* 249 (4968): 505–510. https://doi.org/10.1126/science.2200121.

Yin, K., Wang, Q., Lv, M., and Chen, L. (2019). Microorganism remediation strategies towards heavy metals. *Chemical Engineering Journal* 360: 1553–1563. https://doi.org/10.1016/j.cej.2018.10.226.

Zaghloul, A. (2014). A kinetic approach for remediating ptes in sewaged soils using novel biotechnology. In: 한국토양비료학회 학술발표회 초록집, 62.

Zancan, S., Trevisan, R., and Paoletti, M.G. (2006). Soil algae composition under different agro-ecosystems in North-Eastern Italy. *Agriculture, Ecosystems & Environment* 112 (1): 1–12.

Zhang, J., Barajas, J.F., Burdu, M. et al. (2017). Development of a transcription factor-based lactam biosensor. *ACS Synthetic Biology* 6 (3): 439–445.

# 26

## PGPRs: Toward a Better Greener Future in Sustainable Agriculture

*Soham Das*[1], *V.H.S. Vaishnavee*[1], *Anshika Dedha*[1], *Priya Yadav*[2], *Rahul Prasad Singh*[2], *and Ajay Kumar*[1]

[1] *Amity Institute of Biotechnology, Amity University, Noida, India*
[2] *Laboratory of Algal Research, Department of Botany, Institute of Science, Banaras Hindu University, Varanasi, India*

### 26.1 Introduction

The global population is projected to reach 9 billion by 2050, resulting in a 70% increase in food demand. This poses a significant threat to food security, as agricultural land is limited (Alexandratos and Bruinsma 2012). Facing the negative impacts on environment due to chemical fertilizers and lack of public awareness, there is a pressing need to develop alternative soil fertilization strategies that can support the sustainable production of food for a growing global population. Current agricultural practices, which rely heavily on synthetic chemical fertilizers, may be able to meet the growing demand for food, but they are not sustainable in the long term (Bhardwaj et al. 2014; Kumar et al. 2016a,b). While agricultural production has increased significantly since the Green Revolution, farmers have become dependent on chemical fertilizers, which can lead to environmental problems and economic instability (Nath Bhowmik and Das 2018). Chemical fertilizers are showing detrimental effects on both human health and the environment. While they can promote plant growth in the short term, they do not contribute to improving soil properties and can actually degrade soil quality over time (Choudhary et al. 2018; Etesami et al. 2017; Kumar et al. 2017a,b). This can lead to a number of problems, including reduced water retention capacity, increased salinity, and imbalances in soil nutrients. Excessive use of chemical fertilizers can also lead to environmental pollution through the accumulation and magnification of toxic compounds in soil and water (Margenat et al. 2018; Das et al. 2015). Additionally, organic farming and its inputs offer beneficial alternatives for promoting plant growth and yield, along with soil fertility by using the beneficial interactions between these organisms and plants (Pathan et al. 2020). Plant growth-promoting rhizobacteria (PGPRs) constitute a group of beneficial microbes associated with plants that enhance abiotic and biotic stress tolerance, facilitate nutrient uptake, and promote overall plant health. Sustainable agriculture is essential for meeting the growing global food demand while conserving natural resources and minimizing environmental impacts (Kumar et al. 2016a,b). One promising approach to sustainable agriculture involves harnessing the potential of PGPRs. These helpful microorganisms establish themselves in plant roots and enhance the well-being and growth of plants using different methods. PGPRs have become increasingly popular because they can boost soil quality, raise crop production, and lessen the need for chemical substances. This introduction explores the multifaceted role of PGPRs in sustainable agriculture, highlighting their mechanisms of action and the potential benefits they offer to agricultural systems (Lugtenberg and Kamilova 2009).

In an era where environmental problems, resource limitations, and food security are critical concerns, harnessing the full potential of PGPRs have emerged as a promising endeavor to enhance crop productivity and ecosystem health (Vocciante et al. 2022; Verma et al. 2021; Singh et al. 2022). PGPRs constitute a diverse group of beneficial bacteria that inhabit the rhizosphere, the root zone of plants, and establish complex interactions that can enhance plant growth, nutrient acquisition, disease resistance, and soil structure (Bhanse et al. 2022). By capitalizing on these natural interactions found in nature, PGPRs offer a multifaceted approach to address key issues in modern agricultural methods. The mechanisms through which PGPRs demonstrate their beneficial effects are diverse and complex. These include activities such as $N_2$-fixation, phosphate solubilization and mobilization, the promotion of growth-promoting hormones, and the

*Microbes Based Approaches for the Management of Hazardous Contaminants*, First Edition. Edited by Ajay Kumar, Livleen Shukla, Joginder Singh, and Luiz Fernando R. Ferreira.

production of various biocontrol agents, all of which contribute to enhancing plant health and yield (Tuzun and Bent 2006; Singh et al. 2017a,b; Singh et al. 2018). PGPR's ability to establish symbiotic associations with plants, enhance nutrient availability, and stimulate root growth makes them valuable partners in the pursuit of sustainable agriculture (Gouda et al. 2018). This chapter delves into the multifaceted roles of PGPR in promoting sustainable agriculture. It explores the mechanisms that underlie their beneficial effects, enhances their ability to reduce the environmental impact of farming, and also addresses the challenges and opportunities associated with their integration into modern agricultural practices (Ashkenazy et al. 2018). By emphasizing the transformative potential of PGPR, this chapter underscores the importance of harnessing the capabilities of these beneficial bacteria as catalysts for agricultural sustainability in an era marked by complex challenges faced by all (Solanki et al. 2023).

## 26.2 Brief Introduction of PGPRs

PGPRs play a vital role in transforming sustainable agricultural practices by offering innovative solutions to the challenges faced by conventional farming practices (Perez-Montano et al. 2014). These beneficial microorganisms are commonly found in the rhizosphere and serve to enhance soil productivity, stimulate plant growth, and suppress plant pathogens. Among the most promising and frequently mentioned genera of PGPRs are *Aeromonas, Pseudomonas, Klebsiella, Azoarcus, Azospirillum, Enterobacter, Clostridium, Arthrobacter, Serratia, Azotobacter, Rhizobium, Gluconacetobacter,* and *Bacillus* (Bhattacharyya and Jha 2012). The diversity of PGPRs vary depending on factors such as soil type, environmental conditions, plant species, and nutrient availability in the rhizosphere (Tian et al. 2020). However, it's worth noting that not all strains within a specific genus and species have the same genetic makeup and metabolic capabilities (Huang et al. 2013). In the commercial world, PGPRs are transformed into biofertilizers, which are microbial mixtures. These biofertilizers can be applied directly to seeds or introduced into the area around the plant's roots, where they live and provide essential nutrients to the plant (Malusa and Vassilev 2014). They serve as a safe and eco-friendly substitute for traditional chemical fertilizers, helping enhance global agricultural productivity (Dal Cortivo et al. 2020; Kumar et al. 2024). The application of PGPRs have been shown to lead to enhanced plant growth, including increased seed germination, improved shoot and root development, greater biomass, and a reduction in disease incidence (Table 26.1).

## 26.3 Role of PGPRs

One of the most significant primary contributions of PGPRs in sustainable agriculture is their capacity to enhance nutrient availability within the soil–plant system (Lloret and Martinez-Romero 2005). The process of integrating PGPRs into agricultural practices offers a transformative avenue for reducing reliance on excessive chemical inputs. This approach not only addresses environmental concerns but also contributes to the development of sustainable agricultural systems with increased production capabilities and reduced ecological impact (Figure 26.1) (Raymond et al. 2004). A brief description of these characteristics is provided in the following sections.

### 26.3.1 Nitrogen Fixation

The nif gene is a critical component involved in the biological nitrogen fixation process, containing a structural gene responsible for activating iron proteins and various regulatory genes that coordinate the synthesis of nitrogen-fixing systems in both symbiotic and free-living organisms (Mus et al. 2016). Biological $N_2$-fixation is a crucial ecological process carried out by specific microorganisms, including both free-living and symbiotic $N_2$-fixing bacteria. Notable examples of these bacteria include *Bradyrhizobium, Azotobacter, Herbaspirillum, Mesorhizobium, Achromobacter, Rhizobium, Frankia, Bacillus, and Pseudomonas,* (Mustafa et al. 2019). According to a study, researchers found that when they co-cultivated *Gluconacetobacter diazotrophicus* with young sugarcane plants, the GreA transcription factor exhibited heightened activity. The authors of the research speculate that GreA and ilvA, a gene responsible for producing threonine dehydratase, might be closely located in the genome and subject to similar regulatory mechanisms. What makes this intriguing is that ilvA is an enzyme involved in the production of isoleucine, a compound vital for various plant processes. Isoleucine also interacts with jasmonic acid in a chemical reaction, impacting plant growth and its interactions with microorganisms in the environment. Hence, the researchers suspect that GreA and ilvA may collaborate to influence these crucial plant–microbe

**Table 26.1** List of some PGPRs with their role and future directions.

| S. No. | Bacteria name | Plant | Future directions | Role | References |
|---|---|---|---|---|---|
| 1. | *Rhizobium leguminosarum* | Leguminous crops (e.g., soybeans, peas) | Improved nitrogen fixation efficiency, expanded use in crop rotation systems. | Fixes atmospheric nitrogen, enhancing soil fertility and reducing the need for synthetic fertilizers. | Iqbal et al. (2012) |
| 2. | *Pseudomonas fluorescens* | Leguminous crops | Biocontrol agents against emerging plant pathogens, sustainable disease management. | Suppresses pathogens, induces systemic resistance, and promotes nutrient availability. | Sivasakthi et al. (2014) |
| 3. | *Bradyrhizobium* sp. | Leguminous crops (e.g., beans, lentils) | Symbiotic nitrogen fixation in diverse legume varieties. | Forms nitrogen-fixing nodules on roots, enriching soil nitrogen content. | Ming (2017) |
| 4. | *Gluconacetobacter diazotrophicus* | Sugarcane | Enhanced biofuel production, sustainable sugarcane cultivation. | Fixes nitrogen and produces plant growth regulators. | Oliveira et al. (2018) |
| 5. | *Frankia* sp. | Actinorhizal plants (e.g., alder, bayberry) | Restoration of degraded ecosystems, agroforestry systems. | Fixes nitrogen and forms root nodules in actinorhizal plants. | Gupta et al. (2020) |
| 6. | *Azotobacter* sp. | Leguminous crops | Enhanced soil fertility in low-input systems, sustainable organic farming. | Fixes nitrogen, produces growth-promoting substances. | Chennappa et al. (2014) |
| 7. | Bacillus amyloliquefaciens | Leguminous crops | Biocontrol of plant pathogens, integration into precision agriculture. | Produces antimicrobial compounds, enhances nutrient solubilization. | Gowtham et al. (2018) |
| 8. | *Achromobacter xylosoxidans* | Leguminous crops | Improved soil health and nutrient cycling in degraded soils. | Enhances nutrient availability, contributes to plant growth promotion. | Vyas et al. (2018) |
| 9. | *Serratia marcescens* | Leguminous crops | Biocontrol agents against foliar pathogens, sustainable disease management. | Produces growth hormones, antagonistic compounds against pathogens. | Singh and Jha (2016) |
| 10. | Actinobacteria (general category) | Leguminous crops | Bioprospecting for novel compounds, sustainable pest management. | Produces antibiotics, enzymes, and plant growth regulators. | Singh and Jha (2016) |
| 11. | *Azospirillum brasilense* | Various crops, including cereals and grasses | Biofertilizer development for non-leguminous crops, drought-resistant varieties. | Promotes nutrient uptake, produces growth hormones, and enhances stress tolerance. | Rothballer et al. (2003) |
| 12. | *Bacillus subtilis* | Leguminous crops | Organic farming systems, production of sustainable biopesticides. | Produces antimicrobial compounds, hormones, and enzymes that benefit plant growth and health. | Kumar et al. (2011) |
| 13. | *Methylobacterium* sp. | Leguminous crops | Development of drought-resistant crops, adaptation to changing climate conditions. | Enhances stress tolerance, produces growth-promoting compounds. | Grossi et al. (2020) |
| 14. | *Enterobacter* sp. | Leguminous crops | Phytoremediation in contaminated soils, enhancement of nutrient use efficiency. | Promotes nutrient uptake, produces growth-promoting substances. | Pramanik et al. (2018) |
| 15. | *Streptomyces* sp. | Leguminous crops | Production of novel bioactive compounds, organic pest management. | Produces antibiotics, suppresses pathogens, and contributes to soil health. | Gowdar et al. (2018) |
| 16. | *Burkholderia* sp. | Leguminous crops | Bioremediation, enhanced nutrient cycling in degraded soils. | Supports nutrient uptake, enhances plant growth, and contributes to soil health. | Hassen et al. (2021) |

*(Continued)*

**Table 26.1** (Continued)

| S. No. | Bacteria name | Plant | Future directions | Role | References |
|---|---|---|---|---|---|
| 17. | Cyanobacteria (e.g., *Anabaena* sp., Nostoc) | Rice paddies, various crops | Sustainable rice cultivation, agroecological practices. | Fixes nitrogen, enhances soil fertility, contributes to carbon sequestration. | Prasanna et al. (2012) |
| 18. | *Micrococcus luteus* | Leguminous crops | Soil restoration, revegetation of degraded lands. | Enhances nutrient cycling, contributes to disease suppression. | Pinter et al. (2017) |
| 19. | *Acetobacter diazotrophicus* | Leguminous crops | Biofertilizer development for non-leguminous crops, enhancement of nutrient use efficiency. | Fixes nitrogen, produces growth-promoting substances. | Sevilla et al. (2001) |
| 20. | Mycorrhiza-associated bacteria (MABs) | Various Leguminous crops associated with mycorrhizal fungi | Optimization of mycorrhizal associations for sustainable agriculture. | Enhances nutrient uptake, improves disease resistance in mycorrhizal plants. | Pathak et al. (2017) |

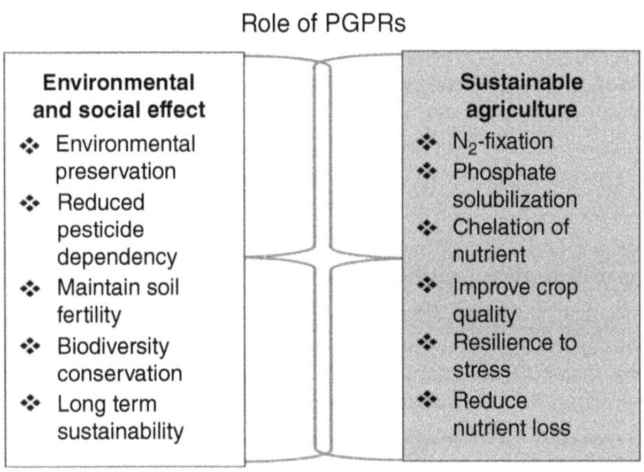


**Figure 26.1** Role of PGPRs in sustainable agriculture and their environmental and social effect.

interactions and development (dos Santos et al. 2010). In another experiment, researchers treated rice plant roots with two different substances, jasmonic acid and a specific strain of *Azoarcus* sp. called BH72. They observed an intriguing phenomenon. When roots were treated with jasmonic acid, there was an increase in the activity of a specific enzyme known as ACC oxidase. However, in the roots treated with the *Azoarcus* sp. strain BH72, this enzyme did not exhibit the same increase. What's significant here is that it indicates a connection between two crucial plant signaling pathways, ethylene and jasmonic acid. These pathways appear to interact with each other, and the activation of ACC oxidase in response to jasmonic acid treatment provides a hint about this interaction. This interaction could have significant implications for how plants respond to various environmental factors and stresses (Miché et al. 2006). Although most reactions controlled by ethylene and jasmonic acid operate separately, there are certain cases, such as when activating the plant defensin gene PDF1.2, where both hormones have to work together. These various microorganisms are vital in transforming atmospheric nitrogen into a form that plants can actually use to grow and thrive (Wang et al. 2002). Their nitrogen-fixing ability holds immense agricultural and ecological significance, as it enhances soil fertility and reduces the dependence on synthetic nitrogen-based fertilizers, ultimately promoting sustainable and environmentally friendly agricultural practices (Xu et al. 2017).

### 26.3.2 Phosphate Solubilization

In the soil, phosphorus exists in two primary forms, inorganic and organic, and neither of these forms are readily accessible to plants for their nutritional needs. However, the intriguing process of phosphate solubilization, facilitated by PGPRs, play a vital role in rendering phosphorus available for plant utilization (Gupta et al. 2015). PGPRs, which are beneficial bacteria residing in the root zone of plants, possess the capability to solubilize phosphorus from these forms. They achieve this by releasing specific enzymes and compounds that transform insoluble phosphorus compounds into a soluble form that plants can uptake through their root systems (Etesami and Adl 2020). In an experiment, when the bacterium *Burkholderia cenocepacia* CR318 was combined with a substance known as $Ca_3(PO_4)_2$, and used as a nutrient source for plants over a period of six weeks, notable effects were observed. In comparison to using only $Ca_3(PO_4)_2$ without the bacterium, the plants exhibited significant improvements. Nevertheless, the growth in shoot length showed a minor 8% rise, but this change didn't have enough statistical support, suggesting that it might have happened by random chance (with a p-value greater than 0.05) (You et al. 2020).

Interestingly, when either *B. cenocepacia* CR318 or $Ca_3(PO_4)_2$ was used individually, or when no treatment was applied at all, there were no discernible differences in the growth of the plants. This implies that the combination of the bacterium and $Ca_3(PO_4)_2$ had a unique and advantageous impact on the plants' growth and overall health. Numerous microorganisms are recognized for their involvement in the process of phosphate solubilization, and these include *Pseudomonas* sp., *Agrobacterium* sp., *Rhizobium* sp., *Bacillus* sp., and *Enterobacter* sp. These bacteria have an important job in breaking down complicated phosphorus compounds in the soil, converting them into a type that plants can easily use. This not only improves plant growth but also cuts down on the need for phosphorus-based fertilizers, which is better for the environment and makes agriculture more sustainable (Alori et al. 2017).

### 26.3.3 Chelation of Nutrients

The presence of heavy metal toxicity in soil represents a substantial environmental threat, resulting from a combination of natural processes and human activities. These heavy metals, including substances such as cadmium, mercury, lead, and cobalt can significantly modify the soil's characteristics. This alteration, in turn, has direct and indirect repercussions on agricultural systems and, ultimately, on all organisms within the environment (Artiola et al. 2019). A promising and environmentally friendly solution for addressing the issue of heavy metal contamination involves the use of PGPRs. They play a vital role in a process known as bioremediation, which aims to remove or reduce heavy metals from the soil (Tak et al. 2012).

The bacterium *Azospirillum brasilense* was discovered to enhance iron uptake mechanisms in cucumber plants, both in conditions with sufficient iron and in iron-deficient conditions. This study identified several genes responsible for this process and found that one of them, $CsFRO_3$, exhibited a positive response to the bacterium in iron-rich environments. Moreover, the bacterium appeared to alleviate copper-related stress in the plants, impacting their growth and nutrient absorption positively (Marastoni et al. 2019). In PGPRs-assisted bioremediation, these bacteria interact with heavy metals in the soil, either immobilizing them or converting them into less toxic forms. This environmentally friendly approach not only aids in the remediation of contaminated soils but also promotes environmental sustainability. It represents a significant step toward making our agricultural practices safer and more environmentally conscious by mitigating the risks associated with heavy metal contamination in soils (Gupta et al. 2023).

### 26.3.4 Prevent Nutrient Losses

The effectiveness of seed inoculation with PGPRs in reducing nutrient losses is not dependent on a specific type of rhizobacteria. This holds true across various scenarios, even when crops are grown in soils with different mineral fertilizer histories (Dasgupta and Desai 2022). To illustrate this idea, let's look at a study where spring barley and winter rye seeds were treated with nitrogen-fixing bacteria, *A. brasilense* 18-2 and *A. brasilense* 410, respectively. In addition, maize seeds were given a dose of phosphate-mobilizing bacterium called *Paenibacillus polymyxa* KB (Adesemoye et al. 2008). The significant discovery here is that, regardless of the specific rhizobacteria strain used or the type and amount of mineral fertilizers employed, there was a consistent reduction in nutrient losses. In simpler terms, the positive effects of PGPRs on nutrient retention and availability were observed across diverse crop varieties and varying soil conditions (Adesemoye and Egamberdieva 2013). To provide an example, a study investigated the potential synergy between a beneficial soil bacterium known as PGPR, specifically *Paraburkholderia* sp. $SOS_3$, and various nitrogen fertilizers with different compositions and concentrations to enhance the growth of *Macadamia integrifolia* seedlings. This study assessed aspects such as plant growth, soil nutrient levels, and various soil characteristics. The main findings of the study revealed that when seedlings were subjected to a combination of inorganic nitrogen fertilizer and organic manure-based fertilizer at high levels (iNoN 100), the inoculation of PGPRs led to a significant increase in both total plant biomass and $N_2$-uptake. However, when exposed to either low or high levels of inorganic nitrogen alone, the presence of PGPRs did not result in noticeable growth effects. Notably, PGPRs played a crucial role in reducing nitrogen loss from the soil while simultaneously enhancing nitrogen availability to the plants. The study's outcomes also highlighted that the influence of PGPRs were contingent upon the type and quantity of nitrogen applied, as demonstrated by Gallart et al. (2021). This underscores the adaptability and efficacy of PGPRs in improving nutrient utilization by plants and mitigating nutrient losses. This makes PGPRs a valuable asset in sustainable agricultural practices, regardless of the specific crop or soil type (Volkogon et al. 2023).

### 26.3.5 Improve Crop Quality

PGPR is known to provide a vast range of functions that promotes plant growth. These remarkable microorganisms serve as a source of various valuable substances, such as metabolites and enzymes, which play an important role in enhancing plant development and health (Du Jardin 2015). One of their key roles is nutrient mobilization, where PGPRs assist in making essential nutrients more accessible to plants. They achieve this through various mechanisms, including phosphate solubilization and $N_2$-fixation, which increase the availability of these vital elements to plants (Etesami and Adl 2020). Biostimulants are becoming increasingly important in sustainable agriculture. The excessive use of chemical products on crops has raised concerns about environmental hazards and soil contamination.

### 26.3.6 Plant Hormones Regulation

Phytohormones are natural molecules made by plants to control different aspects of their growth, such as flowering, handling stress, and ripening fruits. There are five main types of plant hormones, gibberellin, auxin, cytokinin, ethylene, and abscisic acid. Auxin, for instance, is identified by a carboxyl group attached to the third carbon of the indole ring. It mainly affects plant growth by modulating cell division, promoting cell elongation, starting from the growth of main root. In addition also encourage development of the plant's water-conducting tissues (Glick 2012).

Surprisingly, these plant hormones can also be produced by microorganisms such as bacteria, cyanobacteria, and algae. These microorganisms assist in promoting plant growth by softening plant cell walls. In return, they benefit from the nutrients released by plant roots. Therefore, the auxin made by bacteria is quite significant in fostering interactions between plants and bacteria, covering a wide range, from stimulating plant growth to potentially causing diseases (Spaepen et al. 2007).

Furthermore, it has been proposed that enhanced auxin production can facilitate the formation of nodules in plants. This hypothesis gained support when co-inoculating soybean with *R.* sp. NGR234, a low-indole-3-acetic acid (IAA) producing mutant, resulted in less IAA production compared to inoculation with the wild-type strain (Spaepen and Vanderleyden 2011). Additionally, further validation was obtained through the inoculation of beans (*Phaseolus vulgaris* L.; cultivar DOR364) with *Rhizobium etli* CNPAF512 and *A. brasilense* Sp245, which led to an increased number of nodules. It is worth noting that approximately 80% of bacteria isolated from the rice rhizosphere have shown the capability to produce IAA. Other studies have also demonstrated that bacteria belonging to the Enterobacteriaceae family (such as *Escherichia, Enterobacter, Grimontella, Pantoea, Rahnella,* and *Klebsiella*) can synthesize auxins and rely on precursors found in root exudates (Costa et al. 2014). Inoculation with *Azospirillum* sp. has been shown to promote the development and growth of wheat plants due to its ability to produce IAA (Baudoin et al. 2010). PGPRs also influence auxin signaling through nitrite reductase activity, leading to the production of nitric oxide (NO). The production of NO has a role in regulating the formation of lateral roots, which has been observed in *A. brasilense* (Lanteri et al. 2008). Furthermore, the creation of 2,4-diacetylphloroglucinol (DAPG), an antimicrobial substance made by fluorescent pseudomonads, has been found to serve as a signal that influences the structure of plant roots, their branching, what they release into the soil, and their overall resistance to diseases (Vacheron et al. 2013).

In addition to being able to make a plant hormone called IAA, some bacteria such as *Pseudomonas putida* 1290 and *Bradyrhizobium japonicum* have shown that they can break down IAA and use it as a source of carbon (Leveau and Lindow 2005). Research findings on radish co-inoculation with strain 1290 have shown that it can mitigate the harmful effects of pathogenic bacteria such as *Rahnella aquatilis* and *Pseudomonas syringae* by reducing the high concentrations of IAA produced by these bacteria. Hence, organisms in the rhizosphere boost plant growth not just through auxin production but also through auxin degradation (Leveau and Lindow 2005).

Cytokinin, a crucial plant hormone, has a major impact on how plant cells develop and grow. It also has a notable impact on yield and productivity, as it affects apical dominance. Additionally, cytokinins contribute to leaf growth by delaying the aging or senescence of leaves. Therefore, it can be asserted that the balance between auxin and cytokinin levels is a crucial factor that controls organogenesis and overall plant growth (Aloni et al. 2006). PGPRs also exert a substantial influence on the ratio of these hormones due to their ability to synthesize cytokinin's or produce secondary metabolites that can modulate hormonal signaling. Numerous bacteria, such as *Arthrobacter giacomelloi, Pseudomonas fluorescens, A. brasilense, P. polymyxa, Bacillus licheniformis,* and *B. japonicum,* have been identified as capable of producing cytokinin's (Vacheron et al. 2013). While in silico analysis has identified several genes with putative functions related to cytokinin biosynthesis, further experimental validation is required to confirm these hypotheses (Frébort et al. 2011). Planting PGPRs that can make cytokinin has been widely observed to have a beneficial effect on the growth of plant shoots, as indicated by Vacheron et al. (2013). In these interactions, an enzyme called isopentenyl transferase is responsible for catalyzing the conversion of 5'-AMP to isopentenyl adenosine 5'-phosphate. This enzyme has been identified in *Agrobacterium tumefaciens* (ipt) and *P. syringae* pv. *savastanoi*. Interestingly, both enzymes share significant sequence homology and the ability to induce tumor formation in plants. Another crucial plant hormone, gibberellin, plays a role in controlling seed germination and starch hydrolysis by inducing alpha-amylase activity, thereby promoting cellular respiration. Gibberellins also have diverse functions in various physiological processes in plants, including breaking seed dormancy, root development, stem elongation, leaf senescence, and flowering (Yaxley et al. 2001).

Moreover, research has shown that gibberellin produced by plants can aid in the formation of nodules on roots, which is beneficial for certain plants. However, this positive effect only occurs up to a certain point, and beyond that, it can have a negative impact. This idea has been confirmed through experiments with hormone-deficient pea plant mutants, which displayed decreased nodule formation (Vacheron et al. 2013).

Numerous types of bacteria, including *Acinetobacter calcoaceticus*, *Achromobacter xylosoxidans*, various species of *Azotobacter* and *Azospirillum*, various *Bacillus* species, *rhizobia*, *Gluconobacter diazotrophicus*, and *Herbaspirillum seropedicae*, have the ability to produce gibberellin. For instance, when maize is treated with *Azospirillum* bacteria, it promotes root growth by increasing gibberellin levels.

### 26.3.7 Soil Heath Improvement

Beneficial soil microorganisms (PGPRs) have a significant impact on enriching soil quality and promoting the general well-being of the soil. These connections between plants and the soil environment have led to various enhancements in the chemical, biological, and physical aspects of the soil, ultimately leading to more effective and eco-friendly agricultural practices (Adesemoye et al. 2008). Through the enhancement of soil structure and health, PGPRs have provided a sustainable approach to maintaining and improving factors such as soil fertility and productivity (Figure 26.2). The beneficial impact of these bacteria on soil structure extends beyond individual plants, contributing to the overall health of ecosystems and supporting long-term agricultural sustainability consistently (Adesemoye and Kloepper 2009). The treatment with PGPRs yielded some intriguing effects on the composition of microorganisms in the rhizosphere soil. Specifically, it resulted in an increase in the relative abundances of *Pseudomonas* and *Lysobacter*, two types of beneficial bacteria well-known for their positive impact on plant growth and soil health (Jin et al. 2023). Conversely, the application of PGPRs led to reduced levels of *Erwinia* and *Ralstonia*, which are often associated with plant diseases and can have detrimental effects on crops. The rhizosphere, the region where plant roots interact with the soil, plays a crucial role in the ecosystem. It serves as a hub where organic substances and signaling molecules migrate into deeper soil layers, contributing to long-term processes. PGPRs are friendly bacteria that, when introduced into the rhizosphere, exert a positive influence on plants. They excel at competing with other microorganisms in the rhizosphere and do not harm plants. Instead, they establish a robust presence on and

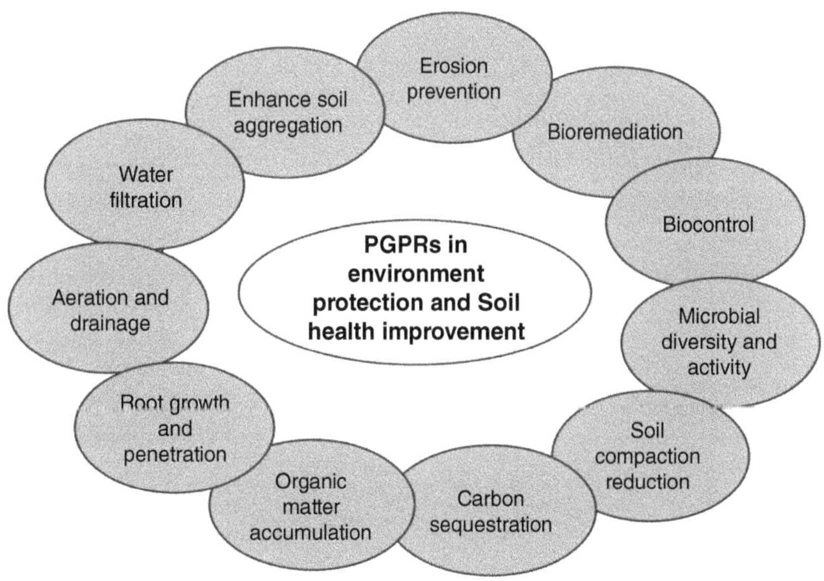


**Figure 26.2** Role of PGPRs in environmental protection and soil health improvement.

around plant roots, thereby aiding in improved plant growth, increased yields, and enhanced plant resilience to various stresses. In essence, PGPRs serve as valuable allies for plant health and growth (Adedeji et al. 2020). These changes in the soil's nutrient content and microbial communities were consistently observed, irrespective of whether the agricultural system used was organic or sustainable ("green"). This suggests that the application of these beneficial PGPRs had a consistent and positive influence on the soil ecosystem, regardless of the specific farming approach employed (Singh et al. 2011). These findings underscore the potential of PGPRs to create a more favorable soil environment for plant growth, regardless of whether farming practices are organic or follow sustainable methods (Yu et al. 2023). They highlight the versatility of PGPRs in contributing to healthier soils and, ultimately, more successful and sustainable agricultural systems.

### 26.3.8 Environmental Protection

Numerous genera of rhizobacteria, including *Rhizobium*, are well-known for their dual roles in promoting plant growth and acting as biocontrol agents. The primary objective is to investigate the specific type of PGPRs and the timing of watering that can have the most significant impact on the vegetative growth of soybean (*Glycine max* L.) (Seneviratne et al. 2011). Soybean is a crucial global crop, and optimizing its growth is vital for agricultural productivity. PGPRs, such as *Rhizobium*, have the potential to enhance soybean growth by facilitating nutrient uptake, improving soil health, and providing protection against specific plant diseases (Khatoon et al. 2020). PGPRs have a significant impact on plant health and growth. They achieve this by producing various compounds, including lipopolysaccharides, lipopeptides, siderophores, pyocyanin, and antibiotics (such as 2,4-diacetylphoroglucinol), as well as volatile substances such as iron-regulated compounds, 2,3-butanediol, and *N*-alkylated benzylamine. Many PGPRs products are commercially available, helping to boost crop yields while safeguarding plants from diseases (Singh et al. 2017a,b). Additionally, some PGPRs products function as biofertilizers, benefiting both native plants and crops, especially in challenging environments. They employ multiple mechanisms to promote plant growth (Figure 26.2) (Meena et al. 2020). Ultimately, this contributes to increase crop yields and promotes sustainable agricultural practices. By understanding the best practices for utilizing PGPRs in soybean cultivation, we can harness the potential of these beneficial rhizobacteria to optimize agricultural outcomes (Kusumastuti et al. 2023).

## 26.4 Social and Economic Impact of PGPRs

The integration of PGPRs into agricultural practices not only brings environmental benefits but also yields significant economic and social outcomes. These impacts extend from the economic viability and growth of farming operations to the improvement of livelihoods in rural communities. This integration creates economic opportunities, enhances rural livelihoods, and promotes sustainable agricultural practices. By incorporating PGPRs into farming systems can contribute to the widespread socioeconomic well-being of communities and societies (Figure 26.3). The overall impact of these economic and social outcomes underscores the significance of PGPRs as a powerful tool for advancing sustainable development for humanity (Ahmed et al. 2019).

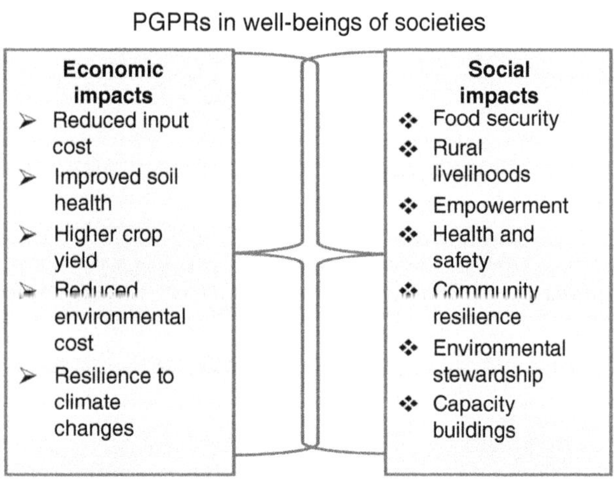


**Figure 26.3** Role of PGPRs in well-beings of societies.

PGPRs play an important role in enhancing soil structure and improving overall soil health. These interactions between plants and the soil ecosystem have resulted in multiple improvements in the biological, chemical, and physical properties of the soil, ultimately leading to more productive and sustainable agricultural systems (Adesemoye et al. 2008). Through the enhancement of soil structure and health, PGPRs have offered a sustainable approach to maintaining and improving factors such as soil fertility and productivity. The beneficial impact of these bacteria on soil structure extends beyond individual plants, contributing to the overall health of ecosystems and supporting long-term agricultural sustainability consistently (Adesemoye and Kloepper 2009). Future research endeavors in the field of PGPRs would also emphasize delving into the intricacies of various types of plant–microbe interactions. This exploration aims to optimize their application methods and develop comprehensive strategies that integrate PGPRs with other sustainable agricultural practices effectively (Fascella et al. 2015). Overcoming these challenges will enable us to refine our understanding of PGPRs and elevate them from a promising and beneficial concept to a powerful tool that paves the way for transformation, fostering greater resilience and sustainability in agriculture. However, unlocking the full potential of PGPRs comes with its share of challenges. From complex scientific aspects to practical implementation issues, a comprehensive approach is needed to translate promising laboratory findings into successful on-field applications. These challenges, although formidable, also present opportunities for innovation, collaboration, and interdisciplinary research. By fostering strong collaboration between scientists, policymakers, farmers, and industry players, we can overcome these obstacles and promote widespread adoption of PGPRs (Katiyar and Goel 2004). In conclusion, the journey toward sustainable agriculture with PGPRs is a motivating narrative of science, responsible stewardship, and optimism. It reminds us that in our mission to feed the world, we also need to take care of the Earth. Incorporating PGPRs into farming practices represents a delicate balance between progress and preservation, innovation and tradition, securing not only a bountiful harvest today but also a prosperous future for generations to come (Basu et al. 2021).

The global biostimulants market was valued at $1.74 billion in 2016, with a projected annual growth rate of 10.2% from 2017 to 2025. This growth is driven by the need for higher productivity and addressing soil degradation issues (Hamid et al. 2021). In 2019, the market was estimated at $2.30 billion and was expected to reach $2.53 billion in 2020. It is predicted to continue growing at a rate of 10.2% annually, reaching $4.14 billion by 2025. While not all biostimulants are biological, microbes, particularly PGPRs, have a long history of association with plants. Interestingly, only a small fraction (less than 25%) of commercial biostimulants are microbial-based. PGPRs also play a crucial role in biological pest control, helping to manage and reduce the impact of harmful pests on crops. Moreover, they have the ability to induce disease resistance in plants, making them more resilient to pathogens and infections. In addition to these functions, PGPRs possess bioremediation potential. They can aid in the detoxification of heavy metals, pollutants, and pesticides present in the soil, contributing to the overall improvement of soil quality and reducing environmental contamination. Phytoextraction, a process in which plants absorb and accumulate contaminants, is one method through which PGPRs can assist in this detoxification process (Nagrale et al. 2023).

## 26.5 Challenges, Future Perspectives and Conclusion

PGPR has become a crucial component of sustainable farming, offering a versatile solution to the complex challenges facing our global food systems. To maintain soil fertility and enhance crop productivity, it is essential to implement alternative strategies, particularly those involving the use of PGPRs. The multifaceted contribution of PGPRs to the sustainability of agriculture is unquestionable in the present day. The application of PGPRs in various agroclimatic conditions across a wide range of crops has consistently demonstrated improved productivity, resulting in reduced global reliance on harmful agrochemicals that disrupt agroecosystems. Figure 26.4 represents the challenges and future direction of PGRPs.

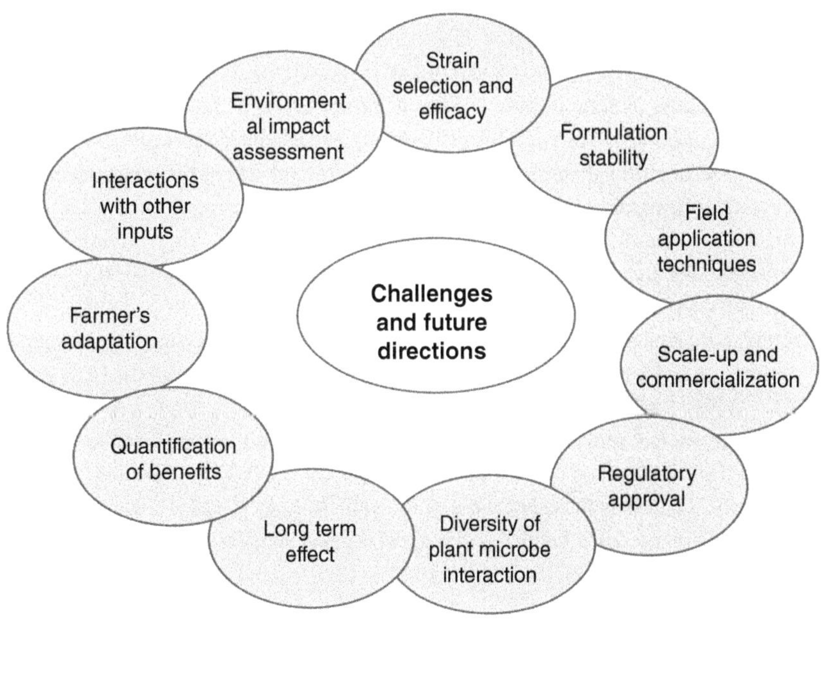


**Figure 26.4** Challenges and future directions in the field of PGPRs application.

However, despite the proven effectiveness of various PGPR strains as biofertilizers, there are still gaps in our understanding of the roles and mechanisms of PGPRs, highlighting the need for greater emphasis on closing these knowledge gaps. As we confront pressing issues such as preserving our environment, conserving resources, and ensuring food security, the importance of PGPRs become increasingly evident. This chapter has explored the myriad ways in which PGPR benefits sustainable agriculture, including increasing nutrient availability, reducing the need for harmful chemicals, enhancing soil quality, and improving economic and social well-being. The potential of PGPR to transform agriculture extends beyond higher crop yields and healthier soil. They also possess the capability to reshape the relationship between farming and the environment. PGPR can contribute to reducing the environmental impact of farming, mitigating chemical pollution, and supporting resilient ecosystems. This vision of agriculture not only meets society's nutritional needs but also maintains the delicate balance of our planet.

## References


Adedeji, A.A., Häggblom, M.M., and Babalola, O.O. (2020). Sustainable agriculture in Africa: plant growth-promoting rhizobacteria (PGPR) to the rescue. *Scientific African* 9: e00492. https://doi.org/10.1016/j.sciaf.2020.e00492.

Adesemoye, A., Torbert, H., and Kloepper, J. (2008). Enhanced plant nutrient use efficiency with PGPR and AMF in an integrated nutrient management system. *Canadian Journal of Microbiology* 54: 876–886. https://doi.org/10.1139/w08-081.

Adesemoye, A.O. and Egamberdieva, D. (2013). Beneficial effects of plant growth-promoting rhizobacteria on improved crop production: prospects for developing economies. In: *Bacteria in Agrobiology: Crop Productivity* (ed. D.K. Maheshwari, M. Saraf, and A. Aeron), 45–63. Berlin, Heidelberg: Springer Berlin Heidelberg https://doi.org/10.1007/978-3-642-37241-4_2.

Adesemoye, A.O. (2009). *Harnessing the Capacity of PGPR with or Without AMF for Improved Plant use Efficiency of Chemical Fertilizers and Organic Manure*. Auburn University.

Alexandratos, N. and Bruinsma, J. (2012). World agriculture towards 2030/2050: the 2012 revision. *Agricultural Economics* https://doi.org/10.22004/ag.econ.288998.

Aloni, R., Aloni, E., Langhans, M., and Ullrich, C.I. (2006). Role of cytokinin and auxin in shaping root architecture: regulating vascular differentiation, lateral root initiation, root apical dominance and root gravitropism. *Annals of Botany* 97 (5): 883–893. https://doi.org/10.1093/aob/mcl027.

Alori, E.T., Dare, M.O., and Babalola, O.O. (2017). Microbial inoculants for soil quality and plant health. *Sustainable Agriculture Reviews* 281–307. https://doi.org/10.1007/978-3-319-48006-0_9.

Artiola, J.F., Walworth, J.L., Musil, S.A., and Crimmins, M.A. (2019). Soil and land pollution. In: *Environmental and Pollution Science* (ed. M.L. Brusseau, I.L. Pepper, and C.P. Gerba), 219–235. Academic Press https://doi.org/10.1016/B978-0-12-814719-1.00014-8.

Ashkenazy, A., Chebach, T.C., Knickel, K. et al. (2018). Operationalising resilience in farms and rural regions–findings from fourteen case studies. *Journal of Rural Studies* 59: 211–221. https://doi.org/10.1016/j.jrurstud.2017.07.008.

Basu, A., Prasad, P., Das, S.N. et al. (2021). Plant growth promoting rhizobacteria (PGPR) as green bioinoculants: recent developments, constraints, and prospects. *Sustainability* 13 (3): 1140. https://doi.org/10.3390/su13031140.

Baudoin, E., Lerner, A., Mirza, M.S. et al. (2010). Effects of *Azospirillum brasilense* with genetically modified auxin biosynthesis gene ipdC upon the diversity of the indigenous microbiota of the wheat rhizosphere. *Research in Microbiology* 161 (3): 219–226. https://doi.org/10.1016/j.resmic.2010.01.005.

Bhanse, P., Kumar, M., Singh, L. et al. (2022). Role of plant growth-promoting rhizobacteria in boosting the phytoremediation of stressed soils: opportunities, challenges, and prospects. *Chemosphere* 303: 134954. https://doi.org/10.1016/j.chemosphere.2022.134954.

Bhardwaj, D., Ansari, M.W., Sahoo, R.K., and Tuteja, N. (2014). Biofertilizers function as key player in sustainable agriculture by improving soil fertility, plant tolerance and crop productivity. *Microbial Cell Factories* 13: 1–10. https://doi.org/10.1186/1475-2859-13-66.

Bhattacharyya, P.N. and Jha, D.K. (2012). Plant growth-promoting rhizobacteria (PGPR): emergence in agriculture. *World Journal of Microbiology and Biotechnology* 28: 1327–1350. https://doi.org/10.1007/s11274-011-0979-9.

Chennappa, G., Adkar-Purushothama, C.R., Naik, M.K. et al. (2014). Impact of pesticides on PGPR activity of *Azotobacter* sp. isolated from pesticide flooded paddy soils. *Greener Journal of Agricultural Sciences* 4 (4): 117–129. https://doi.org/10.15580/GJAS.2014.4.010314003.

Choudhary, M., Panday, S.C., Meena, V.S. et al. (2018). Long-term effects of organic manure and inorganic fertilization on sustainability and chemical soil quality indicators of soybean-wheat cropping system in the Indian mid-Himalayas. *Agriculture, Ecosystems and Environment* 257: 38–46. https://doi.org/10.1016/j.agee.2018.01.029.

Costa, P.B.D., Granada, C.E., Ambrosini, A. et al. (2014). A model to explain plant growth promotion traits: a multivariate analysis of 2,211 bacterial isolates. *PLoS One* 9 (12): e116020. https://doi.org/10.1371/journal.pone.0116020.

Dal Cortivo, C., Ferrari, M., Visioli, G. et al. (2020). Effects of seed-applied biofertilizers on rhizosphere biodiversity and growth of common wheat (*Triticum aestivum* L.) in the field. *Frontiers in Plant Science* 11: 72. https://doi.org/10.3389/fpls.2020.00072.

Das, N.P., Kumar, A., and Singh, P.K. (2015). Cyanobacteria, pesticides and rice interaction. *Biodiversity and Conservation* 24 (4): 995–1005.

Dasgupta, R.B.D.S. and Desai, D.P. (2022). Study on plant growth-promoting rhizobacteria (PGPR): and it's various application. In: *Environment Conservation, Challenges Threats in Conservation of Biodiversity* (ed. V.K. Manam). Scieng Publications p. 198.

Du Jardin, P. (2015). Plant biostimulants: definition, concept, main categories and regulation. *Scientia Horticulturae* 196: 3–14. https://doi.org/10.1016/j.scienta.2015.09.021.

Etesami, H. and Adl, S.M. (2020). Plant growth-promoting rhizobacteria (PGPR) and their action mechanisms in availability of nutrients to plants. In: *Phyto-Microbiome in Stress Regulation*, 147–203. https://doi.org/10.1007/978-981-15-2576-6_9.

Etesami, H., Emami, S., and Alikhani, H.A. (2017). Potassium solubilizing bacteria (KSB): mechanisms, promotion of plant growth, and future prospects a review. *Journal of Soil Science and Plant Nutrition* 17 (4): 897–911. https://doi.org/10.4067/S0718-95162017000400005.

Fascella, G., Montoneri, E., Ginepro, M., and Francavilla, M. (2015). Effect of urban biowaste derived soluble substances on growth, photosynthesis and ornamental value of Euphorbia × lomi. *Scientia Horticulturae* 197: 90–98. https://doi.org/10.1016/j.scienta.2015.10.042.

Frebort, I., Kowalska, M., Hluska, T. et al. (2011). Evolution of cytokinin biosynthesis and degradation. *Journal of Experimental Botany* 62 (8): 2431–2452.

Gallart, M., Paungfoo-Lonhienne, C., Gonzalez, A., and Trueman, S.J. (2021). Nitrogen source influences the effect of plant growth-promoting rhizobacteria (PGPR) on macadamia integrifolia. *Agronomy* 11 (6): 1064. https://doi.org/10.3390/agronomy11061064.

Glick, B.R. (2012). Plant growth-promoting bacteria: mechanisms and applications. *Scientifica* https://doi.org/10.6064/2012/963401.

Gouda, S., Kerry, R.G., Das, G. et al. (2018). Revitalization of plant growth promoting rhizobacteria for sustainable development in agriculture. *Microbiological Research* 206: 131–140. https://doi.org/10.1016/j.micres.2017.08.016.

Gowdar, S.B., Deepa, H., and Amaresh, Y.S. (2018). A brief review on biocontrol potential and PGPR traits of *Streptomyces* sp. for the management of plant diseases. *Journal of Pharmacognosy and Phytochemistry* 7 (5): 03–07.

Gowtham, H.G., Murali, M., Singh, S.B. et al. (2018). Plant growth promoting rhizobacteria-*Bacillus amyloliquefaciens* improves plant growth and induces resistance in chilli against anthracnose disease. *Biological Control* 126: 209–217. https://doi.org/10.1016/j.biocontrol.2018.05.022.

Grossi, C.E.M., Fantino, E., Serral, F. et al. (2020). *Methylobacterium* sp. 2A is a plant growth-promoting rhizobacteria that has the potential to improve potato crop yield under adverse conditions. *Frontiers in Plant Science* 11: 71. https://doi.org/10.3389/fpls.2020.00071.

Gupta, G., Parihar, S.S., Ahirwar, N.K. et al. (2015). Plant growth promoting rhizobacteria (PGPR): current and future prospects for development of sustainable agriculture. *Journal of Microbial and Biochemical Technology* 7 (2): 096–102. https://doi.org/10.4172/1948-5948.1000188.

Gupta, R., Khan, F., Alqahtani, F.M. et al. (2023). Plant growth–promoting rhizobacteria (PGPR) assisted bioremediation of heavy metal toxicity. *Biotechnology and Applied Biochemistry* 1–29. https://doi.org/10.3389/fpls.2020.00071.

Gupta, S.M., Kumar, K., Joshi, R.K. et al. (2020). Frankia: a promising N-fixing plant growth promoting rhizobacteria (PGPR) improved drought tolerance in crops at higher altitude. In: *Microbiological Advancements for Higher Altitude Agro-Ecosystems & Sustainability*, 411–431. https://doi.org/10.1007/978-981-15-1902-4_20.

Hamid, B., Zaman, M., Farooq, S. et al. (2021). Bacterial plant biostimulants: a sustainable way towards improving growth, productivity, and health of crops. *Sustainability* 13 (5): 2856. https://doi.org/10.3390/su13052856.

Hassen, A.I., Khambani, L.S., Swanevelder, Z.H. et al. (2021). Elucidating key plant growth–promoting (PGPR) traits in *Burkholderia* sp. Nafp2/4-1b (= SARCC-3049) using gnotobiotic assays and whole-genome-sequence analysis. *Letters in Applied Microbiology* 73 (5): 658–671. https://doi.org/10.1111/lam.13556.

Huang, C.H., Hsiang, T., and Trevors, J.T. (2013). Comparative bacterial genomics: defining the minimal core genome. *Antonie Van Leeuwenhoek* 103: 385–398. https://doi.org/10.1007/s10482-012-9819-7.

Iqbal, M.A., Khalid, M., Shahzad, S.M. et al. (2012). Integrated use of *Rhizobium leguminosarum*, plant growth promoting rhizobacteria and enriched compost for improving growth, nodulation and yield of lentil (Lens culinaris Medik.). *Chilean Journal of Agricultural Research* 72 (1): 104–110.

Jin, X., Rahman, M.K., Ma, C. et al. (2023). Silicon modification improves biochar's ability to mitigate cadmium toxicity in tomato by enhancing root colonization of plant-beneficial bacteria. *Ecotoxicology and Environmental Safety* 249: 114407. https://doi.org/10.1016/j.ecoenv.2022.114407.

Katiyar, V. and Goel, R. (2004). Siderophore-mediated plant growth promotion at low temperature by mutant of fluorescent pseudomonad. *Plant Growth Regulation* 42: 239–244. https://doi.org/10.1023/B:GROW.0000026477.10681.d2.

Khatoon, Z., Huang, S., Rafique, M. et al. (2020). Unlocking the potential of plant growth-promoting rhizobacteria on soil health and the sustainability of agricultural systems. *Journal of Environmental Management* 273: 111118. https://doi.org/10.1016/j.jenvman.2020.111118.

Kumar, A., Prakash, A., and Johri, B.N. (2011). Bacillus as PGPR in crop ecosystem. In: *Bacteria in Agrobiology: Crop Ecosystems* (ed. D.K. Maheshwari, M. Saraf, and A. Aeron), 37–59. Berlin, Heidelberg: Springer Berlin Heidelberg https://doi.org/10.1007/978-3-642-18357-7_2.

Kumar, A., Singh, A.K., Kaushik, M.S. et al. (2017a). Interaction of turmeric (Curcuma longa L.) with beneficial microbes: a review. *3 Biotech* 7 (6).

Kumar, A., Singh, M., Singh, P.P. et al. (2016b). Isolation of plant growth promoting rhizobacteria and their impact on growth and curcumin content in Curcuma longa L. *Biocatalysis and Agricultural Biotechnology* 8: 1–7.

Kumar, A., Singh, R., Yadav, A. et al. (2016a). Isolation and characterization of bacterial endophytes of Curcuma longa L. 3. *Biotech* 6 (1): 1–8.

Kumar, A., Verma, H., Yadav, A. et al. (2017b). Plant Growth Promoting Rhizobacteria of Curcuma amada (Mango ginger). *Journal of Pure and Applied Microbiology* 11 (1): 513–520. https://doi.org/10.22207/JPAM.11.1.67.

Kumar, A., Solanki, M.K., Wang, Z. et al. (2024). Revealing the seed microbiome: Navigating sequencing tools, microbial assembly, and functions to amplify plant fitness. *Microbiological Research* 279: 127549.

Kusumastuti, C.T., Ardiyanta, A., and Basri, A.I. (2023). Vegetative growth of soybean plants (Glycine mark L.) on the type and interval of PGPR watering. In: *AIP Conference Proceedings* (Vol. 2491, No. 1). AIP Publishing https://doi.org/10.1063/5.0105494.

Lanteri, M.L., Laxalt, A.M., and Lamattina, L. (2008). Nitric oxide triggers phosphatidic acid accumulation via phospholipase D during auxin-induced adventitious root formation in cucumber. *Plant Physiology* 147 (1): 188–198. https://doi.org/10.1104/pp.107.111815.

Leveau, J.H. and Lindow, S.E. (2005). Utilization of the plant hormone indole-3-acetic acid for growth by Pseudomonas putida strain 1290. *Applied and Environmental Microbiology* 71 (5): 2365–2371. https://doi.org/10.1128/AEM.71.5.2365-2371.2005.

Lloret, L. and Martinez-Romero, E. (2005). Evolution and phylogeny of rhizobia. *Revista Latinoamericana de Microbiologia* 47: 43–60.

Lugtenberg, B. and Kamilova, F. (2009). Plant-growth-promoting rhizobacteria. *Annual Review of Microbiology* 63: 541–556. https://doi.org/10.1146/annurev.micro.62.081307.162918.

Malusa, E. and Vassilev, N. (2014). A contribution to set a legal framework for biofertilisers. *Applied Microbiology and Biotechnology* 98: 6599–6607. https://doi.org/10.1007/s00253-014-5828-y.

Marastoni, L., Pii, Y., Maver, M. et al. (2019). Role of Azospirillum brasilense in triggering different Fe chelate reductase enzymes in cucumber plants subjected to both nutrient deficiency and toxicity. *Plant Physiology and Biochemistry* 136: 118–126. https://doi.org/10.1016/j.plaphy.2019.01.013.

Margenat, A., Matamoros, V., Díez, S. et al. (2018). Occurrence and bioaccumulation of chemical contaminants in lettuce grown in peri-urban horticulture. *Science of the Total Environment* 637: 1166–1174. https://doi.org/10.1016/j.scitotenv.2018.05.035.

Meena, M., Swapnil, P., Divyanshu, K. et al. (2020). PGPR-mediated induction of systemic resistance and physiochemical alterations in plants against the pathogens: Current perspectives. *Journal of Basic Microbiology* 60 (10): 828–861. https://doi.org/10.1002/jobm.202000370.

Miché, L., Battistoni, F., Gemmer, S. et al. (2006). Upregulation of jasmonate-inducible defense proteins and differential colonization of roots of Oryza sativa cultivars with the endophyte Azoarcus sp. *Molecular Plant-Microbe Interactions* 19: 502–511. https://doi.org/10.1094/MPMI-19-0502.

Ming, X. (2017). Study on potential of *Bacillus* spp. and *Pseudomonas* spp. PGPR in promoting plant growth and controlling disease. *Oxidation Communications* 40 (2).

Mus, F., Crook, M.B., Garcia, K. et al. (2016). Symbiotic nitrogen fixation and the challenges to its extension to nonlegumes. *Applied and Environmental Microbiology* 82 (13): 3698–3710. https://doi.org/10.1128/AEM.01055-16.

Mustafa, S., Kabir, S., Shabbir, U., and Batool, R. (2019). Plant growth promoting rhizobacteria in sustainable agriculture: from theoretical to pragmatic approach. *Symbiosis* 78: 115–123. https://doi.org/10.1007/s13199-019-00602-w.

Nagrale, D.T., Chaurasia, A., Kumar, S. et al. (2023). PGPR: the treasure of multifarious beneficial microorganisms for nutrient mobilization, pest biocontrol and plant growth promotion in field crops. *World Journal of Microbiology and Biotechnology* 39 (4): 100. https://doi.org/10.1007/s11274-023-03536-0.

Nath Bhowmik, S. and Das, A. (2018). Biofertilizers: a sustainable approach for pulse production. In: *Legumes for Soil Health and Sustainable Management* (ed. D.K. Maheshwari, M. Saraf, and A. Aeron), 445–485. Springer https://doi.org/10.1007/978-981-13-0253-4_14.

Oliveira, M.M., Ramos, E.T.A., Drechsel, M.M. et al. (2018). Gluconacin from Gluconacetobacter diazotrophicus PAL5 is an active bacteriocin against phytopathogenic and beneficial sugarcane bacteria. *Journal of Applied Microbiology* 125 (6): 1812–1826. https://doi.org/10.1111/jam.14074.

Pathak, D., Lone, R., and Koul, K.K. (2017). Arbuscular Mycorrhizal fungi (AMF) and plant growth-promoting Rhizobacteria (PGPR) Association in Potato (Solanum tuberosum L.): A brief Review. In: *Probiotics and Plant Health*, 401–420. https://doi.org/10.1007/978-981-10-3473-2_18.

Pathan, S.I., Ceccherini, M.T., Sunseri, F., and Lupini, A. (2020). Rhizosphere as hotspot for plant-soil-microbe interaction. In: *Carbon and Nitrogen Cycling in Soil* (ed. R. Datta, R.S. Meena, S.I. Pathan, and M.T. Ceccherini), 17–43. Springer https://doi.org/10.1007/978-981-13-7264-3_2.

Pérez-Montaño, F., Alías-Villegas, C., Bellogín, R.A. et al. (2014). Plant growth promotion in cereal and leguminous agricultural important plants: From microorganism capacities to crop production. *Microbiological Research* 169: 325–336. https://doi.org/10.1016/j.micres.2013.09.011.

Pinter, I.F., Salomon, M.V., Berli, F. et al. (2017). Characterization of the As (III) tolerance conferred by plant growth promoting rhizobacteria to in vitro-grown grapevine. *Applied Soil Ecology* 109: 60–68. https://doi.org/10.1016/j.apsoil.2016.10.003.

Pramanik, K., Mitra, S., Sarkar, A., and Maiti, T.K. (2018). Alleviation of phytotoxic effects of cadmium on rice seedlings by cadmium resistant PGPR strain Enterobacter aerogenes MCC 3092. *Journal of Hazardous Materials* 351: 317–329. https://doi.org/10.1016/j.jhazmat.2018.03.009.

Prasanna, R., Rana, A., Chaudhary, V. et al. (2012). Cyanobacteria-PGPR interactions for effective nutrient and pest management strategies in agriculture. In: *Microorganisms in Sustainable Agriculture and Biotechnology* (ed. T. Satyanarayana, B.N. Johri, and A. Prakash), 173–195. Springer https://doi.org/10.1007/978-94-007-2214-9_10.

Raymond, J., Siefert, J.L., and Staples, C.R. (2004). The natural history of nitrogen fixation. *Molecular Biology and Evolution* 21 (3): 541–554. https://doi.org/10.1093/molbev/msh047.

Rothballer, M., Schmid, M., and Hartmann, A. (2003). In situ localization and PGPR-effect of Azospirillum brasilense strains colonizing roots of different wheat varieties. *Symbiosis*.

Seneviratne, G., Weerasekara, M.L.M.A.W., Seneviratne, K.A.C.N. et al. (2011). Importance of biofilm formation in plant growth promoting rhizobacterial action. *Plant Growth and Health Promoting Bacteria* 81–95. https://doi.org/10.1007/978-3-642-13612-2_4.

Sevilla, M., Burris, R.H., Gunapala, N., and Kennedy, C. (2001). Comparison of benefit to sugarcane plant growth and 15N2 incorporation following inoculation of sterile plants with Acetobacter diazotrophicus wild-type and NIF mutant strains. *Molecular Plant-Microbe Interactions* 14 (3): 358–366. https://doi.org/10.1094/MPMI.2001.14.3.358.

Singh, J.S., Pandey, V.C., and Singh, D.P. (2011). Efficient soil microorganisms: a new dimension for sustainable agriculture and environmental development. *Agriculture, Ecosystems and Environment* 140 (3-4): 339–353. https://doi.org/10.1016/j.agee.2011.01.017.

Singh, M., Kumar, A., Singh, R., and Pandey, K.D. (2017a). Endophytic bacteria: a new source of bioactive compounds. 3. *Biotech* 7 (5): 1–14.

Singh, R., Pandey, K.D., Singh, M. et al. (2022). Isolation and characterization of endophytes bacterial strains of momordica Charantia L. and their possible approach in stress management. *Microorganisms* 10 (2): 290.

Singh, R.P. and Jha, P.N. (2016). The multifarious PGPR Serratia marcescens CDP-13 augments induced systemic resistance and enhanced salinity tolerance of wheat (Triticum aestivum L.). *PLoS One* 11 (6): e0155026. https://doi.org/10.1371/journal.pone.0155026.

Singh, V.K., Singh, A.K., and Kumar, A. (2017b). Disease management of tomato through PGPB: current trends and future perspective. 3. *Biotech* 7 (4): 1–10.

Singh, V.K., Singh, A.K., Singh, P.P., and Kumar, A. (2018). Interaction of plant growth promoting bacteria with tomato under abiotic stress: a review. *Agriculture, Ecosystems & Environment* 267: 129–140.

Sivasakthi, S., Usharani, G., and Saranraj, P. (2014). Biocontrol potentiality of plant growth promoting bacteria (PGPR)-Pseudomonas fluorescens and Bacillus subtilis: A review. *African Journal of Agricultural Research* 9 (16): 1265–1277. https://doi.org/10.5897/AJAR2013.7914.

Solanki, M.K., Joshi, N.C., Singh, P.K. et al. (2023). From concept to reality: transforming agriculture through innovative rhizosphere engineering for plant health and productivity. *Microbiological Research* 127553. https://doi.org/10.1016/j.micres.2023.127553.

Spaepen, S. and Vanderleyden, J. (2011). Auxin and plant-microbe interactions. *Cold Spring Harbor Perspectives in Biology* 3 (4): a001438. https://doi.org/10.1101/cshperspect.a001438.

Spaepen, S., Vanderleyden, J., and Remans, R. (2007). Indole-3-acetic acid in microbial and microorganism-plant signaling. *FEMS Microbiology Reviews* 31 (4): 425–448. https://doi.org/10.1111/j.1574-6976.2007.00072.x.

Tak, H.I., Ahmad, F., and Babalola, O.O. (2012). Advances in the application of plant growth-promoting rhizobacteria in phytoremediation of heavy metals. *Reviews of Environmental Contamination and Toxicology* 223: 33–52. https://doi.org/10.1007/978-1-4614-5577-6_2.

Tian, P., Razavi, B.S., Zhang, X. et al. (2020). Microbial growth and enzyme kinetics in rhizosphere hotspots are modulated by soil organics and nutrient availability. *Soil Biology and Biochemistry* 141: 107662. https://doi.org/10.1016/j.soilbio.2019.107662.

Tuzun, S. and Bent, E. (2006). *Multigenic and Induced Systemic Resistance in Plants*. New York: Springer https://doi.org/10.1007/b101085.

Vacheron, J., Desbrosses, G., Bouffaud, M.L. et al. (2013). Plant growth-promoting rhizobacteria and root system functioning. *Frontiers in Plant Science* 4: 356. https://doi.org/10.3389/fpls.2013.00356.

Verma, H., Kumar, D., Kumar, V. et al. (2021). The potential application of endophytes in management of stress from drought and salinity in crop plants. *Microorganisms* 9 (8): 1729.

Vocciante, M., Grifoni, M., Fusini, D. et al. (2022). The role of plant growth-promoting rhizobacteria (PGPR) in mitigating plant's environmental stresses. *Applied Sciences* 12 (3): 1231. https://doi.org/10.3390/app12031231.

Volkogon, V.V., Potapenko, L.V., and Volkogon, M.V. (2023). Vertical migration of nutrients and water-soluble organic matter in the soil profile under pre-sowing seed treatment with plant growth promoting rhizobacteria. *Frontiers in Sustainable Food Systems* 6: 1054113. https://doi.org/10.3389/fsufs.2022.1054113.

Vyas, P., Kumar, D., Dubey, A., and Kumar, A. (2018). Screening and characterization of Achromobacter xylosoxidans isolated from rhizosphere of Jatropha curcas L.(energy crop) for plant-growth-promoting traits. *Journal of Advanced Research in Biotechnology* 3 (1): 1–8.

Wang, K.L.C., Li, H., and Ecker, J.R. (2002). Ethylene biosynthesis and signaling networks. *The Plant Cell* 14 (suppl_1): S131–S151. https://doi.org/10.1105/tpc.001768.

Xu, P., Han, L., He, J. et al. (2017). Research advance on molecular ecology of asymbiotic nitrogen fixation microbes. *Journal of Applied Ecology* 28 (10): 3440–3450. https://doi.org/10.13287/j.1001-9332.201710.035.

Yaxley, J.R., Ross, J.J., Sherriff, L.J., and Reid, J.B. (2001). Gibberellin biosynthesis mutations and root development in pea. *Plant Physiology* 125 (2): 627–633. https://doi.org/10.1104/pp.125.2.627.

You, M., Fang, S., MacDonald, J. et al. (2020). Isolation and characterization of Burkholderia cenocepacia CR318, a phosphate solubilizing bacterium promoting corn growth. *Microbiological Research* 233: 126395. https://doi.org/10.1016/j.micres.2019.126395.

Yu, Y.Y., Xu, J.D., Gao, M.Z. et al. (2023). Exploring plant growth promoting rhizobacteria potential for green agriculture system to optimize sweet potato productivity and soil sustainability in northern Jiangsu, China. *European Journal of Agronomy* 142: 126661. https://doi.org/10.1016/j.eja.2022.126661.

# 27

## Role of MATE Transporters in Xenobiotics Tolerance

*Arathi Radhakrishnan[1], Shakshi[1], Raj Nandini[1], Ajay Kumar[1], Raj Kishor Kapardar[2], and Rajpal Srivastav[1]*

[1] *Amity Institute of Biotechnology, Amity University, Noida, India*
[2] *Microbial Biotechnology Division, The Energy and Resources Institute, New Delhi, India*

## 27.1 Introduction

The impact of xenobiotics on the environment is a serious concern, and there are multiple factors responsible for these effects, including urbanization, population expansion, industrialization, and globalization. All these have an impact on how we live, both positively and negatively, but they are undoubtedly catalysts for change. Xenobiotics are not naturally produced, and they are introduced from outside sources as drugs, environmental pollutants, toxins, or synthetic chemicals. Chemically, xenobiotics are very diverse and encompass a broad range of chemical compounds, including drugs, pesticides, industrial chemicals, pollutants, and toxins. They can have diverse structures, such as organic compounds, inorganic compounds, polycyclic aromatic hydrocarbons (PAHs), metals, antibiotics, pyrethroids, and glyphosate, dioxins. Xenobiotics can undergo biotransformation within human cells. Enzymes in the liver, such as cytochrome $P_{450}$ enzymes, can modify the chemical structure of xenobiotics to facilitate their elimination from the body. Xenobiotics can enter ecosystems through various pathways, including air, water, soil, and food chain. They can have harmful effects on the environment, disrupt ecological balance, and contribute to environmental pollution. The role of export systems, particularly MATE (multidrug and toxic compound extrusion) transporters, becomes important in tolerating the effects of xenobiotics in plants (Figure 27.1).

All forms of life require the extrusion of foreign, harmful substances to preserve cellular equilibrium. Xenobiotic exporting transporters enable the extrusion process. There are six families of transporters, including ATP-binding cassette (ABC), Resistance Nodulation Cell Division (RND), major facilitator superfamily (MFS), small multidrug resistance (SMR), proteobacterial antimicrobial compound efflux (PACE), and multidrug and toxic compound extrusion (MATE) families. The basic membrane transport proteins, known as ABC multidrug transporters, are propelled by ATP hydrolysis. The other types of multidrug transporters are secondary transporters that export medicines using the ionic gradient across the membrane.

The MATE transporters are functionally versatile and employ either the $H^+$ or $Na^+$ electrochemical gradient to export medications, in contrast to the other transporters. The association of MATE transporters with xenobiotics extrusion is crucial for cellular homeostasis in plants (Hvorup et al. 2003; Tanaka et al. 2017).

## 27.2 Degradation and Management of Xenobiotics

Xenobiotic compounds enter into the environment directly or indirectly, as industrial, community, hospital discharge, etc. Additionally, xenobiotics may be released during the process and as a component of the finished product during product production in industries. They might come from mobile or stationary sources and can have various impacts on the environment (Eapen and D'Souza 2005). Therefore, there are directives and regulations from authorities for releasing these compounds, allowing sources to be controlled. The xenobiotics coming from households are challenging to control, so it is

*Microbes Based Approaches for the Management of Hazardous Contaminants*, First Edition. Edited by Ajay Kumar, Livleen Shukla, Joginder Singh, and Luiz Fernando R. Ferreira.

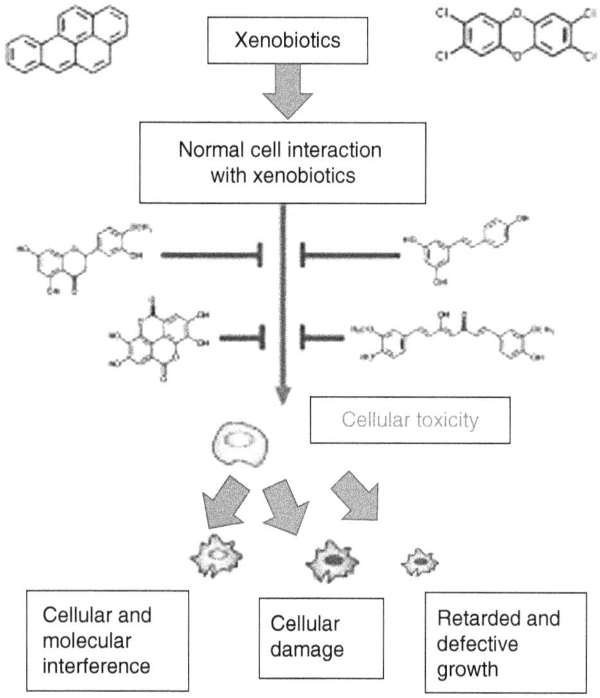


**Figure 27.1** Schematic representation of potential effects of xenobiotics on cellular physiology.

important to sensitize people about the harmful effects of xenobiotics on the environment and human lives. People must be informed about the difficult and time-consuming process of degrading xenobiotics in the environment.

The degradation of xenobiotics involves several enzymatic reactions, which help to reduce the persistence and toxicity of these foreign compounds in ecosystems. Moreover, bioremediation involves introducing suitable microorganisms to contaminated sites to degrade xenobiotics and restore environmental quality. Further, the xenobiotic compounds can be degraded by physiochemical processes like coagulation, filtration, adsorption, chemical precipitation, electrolysis, etc. The waste-water treatment plants are generally not completely sufficient for xenobiotics removal, which allows xenobiotics to enter water bodies and food chain, impact humans, and contribute to the contamination of aquatic bodies with xenobiotics micropollutants. The role of microbial degradation becomes useful in breaking down specific xenobiotics, which is also associated with activated sludge treatment. The operational parameters and xenobiotics in wastewater that are commercially unviable in traditional plants would govern the degradation of xenobiotics. The advanced technology was found to have a significant impact on the removal efficiency of xenobiotics, including pharmaceutical and industrial wastes. The downstream treatment methods were found to vary in their ability to effectively remove xenobiotics.

International environment protection organizations have shown significant concerns about the harmful effects of xenobiotics, encouraged various methods to detect harmful compounds to prevent the harmful effects of xenobiotics, and listed the priority pollutants. Routinely monitoring xenobiotics levels may help in mitigating the problems, designing directives, and upgrading regulations to prevent environmental pollution (Radhakrishnan et al. 2023).

The complex structures of xenobiotics render their degradation a difficult process, and to add to that, biomagnification leads to various harmful effects (Figure 27.2). So, xenobiotic degradation method for analysis should be fast, accurate, and affordable. Xenobiotic contamination in water can be removed by passing the contaminated water through filters, such as activated carbon filters or membrane filters. These filters can trap and adsorb the compounds, allowing clean water to pass through. Adsorption techniques involve the use of adsorbents, such as activated carbon or zeolites, to attract and bind xenobiotic compounds. The chemicals, such as lime or alum, can be added to the contaminated water to induce precipitation of the xenobiotic compounds. The resulting precipitates can then be separated from the water through sedimentation or filtration. The electrolytic methods involve the use of electric current to degrade xenobiotic compounds. The process can generate reactive species, such as hydroxyl radicals, which can break down the compounds into simpler and less harmful substances. Further, ozonation can be used to degrade xenobiotics. Ozone ($O_3$) is a powerful oxidizing agent that can be

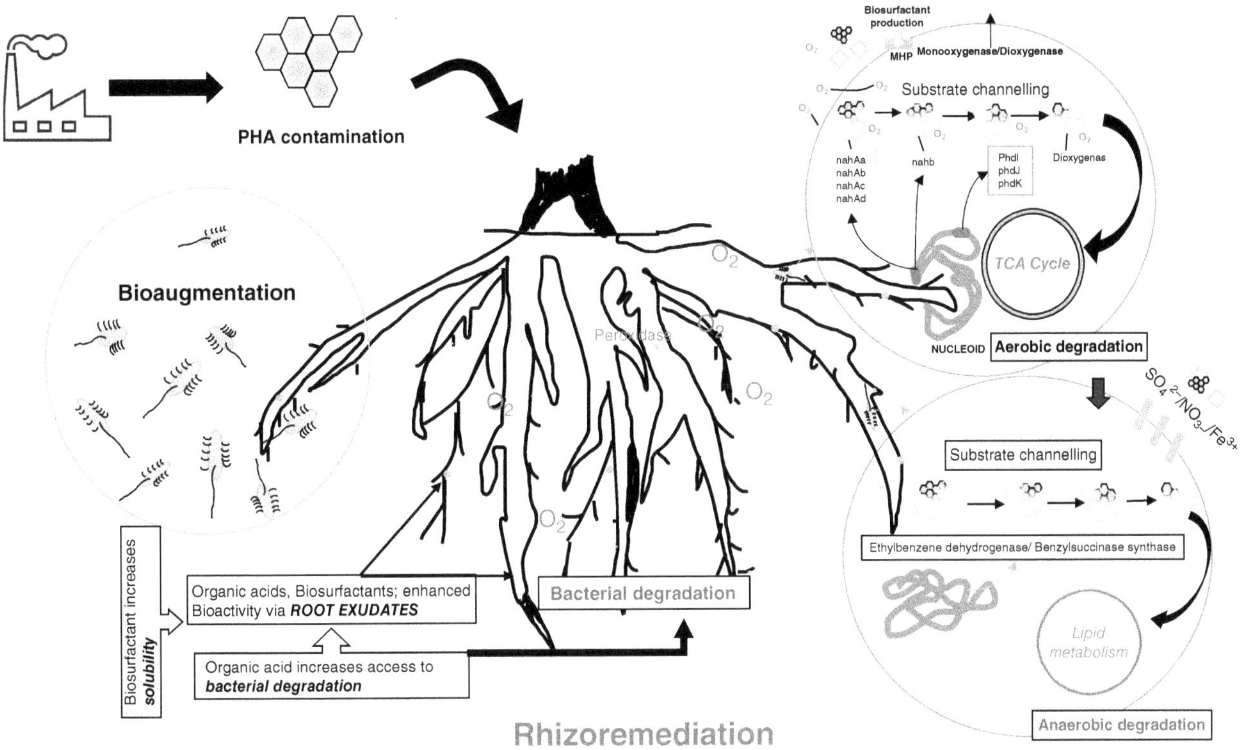


**Figure 27.2** Biodegradation of hydrocarbons involving plants and soil microbiota.

used to degrade xenobiotic compounds. It reacts with the compounds, breaking them down into smaller, less toxic forms. In addition, coagulation can also be useful when it involves the addition of coagulants, such as aluminum sulfate or ferric chloride, to destabilize and aggregate the xenobiotic compounds. The resulting flocs can be easily separated through sedimentation and filtration.

## 27.3 Role of MATE in Xenobiotics' Extrusion and Metabolism

The genomic and functional analysis helped in the identification of the MATE proteins/transporters in plants. MATE transporters are associated with the uptake of nutrients, export of substances, various associated solutes, etc., thus helping in normal plant growth. Proteins like catalytic enzymes and MATE transporters are divided into various classes as per sequence homology analysis (Radchenko et al. 2015; Srivastav et al. 2014). The eukaryotic MATE transporters appear to be proton – dependent, although another transporter can use different ionic gradients for the export of solutes and compounds. Despite having members that are both eubacterial and archaeal, the NorM and DinF subfamilies typically have very little in common in terms of amino acid sequence similarity. These transporters are well characterized in plants and other eukaryotes. These are integral membrane proteins, whose molecular weight varies as per the type of transporter and have different modes of action. They perform multiple functions like growth, senescence, xenobiotic detoxification, disease resistance, phytohormone transport, and stress tolerance, including metal toxicity. These channels can extrude diverse types of compounds that exhibit varied chemical structures. Interestingly, MATE transporters are present ubiquitously across various phyla (Hvorup et al. 2003; Upadhyay et al. 2019). During evolution, the genes of these transporters got dupli cated and amplified in numbers across various species of plants. There has been a greater diversity of these proteins across plant kingdoms, indicating the usefulness of these transporters in stress tolerance and survival in plants. These MATE proteins are also called as "Detoxification" (DTX) proteins because they detoxify cellular environment. The numbers of these transporters vary from plant to plant as well as the cellular and external environments. It is suggested that the number of these transporters be determined by their efficiency and stringency to tolerate xenobiotics.

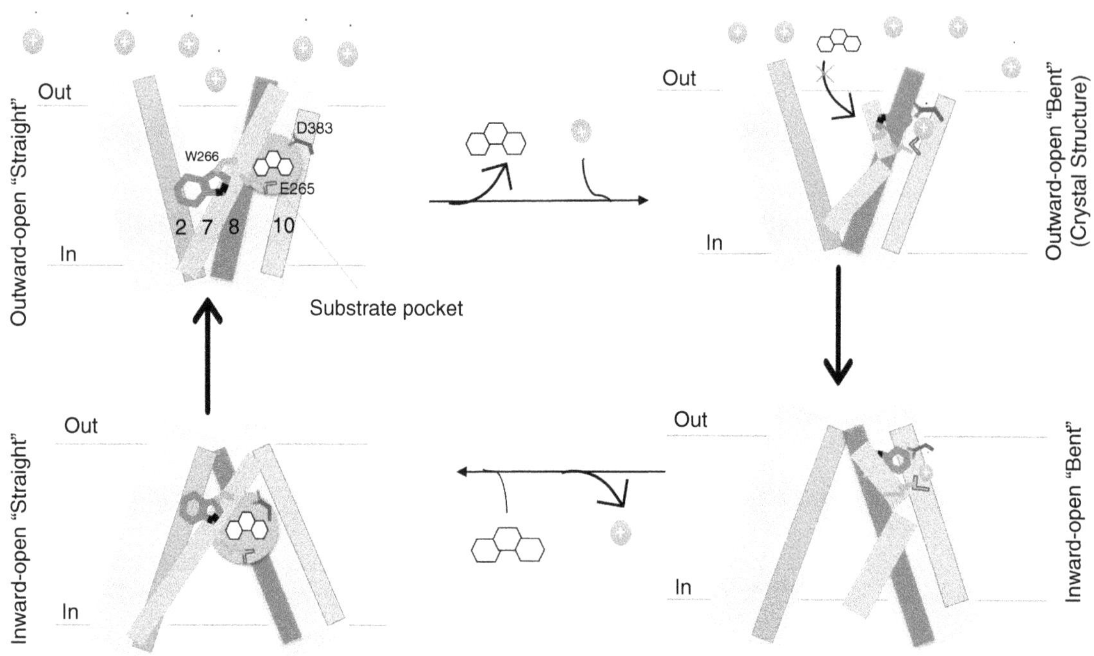


**Figure 27.3** Possible mechanistic extrusion of xenobiotics by plant MATE transporter with associated structural changes for directed movement of substrate (Adaptation: Adapted from Miyauchi et al., 2017).

AtDTX1 is a MATE transporter associated with the tolerance to a wide variety of compounds in plants. This gene was first reported in complementation analysis. The mutant for AtDTX1 gene was sensitive to various compounds in multidrug analysis (Li et al. 2002). This transporter is found to help in survival by detoxifying metal ions, EtBr, berberine alkaloids, and benzene derivatives as xenobiotics. The availability of the crystal structure helps in understanding the mechanism of working of proteins and catalysts (Srivastav et al. 2014). The crystal structure of AtDTX14 helped to decipher the transport mechanism of the MATE transporters in plants. The functional aspects of the substrate or ligand interaction, catalysis, and transport could be decoded by the crystal structure of AtDTX14 (Tanaka et al. 2017). Tanaka et al. described the crystal structure of the plant *Camelina sativa* at 2.9 Å resolution. In these transporters or channel proteins, the role of hydrophobic residues in the ligand or substrate-interacting site becomes relevant as the hydrophobicity of the membrane itself favors the efficient functioning of the transporter. The interaction of the ligand should not be too tight, as it may reduce the affinity for the catalysis (Radchenko et al. 2015).

The hydrogen bonding with conserved residues (acidic residues) plays an important part in the functioning of the transporter. There is extensive hydrogen bonding observed in the carboxy-terminal domain of this MATE transporter, AtDTX14. The transmembrane helix 7 involves major structural changes during the transport process. The protonation–deprotonation defines these structural changes in the transmembrane helix 7. The substrate binding site involves transmembrane helix 7, transmembrane helix 8, and transmembrane helix 10 during straight conformation (Figure 27.3). The binding of the substrate occurs in the carboxy terminus of the transporter. When binding of the substrate takes place, the conformational changes in carboxy-terminal lobe site occur in such a way that it avoids re-binding of substrate. This further guides the directed movement of the substrate across the membrane (Miyauchi et al. 2017).

These MATE transporters are believed to be important contributing factor in bioremediation. The overexpression of transporters like AtDTX1 and AtDTX3 can lead to increased tolerance to xenobiotics. Further, there is differential and spatial expression of MATE transporters in different parts of the plants, like seed and fruit. For example, VcMATE1 and VcMATE4 are differentially expressed in different parts of the blueberry plants (Chen et al., 2015). Therefore, it is thought that these proteins avoid oxidative stress, and they are regulated differentially in different species (Ramel et al. 2012). The differential expression of the genes associated with a cluster in the genome becomes important in regulating the stress response (Srivastav et al. 2014).

Plants can mineralize, detoxify, alter, and compartmentalize the xenobiotics (Chen et al. 2015). In general, plants are capable of metabolizing xenobiotics in three stages: Lipophilic xenobiotic chemicals are modified or activated (via oxido reduction) in phase I (Komives and Gullner 2005). Then, during phase II, xenobiotic metabolites are coupled to an endogenous hydrophilic molecule like glutathione, carbohydrates, or amino acids. In phase III, xenobiotics are attached to lignin/hemicellulose found in cell walls or compartmentalized in vacuoles.

These transporters play a very important role in tolerance to abiotic and biotic stresses in plants. Abiotic stress can be induced by the presence of various xenobiotics like PAHs, polyethylene terephthalate (PET), triclosan, etc. PAHs are diverse types of compounds with fused aromatic rings, pose a significant health risk as mutagens. These PAH xenobiotics are introduced by industrial combustion of organic materials like oil, gases, etc. PET is a common type of plastic used in the production of bottles and containers. PET is highly resistant to degradation in the environment, which leads to its accumulation as plastic waste. However, through metagenomics, the researchers were able to analyze the microbial communities present in soil samples and identify a bacterium named *Ideonella sakaiensis*, which had the ability to degrade PET. PETase is an enzyme capable of breaking down PET. Now, PETase genes can be engineered or modified to enhance their catalytic activity, stability, or specificity for PET degradation. Further investigations are required to understand the ecological distribution of PETase genes in different environments and to understand the microbiome involved with PET degradation for effective bioremediation applications.

Another synthetic antimicrobial agent commonly used in various personal care products is "Triclosan," which eventually leaches into plant systems (Bhargava and Leonard 1996; McMurry et al. 1998). Transcriptomics approaches have been instrumental in elucidating the molecular mechanisms underlying triclosan toxicity and providing valuable insights into its potential impact on the environment. Similarly, resistance to these xenobiotics has been associated with nonspecific efflux systems and cell wall attributes. This is also the basis of cross-antibiotic resistance in microbes. Bisphenol A (BPA) is a synthetic compound that is used primarily in the production of resins, plastics, and derivatives, including toys, storage media, dental composites (Drozdz et al. 2011), medical equipment (Vandentorren et al. 2011) and window foils (Arnold et al. 2013).

It has been found to have potential harmful effects on environment. Another xenobiotics, 2,4-Dichlorophenoxyacetic acid (2,4-D), which is a useful synthetic herbicide, disrupts the normal growth pattern of plants, leading to cellular death. Another widely used herbicide is "Atrazine," which has been found to persist in soil and water systems, posing potential risks to environment. Metagenomics helps in the identification of microbes that can break down atrazine, providing insights into potential strategies for bioremediation of contaminated sites. Exposure to atrazine can induce the expression of AtDTX21, a MATE transporter, indicating its role in xenobiotics tolerance. MATE transporters are helpful to the plant in tolerating ROS levels and fighting the oxidative stress induced by internal and external factors (Ramel et al. 2007).

The integration of the catabolic pathways, or genes, in transgenic plants is an important tool for improving xenobiotic bioremediations. Further, the development of genome-wide analysis can provide knowledge about gene modulations and modifications that is expected to control plant responses to xenobiotics (Weisman et al. 2010). Further, it should be noted that xenobiotic impacts may also involve signaling disruption, in addition to biochemical and physiological disturbance (Unver et al., 2010).

Other alkaloids are thought to be transported by plant MATE transporters. The H+/berberine antiporter may have been used to transfer berberine through the tonoplast, according to biochemical examination of *Coptis japonica* vacuoles (Otani et al. 2005). In *Catharanthus roseus*, two monoterpene indole alkaloids (MIAs), vindoline and catharanthine, are absorbed by mesophyll vacuoles via an H+/MIAs antiport system (Carqueijeiro et al. 2013). The role of MATE transporter, AtDTX51, has been found to be associated with senescence. The process of lateral organ initiation and size of the seed are controlled by the MATE transporters AtDTX54 and AtDTX55 (Suzuki et al. 2015). Under various $CO_2$ circumstances, AtDTX56 modulates stomatal opening (Tian et al. 2015). Thus, these transporters are crucial for controlling different processes in plants.

Secondary metabolites are crucial for disease defense, UV protection, and the color of different organs to draw pollinators. Even though their necessity may be in separate tissues, secondary metabolites are created in specialized tissues. For instance, nicotine manufacturing takes place in the tissues of the roots, but it is necessary in the aerial regions of plants, where it acts as a chemical defense against herbivores. Secondary metabolite transporters are necessary for this translocation across different membranes. It is important to note that plants have evolved to have mechanisms to avoid the cytotoxicity of secondary metabolites by MATE transporters. Toxicology monitoring uses various methods, including biotests with bioluminescent enzyme system, which measure the relationship between the substance's toxicity and its impact on light intensity. Xenobiotics typically reduce the intensity of bioluminescence. Inhibition in bioluminescence is often categorized by traditional biological inhibitors. This method assumes that xenobiotics impair enzyme activity by attaching to substrates/enzyme groups. The rationale is that there are numerous ways for xenobiotic molecules to influence bioluminescence in a complex bioluminescent system. The interaction of the molecules with the main physicochemical processes in the bioluminescence system may modulate their ability to inhibit the process. It has also been reported that MATE transporters are crucial for maintaining Fe homeostasis and absorption by secreting primary metabolites like citrate into the xylem and rhizosphere. In root pericycle cells of several crop plants, such as rice and soybean (*Glycine max*), the MATE transporter FRD3 is highly expressed. Chelation facilitates in the solubilization of iron in the apoplast that surrounds stele cells and permits correct transportation (Kryvoruchko et al. 2018).

In summary, MATE transporters play an important role in regulating the concentrations of various chemicals in cellular environments, including xenobiotics. Further, phytoremediation offers the advantage of cleaning up the environment, particularly soils and water bodies. As future directives, the analysis of xenobiotic degradation through the study of catabolic genes and degradative enzymes can be significant and time-saving through OMIC-based approaches, which have been discussed in the next section.

## 27.4 OMIC-Based Analysis for Xenobiotics Degradation and Metabolism

The study of xenobiotic degradation and its analysis through OMIC-based approaches include genomics, metagenomics, transcriptomics, and metabolomics. Advanced genomics approaches have significantly advanced our understanding of xenobiotic degradation and stress induced by these compounds (Teixeira et al. 2007; Jennings et al. 2009; Srivastav et al. 2019). Genomic approaches have helped in identifying the genes and enzymes involved in the metabolism and detoxification of xenobiotics in various organisms, including bacteria, plants, and animals (Dias et al. 2010; Kretschmann et al. 2011). There are limitations, as some microbes are unculturable in normal laboratory conditions. In such cases, metagenomics becomes an important tool of analysis. The metagenomic analysis provides understanding of the microbial community interacting with plants without the need to culture them.

In a typical OMIC-based analysis, the genomic DNA extraction method would depend on the sample types and the downstream analysis objectives. There are different sequencing platforms available, such as Illumina, PacBio, and Oxford Nanopore, each with its own capabilities, advantages, and limitations (Eapen and D'Souza 2005; Srivastav and Suneja 2019). QC checks the sequencing data quality and removes any low-quality reads or sequencing artifacts. Then sequencing is performed on DNA, assembling the short reads into longer contigs or scaffolds to reconstruct the genomes or metagenomes. Annotation is performed on the assembled genomic/metagenomic contigs or transcripts to identify genes, functional elements, and metabolic pathways, followed by taxonomic profiling. Differential expression analysis is performed to compare the expression levels of genes between different conditions (e.g., PAH-contaminated and uncontaminated samples) to identify differentially expressed genes. Functional Analysis: Assess the functional potential of the microbial communities by predicting the functions of genes or metagenomic contigs. At last, validation of the findings of the genomic analysis through targeted experiments, such as quantitative PCR (qPCR) or functional assays, will confirm the presence and activity of the identified genes or metabolic pathways. At last, interpret the results in the context of the microbial community structure, environmental conditions, and potential xenobiotic degradation or metabolism pathways. Also, we must compare the findings with existing knowledge to gain insights into the role of the microbial community in xenobiotics degradation.

Proteomics/meta-proteomics analysis can focus on the analysis of proteins and their interactions within biological systems. These integrated approaches are preferred to decipher the complex biological ecosystem pertaining to plant–microbial interactions (Figure 27.4). The sensing system in response to xenobiotics has been reported in various organisms, including lower and higher eukaryotes (Heijne et al. 2003; Ulrich 2003; Baker 2005; McMillan and Bradfield 2007; Teixeira et al. 2006; Misra et al. 2011). The utilization of such information via toxico-genomics and chemo-genomics approaches can provide valuable data for xenobiotic response (Heijne et al. 2003; Dias et al. 2010).

## 27.5 Conclusive Remarks

Bioremediation has become an important tool for detoxifying xenobiotic contaminants due to the high costs associated with physical and chemical approaches. Xenobiotic exporting transporters like MATE utilize electrochemical potential to transport xenobiotic compounds out of the cell, thus preventing their toxicity. The complexity of cellular stress induced also determines the need of differential expression of these extrusion transporters. Rhizoremediation also plays an important role in tolerating xenobiotics stress. The molecular, biochemical, and other OMIC-based approaches

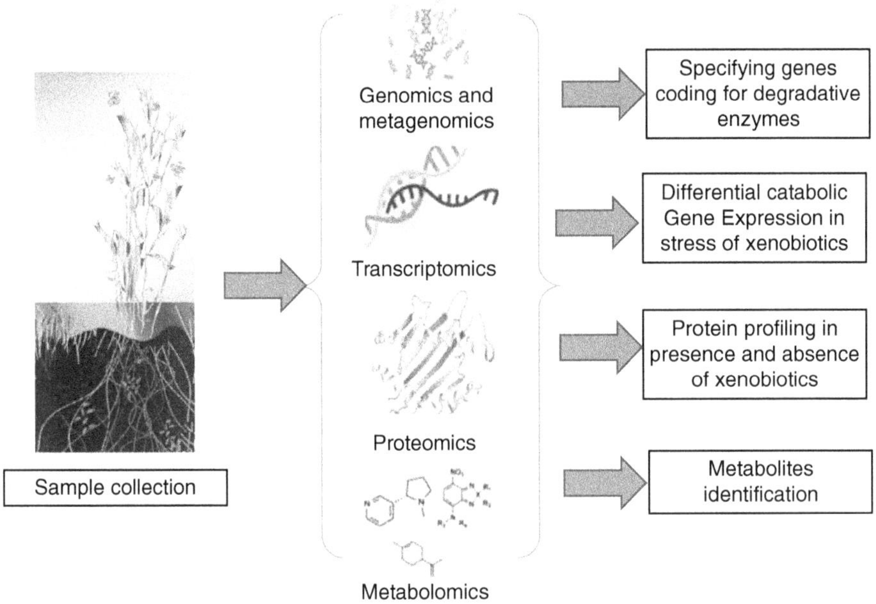


**Figure 27.4** Applications of OMIC-based approaches for identification, analysis, and mitigation of xenobiotics.

help in the analysis of xenobiotic degradation through the study of catabolic genes, degradative enzymes, and associated metabolic pathways. The integrated approaches can help us develop better tools for efficient bioremediation of xenobiotics.

## Acknowledgments

A.R. and R.S. acknowledge INSPIRE, Department of Science and Technology, Ministry of Science and Technology, Government of India, New Delhi, India, for fellowship.

## References

Arnold, S.M., Clark, K.E., Staples, C.A. et al. (2013). Relevance of drinking water as a source of human exposure to bisphenol A. *Journal of Exposure Science & Environmental Epidemiology* 23 (2): 137–144.

Baker, M.E. (2005). Xenobiotics and the evolution of multicellular animals: emergence and diversification of ligand-activated transcription factors. *Integrative and Comparative Biology* 45: 172–178.

Bhargava, H.N. and Leonard, P.A. (1996). Triclosan: applications and safety. *American Journal of Infection Control* 24: 209–218.

Carqueijeiro, I., Noronha, H., Duarte, P. et al. (2013). Vacuolar transport of the medicinal alkaloids from Catharanthus roseus is mediated by a proton-driven antiport. *Plant Physiology* 162 (3): 1486–1496. https://doi.org/10.1104/pp.113.220558. Epub 2013 May 17. PMID: 23686419; PMCID: PMC3707533.

Chen, L., Liu, Y., Hongdi, L. et al. (2015). Identification and expression analysis of MATE genes involved in flavonoid transport in blueberry plants. *PLoS One* 10: e0118578. https://doi.org/10.1371/journal.pone.0118578.

Dias, P.J., Teixeira, M.C., Telo, J.P., and Sá-Correia, I. (2010). Insights into the mechanisms of toxicity and tolerance to the agricultural fungicide mancozeb in yeast, as suggested by the chemogenomic approach. *OMICS* 14: 211–227.

Drozdz, K., Wysokinski, D., Krupa, R., and Woźniak, K. (2011). Bisphenol A-glycidyl methacrylate induces a broad spectrum of DNA damage in human lymphocytes. *Archives of Toxicology* 85 (11): 1453–1461.

Eapen, S. and D'Souza, S.F. (2005). Prospects of genetic engineering of plants for phytoremediation of toxic metals. *Biotechnology Advances* 23: 97–114.

Heijne, W.H., Stierum, R.H., Slijper, M. et al. (2003). Toxicogenomics of bromobenzene hepatotoxicity: a combined transcriptomics and proteomics approach. *Biochemical Pharmacology* 65: 857–875.

Hvorup, R.N., Winnen, B., Chang, A.B. et al. (2003). The multidrug/oligosaccharidyl-lipid/polysaccharide (MOP) exporter superfamily. *European Journal of Biochemistry* 270: 799–813. https://doi.org/10.1046/j.1432-1033.2003.03418.x.

Jennings, L.K., Chartrand, M.M., Lacrampe-Couloume, G. et al. (2009). Proteomic and transcriptomic analyses reveal genes upregulated by cis-dichloroethene in Polaromonas JS666. *Applied and Environmental Microbiology* 11: 3733–3744.

Komives, T. and Gullner, G. (2005). Phase I xenobiotic metabolic systems in plants. *Zeitschrift für Naturforschung. Section C* 60: 179–185.

Kretschmann, A., Ashauer, R., Hitzfeld, K. et al. (2011). Mechanistic toxicodynamic model for receptor-mediated toxicity of diazoxon, the active metabolite of diazinon, in *Daphnia magna*. *Environmental Science and Technology* 45: 4980–4987.

Kryvoruchko, I.S., Routray, P., Sinharoy, S. et al. (2018). An iron-activated citrate transporter, MtMATE67, is required for symbiotic nitrogen fixation. *Plant Physiology* 176 (3): 2315–2329. https://doi.org/10.1104/pp.17.01538. Epub 2017.

Li, L., He, Z., Pandey, G.K. et al. (2002). Functional cloning and characterization of a plant efflux carrier for multidrug and heavy metal detoxification. *Journal of Biological Chemistry* 277 (7): 5360–5368. https://doi.org/10.1074/jbc.M108777200. Epub 2001 Dec 5. PMID: 11739388.

McMillan, B.J. and Bradfield, C.A. (2007). The aryl hydrocarbon receptor sans xenobiotics: endogenous function in genetic model systems. *Molecular Pharmacology* 72: 487–498.

McMurry, L.M., Oethinger, M., and Levy, S.B. (1998). Triclosan targets lipid synthesis. *Nature* 394: 531–532.

Misra, J.R., Horner, M.A., Lam, G., and Thummel, C.S. (2011). Transcriptional regulation of xenobiotic detoxification in *Drosophila*. *Genes and Development* 25: 1796–1806.

Miyauchi, H., Moriyama, S., Kusakizako, T. et al. (2017). Structural basis for xenobiotic extrusion by eukaryotic MATE transporter. *Nature Communications* 8 (1): 1633. https://doi.org/10.1038/s41467-017-01541-0.

Otani, M., Shitan, N., Sakai, K. et al. (2005). Characterization of vacuolar transport of the endogenous alkaloid berberine in Coptis japonica. *Plant Physiology* 138 (4): 1939–1946. https://doi.org/10.1104/pp.105.064352. Epub 2005 Jul 15. PMID: 16024684; PMCID: PMC1183385.

Radchenko, M., Symersky, J., Nie, R., and Lu, M. (2015). Structural basis for the blockade of MATE multidrug efflux pumps. *Nature Communications* 6 (6): 7995. https://doi.org/10.1038/ncomms8995. PMID: 26246409; PMCID: PMC4866600.

Radhakrishnan, A., Balaganesh, P., Vasudevan, M. et al. (2023). Bioremediation of hydrocarbon pollutants: recent promising sustainable approaches, scope, and challenges. *Sustainability* 15 (7): 5847.

Ramel, F., Sulmon, C., Cabello-Hurtado, F. et al. (2007). Genome-wide interacting effects of sucrose and herbicide-mediated stress in *Arabidopsis thaliana*: novel insights into atrazine toxicity and sucrose-induced tolerance. *BMC Genomics* 8: 450.

Ramel, F., Sulmon, C., Serra, A.-A. et al. (2012). Xenobiotic sensing and signalling in higher plants. *Journal of Experimental Botany* 63 (11): 3999–4014. https://doi.org/10.1093/jxb/ers102.

Srivastav, R. and Suneja, G. (2019). Recent advances in microbial genome sequencing. In: *Microbial Genomics in Sustainable Agroecosystems, Vol 2*, 1e (ed. V. Tripathi, P. Kumar, P. Tripathi, et al.). Springer https://doi.org/10.1007/978-981-32-9860-6.

Srivastav, R., Kumar, D., Grover, A. et al. (2014). Unique subunit packing in mycobacterial nanoRNase leads to alternate substrate recognitions in DHH phosphodiesterases. *Nucleic Acids Research* 42 (12): 7894–7910. https://doi.org/10.1093/nar/gku425.

Srivastav, R., Sharma, R., Tandon, S., and Tandon, C. (2019). Role of DHH superfamily proteins in nucleic acids metabolism and stress tolerance in prokaryotes and eukaryotes. *International Journal of Biological Macromolecules* 127: 66–75.

Suzuki, M., Sato, Y., Wu, S. et al. (2015). Conserved functions of the MATE transporter BIG EMBRYO1 in regulation of lateral organ size and initiation rate. *Plant Cell* 27 (8): 2288–2300. https://doi.org/10.1105/tpc.15.00290. Epub 2015 Aug 14. PMID: 26276834; PMCID: PMC4568504.

Tanaka, Y., Iwaki, S., and Tsukazaki, T. (2017). Crystal structure of a plant multidrug and toxic compound extrusion family protein. *Structure* 25: 1455–1460. e2. https://doi.org/10.1016/j.str.2017.07.009.

Teixeira, M.C., Fernandes, A.R., Mira, N.P. et al. (2006). Early transcriptional response of *Saccharomyces cerevisiae* to stress imposed by the herbicide 2,4-dichlorophenoxyacetic acid. *FEMS Yeast Research* 6: 230–248.

Teixeira, M.C., Duque, P., and Sá-Correia, I. (2007). Environmental genomics: mechanistic insights into toxicity of and resistance to the herbicide 2,4-D. *Trends in Biotechnology* 25: 363–370.

Tian, W., Hou, C., Ren, Z. et al. (2015). A molecular pathway for $CO_2$ response in Arabidopsis guard cells. *Nature Communications* 6: 6057. https://doi.org/10.1038/ncomms7057.

Ulrich, R.G. (2003). The toxicogenomics of nuclear receptor agonists. *Current Opinion in Chemical Biology* 7: 505–510.

Unver, T., Bakar, M., Shearman, R.C., and Budak, H. (2010). Genome-wide profiling, and analysis of *Festuca arundinacea* miRNAs and transcriptomes in response to foliar glyphosate application. *Molecular Genetics and Genomics* 283: 397–413.

Upadhyay, N., Kar, D., Deepak Mahajan, B. et al. (2019). The multitasking abilities of MATE transporters in plants. *Journal of Experimental Botany* 70 (18): 4643–4656.

Vandentorren, S., Zeman, F., Morin, L. et al. (2011). Bisphenol-A and phthalates contamination of urine samples by catheters in the Elfe pilot study: implications for large-scale biomonitoring studies. *Environmental Research* 111 (6): 761–764.

Weisman, D., Alkio, M., and Colón-Carmona, A. (2010). Transcriptional responses to polycyclic aromatic hydrocarbon-induced stress in *Arabidopsis thaliana* reveal the involvement of hormone and defense signalling pathways. *BMC Plant Biology* 10: 59.

# Index



*Microbes Based Approaches for the Management of Hazardous Contaminants*, First Edition. Edited by Ajay Kumar, Livleen Shukla, Joginder Singh, and Luiz Fernando R. Ferreira.

www.ingramcontent.com/pod-product-compliance
Ingram Content Group UK Ltd.
Pitfield, Milton Keynes, MK11 3LW, UK
UKHW050952290926
15470UKWH00005B/1781

9 781119 85112